ON THE QUANTUM POTENTIAL

ROBERT CARROLL

Published 2007 by arima publishing

www.arimapublishing.com

ISBN 978-1-84549-211-3

Printed and bound in the United Kingdom

Abramis is an imprint of arima publishing

www.abramis.co.uk

arima publishing
ASK House, Northgate Avenue
Bury St Edmunds, Suffolk IP32 6BB
t: (+44) 01284 700321

ON THE QUANTUM POTENTIAL

Robert Carroll
University of Illinois, Urbana, IL 61801
email: rcarroll@math.uiuc.edu

Contents

PREFACE

The main theme of the book is to relate ideas of quantum fluctuations (expressed via Fisher information for example, or diffusion processes, or fractal structures, or particle creation and annihilation, or whatever) in terms of the so called quantum potential (which arises most conspicuously in deBroglie-Bohm theories). This quantum potential can be directly connected to the diffusion version of the Schrödinger equation as in [**908, 909, 910, 926**] and is also related to the Ricci-Weyl curvature of Dirac-Weyl theory ([**234, 235, 236, 237, 270, 1159, 1160**]) and to the Wheeler-deWittt (WDW) equation (cf. [**255, 258, 259, 599, 1163, 1164**] Further, work of Israelit-Rosen on Weyl-Dirac theory leads to matter production by geometry (cf. [**668, 669, 670, 671**]) which could also be related to the quantum potential via the Dirac field as a matter field. In following this theme we have also been led to developments in scale relativity (cf. [**941, 942, 944**]) where the Schrödinger equation arises from fractal structure and the quantum potential clearly determines a quantization. Quantum potentials also arise in the (x, ψ) duality theme of Faraggi and Matone as developed by Vancea et al; we give a variation on this related to the massless KG equation (or putative aether equation - cf also [**256, 662**]). The quantum fluctuation theme then leads also to stochastic electrodynamics (SED) and the energy of the vacuum (zero point field - ZPF) and thence to further examination of electrodynamics, massless particles, etc. (cf. [**879, 1004**]). There is throughout the book an involvement with quantum field theory (QFT), where in particular we extract from the extensive recent work of Nikolić (see e.g. [**930, 931, 932, 933, 934, 935, 936, 937, 938**]), and there is considerable material devoted to entropy and information. In a sense the magical structure of quantum mechanics (QM) à la von Neumann and others is too perfect; one cannot see what is "really" going on and this makes the deBroglie-Bohm theory attractive, where one has at least the illusion of having particle trajectories, etc. In fact we develop the theme that the Schrödinger mechanics of Hilbert space etc. is inherently incomplete (or uncertain); in particular the uncertainty can arise from its inability to see a third "initial" condition in the microstate trajectories (cf. [**244, 261, 490, 446, 447, 448, 449, 478, 479**]). We show in Remarks 2.2.2 and 3.2.1 how this is related to the Heisenberg uncertainty principle and how it is consequently natural to consider ensembles of particles and probability densities for the Schrödinger picture (cf. also [**241, 244, 250, 445, 446, 478, 479**]). The Hilbert space uncertainty is in fact automatic, due to the operator approach in Hilbert space, and is thus independent of quantum fluctuations, diffusion, fractals, microstates, or whatever, so the Hilbert space formulation is indifferent to the interpretation of uncertainty. However, we are interested in microcausality here and, furthermore, quantum fluctuations generating a quantum potential also have a direct relation to uncertainty via the exact uncertainty principle of Hall-Reginatto; this is discussed at some length.

We want to make a few remarks about writing style. With a mathematical background my writing style has acquired a certain flavor based on equations and sparse in physical motivation or insights. In any case the "meaning" of physics for me lies largely in beautiful equations (Einstein, Maxwell, Schrödinger, Dirac, Hamilton-Jacobi, etc.) and the revelations about the universe presumably therein contained. Of course beauty need not represent "truth" but ugly truth seems meaningless until properly organized and "clothed" conceptually. I hope to convey this spirit in writing and perhaps validate somewhat this approach. Physics is a vast garden of delights and we can only gape at some of the wonders (as expressed here via equations in directions of personal esthetic appeal). We refrain (with some reluctance) from rashly suggesting that nature is simply a manifestation of underlying mathematical structure (i.e. symmetry, combinatorics, topology, geometry, category theory, etc.) played out on a global stage with time, energy and matter as actors. We are also aware that much of physics, both experimental and theoretical, apparently has little to do with equations. Many phenomena are now recognized as emergent (cf. [**801, 1146**]) and one has to deal with phase transitions, self organized criticality, chaos, condensed matter physics, superfluids, etc. (about which I know very little). Such a book as this can probably never be finished since there is new material appearing virtually every weekday on the electronic bulletin boards.

As to the book itself we mainly develop the theme of the quantum potential and with it a a sketch of some aspects of the Bohmian trajectory representation of QM. The quantum potential arises most innocently in the Bohmian theory and the Schrödinger equation (SE) as an expression $Q = -(\hbar^2/2m)(\Delta|\psi|/|\psi|)$ where ψ is the wave function and Q appears then in the corresponding Hamilton-Jacobi (HJ) equation as a potential term. It is possible to relate this term to Fisher information, entropy, and quantum fluctuations in a natural manner and further to hydrodynamics, stress tensors, diffusion, Weyl-Ricci scalar curvatures, fractal velocities, osmotic pressure, etc. It arises in relativistic form via $Q = (\hbar^2/m^2c^2))\Box|\psi|/|\psi|)$ and in field theoretic models as e.g. $Q = -(\hbar^2/2|\Psi|)(\delta^2|\Psi|/\delta|\Psi|^2)$. In terms of lapse and shift functions one has a WDW version $Q = h^2 N q G_{ijk\ell}(\delta^2|\Psi|/\delta q_{ij}\delta q_{k\ell})$ (q is a surface metric). There is also a way in which the quantum potential can be considered as a mass generation term and this is surely related to its role in Weyl-Dirac geometry in determining the matter field. In purely Weyl geometry one can use the quantum matter field (determined by Q) as a metric multiplier to create the conformal geometry. We also exhibit relations between the quantum potential and Ricci flow. We have had recourse to quantum field theory (QFT) at several places in the book, referring to [**38, 149, 202, 616, 677, 1291, 1336**] for example, and we refer to [**254**] for a treatment in terms of Hopf algebras, tau functions, and vertex operators following [**190, 191, 192, 193, 194, 195, 196, 239, 471, 838, 839, 894, 949, 970**] where in particular one to some extent "tames" the combinatorics of QFT via quantum groups for example (cf. also [**323, 324, 426, 778, 779, 780**]). The book represents an expansion and updating of [**254**], which was removed from the market by Springer (about 6 months after publication) due to a complaint by a well known figure who seems to have been offended somehow by my description of some work he had put on the electronic bulletin boards. Although he was explicitly specified as the source of the material he felt this was insufficient referencing (honi soit qui mal y pense). In any case I do have a copyright release from Springer but naturally I have not included the disputed material in this book. It should be emphasized that I have tried to meticulously provide accurate references for all material. I have mentioned names, as well as numbers, when developing material based specifically on someone's work; however with over 1300 references I felt compelled to mainly use number references. Further I have left many references in the form of archive listing numbers (in http://www.arxiv.org) where the updated publication information can usually be found. This should make it easier for everyone and it makes the bibliography considerably shorter and more accessible. Thus Chapters 1-4 from [**254**] are basically unchanged (with some updating), some material is added to Chapters 5 and 6, Chapter 8 is removed, and Chapter 7 is revised and expanded. New Chapters 8-10 are added with contents as indicated. Appendix A seemed ungainly and is replaced by references only while a short Appendix F is added with a few simple formulas.

In a sense we are trying to write a book of a perhaps unusual type, lying between a graduate textbook and a research monograph, which will provide some perspective about ongoing research as well as background. Thus we try to give enough detail so that a graduate student could fill in the details (or be motivated to modify some of the theory) while not focusing on any one topic to the greatest possible depth. The theme of Bohmian mechanics and the quantum potential is of course woven throughout but we do not explore all of the fundamentals or ramifications. Along with others we believe these topics to be a link between classical and quantum mechanics but we do not belabor the point. Further exploration is suggested and copious references are indicated (which in addition give the source of much material); I apologize for any omissions. When describing the work of a referenced author I frequently use the word "one" to refer to the author; this is much shorter than most author's names and avoids a slavish repetition of a given name.

In addition to the people mentioned in [**254**] I would like to express gratitude, for conversations and correspondence on various topics, to F. Calogero, J. Chang, A. Chatzidimitriou-Dreismann, L. Crowell, G. Goldin, A. Khrennikov, M. Kruskal, J. Lee, F. Magri, F. Müller-Hoissen, C. Nucci, R. Parwani, O. Pashaev, M. Pavlov, D. Schuch, and V. Zakharov. As in [**254**] this book is dedicated as before to my beloved wife Denise Rzewska-Bredt-Carroll.

CHAPTER 1

THE SCHRÖDINGER EQUATION

Perhaps no subject has been the focus of as much mystery as "classical" quantum mechanics (QM) even though the standard Hilbert space framework provides an eminently satisfactory vehicle for determining accurate conclusions in many situations (for example seven decimal place accuracy in quantum electrodynamics (QED)). So why all the fuss? The erection of the Hilbert space edifice and the subsequent development of operator algebras (extending now into noncommutative (NC) geometry) has an air of magic. It works but exactly why it works and what it really represents remain shrouded in ambiguity. Also geometrical connections of QM and classical mechanics (CM) are still a source of new work and a modern paradigm focuses on the emergence of CM from QM (or below). Below could mean here a microstructure of space time, or quantum foam, or whatever. Hence we focus on other approaches to QM and will recall any needed Hilbert space ideas as they arise.

1. DIFFUSION AND STOCHASTIC PROCESSES

There are some beautiful stochastic theories for diffusion and QM mainly concerned with origins of the Schrödinger equation (SE). In terms of background information in book form we mention here e.g. [**40, 122, 238, 239, 304, 305, 323, 637, 756, 797, 874, 908, 910, 926, 941, 945, 954, 1069, 1146**]. The present development focuses on certain aspects of the SE involving the wave function form $\psi = Rexp(iS/\hbar)$, hydrodynamical versions, diffusion processes, quantum potentials, and fractal methods. The aim is to envision "structure", both mathematical and physical, and we sometimes avoid detailed technical discussion of mathematical fine points (cf. [**304, 305, 306, 335, 395, 444, 712, 797, 799, 821, 832, 908, 910, 1069, 1209**] for various delicate matters). For example, rather than looking at such topics as Markov processes with jumps we prefer to seek "meaning" for the Schrödinger equation via microstructure and fractals in connection with diffusion processes and kinetic theory.

First consider the SE in the form $-(\hbar^2/2m)\psi'' + V\psi = i\hbar\psi_t$ so that for $\psi = Rexp(iS/\hbar)$ one obtains

$$S_t + \frac{S_X^2}{2m} + V - \frac{\hbar^2 R''}{2mR} = 0;\ \partial_t(R^2) + \frac{1}{m}(R^2 S')' = 0 \tag{1.1}$$

where $S' \sim \partial S/\partial X$. Writing $P = R^2$ (probability density $\sim |\psi|^2$) and $Q = -(\hbar^2/2m)(R''/R)$ (quantum potential) this becomes

$$S_t + \frac{(S')^2}{2m} + Q + V = 0; \; P_t + \frac{1}{m}(PS')' = 0 \tag{1.2}$$

and this has some hydrodynamical interpretations in the spirit of Madelung. Indeed going to Delphenich [**366**] for example we take $p = S'$ with $p = m\dot{q}$ for $\dot{q}$ a velocity (or "collective" velocity - unspecified). Then (1.2) can be written as ($\rho = mP$ is an unspecified mass density)

$$S_t + \frac{p^2}{2m} + Q + V = 0; \; P_t + \frac{1}{m}(Pp)' = 0; \; p = S'; \; P = R^2; \tag{1.3}$$

$$Q = -\frac{\hbar^2}{2m}\frac{R''}{R} = -\frac{\hbar^2}{2m}\frac{\partial^2\sqrt{\rho}}{\sqrt{\rho}}$$

(Q is the quantum potential). Note here

$$\frac{\partial^2\sqrt{\rho}}{\sqrt{\rho}} = \frac{1}{4}\left[\frac{2\rho''}{\rho} - \left(\frac{\rho'}{\rho}\right)^2\right] \tag{1.4}$$

Now from $S' = p = m\dot{q} = mv$ one has

$$P_t + (P\dot{q})' = 0 \equiv \rho_t + (\rho\dot{q})' = 0; \; S_t + \frac{p^2}{2m} + V - \frac{\hbar^2}{2m}\frac{\partial^2\sqrt{\rho}}{\sqrt{\rho}} = 0 \tag{1.5}$$

Differentiating the second equation in X yields ($\partial \sim \partial/\partial X, \; v = \dot{q}$)

$$mv_t + mvv' + \partial V - \frac{\hbar^2}{2m}\partial\left(\frac{\partial^2\sqrt{\rho}}{\sqrt{\rho}}\right) = 0 \tag{1.6}$$

Consequently, multiplying by $p = mv$ and ρ respectively in (1.5) and (1.6), we obtain

$$m\rho v_t + m\rho vv' + \rho\partial V - \frac{\hbar^2}{2m}\rho\partial\left(\frac{\partial^2\sqrt{\rho}}{\sqrt{\rho}}\right) = 0; \; mv\rho_t + mv(\rho' v + \rho v') = 0 \tag{1.7}$$

Then adding in (1.7) we get

$$\partial_t(\rho v) + \partial(\rho v^2) + \frac{\rho}{m}\partial V - \frac{\hbar^2}{2m^2}\rho\partial\left(\frac{\partial^2\sqrt{\rho}}{\sqrt{\rho}}\right) = 0 \tag{1.8}$$

This is similar to an equation in [**366**] (called an "Euler" equation) and it definitely has a hydrodynamic flavor (cf. also Grössing [**567**]).

Now go to [**987**] and write (1.6) in the form ($mv = p = S'$)

$$\frac{\partial v}{\partial t} + (v \cdot \nabla)v = -\frac{1}{m}\nabla(V + Q); \; v_t + vv' = -(1/m)\partial(V + Q) \tag{1.9}$$

The higher dimensional form is not considered here but matters are similar there. This equation (and (1.8)) is incomplete as a hydrodynamical equation as a consequence of a missing term $-\rho^{-1}\nabla\mathfrak{p}$ where $\mathfrak{p}$ is the pressure (cf. [**821**]). Hence one "completes" the equation in the form

$$m\left(\frac{\partial v}{\partial t} + (v \cdot \nabla)v\right) = -\nabla(V + Q) - \nabla F; \; mv_t + mvv' = -\partial(V + Q) - F' \tag{1.10}$$

where $\nabla F = (1/R^2)\nabla\mathfrak{p}$ (or $F' = (1/R^2)\mathfrak{p}'$). By the derivations above this would then correspond to an extended SE of the form

$$i\hbar\frac{\partial\psi}{\partial t} = -\frac{\hbar^2}{2m}\Delta\psi + V\psi + F\psi \tag{1.11}$$

provided one can determine F in terms of the wave function ψ. One notes that it a necessary condition here involves $curlgrad(F) = 0$ or $curl(R^{-2}\nabla\mathfrak{p}) = 0$ which enables one to take e.g. $\mathfrak{p} = -bR^2 = -b|\psi|^2$. For one dimension one writes $F' = -b(1/R^2)\partial|\psi|^2 = -(2bR'/R) \Rightarrow F = -2blog(R) = -blog(|\psi|^2)$. Consequently one has a corresponding SE

$$i\hbar\frac{\partial\psi}{\partial t} = -\frac{\hbar^2}{2m}\psi'' + V\psi - b(log|\psi|^2)\psi \tag{1.12}$$

This equation has a number of nice features discussed in Pardy [**987**] (but serious drawbacks as indicated by Castro, et al in [**275**] - cf. also [**348, 363, 389, 390, 538, 918, 919, 921**] and Chapter 8). For example $\psi = \beta G(x - vt)exp(ikx - i\omega t)$ is a solution of (1.12) with $V = 0$ and for $v = \hbar k/m$ one gets $\psi = cexp[-(B/4)(x - vt + d)^2]exp(ikx - i\omega t)$ where $B = 4mb/\hbar^2$. Normalization $\int_{-\infty}^{\infty}|\psi|^2 = 1$ is possible with $|\psi|^2 = \delta_m(\xi) = \sqrt{m\alpha/\pi}exp(-\alpha m\xi^2)$ where $\alpha = 2b/\hbar^2,\ d = 0$, and $\xi = x - vt$. For $m \to \infty$ we see that δ_m becomes a Dirac delta and this means that motion of a particle with big mass is strongly localized. This is impossible for ordinary QM since $exp(ikx - i\omega t)$ cannot be localized as $m \to \infty$. Such behavior helps to explain the so-called collapse of the wave function and since superposition does not hold Schrödinger's cat is either dead or alive. Further $v = k\hbar/m$ is equivalent to the deBroglie relation $\lambda = h/p$ since $\lambda = (2\pi/k) = 2\pi(\hbar/mv) = 2\pi(h/2\pi)(1/p)$.

REMARK 1.1.1. We go now to Kälbermann [**711**] and the linear SE in the form $i(\partial\psi/\partial t) = -(1/2m)\Delta\psi + U(\vec{r})\psi$; such a situation leads to the Ehrenfest equations which have the form

$$<\vec{v}> = (d/dt)<\vec{r}>;\ <\vec{r}> = \int d^3x|\psi(\vec{r},t)|^2\vec{r};\ m(d/dt)<\vec{v}> = \tag{1.13}$$

$$= \vec{F}(t) = -\int d^3x|\psi(\vec{r},t)|^2\vec{\nabla}U(\vec{r})$$

Thus the quantum expectation values of position and velocity of a suitable quantum system obey the classical equations of motion and the amplitude squared is a natural probability weight. The result tells us that besides the statistical fluctuations quantum systems posess an extra source of indeterminacy, regulated in a very definite manner by the complex wave function. The Ehrenfest theorem can be extended to many point particle systems and in [**711**] one singles out the kind of nonlinearities that violate the Ehrenfest theorem. A theorem is proved that connects Galilean invariance, and the existence of a Lagrangian whose Euler-Lagrange equation is the SE, to the fulfillment of the Ehrenfest theorem. ∎

REMARK 1.1.2. There are many problems with the quantum mechanical theory of derived nonlinear SE (NLSE) but many examples of realistic NLSE arise in the study of superconductivity, Bose-Einstein condensates, stochastic models of quantum fluids, etc. and pick this up again in Chapter 8. We make no attempt to survey this here but rather give an interesting example later from [**275**] related to fractal structures where a number of the difficulties are resolved. For further information, in addition to the references above, we refer to [**124, 349, 502, 538, 711, 722, 724, 725, 1043, 1044, 1045, 1254, 1255**] for some typical situations (see Chapter 8 for a more systematic approach). Let us mention a few cases.

- The program of [**711**] introduces a Schrödinger Lagrangian for a free particle including self-interactions of any nonlinear nature but no explicit dependence on the space of time coordinates. The corresponding action is then invariant under spatial coordinate transformations and by Noether's theorem there arises a conserved current and the physical law of conservation of linear momentum. The Lagrangian is also required to be a real scalar depending on the phase of the wave function only through its derivatives. Phase transformations will then induce the law of conservation of probability identified as the modulus squared of the wave function. Galilean invariance of the Lagrangian then determines a connection between the probability current and the linear momentum which insures the validity of the Ehrenfest theorem.
- We turn next to Kaniadakis [**724**] for a statistical origin for QM (cf. also [**238, 349, 722, 724, 725, 926, 954, 955, 956, 957, 1067, 1127**]). The idea is to build a program in which the microscopic motion, underlying QM, is described by a rigorous dynamics different from Brownian motion (thus avoiding unnecessary assumptions about the Brownian nature of the underlying dynamics). The Madelung approach gives rise to fluid dynamical type equations with a quantum potential, the latter being capable of interpretation in terms of a stress tensor of a quantum fluid. Thus one shows in [**724**] that the quantum state corresponds to a subquantum statistical ensemble whose time evolution is governed by classical kinetics in the phase space. The equations take the form

(1.14)
$$\rho_t + \partial_x(\rho u) = 0;\ \partial_t(\mu\rho u_i) + \partial_j(\rho\phi_{ij}) + \rho\partial_{x_i}V = 0;$$
$$\partial_t(\rho E) + \partial_x(\rho S) - \rho\partial_t V = 0$$

(1.15)
$$\frac{\partial S}{\partial t} + \frac{1}{2\mu}\left(\frac{\partial S}{\partial x}\right)^2 + \mathcal{W} + V = 0$$

for two scalar fields ρ, S determining a quantum fluid. These can be rewritten as

(1.16)
$$\frac{\partial \xi}{\partial t} + \frac{1}{\mu}\frac{\partial^2 S}{\partial x^2} + \frac{1}{\mu}\frac{\partial \xi}{\partial x}\frac{\partial S}{\partial x} = 0;$$
$$\frac{\partial S}{\partial t} - \frac{\eta^2}{4\mu}\frac{\partial^2 \xi}{\partial x^2} - \frac{\eta^2}{8\mu}\left(\frac{\partial \xi}{\partial x}\right)^2 + \frac{1}{2\mu}\left(\frac{\partial S}{\partial x}\right)^2 + V = 0$$

where $\xi = log(\rho)$ and for $\Omega = (\xi/2) + (i/\eta)S = log\Psi$ with $m = N\mu$, $\mathcal{V} = NV$, and $\hbar = N\eta$ one arrives at a SE

(1.17)
$$i\hbar\frac{\partial \Psi}{\partial t} = -\frac{\hbar^2}{2m}\frac{\partial^2 \Psi}{\partial x^2} + \mathcal{V}\Psi$$

Further one can write $\Psi = \rho^{1/2}exp(i\mathfrak{S}/\hbar)$ with $\mathfrak{S} = NS$ and here $N = \int |\Psi|^2 d^n x$. The analysis is very interesting.

We will return to this later. ■

REMARK 1.1.3. Now in Dürr, et al [**412**] one is obliged to use the form $\psi = Rexp(iS/\hbar)$ to make sense out of the constructions (this is no problem with suitable provisos, e.g. that S is not constant - cf. [**238, 244, 245, 254, 445, 446**]). Thus note $\psi'/\psi = (R'/R) + i(S'/\hbar)$ with $\Im(\psi'/\psi) = (1/m)S' \sim p/m$ (see also (1.22) below). Note also $J = (\hbar/m)\Im\psi^*\psi'$ and $\rho = R^2 = |\psi|^2$ represent a current and a density respectively. Then using $p = mv = m\dot{q}$ one can write

$$v = (\hbar/m)\Im(\psi'/\psi);\ J = (\hbar/m)\Im|\psi|^2(\psi^*\psi'/|\psi|^2) = (\hbar/m)\Im(\rho v) \tag{1.18}$$

Then look at the SE in the form $i\hbar\psi_t = -(\hbar^2/2m)\psi'' + V\psi$ with $\psi_t = (R_t + iS_t R/\hbar)exp(iS/\hbar)$ and

$$\psi_{xx} = [(R' + (iS'R/\hbar)exp(iS/\hbar)]' = \tag{1.19}$$

$$[R'' + (2iS'R'/\hbar) + (iS''R/\hbar) + (iS'/\hbar)^2 R]exp(iS/\hbar)$$

which means

$$-\frac{\hbar^2}{2m}\left[R'' - \left(\frac{S'}{\hbar}\right)^2 + \frac{2iS'R'}{\hbar} + \frac{iS''R}{\hbar}\right] + VR = i\hbar\left[R_t + \frac{iS_tR}{\hbar}\right] \Rightarrow \tag{1.20}$$

$$\Rightarrow \partial_t R^2 + \frac{1}{m}(R^2S')' = 0;\ S_t + \frac{(S')^2}{2mR} - \frac{\hbar^2 R''}{2mR} + V = 0$$

This can also be written as (cf. (1.3))

$$\partial_t\rho + \frac{1}{m}\partial(p\rho) = 0;\ S_t + \frac{p^2}{2m} + Q + V = 0 \tag{1.21}$$

where $Q = -\hbar^2 R''/2mR$. Now we sketch the philosophy of [**412, 413**] in part. Most of such aspects are omitted here and we try to isolate the essential mathematical features (see Section 1.2 for more). First one emphasizes configurations based on coordinates whose motion is choreographed by the SE according to the rule (1-D only here)

$$\dot{q} = v = \frac{\hbar}{m}\Im\frac{\psi^*\psi'}{|\psi|^2} \tag{1.22}$$

where $i\hbar\psi_t = -(\hbar^2/2m)\psi'' + V\psi$. The argument for (1.22) is based on obtaining the simplest Galilean and time reversal invariant form for velocity, transforming correctly under velocity boosts. This leads directly to (1.22) (cf. (1.18)) so that Bohmian mechanics (BM) is governed by (1.22) and the SE. It's a fairly convincing argument and no recourse to Floydian time seems possible (cf. [**238, 446, 478, 479**]). Note however that if $S = c$ then $\dot{q} = v = (\hbar/m)\Im(R'/R) = 0$ while $p = S' = 0$ so perhaps this formulation avoids the $S = 0$ problems indicated in [**238, 446, 478, 479**]. One notes also that BM depends only on the Riemannian structure $g = (g_{ij}) = (m_i\delta_{ij})$ in the form

$$\dot{q} = \hbar\Im(grad\psi/\psi);\ i\hbar\psi_t = -(\hbar^2/2)\Delta\psi + V\psi \tag{1.23}$$

What makes the constant $\hbar/m$ in (1.22) important here is that with this value the probability density $|\psi|^2$ on configuration space is equivariant. This means that via the evolution of probability densities $\rho_t + div(v\rho) = 0$ (as in (1.21) with $v \sim p/m$) the density $\rho = |\psi|^2$ is stationary relative to ψ, i.e. $\rho(t)$ retains the form $|\psi(q,t)|^2$. One calls $\rho = |\psi|^2$ the quantum equilibrium density (QED) and says that a system is in quantum equilibrium when its coordinates are randomly distributed according to the QED. The quantum equilibrium hypothesis (QHP) is the assertion that when a system has wave function ψ the distribution ρ of its coordinates satisfies $\rho = |\psi|^2$. ■

REMARK 1.1.4. We extract here from Hall, Reginatto, and Kumar [**594, 596, 598**] (cf. also the references there for background and [**490, 491, 703**] for some information geometry). There are a number of interesting results connecting uncertainty, Fisher information, and QM and we make no attempt to survey the matter. Thus first recall that the classical Fisher information associated with translations of a 1-D observable X with probability density $P(x)$ is

$$F_X = \int dx\, P(x)([log(P(x)]')^2 > 0 \tag{1.24}$$

Recall now the Cramer-Rao inequality $Var(X) \geq F_X^{-1}$ where $Var(X) \sim$ variance of X. A Fisher length for X is defined via $\delta X = F_X^{-1/2}$ and this quantifies the length scale over which $p(x)$ (or better $log(p(x))$) varies appreciably. Then the root mean square deviation ΔX satisfies $\Delta X \geq \delta X$. Let now P be the momentum observable conjugate to X, and P_{cl} a classical momentum observable corresponding to the state ψ given via $p_{cl}(x) = (\hbar/2i)[(\psi'/\psi) - (\bar{\psi}'/\bar{\psi})]$ (cf. (1.22)). One has then the identity $< p >_\psi = < p_{cl} >_\psi$ via integration by parts. Now define the nonclassical momentum by $p_{nc} = p - p_{cl}$ and one shows that $\Delta X \Delta p \geq \delta X \Delta p \geq \delta X \Delta p_{nc} = \hbar/2$. Then go to [**596**] now where two proofs are given for the derivation of the SE from the exact uncertainty principle ($\delta X \Delta p_{nc} = \hbar/2$). Thus consider a classical ensemble of n-dimensional particles of mass m moving under a potential V. The motion can be described via the HJ and continuity equations

$$\frac{\partial s}{\partial t} + \frac{1}{2m}|\nabla s|^2 + V = 0; \quad \frac{\partial P}{\partial t} + \nabla \cdot \left[P\frac{\nabla s}{m}\right] = 0 \tag{1.25}$$

for the momentum potential s and the position probability density P (note that we have interchanged p and P from [**596**] - note also there is no quantum potential and this will be supplied by the information term). These equations follow from the variational principle $\delta L = 0$ with Lagrangian

$$L = \int dt\, d^n x\, P\left[\frac{\partial s}{\partial t} + \frac{1}{2m}|\nabla s|^2 + V\right] \tag{1.26}$$

It is now assumed that the classical Lagrangian must be modified due to the existence of random momentum fluctuations. The nature of such fluctuations is immaterial for (cf. [**596**] for discussion) and one can assume that the momentum associated with position x is given by $p = \nabla s + N$ where the fluctuation term N vanishes on average at each point x. Thus s changes to being an average momentum potential. It follows that the average kinetic energy $<|\nabla s|^2>/2m$ appearing in (1.26) should be replaced by $<|\nabla s+N|^2>/2m$ giving rise to

$$L' = L + (2m)^{-1}\int dt < N\cdot N >= L + (2m)^{-1}\int dt(\Delta N)^2 \tag{1.27}$$

where $\Delta N =< N\cdot N >^{1/2}$ is a measure of the strength of the fluctuations. The additional term is specified uniquely, up to a multiplicative constant, by the following three assumptions

(1) Action principle: L' is a scalar Lagrangian with respect to the fields P and s where the principle $\delta L' = 0$ yields causal equations of motion. Thus $(\Delta N)^2 = \int d^n x\, pf(P, \nabla P, \partial P/\partial t, s, \nabla s, \partial s/\partial t, x, t)$ for some scalar function f.
(2) Additivity: If the system comprises two independent noninteracting subsystems with $P = P_1P_2$ then the Lagrangian decomposes into additive subsystem contributions; thus $f = f_1 + f_2$ for $P = P_1P_2$.
(3) Exact uncertainty: The strength of the momentum fluctuation at any given time is determined by and scales inversely with the uncertainty in position at that time. Thus $\Delta N \to k\Delta N$ for $x \to x/k$. Moreover since position uncertainty is entirely characterized by the probability density P at any given time the function f cannot depend on s, nor explicitly on t, nor on $\partial P/\partial t$.

The following theorem is then asserted (see [**596**] for the proofs).

THEOREM 1.1. The above 3 assumptions imply the relation $(\Delta N)^2 = c\int d^n x\, P|\nabla log(P)|^2$ where c is a positive universal constant.

COROLLARY 1.1. It follows from (1.27) that the equations of motion for p and s corresponding to the principle $\delta L' = 0$ are

$$i\hbar\frac{\partial\psi}{\partial t} = -\frac{\hbar^2}{2m}\nabla^2\psi + V\psi \tag{1.28}$$

where $\hbar = 2\sqrt{c}$ and $\psi = \sqrt{P}exp(is/\hbar)$. ■

REMARK 1.1.5. We sketch here for simplicity and clarity another derivation of the SE along similar ideas following Reginatto [**1062**]. Let $P(y^i)$ be a probability density and $P(y^i + \Delta y^i)$ be the density resulting from a small change in the y^i. Calculate the cross entropy via

$$J(P(y^i + \Delta y^i) : P(y^i)) = \int P(y^i + \Delta y^i) log\frac{P(y^i + \Delta y^i)}{P(y^i)} d^n y \simeq \tag{1.29}$$

$$\simeq \left[\frac{1}{2}\int \frac{1}{P(y^i)}\frac{\partial P(y^i)}{\partial y^i}\frac{\partial P(y^i)}{\partial y^k)} d^n y\right]\Delta y^i\Delta y^k = I_{jk}\Delta y^i\Delta y^k$$

The I_{jk} are the elements of the Fisher information matrix. The most general expression has the form

$$I_{jk}(\theta^i) = \frac{1}{2}\int \frac{1}{P(x^i|\theta^i)}\frac{\partial P(x^i|\theta^i)}{\partial\theta^j}\frac{\partial P(x^i|\theta^i)}{\partial\theta^k} d^n x \tag{1.30}$$

where $P(x^i|\theta^i)$ is a probability distribution depending on parameters θ^i in addition to the x^i. For $P(x^i|\theta^i) = P(x^i + \theta^i)$ one recovers (1.29) (straightforward - cf. [**1062**]). If P is defined over an n-dimensional manifold with positive inverse metric g^{ik} one obtains a natural definition of the information associated with P via

$$I = g^{ik} I_{ik} = \frac{g^{ik}}{2} \int \frac{1}{P} \frac{\partial P}{\partial y^i} \frac{\partial P}{\partial y^k} d^n y \tag{1.31}$$

Now in the HJ formulation of classical mechanics the equation of motion takes the form

$$\frac{\partial S}{\partial t} + \frac{1}{2} g^{\mu\nu} \frac{\partial S}{\partial x^\mu} \frac{\partial S}{\partial x^\nu} + V = 0 \tag{1.32}$$

where $g^{\mu\nu} = diag(1/m, \cdots, 1/m)$. The velocity field u^μ is given by $u^\mu = g^{\mu\nu}(\partial S/\partial x^\nu)$. When the exact coordinates are unknown one can describe the system by means of a probability density $P(t, x^\mu)$ with $\int P d^n x = 1$ and

$$(\partial P/\partial t) + (\partial/\partial x^\mu)(P g^{\mu\nu}(\partial S/\partial x^\nu)) = 0 \tag{1.33}$$

These equations completely describe the motion and can be derived from the Lagrangian

$$L_{CL} = \int P \left\{ \frac{\partial S}{\partial t} + \frac{1}{2} g^{\mu\nu} \frac{\partial S}{\partial x^\mu} \frac{\partial S}{\partial x^\nu} + V \right\} dt d^n x \tag{1.34}$$

using fixed endpoint variation in S and P. Quantization is obtained by adding a term proportional to the information I defined in (1.31). This leads to
(1.35)

$$L_{QM} = L_{CL} + \lambda I = \int P \left\{ \frac{\partial S}{\partial t} + \frac{1}{2} g^{\mu\nu} \left[\frac{\partial S}{\partial x^\mu} \frac{\partial S}{\partial x^\nu} + \frac{\lambda}{P^2} \frac{\partial P}{\partial x^\mu} \frac{\partial P}{\partial x^\nu} \right] + V \right\} dt d^n x$$

Fixed endpoint variation in S leads again to (1.33) while variation in P leads to

$$\frac{\partial S}{\partial t} + \frac{1}{2} g^{\mu\nu} \left[\frac{\partial S}{\partial x^\mu} \frac{\partial S}{\partial x^\nu} + \lambda \left(\frac{1}{P^2} \frac{\partial P}{\partial x^\mu} \frac{\partial P}{\partial x^\nu} - \frac{2}{P} \frac{\partial^2 P}{\partial x^\mu \partial x^\nu} \right) \right] + V = 0 \tag{1.36}$$

These equations are equivalent to the SE if $\psi = \sqrt{P} exp(iS/\hbar)$ with $\lambda = (2\hbar)^2$. ■

REMARK 1.1.6. In Remarks 1.1.6 - 1.1.8 one uses $Q = \pm(1/2m)$ times the standard $Q = -(\hbar^2/2m)(\Delta\sqrt{\rho}/\sqrt{\rho}$. The SE gives to a probability distribution $\rho = |\psi|^2$ (with suitable normalization) and to this one can associate an information entropy $S(t)$ (actually configuration information entropy) $S = -\int \rho log(\rho) d^3x$ which is typically not a conserved quantity (S is an unfortunate notation here but we retain it momentarily since no confusion should arise). The rate of change in time of S can be readily found by using the continuity equation $\partial_t \rho = -\nabla \cdot (v\rho)$ where v is a current velocity field Note here (cf. also [**1003**])

$$\frac{\partial S}{\partial t} = -\int \rho_t (1 + log(\rho)) dx = \int (1 + log(\rho)) \partial(v\rho) \tag{1.37}$$

Note that a formal substitution of $v = -u$ in the contintuity equation implies the standard free Browian motion outcome $dS/dt = D \cdot \int [(\nabla\rho)^2/\rho] d^3x = D \cdot Tr\mathfrak{F} \geq 0$ - use here $u = D\nabla log(\rho)$ with $D = \hbar/2m)$ and (1.37) with $\int (1 + log(\rho))\partial(v\rho) =$

$-\int v\rho\partial log(\rho) = -\int v\rho' \sim \int((\rho')^2/\rho)$ modulo constants involving D etc. Recall here $\mathfrak{F} \sim -(2/D^2)\int \rho Q dx = \int dx[(\nabla\rho)^2/\rho]$ is a functional form of Fisher information. A high rate of information entropy production corresponds to a rapid spreading (flattening down) of the probablity density. This delocalization feature is concomitant with the decay in time property quantifying the time rate at which the far from equilibrium system approaches its stationary state of equilibrium $(d/dt)Tr\mathfrak{F} \leq 0$. ■

REMARK 1.1.7. Now going back to the quantum context one admits general forms of the current velocity v. For example consider a gradient field $v = b - u$ where the so-called forward drift $b(x,t)$ of the stochastic process depends on a particular diffusion model. Then one can rewrite the continuity equation as a standard Fokker-Plank equation $\partial_t\rho = D\Delta\rho - \nabla\cdot(b\rho)$. Boundary restrictions requiring ρ, $v\rho$, and $b\rho$ to vanish at spatial infinities or at boundaries yield the general entropy balance equation

$$(1.38)\quad \frac{dS}{dt} = \int \left[\rho(\nabla\cdot b) + D\cdot\frac{(\nabla\rho)^2}{\rho}\right] d^3x \equiv -D\frac{dS}{dt} = \int \rho(v\cdot u)d^3x =< v\cdot u >$$

The first term in the first equation is not positive definite and can be interpreted as an entropy flux while the second term refers to the entropy production proper. The flux term represents the mean value of the drift field divergence $\nabla\cdot b$ which by itself is a local measure of the flux incoming to or outgoing from an infinitesimal surrounding of x at time t. If locally $(\nabla\cdot b)(x,t) > 0$ on an infinitesimal time scale we would encounter a local entropy increase in the system (increasing disorder) while in case $(\nabla\cdot b)(x,t) < 0$ one thinks of local entropy loss or restoration or order. Only in the situation $< \nabla\cdot b >= 0$ is there no entropy production. Quantum dynamics permits more complicated behavior. One looks first for a general criterion under which the information entropy S is a conserved quantity. Consider (1.8) and invoke the diffusion current to write (recall $u = D(\nabla\rho)/\rho$)

$$(1.39)\qquad D\frac{dS}{dt} = -\int [\rho^{-1/2}(\rho v)]\cdot[\rho^{-1/2}(D\nabla\rho)]d^3x$$

Then by means of the Schwarz inequality one has $D|dS/dt| \leq < v^2 >^{1/2} < u^2 >^{1/2}$ so a necessary (but insufficient) condition for $dS/dt \neq 0$ is that both $< v^2 >$ and $< u^2 >$ are nonvanishing. On the other hand a sufficient condition for $dS/dt = 0$ is that either one of these terms vanishes. Indeed in view of $< u^2 >= D^2\int[(\nabla\rho)^2/\rho]d^3x$ the vanishing information entropy production implies $dS/dt = 0$; the vanishing diffusion current does the same job. ■

REMARK 1.1.8. We develop a little more perspective now (following Garbaczewski [**508**] - first paper). Recall Q written out as

$$(1.40)\qquad Q = 2D^2\frac{\Delta\rho^{1/2}}{\rho^{1/2}} = D^2\left[\frac{\Delta\rho}{\rho} - \frac{1}{2\rho^2}(\nabla\rho)^2\right] = \frac{1}{2}u^2 + D\nabla\cdot u$$

where $u = D\nabla log(\rho)$ is called an osmotic velocity field. The standard Brownian motion involves $v = -u$, known as the diffusion current velocity and (up to a dimensional factor) is identified with the thermodynamic force of diffusion which drives the irreversible process of matter exchange at the macroscopic level. On the other hand, even while the thermodynamic force is a concept of purely statistical origin associated with a collection of particles, in contrast to microscopic forces which have a direct impact on individual particles themselves, it is well known that this force manifests itself as a Newtonian type entry in local conservation laws describing the momentum balance; in fact it pertains to the average (local average) momentum taken over by the particle cloud, a statistical ensemble property quantified in terms of the probability distribution at hand. It is precisely the (negative) gradient of the above potential Q in (1.40) which plays the Newtonian force role in the momentum balance equations. The second analytical expression of interest here involves

$$(1.41) \qquad -\int Q\rho dx = (1/2)\int u^2\rho dx = (1/2)D^2 \cdot F_X;\ F_X = \int \frac{(\nabla\rho)^2}{\rho}dx$$

where F_X is the Fisher information, encoded in the probability density ρ which quantifies its gradient content (sharpness plus localization/disorder) Note that

$$(1.42) \qquad -\int Q\rho = -\int [(1/2)u^2\rho + D\rho u'] = -\int (1/2)u^2\rho + \int Du\rho' =$$

$$= -(1/2)\int D^2(\rho'/\rho)^2\rho + D^2\int \rho'(\rho'/\rho) = (D^2/2)\int (\rho')^2/\rho = (1/2)\int u^2\rho$$

On the other hand the local entropy production inside the system sustaining an irreversible process of diffusion is given via

$$(1.43) \qquad \frac{dS}{dt} = D\cdot\int \frac{(\nabla\rho)^2}{\rho}dx = D\cdot F_X \geq 0$$

This stands for an entropy production rate when the Fick law induced diffusion current (standard Brownian motion case) $j = -D\nabla\rho$, obeying $\partial_t\rho + \nabla j = 0$, enters the scene. Here $S = -\int \rho log(\rho)dx$ plays the role of (time dependent) information entropy in the nonequilibrium statistical mechanics framework for the thermodynamics of irreversible processes. It is clear that a high rate of entropy increase coresponds to a rapid spreading (flattening) of the probability density. This explicitly depends on the sharpness of density gradients. The potential Q(x,t), the Fisher information F_X, the nonequilibrium measure of entropy production dS/dt, and the information entropy $S(t)$ are thus mutually entangled quantities, each being exclusively determined in terms of ρ and its derivatives.

In the standard statistical mechanics setting the Euler equation gives a prototypical momentum balance equation in the (local) mean

$$(1.44) \qquad (\partial_t + v\cdot\nabla)v = \frac{F}{m} - \frac{\nabla P}{\rho}$$

where $F = -\nabla F$ represents normal Newtonian force and P is a pressure term. Q appears in the hydrodynamical formalism of QM via

$$(1.45) \qquad (\partial_t + v\cdot\nabla)v = \frac{1}{m}F - \nabla Q = \frac{1}{m}F + \frac{\hbar^2}{2m^2}\nabla\frac{\Delta\rho^{1/2}}{\rho^{1/2}}$$

Another spectacular example pertains to the standard free Brownian motion in the strong friction regime (Smoluchowski diffusion), namely

$$(\partial_t + v \cdot \nabla)v = -2D^2\nabla\frac{\Delta\rho^{1/2}}{\rho^{1/2}} = -\nabla Q \tag{1.46}$$

where $v = -D(\nabla\rho/\rho)$ (formally $D = \hbar/2m$). ■

REMARK 1.1.9. The papers in Davidson [**362, 363**] contain very interesting derivations of Schrödinger equations via diffusion ideas à la Nelson, Markov wave equations, and suitable "applied" forces (e.g. radiative reactive forces). ■

We go now to Nagasawa [**906, 907, 908, 909, 910**] to see how diffusion and the SE are really connected (cf. also [**17, 173, 275, 555, 913, 914, 926, 954, 955, 956, 957, 964, 965, 966, 967, 968**] for related material, some of which is discussed later in detail); for now we simply sketch some formulas for a simple Euclidean metric where $\Delta = \sum(\partial/\partial x^i)^2$. Then $\psi(t,x) = exp[R(t,x) + iS(t,x)]$ satisfies a SE $i\partial_t\psi + (1/2)\Delta\psi + ia(t,x)\cdot\nabla\psi - V(t,x)\psi = 0$ ($\hbar$ and m omitted with $a(t,x)$ a drift coefficient) if and only if

$$V = -\frac{\partial S}{\partial t} + \frac{1}{2}\Delta R + \frac{1}{2}(\nabla R)^2 - \frac{1}{2}(\nabla S)^2 - a\cdot\nabla S; \tag{1.47}$$

$$0 = \frac{\partial R}{\partial t} + \frac{1}{2}\Delta S + (\nabla S)\cdot(\nabla R) + a\cdot\nabla R$$

in the region $D = \{(s,x) : \psi(s,x) \neq 0\}$ (a harmless gauge factor in the divergence is also being omitted). Solutions are often referred to as weak or distributional but we do not belabor this point. From [**907, 908, 909**] there results

THEOREM 1.2. Let $\psi(t,x) = exp[R(t,x) + iS(t,x)]$ be a solution of the SE above; then $\phi(t,x) = exp[R(t,x) + S(t,x)]$ and $\hat{\phi} = exp[R(t,x) - S(t,x)]$ are solutions of

$$\frac{\partial\phi}{\partial t} + \frac{1}{2}\Delta\phi + a(t,x)\cdot\nabla\phi + c(t,x,\phi)\phi = 0; \tag{1.48}$$

$$-\frac{\partial\hat{\phi}}{\partial t} + \frac{1}{2}\Delta\hat{\phi} - a(t,x)\cdot\nabla\hat{\phi} + c(t,x,\phi)\hat{\phi} = 0$$

where the creation and annihilation term $c(t,x,\phi)$ is given via

$$c(t,x,\phi) = -V(t,x) - 2\frac{\partial S}{\partial t}(t,x) - (\nabla S)^2(t,x) - 2a\cdot\nabla S(t,x) \tag{1.49}$$

Conversely given $(\phi,\hat{\phi})$ as in Theorem 1.2 satisfying (1.48) it follows that ψ satisfies the SE with V as in (1.49) (note $R = (1/2)log(\hat{\phi}\phi)$ and $S = (1/2)log(\phi/\hat{\phi})$ with $exp(R) = (\hat{\phi}\phi)^{1/2}$). ■

We note that the equations (1.48) are not imaginary time SE and from all this one can conclude that nonrelativistic QM is diffusion theory in terms of Schrödinger processes (described by $(\phi,\hat{\phi})$ - more details later). Further it is shown that certain key postulates in Nelson's stochastic mechanics or Zambrini's Euclidean QM (cf. [**1333**]) can both be avoided in connecting the SE to diffusion processes (since they are automatically valid). Look now at Theorem 1.2 for one dimension and write $T = \hbar t$ with $X = (\hbar/\sqrt{m})x$ and $A = a\hbar/\sqrt{m}$; then the SE becomes

$$i\hbar\psi_T = -(\hbar^2/2m)\psi_{XX} - iA\psi_X + V\psi; \tag{1.50}$$
$$i\hbar R_T + (\hbar^2/m^2)R_X S_X + (\hbar^2/2m^2)S_{XX} + AR_X = 0;$$
$$V = -i\hbar S_T + (\hbar^2/2m)R_{XX} + (\hbar^2/2m^2)R_X^2 - (\hbar^2/2m^2)S_X^2 - AS_X$$

Hence

PROPOSITION 1.1. The SE of Theorem 1.2, written in the variables $X = (\hbar/\sqrt{m})x$, $T = \hbar t$, with $A = (\sqrt{m}/\hbar)a$ and $V = V(X,T) \sim V(x,t)$ is equivalent to (2.2).

Making a change of variables in (1.48) now, as in Proposition 1.1, yields

COROLLARY 1.2. Equation (1.48), written in the variables of Proposition 1.2, becomes

$$\hbar\phi_T + \frac{\hbar^2}{2m}\phi_{XX} + A\phi_X + \tilde{c}\phi = 0; \; -\hbar\hat{\phi}_T + \frac{\hbar^2}{2m}\hat{\phi}_{XX} - A\hat{\phi}_X + \tilde{c}\hat{\phi} = 0; \tag{1.51}$$
$$\tilde{c} = -\tilde{V}(X,T) - 2\hbar S_T - \frac{\hbar^2}{m}S_X^2 - 2AS_X$$

Thus the diffusion processes pick up factors of $\hbar$ and $\hbar/\sqrt{m}$. ∎

REMARK 1.1.10. We extract here from the Appendix to [**908**] for some remarks on competing points of view regarding diffusion and the the SE. First some work of Fenyes [**459**] is cited where a Lagrangian is taken as

$$L(t) = \int \left[\frac{\partial S}{\partial t} + \frac{1}{2}(\nabla S)^2 + V + \frac{1}{2}\left(\frac{1}{2}\frac{\nabla\mu}{\mu}\right)^2\right]\mu dx \tag{1.52}$$

where $\mu_t(x) = exp(2R(t,x))$ denotes the distribution density of a diffusion process and V is a potential function. The term $\Pi(\mu) = (1/2)[(1/2)(\nabla\mu/\mu)]^2$ is called a diffusion pressure and since $(1/2)(\nabla\mu/\mu) \sim \nabla R$ the Lagrangian can be written as

$$L = \int \left[\frac{\partial S}{\partial t} + \frac{1}{2}(\nabla S)^2 + \frac{1}{2}(\nabla R)^2 + V\right]\mu dx \tag{1.53}$$

Applying the variational principle $\delta\int_a^b L(t)dt = 0$ one arrives at

$$\frac{\partial S}{\partial t} + \frac{1}{2}\left[(\nabla(R+S))\right]^2 - (\nabla(R+S))\cdot\left(\frac{1}{2}\frac{\nabla\mu}{\mu}\right) + \left(\frac{1}{2}\frac{\nabla\mu}{\mu}\right)^2 - \frac{1}{4}\frac{\Delta\mu}{\mu} + V = 0 \tag{1.54}$$

which is called a motion equation of probability densities. From this he shows that the function $\psi = exp(R + iS)$ satisfies the SE $i\partial_t + (1/2)\Delta\psi - V(t,x)\psi = 0$. Indeed putting $\Pi(\mu)$ and the formula $(1/2)(\Delta\mu/\mu) + (1/2)\Delta R + (\nabla R)^2$ into (1.53) one obtains

$$\frac{\partial S}{\partial t} + \frac{1}{2}(\nabla S)^2 - \frac{1}{2}(\nabla R)^2 - \frac{1}{2}\Delta R + V = 0 \tag{1.55}$$

which goes along with the duality relation $R_t + (1/2)\Delta S + \nabla S\cdot\nabla R + b\cdot\nabla R = 0$ where $u = (1/2)(a + \hat{a}) = \nabla R$ and $v = (1/2)(a - \hat{a}) = \nabla S$ as derived in the

Nagasawa theory. Hence $\psi = exp(R+iS)$ satisfies the SE by previous calculations. One can see however that the equation (1.53) is not needed since the SE and diffusion equations are equivalent and in fact the equations of motion are the diffusion equations. Moreover it is shown in [**908**] that (1.53) is an automatic consequence in diffusion theory with $V = -c - 2S_t - (\nabla S)^2$ and therefore it need not be postulated or derived by other means. This is a simple calculation from the theory developed above. ■

REMARK 1.1.11. Nelson's important work in stochastic mechanics [**926**] produced the SE from diffusion theory but involved a stochastic Newtonian equation which is shown in [**908**] to be automatically true. Thus Nelson worked in a general context which for our purposes here can be considered in the context of Brownian motions

$$B(t) = \partial_t + (1/2)\Delta + b\cdot\nabla + a\cdot\nabla;\ \hat{B}(t) = -\partial_t + (1/2)\Delta - b\cdot\nabla + \hat{a}\cdot\nabla \tag{1.56}$$

and used a mean acceleration $\alpha(t,x) = -(1/2)[B(t)\hat{B}(t)x + \hat{B}(t)B(t)x]$. Assuming the duality relations after (1.55) he obtains a formula

$$\alpha(t,x) = -\frac{1}{2}[B(t)(-b+\hat{a}) + \hat{B}(b+a)] = b_t + (1/2)\nabla(b)^2 - (b+v)\times curl(b) - \tag{1.57}$$

$$-[-v_t + (1/2)\Delta u + (1/2)(\hat{a}\cdot\nabla)a + (1/2)(a\cdot\nabla)\hat{a} - (b\cdot\nabla)v - (v\cdot\nabla)b - v\times curl(b)]$$

Then it is shown that the SE can be deduced from the stochastic Newton's equation

$$\alpha(t,x) = -\nabla V + \frac{\partial b}{\partial t} + \frac{1}{2}\nabla(b^2) - (b+v)\times curl(b) \tag{1.58}$$

Nagasawa shows that this serves only to reproduce a known formula for V yielding the SE; he also shows that (1.57) also is an automatic consequence of the duality formulation of diffusion equations above. This equation (1.57) is often called stochastic quantization since it leads to the SE and it is in fact correct with the V specified there. However the SE is more properly considered as following directly from the diffusion equations in duality and is not correctly an equation of motion. There is another discussion of Euclidean QM developed by Zambrini [**1333**]. This involves $\tilde{\alpha}(t,x) = (1/2)[B(t)B(t)x + \hat{B}(t)\hat{B}(t)x]$ (with $(\sigma\sigma^T)^{ij} = \delta^{ij}$). It is postulated that this equals $-\nabla c + b_t + (1/2)\nabla(b)^2 - b + v) \times curl(b)$ which in fact leads to the same equation for V as above with $V = -c - 2S_t - (\nabla S)^2 - 2b\cdot\nabla S$ so there is nothing new. Indeed it is shown in [**908**] that the postulated equivalence holds automatically as a simple consequence of time reversal of diffusion processes. ■

2. SCALE RELATIVITY

Scale relativity (SR) is due to L. Nottale (cf. [**941, 942, 943, 944, 945, 946, 947**]) and somehow has not been accorded any real recognition by the "establishment". We only touch here on derivations of the SE and will develop further aspects later; the arguments are evidently heuristic but have a compelling interest. More general relativistic and cosmological features are discussed in Chapter 2 where further discussion is given. The ideas involve spacetime having a fractal microstructure containing in particular continuous (self-similar) nondifferentiable paths which serve as geodesic quantum paths of Hausdorff dimension $D = 2$. This is in fact a good notion of quantum path (following Feynman for example - cf. [**1**]) and we will see how it leads to a lovely (heuristic) derivation of the SE which automatically creates a complex wave function.

REMARK 1.2.1. One considers quantum paths à la Feynman so that (**E1**) $lim_{t\to t'}[X(t)-X(t')]^2/(t-t')$ exists. This implies $X(t)\in H^{1/2}$ where H^α means $c\epsilon^\alpha \le |X(t)-X(t')| \le C\epsilon^\alpha$ and from [**444**] for example this means $dim_H X[a,b]=1/2$. Now one "knows" (see e.g. [**1**]) that quantum and Brownian motion paths (in the plane) have H-dimension 2 and some clarification is needed here. We refer to [**849**] where there is a paper on Wiener Brownian motion (WBM), random walks, etc. discussing Hausdorff and other dimensions of various sets. Thus given $0<\lambda<1/2$ with probability 1 a Browian sample function X satisfies $|X(t+h)-X(t)|\le b|h|^\lambda$ for $|h|\le h_0$ where $b=b(\lambda)$. This leads to the result that with probability 1 the graph of a Brownian sample function has Hausdorff and box dimension 3/2. On the other hand a Browian trail (or path) in 2 dimensions has Hausdorff and box dimension 2 (note a quantum path can have self intersections, etc.). ■

There are now several excellent approaches. The method of Nottale [**929, 941, 942, 943, 944, 945, 946, 947**] is preeminent (cf. also Ord et al [**964, 965, 966, 967, 968**]) and there is also a nice derivation of a nonlinear SE via fractal considerations in [**275**] (indicated below). The most elaborate and rigorous approach is due to Cresson [**336, 337, 338**] (cf. also [**3, 254**]. There are various derivations of the SE and we follow [**941**] here (cf. also [**1098**]). The philosophy of scale relativity will be discussed later and we just write down equations here pertaining to the SE. First a bivelocity structure is defined (recall that one is dealing with fractal paths). One defines first

$$\frac{d_+}{dt}y(t)=lim_{\Delta t\to 0_+}\left\langle\frac{y(t+\Delta t)-y(t)}{\Delta t}\right\rangle; \tag{2.1}$$

$$\frac{d_-}{dt}y(t)=lim_{\Delta t\to 0_+}\left\langle\frac{y(t)-y(t-\Delta t)}{\Delta t}\right\rangle$$

Applied to the position vector x this yields forward and backward mean velocities, namely $(d_+/dt)x(t)=b_+$ and $(d_-/dt)x(t)=b_-$. Here these velocities are defined as the average at a point q and time t of the respective velocities of the outgoing and incoming fractal trajectories; in stochastic QM this corresponds to an average on the quantum state. The position vector $x(t)$ is thus "assimilated" to a stochastic process which satisfies respectively after $(dt>0)$ and before $(dt<0)$ the instant t a relation $dx(t)=b_+[x(t)]dt+d\xi_+(t)=b_-[x(t)]dt+d\xi_-(t)$ where $\xi(t)$ is a Wiener process (cf. [**926**]). It is in the description of ξ that the $D=2$ fractal character of trajectories is inserted; indeed that ξ is a Wiener process means that the $d\xi$'s are assumed to be Gaussian with mean 0, mutually independent, and such that

$$<d\xi_{+i}(t)d\xi_{+j}(t)>=2\mathcal{D}\delta_{ij}dt;\ <d\xi_{-i}(t)d\xi_{-j}(t)>=-2\mathcal{D}\delta_{ij}dt \tag{2.2}$$

where $<>$ denotes averaging ($\mathcal{D}$ is now the diffusion coefficient). Nelson's postulate (cf. [**926**]) is that $\mathcal{D}=\hbar/2m$ and this has considerable justification (cf.

[**941**]). Note also that (2.2) is indeed a consequence of fractal (Hausdorff) dimension 2 of trajectories follows from $< d\xi^2 > /dt^2 = dt^{-1}$, i.e. precisely Feynman's result $< v^2 >^{1/2} \sim \delta t^{-1/2}$ (the discussion here in [**941**] is unclear however - cf. [**34, 35**]). Note also that Brownian motion (used in Nelson's postulate) is known to be of fractal (Hausdorff) dimension 2. Note also that any value of $\mathcal{D}$ may lead to QM and for $\mathcal{D} \to 0$ the theory becomes equivalent to the Bohm theory. Now expand any function $f(x,t)$ in a Taylor series up to order 2, take averages, and use properties of the Wiener process ξ to get

$$\frac{d_+ f}{dt} = (\partial_t + b_+ \cdot \nabla + \mathcal{D}\Delta)f; \quad \frac{d_- f}{dt} = (\partial_t + b_- \cdot \nabla - \mathcal{D}\Delta)f \tag{2.3}$$

Let $\rho(x,t)$ be the probability density of $x(t)$; it is known that for any Markov (hence Wiener) process one has $\partial_t \rho + div(\rho b_+) = \mathcal{D}\Delta\rho$ (forward equation) and $\partial_t \rho + div(\rho b_-) = -\mathcal{D}\Delta\rho$ (backward equation). These are called Fokker-Planck equations and one defines two new average velocities $V = (1/2)[b_+ + b_-]$ and $U = (1/2)[b_+ - b_-]$. Consequently adding and subtracting one obtains $\rho_t + div(\rho V) = 0$ (continuity equation) and $div(\rho U) - \mathcal{D}\Delta\rho = 0$ which is equivalent to $div[\rho(U - \mathcal{D}\nabla log(\rho))] = 0$. One can show, using (2.3) that the term in square brackets in the last equation is zero leading to $U = \mathcal{D}\nabla log(\rho)$. Now place oneself in the (U,V) plane and write $\mathcal{V} = V - iU$. Then write $(d_{\mathcal{V}}/dt) = (1/2)(d_+ + d_-)/dt$ and $(d_{\mathcal{U}}/dt) = (1/2)(d_+ - d_-)/dt$. Combining the equations in (2.3) one defines $(d_{\mathcal{V}}/dt) = \partial_t + V \cdot \nabla$ and $(d_{\mathcal{U}}/dt) = \mathcal{D}\Delta + U \cdot \nabla$; then define a complex operator $(d'/dt) = (d_{\mathcal{V}}/dt) - i(d_{\mathcal{U}}/dt)$ which becomes

$$\frac{d'}{dt} = \left(\frac{\partial}{\partial t} - i\mathcal{D}\Delta\right) + \mathcal{V} \cdot \nabla \tag{2.4}$$

One now postulates that the passage from classical mechanics to a new nondifferentiable process considered here can be implemented by the unique prescription of replacing the standard d/dt by d'/dt. Thus consider $\mathfrak{S} = \left\langle \int_{t_1}^{t_2} \mathcal{L}(x, \mathcal{V}, t)dt \right\rangle$ yielding by least action $(d'/dt)(\partial\mathcal{L}/\partial\mathcal{V}_i) = \partial\mathcal{L}/\partial x_i$. Define then $\mathcal{P}_i = \partial\mathcal{L}/\partial\mathcal{V}_i$ leading to $\mathcal{P} = \nabla\mathfrak{S}$ (recall the classical action principle with $dS = pdq - Hdt$). Now for Newtonian mechanics write $L(x,v,t) = (1/2)mv^2 - \mathbf{U}$ which becomes $\mathcal{L}(x,\mathcal{V},t) = (1/2)m\mathcal{V}^2 - \mathfrak{U}$ leading to $-\nabla\mathfrak{U} = m(d'/dt)\mathcal{V}$. One separates real and imaginary parts of the complex acceleration $\gamma = (d'\mathcal{V}/dt$ to get

$$d'\mathcal{V} = (d_{\mathcal{V}} - id_{\mathcal{U}})(V - iU) = (d_{\mathcal{V}}V - d_{\mathcal{U}}U) - i(d_{\mathcal{U}}V + d_{\mathcal{V}}U) \tag{2.5}$$

The force $F = -\nabla\mathfrak{U}$ is real so the imaginary part of the complex acceleration vanishes; hence

$$\frac{d_{\mathcal{U}}}{dt}V + \frac{d_{\mathcal{V}}}{dt}U = \frac{\partial U}{\partial t} + U \cdot \nabla V + V \cdot \nabla U + \mathcal{D}\Delta V = 0 \tag{2.6}$$

from which $\partial U/\partial t$ may be obtained. This is a weak point in the derivation since one has to assume e.g. that $U(x,t)$ has certain smoothness properties (see below for refinements). Differentiating the expression $U = \mathcal{D}\nabla log(\rho)$ and using the continuity equation yields another expression $(\partial U/\partial t) = -\mathcal{D}\nabla(divV) - \nabla(V \cdot U)$. Comparison of these relations yields $\nabla(divV) = \Delta V - U \wedge curlV$ where the $curlU$ term vanishes since U is a gradient. However in the Newtonian case $\mathcal{P} = m\mathcal{V}$

so $\mathcal{PV}\mathfrak{S}$ implies that $\mathcal{V}$ is a gradient and hence a generalization of the classical action S can be defined. Recall $V = 2\mathcal{D}\nabla S$ and $\nabla(divV) = \Delta V$ with $curlV = 0$; combining this with the expression for U one obtains $\mathfrak{S} = log(\rho^{1/2}) + iS$. One notes that this is compatible with [**926**] for example. Finally set $\psi = \sqrt{\rho}exp(iS) = exp(i\mathfrak{S})$ with $\mathcal{V} = -2i\mathcal{D}\nabla(log\psi)$ and note

$$U = \mathcal{D}\nabla log(\rho);\ V = 2\mathcal{D}\nabla S; \tag{2.7}$$

$$\mathcal{V} = -2i\mathcal{D}\nabla log\psi = -i\mathcal{D}\nabla log(\rho) + 2\mathcal{D}\nabla S = V - iU$$

Thus for $\mathcal{P} = m\mathcal{V}$ the relation $\mathcal{P} \sim -i\hbar\nabla$ or $\mathcal{P}\psi = -i\hbar\nabla\psi$ has a natural interpretation. Putting ψ in the equation $-\nabla\mathfrak{U} = m(d'/dt)\mathcal{V}$, which generalizes Newton's law to fractal space the equation of motion takes the form $\nabla\mathfrak{U} = 2i\mathcal{D}m(d'/dt)(\nabla log(\psi))$. Then noting that d' and ∇ do not commute one replaces d'/dt by (2.4) to obtain

$$\nabla\mathfrak{U} = 2i\mathcal{D}m\left[\partial_t\nabla log(\psi) - i\mathcal{D}\Delta(\nabla log(\psi)) - 2i\mathcal{D}(\nabla log(\psi)\cdot\nabla)(\nabla log(\psi)\right] \tag{2.8}$$

This expression can be simplified via

$$\nabla\Delta = \Delta\nabla;\ (\nabla f\cdot\nabla)(\nabla f) = (1/2)\nabla(\nabla f)^2;\ \frac{\Delta f}{f} = \Delta log(f) + (\nabla log(f))^2 \tag{2.9}$$

which implies

$$\frac{1}{2}\Delta(\nabla log(\psi)) + (\nabla log(\psi)\cdot\nabla)(\nabla log(\psi)) = \frac{1}{2}\nabla\frac{\Delta\psi}{\psi} \tag{2.10}$$

Integrating this equation yields $\mathcal{D}^2\Delta\psi + i\mathcal{D}\partial_t\psi - (\mathfrak{U}/2m)\psi = 0$ up to an arbitrary phase factor $\alpha(t)$ which can be set equal to 0 by a suitable choice of phase S. Replacing $\mathcal{D}$ by $\hbar/2m$ one arrives at the SE $i\hbar\psi_t = -(\hbar^2/2m)\Delta\psi + \mathfrak{U}\psi$ and this suggests an interpretation of QM as mechanics in a nondifferentiable (fractal) space.

In fact (using one space dimension for convenience) we see that if $\mathfrak{U} = 0$ then the free motion $m(d'/dt)\mathcal{V} = 0$ yields the SE $i\hbar\psi_t = -(\hbar^2/2m)\psi_{xx}$ as a geodesic equation in "fractal" space. Further from $U = (\hbar/m)(\partial\sqrt{\rho}/\sqrt{\rho})$ and $Q = -(\hbar^2/2m)(\Delta\sqrt{\rho}/\sqrt{\rho})$ one arrives at a lovely relation, namely

PROPOSITION 2.1. The quantum potential Q can be written in the form $Q = -(m/2)U^2 - (\hbar/2)\partial U$ (cf. (1.40) multiplied by $-m$). Hence the quantum potential arises directly from the fractal nonsmooth nature of the quantum paths. Since Q can be thought of as a quantization of a classical motion we see that the quantization corresponds exactly to the existence of nonsmooth paths. Consequently smooth paths imply no quantum mechanics.

REMARK 1.2.2. In Agop et al [**17**] (to be discussed later) one writes again $\psi = Rexp(iS/\hbar)$ with field equations in the hydrodynamical picture (1-D for convenience)

$$d_t(m_0\rho v) = \partial_t(m_0\rho v) + \nabla(m_0\rho v) = -\rho\nabla(u+Q);\ \partial_t\rho + \nabla\cdot(\rho v) = 0 \tag{2.11}$$

where $Q = -(\hbar^2/2m_0)(\Delta\sqrt{\rho}/\sqrt{\rho})$. The Nottale approach is used as above with $d_v \sim d_{\mathcal{V}}$ and $d_u \sim d_{\mathcal{U}}$. One assumes that the velocity field from the hydrodynamical model agrees with the real part v of the complex velocity $V = v - iu$ so

$v = (1/m_0)\nabla s \sim 2\mathcal{D}\partial s$ and $u = -(1/m_0)\nabla\sigma \sim \mathcal{D}\partial log(\rho)$ where $\mathcal{D} = \hbar/2m_0$. In this context the quantum potential $Q = -(\hbar^2/2m_0)\Delta\mathcal{D}\sqrt{\rho}/\sqrt{\rho}$ becomes

$$Q = -m_0\mathcal{D}\nabla\cdot u - (1/2)m_0u^2 \sim -(\hbar/2)\partial u - (1/2)m_0u^2 \tag{2.12}$$

Consequently Q arises from the fractal derivative and the nondifferentiability of spacetime again, as in Proposition 2.1. Further one can relate u (and hence Q) to an internal stress tensor whereas the v equations correspond to systems of Navier-Stokes type.

REMARK 1.2.3. Some of the relevant equations for dimension one are collected together later. We note that it is the presence of $\pm$ derivatives that makes possible the introduction of a complex plane to describe velocities and hence QM; one can think of this as the motivation for a complex valued wave function and the nature of the SE. ∎

We go now to Castro et al [**275**] and will sketch some of the material. Here one extends ideas of Nottale and Ord in order to derive a nonlinear Schrödinger equation (NLSE) (cf. Chapter 7 for more on NLSE). Using the hydrodynamic model in [**987**] one added a hydrostatic pressure term to the Euler-Lagrange equations and another possibility is to add instead a kinematic pressure term. The hydrostatic pressure is based on an Euler equation $-\nabla p = \rho g$ where ρ is density and g the gravitational acceleration (note this gives $p = \rho g x$ in 1-D). In [**987**] one took $\rho = \psi^*\psi$, b a mass-energy parameter, and $p = \rho$; then the hydrostatic potential is (for $\rho_0 = 1$)

$$b\int g(x)\cdot dr = -b\int\frac{\nabla p}{\rho}\cdot dr = -blog(\rho/\rho_0) = -blog(\psi^*\psi) \tag{2.13}$$

Here $-blog(\psi^*\psi)$ has energy units and explains the nonlinear term of [**139**] which involved

$$i\hbar\frac{\partial\psi}{\partial t} = -\frac{\hbar^2}{2m}\nabla^2\psi + U\psi - b[log(\psi^*\psi)]\psi \tag{2.14}$$

A derivation of this equation from the Nelson stochastic QM was given by Lemos (cf. [**811**]). There are however some problems since this equation does not obey the homogeneity condition saying that the state $\lambda|\psi>$ is equivalent to $|\psi>$; however (2.14) is not invariant under $\psi \to \lambda\psi$. Further, plane wave solutions to (2.14) do not seem to have a physical interpretion due to extraneous dispersion relations. Finally one would like to have a SE in terms of ψ alone. Note that another NLSE could be obtained by adding kinetic pressure terms $(1/2)\rho v^2$ and taking $\rho = a\psi^*\psi$ where $v = p/m$. Now using the relations from HJ theory $(\psi/\psi^*) = exp[2i\mathfrak{S}(x)/\hbar]$ and $p = \nabla\mathfrak{S}(x) = mv$ one can write $v = -i(\hbar/2m)\nabla log(\psi/\psi^*)$ so that the energy density becomes

$$(1/2)\rho|v|^2 = (a\hbar^2/8m^2)\psi\psi^*\nabla log(\psi/\psi^*)\cdot\nabla log(\psi^*/\psi) \tag{2.15}$$

This leads to a corresponding nonlinear potential associated with the kinematical pressure via $(a\hbar^2/8m^2)\nabla log(\psi/\psi^*)\cdot\nabla log(\psi^*/\psi)$. Hence a candidate NLSE is

$$i\hbar\partial_t = -\frac{\hbar^2}{2m}\nabla^2\psi + U\psi - b[log(\psi^*\psi)]\psi + \frac{a\hbar^2}{8m^2}\left(\nabla log\frac{\psi}{\psi^*}\cdot\nabla log\frac{\psi^*}{\psi}\right) \tag{2.16}$$

Here the Hamiltonian is Hermitian and $a \neq b$ are both mass-energy parameters to be determined experimentally. The new term can also be written in the form $\nabla log(\psi/\psi^*) \cdot \nabla log(\psi^*/\psi) = -[\nabla log(\psi/\psi^*)]^2$. The goal now is to derive a NLSE directly from fractal space time dynamics for a particle undergoing Brownian motion. This does not require a quantum potential, a hydrodynamic model, or any pressure terms as above.

REMARK 1.2.4. One should make some comments about the kinematic pressure terms $(1/2)\rho v^2 \sim (\hbar^2/2m)(a/m)|\nabla log(\psi)|^2$ versus hydrostatic pressure terms of the form $\int(\nabla p/\rho) \sim -b log(\psi^*\psi)$. The hydrostatic term breaks homogeneity whereas the kinematic pressure term preserves homogeneity (scaling with a λ factor). The hydrostatic pressure term is also not compatible with the motion kinematics of a particle executing a fractal Brownian motion. The fractal formulation will enable one to relate the parameters a, b to $\hbar$. ■

Following Nottale, nondifferentiability implies a loss of causality and one is thinking of Feynmann paths with $< v^2 > \propto (dx/dt)^2 \propto dt^{2[(1/D)-1)}$ with $D = 2$. Now a fractal function $f(x,\epsilon)$ could have a derivative $\partial f/\partial\epsilon$ and renormalization group arguments lead to $(\partial f(x,\epsilon)/\partial log\epsilon) = a(x) + bf(x,\epsilon)$ (cf. [**941**]). This can be integrated to give $f(x,\epsilon) = f_0(x)[1 - \zeta(x)(\lambda/\epsilon)^{-b}]$. Here $\lambda^{-b}\zeta(x)$ is an integration constant and $f_0(x) = -a(x)/b$. This says that any fractal function can be approximated by the sum of two terms, one independent of the resolution and the other resolution dependent; $\zeta(x)$ is expected to be a flucuating function with zero mean. Provided $a \neq 0$ and $b < 0$ one has two interesting cases (i) $\epsilon << \lambda$ with $f(x,\epsilon) \sim f_0(x)(\lambda/\epsilon)^{-b}$ and (ii) $\epsilon >> \lambda$ with f independent of scale. Here λ is the deBroglie wavelength. Now one writes

$$r(t+dt,dt) - r(t,dt) = b_+(r,t)dt + \xi_+(t,dt)\left(\frac{dt}{\tau_0}\right)^\beta; \tag{2.17}$$

$$r(t,dt) - r(t-dt,dt) - b_-(r,t)dt + \xi_-(t,dt)\left(\frac{dt}{\tau_0}\right)^\beta$$

where $\beta = 1/D$ and $b_\pm$ are average forward and backward velocities. This leads to $v_\pm(r,t,dt) = b_\pm(r,t) + \xi_\pm(t,dt)(dt/\tau_0)^{\beta-1}$. In the quantum case $D = 2$ one has $\beta = 1/2$ so $dt^{\beta-1}$ is a divergent quantity (i.e. nondifferentiability ensues). Following [**791, 941, 926**] one defines

$$\frac{d_\pm r(t)}{dt} = lim_{\Delta t \to \pm 0}\left\langle \frac{r(t+\Delta t) - r(t)}{\Delta t}\right\rangle \tag{2.18}$$

from which $d_\pm r(t)/dt = b_\pm$. Now following Nottale one writes

$$\frac{\delta}{dt} = \frac{1}{2}\left(\frac{d_+}{dt} + \frac{d_-}{dt}\right) - \frac{i}{2}\left(\frac{d_+}{dt} - \frac{d_-}{dt}\right) \tag{2.19}$$

which leads to $(\delta/dt) = (\partial/\partial t) + v\cdot\nabla - iD\nabla^2$. Here in principle $\mathcal{D}$ is a real valued diffusion constant to be related to $\hbar$. and $< d\xi_{\pm i} d\xi_{\pm j} >= \pm 2\mathcal{D}\delta_{ij}dt$. Now for the complex time dependent wave function we take $\psi = exp[i\mathfrak{S}/2m\mathcal{D}]$ with $p = \nabla\mathfrak{S}$ so that $v = -2iD\nabla log(\psi)$. The SE is obtained from the Newton equation ($F = ma$) via $-\nabla U = m(\delta/dt)v = -2im\mathcal{D}(\delta/dt)\nabla log(\psi)$ which yields

$$-\nabla U = -2im[\mathcal{D}\partial_t \nabla log(\psi)] - 2\mathcal{D}\nabla\left(\mathcal{D}\frac{\nabla^2\psi}{\psi}\right) \tag{2.20}$$

(see [**941**] for identities involving ∇). Integrating yields $\mathcal{D}^2\nabla^2\psi + i\mathcal{D}\partial_t\psi - (U/2m)\psi = 0$ up to an arbitrary phase factor which may be set equal to zero. Now replacing D by $\hbar/2m$ one gets the SE $i\hbar\partial_t\psi + (\hbar^2/2m)\nabla^2\psi = U\psi$. Here the Hamiltonian is Hermitian, the equation is linear, and the equation is homogeneous of degree 1 under the substitution $\psi \to \lambda\psi$.

Next one generalizes this by relaxing the assumption that the diffusion coefficient is real. Some comments on complex energies are needed - in particular constraints are often needed (cf. [**1043**]). However complex energies are not alien in ordinary QM (cf. [**275**] for references). Now the imaginary part of the linear SE yields the continuity equation $\partial_t\rho + \nabla\cdot(\rho v) = 0$ and with a complex potential the imaginary part of the potential will act as a source term in the continuity equation. Instead of $< d\zeta_\pm d\zeta_\pm >= \pm 2\mathcal{D}dt$ with $\mathcal{D}$ and $2mD = \hbar$ real one sets

$$< d\zeta_\pm d\zeta_\pm >= \pm(\mathcal{D}+\mathcal{D}^*)dt;\ 2m\mathcal{D} = \hbar = \alpha + i\beta \tag{2.21}$$

The complex time derivative operator becomes $(\delta/dt) = \partial_t + v\cdot\nabla - (i/2)(\mathcal{D}+\mathcal{D}^*)\nabla^2$. Writing again $\psi = exp[i\mathfrak{S}/2mD] = exp(i\mathfrak{S}/\hbar)$ one obtains $v = -2i\mathcal{D}\nabla log(\psi)$. The NLSE is then obtained (via the Newton law) via the relation $-\nabla U = m(\delta/dt)v = -2im\mathcal{D}(\delta/dt)\nabla log(\psi)$. Combining equations yields then

$$\nabla U = 2im[\mathcal{D}\partial_t\nabla log(\psi) - 2i\mathcal{D}^2(\nabla log(\psi)\cdot\nabla)(\nabla log(\psi) - \\ -\frac{i}{2}(\mathcal{D}+\mathcal{D}^*)\mathcal{D}\nabla^2(\nabla log(\psi)] \tag{2.22}$$

Now using the identities (i) $\nabla\nabla^2 = \nabla^2\nabla$, (ii) $2(\nabla log(\psi)\cdot\nabla)(\nabla log(\psi) = \nabla(\nabla log(\psi))^2$ and (iii) $\nabla^2 log(\psi) = \nabla^2\psi/\psi - (\nabla log(\psi))^2$ leads to a NLSE with nonlinear (kinematic pressure) potential, namely

$$i\hbar\partial_t\psi = -\frac{\hbar^2}{2m}\frac{\alpha}{\hbar}\nabla^2\psi + U\psi - i\frac{\hbar^2}{2m}\frac{\beta}{\hbar}(\nabla log(\psi))^2\psi \tag{2.23}$$

Note the crucial minus sign in front of the kinematic pressure term and also that $\hbar = \alpha + i\beta = 2m\mathcal{D}$ is complex. When $\beta = 0$ one recovers the linear SE. The nonlinear potential is complex and one defines $W = -(\hbar^2/2m)(\beta/\hbar)(\nabla log(\psi))^2$ with U the ordinary potential; then the NLSE is

$$i\hbar\partial_t\psi = [-(\hbar^2/2m)(\alpha/\hbar)\nabla^2 + U + iW]\psi \tag{2.24}$$

This is the fundamental result of [**275**]; it has the form of an ordinary SE with complex potential $U+iW$ and complex $\hbar$. The Hamiltonian is no longer Hermitian and the potential itself depends on ψ. Nevertheless one can have meaningful physical solutions with real valued energies and momenta; the homogeneity breaking hydrostatic pressure term $-b(log(\psi^*\psi)\psi$ is not present (it would be meaningless) and the NLSE is invariant under $\psi \to \lambda\psi$.

REMARK 1.2.5. One could ask why not simply propose as a valid NLSE an equation

$$i\hbar\partial_t\psi = -\frac{\hbar^2}{2m}\nabla^2\psi + \frac{\hbar^2}{2m}\frac{a}{m}|\nabla log(\psi)|^2\psi$$

Here one has a real Hamiltonian satisfying the homogeneity condition and the equation admits soliton solutions of the form $\psi = CA(x-vt)exp[i(kx-\omega t)]$ where $A(x-vt)$ is to be determined by solving the NLSE. The problem here is that the equation suffers from an extraneous dispersion relation. Thus putting in the plane wave solution $\psi \sim exp[-i(Et-px)]$ one gets an extraneous energy momentum (EM) relation (after setting $U=0$), namely $E = (p^2/2m)[1+(a/m)]$ instead of the usual $E = p^2/2m$ and hence $E_{QM} \neq E_{FT}$ where FT means field theory. ■

REMARK 1.2.6. It has been known since e.g. Puszkarz [**1043**] that the expression for the energy functional in nonlinear QM does not coincide with the QM energy functional, nor is it unique. To see this write down the NLSE of [**139**] in the form $i\hbar\partial_t\psi = \partial H(\psi,\psi^*)/\partial\psi^*$ where the real Hamiltonian density is

$$H(\psi,\psi^*) = -\frac{\hbar^2}{2m}\psi^*\nabla^2\psi + U\psi^*\psi - b\psi^* log(\psi^*\psi)\psi + b\psi^*\psi$$

Then using $E_{FT} = \int Hd^3r$ we see it is different from $<\hat{H}>_{QM}$ and in fact $E_{FT} - E_{QM} = \int b\psi^*\psi d^3r = b$. This problem does not occur in the fractal based NLSE since it is written entirely in terms of ψ. ■

REMARK 1.2.7. In the fractal based NLSE there is no discrepancy between the QM energy functional and the FT energy functional. Both are given by

$$N_{fractal}^{NLSE} = -\frac{\hbar^2}{2m}\frac{\alpha}{\hbar}\psi^*\nabla^2\psi + U\psi^*\psi - i\frac{\hbar^2}{2m}\frac{\beta}{\hbar}\psi^*(\nabla log(\psi)^2\psi$$

The NLSE is unambiguously given by in Remark 1.2.5 and $H(\psi,\psi^*)$ is homogeneous of degree 1 in λ. Such equations admit plane wave solutions with dispersion relation $E = p^2/2m$; indeed, inserting the plane wave solution into the fractal based NLSE one gets (after setting $U=0$)

$$E = \frac{\hbar^2}{2m}\frac{\alpha}{\hbar}\frac{p^2}{2m} + i\frac{\beta}{\hbar}\frac{p^2}{2m} = \frac{p^2}{2m}\frac{\alpha+i\beta}{\hbar} = \frac{p^2}{2m} \tag{2.25}$$

since $\hbar = \alpha + i\beta$. The remarkable feature of the fractal approach versus all other NLSE considered sofar is that the QM energy functional is precisely the FT one. The complex diffusion constant represents a truly new physical phenomenon insofar as a small imaginary correction to the Planck constant is the hallmark of nonlinearity in QM (see [**275**] for more on this). ■

REMARK 1.2.8. Some refinements of the Nottale derivation are given in [**336**] and we consider $x \to f(x(t),t) \in C^n$ with $X(t) \in H^{1/n}$ (i.e. $c\epsilon^{1/n} \leq |X(t') - X(t)| \leq C\epsilon^{1/n}$). Define ($f$ real valued)

$$\nabla_\pm^\epsilon f(t) = \frac{f(t\pm\epsilon)-f(t)}{\pm\epsilon};\quad \frac{\Box_\epsilon f}{\Box t}(f) = \frac{1}{2}(\nabla_+^\epsilon + \nabla_-^\epsilon)f - \frac{i}{2}(\nabla_+^\epsilon - \nabla_-^\epsilon f); \tag{2.26}$$

$$a_{\epsilon,j}(t)=\frac{1}{2}[(\Delta_+^\epsilon x)^j-(-1)^j(\Delta_-^\epsilon x)^j]-\frac{i}{2}[(\Delta_+^\epsilon x)^j+(-1)^j(\Delta_-^\epsilon x)^j]$$

Here one assumes $h>0$ and $\epsilon(f,h)\geq\epsilon>0$ where $\epsilon(f,h)$ is the minimal resolution defined via $inf_\epsilon\{a_\epsilon(f)<h\}$ for $a_\epsilon f(t)=|[f(t+\epsilon)+f(t-\epsilon)-2f(t)]/\epsilon|$. If $\epsilon(f,h)$ is not 0 then f is not differentiable (but not conversely). Now assume some minimal control over the lack of differentiability (cf. [**336**]) and then for f now complex valued with $\Box_\epsilon f/\Box t=(\Box_\epsilon f_{\Re}/\Box t)+i(\Box_\epsilon f_{\Im}/\Box t)$ (note the mixing of i terms is not trivial) one has

$$\frac{\Box_\epsilon f}{\Box t}=\frac{\partial f}{\partial t}+\frac{\Box_\epsilon x}{\Box t}\frac{\partial f}{\partial x}+\sum_2^n\frac{1}{j!}a_{\epsilon,j}(t)\frac{\partial^j f}{\partial x^j}\epsilon^{j-1}+o(\epsilon^{1/n}) \tag{2.27}$$

We sketch now the derivation of a SE in the spirit of Nottale but with more mathematical polish. Going to Cresson [**336**] one defines (for a nondifferentiable function f)

$$f_\epsilon(t)=\frac{1}{2\epsilon}\int_{t-\epsilon}^{t+\epsilon}f(s)ds; \tag{2.28}$$

$$f_\epsilon^+(t)=\frac{1}{2\epsilon}\int_t^{t+\epsilon}f(s)ds;\ f_\epsilon^-(t)=\frac{1}{2\epsilon}\int_{t-\epsilon}^t f(s)ds$$

One considers quantum paths à la Feynman so that $lim_{t\to t'}[X(t)-X(t')]^2/(t-t')$ exists. This implies $X(t)\in H^{1/2}$ where H^α means $c\epsilon^\alpha\leq|X(t)-X(t')|\leq C\epsilon^\alpha$ and from Remark 1.2.1 for example this means $dim_H X[a,b]=1/2$. Next, thinking of classical Lagrangians $L(x,v,t)=(1/2)mv^2+\mathbf{U}(x,t)$, one defines an operator Q via ($(x,t,v)\sim$ classical variables)

$$Q(t)=t;\ Q(x(t))=X(t);\ Q(v(t))=\mathcal{V}(t);\ Q\left(\frac{df}{dt}\right)=Q\left(\frac{d}{dt}\right)\cdot Q(f) \tag{2.29}$$

where $Q(d/dt)=d/dt$ if $Q(f)(t)$ is differentiable and $Q(d/dt)=\Box_\epsilon/\Box t$ where $\epsilon(x,h)>\epsilon>0$ if $Q(f)(t)$ is nondifferentiable. Note $\mathcal{V}(t)=Q(d/dt)[X(t)]$ so regularity of X determines the form of Q here and for $Q(x)=X\in H^{1/2}$ one has $\mathcal{V}=\Box_\epsilon X/\Box t$. The scalar Euler-Lagrange (EL) equation associated to $\mathcal{L}(X(t),\mathcal{V}(t),t)=Q(L(x(t),v(t),t)$ is

$$\frac{\Box_\epsilon}{\Box t}\left(\frac{\partial\mathcal{L}}{\partial\mathcal{V}}(X(t),\mathcal{V}(t),t)\right)=\frac{\partial\mathcal{L}}{\partial X}(X(t),\mathcal{V}(t),t) \tag{2.30}$$

Now given a classical $v\sim(1/m)(\partial S/\partial x)$ one gets $\mathcal{V}=(1/m)(\partial\mathfrak{S}/\partial X)$ and $\mathcal{L}=(\partial\mathfrak{S}/\partial t)$ with $\psi(X,t)=exp[i\mathfrak{S}(X,t)/2m\gamma]$. For $\mathcal{L}\sim(1/2)m\mathcal{V}^2+\mathfrak{U}$ then the quantum (EL) equation is $m(\Box_\epsilon\mathcal{V}/\Box t)=(\partial\mathfrak{U}/\partial X)$ leading to
(2.31)

$$2i\gamma m\left[-\frac{\psi_X^2}{\psi}\left(i\gamma+\frac{a_\epsilon(t)}{2}\right)+\partial_t\psi+\frac{a_\epsilon(t)}{2}\frac{\partial^2\psi}{\partial X^2}\right]=(\mathfrak{U}(X)+\alpha(X))\psi+o(\epsilon^{1/2})$$

where

$$a_\epsilon(t)=\frac{1}{2}\{[(\nabla_+^\epsilon X(t))^2-(\nabla_-^\epsilon X(t))^2]-i[(\nabla_+^\epsilon X(t))^2+(\nabla_-^\epsilon X(t))^2]\} \tag{2.32}$$

Then (2.32) is called the generalized SE and the nonlinear character of such equations is discussed in [**239, 275**] for example. In [**336**] one then arrives at a conventional looking SE under the assumption $a_\epsilon = -2i\gamma$, leading to

$$\gamma^2 \frac{\partial^2 \psi}{\partial X^2} + i\gamma \frac{\partial \psi}{\partial t} = [\mathfrak{U}(X,t) + \alpha(X)]\frac{\psi}{2m} + o(\epsilon^{1/2}) \tag{2.33}$$

One can then always take $\alpha(X) = 0$ and choosing $\gamma = \hbar/2m$ one arrives at $i\hbar\psi_t + (\hbar^2/2m)(\partial^2\psi/\partial t^2) = \mathfrak{U}\psi$. However the requirement $a_\epsilon(t) = -2i\gamma$ seems quite restrictive.

- Note here that the argument using $a_\pm$ is rigorous via [**336**]. $a_\epsilon = -i\hbar/m$ is permissible and in fact can have solutions of $\nabla_\sigma^\epsilon X(t) = constant$ via $X_c(t) = \pm\sqrt{\hbar/2m)}(t - c - (\epsilon/2)) + P_\epsilon(t)$ where $P_\epsilon \in H^{1/2}$ is an arbitrary periodic function.

Referring back to Example 1.2.3 we have $b_\pm(t)(t) \sim \Box_\pm x(t)$ and $V = (/2)(\Box_+ x + \Box_- x)(t)$ with $U = (1/2)(\Box_+ x - \Box_- x)(t)$. The relation between U and the quantum potential Q will formally still hold (cf. also [**337, 338**] on nondifferentiable variational principles) and one can rewrite this as $\sqrt{\rho}U = (\hbar/m)\partial\sqrt{\rho}$; $\sqrt{\rho}Q = -(\hbar^2/2m)\partial^2\sqrt{\rho}$ along with $\partial(\sqrt{\rho}U) = -(2/\hbar)\sqrt{\rho}Q$. If U is not differentiable one could also look at $\sqrt{\rho}U = -(2/\hbar)\int_0^X \sqrt{\rho}Q dX' + f(t)$ with $f(t)$ possibly determinable via the term $(\sqrt{\rho}U)(0,t)$. ■

3. REMARKS ON FRACTAL SPACETIME

There have been a number of articles and books involving fractal methods in spacetime or fractal spacetime itself with impetus coming from quantum physics and relativity. We refer here especially to [**1, 234, 235, 254, 272, 273, 278, 556, 912, 913, 914**] for background to this paper. Many related papers are omitted here and we refer in particular to the journal Chaos, Solitons, and Fractals CSF) for further information. For information on fractals and stochastic processes we refer for example to [**40, 104, 304, 305, 306, 444, 557, 756, 764, 799, 849, 850, 867, 925, 953, 994, 1016, 1069, 1209, 1239, 1293**]. We discuss here a few background ideas and constructions in order to indicate the ingredients for El Naschie's Cantorian spacetime $\mathfrak{E}^\infty$, whose exact nature is elusive. Suitable references are given but there are many more papers in the journal CSF by El Naschie (and others) based on these fundamental ideas and these are either important in a revolutionary sense or a fascinating refined form of science fiction. In what appears at times to be pure numerology one manages to (rather hastily) produce amazingly close numerical approximations to virtually all the fundamental constants of physics (including string theory). The key concepts revolve around the famous golden ratio $(\sqrt{5} - 1)/2$ and a strange Cantorian space $\mathfrak{E}^\infty$ which we try to describe below. It is very tempting to want all of these (heuristic) results to be true and the approach seems close enough and universal enough to compel one to think something very important must be involved. Moreover such scope and accuracy cannot be ignored so we try to examine some of the constructions in a didactic manner in order to possibly generate some understanding.

3.1. COMMENTS ON CANTOR SETS.

EXAMPLE 3.1. In the paper [**867**] one discusses random recursive constructions leading to Cantor sets, etc. Associated with each such construction is a universal number α such that almost surely the random object has Hausdorff dimension α (we assume that ideas of Hausdorff and Minkowski-Bouligand (MB) or upper box dimension are known - cf. [**104, 234, 444, 799**]). One construction of a Cantor set goes as follows. Choose x from $[0,1]$ according to the uniform distribution and then choose y from $[x,1]$ according to the uniform distribution on $[x,1]$. Set $J_0=[0,x]$ and $J_1=[y,1]$ and recall the standard $1/3$ construction for Cantor sets. Continue this procedure by rescaling to each of the intervals already obtained. With probability one one then obtains a Cantor set S_c^0 with Hausdorff dimension $\alpha=\phi=(\sqrt{5}-1)/2\sim .618$. Note that this is just a particular random Cantor set; there are others with different Hausdorff dimensions (there seems to be some - possibly harmless - confusion on this point in the El Naschie papers). However the golden ratio ϕ is a very interesting number whose importance rivals that of π or e. In particular (cf. [**1**]) ϕ is the hardest number to approximate by rational numbers and could be called the most irrational number. This is because its continued fraction represention involves all $1's$. ∎

EXAMPLE 3.2. From El Naschie [**913**] the Hausdorff (H) dimension of a traditional triadic Cantor set is $d_c^{(0)}=log(2)/log(3)$. To determine the equivalent to a triadic Cantor set in 2 dimensions one looks for a set which is triadic Cantorian in all directions. The analogue of an area $A=1\times 1$ is a quasi-area $A_c=d_c^{(0)}\times d_c^{(0)}$ and to normalize A_c one uses $\rho_2=(A/A_c)_2=1/(d_c^{(0)})^2$ (for n-dimensions $\rho_n=1/(d_c^{(0)})^{n-1}$). Then the nth Cantor like H dimension $d_c^{(n)}$ will have the form $d_c^{(n)}=\rho_n d_c^{(0)}=1/(d_c^{(0)})^{n-1}$. Note also that the H dimension of a Sierpinski gasket is $d_c^{(n+1)}/d_c^{(n)}=1/d_c^{(0)}=log(3)/log(2)$ and in any event the straight-forward interpretation of $d_c^{(2)}=log(3)/log(2)$ is a scaling of $d_c^{(0)}=log(2)/log(3)$ proportional to the ratio of areas $(A/A_c)_2$. One notes that $d_c^{(4)}=1/(d_c^{(0)})^3=(log(3)/log(2))^3\simeq 3.997\sim 4$ so the 4-dimensional Cantor set is essentially "space filling".

Another derivation goes as follows. Define probability quotients via $\Omega=dim(subset)/dim(set)$. For a triadic Cantor set in 1-D $\Omega^{(1)}=d_c^{(0)}/d_c^{(1)}=d_c^{(0)}$ ($d_c^{(1)}=1$). To lift the Cantor set to n-dimensions look at the multiplicative probability law $\Omega^{(n)}=(\Omega^{(1)})^n=(d_c^{(0)})^n$. However since $\Omega^{(1)}=d_c^{(0)}/d_c^{(n)}$ we get

$$d_c^{(0)}/d_c^{(n)}=(d_c^{(0)})^n\Rightarrow d_c^{(n)}=1/(d_c^{(0)})^{n-1} \tag{3.1}$$

Since $\Omega^{(n-1)}$ is the probability of finding a Cantor point (Cantorian) one can think of the H dimension $d_c^{(n)}=1/\Omega^{(n-1)}$ as a measure of ignorance. One notes here also that for $d_c^{(0)}=\phi$ (the Cantor set $S_c^{(0)}$ of Example 3.1) one has $d_c^{(4)}=1/\phi^3=4+\phi^3\simeq 4.236$ which is surely space filling. ∎

Based on these ideas one proves in [**913, 914**] a number of theorems and we sketch some of this here. One picks a "backbone" Cantor set with H dimension

$d_c^{(0)}$ (the choice of $\phi = d_c^{(0)}$ will turn out to be optimal for many arguments). Then one imagines a Cantorian spacetime $\mathfrak{E}^\infty$ built up of an infinite number of spaces of dimension $d_c^{(n)}$ $(-\infty \leq n < \infty)$. The exact form of embedding etc. here is not specified so one imagines e.g. $\mathfrak{E}^\infty = \cup\mathfrak{E}^{(n)}$ (with unions and intersections) in some amorphous sense. There are some connections of this to vonNeumann's continuous geometries indicated in [**913, 914**]. In this connection we remark that only $\mathfrak{E}^{(-\infty)}$ is the completely empty set ($\mathfrak{E}^{(-1)}$ is not empty). First we note that $\phi^2 + \phi - 1 = 0$ leading to

$$(3.2)\quad 1+\phi = 1/\phi,\ \phi^3 = (2+\phi)/\phi,\ (1+\phi)/(1-\phi) = 1/\phi(1-\phi) = 4+\phi^3 = 1/\phi^3$$

(a very interesting number indeed). Then one asserts that

THEOREM 3.1. Let $(\Omega^{(1)})^n$ be a geometrical measure in n-dimensional space of a multiplicative point set process and $\Omega^{(1)}$ be the Hausdorff dimension of the backbone (generating) set $d_c^{(0)}$. Then $< d >= 1/d_c^{(0)}(1 - d_c^{(0)})$ (called curiously an average Hausdorff dimension) will be exactly equal to the average space dimension $\tilde{}< n >= (1 + d_c^{(0)})(1 - d_c^{(0)})$ and equivalent to a 4-dimensional Cantor set with H-dimension $d_c^{(4)} = 1/(d_c^{(0)})^3$ if and only if $d_c^{(0)} = \phi$.

To see this take $\Omega^{(n)} = (\Omega^{(1)})^n$ again and consider the total probability of the additive set described by the $\Omega^{(n)}$, namely $Z_0 = \sum_0^\infty (\Omega^{(1)})^n = 1/(1 - \Omega^{(1)})$. It is conceptually easier here to regard this as a sum of weighted dimensions (since $d_c^{(n)} = 1/(d_c^{(0)})^{n-1}$) and consider $w_n = n(d_c^{(0)})^n$. Then the expectation of n becomes (note $d_c^{(n)} \sim 1/(d_c^{(0)})^{n-1} \sim 1/\Omega^{(n-1)}$ so $n(d_c^{(0)})^{n-1} \sim n/d_c^{(n)}$)

$$(3.3)\qquad E(n) = \frac{\sum_1^\infty n^2 (d_c^{(0)})^{n-1}}{\sum_1^\infty n(d_c^{(0)})^{n-1}} = \tilde{}< n >= \frac{1 + d_c^{(0)}}{1 - d_c^{(0)}}$$

Another average here is defined via (blackbody gamma distribution)

$$(3.4)\qquad < n >= \frac{\int_0^\infty n^2 (\Omega^{(1)})^n dn}{\int_0^\infty n(\Omega^{(1)})^n dn} = \frac{-2}{log(\Omega^{(1)})}$$

which corresponds to $\tilde{}< n >$ after expanding the logarithm and omitting higher order terms. However $\tilde{}< n >$ seems to be the more valid calculation here. Similarly one defines (somewhat ambiguously) an expected value for $d_c^{(n)}$ via

$$(3.5)\qquad < d >= \frac{\sum_1^\infty n(d_c^{(0)})^{n-1}}{\sum_1^\infty (d_c^{(0)})^n} = \frac{1}{d_c^{(0)}(1 - d_c^{(0)})}$$

This is contrived of course (and cannot represent $E(d_c^{(n)})$ since one is computing reciprocals $\sum(n/d_c^{(n)})$ but we could think of computing an expected ignorance and identifying this with the reciprocal of dimension. Thus the label $< d >$ does not seem to represent an expected dimension but if we accept it as a symbol then for $d_c^{(0)} = \phi$ one has

$$(3.6)\qquad \tilde{}< n >= \frac{1+\phi}{1-\phi} =< d >= \frac{1}{\phi(1-\phi)} = d_c^{(4)} = 4 + \phi^3 = \frac{1}{\phi^3} \sim 4.236$$

REMARK 1.3.1. We note that the normalized probability $N = \Omega^{(1)}/Z_0 = \Omega^{(1)}(1-\Omega^{(1)}) = 1/<d>$ for any $d_c^{(0)}$. Further if $<d>= 4 = 1/d_c^{(0)}(1-d_c^{(0)})$ one has $d_c^{(0)} = 1/2$ while $\sim <n>= 3 < 4 =<d>$. One sees also that $d_c^{(0)} = 1/2$ is the minimum (where $d<d>/d(d_c^{(0)}) = 0$). ■

REMARK 1.3.2. The results of Theorem 3.1 should really be phrased in terms of $\mathfrak{E}^\infty$ (cf. [**913, 914**]). thus ($H \sim$ Hausdorff dimension and $T \sim$ topological dimension)

$$dim_H \mathfrak{E}^{(n)} = d_c^{(n)} = \frac{1}{(d_c^{(0)})^{n-1}}; \tag{3.7}$$

$$<d>= \frac{1}{d_c^{(0)}(1-d_c^{(0)})}; \sim< dim_T \mathfrak{E}^\infty >= \frac{1+d_c^{(0)}}{1-d_c^{(0)}} =\sim <n>$$

In any event $\mathfrak{E}^\infty$ is formally infinite dimensional but effectively it is $4\pm$ dimensional with an infinite number of internal dimensions. We emphasize that $\mathfrak{E}^\infty$ appears to be constructed from a fixed backbone Cantor set with H dimension $1/2 \le d_c^{(0)} < 1$; thus each such $d_c^{(0)}$ generates an $\mathfrak{E}^\infty$ space. Note that in [**913, 914**] $\mathfrak{E}^\infty$ is looked upon as a transfinite discretum underpinning the continuum (whatever that means). ■

REMARK 1.3.3. An interesting argument from El Naschie [**913**] goes as follows. Thinking of $d_c^{(0)}$ as a geometrical probability one could say that the spatial (3-dimensional) probability of finding a Cantorian "point" in $\mathfrak{E}^\infty$ must be given by the intersection probability $P = (d_c^{(0)})^3$ where $3 \sim 3$ topological spatial dimension. P could then be regarded as a Hurst exponent (cf. [**1, 941, 1293**]) and the Hausdorff dimension of the fractal path of a Cantorian would be $d_{path} = 1/H = 1/P = 1/(d_c^{(0)})^3$. Given $d_c^{(0)} = \phi$ this means $d_{path} = 4 + \phi^3 \sim 4^+$ so a Cantorian in 3-D would sweep out a 4-D world sheet; i.e. the time dimension is created by the Cantorian space $\mathfrak{E}^\infty$ (! - ?). Conjecturing further (wildly) one could say that perhaps space (and gravity) is created by the fractality of time. This is a typical form of conjecture to be found in the El Naschie papers - extremely thought provoking but ultimately heuristic. Regarding the Hurst exponent one recalls that for Feynmann trajectories in $1+1$ dimensions $d_{path} = 1/H = 1/d_c^{(0)} = d_c^{(2)}$. Thus we are concerned with relating the two determinations of d_{path} (among other matters). Note that path dimension is often thought of as a fractal dimension (M-B or box dimension), which is not necessarily the same as the Hausdorff dimension. However in [**34**] one shows that quantum mechanical free motion produces fractal paths of Hausdorff dimension 2 (cf. also [**781**]). ■

REMARK 1.3.4. Following [**279**] let $S_c^{(0)}$ correspond to the set with dimension $d_c^{(0)} = \phi$. Then the complementary dimension is $\tilde{d}_c^{(0)} = 1-\phi = \phi^2$. The path dimension is given as in Remark 1.3.3 by $d_{path} = d_c^{(2)} = 1/\phi = 1+\phi$ and $\tilde{d}_{path} = \tilde{d}_c^{(2)} = 1/(1-\phi) = 1/\phi^2 = (1+\phi)^2$. Following El Naschie for an equivalence between unions and intersections in a given space one requires (in the present situation) that

(3.8) $$d_{crit} = d_c^{(2)} + \tilde{d}_c^{(2)} = \frac{1}{\phi} + \frac{1}{\phi^2} = \frac{\phi(1+\phi)}{\phi^3} = \frac{1}{\phi^3} = \frac{1}{\phi}\cdot\frac{1}{\phi^2} = d_c^{(2)}\cdot\tilde{d}_c^{(2)} = 4+\phi^3$$

where $d_{crit} = 4 + \phi^3 = d_c^{(4)} \sim 4.236$. Thus the critical dimension coincides with the Hausdorff dimension of $S_c^{(4)}$ which is embedded densely into a smooth space of topological dimension 4. On the other hand the backbone set of dimension $d_c^{(0)} = \phi$ is embedded densely into a set of topological dimension zero (a point). Thus one thinks in general of $d_c^{(n)}$ as the H dimension of a Cantor set of dimension ϕ embedded into a smooth space of integer topological dimension n. ■

REMARK 1.3.5. In Castro [**279**] it is also shown that realization of the spaces $\mathfrak{E}^{(n)}$ comprising $\mathfrak{E}^{\infty}$ can be expressed via the fractal sprays of Lapidus-van Frankonhuyoon (cf. [**700**]). Thus we refer to [**799**] for graphics and details and simply sketch some ideas here (with apologies to M. Lapidus). A fractal string is a bounded open subset of **R** which is a disjoint union of an infinite number of open intervals $\mathfrak{L} = \ell_1, \ell_2, \cdots$. The geometric zeta function of $\mathfrak{L}$ is $\zeta_{\mathfrak{L}}(s) = \sum_1^{\infty} \ell_j^{-s}$. One assumes a suitable meromorphic extension of $\zeta_{\mathfrak{L}}$ and the complex dimensions of $\mathfrak{L}$ are defined as the poles of this meromorphic extension. The spectrum of $\mathfrak{L}$ is the sequence of frequencies $f = k\cdot\ell_j^{-1}$ $(k = 1, 2, \cdots)$ and the spectral zeta function of $\mathfrak{L}$ is defined as $\zeta_\nu(s) = \sum_f f^{-s}$ where in fact $\zeta_\nu(s) = \zeta_{\mathfrak{L}}(s)\zeta(s)$ (with $\zeta(s)$ the classical Riemann zeta function). Fractal sprays are higher dimensional generalizations of fractal strings. As an example consider the spray Ω obtained by scaling an open square B of size 1 by the lengths of the standard triadic Cantor string CS. Thus Ω consists of one open square of size 1/3, 2 open squares of size 1/9, 4 open squares of size 1/27, etc. (see [**799**] for pictures and explanations). Then the spectral zeta function for the Dirichlet Laplacian on the square is $\zeta_B(s) = \sum_{n_1,n_2=1}^{\infty}(n_1^2+n_2^2)^{s/2}$ and the spectral zeta function of the spray is $\zeta_\nu(s) = \zeta_{CS}(s)\cdot\zeta_B(s)$. Now $\mathfrak{E}^{\infty}$ is composed of an infinite hierarchy of sets $\mathfrak{E}^{(j)}$ with dimension $(1+\phi)^{j-1} = 1/\phi^{j-1}$ $(j = 0, \pm 1, \pm 2, \cdots)$ and these sets correspond to a special case of boundaries $\partial\Omega$ for fractal sprays Ω whose scaling ratios are suitable binary powers of $2^{-\phi^{j-1}}$. Indeed for $n = 2$ the spectral zeta function of the fractal golden spray indicated above is $\zeta_\nu(s) = (1/(1-2\cdot 2^{s\phi})\zeta_B(s)$. The poles of $\zeta_B(s)$ do not coincide with the zeros of the denominator $1 - 2\cdot 2^{-s\phi}$ so the (complex) dimensions of the spray correspond to those of the boundary $\partial\Omega$ of Ω. One finds that the real part $\Re s$ of the complex dimensions coincides with $dim\,\mathfrak{E}^{(2)} = 1+\phi = 1/\phi^2$ and one identifies then $\partial\Omega$ with $\mathfrak{E}^{(2)}$. The procedure generalizes to higher dimensions (with some stipulations) and for dimension n there results $\Re s = 1/\phi^{n-1} = dim\,\mathfrak{E}^{(n)}$. This produces a physical model of the Cantorian fractal space from the boundaries of fractal sprays (see [**279**] for further details and [**799**] for precision). Other (putative) geometric realizations of $\mathfrak{E}^{\infty}$ are indicated in [?] in terms of wild topologies, etc. ■

3.2. COMMENTS ON HYDRODYNAMICS. We sketch first some material from Agop et al [**17**] (see also [**234, 941, 944, 946**] and Sections 1-2 for background). Thus let ψ be the wave function of a test particle of mass m_0 in a force field $U(r,t)$ determined via $i\hbar\partial_t\psi = U\psi - (\hbar^2/2m)\nabla^2\psi$ where $\nabla^2 = \Delta$. One writes $\psi(r,t) = R(r,t)exp(iS(r,t))$ with $v = (\hbar/2m)\nabla S$ and $\rho = R \cdot R$ (one assumes $\rho \neq 0$ for physical meaning). Thus the field equations of QM in the hydrodynamic picture are

$$d_t(m_0\rho v) = \partial_t(m_0\rho v) + \nabla(m_0\rho v) = -\rho\nabla(U+Q);\ \partial_t\rho + \nabla\cdot(\rho v) = 0 \tag{3.9}$$

where $Q = -(\hbar^2/2m_0)(\Delta\sqrt{\rho}/\sqrt{\rho})$ is the quantum potential (or interior potential). Now because of the nondifferentiability of spacetime an infinity of geodesics will exist between any couple of points A and B. The ensemble will define the probability amplitude (this is a nice assumption but geodesics should be defined here). At each intermediate point C one can consider the family of incoming (backward) and outgoing (forward) geodesics and define average velocities $b_+(C)$ and $b_-(C)$ on these families. These will be different in general and following Nottale this doubling of the velocity vector is at the origin of the complex nature of QM. Even though Nottale reformulates Nelson's stochastic QM the former's interpretation is profoundly different. While Nelson (cf. [**926**]) assumes an underlying Brownian motion of unknown origin which acts on particles in Minkowskian spacetime, and then introduces nondifferentiability as a byproduct of this hypothesis, Nottale assumes as a fundamental and universal principle that spacetime itself is no longer Minkowskian nor differentiable. An interesting comment here from [**17**] is that with Nelson's Browian motion hypothesis, nondifferentiability is but an approximation which expected to break down at the scale of the underlying collisions, where a new physics should be introduced, while Nottale's hypothesis of nondifferentiability is essential and should hold down to the smallest possible length scales. Following Nelson one defines now the mean forward and backward derivatives

$$\frac{d_\pm}{dt}y(t) = lim_{\Delta t\to 0\pm}\left\langle \frac{y(t+\Delta t)-y(t)}{\Delta t}\right\rangle \tag{3.10}$$

This gives forward and backward mean velocities $(d_+/dt)x(t) = b_+$ and $(d_-/dt)x(t) = b_-$ for a position vector x. Now in Nelson's stochastic mechanics one writes two systems of equations for the forward and backward processes and combines them in the end in a complex equation, Nottale works from the beginning with a complex derivative operator

$$\frac{\delta}{dt} = \frac{(d_+ + d_-) - i(d_+ - d_-)}{2dt} \tag{3.11}$$

leading to $V = (\delta/dt)x(t) = v - iu = (1/2)(b_+ + b_-) - (i/2)(b_+ - b_-)$. One defines also $(d_v/dt) = (1/2)(d_+ + d_-)/dt$ and $(d_u/dt) = (1/2)(d_+ - d_-)/dt$ so that $d_vx/dt = v$ and $d_ux/dt = u$. Here v generalizes the classical velocity while u is a new quantity arising from nondifferentiability. This leads to a stochastic process satisfying (respectively for the forward ($dt > 0$) and backward ($dt < 0$) processes) $dx(t) = b_+[x(t)] + d\xi_+(t) = b_-[x(t)] + d\xi_-(t)$. The $d\xi(t)$ terms can be seen as fractal functions and they amount to a Wiener process when the fractal dimension $D = 2$. Then the $d\xi(t)$ are Gaussian with mean zero, mutually independent, and satisfy $< d\xi_{\pm i}d\xi_{\pm j} >= \pm 2\mathcal{D}\delta_{ij}dt$ where $\mathcal{D}$ is a diffusion coefficient determined as $\mathcal{D} = \hbar/2m_0$ when $\tau_0 = \hbar/(m_0c^2)$ (deBroglie time scale in the rest frame (cf.

[**17**]). This allows one to give a general expression for the complex time derivative, namely

$$df = \frac{\partial f}{\partial t} + \nabla f \cdot dx + \frac{1}{2}\frac{\partial^2 f}{\partial x_i \partial x_j} dx_i dx_j \tag{3.12}$$

Next compute the forward and backward derivatives of f; then one will arrive at $< dx_i dx_j > \to < d\xi_{\pm i} d\xi_{\pm j} >$ so the last term in (3.12) amounts to a Laplacian and one obtains $(d_{\pm}f/dt) = [\partial_t + b_{\pm} \cdot \nabla \pm \mathcal{D}\Delta]f$ which is an important result. Thus assume the fractal dimension is not 2 in which case there is no longer a cancellation of the scale dependent terms in (3.12) and instead of $\mathcal{D}\Delta f$ one would obtain an explicitly scale dependent behavior proportional to $\delta t^{(2/D)-1}\Delta f$. In other words the value $D = 2$ for the fractal dimension implies that the scale symmetry becomes hidden in the operator formalism. One obtains the complex time derivative operator in the form $(\delta/dt) = \partial_t + V \cdot \nabla - iD\Delta$ (V as above). Nottale's prescription is then to replace d/dt by δ/dt. In this spirit one can write now $\psi = exp(i(\mathfrak{S}/2m_0\mathcal{D}))$ so that $V = -2i\mathcal{D}\nabla(log(\psi))$ and then the generalized Newton equation $-\nabla U = m_0(\delta/dt)V$ reduces to the SE $(L = (1/2)mv^2 - U)$.

Now assume the velocity field from the hydrodynamic model agrees with the real part v of the complex velocity V and equate the wave functions from the two models $\psi = exp(i\mathfrak{S}/2m_0\mathcal{D})$ and $\psi = Rexp(iS)$ with $m = m_0$; one obtains for $\mathfrak{S} = s + i\sigma$ the formulas $s = 2m_0\mathcal{D}S$, $\mathcal{D} = (\hbar/2m_0)$, and $\sigma = -m_0\mathcal{D}log(\rho)$. Using the definition $V = (1/m_0)\nabla\mathfrak{S} = (1/m_0)\nabla s + (i/m_0)\nabla\sigma = v - iu$ (which results from the above equations) we get

$$v = (1/m_0)\nabla s = 2D\nabla S;\ u = -(1/m_0)\nabla\sigma = \mathcal{D}\nabla log(\rho) \tag{3.13}$$

Note that the imaginary part of the complex velocity coincides with Nottale. Dividing the time dependent SE $i\hbar\psi_t = U\psi - (\hbar^2/2m_0)\Delta\psi$ by $2m_0$ and taking the gradient gives $\nabla U/m_0 = 2\mathcal{D}\nabla[i\partial_t log(\psi) + \mathcal{D}(\Delta\psi/\psi)]$ where $\hbar/2m_0$ has been replaced by $\mathcal{D}$. Then consider the identities

$$\Delta\nabla = \nabla\Delta;\ (\nabla f \cdot \nabla)(\nabla f) = (1/2)\nabla(\nabla f)^2;\ \frac{\Delta f}{f} = \Delta log(f) + (\nabla log(f))^2 \tag{3.14}$$

Then the second term in the right of the equation for $\nabla U/m_0$ becomes $\nabla(\Delta\psi/\psi) = \Delta(\nabla log(\psi)) + 2(\nabla log(\psi) \cdot \nabla)(\nabla log(\psi))$ so we obtain

$$\nabla U = 2iDm_0[\partial_t \nabla log(\psi) - i\mathcal{D}\Delta(\nabla log(\psi) - 2i\mathcal{D}(\nabla log(\psi) \cdot \nabla)(\nabla log(\psi))] \tag{3.15}$$

One can show that this is nothing but the generalized Newton equation $-\nabla U = m_0(\delta/dt)V$. Now replacing the complex velocity $V = -2i\mathcal{D}\nabla log(\psi)$ and taking into account the form of V, we get

$$-\nabla U = m_0\{\partial_t(v - i\mathcal{D}\nabla log(\rho) + [i(v - i\mathcal{D}\nabla log(\rho) \cdot \nabla](v - i\mathcal{D}\nabla log(\rho)) - \\ -i\mathcal{D}\Delta(v - i\mathcal{D}\nabla log(\rho))\} \tag{3.16}$$

Equation (3.16) is a complex differential equation and reduces to

$$m_0[\partial_t v + (v \cdot \nabla)v] = -\nabla\left(U - 2m_0\mathcal{D}^2\frac{\Delta\sqrt{\rho}}{\sqrt{\rho}}\right);\ \nabla\left\{\frac{1}{\rho}\left[\partial_t\rho + \nabla \cdot (\rho v)\right]\right\} \tag{3.17}$$

The last equation in (3.17) reduces to the continuity equation up to a phase factor $\alpha(t)$ which can be set equal to zero (note again that $\rho \neq 0$ is posited). Thus (3.17) is nothing but the fundamental equations (3.9) of the hydrodynamic model. Further combining the imaginary part of the complex velocity with the quantum potential, and using (3.14), one gets $Q = -m_0\mathcal{D}\nabla \cdot u - (1/2)m_0u^2$ (as indicated in Remark 1.2.2). Since u arises from nondifferentiability according to our nondifferentiable space model of QM it follows that the quantum potential comes from the nondifferentiability of the quantum spacetime (note that the x derivatives should be clarified and $\mathfrak{E}^\infty$ has not been utilized).

Putting $U = 0$ in the first equation of (3.17), multiplying by ρ, and taking the second equation into account yields

$$\text{(3.18)}\qquad \partial_t(m_0\rho\nu_k) + \frac{\partial}{\partial x_i}(m_0\rho\nu_i\nu_k) = -\rho\frac{\partial}{x_k}\left[2m_0\mathcal{D}^2\frac{1}{\sqrt{\rho}}\frac{\partial}{\partial x_i}\frac{\partial}{\partial x_i}(\sqrt{\rho})\right]$$

(here $\nu_k \sim v_k$ seems indicated). Now set $\Pi_{ik} = m_0\rho\nu_i\nu_k - \sigma_{ik}$ along with $\sigma_{ik} = m_0\rho D^2(\partial/\partial x_i)(\partial/\partial x_k)(log(\rho))$. Then (3.18) takes the simple form

$$\text{(3.19)}\qquad \partial_t(m_0\rho\nu_k) = -\partial\Pi_{ik}/\partial x_i$$

The analogy with classical fluid mechanics works well if one introduces the kinematic $\mu = D/2$ and dynamic $\eta = (1/2)m_0\mathcal{D}\rho$ viscosities. Then Π_{ik} defines the momentum flux density tensor and σ_{ik} the internal stress tensor $\sigma_{ik} = \eta[(\partial u_i/\partial x_k) + (\partial u_k/\partial x_i)]$. One can see that the internal stress tensor is build up using the quantum potential while the equations (3.18) or (3.19) are nothing but systems of Navier-Stokes type for the motion where the quantum potential plays the role of an internal stress tensor. In other words the nondifferentiability of the quantum spacetime manifests itself like an internal stress tensor. For clarity in understanding (3.19) we put this in one dimensional form so (3.18) becomes

$$\text{(3.20)}\qquad \partial_t(m_0\rho v) + \partial_x(m_0\rho v^2) = -\rho\partial\left(2m_0\mathcal{D}^2\frac{1}{\sqrt{\rho}}\partial^2\sqrt{\rho}\right) = \rho\partial Q$$

and $\Pi = m_0\rho v^2 - \sigma$ with $\sigma = m_0\rho D^2\partial^2 log(\rho)$ which agrees with standard formulas. Now note $\partial\sqrt{\rho} = (1/2)\rho^{-1/2}\rho'$ and $\partial^2\sqrt{\rho} = (1/2)[-(1/2)\rho^{-3/2}(\rho')^2 + \rho^{-1/2}\rho'']$ with $\partial^2 log(\rho) = \partial(\rho'/\rho) = (\rho''/\rho) - (\rho'/\rho)^2$ while

(3.21)

$$-\rho\partial\left[2m_0\mathcal{D}^2\frac{1}{\sqrt{\rho}}(\partial^2\sqrt{\rho})\right] = -2m_0\mathcal{D}^2\rho\partial\left[\frac{1}{2\sqrt{\rho}}\left(-\frac{1}{2}\rho^{-3/2}(\rho')^2 + \rho^{-1/2}\rho''\right)\right] =$$

$$= -2m_0\mathcal{D}^2\rho\partial\left[\frac{\rho''}{2\rho} - \frac{1}{4}\left(\frac{(\rho'}{\rho}\right)^2\right] = -m_0\mathcal{D}^2\rho\partial\left[\frac{\rho''}{\rho} - \frac{1}{2}\left(\frac{\rho'}{\rho}\right)^2\right]$$

One wants to show then that (3.19) holds or equivalently $-\partial\sigma = (3.21)$. However

(3.22)

$$-\partial\sigma = -\partial[m_0\rho D^2\partial^2 log(\rho)] = -m_0\mathcal{D}^2\left[\rho'\left(\frac{\rho''}{\rho} - \left(\frac{\rho'}{\rho}\right)^2\right) + \rho\partial\left(\frac{\rho''}{\rho} - \frac{(\rho')^2}{\rho}\right)\right]$$

so we want $(3.22) = (3.21)$ which is easily verified.

4. REMARKS ON FRACTAL CALCULUS

We sketch first (in summary form) some material from Parvate and Gangal [**994**] where a calculus based on fractal subsets of the real line is formulated. A local calculus based on renormalizing fractional derivatives à la [**764**] is subsumed and embellished. Consider first the concept of content or α-mass for a (generally fractal) subset $F \subset [a,b]$ (in what follows $0 < \alpha \leq 1$). Then define the flag function for a set F and a closed interval I as $\theta(F,I) = 1$ ($F \cap I \neq \emptyset$ and otherwise $\theta = 0$. Then a subdivision $P_{[a,b]} \sim P$ of $[a,b]$ $(a < b)$ is a finite set of points $\{a = x_0, x_1, \cdots, x_n = b\}$ with $x_i < x_{i+1}$. If Q is any subdivision with $P \subset Q$ it is called a refinement and if $a = b$ the set $\{a\}$ is the only subdivision. Define then

$$\sigma^\alpha[F,p] = \sum_0^{n-1} \frac{(x_{i+1}-x_i)^\alpha}{\Gamma(\alpha+1)} \theta(F,[x_i,x_{i+1}]) \tag{4.1}$$

For $a = b$ one defines $\sigma^\alpha[F,P] = 0$. Next given $\delta > 0$ and $a \leq b$ the coarse grained mass $\gamma_\delta^\alpha(F,a,b)$ of $F \cap [a,b]$ is given via

$$\gamma_\delta^\alpha(F,a,b) = inf_{|P| \leq \delta} \sigma^\alpha[F,P] \;\; (|P| = max_{0 \leq i \leq n-1}(x_{i+1}-x_i)) \tag{4.2}$$

where the infimum is over P such that $|P| \leq \delta$. Various more or less straightforward properties are:

- For $a \leq b$ and $\delta_1 < \delta_2$ one has $\gamma_{\delta_1}^\alpha(F,a,b) \geq \gamma_{\delta_2}^\alpha(F,a,b)$.
- For $\delta > 0$ and $a < b < c$ one has $\gamma_\delta^\alpha(F,a,b) \leq \gamma_\delta^\alpha(F,a,c)$ and $\gamma_\delta^\alpha(F,b,c) \leq \gamma_\delta^\alpha(F,a,c)$.
- γ_δ^α is continuous in b and a.

Now define the mass function $\gamma^\alpha(F,a,b)$ via $\gamma^\alpha(F,a,b) = lim_{\delta \to 0} \gamma_\delta^\alpha(F,a,b)$. The following results are proved

(1) If $F \cap (a,b) = \emptyset$ then $\gamma^\alpha(F,a,b) = 0$.
(2) Let $a < b < c$ and $\gamma^\alpha(F,a,c) < \infty$. Then $\gamma^\alpha(F,a,c) = \gamma^\alpha(F,a,b) + \gamma^\alpha(F,b,c)$. Hence $\gamma^\alpha(F,a,b)$ is increasing in b and decreasing in a.
(3) Let $a < b$ and $\gamma^\alpha(F,a,b) \neq 0$ be finite. If $0 < y < \gamma^\alpha(F,a,b)$ then there exists c, $a < c < b$ such that $\gamma^\alpha(F,a,c) = y$. Further if $\gamma^\alpha(F,a,b)$ is finite then $\gamma^\alpha(F,a,x)$ is continuous for $x \in (a,b)$.
(4) For $F \subset \mathbf{R}$ and $\lambda \in \mathbf{R}$ let $F + \lambda = \{x + \lambda;\, x \in F\}$. Then $\gamma^\alpha(F+\lambda, a+\lambda, b+\lambda) = \gamma^\alpha(F,a,b)$ and $\gamma^\alpha(\lambda F, \lambda a, \lambda b) = \lambda^\alpha \gamma^\alpha(F,a,b)$.

Now for a_0 an arbitrary fixed real number one defines the integral staircase function of order α for F is

$$S_F^\alpha(x) = \begin{cases} \gamma^\alpha(F,a_0,x) & x \geq a_0 \\ -\gamma^\alpha(F,x,a_0) & otherwise \end{cases} \tag{4.3}$$

The following properties of S_F are restatements of properties for γ^α. thus

- $S_F^\alpha(x)$ is increasing in x.
- If $F \cap (x,y) = 0$ then S_F^α is constant in $[x,y]$.
- $S_F^\alpha(y) - S_F^\alpha(x) = \gamma^\alpha(F,x,y)$.
- S_F^α is continuous on (a,b).

Now one considers the sets F for which the mass function $\gamma^\alpha(F,a,b)$ gives the most useful information. Indeed one can use the mass function to define a fractal dimension. If $0<\alpha<\beta\leq 1$ one writes

(4.4)
$$\sigma^\beta[F,P]\leq |P|^{\beta-\alpha}\sigma^\alpha[F,P]\frac{\Gamma(\alpha+1)}{\Gamma(\beta+1)};\ \gamma^\beta_\delta(F,a,b)\leq \delta^{\beta-\alpha}\gamma^\alpha_\delta(F,a,b)\frac{\Gamma(\alpha+1)}{\Gamma(\beta+1)}$$

Thus in the limit $\delta\to 0$ one gets $\gamma^\beta(F,a,b)=0$ provided $\gamma^\alpha(F,a,b)<\infty$ and $\alpha<\beta$. It follows that $\gamma^\alpha(F,a,b)$ is infinite up to a certain value α_0 and then jumps down to zero for $\alpha>\alpha_0$ (if $\alpha_0<1$). This number is called the γ-dimension of F; $\gamma^{\alpha_0}(F,a,b)$ may itself be zero, finite, or infinite. To make the definition precise one says that the γ-dimension of $F\cap[a,b]$, denoted by $dim_\gamma(F\cap[a,b])$, is

$$dim_\gamma(F\cap[a,b])=\left\{\begin{array}{l} inf\{\alpha;\ \gamma^\alpha(F,a,b)=0\}\\ sup\{\alpha;\ \gamma^\alpha(F,a,b)=\infty\}\end{array}\right. \tag{4.5}$$

One shows that $dim_H(F\cap[a,b])\leq dim_\gamma(F\cap[a,b])$ where dim_H denotes Hausdorff dimension. Further $dim_\gamma(F\cap[a,b])\leq dim_B(F\cap[a,b])$ where dim_B is the box dimension. Some further analysis shows that for $F\subset\mathbf{R}$ compact $dim_\gamma F=dim_H F$.

Next one notes that the correspondence $F\to S^\alpha_F$ is many to one (examples from Cantor sets) and one calls the sets giving rise to the same staircase function "staircasewise congruent". The equivalence class of congruent sets containing F is denoted by $\mathcal{E}_F$; thus if $G\in\mathcal{E}_F$ it follows that $S^\alpha_G=S^\alpha_F$ and $\mathcal{E}^\alpha_G=\mathcal{E}^\alpha_F$. One says that a point x is a point of change of f if f is not constant over any open interval (c,d) containing x. The set of all points of change of f is denoted by $Sch(f)$. In particular if $G\in\mathcal{E}^\alpha_F$ then $S^\alpha_G(x)=S^\alpha_F(x)$ so $Sch(S^\alpha_G)=Sch(S^\alpha_F)$. Thus if $F\subset\mathbf{R}$ is such that $S^\alpha_F(x)$ is finite for all x ($\alpha=dim_\gamma F$) then $H=Sch(S^\alpha_F)\in\mathcal{E}^\alpha_F$. This takes some proving which we omit (cf. [**994**]). As a consequence let $F\subset\mathbf{R}$ be such that $S^\alpha_F(x)$ is finite for all $x\in\mathbf{R}$ ($\alpha=dim_\gamma F$). Then the set $H=Sch(S^\alpha_F)$ is perfect (i.e. H is closed and every point is a limit point). Hence given $S^\alpha_F(x)$ finite for all x ($\alpha=dim_\gamma F$) one calls $Sch(S^\alpha_F)$ the α-perfect representative of $\mathcal{E}^\alpha_F$ and one proves that it is the minimal closed set in $\mathcal{E}^\alpha_F$. Indeed an α-perfect set in $\mathcal{E}^\alpha_F$ is the intersection of all closed sets G in $\mathcal{E}^\alpha_F$. One can also say that if $F\subset\mathbf{R}$ is α-perfect and $x\in F$ then for $y<x<z$ either $S^\alpha_F(y)<S^\alpha_F(x)$ or $S^\alpha_F(x)<S^\alpha_F(z)$ (or both). Thus for an α-perfect set it is assured that the values of $S^\alpha_F(y)$ must be different from $S^\alpha_F(x)$ at all points y on at least one side of x. As an example one shows that the middle third Cantor set $C=E_{1/3}$ is α-perfect for $\alpha=log(2)/log(3)=d_H(C)$ so $C=Sch(S^\alpha_C)$.

Now look at F with the induced topology from $\mathbf{R}$ and consider the idea of F-continuity.

DEFINITION 4.1. Let $F\subset\mathbf{R}$ and $f:\mathbf{R}\to\mathbf{R}$ with $x\in F$. A number ℓ is said to be the limit of f through the points of F, or simply F-limit, as $y\to x$ if given $\epsilon>0$ there exists $\delta>0$ such that $y\in F$ and $|y-x|<\delta\Rightarrow|f(y)-\ell|<\epsilon$. In such a case one writes $\ell=F-limit_{y\to x}f(y)$. A function f is F-continuous at $x\in F$ if $f(x)=F-limit_{y\to x}f(y)$ and uniformly F-continuuous on $E\subset F$ if for $\epsilon>0$ there exists $\delta>0$ such that $x\in F$, $y\in E$ and $|y-x|<\delta\Rightarrow|f(y)-f(x)|<\epsilon$.

One sees that if f is F-continuous on a compact set $E \subset F$ then it is uniformly F-continuous on E. ■

DEFINITION 4.2. The class of functions $f : \mathbf{R} \to \mathbf{R}$ which are bounded on F is denoted by $B(F)$. Define for $f \in B(F)$ and I a closed interval

$$M[f,F,I] = \begin{cases} sup_{x \in F \cap I} f(x) & F \cap I \neq \emptyset \\ 0 & otherwise \end{cases} \tag{4.6}$$

$$m[f,F,I] = \begin{cases} inf_{x \in F \cap I} f(x) & F \cap I \neq \emptyset \\ 0 & otherwise \end{cases}$$

■

DEFINITION 4.3. Let $S_F^\alpha(x)$ be finite for $x \in [a,b]$ and P be a subdivision with points $x_0, \cdots, x_n$. The upper F^α and lower F^α sums over P are given respectively by

$$U^\alpha[f,F,P] = \sum_0^{n-1} M[f,F,[x_i,x_{i+1}]](S_F^\alpha(x_{i+1}) - S_F^\alpha(x_i)); \tag{4.7}$$

$$L^\alpha[f,F,P] = \sum_0^{n-1} m[f,F,[x_i,x_{i+1}]](S_F^\alpha(x_{i+1}) - S_F^\alpha(x_i))$$

This is sort of like Riemann-Stieltjes integration and in fact one shows that if Q is a refinement of P then $U^\alpha[f,F,Q] \leq U^\alpha[f,F,P]$ and $L^\alpha[f,F,Q] \geq L^\alpha[f,F,P]$. Further $U^\alpha[f,F,P] \geq L^\alpha[f,F,Q]$ for any subdivisions of $[a,b]$ and this leads to the idea of F-integrability. Thus assume S_F^α is finite on $[a,b]$ and for $f \in B(F)$ one defines lower and upper F^α-integrals via

$$\underline{\int_a^b} f(x) d_F^\alpha x = sup_P L^\alpha[f,F,P]; \quad \overline{\int_a^b} f(x) d_F^\alpha x = inf_P U^\alpha[f,F,P] \tag{4.8}$$

One then says that f is F^α-integrable if **(D15)** $\underline{\int_a^b} f(x) d_F^\alpha x = \overline{\int_a^b} f(x) d_F^\alpha x = \int_a^b f(x) d_F^\alpha x$. ■

One shows then

(1) $f \in B(F)$ is F^α-integrable on $[a,b]$ if and only if for any $\epsilon > 0$ there is a subdivision P of $[a,b]$ such that $U^\alpha[f,F,P] < L^\alpha[f,F,P] + \epsilon$.
(2) Let $F \cap [a,b]$ be compact with S_F^α finite on $[a,b]$. Let $f \in B(F)$ and $a < b$; then if f is F-continuous on $F \cap [a,b]$ it follows that f is F^α-integrable on $[a,b]$.
(3) Let $a < b$ and f be F^α-integrable on $[a,b]$ with $c \in (a,b)$. Then f is F^α-integrable on $[a,c]$ and $[c,b]$ with $\int_a^b f(x) d_F^\alpha x = \int_a^c f(x) d_F^\alpha x + \int_c^b f(x) d_F^\alpha x$.
(4) If f is F^α-integrable then $\int_a^b \lambda f(x) d_F^\alpha x = \lambda \int_a^b f(x) d_F^\alpha x$ and, for g also F^α-integrable, $\int_a^b (f(x) + g(x)) d_F^\alpha x = \int_a^b f(x) d_F^\alpha x + \int_a^b g(x) d_F^\alpha x$.
(5) If f, g are F^α-integrable and $f(x) \geq g(x)$ for $x \in F \cap [a,b]$ then $\int_a^b f(x) d_F^\alpha x \geq \int_a^b g(x) d_F^\alpha x$.

One specifies also $\int_b^a f(x)d_F^\alpha x = -\int_a^b f(x)d_F^\alpha x$ and it is easily shown that if $\chi_F(x)$ is the characteristic function of F then $\int_a^b \chi_F(x)d_F^\alpha x = S_F^\alpha(b) - S_F^\alpha(a)$. Now for differentiation one writes

$$\mathcal{D}_F^\alpha f(x) = \begin{cases} F - lim_{y\to x}\frac{f(y)-f(x)}{S_F^\alpha(y)-S_F^\alpha(x)} & x \in F \\ 0 & otherwise \end{cases} \tag{4.9}$$

if the limit exists. One shows then

(1) If $\mathcal{D}_F^\alpha f(x)$ exists for all $x \in (a,b)$ then $f(x)$ is F-continuous in (a,b).
(2) With obvious hypotheses $\mathcal{D}_F^\alpha(\lambda f(x)) = \lambda\mathcal{D}_F^\alpha f(x)$ and $\mathfrak{D}_F^\alpha(f+g)(x) = \mathcal{D}_F^\alpha f(x) + \mathcal{D}_F^\alpha g(x)$. Further if f is constant then $\mathcal{D}_F^\alpha f = 0$.
(3) $\mathfrak{D}_F^\alpha(S_F^\alpha(x)) = \chi_F(x)$.
(4) (Rolle's theorem) Let $f : \mathbf{R} \to \mathbf{R}$ be continuous with $Sch(f) \subset F$ where F is α-perfect and assume $\mathcal{D}_F^\alpha f(x)$ is defined for all $x \in [a,b]$ with $f(a) = f(b) = 0$. Then there is a point $c \in F \cap [a,b]$ such that $\mathcal{D}_F^\alpha f(c) \geq 0$ and a point $d \in F \cap [a,b]$ where $\mathcal{D}_F^\alpha f(d) \leq 0$.

EXAMPLE 4.1. This is the best that can be done with Rolle's theorem since for C the Cantor set $E_{1/3}$ take $f(x) = S_C^\alpha(x)$ for $0 \leq x \leq 1/2$ and $f(x) = 1 - S_C^\alpha(x)$ for $1/2 < x \leq 1$. This function is continuous with $f(0) = f(1) = 0$ and the set of change $(Sch(f))$ is C. The C^α-derivative is given by $\mathcal{D}_C^\alpha f(x) = \chi_C(x)$ for $0 \leq x \leq 1/2$ and by $-\chi_C(x)$ for $1/2 < x \leq 1$. Thus $x \in C$ which implies $\mathcal{D}_C^\alpha f(x) = \pm 1 \neq 0$. ∎

As a corollary one has the following result: Let f be continuous with $Sch(f) \subset F$ where F is α-perfect; assume $\mathcal{D}_F^\alpha f(s)$ exists at all points of $[a,b]$ and that $S_F^\alpha(b) \neq S_F^\alpha(a)$. Then there are points $c, d \in F$ such that

$$\mathcal{D}_F^\alpha f(c) \geq \frac{f(b)-f(a)}{S_F^\alpha(b) - S_F^\alpha(b)};\ \mathcal{D}_F^\alpha f(d) \leq \frac{f(b)-f(a)}{S_F^\alpha(b) - S_F^\alpha(a)} \tag{4.10}$$

Similarly if f is continuous with $Sch(f) \subset F$ and $\mathcal{D}_F^\alpha f(x) = 0\ \forall x \in [a,b]$ then $f(x)$ is constant on $[a,b]$. There are also other fundamental theorems as follows

(1) (Leibniz rule) If $u, v : \mathbf{R} \to \mathbf{R}$ are F^α-differentiable then $\mathcal{D}_F^\alpha(uv)(x) = (\mathcal{D}_F^\alpha u(x))v(x) + u(x)\mathcal{D}_F^\alpha v(x)$.
(2) Let $F \subset \mathbf{R}$ be α-perfect. If $f \in B(F)$ is F-continuous on $F \cap [a,b]$ with $g(x) = \int_a^x f(y)d_F^\alpha y$ for all $x \in [a,b]$ then $\mathcal{D}_F^\alpha g(x) = f(x)\chi_F(x)$.
(3) Let $f : \mathbf{R} \to \mathbf{R}$ be continuous and F^α-differentiable with $Sch(f)$ contained in an α-perfect set F; let also $h : \mathbf{R} \to \mathbf{R}$ be F-continuous such that $h(x)\chi_F(x) = \mathcal{D}_F^\alpha f(x)$. Then $\int_a^b h(x)d_F^\alpha x = f(b) - f(a)$.
(4) (Integration by parts) Assume: (i) u is continuous on $[a,b]$ and $Sch(u) \subset F$. (ii) $\mathcal{D}_F^\alpha u(x)$ exists and is F-continuous on $[a,b]$. (iii) v is F-continuous on $[a,b]$. Then

$$\int_a^b uvd_F^\alpha x = \left[u(x)\int_a^x v(x')d_F^\alpha x'\right]\Big|_a^b - \int_a^b \mathcal{D}_F^\alpha u(x)\int_a^x v(x')d_F^\alpha x' d_F^\alpha x \tag{4.11}$$

Some examples are given relative to applications and we mention e.g.

EXAMPLE 4.2. Following [**764**] one has a local fractal diffusion equation

$$\mathcal{D}^{\alpha}_{F,t}(W(x,t)) = \frac{\chi_F(t)}{2}\frac{\partial^2}{\partial x^2}W(x,t) \tag{4.12}$$

with solution

$$W(x,t) = \frac{1}{(2\pi S^{\alpha}_F(t))^{1/2}} exp\left(\frac{-x^2}{2S^{\alpha}_F(t)}\right) \tag{4.13}$$

The appendix to [**994**] also gives some formulas for repeated integration and differentiation. For example it is shown that

$$(\mathcal{D}^{\alpha}_F)^2(S^{\alpha}_F(x))^2 = 2\chi_F(x);\ \int_a^{x'}(S^{\alpha}_F(x))^n d^{\alpha}_F x = \frac{1}{n+1}(S^{\alpha}_F(x'))^{n+1} \tag{4.14}$$

We refer to [**764, 994**] for other interesting material. ∎

5. A BOHMIAN APPROACH TO QUANTUM FRACTALS

The powerful exact uncertainty method of Hall and Reginatto for passing from classical to quantum mechanics has been further embellished and deepened in recent years (see e.g. [**234, 235, 237, 250, 254, 255, 258, 259, 261, 1065, 592, 593, 594, 596, 598, 599, 600, 995, 996, 1062, 1063, 1064, 1122, 1123**] and Sections 1.1, 3.1, and 4.7 in this book. In [**593**] one finds an apparent incompleteness in the traditional trajectory based Bohmian mechanics when dealing with a quantum particle in a box. It turns out that there is no suitable HJ equation for describing the motion which in fact has a fractal character. After reviewing the material on scale relativity in Section 1.2 for example it is not surprising to encounter such situations and in Sanz [**1122**] the Bohmian point of view is reinstated for fractal trajectories. One should also remark in passing that there is much material available on weak or distribution solutions of HJ type equations and some of this should come into play here (cf. [**251**]). The main issue here however is that in order to treat wave functions displaying fractal features (quantum fractals) one needs to enlarge the picture via limiting processes. One derives the quantum trajectories by means of limiting procedures that involve the expansion of the wave function in a series of eigenvectors of the Hamiltonian.

Consider first the quantum analogue of the Weierstrass function

$$W(x) = \sum_0^{\infty} b^r Sin(a^r x);\ a > 1 > b > 0;\ ab \geq 1 \tag{5.1}$$

Then in the problem of a particle in a 1-dimensional box of length L (with $0 < x < L$) one can construct wave functions of the form

$$\Phi_t(x;R) = A\sum_{r=0}^{R} n^{r(s-2)} Sin(p_{n,r}x/\hbar)e^{-iE_{n,r}t/\hbar} \tag{5.2}$$

with $2 > s > 0$ and $n \geq 2$. Here $p_{n,r} = n^r\pi\hbar/L$ is the quantized momentum (with integer quantum number given by $n' = n^r$). $E_{n,r} = p^2_{n,r}/2m$ is the eigenenergy and a is a normalization constant. This wave function, which is a solution of the time dependent SE, is continuous and differentiable everywhere. However the wave function resulting in the limit, namely $\Phi_t(x) = lim_{R\to\infty}\Phi_t(x;R)$ is a fractal object in both space and time (cf. Mandelbrot [**850**]). This method for generating quantum fractals basically involves (given s) choosing a quantum number, say n, and then considering the series that contains its powers $n' = n^r$. There is also another related method (cf. [**174**]) of generating quantum fractals based on the presence of discontinuities in the wave function. The emergence of fractal features arises from the perturbations that such discontinuities cause in the wave function during propagation. This generating process can be easily understood by considering a wave function initially uniform along a certain interval $\ell = x_2 - x_1 \leq L$ inside the box

$$\Psi_0(x) = \begin{cases} \frac{1}{\sqrt{\ell}} & x_1 < x < x_2 \\ 0 & otherwise \end{cases} \tag{5.3}$$

The Fourier decomposition of this wave function is

$$\Psi_0(x) = \frac{2}{\pi\sqrt{\ell}}\sum_1^\infty \frac{1}{n}[Cos(p_n x_1/\hbar) - Cos(p_n x_2/\hbar)]Sin(p_n x/\hbar) \tag{5.4}$$

whose time evolved form is

$$\Psi_t(x) = \frac{2}{\pi\sqrt{\ell}}\sum_1^\infty \frac{1}{n}[Cos(p_n x_1/\hbar) - Cos(p_n x_2/\hbar)]Sin(p_n x/\hbar)e^{-iE_n t/\hbar} \tag{5.5}$$

It is equivalent to consider $r = R = 1$ in (5.2) and sum over n from 1 to N; the quantum fractal is then obtained in the limit $N \to \infty$. This equivalence is based on the fact that the Fourier decomposition of Ψ_0 gives precisely its expansion in terms of the eigenvectors of the Hamiltonian in the problem of a particle in a box (this is not a general situation).

EXAMPLE 5.1. The fractality of wave functions like $\Phi_t(x)$ or $\Psi_t(x)$ can be analytically estimated (cf. Boyarsky et al [**174**]) by taking advantage of a result for Fourier series. Thus given an arbitrary function $f(x) = \sum_1^N a_n exp(-inx)$ its real and imaginary pars are fractals (and also $|f(x)|^2$) with dimension $D_f = (5-\beta)/2$ if its power spectrum has the asymptotic form $|a_n|^2 \sim n^{-\beta}$ for $N \to \infty$ with $1 < \beta \leq 3$. Alternatively the fractality of $f(x)$ can also be calculated by measuring the length $\mathfrak{L}$ of its real and imaginary parts (or $|f(x)|^2$) as a function of the number of terms N considered in the generating sries. Asymptotically the relation bewteen $\mathfrak{L}$ and N can be expressed as $\mathfrak{L}(N) \propto N^{D_f - 1}$ which diverges if $f(x)$ is a fractal object. One notes that to increase the number of terms contributing to $f(x)$ is analogous to measuring its length with more precision. ■

It is known that for quantum fractals the corresponding expected value of the energy $<\hat{H}>$ becomes infinite. This is related to the fact that the familiar form of the SE

$$i\hbar\partial_t\Psi_t(x) = \hat{H}\Psi_t(x) \tag{5.6}$$

does not hold in general (cf. [**593, 1316**]). In this case neither the left side of (5.6) nor the right side belong to the Hilbert space; however the identity

$$[\hat{H} - i\hbar\partial_t]\Psi_t(x) = 0 \tag{5.7}$$

still remains valid. In this situation one says that $\Psi_t(x)$ is a weak solution of the SE (note weak solutions have many meanings and have been extensively studied in PDE - cf. [**251**]).

The formal basis of Bohmian mechanics (BM) is usually established via

$$\Psi_t(x) = \rho_t^{1/2}(x)e^{iS_t(x)/\hbar};\ \frac{\partial\rho_t}{\partial t} + \nabla\cdot\left(\rho_t\frac{\nabla S_t}{m}\right) = 0; \tag{5.8}$$

$$\frac{\partial S_t}{\partial t} + \frac{(\nabla S_t)^2}{2m} + V + Q_t = 0;\ Q_t = -\frac{\hbar^2}{2m}\frac{\nabla^2\rho_t^{1/2}}{\rho_t^{1/2}}$$

One postulates also the trajectory velocity as

$$\dot{x} = \frac{\nabla S_t}{m} = \frac{\hbar}{m}\Im[\Psi_t^{-1}\nabla\Psi_t] \tag{5.9}$$

Now Q_t in (5.8) is well defined provided that the quantum state is also well defined (i.e. continuous and differentiable). However this is not the case for quantum fractals and the theory seems incomplete; the solution is to take into account the decomposition of the quantum fractal in terms of differentiable eigenvectors and redefining Q_t in (5.8). Thus any wave function Ψ_t is expressible as

$$\Psi_t(x:N) = \sum_1^N c_n\xi_n(x)e^{-iE_nt/\hbar} \tag{5.10}$$

in the limit $N\to\infty$ (cf. Φ_t above and (5.5)) where the $\xi_n(x)$ are eigenvectors with eigenvalues E_n of the corresponding Hamiltonian. One can then define the quantum trajectories evolving under the guidance of this wave as

$$x_t = lim_{N\to\infty}x_N(t);\ \dot{x}_N = \frac{\hbar}{m}\Im\left[\Psi_t^{-1}(x;N)\frac{\partial\Psi_t(x;N)}{\partial x}\right] \tag{5.11}$$

Note the calculation of trajectories is not based on S_t, which has no trivial decomposition in a series of nice functions, but this kind of velocity formulation is common in e.g. [**412, 413, 414, 415, 416, 417, 418, 544, 545, 546, 548, 549, 550**] where one modern version of BM is being developed.

EXAMPLE 5.2. A numerical example is given in [**1122**] and we only mention a few features here. Thus one considers a highly delocalized particle in a box with wave function (5.4) and $x_1 = 0$ with $x_2 = L$. Then (5.5) becomes

$$\Psi_t(x) = \frac{4}{\pi\sqrt{L}}e^{-iE_1t/\hbar}\sum_{n\ odd}\frac{1}{n}Sin(p_nx/\hbar)e^{i\omega_{n,1}t} \tag{5.12}$$

where $\lambda_{n,1} = (E_n - E_1)/\hbar$ (in the numerical calculations one uses $L = m = \hbar = 1$). Here the probability density ρ_t is periodic in time but the wave function is not periodic (this does not affect (5.11)). Various features are observed (e.g. Cantor set structures, Gibbs phenomena, etc.) and graphs are displayed - we omit any further discussion here. ■

In summary, although the SE is not satisfied by quantum fractals as a whole, it is when one considers its decomposition in terms of the eigenvectors of the Hamiltonian. The contributing eigenvectors are continuous and differentiable and any wave function (regular or not) admits a decomposition in terms of eigenvectors. Correspondingly the Bohmian equation of motion must be reformulated in terms of such decompositions via (5.11) and this can be regarded also as a generalization of (5.8). We mention in passing that from time to time there are papers claiming contradictions between BM and QM and we refer here to [**580, 852**] for some refutations.

REMARK 5.3. Let us mention here a suggestion of 't Hooft [**641**] about establishing the physical link between classical and quantum mechanics by employing the underlying equations of classical mechanics and including into them a specially chosen dissipative function. The wave like QM turns out to follow from the particle like classical mechanics due to embedding in the latter a dissipation "device" responsible for loss of information. Thus the initial precise information about the classical trajectory is lost in QM due to the "dissipative spread" of the trajectory and its transformation into a fuzzy object such as the fractal Hausdorff path of dimension 2 in a simple case of a spinless particle. Some rough calculations in this direction appear in [**561**] and we refer also to [**151, 562, 995**]. ∎

CHAPTER 2

DEBROGLIE-BOHM IN VARIOUS CONTEXTS

The quantum potential arises in various forms, some of which were discussed in Chapter 1. We return to this now in a somewhat more systematic manner. The original theory goes back to deBroglie and D. Bohm (see e.g. [**118, 119, 160, 161, 187, 637, 638, 713**]) and in its modern version the dominant themes seem to be contained in [**109, 85, 126, 127, 128, 129, 130, 207, 357, 367, 412, 413, 414, 415, 416, 417, 418, 497, 520, 544, 545, 546, 547, 548, 549, 550, 1218, 1245**] with variations as in [**133, 234, 235, 236, 237, 238, 241, 244, 245, 445, 446, 447, 448, 449, 478, 479, 480, 481**] based on work of Bertoldi, Faraggi, Floyd, and Matone (cf. also [**85, 168, 178, 199, 294, 382, 383, 637, 638, 639, 640, 699, 773, 774, 775, 1160, 1170, 1251, 1252**]) and cosmology following [**153, 236, 237, 270, 668, 669, 670, 671, 770, 931, 934, 935, 1115, 1116, 1158, 1159, 1160, 1162, 1163, 1181, 1299, 1300**]. In any event the quantum potential does enter into any trajectory theory of deBroglie-Bohm (dBB) type. The history is discussed for example in Holland [**637**] (cf. also [**85, 158, 159, 161, 187**]) and we have seen how this quantum potential idea can be formulated in various ways in terms of statistical mechanics, hydrodynamics, information and entropy, etc. when dealing with different versions and origins of the SE. Given the existence of particles we finds the pilot wave of thinking very attractive, with the wave function serving to choreograph the particle motion (or perhaps to "create" particles and/or spacetime paths). However the existence of particles itself is not such an assured matter and in field theory approaches for example one will deal with particle currents (cf. Nikolić [**930, 931**] and see also e.g. [**118, 119, 414, 520, 1245**]). The whole idea of quantum particle path seems in any case to be either fractal (cf. [**1, 3, 16, 17, 234, 275, 287, 337, 913, 941, 943, 946, 965, 966, 967, 968**], stochastic (see e.g. [**85, 178, 234, 490, 491, 594, 596, 598, 599, 722, 725, 907, 910, 926, 1062, 1063**], or field theoretic (cf. [**118, 119, 414, 520, 930, 931, 932, 934, 935, 1245**]. The fractal approach sometimes imagines an underlying micro-spacetime where paths are perhaps fractals with jumps, etc. and one possible advantage of a field theoretic approach would be to let the fields sense the ripples, which as e.g. operator valued Schwartz type distributions, they could well accomplish. In fact what comes into question here is the structure of the vacuum and/or of spacetime itself. One can envision microstructures as in [**234, 556, 913**] for example, textures (topological defects) as in [**88, 96, 97, 209, 1282**], Planck scale structure and QFT, along with space-time uncertainty relations as in [**88, 396, 397, 818, 1326**], vacuum structures and conformal invariance as in [**901, 902, 1104, 1108, 1110, 1111, 1117**], pilot wave cosmology as in [**1107, 1170**], ether theories as in [**1129, 1210**], etc. Generally there seems to be a sense in which particles cannot be measured as such and hence the idea of particle currents (perhaps corresponding to fuzzy particles or ergodic clumps) should prevail perhaps along with the idea of probability packets. A number of arguments work with a (representative) trajectory as if it were a single particle but there is no reason to take this too seriously; it could be thought of perhaps as a "typical" particle in a cloud but conclusions should perhaps always be constructed from an ensemble point of view. We will try to develop some of this below. The sticky point as we see it now goes as follows. Even though one can write stochastic equations for (typical) particle motion as in the Nelson theory for example one runs into the problem of ever actually being able to localize a particle. Indeed as indicated in [**396, 397**] (working in a relativistic context but this should hold in general) one expects space time uncertainty relations even at a semiclassical level since any localization experiment will generate a gravitational field and deform spacetime. Thus there are relations $[q_\mu, q_\nu] = i\lambda_P^2 Q_{\mu\nu}$ where λ_P is the Planck length and the picture of spacetime as a local Minkowski manifold should break down at distances of order λ_P. One wants the localization experiment to avoid creating a black hole (putting the object out of "reach") for example and this suggests $\Delta x_0(\sum_1^3 \Delta x_i) \gtrsim \lambda_P^2$ with $\Delta x_1\Delta x_2 + \Delta x_2\Delta x_3 + \Delta x_3\Delta x_1 \gtrsim \lambda_P^2$ (cf. [**396, 397**]). On the other hand in [**930, 931**] it is shown that in a relativistic bosonic field theory for example one can speak of currents and n-particle wave functions can have particles attributed to them with well defined trajectories, even though the probability of their experimental detection is zero. Thus one enters an arena of perfectly respectible but undetectible particle trajectories. The discussion in [**320, 414, 1211, 1251, 1262**] is also relevant here; some recourse to the idea of beables, reality, and observables as beables, etc. is also involved (cf. [**118, 119, 320, 1262**]). We will have something to say about all these matters.

The dominant approach as in Dürr et al [**412, 413, 414, 415, 520, 1245**] will be discussed as needed (a thorough discussion would take a book in itself) and we only note here that one is obliged to use the form $\psi = Rexp(iS/\hbar)$ to make sense out of the constructions (this is no problem with suitable provisos, e.g. that S is not constant - cf. [**238, 445, 446, 478, 479**] and comments later). This leads to

$$(\bigstar)\ S_t + \frac{(S')^2}{2m} - \left(\frac{\hbar^2}{2m}\right)\left(\frac{R''}{R}\right) + V = 0;\ \partial_t R^2 + \partial\left(\frac{R^2 S'}{m}\right) = 0$$

(cf. (1.1.1)) where $Q = -\hbar^2 R''/2mR$ arising from a SE $i\hbar\partial_t\psi = -(\hbar^2/2m)\psi_{xx} + V\psi$ (we use 1-D for simplicity here). In [**412**] one emphasizes configurations based on coordinates whose motion is choreographed by the SE according to the rule

$$(\bigstar\bigstar)\ \dot{q} = v = \frac{\hbar}{m}\Im\frac{\psi^*\psi'}{|\psi|^2} = \frac{\hbar}{m}\Im\left(\frac{\psi'}{\psi}\right)$$

The argument for (★★) is based on obtaining the simplest Galilean and time reversal invariant form for velocity, transforming correctly under velocity boosts. This leads directly to (★★) so that Bohmian mechanics (BM) is governed by (★★) and the SE. It's a fairly convincing argument and no recourse to Floydian time need be involved (cf. [**238, 446, 478, 479**]). Note however that if $S = c$

then $\dot{q} = v = (\hbar/m)\Im(R'/R) = 0$ while $p = S' = 0$ so this formulation seems to avoid the $S = constant$ problems indicated in [**133, 238, 446, 478, 479**]. What makes the constant $\hbar/m$ in (★★) important here is that with this value the probability density $|\psi|^2$ on configuration space is equivariant. This means that via the evolution of probability densities $\rho_t + div(v\rho) = 0$ (as in (1.1.5)) the density $\rho = |\psi|^2$ is stationary relative to ψ, i.e. $\rho(t)$ retains the form $|\psi(q,t)|^2$. One calls $\rho = |\psi|^2$ the quantum equilibrium density (QEDY) and says that a system is in quantum equilibrium when its coordinates are randomly distributed according to the QEDY. The quantum equilibrium hypothesis (QEHP) is the assertion that when a system has wave function ψ the distribution ρ of its coordinates satisfies $\rho = |\psi|^2$.

1. THE KLEIN-GORDON AND DIRAC EQUATIONS

Before embarking on further discussion of QM it is necessary to describe some aspects of quantum field theory (QFT) and in particular to give some foundation for the Klein-Gordon (KG) and Dirac equations. For QFT we rely on [**149, 616, 709, 1017, 1096, 1291, 1336**] and concentrate on aspects of general quantum theory that are expressed through such equations. We alternate between signature $(-,+,+,+)$ and $(+,-,-,-)$ in Minkowski space, depending on the source. It is hard to avoid using units $\hbar = c = 1$ when sketching theoretical matters (which is personally repugnant) but we will set $\hbar = c = 1$ and shift to the general notation whenever any real meaning is desired. Thus $|length| \sim |time| \sim |energy|^{-1} \sim |mass|^{-1}$ and $m =$ the inverse Compton wavelength ($mc/\hbar = \ell_C^{-1}$). The approaches in [**616, 1017**] seem best adapted to our needs and in particular [**616**] gives a nice discussion motivating second quantization of a nonrelativistic SE. The resulting second quantization would be Galiean invariant but not Lorentz invariant so we go directly to the KG equation as follows. Note that there are often notational differences in various treatments of QFT and we use that of [**616**] in general. Start now from $E^2 = p^2 + m^2$ (which is the relativistic form of $E = p^2/2m$) to arrive, via $E \to i\partial_t$ and $p_j \to -i\partial_j = -i\partial/\partial x^j$, at the KG equation

$$(\partial_t^2 - \nabla^2)\phi + m^2\phi = 0 \tag{1.1}$$

where $\phi = \phi(\mathbf{x}, t)$ is a scalar wave function. This can also be derived from an action

$$S(\phi) = \int d^4x \mathfrak{L}(\phi, \partial_\mu\phi) = \frac{1}{2}\int d^4x(\partial^\mu\partial_\mu\phi - m^2\phi) \tag{1.2}$$

($x^0 = t$, $x = (\mathbf{x}, t)$), provided ϕ transforms as a Lorentz scalar (required also in (1.1)). The first problems arise from negative energy solutions (e.g. $exp[i(\mathbf{k}\cdot\mathbf{x}+\omega t)]$ is a solution of (1.1) with $E = -\omega = -(\mathbf{k}^2 + m^2)^{1/2}$). Secondly the energy spectrum is not bounded below (i.e. one could extract an arbitrary amount of energy from a single particle system). Further, using a positve square root of $E^2 = p^2 + m^2$ would involve a square root of a differential operator and nonlocal terms. Next observe that conserved currents j_μ (with $\partial^\mu j_\mu = 0$) arise à la E.

Noether in the form

$$j_0 = \rho = \frac{i}{2m}(\phi^*\phi_t - \phi_t^*\phi);\ j_i = \frac{1}{2im}(\phi^*\partial_i\phi - (\partial_i\phi^*)\phi) \tag{1.3}$$

(where normal ordering is implicit here in order to avoid dealing with a vacuum energy term - to be discussed later). We note that for the plane wave solution above $\rho = -\omega/m = -(1/m)(\mathbf{k}^2 + m^2)^{1/2}$ and this is not a good probability density. The difficulties are resolved by giving up the idea of a one particle theory; it is not compatible with Lorentz invariance and the solution is to quantize the field ϕ.

Thus take $S(\phi)$ as in (1.2) with $\pi = \partial\mathfrak{L}/\partial(\partial_\mu\phi) = \partial_t\phi = \dot{\phi}$ and construct a Hamiltonian

$$H = \frac{1}{2}\int d^3x[\pi^2(x) + |\nabla\phi(x)|^2 + m^2\phi^2(x)] \tag{1.4}$$

In analogy with QM where $[x,p] = i$ one stipulates

$$[\phi(\mathbf{x},t), \pi(\mathbf{y},t)] = i\delta(\mathbf{x}-\mathbf{y});\ [\phi(\mathbf{x},t), \phi(\mathbf{y},t)] = [\pi(\mathbf{x},t), \pi(\mathbf{y},t)] = 0 \tag{1.5}$$

The operator equation $\dot{\phi} = i[H,\phi]$ then yields $\pi = \dot{\phi}$ and $\dot{\pi} = i[H,\pi]$ reproduces the KG equation. This is now a quantum field theory and for a particle interpretation one expands $\phi(\mathbf{x},t)$ in terms of classical solutions of the KG equation via

$$\phi(\mathbf{x},t) = \sum a(\mathbf{k})\phi_k^+(x) + b(\mathbf{k})\phi_k^-(x) = \tag{1.6}$$

$$= \int \frac{d^3\mathbf{k}}{(2\pi)^3}\frac{1}{2\omega_k}(a(\mathbf{k})e^{-i[\omega_k t - \mathbf{k}\cdot\mathbf{x}]} + b(\mathbf{k})e^{i[\omega_k t - \mathbf{k}\cdot\mathbf{x}]})$$

where $\omega_k = (\mathbf{k}^2 + m^2)^{1/2}$ and $\phi_k^\pm$ denotes a classical positive (resp. negative) energy plane wave solution of (1.1) ($k\cdot x = k_0x_0 - \mathbf{k}\cdot\mathbf{x} = \omega_k t - \mathbf{k}\cdot\mathbf{x}$). With $\phi(\mathbf{x},t)$ an operator one has operators $a(\mathbf{k})$ and $b(\mathbf{k})$; further since $\phi(\mathbf{x},t)$ is classically a real field we must have a Hermitian operator here and hence $b(\mathbf{k}) = a^\dagger(\mathbf{k})$. The normalization factor $1/2\omega_k$ is chosen for Lorentz invariance (cf. Hatfield [**616**] for details). It follows immediately from $\pi = \partial_t\phi$ that

$$\pi(\mathbf{x},t) = \int \frac{k^3k}{(2\pi)^3}\frac{1}{2\omega_k}(-i\omega_k a(\mathbf{k})e^{-ik\cdot x} + i\omega_k a^\dagger(\mathbf{k})e^{ik\cdot x}) \tag{1.7}$$

Some calculation (via Fourier formulas) leads then to

$$a(\mathbf{k}) = \int d^3x e^{ik\cdot x}[\omega_k\phi(\mathbf{x},t) + i\pi(\mathbf{x},t)] \tag{1.8}$$

and the algebra of a, $a^\dagger$ is then determined by $[a(\mathbf{k}), a^\dagger(\mathbf{k}')] = (2\pi)^3 2\omega_k\delta^3(\mathbf{k}-\mathbf{k}')$. The Hamiltonian (1.4) yields

$$H = \frac{1}{2}\int \frac{d^3k}{(2\pi)^3}\frac{1}{2\omega_k}\omega_k[a^\dagger(\mathbf{k})a(\mathbf{k}) - a(\mathbf{k})a^\dagger(\mathbf{k})] \tag{1.9}$$

There is a bit of hocus-pocus here since the calculation gives $a^\dagger a + (1/2)[a, a^\dagger]$ ($= (1/2)(a^\dagger a + aa^\dagger)$ formally) but $[a, a^\dagger] \sim \delta(0)$ corresponds to the sum oveer all modes of zero point energies $\omega_k/2$. This infinite energy cannot be detected experimaentally since experiments only measure differences from the ground state of H. In any event the zero point field (ZPF) will be discussed in some detail later.

Now the ground state is defined via $a(\mathbf{k})|0>=0$ with $<0|0>=1$ ($a^\dagger(\mathbf{k})|0>$ is a one particle state with energy ω_k and momentum $\mathbf{k}$ while $(a^\dagger(\mathbf{k}))^2|0>$ contains two such particles, etc.). One notes however that the state $a^\dagger(\mathbf{k})|0>$ is not normalizable since $<0|a(\mathbf{k})a^\dagger(\mathbf{k})|0>=\delta(0)$ is not normalizable. This is not surprising since $a^\dagger(\mathbf{k})$ creates a particle of definite energy and momentum and by the uncertainty principle its location is unknown. Thus its wave function is a plane wave and such states are not normalizable. In fact $a^\dagger(\mathbf{k})$ is an operator valued distribution and one can do calculations by "smearing" and considering states $\int d^3k f(\mathbf{k})a^\dagger|0>$ for functions f such that $\int d^3k|f(\mathbf{k})|^2 < \infty$ for example. One sees also that the bare vacuum $|0>$ is an eigenstate of the Hamiltonian but its energy is divergent via $<0|H|0>=(1/2)\int d^3k\omega_k\delta^3(0)$ (where $(2\pi)^3\delta^3(0)\sim\int d^3x$). To deal with such infinities one subtracts them away, i.e. $H\to H-<0|H|0>$ and this corresponds to normal ordering the Hamiltonian via $:aa^\dagger:=:a^\dagger a:=a^\dagger a$ leading to

$$:H:=\int\frac{d^3k}{(2\pi)^3}\frac{1}{2\omega_k}\omega_k a^\dagger(\mathbf{k})a(\mathbf{k}) \tag{1.10}$$

with vanishing vacuum expectation.

REMARK 2.1.1. Regarding Lorentz invariance one recalls that the Lorentz group $O(3,1)$ is the set of 4×4 matrices leaving the form $s^2=(x^0)^2-\sum(x^i)^2=x^\mu g_{\mu\nu}x^\nu$ invariant. One writes $(x')^\mu=\Lambda^\mu_\nu x^\nu$ and notes that $g_\mu=\Lambda^\rho_\mu g_{\rho\sigma}\Lambda^\sigma_\nu\sim g=\Lambda^T g\Lambda$. Since s^2 can be plus or minus there is a splitting into regions $(x-y)^2>0$ (time-like), $(x-y)^2<0$ (space-like), and $(x-y)^2=0$ (light-like). A standard parametrization for Lorentz boosts involves ($x^0=ct$)

$$x'=\frac{x+vt}{\sqrt{1-(v/c)^2}};\ y'=y;\ z'=z;\ t'=\frac{t+(vx/c^2)}{\sqrt{1-(v/c)^2}} \tag{1.11}$$

One writes e.g. $\gamma=1/\sqrt{1-(v/c)^2}=cosh(\phi)$ with $sinh(\phi)=\beta\gamma=v\gamma/c$. ∎

REMARK 2.1.2. The total 4-momentum operator is

$$P^\mu=\int\frac{d^3k}{(2\pi)^3}\frac{1}{2\omega_k}k^\mu a^\dagger(\mathbf{k})a(\mathbf{k}) \tag{1.12}$$

and the total angular momentum operator is

$$M^{\mu\nu}=\int d^3x(x^\mu p^\nu-x^\nu p^\mu) \tag{1.13}$$

The Lorentz algebra (for infinitesimal Lorentz transformations) is

$$[M^{\mu\nu},M^{\lambda\sigma}]=i(\eta^{\mu\lambda}M^{\nu\sigma}-\eta^{\nu\lambda}M^{\mu\sigma}-\eta^{\mu\sigma}M^{\nu\lambda}+\eta^{\nu\sigma}M^{\mu\lambda}) \tag{1.14}$$

where $\eta^{\mu\nu}=diag(1,-1,-1,-1)$. ∎

REMARK 2.1.3. The commutator rules (1.5) are not manifestly Lorentz covariant. However one can verify that the same quantum theory is obtained regardless of what Lorentz frame is chosen; to do this one shows that the QM operator forms of the Lorentz generators satisfy the Lorentz algebra after quantization (this is given as an exercise in [**616**]). ∎

EXAMPLE 1.1. Quantum fields are also discussed briefly by Holland in [**637**] and we extract here from this source. The approach follows [**160**] and one takes $\mathcal{L} = (1/2)\partial_\mu\psi\partial^\mu\psi = (1/2)[\dot{\psi}^2 - (\nabla\psi)^2]$ as Lagrangian where $\dot{\psi} = \partial_t\psi$ and variational technique yields the wave equation $\Box\psi = 0$ ($\hbar = c = 1$). Define conjugate momentum as $\pi = \partial\mathcal{L}/\partial\dot{\psi}$, the Hamiltonian via $\mathcal{H} = \pi\dot{\psi} - \mathcal{L} = (1/2)[\pi^2 + (\nabla\psi))^2]$, and the field Hamiltonian by $\mathfrak{H} = \int \mathcal{H}d^3x$. Replacing π by $\delta S/\delta\psi$ where $S[\psi]$ is a functional the classical HJ equation of the field $\partial_t S + H = 0$ becomes

$$\frac{\partial S}{\partial t} + \frac{1}{2}\int d^3x \left[\left(\frac{\delta S}{\delta\psi}\right)^2 + (\nabla S)^2\right] = 0 \tag{1.15}$$

The term $(1/2)\int d^3x(\nabla\psi)^2$ plays the role of an external potential. To quantize the system one treats $\psi(\mathbf{x})$ and $\pi(\mathbf{x})$ as Schrödinger operators with $[\psi(\mathbf{x}), \psi(\mathbf{x}')] = [\pi(\mathbf{x}), \pi(\mathbf{x}')] = 0$ and $[\psi(\mathbf{x}), \pi(\mathbf{x}')] = i\delta(\mathbf{x}-\mathbf{x}')$. Then one works in a representation $|\psi(\mathbf{x}) >$ in which the Hermitian operator $\psi(\mathbf{x})$ is diagonal. The Hamiltonian becomes an operator $\hat{H}$ acting on a wavefunction $\Psi[\psi(\mathbf{x}), t) =< \psi(\mathbf{x})|\Psi(t) >$ which is a functional of the real field ψ and a function of t. This is not a point function of $\mathbf{x}$ since Ψ depends on the variable ψ for all $\mathbf{x}$. Now the SE for the field is $i\partial_t\Psi = \hat{H}\Psi$ or explicitly

$$i\frac{\partial\Psi}{\partial t} = \int d^3x \frac{1}{2}\left[-\frac{\delta^2}{\delta\psi^2} + (\nabla\psi)^2\right]\Psi \tag{1.16}$$

Thus ψ is playing the role of the space variable $\mathbf{x}$ in the particle SE and the continuous index $\mathbf{x}$ here is analogous to a discrete index n in the many particle theory. To arrive at a causal interpretation now one writes $\Psi = Rexp(iS)$ for $R, S[\psi, t]$ real functionals and decomposes (1.16) as

$$\frac{\partial S}{\partial t} + \frac{1}{2}\int d^3x \left[\left(\frac{\delta S}{\delta\psi}\right)^2 + (\nabla\psi)^2\right] + Q = 0;\ \frac{\partial R^2}{\partial t} + \int d^3x \frac{\delta}{\delta\psi}\left(R^2\frac{\delta S}{\delta\psi}\right) = 0 \tag{1.17}$$

where the quantum potential is now $Q[\psi, t] = -(1/2R)\int d^3x(\delta^2 R/\delta\psi^2)$. (1.17) now gives a conservation law wherein, at time t, $R^2 D\psi$ is the probability for the field to lie in an element of volume $D\psi$ around ψ, where $D\psi$ means roughly $\prod_{\mathbf{x}} d\psi$ and there is a normalization $\int |\Psi|^2 D\psi = 1$. Now introduce the assumption that at each instant t the field ψ has a well defined value for all $\mathbf{x}$ as in classical field theory, whatever the state Ψ. Then the time evolution is obtained from the solution of the "guidance" formula

$$\frac{\partial\psi(\mathbf{x}, t)}{\partial t} = \left.\frac{\delta S[\psi(\mathbf{x}), t]}{\delta\psi(\mathbf{x})}\right|_{\psi(\mathbf{x})=\psi(\mathbf{x},t)} \tag{1.18}$$

(analogous to $m\ddot{\mathbf{x}} = \nabla S$) once one has specified the initial function $\psi_0(\mathbf{x})$ in the HJ formalism. To find the equation of motion for the field coordinates apply $\delta/\delta\psi$ to the HJ equation (1.17) to get formally ($\dot{\psi} \sim \delta S/\delta\psi$)

$$\frac{d}{dt}\dot{\psi} = -\frac{\delta}{\delta\psi}\left[Q + \frac{1}{2}\int d^3x(\nabla\psi)^2\right];\ \frac{d}{dt} = \frac{\partial}{\partial t} + \int d^3x \frac{\partial\psi}{\partial t}\frac{\delta}{\delta\psi} \tag{1.19}$$

This is analogous to $m\ddot{\mathbf{x}} = -\nabla(V+Q)$ and, noting that $d\dot{\psi}/dt = \partial\dot{\psi}/dt$ and taking the classical external force term to the right one arrives, via standard variational methods, at

$$\Box\psi(\mathbf{x}),t) = -\left.\frac{\delta Q[\psi(\mathbf{x},t]}{\delta\psi(\mathbf{x})}\right|_{\psi(\mathbf{x})=\psi(\mathbf{x},t)} \tag{1.20}$$

(note $(\delta/\delta\psi)\int d^3x(\nabla\psi)^2 \sim -2\Delta\psi$ and $(\delta/\delta\psi 0\partial_t\psi = \partial_t(\delta/\delta\psi)\psi = 0)$. The quantum force term on the right side is responsible for all the characteristic effects of QFT. In particular comparing to a classical massive KG equation $\Box\psi + m^2\psi = 0$ with suitable initial conditions one can argue that the quantum force generates mass in the sense that the massless quantum field acts as if it were a classical field with mass given via the quantum potential (cf. Remark 2.2.1 below). ■

1.1. ELECTROMAGNETISM AND THE DIRAC EQUATION. It will be useful to have a differential form discription of EM fields and we supply this via Ohanian-Ruffini [**950**]. Thus one thinks of tensors $T = T^{\sigma}_{\mu\nu}\partial_\sigma \otimes dx^\mu \otimes dx^\nu$ with contractions of the form $T(dx^\sigma, \partial_\sigma) \sim T_\nu dx^\nu$. For $\eta = \eta_{\mu\nu}dx^\mu \otimes dx^\nu$ one has $\eta^{-1} = \eta^{\mu\nu}\partial_\mu \otimes \partial_\nu$ and $\eta\eta^{-1} = 1 \sim diag(\delta^\mu_\mu)$. Note also e.g.

$$\begin{aligned} \eta_{\mu\nu}dx^\mu \otimes dx^\nu(\mathbf{u},\mathbf{w}) &= \eta_{\mu\nu}dx^\mu(\mathbf{u})dx^\nu(\mathbf{w}) = \\ &= \eta_{\mu\nu}dx^\mu(u^\alpha\partial_\alpha)dx^\nu(w^\tau\partial_\tau) = \eta_{\mu\nu}u^\mu w^\nu \end{aligned} \tag{1.21}$$

$$\begin{aligned} \eta(\mathbf{u}) &= \eta_{\mu\nu}dx^\mu \otimes dx^\nu(\mathbf{u}) = \eta_{\mu\nu}dx^\mu(\mathbf{u})dx^\nu = \\ &= \eta_{\mu\nu}dx^\mu(u^\alpha\partial_\alpha)dx^\nu = \eta_{\mu\nu}u^\mu dx^\nu = u_\nu dx^\nu \end{aligned} \tag{1.22}$$

for a metric η. Recall $\alpha\wedge\beta = \alpha\otimes\beta - \beta\otimes\alpha$ and

$$\alpha\wedge\beta = \alpha_\mu dx^\mu \wedge \beta_\nu dx^\nu = (1/2)(\alpha_\mu\beta_\nu - \alpha_\nu\beta_\mu)dx^\mu\wedge dx^\nu \tag{1.23}$$

The EM field tensor is $F = (1/2)F_{\mu\nu}dx^\mu\wedge dx^\nu$ where

$$F_{\mu\nu} = \begin{pmatrix} 0 & E_x & E_y & E_z \\ -E_x & 0 & -B_z & B_y \\ -E_y & B_z & 0 & -B_x \\ -E_z & -B_y & B_x & 0 \end{pmatrix}; \tag{1.24}$$

$$\begin{aligned} F = E_x dx^0\wedge dx^1 + E_y dx^0\wedge dx^2 + E_z dx^0 \wedge dx^3 - B_z dx^1\wedge dx^2 + \\ +B_y dx^1\wedge dx^3 - B_x dx^2\wedge dx^3 \end{aligned}$$

The equations of motion of an electric charge is then $d\mathbf{p}/d\tau = (e/m)\mathbf{F}(\mathbf{p})$ where $\mathbf{p} = p^\mu\partial_\mu$. There is only one 4-form, namely $\epsilon = dx^0\wedge dx^1\wedge dx^2\wedge dx^3 = (1/4!)\epsilon_{\mu\nu\sigma\tau}dx^\mu\wedge dx^\nu\wedge dx^\sigma\wedge dx^\tau$ where $\epsilon_{\mu\nu\sigma\tau}$ is totally antisymmetric. Recall also for $\alpha = \alpha_{\mu\nu\ldots}dx^\mu\wedge dx^\nu\cdots$ one has $d\alpha = d\alpha_{\mu\nu\ldots}\wedge dx^\mu\wedge dx^\nu\cdots = \partial_\alpha\alpha_{\mu\nu\ldots}dx^\sigma\wedge dx^\mu\wedge dx^\nu\cdots$ and $dd\alpha = 0$. Define also the Hodge star operator on F and j via ${}^*F = (1/4)\epsilon_{\mu\nu\sigma\tau}F^{\sigma\tau}dx^\mu\wedge dx^\nu$ and ${}^*j = (1/3!)\epsilon_{\mu\nu\sigma\tau}j^\tau dx^\mu\wedge dx^\nu\wedge dx^\sigma$; these are called dual tensors. Now the Maxwell equations are

$$\partial_\mu F^{\mu\nu} = \frac{4\pi}{c}j^\nu;\ \partial^\alpha F^{\mu\nu} + \partial^\mu F^{\nu\alpha} + \partial^\nu F^{\alpha\mu} = 0 \tag{1.25}$$

and this can now be written in the form

$$dF = 0;\ d^*F = \frac{4\pi}{c} * j \tag{1.26}$$

and $0 = d^*j = 0$ is automatic. In terms of $A = A_\mu dx^\mu$ where $F = dA$ the relation $dF = 0$ is an identity $ddA = 0$.

A few remarks about the tensor nature of j^μ and $F^{\mu\nu}$ are in order and we write $n = n(x)$ and $\mathbf{v} = \mathbf{v}(x)$ for number density and velocity with charge density $\rho(x) = qn(x)$ and current density $\mathbf{j} = qn(x)\mathbf{v}(x)$. The conservation of particle number leads to $\nabla \cdot \mathbf{j} + \rho_t = 0$ and one writes

(1.27) $$j^\nu = (c\rho, j_x, j_y, j_z) = (c\rho n, qnv_x, qnv_y, qnv_z) \equiv j^\nu = n_0 q u^\nu \equiv j^\nu = \rho_0 u^\nu$$

where $n_0 = n\sqrt{1-(v^2/c^2)}$ and $\rho_0 = qn_0$ (ρ_0 here is charge density). Since j^ν consists of u^ν multiplied by a scalar it must have the transformation law of a 4-vector $j'^\beta = a^\beta_{\ \nu} j^\nu$ under Lorentz transformations ($a^\beta_\nu \sim \Lambda^\beta_\nu$). Then the conservation law can be written as $\partial_\nu j^\nu = 0$ with obvious Lorentz invariance. After some argument one shows also that $F^{\mu\nu} = a^{\ \nu}_\beta a^{\ \mu}_\alpha F'^{\alpha\beta}$ under Lorentz transformations so $F^{\mu\nu}$ is indeed a tensor. The equation of motion for a charged particle can be written now as

(1.28) $$(d\mathbf{p}/dt) = q\mathbf{E} + (q/c)\mathbf{v} \times \mathbf{B};\ \mathbf{p} = m\mathbf{v}/\sqrt{1-(v^2/c^2)}$$

This is equivalent to $dp^\mu/dt = (q/m)p_\nu F^{\mu\nu}$ with obvious Lorentz invariance. The energy momentum tensor of the EM field is

(1.29) $$T^{\mu\nu} = -(1/4\pi)[F^{\mu\alpha}F^\nu_{\ \alpha} - (1/4)\eta^{\mu\nu}F^{\alpha\beta}F_{\alpha\beta}]$$

(cf. [**950**] for details) and in particular $T^{00} = (1/8\pi)(\mathbf{E}^2 + \mathbf{B}^2)$ while the Poynting vector is $T^{0k} = (1/4\pi)(\mathbf{E} \times \mathbf{B})^k$.

One can equally well work in a curved space where e.g. covariant derivatives are defined via $\nabla_n T = lim_{d\lambda \to 0}[(T(\lambda + d\lambda) - T(\lambda) - \delta T]/d\lambda$ where δT is the change in T produced by parallel transport. One has then the usual rules $\nabla_u(T \otimes R) = \nabla_u T \otimes R + T \otimes \nabla_u R$ and for $\mathbf{v} = v^\nu \partial_\nu$ one finds $\nabla_\mu \mathbf{v} = \partial_\mu v^\nu \partial_\nu + v^\nu \nabla_\mu \partial_\nu$. Now if $\mathbf{v}$ was constructed by parallel transport its covariant derivative is zero so, acting with the dual vector dx^α gives

(1.30) $$\frac{\partial x^\nu}{\partial x^\mu} dx^\alpha(\partial_\nu) + v^\nu dx^\alpha(\nabla_\mu \partial_\nu) = 0 \equiv \partial_\mu v^\alpha + v^\nu dx^\alpha(\nabla_\mu \partial_\nu) = 0$$

Comparing this with the standard $\partial_\mu v^\alpha + \Gamma^\alpha_{\ \mu\nu} v^\nu = 0$ gives $dx^\alpha(\nabla_\mu \partial_\nu) = \Gamma^\alpha_{\ \mu\nu}$. One can show also for vectors u, v, w (boldface omitted) and a 1-form α

(1.31) $$(\nabla_u \nabla_v - \nabla_v \nabla_u - uv + vu)\alpha(w) = R(\alpha, u, v, w);$$
$$R = F^\sigma_{\ \beta\mu\nu} \partial_\sigma \otimes dx^\beta \otimes dx^\mu \otimes dx^\nu$$

so R represents the Riemann tensor.

For the nonrelativistic theory we recall from [**876**] that one can define a transverse and longitudinal component of a field F via

(1.32) $$F^{||}(r) = -\frac{1}{4\pi}\int d^3r' \frac{\nabla' \cdot F(r')}{|r - r'|};\ F^\perp(r) = \frac{1}{4\pi}\nabla \times \nabla \times \int d^3r' \frac{F(r')}{|r - r'|}$$

For a point particle of mass m and charge e in a field with potentials A and ϕ one has nonrelativistic equations $m\ddot{x} = eE + (e/c)v \times B$ (boldface is suppressed here) where one recalls $B = \nabla \times A$, $v = \dot{x}$, and $E = -\nabla\phi - (1/c)A_t$ with $H = (1/2m)(p - (e/c)A)^2 + e\phi$ leading to

$$\dot{x} = \frac{1}{2m}\left(p - \frac{e}{c}A\right);\ \dot{p} = \frac{e}{c}[v \times B + (v \cdot \nabla)A] - e\nabla\phi \tag{1.33}$$

Recall here also

$$B = \nabla \times A;\ \nabla \cdot E = 0;\ \nabla \cdot B = 0;\ \nabla \times E = -(1/c)B_t; \quad \nabla \times B = (1/c)E_t;\ E = -(1/c)A_t - \nabla\phi \tag{1.34}$$

(the Coulomb gauge $\nabla \cdot A = 0$ is used here). One has now $E = E^{\perp} + E^{||} \sim E^T + E^L$ with $\nabla \cdot E^{\perp} = 0$ and $\nabla \times E^{||} = 0$ and in Coulomb gauge $E^{\perp} = -(1/c)A_t$ and $E^{||} = -\nabla\phi$. Further

$$H \sim \frac{1}{2m}\left(p - \frac{e}{c}A\right)^2 + e\phi + \frac{1}{8\pi}\int d^3r((E^{\perp})^2 + B^2) \tag{1.35}$$

(covering time evolution of both particle and fields).

For the relativistic theory one goes to the Dirac equation

$$i(\partial_t + \alpha \cdot \nabla)\psi = \beta m\psi \tag{1.36}$$

which, to satisfy $E^2 = \mathbf{p}^2 + m^2$ with $E \sim i\partial_t$ and $\mathbf{p} \sim -i\nabla$, implies $-\partial_t^2\psi = (-i\alpha \cdot \nabla + \beta m)^2\psi$ and ψ will satisfy the KG equation if $\beta^2 = 1$, $\alpha_i\beta + \beta\alpha_i \equiv \{\alpha_i, \beta\} = 0$, and $\{\alpha_i, \alpha_j\} = 2\delta_{ij}$ (note $c = \hbar = 1$ here with $\alpha \cdot \nabla \sim \sum \alpha_\mu \partial_\mu$ and cf. [**874, 879**] for notations and background). This leads to matrices

$$\sigma_1 = \begin{pmatrix} 0 & 1 \\ 1 & 0 \end{pmatrix};\ \sigma_2 = \begin{pmatrix} 0 & -i \\ i & 0 \end{pmatrix}; \tag{1.37}$$

$$\sigma_3 = \begin{pmatrix} 1 & 0 \\ 0 & -1 \end{pmatrix};\ \alpha_i = \begin{pmatrix} 0 & \sigma_i \\ \sigma_i & 0 \end{pmatrix};\ \beta = \begin{pmatrix} 1 & 0 \\ 0 & -1 \end{pmatrix}$$

where α_i and β are 4×4 matrices. Then for convenience take $\gamma^0 = \beta$ and $\gamma^i = \beta\alpha_i$ which satisfy $\{\gamma^\mu, \gamma^\nu\} = 2g^{\mu\nu}$ (Lorentz metric) with $(\gamma^i)^\dagger = -\gamma^i$, $(\gamma^i)^2 = -1$, $(\gamma^0)^\dagger = \gamma^0$, and $(\gamma^0)^2 = 1$. The Dirac equation for a free particle can now be written

$$\left(i\gamma^\mu \frac{\partial}{\partial x^\mu} - m\right)\psi = 0 \equiv (i\partial\!\!\!/ - m)\psi = 0 \tag{1.38}$$

where $A\!\!\!/ = g_{\mu\nu}\gamma^\mu A^\nu = \gamma^\mu A_\mu$ and $\partial\!\!\!/ = \gamma^\mu \partial_\mu$. Taking Hermitian conjugates in (1.36), noting that α and β are Hermitian, one gets $\bar{\psi}(i\overleftarrow{\partial\!\!\!/} + m) = 0$ where $\bar{\psi} = \psi^\dagger\beta$. To define a conserved current one has an equation $\bar{\psi}\gamma^\mu\partial_\mu\psi + \gamma^\mu\bar{\psi}_\mu\psi = \partial_\mu(\bar{\psi}\gamma^\mu\psi) = 0$ leading to the conserved current $j^\mu = \bar{\psi}\gamma^\mu\psi = (\psi^\dagger\psi, \psi^\dagger\alpha\psi)$ (this means $\rho = \psi^\dagger\psi$ and $\mathbf{j} = \psi^\dagger\alpha\psi$ with $\partial_t\rho + \nabla \cdot \mathbf{j} = 0$). The Dirac equation has the Hamiltonian form

$$i\partial_t\psi = -i\alpha \cdot \nabla\psi + \beta m\psi = (\alpha \cdot \mathbf{p} + \beta m)\psi \equiv H\psi \tag{1.39}$$

($\alpha \cdot \mathbf{p} \sim \sum \alpha_\mu p_\mu$). To obtain a Dirac equation for an electron coupled to a prescribed external EM field with vector and scalar potentials A and ϕ one substitutes $p^\mu \to p^\mu - eA^\mu$, i.e. $\mathbf{p} \to \mathbf{p} - e\mathbf{A}$ and $p^0 = i\partial_t \to i\partial_t - e\Phi$, to obtain

$$i\partial_t\psi = [\alpha \cdot (\mathbf{p} - e\mathbf{A}) + e\Phi + \beta m]\psi \tag{1.40}$$

This identifies the Hamiltonian as $H = \alpha\cdot(\mathbf{p}-e\mathbf{A})+e\Phi+\beta m = \alpha\cdot\mathbf{p}+\beta m+H_{int}$ where $H_{int} = -e\alpha\cdot\mathbf{A}+e\Phi$, suggesting α as the operator corresponding to the velocity v/c; this is strengthened by the Heisenberg equations of motion

$$(1.41)\qquad \dot{\mathbf{r}} = \left(\frac{1}{i\hbar}\right)[\mathbf{r},H] = \alpha;\ \dot{\pi} = \left(\frac{1}{i\hbar}\right)[\pi,H] = e(\mathbf{E}+\alpha\times\mathbf{B})$$

Another bit of notation now from [**879**] is useful. Thus (again with $c=\hbar=1$) one can define e.g.

$$(1.42)\qquad \sigma_z = -i\alpha_x\alpha_y;\ \sigma_x = -i\alpha_y\alpha_z;\ \sigma_y = -i\alpha_z\alpha_x;\ \rho_3 = \beta;$$
$$\rho_1 = \sigma_z\alpha_z = -i\alpha_x\alpha_y\alpha_z;\ \rho_2 = i\rho_1\rho_3 = \beta\alpha_x\alpha_y\alpha_z$$

so that $\beta = \rho_3$ and $\alpha^k = \rho_1\sigma^k$. Recall also that the angular momentum $\vec{\ell}$ of a particle is $\vec{\ell} = \mathbf{r}\times\mathbf{p}$ ($\sim(-i)\mathbf{r}\times\nabla$) with components ℓ_k satisfying $[\ell_x,\ell_y] = i\ell_z$, $[\ell_y,\ell_z] = i\ell_x$, and $[\ell_z,\ell_x] = i\ell_y$. Any vector operator L satisfying such relations is called an angular momentum. Next one defines $\sigma_{\mu\nu} = (1/2)i[\gamma_\mu,\gamma_\nu] = i\gamma_\mu\gamma_\nu$ $(\mu\neq\nu)$ and $S_{\alpha\beta} = (1/2)\sigma_{\alpha\beta}$. Then the 6 components $S_{\alpha\beta}$ satisfy

$$(1.43)\qquad S_{10} = (i/2)\alpha_x;\ S_{20} = (i/2)\alpha_y;\ S_{30} = (i/1)\alpha_z;$$
$$S_{23} = (1/2)\sigma_x;\ ,\ S_{31} = (1/2)\sigma_y;\ S_{12} = (1/2)\sigma_z$$

The $S_{\alpha\beta}$ arise in representing infinitesimal rotations for the orthochronous Lorentz group via matrices $I+i\epsilon S_{\alpha\beta}$. Further one can represent total angular momentum J in the form $J = L+S$ where $L = \mathbf{r}\times\mathbf{p}$ and $S = (1/2)\sigma$ (L is orbital angular momentum and S represents spin). We recall that the gamma matrices are given via $\gamma = \beta\alpha$. Finally $[(i\partial_t - e\phi)-\alpha\cdot(-i\nabla - e\mathbf{A})-\beta m]\psi = 0$ (cf. (1.40)) and one gets

$$(1.44)\qquad [i\gamma^\mu D_\mu - m)\psi = [\gamma^\mu(i\partial_\mu - eA_\mu)-m]\psi = 0$$
$$D_\mu = \partial_\mu + ieA_\mu \equiv (\partial_0 + ie\phi, \nabla - ie\mathbf{A})$$

Working on the left with $(-i\gamma^\lambda D_\lambda - m)$ gives then $[\gamma^\lambda\gamma^\mu D_\lambda D_\mu + m^2]\psi = 0$ where $\gamma^\lambda\gamma^\mu = g^{\lambda\mu}+(1/2)[\gamma^\lambda,\gamma^\mu]$. By renaming the dummy indices one obtains $[\gamma^\lambda,\gamma^\mu]D_\lambda D_\mu = -[\gamma^\lambda,\gamma^\mu]D_\mu D_\lambda = (1/2)[\gamma^\lambda,\gamma^\mu][D_\lambda,D_\mu]$ leading to

$$(1.45)\qquad [D_\lambda,D_\mu] = ie[\partial_\lambda,A_\mu]+ie[A_\lambda,\partial_\mu] = ie(\partial_\lambda A_\mu - \partial_\mu A_\lambda) = ieF_{\lambda\mu}$$

This yields then $\gamma^\lambda\gamma^\mu D_\lambda D_\mu = D_\mu D_\lambda + eS^{\lambda\mu}F_{\lambda\mu}$ where $S^{\lambda\mu}$ represents the spin of the particle. Therefore one can write $[D_\mu D^\mu + eS^{\lambda\mu}F_{\lambda\mu}+m^2]\psi = 0$. Comparing with the standard form of the KG equation we see that this differs by the term $eS^{\lambda\mu}F_{\lambda\mu}$ which is the spin coupling of the particle to the EM field and has no classical analogue.

2. FARAGGI-MATONE THEORY

The equivalence principle (EP) of Faraggi-Matone (cf. [**133, 238, 240, 245, 446, 865**]) is based on the idea that all physical systems can be connected by a coordinate transformation to the free situation with vanishing energy (i.e. all potentials are equivalent under coordinate transformations). This automatically leads to the quantum stationary Hamilton-Jacobi equation (QSHJE) which is a third order nonlinear differential equation providing a trajectory representation of quantum mechanics (QM). The theory transcends in several respects the Bohm theory and in particular utilizes a Floydian time (cf. Floyd [**478, 479**]) leading to $\dot{q} = p/m_Q \neq p/m$ where $m_Q = m(1 - \partial_E Q)$ is the "quantum mass" and Q the "quantum potential" (cf. also Section 7.4). Thus the equivalence principle (EP) is reminscient of the Einstein equivalence of relativity theory. This latter served as a midwife to the birth of relativity but was somewhat inaccurate in its original form. It is better put as saying that all laws of physics should be invariant under general coordinate transformations (cf. [**950**]). This demands that not only the form but also the content of the equations be unchanged. More precisely the equations should be covariant and all absolute constants in the equations are to be left unchanged (e.g. c, $\hbar$, e, m and $\eta_{\mu\nu}$ = Minkowski tensor). Now for the EP, the classical picture with $S^{cl}(q, Q^0, t)$ the Hamilton principal function ($p = \partial S^{cl}/\partial q$) and P^0, Q^0 playing the role of initial conditions involves the classical HJ equation (CHJE) $H(q,p) = (\partial S^{cl}/\partial q), t) + (\partial S^{cl}/\partial t) = 0$. For time independent V one writes $S^{cl} = S_0^{cl}(q, Q^0) - Et$ and arrives at the classical stationary HJ equation (CSHJE) $(1/2m)(\partial S_0^{cl}/\partial q)^2 + \mathfrak{W} = 0$ where $\mathfrak{W} = V(q) - E$. In the Bohm theory one looked at Schrödinger equations $i\hbar\psi_t = -(\hbar^2/2m)\psi'' + V\psi$ with $\psi = \psi(q)exp(-iEt/\hbar)$ and $\psi(q) = R(q)exp(i\hat{W}/\hbar)$ leading to

$$\left(\frac{1}{2m}\right)(\hat{W}')^2 + V - E - \frac{\hbar^2 R''}{2mR} = 0; \; (R^2\hat{W}')' = 0 \tag{2.1}$$

where $\hat{Q} = -\hbar^2 R''/2mR$ was called the quantum potential; this can be written in the Schwartzian form $\hat{Q} = (\hbar^2/4m)\{\hat{W}; q\}$ (via $R^2\hat{W}' = c$). Here $\{f; q\} = (f'''/f') - (3/2)(f''/f')^2$. Writing $\mathfrak{W} = V(q) - E$ as in above we have the quantum stationary HJ equation (QSHJE)

$$(1/2m)(\partial\hat{W}'/\partial q)^2 + \mathfrak{W}(q) + \hat{Q}(q) = 0 \equiv \mathfrak{W} = -(\hbar^2/4m)\{exp(2iS_0/\hbar); q\} \tag{2.2}$$

This was worked out in the Bohm school (without the Schwarzian connections) but $\psi = Rexp(i\hat{W}/\hbar)$ is not appropriate for all situations and care must be taken ($\hat{W} = constant$ must be excluded for example - cf. [**446, 478, 479**]). The technique of Faraggi-Matone (FM) is completely general and with only the EP as guide one exploits the relations between Schwarzians, Legendre duality, and the geometry of a second order differential operator $D_x^2 + V(x)$ (Möbius transformations play an important role here) to arrive at the QSHJE in the form

$$\frac{1}{2m}\left(\frac{\partial S_0^v(q^v)}{\partial q^v}\right)^2 + \mathfrak{W}(q^v) + \mathfrak{Q}^v(q^v) = 0 \tag{2.3}$$

where $v: q \to q^v$ represents an arbitrary locally invertible coordinate transformation. Note in this direction for example that the Schwarzian derivative of the the ratio of two linearly independent elements in $ker(D_x^2 + V(x))$ is twice $V(x)$. In particular given an arbitrary system with coordinate q and reduced action $S_0(q)$ the system with coordinate q^0 corresponding to $V - E = 0$ involves $\mathfrak{W}(q) = (q^0; q)$ where (q^0, q) is a cocycle term which has the form $(q^a; q^b) = -(\hbar^2/4m)\{q^a; q^b\}$. In fact it can be said that the essence of the EP is the cocycle condition

$$(q^a; q^c) = (\partial_{q^c} q^b)^2[(q^a; q^b) - (q^c; q^b)] \tag{2.4}$$

In addition FM developed a theory of (x, ψ) duality (cf. [**445**])) which related the space coordinate and the wave function via a prepotential (free energy) in the form $\mathfrak{F} = (1/2)\psi\bar{\psi} + iX/\epsilon$ for example. A number of interesting philosophical points arise (e.g. the emergence of space from the wave function) and we connected this to various features of dispersionless KdV in [**238, 245**] in a sort of extended WKB spirit. One should note here that although a form $\psi = Rexp(i\hat{W}/\hbar)$ is not generally appropriate it is correct when one is dealing with two independent solutions of the Schrödinger equation ψ and $\bar{\psi}$ which are not proportional. In this context we utilized some interplay between various geometric properties of KdV which involve the Lax operator $L^2 = D_x^2 + V(x)$ and of course this is all related to Schwartzians, Virasoro algebras, and vector fields on S^1 (see e.g. [**238, 239, 245, 247, 248**]). Thus the simple presence of the Schrödinger equation (SE) in QM automatically incorporates a host of geometrical properties of $D_x = d/dx$ and the circle S^1. In fact since the FM theory exhibits the fundamental nature of the SE via its geometrical properties connected to the QSHJE one could speculate about trivializing QM (for 1-D) to a study of S^1 and ∂_x.

We import here some comments based on Bertoldi-Faraggi-Matone [**133, 446**] concerning the Klein-Gordon (KG) equation and the equivalence principle (EP) (details are in [**133**] and cf. also [**198, 199, 197, 201, 371, 638, 639, 640, 644, 645, 646, 899, 900**] for the KG equation which is treated in some detail later at several places in this book). One starts with the relativistic classical Hamilton-Jacobi equation (RCHJE) with a potential $V(q,t)$ given as

$$\frac{1}{2m}\sum_1^D(\partial_k S^{cl}(q,t))^2 + \mathfrak{W}_{rel}(q,t) = 0; \tag{2.5}$$

$$\mathfrak{W}_{rel}(q,t) = \frac{1}{2mc^2}[m^2c^4 - (V(q,t) + \partial_t S^{cl}(q,t))^2]$$

In the time-independent case one has $S^{cl}(q,t) = S_0^{cl}(q) - Et$ and (1.26) becomes

$$\frac{1}{2m}\sum_1^D(\partial_k S_0^{cl})^2 + \mathfrak{W}_{rel} = 0;\ \mathfrak{W}_{rel}(q) = \frac{1}{2mc^2}[m^2c^4 - (V(q) - E)^2] \tag{2.6}$$

In the latter case one can go through the same steps as in the nonrelativistic case and the relativistic quantum HJ equation (RQHJE) becomes

$$(1/2m)(\nabla S_0)^2 + \mathfrak{W}_{rel} - (\hbar^2/2m)(\Delta R/R) = 0;\ \nabla\cdot(R^2\nabla S_0) = 0 \tag{2.7}$$

these equations imply the stationary KG equation

$$-\hbar^2c^2\Delta\psi + (m^2c^4 - V^2 + 2EV - E^2)\psi = 0 \tag{2.8}$$

where $\psi = Rexp(iS_0/\hbar)$. Now in the time dependent case the (D+1)-dimensional RCHJE is ($\eta^{\mu\nu} = diag(-1, 1, \cdots, 1)$)

$$(1/2m)\eta^{\mu\nu}\partial_\mu S^{cl}\partial_\nu S^{cl} + \mathfrak{W}'_{rel} = 0; \tag{2.9}$$

$$\mathfrak{W}'_{rel} = (1/2mc^2)[m^2c^4 - V^2(q) - 2cV(q)\partial_0 S^{cl}(q)]$$

with $q = (ct, q_1, \cdots, q_D)$. Thus (2.9) has the same structure as (2.6) with Euclidean metric replaced by the Minkowskian one. We know how to implement the EP by adding Q via $(1/2m)(\partial S)^2 + \mathfrak{W}_{rel} + Q = 0$ (cf. [**446**] and remarks above). Note now that $\mathfrak{W}'_{rel}$ depends on S^{cl} requires an identification

$$\mathfrak{W}_{rel} = (1/2mc^2)[m^2c^4 - V^2(q) - 2cV(q)\partial_0 S(q)] \tag{2.10}$$

(S replacing S^{cl}) and implementation of the EP requires that for an arbitrary $\mathfrak{W}^a$ state ($q \sim q^a$) one must have

$$\mathfrak{W}^b_{rel}(q^b) = (p^b|p^a)\mathfrak{W}^a_{rel}(q^a) + (q^q;q^b);\ Q^b(q^b) = (p^b|p^a)Q(q^a) - (q^a;q^b) \tag{2.11}$$

where

$$(p^b|p) = [\eta^{\mu\nu}p^b_\mu p^b_\nu/\eta^{\mu\nu}p_\mu p_\nu] = p^T J\eta J^T p/p^T\eta p;\ J^\mu_\nu = \partial q^\mu/\partial q^{b^\nu} \tag{2.12}$$

(J is a Jacobian and these formulas are the natural multidimensional generalization - see [**133**] for details). Furthermore there is a cocycle condition $(q^a;q^c) = (p^c|p^b)[(q^a;q^b) - (q^c;q^b)]$.

Next one shows that $\mathfrak{W}_{rel} = (\hbar^2/2m)[\Box(Rexp(iS/\hbar))/Rexp(iS/\hbar)]$ and hence the corresponding quantum potential is $Q_{rel} = -(\hbar^2/2m)[\Box R/R]$. Then the RQHJE becomes $(1/2m)(\partial S)^2 + \mathfrak{W}_{rel} + Q = 0$ with $\partial\cdot(R^2\partial S) = 0$ (here $\Box R = \partial_\mu\partial^\mu R$) and this reduces to the standard SE in the classical limit $c \to \infty$ (note $\partial \sim (\partial_0, \partial_1, \cdots, \partial_D)$ with $q_0 = ct$, etc. - cf. (2.9)). To see how the EP is simply implemented one considers the so called minimal coupling prescription for an interaction with an electromagnetic four vector A_μ. Thus set $P^{cl}_\mu = p^{cl}_\mu + eA_\mu$ where p^{cl}_μ is a particle momentum and $P^{cl}_\mu = \partial_\mu S^{cl}$ is the generalized momentum. Then the RCHJE reads as $(1/2m)(\partial S^{cl} - eA)^2 + (1/2)mc^2 = 0$ where $A_0 = -V/ec$. Then $\mathfrak{W} = (1/2)mc^2$ and the critical case $\mathfrak{W} = 0$ corresponds to the limit situation where $m = 0$. One adds the standard Q correction for implementation of the EP to get $(1/2m)(\partial S - eA)^2 + (1/2)mc^2 + Q = 0$ and there are transformation properties (here $(\partial S - eA)^2 \sim \sum(\partial_\mu S - eA_\mu)^2$)

$$\mathfrak{W}(q^b) = (p^b|p^a)\mathfrak{W}^a(q^a) + (q^a;q^b);\ Q^b(q^b) = (p^q|p^a)Q^a(q^a) - (q^a;q^b) \tag{2.13}$$
$$(p^b|p) = \frac{(p^b - eA^b)^2}{(p - eA)^2} = \frac{(p-eA)^T J\eta J^T(p-eA)}{(p-eA)^T\eta(p-eA)}$$

Here J is a Jacobian $J^\mu_\nu = \partial q^\mu/\partial q^{b^\nu}$ and this all implies the cocycle condition again. One finds now that (recall $\partial\cdot(R^2(\partial S - eA)) = 0$ - continuity equation)

$$(\partial S - eA)^2 = \hbar^2\left(\frac{\Box R}{R} - \frac{D^2(Re^{iS/\hbar})}{Re^{iS/\hbar}}\right);\ D_\mu = \partial_\mu - \frac{i}{\hbar}eA_\mu \tag{2.14}$$

and it follows that

$$\mathfrak{W} = \frac{\hbar^2}{2m}\frac{D^2(Re^{iS/\hbar})}{Re^{iS/\hbar}};\ Q = -\frac{\hbar^2}{2m}\frac{\Box R}{R};\ D^2 = \Box - \frac{2ieA\partial}{\hbar} - \frac{e^2A^2}{\hbar^2} - \frac{ie\partial A}{\hbar} \tag{2.15}$$

$$(\partial S - eA)^2 + m^2c^2 - \hbar^2\frac{\Box R}{R} = 0;\ \partial\cdot(R^2(\partial S - eA)) = 0 \tag{2.16}$$

Note also that (2.9) agrees with $(1/2m)(\partial S^{cl} - eA)^2 + (1/2)mc^2 = 0$ after setting $\mathfrak{W}_{rel} = mc^2/2$ and replacing $\partial_\mu S^{cl}$ by $\partial_\mu S^{cl} - eA_\mu$. One can check that (2.16) implies the KG equation $(i\hbar\partial + eA)^2\psi + m^2c^2\psi = 0$ with $\psi = Rexp(iS/\hbar)$.

REMARK 2.2.1. We extract now a remark about mass generation and the EP from [**133, 446**]. Thus a special property of the EP is that it cannot be implemented in classical mechanics (CM) because of the fixed point corresponding to $\mathfrak{W} = 0$. One is forced to introduce a uniquely determined piece to the classical HJ equation (namely a quantum potential Q). In the case of the RCHJE the fixed point $\mathfrak{W}(q^0) = 0$ corresponds to $m = 0$ and the EP then implies that all the other masses can be generated by a coordinate transformation. Consequently one concludes that masses correspond to the inhomogeneous term in the transformation properties of the $\mathfrak{W}^0$ state, i.e. $(1/2)mc^2 = (q^0; q)$. Furthermore by (2.13) masses are expressed in terms of the quantum potential $(1/2)mc^2 = (p|p^0)Q^0(q^0) - Q(q)$. In particular in [**446**] the role of the quantum potential was seen as a sort of intrinsic self energy which is reminiscent of the relativistic self energy and this provides a more explicit evidence of such an interpretation. ■

REMARK 2.2.2. In a previous paper [**241**] (working with stationary states and ψ satisfying the Schrödinger equation (SE) $-(\hbar^2/2m)\psi'' + V\psi = E\psi$) we suggested that the notion of uncertainty in quantum mechanics (QM) could be phrased as incomplete information. The background theory here is taken to be the trajectory theory of Faraggi-Matone (and Floyd) as above and the idea in [**241**] goes as follows. First recall that microstates satisfy a third order quantum stationary Hamilton-Jacobi equation (QSHJE)

$$\frac{1}{2m}(S_0')^2 + \mathfrak{W}(q) + Q(q) = 0;\ Q(q) = \frac{\hbar^2}{4m}\{S_0; q\}; \tag{2.17}$$

$$\mathfrak{W}(q) = -\frac{\hbar^2}{4m}\{exp(2iS_0/\hbar); q\} \sim V(q) - E$$

where $\{f; q\} = (f'''/f') - (3/2)(f''/f')^2$ is the Schwarzian and S_0 is the Hamilton principle function. Also one recalls that the EP of Faraggi-Matone can only be implemented when $S_0 \neq const$; thus consider $\psi = Rexp(iS_0/\hbar)$ with $Q = -\hbar^2 R''/2mR$ and $(R^2 S_0')' = 0$ where $S_0' = p$ and $m_Q\dot{q} = p$ with $m_Q = m(1 - \partial_E Q)$ and $t \sim \partial_E S_0$ (Q in (2.17) is the definitive form - cf. [**448**]). Thus microstates require three initial or boundary conditions in general to determine S_0 whereas the SE involves only two such conditions (cf. also [**168, 170, 169, 382, 383, 384, 385, 386, 446, 447, 448, 478, 479, 480, 699**]). Hence in dealing with the SE in the standard QM Hilbert space formulation one is not using complete information about the "particles" described by microstate trajectories. The price of underdetermination is then uncertainty in q, p, t for example. In the present note we will make this more precise and add further discussion. Following [**244**] we now make this more precise and add further discussion. For the stationary SE $-(\hbar^2/2m)\psi'' + V\psi = E\psi$ it is shown in [**446**] that one has a general formula

$$e^{2iS_0(\delta)/\hbar} = e^{i\alpha}\frac{w + i\bar{\ell}}{w - i\ell} \tag{2.18}$$

($\delta \sim (\alpha, \ell)$) with three integration constants, α, ℓ_1, ℓ_2 where $\ell = \ell_1 + i\ell_2$ and $w \sim \psi^D/\psi \in \mathbf{R}$. Note ψ and ψ^D are linearly independent solutions of the SE and one can arrange that $\psi^D/\psi \in \mathbf{R}$ in describing any situation. Here p is determined by the two constants in ℓ and has a form

$$p = \frac{\pm\hbar\Omega\ell_1}{|\psi^D - i\ell\psi|^2} \tag{2.19}$$

(where $w \sim \psi^D/\psi$ above and $\Omega = \psi'\psi^D - \psi(\psi^D)')$. Now let p be determined exactly with $p = p(q,E)$ via the Schrödinger equation and S_0'. Then $\dot{q} = (\partial_E p)^{-1}$ is also exact so $\Delta q = (\partial_E p)^{-1}(\tau)\Delta t$ for some τ with $0 \le \tau \le t$ is exact (up to knowledge of τ). Thus given the wave function ψ satisfying the stationary SE with two boundary conditions at $q = 0$ say to fix uniqueness, one can create a probability density $|\psi|^2(q,E)$ and the function S_0'. This determines p uniquely and hence $\dot{q}$. The additional constant needed for S_0 appears in (2.18) and we can write $S_0 = S_0(\alpha, q, E)$ since from (2.18) one has

$$S_0 - (\hbar/2)\alpha = -(i\hbar/2)log(\beta) \tag{2.20}$$

and $\beta = (w + i\bar{\ell})/(w - i\ell)$ with $w = \psi^D/\psi$ is to be considered as known via a determination of suitable ψ, ψ^D. Hence $\partial_\alpha S_0 = -\hbar/2$ and consequently $\Delta S_0 \sim \partial_\alpha S_0 \delta\alpha = -(\hbar/2)\Delta\alpha$ measures the indeterminacy or uncertainty in S_0.

Let us expand upon this as follows. Note first that the determination of constants necessary to fix S_0 from the QSHJE is not usually the same as that involved in fixing ℓ, $\bar{\ell}$ in (2.18). In paricular differentiating in q one gets

$$S_0' = -\frac{i\hbar\beta'}{\beta};\ \beta' = -\frac{2i\Re\ell w'}{(w - i\ell)^2} \tag{2.21}$$

Since $w' = -\Omega/\psi^2$ where $\Omega = \psi'\psi^D - \psi(\psi^D)'$ we get $\beta' = -2i\ell_1\Omega/(\psi^D - i\ell\psi)^2$ and consequently

$$S_0' = -\frac{\hbar\ell_1\Omega}{|\psi^D - i\ell\psi|^2} \tag{2.22}$$

which agrees with p in (2.19) ($\pm\hbar$ simply indicates direction). We see that e.g. $S_0(x_0) = i\hbar\ell_1\Omega/|\psi^D(x_0) - i\ell\psi(x_0)|^2 = f(\ell_1, \ell_2, x_0)$ and $S_0'' = g(\ell_1, \ell_2, x_0)$ determine the relation between $(p(x_0), p'(x_0))$ and $(\ell_1,\ \ell_2)$ but they are generally different numbers. In any case, taking α to be the arbitrary unknown constant in the determination of S_0, we have $S_0 = S_0(q, E, \alpha)$ with $q = q(S_0, E, \alpha)$ and $t = t(S_0, E, \alpha) = \partial_E S_0$ (emergence of time from the wave function). One can then write e.g.

$$\Delta q = (\partial q/\partial S_0)(\hat{S}_0, E, \alpha)\Delta S_0 = (1/p)(\hat{q}, E)\Delta S_0 = -(1/p)(\hat{q}, E)(\hbar/2)\Delta\alpha \tag{2.23}$$

(for intermediate values $(\hat{S}_0,\ \hat{q})$) leading to

THEOREM 2.1. With p determined uniquely by two "initial" conditions, so that Δp (independent of $\hbar$) can be specified (independently of $\hbar$) and q given via (2.18), we have from (2.23) the inequality $\Delta p \Delta q = O(\hbar)$ which resembles the Heisenberg uncertainty relation.

COROLLARY 2.1. Similarly $\Delta t = (\partial t/\partial S_0)(\hat{S}_0, E, \alpha)\Delta S_0$ for some intermediate value $\hat{S}_0$ and hence as before $\Delta E \Delta t = O(\hbar)$ (ΔE being specified, independent of $\hbar$).

Note that there is no physical argument here; one is simply looking at the number of conditions necessary to fix solutions of a differential equation. In fact (based on some corresondence with E. Floyd) it seems somewhat difficult to produce a viable physical argument. We refer also to Remark 3.1.2 for additional discussion. ■

REMARK 2.2.3. In order to get at the time dependent SE from the FM (Faraggi-Matone) theory we proceed following [**133, 446**]. From the previous discussion on the KG equation one sees that (dropping the EM terms) in the time independent case one has $S^{cl}(q,t) = S_0^{cl}(q) - Et$

$$(2.24)\quad (1/2m)\sum_1^D(\partial_k S_0^{cl})^2 + \mathfrak{W}_{rel} = 0;\ \mathfrak{W}_{rel}(q) = (1/2mc^2)[m^2c^4 - (V(q)-E)^2]$$

leading to a stationary RQHJE

$$(2.25)\qquad (1/2m)(\nabla S_0)^2 + \mathfrak{W}_{rel} - (\hbar^2/2m)(\Delta R/R) = 0;\ \nabla\cdot(R^2\nabla S_0) = 0$$

This implies also the stationary KG equation

$$(2.26)\qquad -\hbar^2c^2\Delta\psi + (m^2c^4 - V^2 + 2VE - E^2)\psi = 0$$

Now in the time dependent case one can write $(1/2m)\eta^{\mu\nu}\partial_\mu S^{cl}\partial_\nu S^{cl} + \mathfrak{W}'_{rel} = 0$ where $\eta \sim diag(-1,1,\cdots,1)$ and

$$(2.27)\qquad \mathfrak{W}'_{rel}(q) = (1/2mc^2)[m^2c^4 - V^2(q) - 2cV(q)\partial_0 S^{cl}(q)]$$

with $q \equiv (ct, q_1, \cdots, q_D)$. Thus we have the same structure as (2.24) with Euclidean metric replaced by a Minkowskian one. To implement the EP we have to modify the classical equation by adding a function to be determined, namely $(1/2m)(\partial S)^2 + \mathfrak{W}_{rel} + Q = 0$ ($(\partial S)^2 \sim \sum(\partial_\mu S)^2$ etc.). Observe that since $\mathfrak{W}'_{rel}$ depends on S^{cl} we have to make the identification $\mathfrak{W}_{rel} = (1/2mc^2)[m^2c^4 - V^2(q) - 2cV(q)\partial_0 S(q)]$ which differs from $\mathfrak{W}'_{rel}$ since S now appears instead of S^{cl}. Implementation of the EP requires that for an arbitrary $\mathfrak{W}^a$ state

$$(2.28)\quad \mathfrak{W}^b_{rel}(q^b) = (p^b|p^a)\mathfrak{W}^a_{rel}(q^a) + (q^q;q^b);\ Q^b(q^b) = (p^b|p^a)Q^a(q^a) - (q^a;q^b)$$

where now $(p^b|p) = \eta^{\mu\nu}p^b_\mu p^b_\nu/\eta^{\mu\nu}p_\mu p_\nu = p^TJ\eta J^Tp/p^T\eta p$ and $J^\mu_\nu = \partial q^\mu/\partial(q^b)^\nu$. This leads to the cocycle condition $(q^a;q^c) = (p^c|p^b)[(q^q;q^b) - (q^c;q^b)]$ as before. Now consider the identity

$$(2.29)\qquad \alpha^2(\partial S)^2 = \Box(Rexp(\alpha S))/Rexp(\alpha S) - (\Box R/R) - (\alpha\partial\cdot(R^2\partial S)/R^2)$$

and if R satisfies the continuity equation $\partial\cdot(R^2\partial S) = 0$ one sets $\alpha = i/\hbar$ to obtain

$$(2.30)\qquad \frac{1}{2m}(\partial S)^2 = -\frac{\hbar^2}{2m}\frac{\Box(Re^{iS/\hbar})}{Re^{iS/\hbar}} + \frac{\hbar^2}{2m}\frac{\Box R}{R}$$

Then it is shown that $\mathfrak{W}_{rel} = (\hbar^2/2m)(\Box(Rexp(iS/\hbar))/Rexp(iS/\hbar)$ so $Q_{rel} = -(\hbar^2/2m)(\Box R/R)$. Thus the RQHJE has the form (cf. (2.14) - (2.16))

$$(2.31)\qquad \frac{1}{2m}(\partial S)^2 + \mathfrak{W}_{rel} - \frac{\hbar^2}{2m}\frac{\Box R}{R} = 0;\ \partial\cdot(R^2\partial S) = 0$$

Now for the time dependent SE one takes the nonrelativistic limit of the RQHJE. For the classical limit one makes the usual substitution $S = S' - mc^2t$

so as $c \to \infty$ $\mathfrak{W}_{rel} \to (1/2)mc^2 + V$ and $-(1/2m)(\partial_0 S)^2 \to \partial_t S' - (1/2)mc^2$ with $\partial(R^2\partial S) = 0 \to m\partial_t(R')^2 + \nabla\cdot((R')^2\nabla S') = 0$. Therefore (removing the primes) (6.2) becomes $(1/2m)(\nabla S)^2 + V + \partial_t S - (\hbar^2/2m)(\Delta R/R) = 0$ with the time dependent nonrelativistic continuity equation being $m\partial_t R^2 + \nabla\cdot(R^2\nabla S) = 0$. This leads then (for $\psi \sim Rexp(iS/\hbar)$) to the SE

$$i\hbar\partial_t\psi = \left(-\frac{\hbar^2}{2m}\Delta + V\right)\psi \tag{2.32}$$

One sees from all this that the FM theory is profoundly governed by the equivalence principle and produces a usable framework for computation. It is surprising that it has not attracted more adherents. ∎

3. FIELD THEORY MODELS

In trying to imagine particle trajectories of a fractal nature or in a fractal medium we are tempted to abandon (or rather relax) the particle idea and switch to quantum fields (QF). Let the fields sense the bumps and fractality; if one can think of fields as operator valued distributions for example then fractal supports for example are quite reasonable. There are other reasons of course since the notion of particle in quantum field theory (QFT) has a rather fuzzy nature anyway. Then of course there are problems with QFT itself (cf. Wallace [**1275**]) as well as arguments that there is no first quantization (except perhaps in the Bohm theory - cf. [**930, 935, 1337**]). We review here some aspects of particles arising from QFT methods, especially in a Bohmian spirit (cf. [**99, 133, 320, 412, 413, 414, 606, 638, 644, 645, 646, 662, 713, 858, 930, 931, 934, 935, 1292**]). We refer to [**606, 1275**] for interesting philosophical discussion about particles and localized objects in a QFT and will extract here from [**99, 320, 414, 930, 931**]; for QFT we refer to [**38, 616, 677, 1201, 1291, 1336**]. Many details are omitted and standard QFT techniques are assumed to be known and we will concentrate here on derivations of KG type equations and the nature of the quantum potential (the Dirac equation will be treated later). There has been a remarkable effusion of important results in Bohmian mechanics, covariant field theory, many fingered time, fundamentals of quantum theory and relativity, etc. by H. Nikolić (a partial summary by Nikolić appears in hep-th 0610138 in [**930**]). We have organized some of this in [**930**]-[**938**] and have sketched some of the material and used some of the ideas in applications to the WDW equation. The interested reader would do well to read all of these papers.

3.1. EMERGENCE OF PARTICLES. The papers of Nikolić [**930, 931**] are impressive in producing a local operator describing the particle density current for scalar and spinor fields in an arbitrary gravitational and electromagnetic background. This enables one to describe particles in a local, general covariant, and gauge invariant manner. The current depends on the choice of a 2-point Wightman function and a most natural choice based on the Green's function à la Schwinger-deWitt leads to local conservation of the current provided that interaction with quantum fields is absent. Interactions lead to local nonconservation of current which describes local particle production consistent with the usual global description based on the interaction picture. The material is quite technical but we feel it is important and will sketch some of the main points; the discussion should provide a good exercise in field theoretic technique. The notation is indicated as we proceed and we make no attempt to be consistent with other notation. Thus let $g_{\mu\nu}$ be a classical background metric, g the determinant, and R the curvature. The action of a Hermitian scalar field ϕ can be written as

$$S = \frac{1}{2}\int d^4x|g|^{1/2}[g^{\mu\nu}(\partial_\mu\phi)(\partial_\nu\phi) - m^2\phi^2 - \xi R\phi^2] \tag{3.1}$$

where ξ is a coupling constant. Writing this as $S = \int d^4x|g|^{1/2}\mathcal{L}$ the canonical mommentum vector is $\pi_\mu = [\partial\mathcal{L}/\partial(\partial^\mu\phi)] = \partial_\mu\phi$ (standard $g_{\mu\nu}$). The corresponding equation of motion is $(\nabla^\mu\partial_\mu + m^2 + \xi R)\phi = 0$ where ∇^μ is the covariant derivative. Let Σ be a spacelike Cauchy hypersurface with unit normal vector n^μ; the canonical momentum scalar is defined as $\pi = n^\mu\pi_\mu$ and the volume element on Σ is $d\Sigma^\mu = d^3x|g^{(3)}|^{1/2}n^\mu$ with scalar product $(\phi_1,\phi_2) = i\int_\Sigma d\Sigma^\mu\phi_1^*\overleftrightarrow{\partial}_\mu\phi_2$ where $a\overleftrightarrow{\partial}_\mu b = a\partial_\mu a - (\partial_\mu a)b$. If ϕ_i are solutions of the equation of motion then the scalar product does not depend on Σ. One chooses coordinates (t,x) such that $t = c$ on Σ so that $n^\mu = g_0^\mu/\sqrt{g_{00}}$ and the canonical commutation relations become

$$[\phi(x),\phi(x')]_\Sigma = [\pi(x),\pi(x')]_\Sigma = 0;\ [\phi(x),\pi(x')]_\Sigma = |g^{(3)}|^{-1/2}i\delta^3(x-x') \tag{3.2}$$

(here x, x' lie on Σ). This can be written in a manifestly covariant form via

$$\int_\Sigma d\Sigma'^\mu[\phi(x),\partial'_\mu\phi(x')]\chi(x') = \int_\Sigma d\Sigma'^\mu[\phi(x'),\partial_\mu\phi(x)]\chi(x') = i\chi(x) \tag{3.3}$$

for an arbitrary test function χ. For practical reasons one writes $\tilde{n}^\mu = |g^{(3)}|^{1/2}n^\mu$ where the tilde indicates that it is not a vector. Then $\nabla_\mu\tilde{n}_\nu = 0$ and in fact $\tilde{n}^\mu = (|g^{(3)}|^{1/2}/\sqrt{g_{00}},0,0,0)$. It follows that $d\Sigma^\mu = d^3x\tilde{n}^\mu$ while (1.2) can be written as $\tilde{n}^0(x')[\phi(x),\partial'_0\phi(x')]_\Sigma = i\delta^3(x-x')$. Consequently

$$[\phi(x),\tilde{\pi}(x')]_\Sigma = i\delta^3(x-x');\ \tilde{\pi} = |g^{(3)}|^{1/2}\pi \tag{3.4}$$

Now choose a particular complete orthonormal set of solutions $\{f_k(x)\}$ of the equation of motion satisfying therefore

$$(f_k,f_{k'}) = -(f_k^*,f_{k'}^*) = \delta_{kk'};\ (f_k^*,f_{k'}) = (f_k,f_{k'}^*) = 0 \tag{3.5}$$

One can then write $\phi(x) = \sum_k a_kf_k(x) + a_k^\dagger f_k^*(x)$ from which we deduce that $a_k = (f_k,\phi)$ and $a_k^\dagger = -(f_k^*,\phi)$ while $[a_k,a_{k'}^\dagger] = \delta_{kk'}$ and $[a_k,a_{k'}] = [a_k^\dagger,a_{k'}^\dagger] = 0$. The lowering and raising operators a_k and $a_k^\dagger$ induce the representation of the field algebra in the usual manner and $a_k|0>= 0$. The number operator is $N = \sum a_k^\dagger a_k$ and one defines a two point function $W(x,x') = \sum f_k(x)f_k^*(x')$ (different definitions appear later). Using the equation of motion one finds that W is a Wightman function $W(x,x') =<0|\phi(x)\phi(x')|0>$ and one has $W^*(x,x') = W(x',x)$. Further, via the equation of motion, for f_k, f_k^* one has

$$(\nabla^\mu\partial_\mu + m^2 + \xi R(x))W(x,x') = 0 = (\nabla'^\mu\partial'_\mu + m^2 + \xi R(x'))W(x,x') \tag{3.6}$$

From the form of W and the commutation relations there results also

$$W(x,x')|_\Sigma = W(x',x)|_\Sigma;\ \partial_0\partial'_0W(x,x')|_\Sigma = \partial_0\partial'_0W(x',x)|_\Sigma; \tag{3.7}$$
$$\tilde{n}^0\partial'_0[W(x,x') - W(x',x)]_\Sigma = i\delta^3(x-x')$$

The number operator given by $N = \sum a_k^\dagger a_k$ is a global quantity. However a new way of looking into the concept of particles emerges when $a_k = (f_k, \phi)$, etc. is put into N; using the scalar product along with the expression for W leads to

$$N = \int_\Sigma d\Sigma^\mu \int_\Sigma d\Sigma'^\nu W(x, s') \overleftrightarrow{\partial_\mu} \overleftrightarrow{\partial'_\nu} \phi(x)\phi(x') \tag{3.8}$$

By interchanging the names of the coordinates x, x' and the names of the indices μ, ν this can be written as a sum of two equal terms

$$N = \frac{1}{2}\int_\Sigma d\Sigma^\mu \int_\Sigma d\Sigma'^\nu W(x, x') \overleftrightarrow{\partial_\mu} \overleftrightarrow{\partial'_\nu} \phi(x)\phi(x') + \tag{3.9}$$
$$\frac{1}{2}\int_\Sigma d\Sigma^\mu \int_\Sigma d\Sigma'^\nu W(x', x) \overleftrightarrow{\partial_\mu} \overleftrightarrow{\partial'_\nu} \phi(x')\phi(x)$$

Using also $W^*(x, x') = W(x', x)$ one sees that (3.9) can be written as $N = \int_\Sigma d\Sigma^\mu j_\mu(x)$ where

$$j_\mu(x) = (1/2)\int_\Sigma d\Sigma'^\nu \{W(x, x') \overleftrightarrow{\partial_\mu} \overleftrightarrow{\partial'_\nu} \phi(x)\phi(x') + h.c.\} \tag{3.10}$$

(where $h.c.$ denotes hermitian conjugate). Evidently the vector $j_\mu(x)$ should be interpreted as the local current of particle density. This representation has three advantages over $N = \sum a_k^\dagger a_k$: (i) It avoids the use of a_k, $a_k^\dagger$ related to a particular choice of modes $f_k(x)$. (ii) It is manifestly covariant. (iii) The local current $j_\mu(x)$ allows one to view the concept of particles in a local manner. If now one puts all this together with the antisymmetry of $\overleftrightarrow{\partial_\mu}$ we find

$$j_\mu = i\sum_{k,k'} f_k^* \overleftrightarrow{\partial_\mu} f_{k'} a_k^\dagger a_{k'} \tag{3.11}$$

From this we see that j_μ is automatically normally ordered and has the property $j_\mu|0>= 0$ (not surprising since $N = \sum a_k^\dagger a_k$ is normally ordered). Further one finds $\nabla^\mu j_\mu = 0$ (covariant conservation law) so the background gravitational field does not produce particles provided that a unique vacuum defined by $a_k|0>= 0$ exists. This also implies global conservation since it provides that $N = \int_\Sigma d\Sigma^\mu j_\mu(x)$ does not depend on time. The extra terms in $\nabla^\mu j_\mu = 0$ originating from the fact that $\nabla_\mu \neq \partial_\mu$ are compensated by the extra terms in $N = \int_\Sigma d\Sigma^\mu j_\mu$ that originate from the fact that $d\Sigma^\mu$ is not written in "flat" coordinates. The choice of vacuum is related to the choice of $W(x, x')$. Note that although $j_\mu(x)$ is a local operator some nonlocal features of the particle concept still remain because (3.10) involves an integration over Σ on which x lies. Since $\phi(x')$ satisfies the equation of motion and $W(x, x')$ satisfies (3.6) this integral does not depend on Σ. However it does depend on the choice of $W(x, x')$. Note also the separation between x and x' in (3.10) is spacelike which softens the nonlocal features because $W(x, x')$ decreases rapidly with spacelike separation - in fact it is negligible when the space like separation is much larger than the Compton wavelength.

We pick this up now in Nikolić [**935**]. Thus consider a scalar Hermitian field $\phi(x)$ in a curved background satisfying the equation of motion and choose a particular complete orthonormal set $\{f_k(x)\}$ having relations and scalar product as before. The field ϕ can be expanded as $\phi(x) = \phi^+(x) + \phi^-(x)$ where $\phi^+(x) = \sum a_k f_k(x)$ and $\phi^-(x) = \sum a_k^\dagger f_k^*(x)$. Introducing the two point function $W^+(x,x') = \sum f_k(x) f_k^*(x')$ with $W^-(x,x') = \sum f_k^*(x) f_k(x')$ one finds the remarkable result that

(3.12)

$$\phi^+(x) = i\int_\Sigma d\Sigma'^\nu W^+(x,x')\overleftrightarrow{\partial}'_\nu \phi(x');\ \phi^-(x) = -i\int_\Sigma d\Sigma'^\nu W^-(x,x')\overleftrightarrow{\partial}'_\nu \phi(x')$$

We see that the extraction of $\phi^\pm(x)$ from $\phi(x)$ is a nonlocal procedure. Note however that the integrals in (3.12) do not depend on the choice of the timelike Cauchy hypersurface Σ because $W^\pm(x,x')$ satisfies the equation of motion with respect to x' just as $\phi(x')$ does. However these integrals do depend on the choice of $W^\pm(x,x')$, i.e. on the choice of the set $\{f_k(x)\}$. Now define normal ordering in the usual way, putting ϕ^- on the left, explicitly : $\phi^+\phi^- := \phi^-\phi^+$ while the ordering of the combinations $\phi^-\phi^+$, $\phi^+\phi^+$, and $\phi^-\phi^-$ leaves these combinations unchanged. Generalize this now by introducing 4 different orderings $N_{(\pm)}$ and $A_{(\pm)}$ defined via

(3.13)
$$N_+\phi^+\phi^- = \phi^-\phi^+;\ N_-\phi^+\phi^- = -\phi^-\phi^+;$$
$$A_+\phi^-\phi^+ = \phi^+\phi^-;\ A_-\phi^-\phi^+ = -\phi^+\phi^-$$

Thus N_+ is normal ordering, N_- will be useful, and the antinormal orderings $A_\pm$ can be used via symmetric orderings $S_+ = (1/2)[N_+ + A_+]$ and $S_- = (1/2)[N_- + A_-]$. When S_+ acts on a bilinear combination of fields it acts as the default ordering, i.e. $S_+\phi\phi = \phi\phi$.

Now the particle current for scalar Hermitian fields can be written as (cf. (3.10))

(3.14)

$$j_\mu(x) = \frac{1}{2}\int_\Sigma d\Sigma'^\nu \left[W^+(x,x')\overleftrightarrow{\partial_\mu}\overleftrightarrow{\partial}'_\nu\phi(x)\phi(x;) + W^-(x,x')\overleftrightarrow{\partial_\mu}\overleftrightarrow{\partial}'_\nu\phi(x')\phi(x)\right]$$

(3.14) can be written in a local form as $j_\mu(x) = (i/2)[\phi(x)\overleftrightarrow{\partial}_\mu\phi^+(x) + \phi^-(x)\overleftrightarrow{\partial}_\mu\phi(x)]$ (via (3.12)). Using the identities $\phi^+\overleftrightarrow{\partial}_\mu\phi^+ = \phi^-\overleftrightarrow{\partial}_\mu\phi^-$ this can be written in the elegant form $j_\mu = i\phi^-\overleftrightarrow{\partial}_\mu\phi^+$. Similarly using (3.13) this can be written in another elegant form without explicit use of $\phi^\pm$, namely $j_\mu = (i/2)N_-\phi\overleftrightarrow{\partial}_\mu\phi$. Note that the expression on the right here without the ordering N_- vanishes identically - this peculiar feature may explain why the particle current was not previously discovered. The normal ordering N_- provides that $j_\mu|0>= 0$ which is related to the fact that the total number of particles is $N = \int_\Sigma d\Sigma^\mu j_\mu = \sum a_k^\dagger a_k$. Alternatively one can choose the symmetric ordering S_- and define the particle current as $j_\mu = (i/2)S_-\phi\overleftrightarrow{\partial}_\mu\phi$. This leads to the total number of particles $N = (1/2)\sum(a_k^\dagger a_k + a_k a_k^\dagger) = \sum[a_k^\dagger a_k + (1/2)]$.

When the gravitational background is time dependent one can introduce a new set of solutions $u_k(x)$ for each time t, such that the $u_k(x)$ are positive frequency modes at that time. This leads to functions with an extra time dependence $u_k(x,t)$

that do not satisfy the equation of motion (cf. Nikolić [**931**]). Define $\phi^{\pm}$ as in (3.12) but with the two point functions

$$W^+(x,x') = \sum u_k(x,t)u_k^*(x',t');\ W^-(x,x') = \sum u_k^*(x,t)u_k(x',t') \tag{3.15}$$

As shown in [**931**] such a choice leads to a local description of particle creation consistent with the conventional global description based on the Bogoliubov transformation. Putting $\phi(x) = \sum a_k f_k(x) + a_k^{\dagger} f_k^*(x)$ in (3.12) with (3.15) yields $\phi^+(x) = \sum A_k(t)u_k(x,t)$ and $\phi^-(x) = \sum A_k^{\dagger}(t)u_k^*(x)$ where

$$A_k(t) = \sum \alpha_{kj}^*(t)a_j - \beta_{kj}^*(t)a_k^{\dagger};\ \alpha_{jk} = (f_j, u_k);\ \beta_{jk}(t) = -(f_j^*, u_k) \tag{3.16}$$

Putting these $\phi^{\pm}$ in $j_\mu = i\phi^- \overleftarrow{\partial}\phi^+$ one finds

$$j_\mu(x) = i\sum_{k,k'} A_k^{\dagger}(t)u_k^*(x,t)\overleftrightarrow{\partial_\mu} A_{k'}(t)u_{k'}(x,t) \tag{3.17}$$

Note that because of the extra time dependence the fields $\phi^{\pm}$ do not satisy the equation of motion $(\nabla^\mu\partial_\mu + m^2 + \xi R)\phi = 0$ and hence the current (3.17) is not conserved, i.e. $\nabla^\mu j_\mu$ is a nonvanishing local scalar function describing the creation of particles in a local and invariant manner as in [**931**]. In [**935**] there follows a discussion about where and when particles are created with conclusion that this happens at the spacetime points where the metric is time dependent. Hawking radiation is then cited as an example. Generally the choice of the 2-point function (3.15) depends on the choice of time coordinate. Therefore in general a natural choice of the 2-point function (3.15) does not exist. In [**931**] an alternative choice is introduced via $W^{\pm}(x,x') = G^{\pm}(x,x')$ where $G^{\pm}(x,x')$ is determined by the Schwinger- deWitt function. As argued in [**931**] this choice seems to be the most natural since the $G^{\pm}$ satisfy the equation of motion and hence the particle current in which $\phi^{\pm}$ are calculated by putting $:\phi^+\phi^- := \phi^-\phi^+$ in (3.12) is conserved; this suggests that classical gravitational backgrounds do not create particles (see below).

A complex scalar field $\phi(x)$ and its Hermitian conjugate $\phi^{\dagger}$ in an arbitrary gravitational background can be expanded as

$$\phi = \phi^{P+} + \phi^{A-};\ \phi^{\dagger} = \phi^{P-} + \phi^{A+};\ \phi^{P+} = \sum a_k f_k(x); \tag{3.18}$$

$$\phi^{P-} = \sum a_k^{\dagger} f_k^*;\ \phi^{A+} = \sum b_k f_k(x);\ \phi^{A-} = \sum b_k^{\dagger} f_k^*(x)$$

In a similar manner to the preceeding one finds also

$$\phi^{P+} = i\int d\Sigma'^{\nu} W^+(x,x')\overleftrightarrow{\partial'_\nu}\phi(x');\ \phi^{A+} = i\int_\Sigma d\Sigma'^{\nu} W^+(x,x')\overleftrightarrow{\partial'_\nu}\phi^{\dagger}(x'); \tag{3.19}$$

$$\phi^{P-} = -i\int_\Sigma d\Sigma'^{\nu} W^-(x,x')\overleftrightarrow{\partial'_\nu}\phi^{\dagger}(x');\ \phi^{A-} = -i\int_\Sigma d\Sigma'^{\nu} W^-(x,x')\overleftrightarrow{\partial'_\nu}\phi(x')$$

The particle current $j_\mu^P(x)$ and the antiparticle current $j_\mu^A(x)$ are then (cf. [**931**])

(3.20)

$$j_\mu^P(x) = \frac{1}{2}\int_\Sigma d\Sigma'^{\nu}\left[W^+(x,x')\overleftrightarrow{\partial_\mu}\overleftrightarrow{\partial'_\nu}\phi^{\dagger}(x)\phi(x') + W^-(x,x')\overleftrightarrow{\partial_\mu}\overleftrightarrow{\partial'_\nu}\phi^{\dagger}(x')\phi(x)\right];$$

$$j^A_\mu(x)=\frac{1}{2}\int_\Sigma d\Sigma'^\nu\left[W^+(x,x')\overleftrightarrow{\partial_\mu}\overleftrightarrow{\partial'_\nu}\phi(x)\phi^\dagger(x')+W^-(x,x')\overleftrightarrow{\partial_\mu}\overleftrightarrow{\partial'_\nu}\phi(x')\phi^\dagger(x)\right]$$

Consequently they can be written in a purely local form as

$$j^P_\mu=i\phi^{P-}\overleftrightarrow{\partial_\mu}\phi^{P+}+j^{mix}_\mu;\ j^A_\mu=i\phi^{A-}\overleftrightarrow{\partial_\mu}\phi^{A+}-j^{mix}_\mu; \tag{3.21}$$

$$j^{mix}_u=\frac{i}{2}\left[\phi^{P-}\overleftrightarrow{\partial_\mu}\phi^{A-}-\phi^{P+}\overleftrightarrow{\partial_\mu}\phi^{A+}\right]$$

The current of charge j^-_μ has the form $j^-_\mu=j^P_\mu-j^A_\mu$ which can be written as (cf. [**931**]) $j^-_\mu=:i\phi^\dagger\overleftrightarrow{\partial_\mu}\phi:=\frac{i}{2}\left[\phi^\dagger\overleftrightarrow{\partial_\mu}\phi-\phi\overleftrightarrow{\partial_\mu}\phi^\dagger\right]$. Using (3.13) this can also be written as

$$j^-_\mu=N_+i\phi^\dagger\overleftrightarrow{\partial_\mu}\phi=(i/2)N_+[\phi^\dagger\overleftrightarrow{\partial_\mu}\phi-\phi\overleftrightarrow{\partial_\mu}\phi^\dagger] \tag{3.22}$$

The current of total number of particles is now defined as $j^+_\mu=j^P_\mu+j^A_\mu$ and it is shown in [**931**] that j^+_μ can be written as $j_\mu=j^1_\mu+j^2_\mu$ where $\phi=(1/\sqrt{2})(\phi_1+i\phi_2)$ (j^l_μ are two currents of the form (3.14). Therefore using $j_\mu=(i/2)N_-\phi\overleftrightarrow{\partial_\mu}\phi$ one can write j_μ as $j^+_\mu=(i/2)N_-[\phi_1\overleftrightarrow{\partial_\mu}\phi_1+\phi_2\overleftrightarrow{\partial_\mu}\phi_2]$. Finally one shows that this can be written in a form analogous to (3.22) as $j^+_\mu=(i/2)N_-[\phi^\dagger\overleftrightarrow{\partial_\mu}\phi+\phi\overleftrightarrow{\partial_\mu}\phi^\dagger]$. This can be summarized by defining currents $q^\pm_\mu=(1/2)[\phi^\dagger\overleftrightarrow{\partial_\mu}\phi\pm\phi\overleftrightarrow{\partial_\mu}\phi^\dagger]$ leading to $j^\pm_\mu=N_\mu q^\pm_\mu$. The current q^+_μ vanishes but $N_-q^+_\mu$ does not vanish. These results can be easily generalized to the case where the field interacts with a backgound EM field (as in [**931**]). The equations are essentially the same but the derivatives ∂_μ are replaced by the corresponding gauge covariant derivatives and the particle 2-point functions $W^{P\pm}$ are not equal to the antiparticle 2-point functions $W^{A\pm}$. As in the gravitational case in the case of interaction with an EM background three different choices for the 2-point functions exist and we refer to Nikolić [**931**] for details.

REMARK 2.3.1 In a classical field theory the energy-momentum tensor (EMT) of a real scalar field is

$$T_{\mu\nu}=(\partial_\mu\phi)(\partial_\nu\phi)-(1/2)g_{\mu\nu}[g^{\alpha\beta}(\partial_\alpha\phi)(\partial_\beta\phi)-m^2\phi^2] \tag{3.23}$$

Contrary to the conventional idea of particles in QFT the EMT is a local quantity. Therefore the relation between the definition of particles and that of EMT is not clear in the conventional approach to QFT in curved spacetime. Here one can exploit the local and covariant description of particles to find a clear relation between particles and EMT. One has to choose some ordering of the operators in (3.23) just as a choice of ordering is needed in order to define the particle current. Although the choice is not obvious it seems natural that the choice for one quantity should determine the choice for the other. Thus if the quantum EMT is defined via : $T_{\mu\nu}:=N_+T_{\mu\nu}$ then the particle current should be defined as $N_-i\phi\overleftrightarrow{\partial_\mu}\phi$. The nonlocalities related to the extraction of ϕ^+ and ϕ^- from ϕ needed for the definitions of the normal orderings N_+ and N_- appear both in the EMT and in the particle current. Similarly if $W^\pm$ is chosen as in $W^+(x,x')=\sum f_k(x)f^*_k(x')$ for one quantity then it should be chosen in the same way for the other. The choices as above lead to a consistent picture in which both the energy and the number of particles vanish in the vacuum $|0>$ defined by $a_k|0>=0$. Alternatively if $W^{\pm}$ is chosen as in (3.15) for the definition of particles it should be chosen in the same way for the definition of the EMT. Assume for simplicity that spacetime is flat at some late time t. Then the normally ordered operator of the total number of particles at t is $N(t)=\sum_q A_q^{\dagger}(t)A_q(t)$ (cf. (3.16)) while the normally ordered operator of energy is $H(t)=\sum_q \omega_q A^{\dagger}(t)A_q(t)$ (note here $q\sim\mathbf{q}$). Owing to the extra time dependence it is clear that both the particle current and the EMT are not conserved in this case. Thus it is clear that the produced energy exactly corresponds to the produced particles. A similar analysis can be caried out for the particle-antiparticle pair creation caused by a classical EM background. Since the energy should be conserved this suggests that $W^{\pm}$ should not be chosen as in (3.15), i.e. that classical backgrounds do not cause particle creation (see Nikolić [**935**] for more discussion). The main point in all this is that particle currents as developed above can be written in a purely local form. The nonlocalities are hidden in the extraction of $\phi^{\pm}$ from ϕ. The formalism also reveals a relation between EM and particles suggesting that it might not be consistent to use semiclassical methods to describe particle creation; it also suggests that the vacuum energy might contribute to dark matter that does not form structures, instead of contributing to the cosmological constant. ■

3.2. BOSONIC BOHMIAN THEORY. We follow here Nikolić [**930, 931**] concerning Bohmian particle trajectories in relativistic bosonic and fermionic QFT. First we recall that there is no objection to a Bohmian type theory for QFT and no contradiction to Bell's theorems etc. (see e.g. [**99, 158, 320, 414**]). Without discussing all the objections to such a theory we simply construct one following Nikolic (cf. also [**226, 603, 791**] for related information). Thus consider first particle trajectories in relativistic QM and posit a real scalar field $\phi(x)$ satisfying the Klein-Gordon equation in a Minkowski metric $\eta_{\mu\nu}=diag(1,-1,-1,-1)$ written as $(\partial_0^2-\nabla^2+m^2)\phi=0$. Let $\psi=\phi^+$ with $\psi^*=\phi^-$ correspond to positive and negative frequency parts of $\phi=\phi^++\phi^-$. The particle current is $j_\mu=i\psi^*\overleftrightarrow{\partial_\mu}\psi$ and $N=\int d^3xj_0$ is the positive definite number of particles (not the charge). This is most easily seen from the plane wave expansion $\phi^+(x)=\int d^3ka(\kappa)exp(-ikx)/\sqrt{(2\pi)^3 2k_0}$ since then $N=\int d^3ka^{\dagger}(\kappa)a(\kappa)$ (see above and [**931, 935**] where it is shown that the particle current and the decomposition $\phi=\phi^++\phi^-$ make sense even when a background gravitational field or some other potential is present). One can write also $j_0=i(\phi^-\pi^+-\phi^+\pi^-)$ where $\pi=\pi^++\pi^-$ is the canonical momentum (cf. Holland [**637**]). Alternatively ϕ may be interpreted not as a field containg an arbitrary number of particles but rather as a one particle wave function. Here we note that contrary to a field a wave function is not an observable and so doing we normalize ϕ here so that $N=1$. The current j_μ is conserved via $\partial_\mu j^\mu=0$ which implies that $N=\int d^3xj_0$ is also conserved, i.e. $dN/dt=0$. In the causal interpretation one postulates that the particle has the trajectory determined by $dx^\mu/d\tau=j^\mu/2m\psi^*\psi$. The affine parameter τ can be eliminated by writing the trajectory equation as $d\mathbf{x}/dt=\mathbf{j}(t,\mathbf{x})/j_0(t,\mathbf{x})$ where $t=x^0$, $\mathbf{x}=(x^1,x^2,x^3)$ and $\mathbf{j}=(j^1,j^2,j^3)$. By writing $\psi=Rexp(iS)$ where R, S) are real one arrives at a Hamilton-Jacobi (HJ) form $dx^\mu/d\tau=-(1/m)\partial^\mu S$

and the KG equation is equivalent to

$$\partial^\mu(R^2\partial_\mu S) = 0;\ \frac{(\partial^\mu S)(\partial_\mu S)}{2m} - \frac{m}{2} + Q = 0 \tag{3.24}$$

Here $Q = -(1/2m)(\partial^\mu\partial_\mu R/R$ is the quantum potential. One has put here $c = \hbar = 1$ and reinserted we would have

$$\frac{(\partial^\mu S)(\partial_\mu S)}{2m} - \frac{c^2 m}{2} - \frac{\hbar^2}{2m}\frac{\partial^\mu\partial_\mu R}{R} = 0 \tag{3.25}$$

From the HJ form and (3.24) plus the identity $d/d\tau = (dx^\mu/dt)\partial_\mu$ one arrives at the equations of motion $m(d^2x^\mu/d\tau^2) = \partial^\mu Q$. A typical trajectory arising from $d\mathbf{x}/dt = \mathbf{j}/j_0$ could be imagined as an S shaped curve in the $t - x$ plane (with t horizontal) and cut with a vertical line through the middle of the S. The velocity may be superluminal and may move backwards in time (at points where $j_0 < 0$). There is no paradox with backwards in time motion since it is physically indistinguishable from a motion forwards with negative energy. One introduces a physical number of particles via $N_{phys} = \int d^3x|j_0|$. Contrary to $N = \int d^3x j_0$ the physical number of particles is not conserved. A pair of particles one with positive and the other with negative energy may be created or annihilated; this resembles the behavior of virtual particles in convential QFT.

Now go to relativistic QFT where in the Heisenberg picture the Hermitian field operator $\hat{\phi}(x)$ satisfies

$$(\partial_0^2 - \nabla^2 + m^2)\hat{\phi} = J(\hat{\phi}) \tag{3.26}$$

where J is a nonlinear function describing the interaction. In the Schrödinger picture the time evolution is determined via the Schrödinger equation (SE) in the form $H[\phi, -i\delta/\delta\phi]\Psi[\phi,t] = i\partial_t\Psi[\phi,t]$ where Ψ is a functional with respect to $\phi(\mathbf{x})$ and a function of t. A normalized solution of this can be expanded as $\Psi[\phi,t] = \sum_{-\infty}^{\infty}\tilde{\Psi}_n[\phi,t]$ where the $\tilde{\Psi}_n$ are unnormalized n-particle wave functionals. Since any (reasonable) $\phi(\mathbf{x})$ can be Fourier expanded one can write

$$\tilde{\Psi}_n[\phi,t] = \int d^3k_1\cdots d^3k_n c_n(\mathbf{k}^{(n)},t)\Psi_{n,\mathbf{k}^{(n)}}[\phi] \tag{3.27}$$

where $\mathbf{k}^{(n)} = \{\mathbf{k}_1,\cdots,\mathbf{k}_n\}$. These functionals in (3.27) constitute a complete orthonormal basis which generalizes the basis of Hermite functions and they satisfy

$$\int \mathcal{D}\phi\Psi_0^*[\phi]\phi(\mathbf{x}_1)\cdots\phi(\mathbf{x}_{n'})\Psi_{n,\mathbf{k}^{(n)}}[\phi] = 0\ (n \neq n) \tag{3.28}$$

For free fields (i.e. when $J = 0$ in (3.26) one has

$$c_n(\mathbf{k}^{(n)},t) = c_n(\mathbf{k}^{(n)})e^{-i\omega_n(\mathbf{k}^{(n)})t};\ \omega_n = E_0 + \sum_1^n\sqrt{\mathbf{k}_j^2 + m^2} \tag{3.29}$$

where E_0 is the vacuum energy. In this case the quantities $|c_n(\mathbf{k}^{(n)},t)|^2$ do not depend on time so the number of particles (corresponding to the quantized version of $N = \int d^3x j_0$) is conserved. In a more general situation with interactions the

SE leads to a more complicated time depenedence of the coefficients c_n and the number of particles is not conserved. Now the n-particle wave function is

$$\psi_n(\mathbf{x}^{(n)},t) =< 0|\hat{\phi}(t,\mathbf{x}_1)\cdots\hat{\phi}(t,\mathbf{x}_n)|\Psi > \tag{3.30}$$

(the multiplication of the right side by $(n!)^{-1/2}$ would lead to a normalized wave function only if $\Psi = \tilde{\Psi}_n$). The generalization of (3.30) to the interacting case is not trivial because with an unstable vacuum it is not clear what is the analogue of $< 0|$. Here the Schrödinger picture is more convenient where (3.30) becomes

$$\psi_n(\mathbf{x}^{(n)},t) = \int \mathcal{D}\phi\Psi_0^*[\phi]exp(-i\phi_0(t))\phi(\mathbf{x}_1)\cdots\phi(\mathbf{x}_n)\Psi[\phi,t] \tag{3.31}$$

where $\phi_0(t) = -E_0 t$. For the interacting case one uses a different phase $\phi_0(t)$ defined via an expansion, namely

$$\hat{U}(t)\Psi_0[\phi] = r_0(t)exp(i\phi_0(t))\Psi_0[\phi] + \sum_1^\infty \cdots \tag{3.32}$$

where $r_0(t) \geq 0$ and $\hat{U}(t) = U(\phi, -i\delta/\delta\phi, t]$ is the unitary time evolution operator. One sees that even in the interacting case only the $\tilde{\Psi}_n$ part of Ψ contributes to (3.31) so $\tilde{\Psi}_n$ can be called the n-particle wave functional. The wave function (3.30) can also be generalized to a nonequaltime wave function $\psi_n(x^{(n)}) = S_{\{x_j\}} < 0|\hat{\phi}(x_1)\cdots\hat{\phi}(x_n)|\Psi >$ (here $S_{\{x_j\}}$ denotes symmetrization over all x_j which is needed because the field operators do not commute for nonequal times. For the interacting case the nonequaltime wave function is defined as a generalization of (3.30) with the replacements

$$\phi(\mathbf{x}_j) \to \hat{U}^\dagger(t_j)\phi(\mathbf{x}_j)\hat{U}(t_j);\ \Psi[\phi,t] \to \hat{U}^\dagger(t)\Psi[\phi,t] = \Psi[\phi]; \tag{3.33}$$
$$e^{-i\phi_0(t)} \to e^{-i\phi_0(t_1)}\hat{U}(t_1)$$

followed by symmetrization.

In the deBroglie-Bohm (dBB) interpretation the field $\phi(x)$ has a causal evolution determined by

$$(\partial_0^2 - \nabla^2 + m^2)\phi(x) = J(\phi(x)) - \left(\frac{\delta Q[\phi,t]}{\delta\phi(\mathbf{x})}\right)_{\phi(\mathbf{x})=\phi(x)}; \tag{3.34}$$
$$Q = -\frac{1}{2|\Psi|}\int d^3x \frac{\delta^2|\Psi|}{\delta\phi^2(\mathbf{x})}$$

where Q is the quantum potential again. However the n particles attributed to the wave function ψ_n also have causal trajectories determined by a generalization of $d\mathbf{x}/dt = \mathbf{j}/j_0$ as

$$\frac{d\mathbf{x}_{n,j}}{dt} = \left(\frac{\psi_n^*(x^{(n)})\overleftrightarrow{\nabla}_j\psi_n(x^{(n)})}{\psi_n^*(x^{(n)})\overleftrightarrow{\partial}_{t_j}\psi_n(x^{(n)})}\right)_{t_1=\cdots=t_n=t} \tag{3.35}$$

These n-particles have well defined trajectories even when the probability (in the conventional interpretation of QFT) of the experimental detection is equal to zero.

In the dBB interpretation of QFT we can introduce a new causally evolving "effectivity" parameter $e_n[\phi, t]$ defined as

(3.36)
$$e_n[\phi, t] = |\tilde{\Psi}_n[\phi, t]|^2 / \sum_{n'}^{\infty} |\tilde{\Psi}_{n'}[\phi, t]|^2$$

The evolution of this parameter is determined by the evolution of ϕ given via (3.34) and by the solution $\Psi = \sum \tilde{\Psi}$ of the SE. This parameter might be interpreted as a probability that there are n particles in the system at time t if the field is equal (but not measured!) to be $\phi(\mathbf{x})$ at that time. However in the dBB theory one does not want a stochastic interpretation. **Hence assume that e_n is an actual property of the particles guided by the wave function ψ_n and call it the effectivity of these n particles.** This is a nonlocal hidden variable attributed to the particles and it is introduced to provide a deterministic description of the creation and destruction of particles. One postulates that the effective mass of a particles guided by ψ_n is $m_{eff} = e_n m$ and similarly for the energy, momentum, charge, etc. This is achieved by postulating that the mass density is $\rho_{mass}(\mathbf{x}, t) = m\sum_1^{\infty} e_n \sum_1^n \delta^3(\mathbf{x} - \mathbf{x}_{n,j}(t))$ and similarly for other quantities. Thus if $e_n = 0$ such particles are ineffective, i.e. their effect is the same as if they didn't exist while if $e_n = 1$ they exist in the usual sense. However the trajectories are defined even for the particles for which $e_n = 0$ and QFT is a theory of an infinite number of particles although some of them may be ineffective (conventionally one would say they are virtual). We will say more about this later.

3.3. FERMIONIC THEORY. This extraction from Nikolić [**930, 931**] (cf. also [**413**]) becomes even more technical but a sketch should be rewarding; there is more detail and discussion in [**930, 931**]. The Dirac equation in Minkowski space $\eta_{\mu\nu} = diag(1, -1, -1, -1)$ is $i\gamma^{\mu}\partial_{\mu} - m)\psi(x) = 0$ where $x = (x^i) = (t, \mathbf{x})$ with $\mathbf{x} \in \mathbf{R}^3$ (cf. Section 2.1.1). A general solution can be written as $\psi(x) = \psi^P(x) + \psi^A(x)$ where the particle and antiparticle parts can be expanded as $\psi^P = \sum b_k u_k(x)$ and $\psi^A = \sum d_k^* v_k(x)$. Here u_k (resp. v_k) are positive (resp. negative) frequency 4-spinors that, together, form a complete orthonormal set of solutions to the Dirac equation. The label k means $(\mathbf{k}, s)$ where $s = \pm 1/2$ is the spin label. Writing $\Omega^P(x, x') = \sum u_k(x)u_k^{\dagger}(x')$ and $\Omega^A(x, x') = \sum v_k(x)v_k^{\dagger}(x')$ one can write

(3.37)
$$\psi^P = \int d^3x'\Omega^P(x, x')\psi(x'); \ \psi^A(x) = \int d^3x'\Omega^A(x, x')\psi(x')$$

where $t = t'$. The particle and antiparticle currents are $j_{\mu}^P = \bar{\psi}^P\gamma_{\mu}\psi^P$ and $j_{\mu}^A = \bar{\psi}^A\gamma_{\mu}\psi^A$ where $\bar{\psi} = \psi^{\dagger}\gamma_0$. Since ψ^P and ψ^A satisfy the Dirac equation the currents j_{μ}^P, j_{μ}^A are separately conserved, i.e. $\partial^{\mu}j_{\mu}^P = \partial^{\mu}j_{\mu}^A = 0$. One postulates then trajectories of the form

(3.38)
$$\frac{d\mathbf{x}^P}{dt} = \frac{\mathbf{j}^P(t, \mathbf{x}^P)}{j_0^P(t, \mathbf{x})}; \ \frac{d\mathbf{x}^A}{dt} = \frac{\mathbf{j}^A(t, \mathbf{x}^A)}{j_0^A(t, \mathbf{x}^A)}$$

where $\mathbf{j} = (j^1, j^2, j^3)$ for a causal interpretation of the Dirac equation. Now in QFT the coefficients b_k and d_k^* become anticommuting operators with $\hat{b}_k^{\dagger}$ and $\hat{d}_k^{\dagger}$ creating particles and antiparticles while $\hat{b}_k$ and $\hat{d}_k$ annihilate them. In the

Schrödinger picture the field opperators $\hat{\psi}(\mathbf{x})$ and $\hat{\psi}^\dagger(\mathbf{x})$ satisfy the commutation relations $\{\hat{\psi}_a(\mathbf{x}), \hat{\psi}^\dagger_{a'}(\mathbf{x}')\} = \delta_{aa'}\delta^3(\mathbf{x}-\mathbf{x}')$ while other commutators vanish (a is the spinor index). These relations can be represented via

$$\hat{\psi}_a(\mathbf{x}) = \frac{1}{\sqrt{2}}\left[\eta_a(\mathbf{x}) + \frac{\delta}{\delta\eta^*_a(\mathbf{x})}\right];\ \hat{\psi}^\dagger_a(\mathbf{x}) = \frac{1}{\sqrt{2}}\left[\eta^*_a(\mathbf{x}) + \frac{\delta}{\delta\eta_z(\mathbf{x})}\right] \tag{3.39}$$

where η_a, η^*_a are anticommuting Grassmann numbers satisfying $\{\eta_a(\mathbf{x}), \eta_{a'}(\mathbf{x}')\} = \{\eta^*_a(\mathbf{x}), \eta^*_{a'}(\mathbf{x}')\} = \{\eta_a(\mathbf{x}), \eta^*_{a'}(\mathbf{x}')\} = 0$. Next introduce a complete orthonormal set of spinors $u_k(\mathbf{x})$ and $v_k(\mathbf{x})$ which are equal to the spinors $u_k(x)$ and $v_k(x)$ at $t=0$. An arbitrary quantum state may then be obtained by acting with creation operators

$$b^\dagger_k = \int d^3x\hat{\psi}^\dagger(\mathbf{x})u_k(\mathbf{x});\ \hat{d}^\dagger_k = \int d^3x v^\dagger_k(\mathbf{x})\hat{\psi}(\mathbf{x}) \tag{3.40}$$

on the vacuum $|0>=|\Psi_0>$ represented by

$$\Psi_0[\eta,\eta^\dagger] = Nexp\{\int d^3x\int d^3x'\eta^\dagger(\mathbf{x})\Omega(\mathbf{x},\mathbf{x}')\eta(\mathbf{x}')\} \tag{3.41}$$

Here $\Omega(\mathbf{x},\mathbf{x}') = (\Omega^A - \Omega^P)(\mathbf{x},\mathbf{x}')$, N is a constant such that $<\Psi_0|\Psi_0>=1$ and the scalar product is $<\Psi|\Psi>=\int\mathcal{D}^2\eta\Psi^*[\eta,\eta^\dagger\Psi'[\eta,\eta^\dagger]$; also $\mathcal{D}^2 = \mathcal{D}\eta\mathcal{D}\eta^\dagger$ and Ψ^* is dual (not simply the complex conjugate) to Ψ. The vacuum is chosen such that $\hat{b}_k\Psi_0 = \hat{d}_k\Psi_0 = 0$. A functional $\Psi[\eta,\eta^\dagger]$ can be expanded as $\Psi[\eta,\eta^\dagger] = \sum c_K\Psi_K[\eta,\eta^\dagger]$ where the set $\{\psi_K\}$ is a complete orthonormal set of Grassmann valued functionals. This is chosen so that each Ψ_K is proportional to a functional of the form $\hat{b}^\dagger_{k_1}\cdots\hat{b}^\dagger_{k_{n_P}}\hat{d}^\dagger_{k'_1}\cdots\hat{d}^\dagger_{k'_{n_A}}\Psi_0$ which means that each Ψ_K has a definite number n_P of particles and n_A of antiparticles. Therefore one can write $\Psi[\eta,\eta^\dagger] = \sum_{n_P,n_A=0}^\infty\tilde{\Psi}_{n_P,n_A}[\eta,\eta^\dagger]$ where the tilde denotes that these functionals, in contrast to Ψ and Ψ_K, do not have unit norm. Time dependent states $\Psi[\eta,\eta^\dagger,t]$ can be expanded as

$$\Psi[\eta,\eta^\dagger,t] = \sum_K c_K(t)\Psi_K[\eta,\eta^\dagger] = \sum_{n_p,n_A=0}^\infty\tilde{\Psi}_{n_P,n_A}[\eta,\eta^\dagger,t] \tag{3.42}$$

The time dependence of the c-number coefficients $c_K(t)$ is governed by the functional SE

$$H[\hat{\psi},\hat{\psi}^\dagger]\Psi[\eta,\eta^\dagger,t] = i\partial_t\Psi[\eta,\eta^\dagger,t] \tag{3.43}$$

Since the Hamiltonian H is a Hermitian operator the norms $<\Psi(t)|\Psi(t)>=\sum|c_K(t)|^2$ do not depend on time. In particular if H is the free Hamiltonian (i.e. the Hamiltonian that generates the second quantized free Dirac equation) then the quantities $|c_K(t)|$ do not depend on time, which means that the average number of particles and antiparticles does not change with time when there are no interactions.

Next introduce the wave function of n_P particles and n_A antiparticles via

$$\psi_{n_P,n_A} \equiv \psi_{b_1\cdots b_{n_P}d_1\cdots d_{n_A}}(\mathbf{x}_1,\cdots,\mathbf{x}_{n_P},\mathbf{y}_1\cdots,\mathbf{y}_{n_A},t) \tag{3.44}$$

It has $n_P + n_A$ spinor indices and for free fields the (unnormalized) wave function can be calculated using the Heisenberg picture as

$$\psi_{n_P,n_A} = < 0|\hat{\psi}^P_{b_1}(t,\mathbf{x}_1)\cdots\hat{\psi}^{A\dagger}_{d_{n_A}}(t,\mathbf{y}_{n_A})|\Psi > \tag{3.45}$$

where $\hat{\psi}^P$ and $\hat{\psi}^A$ are extracted from $\hat{\psi}$ using (3.37). In the general interacting case the wave function can be calculated using the Schrödinger picture as

$$\psi_{n_P,n_A} = \int \mathcal{D}^2\eta\Psi_0^*[\eta,\eta^\dagger]e^{-i\phi_0(t)}\hat{\psi}^P_{b_1}(\mathbf{x}_1)\cdots\hat{\psi}^{A\dagger}_{d_{n_A}}(y_{n_A})\Psi[\eta,\eta^\dagger,t] \tag{3.46}$$

Here the phase $\phi_0(t)$ is defined by an expansion as in (3.42), namely

$$\hat{U}(t)\Psi_0[\eta,\eta^\dagger] = r_0(t)exp(i\phi_0(t))\Psi_0[\eta,\eta^\dagger] + \sum_{(n_P,n_A)\neq(0,0)} \cdots \tag{3.47}$$

where $r_0(t) \geq 0$ and $\hat{U}(t) = U[\hat{\psi},\hat{\psi}^\dagger,t]$ is the unitary time evolution operator that satisfies the SE (3.43). The current attributed to the ith corpuscle (particle or antiparticle) in the wave function ψ_{n_P,n_A} is $j_{\mu(i)} = \bar{\psi}_{n_P,n_A}\gamma_{\mu(i)}\psi_{n_P,n_A}$ where one writes

$$\bar{\psi}\Gamma_i\psi = \bar{\psi}_{a_1\cdots a_i\cdots a_n}(\Gamma)_{a_ia_i'}\psi_{a_1\cdots a_i'\cdots a_n}; \tag{3.48}$$

$$\bar{\psi}_{a_1\cdots a_n} = \psi^*_{a_1'\cdots a_n'}(\gamma_0)_{a_1'a_1}\cdots(\gamma_0)_{a_n'a_n}$$

Hence the trajectory of the ith corpuscle guided by the wave function ψ_{n_P,n_A} is given by the generalization of (3.38), namely $d\mathbf{x}_i/dt = \mathbf{j}_i/j_{0(i)}$.

We now need a causal interpretation of the processes of creation and destruction of particles and antiparticles. For bosonic fields this was achieved by introducing the effectivity parameter in Section 1.3.2 but this cannot be done for the Grassmann fields η, $\eta^\dagger$ because $\Psi^*[\eta,\eta^\dagger,t]\Psi[\eta,\eta^\dagger,t]$ is Grassmann valued and cannot be interpreted as a probability density. Hence another formulation of fermionic states is developed here, more similar to the bosonic states. First the notion of the scalar product can be generalized in such a way that it may be Grassmann valued which allows one to write $\Psi[\eta,\eta^\dagger,t] =< \eta,\eta^\dagger|\Psi(t) >$ and $1 = \int \mathcal{D}^2\eta|\eta,\eta^\dagger >< \eta,\eta^\dagger|$ (cf. Hatfield [**616**]). We can also introduce

$$< \phi,\phi^\dagger|\eta,\eta^\dagger > = \sum_K < \phi,\phi^\dagger|\Psi_K >< \Psi_K|\eta,\eta^\dagger > = \sum_K \Psi_K[\phi,\phi^\dagger]\Psi_K^*[\eta,\eta^\dagger] \tag{3.49}$$

so one sees that the sets $\{\Psi_K[\eta,\eta^\dagger]\}$ and $\{\Psi_K[\phi,\phi^\dagger]\}$ are two representations of the same orthonormal basis $\{|\Psi_K >\}$ for the same Hilbert space of fermionic states. In other words the state $|\Psi(t) >$ can be represented as $\Psi[\phi,\phi^\dagger,t] =< \phi,\phi^\dagger|\Psi(t) >$ which can be expanded as

$$\Psi[\phi,\phi^\dagger,t] = \sum_K c_K(t)\Psi_K[\phi,\phi^\dagger] = \sum_{n_P,n_A=0}^{\infty} \tilde{\psi}_{n_P,n_A}[\phi,\phi^\dagger,t] \tag{3.50}$$

Putting the unit operator $1 = \int \mathcal{D}^2\phi|\phi,\phi^\dagger >< \phi,\phi^\dagger|$ in the expression for $< \Psi(t),\Psi(t) >$ we see that the time independent norm can be written as

$$< \Psi(t)|\Psi(t) > = \int \mathcal{D}^2\phi\Psi^*[\phi,\phi^\dagger,t)]\Psi[\phi,\phi^\dagger,t] \tag{3.51}$$

Therefore the quantity $\rho[\phi,\phi^\dagger,t] = \Psi^*[\phi,\phi^\dagger,t]\Psi[\phi,\phi^\dagger,t]$ can be interpreted as a positive definite probability density for spinors ϕ, $\phi^\dagger$ to have space dependence $\phi(\mathbf{x})$ and $\phi^\dagger(\mathbf{x})$ respectively at time t. The SE (3.43) can also be written in the ϕ-representation as $\hat{H}_\phi\Psi[\phi,\phi^\dagger,t] = i\partial_t\Psi[\phi,\phi^\dagger,t]$ where the Hamiltonian $\hat{H}_\phi$ is defined by its action on wave functionals $\Psi[\phi,\phi^\dagger,t]$ determined via
(3.52)

$$\hat{H}_\phi[\phi,\phi^\dagger,t] = \int \mathcal{D}^2\eta \int \mathcal{D}^2\phi' < \phi,\phi^\dagger|\eta,\eta^\dagger > \hat{H} < \eta,\eta^\dagger|\phi',\phi'^\dagger > \Psi[\phi',\phi'^\dagger,t]$$

where $\hat{H} = H[\hat{\psi},\hat{\psi}^\dagger]$ is the Hamiltonian of (3.43).

One can now obtain a causal interpretation of a quantum system described by a c-number valued wave function satisfying a SE. The material is written for an n-dimensional vector $\vec{\phi}$ but in a form that generalizes to infinite dimensions. The wave function $\psi(\vec{\phi},t)$ satisfies the SE $\hat{H}\psi = i\partial_t\psi$ where $\hat{H}$ is an arbitrary Hermitian Hamiltonian written in the $\vec{\phi}$ representation. The quantity $\rho = \psi^*\psi$ is the probability density for the variables $\vec{\phi}$ and the average velocity is

$$(3.53)\qquad d < \vec{\phi} > (t)/dt = \int d^n\phi\rho(\vec{\phi},t)\vec{u}(\vec{\phi},t);\ \vec{u} = i\psi^*[\hat{H},\vec{\phi}]\psi/\psi^*\psi$$

Introduce a source J via $J = (\partial\rho/\partial t)+\vec{\nabla}(\rho\vec{u})$ (note e.g. for the example of (3.52) J does not vanish even though it frequently will vanish). One wants to find a quantity $\vec{v}(\vec{\phi},t)$ that has the property (3.53) in the form $d < \vec{\phi} > (t)/dt = \int d^n\phi\rho(\vec{\phi},t)\vec{v}(\vec{\phi},t)$ but at the same time satisfies the equivariance property $\partial_t\rho+\vec{\nabla}(\rho\vec{v}) = 0$. These two properties allow one to postulate a consistent causal interpretation of QM in which $\vec{\phi}$ has definite values at each time t determined via $d\vec{\phi}/dt = \vec{v}(\vec{\phi},t)$. In particular the equivariance provides that the statistical distribution of the variables $\vec{\phi}$ is given by ρ for any time t provided that it is given by ρ for some initial time t_0. When $J = 0$ then $\vec{v} = \vec{u}$ which corresponds to the dBB interpretation. The aim now is to generalize this to the general case of $\vec{v}$ in the form $\vec{v} = \vec{u}+\rho^{-1}\vec{\mathcal{E}}$ where $\vec{\mathcal{E}}(\vec{\phi},t)$ is the quantity to be determined. From $\partial_t\rho+\vec{\nabla}(\rho\vec{v}) = 0$ we see that $\vec{\mathcal{E}}$ must be a solution of the equation $\vec{\nabla}\vec{\mathcal{E}} = -J$. Now let $\vec{\mathcal{E}}$ be some particular solution of this equation; then $\vec{\mathcal{E}}(\vec{\phi},t) = \vec{e}(t) + \vec{E}(\vec{\phi},t)$ is also a solution for an arbitrary $\vec{\phi}$ independent function $\vec{e}(t)$. Comparing with (3.53) one sees that $\int d^n\phi\vec{\mathcal{E}} = 0$ is required. This fixes the function $\vec{e}$ to be $\vec{e}(t) = -V^{-1}\int d^n\phi\vec{E}(\vec{\phi},t)$ where $V = \int d^n\phi$. Thus it remains to choose $\vec{E}$ and in Nikolić [**930, 931**] one takes $\vec{E}$ such that $\vec{E} = 0$ when $J = 0$ so that $\vec{\mathcal{E}} = 0$ when $J = 0$ as well; thus $\vec{v} = \vec{u}$ when $J = 0$. There is still some arbitrariness in $\vec{E}$ so take $\vec{E} = \vec{\nabla}\Phi$ where $\vec{\nabla}^2\Phi = -J$, which is solved via $\Phi(\vec{\phi},t)\int d^n\phi' G(\vec{\phi},\vec{\phi}')J(\vec{\phi}',t)$, so that $\vec{\nabla}^2 G(\vec{\phi},\vec{\phi}') = -\delta^n(\vec{\phi}-\vec{\phi}')$. The solution can be expressed as a Fourier transform $G(\vec{\phi},\vec{\phi}') = \int(d^nk/(2\pi)^n)exp[i\vec{k}(\vec{\phi}-\vec{\phi}')]/\vec{k}^2$. To eleminate the factor $1/(2\pi)^n$ one uses a new integration variable $\vec{\chi} = \vec{k}/2\pi$ and we obtain

$$(3.54)\qquad \Phi(\vec{\phi},t) = \int d^n\chi \int d^n\phi'[exp(2i\pi\vec{\chi}(\vec{\phi}-\vec{\phi}'))]J(\vec{\phi}',t)/(2\pi)^2\vec{\chi}^2]$$

Now for a causal interpretation of fermionic QFT one writes first for simplicity $A[\mathbf{x}]$ for functionals of the form $A[\phi, \phi^\dagger, t, \mathbf{x}]$ and introduces

$$u_a[\mathbf{x}] = i\frac{\Psi^*[\hat{H}_\phi, \phi_a(\mathbf{x})]\Psi}{\Psi^*\Psi};\ u_a^*[\mathbf{x}] = i\frac{\Psi^*[\hat{H}_\phi, \phi_a^*(\mathbf{x})]\Psi}{\Psi^*\Psi} \tag{3.55}$$

where $\Psi = \Psi[\phi, \phi^\dagger, t]$. Next introduce the source

$$J = \frac{\partial\rho}{\partial t} + \sum_a \int d^3x \left[\frac{\delta(\rho u_a[\mathbf{x}])}{\delta\phi_z(\mathbf{x})} + \frac{\delta(\rho u_a^*[\mathbf{x}])}{\delta\phi_a^*(\mathbf{x})}\right] \tag{3.56}$$

where $\rho = \Psi^*\Psi$. Introduce now the notation $\alpha \cdot \beta = \sum_a \int d^3x[\alpha_a(\mathbf{x})\beta_a(\mathbf{x}) + \alpha_a^*(\mathbf{x})\beta_a^*(\mathbf{x})]$ and (3.51) generalizes to

$$\Phi[\phi, \phi^\dagger, t] = \int \mathcal{D}^2\chi \int \mathcal{D}^2\phi' \frac{e^{2\pi i\chi\cdot(\phi-\phi')}}{(2\pi)^2\chi\cdot\chi} J[\phi', \phi'^\dagger, t] \tag{3.57}$$

Then write for $V = \int \mathcal{D}^2\phi$

$$E_a[\mathbf{x}] = \frac{\delta\Phi}{\delta\phi_a(\mathbf{x})};\ E_a^*[\mathbf{x}] = \frac{\delta\Phi}{\delta\phi_a^*(\mathbf{x})}; \tag{3.58}$$

$$e_a(t, \mathbf{x}) = -V^{-1}\int \mathcal{D}^2\phi E_a[\phi, \phi^\dagger, t, \mathbf{x}];\ e_a^*(t, \mathbf{x}) = -V^{-1}\int \mathcal{D}^2\phi E_a^*[\phi, \phi^\dagger, t, \mathbf{x}]$$

The corresponding velocities are then

$$v_a[\mathbf{x}] = u_a[\mathbf{x}] + \rho^{-1}(e_a(t, \mathbf{x}) + E_a[\mathbf{x}]);\ v_a^*[\mathbf{x}] = u_a^*[\mathbf{x}] + \rho^{-1}(e_a^*(t, \mathbf{x}) + E_a^*[\mathbf{x}]) \tag{3.59}$$

Next introduce hidden variables $\phi(t, \mathbf{x})$ and $\phi^\dagger(t, \mathbf{x})$ with causal evolution given then by

$$\frac{\partial\phi_a(t, \mathbf{x})}{\partial t} = v_a[\phi, \phi^\dagger, t, \mathbf{x}];\ \frac{\partial\phi_a^*(t, \mathbf{x})}{\partial t} = v_a^*[\phi, \phi^\dagger, t, \mathbf{x}] \tag{3.60}$$

where it is understood that the right sides are calculated at $\phi(\mathbf{x}) = \phi(t, \mathbf{x})$ etc. In analogy with the bosonic fields treated earlier one introduces effectivity parameters guided by the wave function ψ_{n_P, n_A} given by

$$e_{n_P, n_A}[\phi, \phi^\dagger, t] = \frac{|\tilde{\Psi}_{n_P, n_A}[\phi, \phi^\dagger, t]|^2}{\sum_{n'_P, n'_A} |\tilde{\Psi}_{n'_P, n'_A}[\phi, \phi^\dagger, t]|^2} \tag{3.61}$$

REMARK 2.3.2. Concerning the nature of the effectivity parameter we extract from Nikolić [**930**] as follows. In the bosonic theory the analogue of (3.61) is

$$e_n[\{\phi\}, t] = \frac{|\tilde{\Psi}_n[\{\phi\}], t|^2}{\sum_{n'} |\tilde{\Psi}_{n'}[\{\phi\}], t|^2} \tag{3.62}$$

$\{\phi\} = \{\phi_1, \cdots, \phi_{N_s}\}$ where N_s is the number of different particle species. Now the measured effectivity can be any number between 0 and 1 and this is no contradiction since if different $\tilde{\Psi}_n$ in the expansion do not overlap in the ϕ space then they represent a set of nonoverlapping "channels" for the causally evolving field ϕ. The field necessarily enters one and only one of the channels and one sees that $e_n = 1$ for the nonempty channel with $e_{n'} = 0$ for all empty channels. The effect is the same as if the wave functional Ψ "collapsed" into one of the states $\tilde{\psi}_n$ with a definite number of particles. In a more general situation different $\tilde{\Psi}_n$ of the measured particles may overlap. However the general theory of ideal quantum measurements (cf. [**158**]) provides that the total wave functional can be written again as a sum of nonoverlappiing wave functionals in the $\{\phi\}$ space, where one of the fields represents the measured field, while the others represent fields of the measuring apparatus. Thus only one of the $\tilde{\Psi}$ in (3.62) becomes nonempty with the corresponding $e_n = 1$ while all the other $e_{n'} = 0$. The essential point is that from the point of view of an observer who does not know the actual field configuration the probability for such an effective collapse of the wave functional is exactly equal to the usual quantum mechanical probability for such a collapse. Hence the theory has the same statistical properties as the usual theory. In the case when all the effectivities are less than 1 (i.e. the wave functional has not collapsed) the theory does not agree nor disagree with standard theory; effectivity is a hidden variable. This agrees with the Bohmian particle positions which agree with the standard quantum theory only when the wave function effectively collapses into a state with a definite particle position. Similar comments apply to the fermionic picture. In an ideal experiment in which the number of particles is measured, different Ψ_{n_P,n_A} do not overlap in the $(\phi, \phi^\dagger)$ space and the fields ϕ, $\phi^\dagger$ necessarily enter into a unique "channel" $\tilde{\Psi}_{n_P,n_A}$, etc. ∎

REMARK 2.3.3. In Nikolić [**931**] one addresses the question of statistical transparency. Thus the probabilitistic interpretation of the nonrelativistic SE does not work for the relativistic KG equation $(\partial^\mu\partial_\mu + m^2)\psi = 0$ (where $x = (\mathbf{x}, t)$ and $\hbar = c = 1$) since $|\psi|^2$ does not correspond to a probability density. There is a conserved current $j^\mu = i\psi^* \overleftrightarrow{\partial^\mu}\psi$ (where $a\overleftrightarrow{\partial^\mu}b = a\partial^\mu b - b\partial^\mu a$) but the time component j^0 is not positive definite. In [**930, 931**] the equations that determine the Bohmain trajectories of relativistic quantum particles described by many particle wave functions were written in a form requiring a preferred time coordinate. However a preferred Lorentz frame is not necessary (cf. Berndl et al [**129**]) and this is developed in [**931**] following [**129, 930, 931**]. First note that as in [**129, 930, 931**] it appears that particles may be superluminal and the principle of Lorentz covariance does not forbid superluminal velocities and conversly superluminal velocities do not lead to causal paradoxes (cf. [**129, 931**]). As noted in [**129**] the Lorentz-covariant Bohmian interprtation of the many particle KG equation is not statistically transparent. This means that the statistical distribution of particle positions cannot be calculated in a simple way from the wave function alone without the knowledge of particle trajetories. One knows that classcal QM is statistically transparent of course and this perhaps helps to explain why Bohmian mechanics has not attracted more attention. However statistical transparency (ST) may not be a fundamental property of nature as the following facts suggest:

- Classical mechanics, relativistic or nonrelativistic, is not ST.
- Relativistic QM based on the KG equation (or some of its generalizations) is not ST.
- The relativistic Dirac equation is ST but its many particle relativistic generalization is not (unless a preferred time coordinate is determined in an as yet unknown dynamical manner).

- Nonrelativistic QM is ST but not completely so since it distinguishes the time variable (e.g. $\rho(x^1, x^2, t)$ is not a probability density).
- The background independent quantum gravity based on the Wheeler-DeWitt (WDW) equation lacks the notion of time and is not ST.

The upshot is that since statistical probabilities can be calculated via Bohmian trajectories that theory is more powerful than other interpretations of general QM (see [**930, 931**] for discussion on this). Now let $\hat{\phi}(x)$ be a scalar field operator satisfying the KG equation (an Hermitian uncharged field for simplicity so that negative values of the time component of the current cannot be interpreted as negatively charged particles). The corresponding n-particle wave function is (cf. [**930, 931**])

$$\psi(x_1, \cdots, x_n) = (n!)^{-1/2} S_{\{x_a\}} < 0|\hat{\phi}(x_1) \cdots \hat{\phi}(x_n)|n > \tag{3.63}$$

Here $S_{\{x_a\}}$ $(a = 1, \cdots, n)$ denotes the symmetrization over all x_a which is needed because the field operators do not commute for nonequal times. The wave function ψ satisfies n KG equations

$$(\partial_a^\mu \partial_{a\mu} + m^2)\psi(x_1, \cdots, x_n) = 0 \tag{3.64}$$

Although the operator $\hat{\phi}$ is Hermitian the nondiagonal matrix element ψ defined by (3.63) is complex and one can introduce n real 4-currrents $j_a^\mu = i\psi^* \overleftrightarrow{\partial_a^\mu} \psi$ each of which is separately conserved via $\partial_a^\mu j_{a\mu} = 0$. Equation (3.64) also implies

$$\left(\sum_a \partial_a^\mu \partial_{a\mu} + nm^2\right)\psi(x_1, \cdots, x_n) = 0 \tag{3.65}$$

and the separate conservation equations imply that $\sum_a \partial_a^\mu j_{a\mu} = 0$. Now write $\psi = Rexp(iS)$ with R and S real. Then (3.65) is equivalent to a set of two equations

(3.66)

$$\sum_a \partial_a^\mu (^2\partial_{a\mu} S) = 0; \ -\frac{\sum_a (\partial_a^\mu S)(\partial_{a\mu} S)}{2m} + \frac{nm}{2} + Q = 0; \ Q = \frac{1}{2m}\frac{\sum_a \partial_a^\mu \partial_{a\mu} R}{2mR}$$

where Q is the quantum potential. The first equation is equivalent to a current conservation equation while the second is the quantum analogue of the relativistic HJ equation for n particles. The Bohmian interpretation consistists in postulating the existence of particle trajectories $x_a^\mu(s)$ satisfying $dx_a^\mu/ds = -(1/m)\partial_a^\mu s$ where s is an affine parameter along the n curves in the 4-dimensional Minkowski space. This equation has a form identical to the corresponding classical relativistic equation and can also be written as $dx_a^\mu/ds = j_a^\mu/2m\psi^*\psi$. Hence using $d/ds = \sum_a (dx_z^\mu/ds)\partial_{a\mu}$ one finds the equations of motion

$$m\frac{d^2 x_a^\mu}{ds^2} = \partial_a^\mu Q \tag{3.67}$$

Note that the equations above for the particle trajectories are nonlocal but still Lorentz covariant. The Lorentz covariance is a consequence of the fact that the trajectories in spacetime do not depend on the choice of affine parameter s (cf. [**129**]). Instead, by choosing n "initial" spacetime positions x_a, the n trajectories are uniquely determined by the vector fields j_a^μ or $-\partial_a^\mu S$ (i.e. the trajectories are integral curves of these vector fields). The nonlocality is encoded in the fact that the right hand side of (3.67) depends not only on x_a but also on all the other $x_{a'}$. This is a consequence of the fact that $Q(x_1, \cdots, x_n)$ in (3.66) is not of the form $\sum_a Q_a(x_1, \cdots, x_n)$, which in turn is related to the fact that $S(x_1, \cdots, x_n)$ is not of the form $\sum_a S(x_a)$. Note also that the fact that we parametrize all trajectories with the same parameter s is not directly related to the nonlocality, because such a parametrization can be used even in local classical physics. When the interactions are local then one can even use another parameter s_a for each curve but when the interactions are not local one must use a single parameter s; new separate parameters could only be used after the equations are solved. In the nonrelativistic limit all wave function frequencies are (approximately) equal to m so from $j^\mu_a \psi^* \overleftrightarrow{\partial^\mu_a} \psi$ all time components are equal and given by $j^0_a = 2m\psi^*\psi = \tilde{\rho}$ which does not depend on a. Writing then $\rho(\mathbf{x}_1, \cdots, \mathbf{x}_n) = \tilde{\rho}(x_1, \cdots, x_n)|_{t_1=\cdots=t_n=t}$ one obtains $\partial_t \rho + \sum_a \partial^i_a j_{ai} = 0$ and this implies that ρ can be interpreted as a probability density. In the full relativistic there is generally no analogue of such a function ρ. We refer to [**930, 931**] for more discussion. ∎

4. DeDONDER, WEYL, AND BOHM

We go here to a fascinating paper in Nikolić [**931**] which gives a manifestly covariant canonical method of field quantization based on the classical DeDonder-Weyl (DW) formulation of field theory (cf. also Appendix A in [**254**] for some background on DW theory following Kyprianidis [**788**]). The Bohmian formulation is not postulated for intepretational purposes here but derived from purely technical requirements, namely covariance and consistency with standard QM. It arises automatically as a part of the formalism without which the theory cannot be formulated consistently. This together with the results of Nikolić [**930, 931**] suggest that it is Bohmian mechanics that might be the missing bridge between QM and relativity; further (as will be seen later) it should play an important role in cosmology. The classical covariant canonical DeDonder-Weyl formalism is given first following [**788**] and for simplicity one real scalar field in Minkowski spacetime is used. Thus let $\phi(x)$ be a real scalar field described by

$$\mathfrak{A} = \int d^4x \mathfrak{L};\ \mathfrak{L} = \frac{1}{2}(\partial^\mu \phi)(\partial_\mu \phi) - V(\phi) \tag{4.1}$$

As usual one has

$$\pi^\mu = \frac{\partial \mathfrak{L}}{\partial(\partial_\mu \phi)} = \partial^\mu \phi;\ \partial_\mu \phi = \frac{\partial \mathfrak{H}}{\partial \pi^\mu};\ \partial_\mu \pi^\mu = -\frac{\partial \mathfrak{H}}{\partial \phi} \tag{4.2}$$

where the scalar DeDonder-Weyl (DDW) Hamilonian (not related to the energy density) is given by the Legendre transform $\mathfrak{H}(\pi^\mu, \phi) = \pi^\mu \partial_\mu \phi - \mathfrak{L} = (1/2)\pi^\mu \pi_\mu + V$. The equations (4.2) are equivalent to the standard Euler-Lagrange (EL) equations and by introducing the local vector $S^\mu(\phi(x), x)$ the dynamics can also be described by the covariant DDW HJ equation and equations of motion

$$\mathfrak{H}\left(\frac{\partial S^\alpha}{\partial \phi}, \phi\right) + \partial_\mu S^\mu = 0;\ \partial^\mu \phi = \pi^\mu = \frac{\partial S^\mu}{\partial \phi} \tag{4.3}$$

Note here ∂_μ is the partial derivative acting only on the second argument of $S^\mu(\phi(x),x)$; the corresonding total derivative is $d_\mu = \partial_\mu + (\partial_\mu\phi)(\partial/\partial\phi)$. Note that the first equation in (4.3) is a single equation for four quantities S^μ so there is a lot of freedom in finding solutions. Nevertheless the theory is equivalent to other formulations of classical field theory. Now following Kanachikov [**720**] one considers the relation between the covariant HJ equation and the conventional HJ equation; the latter can be derived from the former as follows. Using (4.2), (4.3) takes the form $(1/2)\partial_\phi S_\mu \partial_\phi S^\mu + V + \partial_\mu S^\mu = 0$. Then using the equation of motion in (4.3) write the first term as

$$\frac{1}{2}\frac{\partial S_\mu}{\partial\phi}\frac{\partial S^\mu}{\partial\phi} = \frac{1}{2}\frac{\partial S^0}{\partial\phi}\frac{\partial S^0}{\partial\phi} + \frac{1}{2}(\partial_i\phi)(\partial^i\phi) \tag{4.4}$$

Similarly using (4.3) the last term is $\partial_\mu S^\mu = \partial_0 S^0 + d_i S^i - (\partial_i\phi)(\partial^i\phi)$. Now introduce the quantity $\mathfrak{S} = \int d^3x S^0$ so $[\partial S^0(\phi(x),x)/\partial\phi(x)] = [\delta\mathfrak{S}([\phi(\mathbf{x},t)],t)/\delta\phi(\mathbf{x},t)]$ where $\delta/\delta\phi(\mathbf{x},t) \equiv [\delta/\delta\phi(x)]_{\phi(x)=\phi(\mathbf{x},t)}$ is the space functional derivative. Putting this together gives then

$$\int d^3x\left[\frac{1}{2}\left(\frac{\delta\mathfrak{S}}{\delta\phi(\mathbf{x},t)}\right)^2 + \frac{1}{2}(\nabla\phi)^2 + V(\phi)\right] + \partial_t\mathfrak{S} = 0 \tag{4.5}$$

which is the standard noncovariant HJ equation. The time evolution of $\phi(\mathbf{x},t)$ is given by $\partial_t\phi(\mathbf{x},t) = \delta\mathfrak{S}/\delta\phi(\mathbf{x},t)$ which arises from the time component of (4.3). Note that in deriving (4.5) it was necessary to use the space part of the equations of motion (4.3) (this does not play an important role in classical physics but is important here). Now for the Bohmian formulation look at the SE $\hat{H}\Psi = i\hbar\partial_t\Psi$ where we write

$$\hat{H} = \int d^3x\left[-\frac{\hbar^2}{2}\left(\frac{\delta}{\delta\phi(\mathbf{x})}\right)^2 + \frac{1}{2}(\nabla\phi)^2 + V(\phi)\right]; \tag{4.6}$$

$$\Psi([\phi(\mathbf{x})],t) = \mathfrak{R}([\phi(\mathbf{x})],t)e^{i\mathfrak{S}(([\phi(\mathbf{x})],t)/\hbar}$$

Then the complex SE equation is equivalent to two real equations

$$\int d^3x\left[\frac{1}{2}\left(\frac{\delta\mathfrak{S}}{\delta\phi(\mathbf{x})}\right)^2 + \frac{1}{2}(\nabla\phi)^2 + V(\phi) + Q\right] + \partial_t S = 0; \tag{4.7}$$

$$\int d^3x\left[\frac{\delta\mathfrak{R}}{\delta\phi(\mathbf{x})}\frac{\delta\mathfrak{S}}{\delta\phi(\mathbf{x})} + J\right] + \partial_t\mathfrak{R} = 0;\ Q = -\frac{\hbar^2}{2\mathfrak{R}}\frac{\delta^2\mathfrak{R}}{\delta\phi^2(\mathbf{x})};\ J = \frac{\mathfrak{R}}{2}\frac{\delta^2\mathfrak{S}}{\delta\phi^2(\mathbf{x})}$$

The second equation is also equivalent to

$$\partial_t\mathfrak{R}^2 + \int d^3x\frac{\delta}{\delta\phi(\mathbf{x})}\left(\mathfrak{R}^2\frac{\delta\mathfrak{S}}{\delta\phi(\mathbf{x})}\right) = 0 \tag{4.8}$$

and this exhibits the unitarity of the theory because it provides that the norm $\int[d\phi(\mathbf{x})]^2\Psi^*\Psi = \int[d\phi(\mathbf{x})]\mathfrak{R}^2$ does not depend on time. The quantity $\mathfrak{R}^2([\phi(\mathbf{x})],t)$ represents the probability density for fields to have the configuration $\phi(\mathbf{x})$ at time t. One can take (4.7) as the starting point for quantization of fields (note $exp(i\mathfrak{S}/\hbar)$ should be single valued). Equations (4.7) and (4.8) suggest a Bohmian interpretation with deterministic time evolution given via $\partial_t\phi$. Remarkably the statistical predictions of this deterministic interpretation are equivalent to those of the conventional interpretation. All quantum uncertainties are a consequence of the ignorance of the actual initial field configuration $\phi(\mathbf{x}, t_0)$. The main reason for the consistency of this interpretation is the fact that (4.8) with $\partial_t\phi$ as above represents the continuity equation which provides that the statistical distribution $\rho([\phi(\mathbf{x})], t)$ of field configurations $\phi(\mathbf{x})$ is given by the quantum distribution $\rho = \mathfrak{R}^2$ at any time t, provided that ρ is given by $\mathfrak{R}^2$ at some initial time. The initial distribution is arbitrary in principle but a quantum H theorem explains why the quantum distribution is the most probable (cf. Valentini [**1252**]). Comparing (4.7) with (4.5) we see that the quantum field satisfies an equation similar to the classical one, with the addition of a term resulting from the nonlocal quantum potential Q. The quantum equation of motion then turns out to be

$$\partial^\mu\partial_\mu\phi + \frac{\partial V(\phi)}{\partial\phi} + \frac{\delta\mathfrak{Q}}{\delta\phi(\mathbf{x};t)} = 0 \tag{4.9}$$

where $\mathfrak{Q} = \int d^3xQ$. A priori perhaps the main unattractive feature of the Bohmian formulation appears to be the lack of covariance, i.e. a preferred Lorentz frame is needed and this can be remedied with the DDW presentation to follow.

Thus one wants a quantum substitute for the classical covariant DDW HJ equation $(1/2)\partial_\phi S_\mu\partial_\phi S^\mu + V + \partial_\mu S^\mu = 0$. Define then the derivative

$$\frac{dA([\phi], x)}{d\phi(x)} = \int d^4x'\frac{\delta A([\phi], x')}{\delta\phi(x)} \tag{4.10}$$

where $\delta/\delta\phi(x)$ is the spacetime functional derivative (not the space functional derivative used before in (4.5)). In particular if $A([\phi], x)$ is a local functional, i.e. if $A([\phi], x) = A(\phi(x), x)$ then

$$\frac{dA(\phi(x), x)}{d\phi(x)} = \int d^4x'\frac{\delta A(\phi(x'), x')}{\delta\phi(x)} = \frac{\partial A(\phi(x), x)}{\partial\phi(x)} \tag{4.11}$$

Thus $d/d\phi$ is a generalization of $\partial/\partial\phi$ such that its action on nonlocal functionals is also well defined. An example of interest is a functional nonlocal in space but local in time so that

$$\frac{\delta A([\phi], x')}{\delta\phi(x)} = \frac{\delta A([\phi], x')}{\delta\phi(\mathbf{x}), x^0)}\delta((x')^0 - x^0) \Rightarrow \tag{4.12}$$

$$\Rightarrow \frac{dA([\phi], x)}{d\phi(x)} = \frac{\delta}{\delta\phi(\mathbf{x}, x^0)}\int d^3x' A([\phi], \mathbf{x}', x^0)$$

Now the first equation in (4.3) and the equations of motion become

$$\frac{1}{2}\frac{dS_\mu}{d\phi}\frac{dS^\mu}{d\phi} + V + \partial_\mu S^\mu = 0;\ \partial^\mu\phi = \frac{dS^\mu}{d\phi} \tag{4.13}$$

which is appropriate for the quantum modification. Next one proposes a method of quantization that combines the classical covariant canonical DDW formalism with the standard specetime asymmetric canonical quantization of fields. The starting point is the relation between the noncovariant classical HJ equation (4.5) and its quantum analogue (4.7). Suppressing the time dependence of the field in (4.5) we see that they differ only in the existence of the Q term in the quantum case. This suggests the following quantum analogue of the classical covariant equation (4.13)

$$\frac{1}{2}\frac{dS_\mu}{d\phi}\frac{dS^\mu}{d\phi}+V+Q+\partial_\mu S^\mu=0 \tag{4.14}$$

Here $S^\mu = S^\mu([\phi],x)$ is a functional of $\phi(x)$ so S^μ at x may depend on the field $\phi(x')$ at all points x'. One can also allow for time nonlocalities (cf. Nikolić [**931**]). Thus (4.15) is manifestly covariant provided that Q given by (4.7) can be written in a covariant form. The quantum equation (4.14) must be consistent with the conventional quantum equation (4.7); indeed by using a similar procedure to that used in showing that (4.3) implies (4.5) one can show that (4.14) implies (4.7) provided that some additional conditions are fulfilled. First S^0 must be local in time so that (4.12) can be used. Second S^i must be completely local so that $dS^i/d\phi = \partial S^i/\partial\phi$, which implies

$$d_iS^i=\partial_iS^i+(\partial_i\phi)\frac{dS^i}{d\phi} \tag{4.15}$$

However just as in the classical case in this procedure it is necessary to use the space part of the equations of motion (4.3). Therefore these classical equations of motion must be valid even in the quantum case. Since we want a covariant theory in which space and time play equal roles the validity of the space part of the (4.3) implies that its time part should also be valid. Consequently in the covariant quantum theory based on the DDW formalism one must require the validity of the second equation in (4.13). This requirement is nothing but a covariant version of the Bohmian equation of motion written for an arbitrarily nonlocal S^μ (this clarifies and generalizes results in Kanatchikov [**720**]). The next step is to find a covariant substitute for the second equation in (4.7). One introduces a vector $R^\mu([\phi],x)$ which will generate a preferred foliation of spacetime such that the vector R^μ is normal to the leaves of the foliation. Then define

$$\mathfrak{R}([\phi],\Sigma)=\int_\Sigma d\Sigma_\mu R^\mu;\ \mathfrak{S}([\phi],x)=\int_\Sigma d\Sigma_\mu S^\mu \tag{4.16}$$

where Σ is a leaf (a 3-dimensional hypersurface) generated by R^μ. Hence the covariant version of $\Psi=\mathfrak{R}exp(i\mathfrak{S})$ is $\Psi([\phi],\Sigma)=\mathfrak{R}([\phi],\Sigma)exp(i\mathfrak{S}([\phi],\Sigma)/\hbar)$. For R^μ one postulates the equation

$$\frac{dR^\mu}{d\phi}\frac{dS^\mu}{d\phi}+J+\partial_\mu R^\mu=0 \tag{4.17}$$

In this way a preferred foliation emerges dynamically as a foliation generated by the solution R^μ of the equations (4.17) and (4.14). Note that R^μ does not play any role in classical physics so the existence of a preferred foliation is a purely quantum effect. Now the relation between (4.17) and (4.7) is obtained by assuming that nature has chosen a solution of the form $R^\mu=(R^0,0,0,0)$ where R^0 is local in time. Then integrating (4.17) over d^3x and assuming again that S^0 is local in time one obtains (4.7). Thus (4.17) is a covariant substitute for the second equation in

(4.7). It remains to write covariant versions for Q and J and these are

$$Q = -\frac{\hbar^2}{2\mathfrak{R}}\frac{\delta^2\mathfrak{R}}{\delta_\Sigma\phi^2(x)};\ J = \frac{\mathfrak{R}}{2}\frac{\delta^2\mathfrak{S}}{\delta_\Sigma\phi^2(x)} \tag{4.18}$$

where $\delta/\delta_\Sigma\phi(x)$ is a version of the space functional derivative in which Σ is generated by R^μ. Thus (4.17) and (4.14) with (4.18) represent a covariant substitute for the functional SE equivalent to (4.8). The covariant Bohmain equations (4.13) imply a covariant version of (4.9), namely

$$\partial^\mu\partial_\mu\phi + \frac{\partial V}{\partial\phi} + \frac{dQ}{d\phi} = 0 \tag{4.19}$$

Since the last term can also be written as $\delta(\int d^4xQ)/\delta\phi(x)$ the equation of motion (4.19) can be obtained by varying the quantum action

$$\mathfrak{A}_Q = \int d^4x\mathfrak{L}_Q = \int d^4x(\mathfrak{L} - Q) \tag{4.20}$$

Thus in summary the covariant canonical quantization of fields is given by equations (4.13), (4.14), (4.17), and (4.18). The conventional functional SE corresponds to a special class of solutions for which $R^i = 0$, S^i are local, while R^0 and S^0 are local in time. A multifield generalization is also spelled out by Nikolić, a toy model is considered, and applications to quantum gravity are treated. The main result is that a manifestly covariant method of field quantization based on the DDW formalism is developed which treats space and time on an equal footing. Unlike the conventional canonical quantization it is not formulated in terms of a single complex SE but in terms of two coupled real equations. The need for a Bohmian formulation emerges from the requirement that the covariant method should be consistent with the conventional noncovariant method. This suggests that Bohmian mechanics (BM) might be a part of the formalism without which the covariant quantum theory cannot be formulated consistently.

5. QFT AND STOCHASTIC JUMPS

The most extensive modern treatment of Bohmian theory is due to a group based in Germany, Italy, and the USA consisting of V. Allori, A. Barut, K. Berndl, M. Daumer, D. Dürr, H. Georgi, S. Goldstein, J. Lebowitz, S. Teufel, R. Tumulka, and N. Zanghi (cf. [**27, 28, 109, 126, 127, 128, 129, 130, 357, 412, 413, 414, 415, 416, 417, 418, 544, 545, 546, 547, 548, 545, 546, 547, 548, 549, 550, 1218, 1219, 1245**]). There is also of course the pioneering work of deBroglie and Bohm (see e.g. [**187, 158, 159, 160, 161, 188, 162**]) as well as work of many other people (cf. [**85, 118, 119, 133, 168, 178, 198, 199, 197, 234, 235, 236, 237, 238, 244, 245, 254, 261, 294, 344, 345, 367, 371, 382, 383, 445, 446, 478, 479, 480, 637, 638, 639, 640, 644, 645, 646, 1160, 1196, 1251, 1262**]). We make no attempt to survey the philosophy of Bohmian mechanics (BM), or better deBroglie-Bohm theory (dBB theory), here. This involves many issues, some of them delicate, which are discussed at length in the references cited. There is a lot of associated "philosophy", involving hidden variables, nonlocality, EPR ideas, wave function collapse, pilot waves, implicate order, measurement problems, decoherence, etc., much of which has been resolved or might well be forgotten.

Many matters are indeed clarified already in the literature above (cf. in particular [**118, 119, 418, 545, 637**]) and we will not belabor philosophical matters. It may well be that a completely unified mathematical theory is beyond reach at the moment but thre are already quite accurate and workable models available and the philosophy of dBB theory as developed by the American-German-Italian school mentioned is quite sophisticated. We also feel that the Faraggi-Matone-Bertoldi theory of [**133, 446**] is fundamental and very important and the collection of papers by Nikolić in [**930**]-[**938**] provides a possible framework for even deeper understanding of some matters.

Basically, following Goldstein [**545**], for the nonrelativistic theory, GM for N particles is determined by the two equations

$$i\hbar\psi_t = H\psi; \quad \frac{dq_k}{dt} = \frac{\hbar}{2m_k}\Im\left[\frac{\psi^*\partial_k\psi}{\psi^*\psi}\right] \tag{5.1}$$

The latter equation is called the guidance or pilot equation which choreographs the motion of the particles. If ψ is spinor valued the products in the numerator and denominator are scalar products and if external magnetic fields are present the gradient $\nabla \sim (\partial_k)$ should be understood as the covariant derivative involving the vector potential (thus accomodating some versions of field theory - more on this later). Since the denominator vanishes at nodes of ψ existence and uniqueness of solutions for Bohmian dynamics is nontrivial but this is proved in [**128, 130**]. This formula extends to spin and the right side corresponds to J/ρ which is the ratio of the quantum probability current to the quantum probability density. Further from the quantum continuity condition $\partial_t\rho + div(J) = 0$ (derivable from the SE) it follows that if the configuration of particles is random at the initial time t_0 with probability distribution $\psi^*\psi$ then this remains true for all times (assuming no interaction with the environment). Upon setting $\psi = Rexp(iS/\hbar)$ one identifies $p_k = m_k v_k$ with $\partial_k S$ (which is equivalent to the guiding equation for particles without spin) and this corresponds to particles being acted upon by the force $\partial_k Q$ generated by the quantum potential (in addition to any "classical" forces).

REMARK 2.5.1. Recall from Section 2.3.2 that in the FM theory of Bohmian type $\dot{q} = p/m_Q \neq p/m$ where $m_Q = m(1 - \partial_E Q)$ in stationary situations with energy E. Here one is using a Floydian time and there has been a great deal of discussion, involving e.g. tunneling times (see e.g. [**168, 169, 170, 238, 241, 244, 245, 382, 383, 384, 386, 446, 478, 479, 480, 481, 699**]). We do not attempt to resolve any issues here and refer to the references for up to date information. ■

In any event we proceed with BM or dBB theory in full confidence not only that it works but that it is probably the best way to look at QM. We regard the quantum potential Q as being a quantization vehicle which expresses the influence of quantum fluctuations (cf. Chapters 1,4,5); it also arises in describing Weyl curvature (cf. Chapters 4,5) and thus we regard it as perhaps the fundamental object of QM. Returning now to [**545**] one notes that the predictions of

BM for measurements must agree with those of standard QM provided configurations are random with distributions given by the quantum equilibrium distribution $|\psi|^2$. Then a probability distribution ρ^{ψ} depending on ψ is called equivariant if $(\rho^{\psi})_t = \rho^{\psi_t}$ where the right side comes from the SE and the left from the guiding equation (since $\rho_t + div(J) = 0$ with $v = J/\rho$ arises in (5.1)). This has been studied in detail and we summarize some results below. Further BM can handle spin via (5.1) (as mentioned above) and nonlocality is no problem; however Lorentz invariance, even for standard QM, is tricky and one views it as an emergent symmetry. Further QFT with particle creation and annihilation is a current topic of research (cf. Sections 2.3 and 2.4) and some additional remarks in this direction will follow. The papers [**412, 413**] of Dürr et al are mainly about quantum equilibrium, absolute uncertainty, and the nature of operators. There are two long papers here (75 and 77 pages) and an earlier paper of 35 pages so we make no attempt to cover this here. We mention briefly some results of the two more recent papers however. Thus from the abstract to the second paper of [**412**] the quantum formalism is treated as a measurement formalism, i.e. a phenomenological formalism describing certain macroscopic regularities. One argues that it can be regarded and best be understood as arising from Bohmian mechanics, which is what emerges from the SE for a system of particles when one merely insists that "particles' means particles. BM is a fully deterministic theory of particles in motion, a motion choreographed by the wave function. One finds that a Bohmian universe, although deterministic, evolves in such a manner that an appearance of randomness emerges, precisely as described by the quantum formalism and given by $\rho = |\psi|^2$. A crucial ingredient in the analysis of the origin of this randomness is the notion of the effective wave function of a subsystem. When the quantum formalism is regarded as arising in this way the paradoxes and perplexities so often associated with (nonrelativistic) quantum theory evaporate. A fundamental fact here is that given a SE $i\hbar\psi_t = -(\hbar^2/2)\sum(\Delta_k\psi/m_k)+V\psi$ one can derive a velocity formula $v_k^{\psi} = (\hbar/m_k)\Im(\nabla_k\psi/\psi)$ by general arguments based on symmetry considerations and this yields (5.1) without any recourse to a formula $\psi = Rexp(iS/\hbar)$. Further the continuity equation $\rho_t + div(\rho v^{\psi}) = 0$ holds and this implies the equivariance $\rho(q,t) = |\psi(q,t)|^2$ provided this is true at (t_0, q_0). The distribution $\rho = |\psi|^2$ is called the quantum equilibrium distribution (QELD) and a system is in quantum equilibrium when the QELD is appropriate for its description. The quantum equilibrium hypothesis (QEH) is that if a system has wave function ψ then $\rho = |\psi|^2$. It is necessary to discuss wave functions of systems and subsystems at some length and it is argued that in a universe governed by BM it is impossible to know more about the configuration of any subsystem than what is expressed via $\rho = |\psi|^2$ (despite the fact that for BM the actual configuration is an objective property, beyond the wave function). Moreover, this uncertainty, of an absolute and precise character, emerges with complete ease, the structure of BM being such that it allows for the formulation and clean demonstration of statistical statements of a purely objective character which nontheless imply the claims concerning the irreducible limitations on possible knowledge, whatever this knowledge may precisely mean and however one might attempt to obtain this knowledge, provided it is consistent with BM. This limitation on what can be known is called absolute uncertainty. One proceeds by analysis of systems and subsystems and we refer to [**412**] for details. In [**413**] one shows how the entire quantum formalism, operators as observables, etc. naturally emerges in BM from the analysis of measurments. It is however quite technical, with considerable important and delicate reasoning, and we cannot possibly deal with it in a reasonable number of pages.

We go to Dürr et al [**414**] now where a comprehensive theory is developed for Bohmian mechanics and QFT (cf. also [**418**]). Bohm and subsequently Bell had proposed such models and the latter is model is modified and expanded in [**414, 418**] in the context of what are called Bell models. One will treat the configuration space variables in terms of Markov processes with jumps (which is reminiscent of the diffusion picture in Nagasawa [**909, 910**] (cf. also [**234**]). Roughly one thinks of world lines involving particle creation and annihilation, hence jumps, and writes $\mathfrak{Q} = \cup_0^\infty \mathfrak{Q}^n$ where, taking identical particles, the sector $\mathfrak{Q}^n$ is best defined as $\mathbf{R}^{3n}/S_n$ where $S \sim$ permutations. For several particle species one forms several coples of $\mathfrak{Q}$, one for each species, and obtains a union of sectors $\mathfrak{Q}^{(n)}$ where now $n \sim (n_1, \ldots, n_k)$ for the k species of particles. Note that a path $Q(t)$ will typically have discontinuities, even if there is nothing discontinuous in the world line pattern, because it jumps to a different sector at every creation or annihilation event. One can think of the bosonic Fock space as a space of L^2 functions on $\cup_n \mathbf{R}^{3n}/S_n$ with the fermionic Fock space being L^2 functions on $\cup_n \mathbf{R}^{3n}$, antisymmetric under permutation. A Bell type QFT specifies such world line patterns or histories in configuration space by specifying three sorts of "laws of motion": when to jump, where to jump, and how to move between jumps. One consequence of these laws (to be enumerated) is the property of preservation of $|\Psi_0|^2$ at time t_0 to be equal to $|\Psi_t|^2$ at time t; this is called equivariance (see above and cf. [**413, 415**] for more detail on equivariance for Bohmian mechanics - the same sort of reasoning will apply here). One will use the quantum state vector Ψ to determine the laws of motion and here a state described by the pair (Ψ_t, Q_t) where Ψ evolves according to the SE $i\hbar\partial_t\Psi_t = H\Psi$. Typically $H = H_0 + H_I$ and it is important to note that although there is an actual particle number $N(t) = \#Q(t)$ or $Q(t) \in \mathfrak{Q}^{N(t)}$, Ψ need not be a number eigenstate (i.e. concentrated in one sector). This is similar to the usual double-slit experiment in which the particle passes through only one slit although the wavefunction passes through both. As with this experiment, the part of the wave function that passes through another sector of $\mathfrak{Q}$ (or another slit) may well influence the behavior of $Q(t)$ at a later time. The laws of motion of Q_t depend on Ψ_t (and on H) and the continuous part of the motion is governed by

$$\text{(5.2)}\quad \frac{dQ_t}{dt} = v^{\Psi_t}(Q_t) = \Re\frac{\Psi_t^*(Q_t)(\dot{\hat{q}}\Psi_t)(Q_t)}{\Psi_t^*(Q_t)\Psi_t(Q_t)};\ \dot{\hat{q}} = \frac{d}{d\tau}e^{iH_0\tau/\hbar}\Big|_{\tau=0} = \frac{i}{\hbar}[H_0, \hat{q}]$$

Here $\dot{\hat{q}}$ is the time derivative of the $\mathfrak{Q}$ valued Heisenberg position operator $\hat{q}$ evolved with H_0 alone. One should understand this as saying that for any smooth function $f : \mathfrak{Q} \to \mathbf{R}$

$$\text{(5.3)}\quad \frac{df(Q_t)}{dt} = \Re\frac{\Psi_t^*(Q_t)(i/\hbar)[H_0, \hat{f}]\Psi_t)(Q_t)}{\Psi_t^*(Q_t)\Psi_t(Q_t)}$$

This expression is of the form $v^\Psi \cdot \nabla f(Q_t)$ (as it must be for defining a dynamics for Q_t) if the free Hamiltonian is a differential operator of up to second order (more on this later). **Note that the KG equation is not covered by** (5.2) **or** (5.3). The numerator and denominators above involve, when appropriate, scalar products in spin space. One may view v as a vector field on $\mathfrak{Q}$ and thus as consisting of one vector field v^n on every manifold $\mathfrak{Q}^n$; it is then $v^{N(t)}$ that governs the motion of $Q(t)$ in (5.2). If H_0 were the Schrödinger operator $-\sum_1^n (\hbar^2/2m)\Delta_i + V$ (5.2) yields the Bohm velocities $v_i^\Psi = (\hbar/m_i)\Im[\Psi^* \nabla_i \Psi/\Psi^*\Psi]$. When H_0 is the "second quantization" of a 1-particle Schrödinger operator (5.2) involves equal masses in every sector $\mathfrak{Q}^n$. Similarly in case H_0 is the second quantization of the Dirac opertor $-ic\hbar\vec{\alpha} \cdot \nabla + \beta mc^2$ (5.2) says that a configuration $Q(t)$ (with N particles) moves according to (the N-particle version of) the known variant of Bohm's velocity formula for Dirac wavefunctions $v^\Psi = (\Psi^* \alpha \Psi/\Psi^*\Psi)c$ (cf. [**159**]). The jumps now are stochastic in nature, i.e. they occur at random times and lead to random destinations. In Bell type QFT God does play dice. There are no hidden variables which would fully determine the time and destination of a jump (cf. here Section 2.3 and the effectivity parameters). The probability of jumping, with the next dt seconds to the volume dq in $\mathfrak{Q}$ is $\sigma^\Psi(dq|Q_t)dt$ with

$$\sigma^\Psi(dq|q') = \frac{2}{\hbar}\frac{[\Im\Psi^*(q) < q|H_I|q' > \Psi(q')]^+}{\Psi^*(q')\Psi(q')}dq \tag{5.4}$$

where $x^+ = max(x, 0)$. Thus the jump rate σ^Ψ depends on the present configuration Q_t, on the state vector Ψ_t which has a guiding role similar to that in the Bohm theory, and of course on the overall setup of the QFT as encoded in the interaction Hamiltonian H_I (cf. [**414**] for a simple example). There is a striking similarity between (5.4) and (5.2) in that they are both cases of "minimal" Markov processes associated with a given Hamiltonian (more on this below). When H_0 is replaced by H_I in the right side of (5.3) one obtains an operator on functions $f(q)$ that is naturally associated with the process defined by the jump rates (5.4).

The field operators (operator valued fields on spacetime) provide a connection, the only connection in fact, between spacetime and the abstract Hilbert space containing the quantum states $|\Psi>$, which are usually regarded not as functions but as abstract vectors. What is crucial now is that (i) The field operators naturally correspond to the spatial structure provided by a projection valued (PV) measure on configuration space $\mathfrak{Q}$, and (ii) The process defined here can be efficiently expressed in terms of a PV measure. Thus consider a PV measure P on $\mathfrak{Q}$ acting on $\mathcal{H}$ where for $B \subset \mathfrak{Q}$, $P(B)$ means the projection to the space of states localized in B. Then one can rewrite the formulas above in terms of P and $|\Psi>$ and we get

$$\frac{df(Q_t)}{dt} = \Re\left.\frac{<\Psi|P(dq)\frac{i}{\hbar}[H_0, \hat{f}]|\Psi>}{<\Psi|P(dq)|\Psi>}\right|_{q=Q_t}; \ \hat{f} = \int_{q\in\mathfrak{Q}} f(q)P(dq) \tag{5.5}$$

(for smooth functions $f : \mathfrak{Q} \to \mathbf{R}$) and

$$\sigma^\Psi(dq|q') = \frac{2}{\hbar}\frac{\Im <\Psi|P(dq)H_IP(dq')|\Psi>]^+}{<\Psi|P(dq')|\Psi>} \tag{5.6}$$

Note that $< \Psi|P(dq)|\Psi >$ is the probability distribution analogous to the standard $|\Psi(q)|^2dq$. The next question is how to obtain the PV measure P from the field operators. Such a measure is equivalent to a system of number operators (more on this below); thus an additive operator valued set function $N(R)$, $R \in \mathbf{R}^3$ such that the $N(R)$ commute pairwise and have spectra in the nonnegative integers. By virtue of the canonical commutation and anticommutation relations for the field operators $\phi(\mathbf{x})$ the easiest way to obtain such a system of number operators is via $N(R) = \int_R \phi^*(\mathbf{x})\phi(\mathbf{x})d^3\mathbf{x}$. Thus what one needs from a QFT in order to construct trajectories are: (i) a Hilbert space $\mathcal{H}$ (ii) a Hamiltonian $H = H_0 + H_I$ (iii) a configuration space $\mathfrak{Q}$ (or measurable space), and (iv) a PV measure on $\mathfrak{Q}$ acting on $\mathcal{H}$. This will be done below following [**414**].

We go now to the last paper in Dürr et al [**414**] which is titled quantum Hamiltonians and stochastic jumps. The idea is that for the Hamiltonian of a QFT there is associated a $|\Psi|^2$ distributed Markov process, typically a jump process, on the configuration space of a variable number of particles. A theory is developed generalizing work of J. Bell and the authors of [**414**]. The central formula of the paper is

$$\sigma(dq|q') = \frac{[(2/\hbar)\Im < \Psi|P(dq)HP(dq')|\Psi >]^+}{< \Psi|P(dq')|\Psi >} \tag{5.7}$$

It plays a role similar to that of Bohm's equation of motion

$$\frac{dQ}{dt} = v(Q); \;\; v = \hbar\Im\frac{\Psi^*\nabla\Psi}{\Psi^*\Psi} \tag{5.8}$$

Together these two equations make possible a formulation of QFT that makes no reference to observers or measurements, while implying that observers, when making measurements, will arrive at precisely the results that QFT is known to predict. This formulation takes up ideas from the seminal papers of J. Bell [**118, 119**] and such theories will be referred to as Bell-type QFT's. The aim is to present methods for constructing a canonical Bell type model for more of less any regularized QFT. One assumes a well defined Hamiltonian as given (with cutoffs included if needed). The primary variables of such theories are particle positions and Bell suggested a dynamical law governing the motion of the particles in which the Hamiltonian H and the state vector Ψ determine the jump rates σ. These rates are in a sense the smallest choice possible (explained below) and are called minimal jump rates; they preserve the $|\Psi|^2$ distribution. Bell type QFT's can also be regarded as extensions of Bohmian mechanics which cover particle creation and annihilation; the quantum equilibrium distribution more or less dictates that creation of a particle occurs in a stochastic manner as in the Bell model. We recall that for Bohmian mechanics in addition to (5.8) one has an evolution equation $i\hbar\partial_t\Psi = H\Psi$ for the wave function with $H = -(\hbar^2/2\Delta + V$ for spinless particles ($\Delta = div\nabla$). For particles with spin Ψ takes values in the appropriate spin space $\mathbf{C}^k$, V may be matrix valued, and inner products in (3.6) are understood as involving inner products in spin spaces. The success of the Bohmian method is based on the preservation of $|\Psi|^2$, called equivariance and this follows immediately from comparing the continuity equation for a probability distribution ρ associated with (5.8), namely $\partial_t\rho = -div(\rho v)$, with the equation for $|\Psi|^2$ following from the SE, namely

$$(5.9) \qquad \partial_t|\Psi|^2(q,t) = (2/\hbar)\Im[\Psi^*(q,t)(H\Psi)(q,t)]$$

In fact it follows from the continuity equation that

$$(5.10) \qquad (2/\hbar)\Im[\Psi^*(q,t)(H\Psi)(q,t)] = -div[\hbar\Im\Psi^*(q,t)\nabla\Psi(q,t)]$$

so recalling (5.8), one has $\partial_t|\Psi|^2 = -div(|\Psi|^2 v)$, and hence if $\rho + |\Psi_t|^2$ as some time t there results $\rho = |\psi_t|^2$ for all times. One is led naturaly to the consideration of Markov processes as candidates for the equivariant motion of the configuration Q for more general Hamiltonians (see e.g. [**676, 910, 1069, 1079**] for Markov processes - [**910**] is especially good for Markov processes with jumps and dynamics but we follow [**676, 1079**] for background since the ideas are more or less clearly stated without a deathly deluge of definitions and notation - of course for a good theory much of the verbiage is actually important).

DEFINITION 5.1. Let ($\mathfrak{E}$ be a Borel σ-algebra of subsets of E. For Ω generally a path space (e.g. $\Omega \sim C(\mathbf{R}^+, E)$ with $X_t(\omega) = X(t,\omega) = \omega(t)$ and $\mathfrak{F}_t = \sigma(X_s(\omega),\ s \leq t)$ a filtration by Borel sub σ-algebras) a Markov process **(C11)** $X = (\Omega, \{\mathfrak{F}_t\}, \{X_t\}, \{P_t\}, \{P^x, x\}E\})$ with $t \geq 0$ and state space $(E, \mathfrak{E})$, is an E valued stochastic process adapted to $\{\mathfrak{F}_t\}$ such that for $0 \leq s \leq t$, $f \in b\mathfrak{E}$ ($b\mathfrak{E}$ means bounded $\mathfrak{E}$ measurable functions), and $x \in E$, $E^x[f(X_{s+t})|\mathfrak{F}_t] = (P_t f)(X_s)$, $P^x ae$ (ae means almost everywhere). Here $\{P_t\}$ is a transition function on $(E,\mathfrak{E})$, i.e. a family of kernels $P_t : E \times \mathfrak{E} \to [0,1]$ such that

(1) For $t \geq 0$ and $x \in E$, $P_t(x,\cdot)$ is a measure on $\mathfrak{E}$ with $P_t(x,E) \leq 1$
(2) For $t \geq 0$ and $\Gamma \in \mathfrak{E}$ $P_t(\cdot,\Gamma)$ is $\mathfrak{E}$ measurable
(3) For $x, t \geq 0$, $x \in E$, and $\Gamma \in \mathfrak{E}$ one has $P_{t+s}(x,\Gamma) = \int_E P_s(x,dy)P_t(y,\Gamma)$

The equation in #3 is called the Chapman-Kolmogorov (CK) equation and, thinking of the transition functions as inducing a family $\{P_t\}$ of positive bounded operators or norm less than or equal to 1 on $b\mathfrak{E}$ one has $P_t f(x) = (P_t f)(x) = \int_E P_t(x,dy)f(y)$ in which case the CK equation has the semigroup property $P_s P_t = P_{s+t}$ for $s,t \geq 0$. ■

Under mild regularity conditions if a transition semigroup $\{P_t\}$ is given there will exist on some probability space a Markov process X with suitable paths such that the strong Markov property holds, i.e. $E^x[f(X_{S+t})]|\mathfrak{F}_x] = (P_t f)(X_S)\ P^x\ ae$ whenever S is a finite stopping time (here $S : \Omega \to [0,\infty]$ is a $\mathfrak{E}$ stopping time if $\{S \leq t\} = \{\omega;\ S(\omega) \leq t\} \in \mathfrak{E}_t$ for every $t < \infty$).

EXAMPLE 5.1. A Markov process with countable state space is called a Markov chain. One writes $p_{ij}(t) = P_t(i,\{j\})$ with $P(t) = \{p_{ij}(t);\ i,j \in E\}$. Assume $P_t(i,E) = 1$ and $p_{ij}(t) \to \delta_{ij}$ as $t \downarrow 0$. This will imply that in fact $p'_{ij}(0) = q_{ij}$ exists and the matrix $Q = (q_{ij})$ is called an infinitesimal generator of $\{P_t\}$ with $q_{ij} \geq 0$ $(i \neq j)$ and $\sum_j q_{ik} = 0$ $(i \in E)$. This illustrates some important structure for Markov processes. Thus when $P'(0) = Q$ exists one can write $P'(t) = lim_{\epsilon\to 0}\epsilon^{-1}[P(t+\epsilon) - P(t)] = lim_{\epsilon\to 0}P(t)[P(\epsilon) - I] = P(t)Q$. Then solving this equation one has $P(t) = exp(tQ$ as a semigroup generated by Q. The resolvent is defined via $R_\lambda = \int_0^\infty exp(-\lambda t)P_t dt$ and one can regard it as

$$(\lambda R_\lambda)_{ij} = \int_0^\infty \lambda exp(-\lambda t)p_{ij}(t)dt = \mathbf{P}(X_{\mathbf{T}} = j|X_0 = i) \tag{5.11}$$

where $\mathbf{T}$ is a random variable independent of X with the exponential distribution of rate λ. It follows then that $R_\lambda = (\lambda - Q)^{-1}$ and $R_\lambda - R_\mu = (\mu - \lambda)R_\lambda R_\mu$. The whole subject is full of pathological situations however and we make no attempt to describe this. ■

REMARK 2.5.2. Nagasawa [**910**] is oriented toward diffusion processes and departs from the concept that the kinematics of quantum particles is stochastic calculus (in particular Markov processes) while the kinematics of classical particles is classical differential calculus. The relation between these two calculi must be established. Thus classically $x(t) = x(a) + \int_a^t v(s, x(s))ds$ while for a particle with say Brownian noise B_t and a drift field $a(t, x(t))$ one has

$$X_t = X_a + \int_a^t a(s, X_s)ds + \int_a^t \sigma(s, X_s)dB_s \tag{5.12}$$

We recall $dB_t^2 \sim dt$ so X_t has no velocity and the drift field $a(t, x)$ is not an average speed. However $P[X] = \int_\Omega XdP$ is the expectation (since $P[\sigma(t, X_t)dB_t] = 0$). Now the notation for Markov processes involves nonnegative transition functions $P(s, x; t, B)$ with $a \leq s \leq t \leq b$, $x \in \mathbf{R}^d$, and $B \in \mathfrak{B}(\mathbf{R}^d)$ which are measures in B, measurable in x, and satisfy the CK equation

$$P(s, x; t, B) = \int_{\mathbf{R}^d} P(s, x; r, dy)P(r, y; t, B);\ P(s, x; t, \mathbf{R}^d) = 1 \tag{5.13}$$

If there is a measurable function p such that $P(s, x; t, B) = \int_B p(s, x; t, y)dy$ $(t - s > 0)$ then p is a transition density. One defines a probability measure P on a path space $\Omega = (\mathbf{R}^d)^{[a,b]}$ via finite dimensional distributions
(5.14)

$$P[f(X_a, X_{t_1}, \cdots, X_{t_n}, X_b)] = \int \mu_a(dx_0)P(a, x_0; t_1, dx_1)P(t_1, x_1; t_2, dx_2) \cdots \times$$
$$\times \cdots P(t_{n-1}, x_{n-1}; b, dx_n)f(x_0, \cdots, x_n)$$

Moreover one defines a family $\{X_t;\ t \in [a, b]\}$ on Ω via $X_t(\omega) = \omega(t),\ \omega \in \Omega$. Note one assumes the right continuity of $X_t(\omega)$ ae. This representation can be written as $P = [\mu_a P >>$ and is called the Kolmogorov representation of P. Let now $\{\mathfrak{F}_s^t\}$ be a filtration as before, i.e. a family of σ-fields generated by $\{X_r(\omega);\ s \leq r \leq t\}$. Then we have a Markov process $\{X_t, t \in [a, b], \mathfrak{F}_s^t, P\}$. Replacing μ_a by δ_x and a by s with $s < t_1 < \cdots < t_{n-1} < t_n \leq b$ one defines probability measures $P_{(s,x)}$, $(s, x) \in [a, b] \times \mathbf{R}^d$ from (3.7) via

$$P_{(s,x)}[f(X_{t_1}, \cdots, X_{t_{n-1}}, X_{t_n})] = \tag{5.15}$$
$$= \int P(s, x; t_1, dx_1) \cdots P(t_{n-1}, x_{n-1}; t_n, dx_n)f(x_1, \cdots, x_n)$$

As a special case one has $P_{(s,x)}[f(t, X_t)] = \int P(s, x; t, dy)f(t, y)$ and one can also prove that $P[GF] = P[GP_{(s,X_s)}[F]]$ for any bounded $\mathfrak{F}_a^s$ measurable G and any bounded $\mathfrak{F}_s^b$ measurable F. This is the time inhomogeneous Markov property which can be written in terms of conditional expectations as $P[F|\mathfrak{F}_a^s] = P_{(s,X_s)}[F]$, P $ae.$ There is a great deal of material in [**910**] about Markov processes with jumps but we prefer to stay here with [**414**] for notational convenience. ∎

Going back to [**414**] we consider a Markov process Q_t on configuration space with transition probabilities characterized by the backward generator L_t, a time dependent linear operator acting on functions f via $L_t f(q) = (d/ds)E(f(Q_{t+s}|Q_t = q)$ where d/ds means the right derivative at $s = 0$ and $E(\cdot|\cdot)$ is conditional expectation. Equivalently the transition probabilities are characterized by the forward generator $\mathcal{L}_t$ (or simply generator) which is also a linear operator but acts on (signed) measures on the configuration space. Its defining property is that for every process Q_t with the given transition properties $\partial_t \rho_t = \mathcal{L}_t \rho_t$. Thus $\mathcal{L}$ is dual to L_t in the sense

$$\int f(q)\mathcal{L}_t\rho(dq) = \int L_t f(q)\rho(dq) \tag{5.16}$$

Given equivariance for $|\Psi|^2$, one says that the corresponding transition probabilities are equivariant and this is equivalent to $\mathcal{L}_t|\Psi|^2 = \partial_y|\Psi|^2$ for all t; when this holds one says that $\mathcal{L}_t$ is an equivariant generator (with respect to Ψ_t and H). One says that a Markov process is Q equivariant if and only if for every t the distribution ρ_t of Q_t equals $|\Psi_t|^2$. For this equivariant transition probabilities are necessary but not sufficient; however for equivariant transition probabilities there is a unique equivariant Markov process. The crucial idea here for construction of an equivariant Markov process is to note that (5.9) is completely general and to find a generator $\mathcal{L}_t$ such that the right side of (5.9) can be read as the action of $\mathcal{L}$ on $\rho = |\Psi|^2$ means $(2/\hbar)\Im\Psi^* H\Psi = \mathcal{L}|\Psi|^2$. This will be implemented later. For H of the form $-(\hbar^2/2)\Delta + V$ one has (5.10) and hence

$$\frac{2}{\hbar}\Im\Psi^* H\Psi = -div(\hbar\Im\Psi^*\nabla\Psi) = -div\left(|\Psi|^2\hbar\Im\frac{\Psi^*\nabla\Psi}{|\Psi|^2}\right) \tag{5.17}$$

Since the generator of the (deterministic) Markov process corresponding to the dynamical system $dQ/dt = v(Q)$ is given by a velocity vector field is $\mathcal{L}\rho = -div(\rho v)$ we may recognize the last term of (3.7) as $\mathcal{L}|\Psi|^2$ with $\mathcal{L}$ the generator of the deterministic process defined by (5.8). Thus Bohmian mechanics arises as the natural equivariant process on configuration space associated with H and Ψ. One notes that Bohmian mechanics is not the only solution of $(2/\hbar)\Im\Psi^* H\Psi - \mathfrak{L}|\Psi|^2$; there are alternatives such as Nelson's stochastic mechanics (and hence Nagasawa's theory of [**908, 910**]) and other velocity formulas (cf. DeOtto and Ghirardi [**367**]).

For equivariant jump processes one says that a (pure) jump process is a Markov process on $\mathfrak{Q}$ for which the only motion that occurs is via jumps. Given that $Q_t = q$ the probability for a jump to q' (i.e. into the infinitesimal volume dq' around q') by time $t+dt$ is $\sigma_t(d'q|q)dt$ where σ is called the jump rate. Here σ is a finite measure in the first variable; $\sigma(B|q)$ is the rate (i.e. the probability per unit time) of jumping to somewhere in the set $B \subset \mathfrak{Q}$ given that the present location is q. The overall jump rate is $\sigma(\mathfrak{Q}|q)$ (sometimes one writes $\rho(dq) = \rho(q)dq$). A jump first occurs when a random waiting time T has elapsed, after the time t_0 at which the process was started or at which the most recent previous jump has occured. For purposes of simulating or constructing the process, the destination q' can be chosen at the time of jumping, $t_0 + T$, with probability distribution $\sigma_{t_0+T}(\mathfrak{Q}|q)^{-1}\sigma_{t_0+T}(\cdot|q)$. In case the overall jump rate is time independent T is exponentially distributed with mean of $\sigma(\mathfrak{Q}|q)^{-1}$. When the rates are time dependent (as they will typically in what follows) the waiting time remains such that $\int_{t_0}^{t_0+T} \sigma_t(\mathfrak{Q}|q)dt$ is exponentially distributed with mean 1, i.e. T becomes exponential after a suitable (time dependent) rescaling of time. The generator of a pure jump process can be expressed in terms of the rates

$$\mathcal{L}\rho(dq) = \int_{q'\in\mathfrak{Q}} (\sigma(dq|q')\rho(dq') - \sigma(dq')|q)\rho(dq)) \tag{5.18}$$

which is a balance or master equation expressing $\partial_t\rho$ as the gain due to jumps to dq minus the loss due to jumps away from dq. One says the jump rates are equivariant if $\mathcal{L}_\upsilon$ is an equivariant generator.

Given a Hamiltonian $H = H_0 + H_I$ one obtains

$$(2/\hbar)\Im\Psi^* H_0\Psi + (2/\hbar)\Im\Psi^* H_I\Psi - \mathcal{L}|\Psi|^2 \tag{5.19}$$

This opens the possibility of finding a generator $\mathcal{L} = \mathcal{L}_0 + \mathcal{L}_I$ given $(2/\hbar)\Im\Psi^* H_0\Psi = \mathcal{L}_0|\Psi|^2$ and $(2/\hbar)\Im\Psi^* H_I\Psi = \mathcal{L}_I|\Psi|^2$; this will be called process additivity and correspondingly $L = L_0 + L_I$. If one has two deterministic processes of the form $\mathcal{L}\rho = -div(\rho v)$ then adding generators corresponds to $v = v_+ v_2$. For a pure jump process adding generators corresponds to adding rates σ_i which is equivalent to saying there are two kinds of jumps. Now add generators for a deterministic and a jump process via

$$\mathcal{L}\rho(q) = -div(\rho v)(q) + \int_{q'\in\mathfrak{Q}} (\sigma(q|q')\rho(q') - \sigma(q'|q)\rho(q))\, dq' \tag{5.20}$$

This process moves with velocity $v(q)$ until it jumps to q' where it continues moving with velocity $v(q')$. One can understand (5.20) in terms of gain or loss of probability density due to motion and jumps; the process is piecewise deterministic with random intervals between jumps and random destinations. Note that for a Wiener process the generator is the Laplacian and adding to it the generator of a deterministic process means introducing a drift.

Now consider H_I and note that in QFT's with cutoffs it is usually the case that H_I is an integral operator. Hence one writes here $H \sim H_I$ and thinks of it as an integral operator with $\mathfrak{Q} \sim \mathbf{R}^n$. What characterizes jump processes is that some amount of probability that vanishes at $q \in \mathfrak{Q}$ can reappear in an entirely different region say at $q' \in \mathfrak{Q}$. This suggests that the Hamiltonians for which the expression (5.9) for $\partial_t|\Psi|^2$ is naturally an integral over q' correspond to pure jump processes. Thus when is the left side of $(2/\hbar)\Im|psi^* H\Psi = \mathfrak{L}|\Psi|^2$ an integral over q' or $(H\Psi)(q) = \int dq' < q|H|q' > \Psi(q')$. In this case one should choose the jump rates so that when $\rho = |\Psi|^2$ one has

$$\sigma(q|q')\rho(q') - \sigma(q'|q)\rho(q) = (2/\hbar)\Im\Psi^*(q) < q|H|q' > \Psi(q') \tag{5.21}$$

This suggests, since jump rates are nonnegative and the right side of (5.21) is antisymmetric) that $\sigma(q|q')\rho(q') = [(2/\hbar)\Im\Psi^*(q) < q|H|q' > \Psi(q')]^+$ or

$$\sigma(q|q') = \frac{(2/\hbar)\Im\Psi^*(q) < q|H|q' > \Psi(q')]^+}{\Psi^*(q')\Psi(q')} \tag{5.22}$$

These rates are an instance of what can be called minimal jump rates associated with H (and Ψ). They are actually the minimal possible values given (5.21) and this is discussed further in [**414**]. Minimality entails that at any time t one of the transitions $q_1 \to q_2$ or $q_2 \to q_1$ is forbidden and this will be called a minimal jump process. One summarizes motions via

H	*motion*
integral operator	*jumps*
differential operator	*deterministic continuous motion*
multiplication operator	*no motion* ($\mathcal{L} = 0$)

The reasoning above applies to the more general setting of arbitrary configuration spaces $\mathfrak{Q}$ and generalized observables - POVM's - defining what the "position" representation is to be. One takes the following ingredients from QFT

(1) A Hilbert space $\mathcal{H}$ with scalar product $< \Psi|\Phi >$.
(2) A unitary one parameter group U_t in $\mathcal{H}$ with Hamiltonian H, i.e. $U_t = exp[-(i/\hbar)tH]$, so that in the Schrödinger picture the state Ψ evolves via $i\hbar\partial_t\Psi = H\Psi$. U_t could be part of a representation of the Poincaré group.
(3) A positive operator valued measure (POVM) $P(dq)$ on $\mathfrak{Q}$ acting on $\mathcal{H}$ so that the probability that the system in the state Ψ is localized in dq at time t is $\mathbf{P}_t(dq) =< \Psi_t|P(dq)|\Psi_t >$.

Mathematically a POVM on $\mathfrak{Q}$ is a countably additive set function (measure) defined on measurable subsets of $\mathfrak{Q}$ with values in the positive (bounded self adjoint) operators on a Hilbert space $\mathcal{H}$ such that $P(\mathfrak{Q}) = Id$. Physically for purposes here $P(\cdot)$ represents the (generalized) position observable, with values in $\mathfrak{Q}$. The notion of POVM generalizes the more familiar situation of observables given by a set of commuting self adjoint operators, corresponding by means of the spectral theorm to a projection valued measure (PVM) - the case where the positive operators are projection operators (see Dürr et al [**414**] for discussion). The goal now is to specify equivariant jump rates $\sigma = \sigma^{\Psi,H,P}$ so that $\mathcal{L}_\sigma\mathbf{P} = d\mathbf{P}/dt$. To this end one could take the following steps.

(1) Note that $(d\mathbf{P}_t(dq)/dt) = (2/\hbar)\Im < \Psi_t|P(dq)H|\Psi_t >$.
(2) Insert the resolution of the identity $I = \int_{q'\in\mathfrak{Q}} P(dq')$ and obtain

$$(d\mathbf{P}_t(dq)/dt) = \int_{q'\in\mathfrak{Q}} \mathbf{J}_t(dq, dq'); \tag{5.23}$$

$$\mathbf{J}_t(dq, dq') = (2/\hbar)\Im < \Psi_t|P(dq)HP(dq')|\Psi_t >$$

(3) Observe that $\mathbf{J}$is antisymmetric so since $x = x^+ - (-x)^+$ one has

$$\mathbf{J}(dq, dq') = [(2/\hbar)\Im < \Psi|P(dq)HP(dq')|\Psi]^+ - \tag{5.24}$$
$$-[(2/\hbar)\Im < \Psi|P(dq')HP(dq)|\Psi >]^+$$

(4) Multiply and divide both terms by $\mathbf{P}(\cdot)$ obtaining

$$\int_{q'\in\Omega} \mathbf{J}(dq,dq') = \int_{q'\in\Omega} \Big(\frac{(2/\hbar)\Im < \Psi|P(ddq)HP(dq')|\Psi >]^+}{< \Psi|P(dq')|\Psi >} \mathbf{P}(dq') - \tag{5.25}$$

$$- \frac{[(2/\hbar)\Im < \Psi|P(dq')HP(dq)|\Psi >]^+}{< \Psi|P(dq)|\Psi >} \mathbf{P}(dq) \Big)$$

(5) By comparison with (5.18) recognize the right side of the above equation as $\mathcal{L}_\sigma \mathbf{P}$ with $\mathcal{L}_\sigma$ the generator of a Markov jump process with jump rates (5.7) (minimal jump rates).

Note the right side of (5.7) should be understood as a density (Radon-Nikodym derivative).

When H_0 is made of differential operators of up to second order one can characterize the process associated with H_0 in a particularly succinct manner as follows. Define for any H, P, Ψ an operator L acting on functions $f : \Omega \to \mathbf{R}$ which may or may not be the backward generator of a process via

$$Lf(q) = \Re \frac{< \Psi|P(dq)\hat{L}\hat{f}|\Psi >}{< \Psi|P(dq)\Psi >} = \Re \frac{< \Psi|P(dq)(i/\hbar)[H,\hat{f}]|\Psi >}{< \Psi|P(dq)|\Psi >} \tag{5.26}$$

where [,] means the commutator and $\hat{f} = \int_{q\in\Omega} f(q)P(dq)$ with $\hat{L}$ the generator of the (Heisenberg) evolution $\hat{f}$,

$$\hat{L}\hat{f} = (d/d\tau)exp(iH\tau/\hbar)\hat{f}exp(-iH\tau/\hbar)|_{\tau=0} = (i/\hbar)[H,\hat{f}] \tag{5.27}$$

Note if P is a PVM then $\hat{f} = f(\hat{q})$. (5.26) could be guessed in the following manner: Since Lf is in a certain sense the time derivative of f it might be expected to be related to $\hat{L}\hat{f}$ which is in a certain sense (cf. (5.27)) the time derivative of $\hat{f}$. As a way of turning the operator $\hat{L}\hat{f}$ into a function $Lf(q)$ the middle term in (5.26) is an obvious possibility. Note also that this way of arriving at (5.26) does not make use of equivariance. The formula for the forward generator equivalent to (5.26) reads

$$\mathcal{L}\rho(dq) = \Re < \Psi|(\widehat{d\rho/d\mathbf{P}})(i/\hbar)[H,P(dq)]|\Psi > \tag{5.28}$$

Whenever L is indeed a backward generator we call it the minimal free (backward) generator associated with Ψ, H, P. Then the corresponding process is equivariant and this is the case if (and there is reason to expect, only if) P is a PVM and H is a differential operator of up to second order in the position representation, in which P is diagonal. In that case the process is deterministic and the backward generator has the form $L = v \cdot \nabla$ where v is the velocity field; thus (5.26) directly specifies the velocity in the form of a first order differential operator $v \cdot \nabla$. In case H is the N-particle Schrödinger operator with or without spin (5.26) yields the Bohmian velocity (5.8) and if H is the Dirac operator the Bohm-Dirac velocity emerges. Thus in some cases (5.26) leads to just the right backward generator. In [**414**] there are many examples and mathematical sections designed to prove various assertions but we omit this here..

6. BOHMIAN MECHANICS IN QFT

We extract here from a fascinating paper in [**932**] by H. Nikolić. Quantum field theory (QFT) can be formulated in the Schrödinger picture by using a functional time dependent SE but this requires a choice of time coordinate and the corresponding choice of a preferred foliation of spacetime producing a relativistically noncovariant theory. The problem of noncovariance can be solved by replacing the usual time dependent SE with the many fingered time (MFT) Tomonaga-Schwinger equation, which does not require a preferred foliation and the quantum state is a functional of an arbitrary timelike hypersurface. In a manifestly covariant formulation introduced in [**396**] the hypersurface does not even have to be timelike. In [**932**] one develops a Bohmian interpretation for the MFT theory for QFT and refers to [**99, 108, 319, 639, 644, 789, 931, 1022, 1025, 1027, 1085, 1086, 1163, 1199**] for background and related information. Thus let $x = \{x^\mu\} = (x^0, \mathbf{x})$ be spacetime coordinates. A timelike Cauchy hypersurface Σ can be defined via $x^0 = T(\mathbf{x})$ with $\mathbf{x}$ denoting coordinates on Σ. Let $\phi(\mathbf{x})$ be a dynamical field on Σ (a real scalar field for convenience) and write T, ϕ without an argument for the functions themselves with $\phi = \phi|_\Sigma$ etc. Let $\hat{\mathfrak{H}}(\mathbf{x})$ be the Hamiltonian density operator and then the dynamics of a field ϕ is described by the MFT Tomonaga-Schwinger equation

$$\hat{\mathfrak{H}}\Psi[\phi, T] = i\frac{\delta\Psi[\phi, T]}{\delta T(\mathbf{x})} \tag{6.1}$$

Note $\delta T(\mathbf{x})$ denotes an infinitesimal change of the hypersurface Σ. The quantity $\rho[\phi, T] = |\Psi[\phi, T]|^2$ represents the probability density for the field to have a value ϕ on Σ or equivalently the probability density for the field to have a value ϕ at time T. One can say that ϕ has a definite value φ at some time T_0 if

$$\Psi[\phi, T_0] = \delta(\phi - \varphi) = \prod_{\mathbf{x}\in\Sigma} \delta(\phi(\mathbf{x}) - \varphi(\mathbf{x})) \tag{6.2}$$

[**932**] then provides an important discussion of measurement and contextuality in QM which we largly omit here in order to go directly to the Bohmian formulation.

For simplicity take a free scalar field with

$$\hat{\mathfrak{H}}(\mathbf{x}) = -\frac{1}{2}\frac{\delta^2}{\delta\phi^2(\mathbf{x})} + \frac{1}{2}[(\nabla\phi(\mathbf{x}))^2 + m^2\phi^2(\mathbf{x})] \tag{6.3}$$

Writing $\Psi = Rexp(iS)$ with R and S real functionals the complex equation (6.1) is equivalent to two real equations with

$$\frac{1}{2}\left(\frac{\delta S}{\delta\phi(\mathbf{x})}\right)^2 + \frac{1}{2}[(\nabla\phi(\mathbf{x}))^2 + m^2\phi^2(\mathbf{x})] + \mathfrak{Q}(\mathbf{x}, \phi, T] + \frac{\delta S}{\delta T(\mathbf{x})} = 0; \tag{6.4}$$

$$\frac{\delta\rho}{\delta T(\mathbf{x})} + \frac{\delta}{\delta\phi(\mathbf{x})}\left(\rho\frac{\delta S}{\delta T(\mathbf{x})}\right) = 0;\ \mathfrak{Q}(\mathbf{x}, \phi, T] = -\frac{1}{2R}\frac{\delta^2 R}{\delta\phi^2(\mathbf{x})}$$

The conservation equation shows that it is consistent to interpret $\rho[\phi, T]$ as the probability density for the field to have the value ϕ at the hypersurface determined by the time T. Now let σ_x be a small region around $\mathbf{x}$ and define the derivative

$$\frac{\partial}{\partial T(\mathbf{x})} = \lim_{\sigma_x \to 0} \int_{\sigma_x} d^3x \frac{\delta}{\delta T(\mathbf{x})} \tag{6.5}$$

where $\sigma_x \to 0$ means that the 3-volume goes to zero (note $\partial T(\mathbf{y})/\partial T(\mathbf{x}) = \delta_{xy}$). It is convenient to integrate (6.4) inside a small σ_x leading to

$$\frac{\partial \rho}{\partial T(\mathbf{x})} + \frac{\partial}{\partial \phi(\mathbf{x})}\left(\rho \frac{\delta S}{\delta \phi(\mathbf{x})}\right) = 0 \tag{6.6}$$

where $\partial/\partial\phi(\mathbf{x})$ is defined as in (6.5). The Bohmian interpretation consists now in introducing a deterministic time dependent hidden variable such that the time evolution of this variable is consistent with the probabilistic interpretation of ρ. From (6.6) one sees that this is naturally achieved by introducing a MFT field $\Phi(x, T]$ that satisfies the MFT Bohmian equations of motion

$$\frac{\partial \Phi(\mathbf{x}, T]}{\partial T(\mathbf{x})} = \left.\frac{\delta S}{\delta \phi(\mathbf{x})}\right|_{\phi=\Phi} \tag{6.7}$$

From (6.7) and the quantum MFT HJ equation (6.4) results

$$\left[\left(\frac{\partial}{\partial T(\mathbf{x})}\right)^2 - \nabla_x^2 + m^2\right] \Phi(\mathbf{x}, T] = -\left.\frac{\partial \mathfrak{Q}(\mathbf{x}, \phi, T]}{\partial \psi(\mathbf{x})}\right|_{\phi=\Phi} \tag{6.8}$$

This can be viewed as a MFT KG equation modified with a nonlocal quantum term on the right side. The general solution of (6.7) has the form

$$\Phi_{gen}(\mathbf{x}), T] = F(\mathbf{x}, c(\mathbf{x}, T]; T] \tag{6.9}$$

where F is a function(al) that depends on the right side of (6.7) and $c(\mathbf{x}, T]$ is an arbitrary function(al) with the property

$$\frac{\delta c(\mathbf{x}, T]}{\delta T(\mathbf{x})} = 0 \tag{6.10}$$

This quantity can be viewed as an arbitrary MFT integration constant - it is constant in the sense that it does not depend on $T(\mathbf{x})$, but it may depend on T at other points $\mathbf{x}' \neq \mathbf{x}$. To provide the correct classical limit (indicated below) one restricts $c(\mathbf{x}, T]$ to satisfy

$$c(\mathbf{x}, T] = c(\mathbf{x}) \tag{6.11}$$

where $c(\mathbf{x})$ is an arbitrary function. Here it is essential to realize that $\Phi(\mathbf{x}, T]$ is a function of $\mathbf{x}$ but a functional of T; the field Φ depends not only on $(\mathbf{x}, T(\mathbf{x})) \equiv (\mathbf{x}, x^0) \equiv x$ but also on the choice of the whole hypersurface Σ that contains the point $\mathbf{x}$. Consequently the MFT Bohmian interpretation does not in general assign a value of the field at the point x unless the whole hypersurface containing x is specified. On the other hand if e.g. $\delta S/\delta\phi(\mathbf{x})$ on the right side in (6.7) is a local functional, i.e. of the form $V(\mathbf{x}, \phi(\underline{x}), T(\mathbf{x})$, then the solution of (6.7) is a local functional of the form

$$\Phi(\mathbf{x}, T(\mathbf{x}) = \Phi(\mathbf{x}, x^0) = \Phi(x) \tag{6.12}$$

This occurs for example when the wave functional is a local product $\Psi[\phi, T] = \prod_{\mathbf{x}} \psi_{\mathbf{x}}(\phi(\mathbf{x}, T(\mathbf{x}))$. Interactions with the measuring apparatus can also produce locality. As for the classical limit one can formulate the classical HJ equation as a MFT theory (cf. Rovelli [**1085, 1086**]) without of course the $\mathfrak{Q}$ term. Hence by imposing a restriction similar to (6.11) the solution $S[\phi, T]$ can be chosen so that $\delta S/\delta\phi(\mathbf{x})$ is a local functional; the restriction (6.11) again implies that the classical solution Φ is also a local functional.

The MFT formalism was introduced by Tomonaga and Schwinger to provide the manifest covariance of QFT in the interaction picture. The picture here is so far not manifestly covariant since time is not treated on an equal footing with space. However the MFT formalism can be also phrased in a manifestly covariant manner via [**396, 1085, 1086**]. One starts by introducing a set of 3 real parameters $\{s^1, s^2, s^3\} \equiv \mathbf{s}$ to serve as coordinates on a 3-dimensional manifold (a priori $\mathbf{s}$ is not related to $\mathbf{x}$). The 3-dimensional manifold Σ can be embedded in the 4-dimensional spacetime by introducing 4 functions $X^\mu(\mathbf{s})$ and a 3-dimensional hypersurface is given via $x^\mu = X^\mu(\mathbf{s})$. The 3 parameters s^i can be eliminated leading to an equation of the form $f(x^0, x^1, x^2, x^3) = 0$ and assuming that the background spacetime metric $g_{\mu\nu}(x)$ is given the induced metric $q_{ij}(\mathbf{s})$ on this hypersurface is

$$q_{ij}(\mathbf{s}) = g_{\mu\nu}(X(\mathbf{s}))\frac{\partial X^\mu(\mathbf{s})}{\partial s^i}\frac{\partial X^\nu(\mathbf{s})}{\partial s^j} \tag{6.13}$$

Similarly a normal to the surface is

$$\tilde{n}(\mathbf{s}) = \epsilon_{\mu\alpha\beta\gamma}\frac{\partial X^\alpha}{\partial s^1}\frac{\partial X^\beta}{\partial s^2}\frac{\partial X^\gamma}{\partial s^3} \tag{6.14}$$

and the unit normal transforming as a spacetime vector is

$$n^\mu(\mathbf{s}) = \frac{g^{\mu\nu}\tilde{n}_\nu}{\sqrt{|g^{\alpha\beta}\tilde{n}_\alpha\tilde{n}_\beta|}} \tag{6.15}$$

Now some of the original equations above can be written in a covariant form by making the replacements

$$\mathbf{x} \to \mathbf{s};\quad \frac{\delta}{\delta T(\mathbf{x})} \to n^\mu(\mathbf{s})\frac{\delta}{\delta X^\mu(\mathbf{s})} \tag{6.16}$$

The Tomonaga-Schwinger equation (6.1) becomes

$$\hat{\mathfrak{H}}(\mathbf{s})\Psi[\phi, X] = i n^\mu(\mathbf{s})\frac{\delta\Psi[\phi, X]}{\delta X^\mu(\mathbf{s})} \tag{6.17}$$

For free fields the Hamiltonian density operator in curved spacetime is

$$\hat{\mathfrak{H}} = \frac{-1}{2|q|^{1/2}}\frac{\delta^2}{\delta\phi^2(\mathbf{s})} + \frac{|q|^{1/2}}{2}[-q^{ij}(\partial_i\phi)(\partial_j\phi) + m^2\phi^2] \tag{6.18}$$

The Bohmian equations of motion (6.7) becomes

$$\frac{\partial\Phi(\mathbf{s}, T]}{\partial\tau(\mathbf{s})} = \frac{1}{|q(\mathbf{s})|^{1/2}}\frac{\delta S}{\delta\phi(\mathbf{s})}\Big|_{\phi=\Phi};\quad \frac{\partial}{\partial\tau(\mathbf{s})} \equiv lim_{\sigma_x\to 0}\int_{\sigma_x} d^3s\, n^\mu(\mathbf{s})\frac{\delta}{\delta X^\mu(\mathbf{s})} \tag{6.19}$$

Similarly (6.8) becomes

$$\left[\left(\frac{\partial}{\partial\tau(\mathbf{s})}\right)^2 + \nabla^i\nabla_i + m^2\right]\Phi(\mathbf{s},X] = -\frac{1}{|q(\mathbf{s})|^{1/2}}\frac{\partial\Omega(\mathbf{s},\phi,X]}{\partial\phi\mathbf{s})}\Bigg|_{\phi=\Phi} \tag{6.20}$$

where ∇_i is the covariant derivative with respect to s^i and

$$\Omega(\mathbf{s},\phi,X] = -\frac{1}{|q(\mathbf{s})|^{1/2}}\frac{1}{2R}\frac{\delta^2 R}{\delta\phi^2(\mathbf{s})} \tag{6.21}$$

corresponding to a quantum potential. A given hypersurface Σ can be parametrized by different sets of 4 functions $X^\mu(\mathbf{s})$ of course but quantities such as $\Psi[\phi,X]$ and $\Phi(\mathbf{s},X]$ depend on Σ, not on the parametrization. The freedom in choosing functions $X^\mu(\mathbf{s})$ is sort of a gauge freedom related to the covariance. Now to find a solution of the covariant equations above it is convenient to fix a gauge and for a timelike surface the simplest choice is $X^i(\mathbf{s}) = s^i$. This implies $\delta X(\mathbf{s}) = 0$ which leads to equations similar to those obtained previously. For example (6.19) becomes

$$(g^{00}(\mathbf{x}))^{1/2}\frac{\partial\Phi(\mathbf{x},X^0]}{\partial X^0(\mathbf{x})} = \frac{1}{|q(\mathbf{x})|^{1/2}}\frac{\delta S}{\delta\phi(\mathbf{x})}\Bigg|_{\phi=\Phi} \tag{6.22}$$

which is the curved spacetime version of (6.7).

CHAPTER 3

GRAVITY AND THE QUANTUM POTENTIAL

Just as we plunged into QM in Chapters 1 and 2 we plunge again into general relativity (GR), Weyl geometry, Dirac-Weyl (DW) theory, and deBroglie-Bohm-Weyl (dBBW) theory. There are many good books available for background in general relativity, especially Baez-Muniain [**86**] (marvelous for conceptual purposes and for a modern perspective) and Adler et al [**12**] (a classic masterpiece with all the indices in place). In addition we mention some excellent books and papers which will arise in references later, namely [**66, 150, 263, 451, 618, 668, 749, 884, 941, 950, 1085, 1199, 1274**]. To develop all the background differential geometry requires a book in itself and the presentation adopted here will in fact include much of this implicitly since the topics range over a fairly wide field (see also Chapter 5, 7, and 8).

1. INTRODUCTION

A complete description of necessary geometric ideas appears in Misner et al [**884**] for example and we only make some definitions and express some relations here, usng the venerable tensor notation of indices, etc., since even today much of the physics literature appears in this form. For differential geometry one can refer to [**165, 341, 1309**]. First we give some background on Weyl geometry and Brans-Dicke theory following [**12**]; for differential geometry we use the tensor notation of [**12**] and refer to e.g. [**150, 457, 618, 668, 950, 962, 1274, 1309**] for other notation (see also [**1300**] for an interesting variation). One thinks of a differential manifold $M = \{U_i, \phi_i\}$ with $\phi : U_i \to \mathbf{R}^4$ and metric $g \sim g_{ij}dx^i dx^j$ satisfying $g(\partial_k, \partial_\ell) = g_{k\ell} =< \partial_k, \partial_\ell >= g_{\ell k}$. This is for the bare essentials; one can also imagine tangent vectors $X_i \sim \partial_i$ and dual cotangent vectors $\theta^i \sim dx^i$, etc. Given a coordinate change $\tilde{x}^i = \tilde{x}^i(x^j)$ a vector ξ^i transforming via $\tilde{\xi}^i = \sum \partial_i \tilde{x}^j \xi^j$ is called contravariant (e.g. $d\tilde{x}^i = \sum \partial_j \tilde{x}^i dx^j$). On the other hand $\partial\phi/\partial\tilde{x}^i = \sum(\partial\phi/\partial x^j)(\partial x^j/\partial\tilde{x}^i$ leads to the idea of covariant vectors $A_j \sim \partial\phi/\partial x^j$ transforming via $\tilde{A}_i = \sum(\partial x^j/\partial\tilde{x}^i)A_j$ (i.e. $\partial/\partial\tilde{x}^i \sim (\partial x^j/\partial\tilde{x}^j)\partial/\partial x^j$). Now define connection coefficients or Christoffel symbols via (strictly one writes $T^\gamma{}_\alpha = g_{\alpha\beta}T^{\gamma\beta}$ and $T_\alpha{}^\gamma = g_{\alpha\beta}T^{\beta\gamma}$ which are generally different - we use that notation here but it is sometimes not used later when it is unnecessary due to symmetries, etc.)

$$(1.1) \qquad \Gamma^r{}_{ki} = -\left\{ \begin{array}{c} r \\ k\ i \end{array} \right\} = -\frac{1}{2}\sum(\partial_i g_{k\ell} + \partial_k g_{\ell i} - \partial_\ell g_{ik})g^{\ell r} = \Gamma^r{}_{ik}$$

(note this differs by a minus sign from some other authors). Note also that (1.1) follows from equations

$$\partial_\ell g_{ik} + g_{rk}\Gamma^r{}_{i\ell} + g_{ir}\Gamma^r{}_{\ell k} = 0 \tag{1.2}$$

and cyclic permutation; the basic definition of $\Gamma^i{}_{mj}$ is found in the transplantation law $d\xi^i = \Gamma^i{}_{mj}dx^m\xi^j$. Next for tensors $T^\alpha_{\beta\gamma}$ define derivatives $T^\alpha_{\beta\gamma|k} = \partial_k T^\alpha_{\beta\gamma}$ and

$$T^\alpha_{\beta\gamma||\ell} = \partial_\ell T^\alpha_{\beta\gamma} - \Gamma^\alpha{}_{\ell s}T^s_{\beta\gamma} + \Gamma^s{}_{\ell\beta}T^\alpha_{s\gamma} + \Gamma^s{}_{\ell\gamma}T^\alpha_{\beta s} \tag{1.3}$$

In particular covariant derivatives for contravariant and covariant vectors respectively are defined via

$$\xi^i_{||k} = \partial_k\xi^i - \Gamma^i{}_{k\ell}\xi^\ell = \nabla_k\xi^i;\ \eta_{m||\ell} = \partial_\ell\eta_m + \Gamma^r{}_{m\ell}\eta_r = \nabla_\ell\eta_m \tag{1.4}$$

respectively. Now to describe Weyl geometry one notes first that for Riemannian geometry transplantation holds along with

$$\ell^2 = \|\xi\|^2 = g_{\alpha\beta}\xi^\alpha\xi^\beta;\ \xi^\alpha\eta_\alpha = g_{\alpha\beta}\xi^\alpha\eta^\beta \tag{1.5}$$

For Weyl geometry however one does not demand conservation of lengths and scalar products under affine transplantation as above. Thus assume $d\ell = (\phi_\beta dx^\beta)\ell$ where the covariant vector ϕ_β plays a role analogous to $\Gamma^\alpha{}_{\beta\gamma}$ and one obtains

$$d\ell^2 = 2\ell^2(\phi_\beta dx^\beta) = d(g_{\alpha\beta}\xi^\alpha\xi^\beta) = \tag{1.6}$$

$$= g_{\alpha\beta|\gamma}\xi^\alpha\xi^\beta dx^\gamma + g_{\alpha\beta}\Gamma^\alpha{}_{\rho\gamma}\xi^\rho\xi^\beta dx^\gamma + g_{\alpha\beta}\Gamma^\beta{}_{\rho\gamma}\xi^\alpha\xi^\rho dx^\gamma$$

Rearranging etc. and using (1.5) again gives

$$(g_{\alpha\beta|\gamma} - 2g_{\alpha\beta}\phi_\gamma) + g_{\sigma\beta}\Gamma^\sigma{}_{\alpha\gamma} + g_{\sigma\alpha}\Gamma^\sigma{}_{\beta\gamma} = 0; \tag{1.7}$$

$$\Gamma^\alpha{}_{\beta\gamma} = -\left\{\begin{array}{c}\alpha\\ \beta\ \gamma\end{array}\right\} + g^{\sigma\alpha}[g_{\sigma\beta}\phi_\gamma + g_{\sigma\gamma}\phi_\beta - g_{\beta\gamma}\phi_\sigma]$$

Thus we can prescribe the metric $g_{\alpha\beta}$ and the covariant vector field ϕ_γ and determine by (1.7) the field of connection coefficients $\Gamma^\alpha{}_{\beta\gamma}$ which admits the affine transplantation law as above. If one takes $\phi_\gamma = 0$ the Weyl geometry reduces to Riemannian geometry. This leads one to consider new metric tensors via a metric change $\hat{g}_{\alpha\beta} = f(x^\lambda)g_{\alpha\beta}$ and it turns out that $(1/2)\partial log(f)/\partial x^\lambda$ plays the role of ϕ_λ. Here the metric change is called a gauge transformation and the ordinary connections $\left\{\begin{array}{c}\alpha\\ \beta\ \gamma\end{array}\right\}$ constructed from $g_{\alpha\beta}$ are equal to the more general connections $\hat{\Gamma}^\alpha{}_{\beta\gamma}$ constructed according to (1.7) from $\hat{g}_{\alpha\beta}$ and $\hat{\phi}_\lambda = (1/2)\partial log(f)/\partial x^\lambda$. The generalized differential geometry is conformal in that the ratio

$$\frac{\xi^\alpha\eta_\alpha}{\|\xi\|\|\eta\|} = \frac{g_{\alpha\beta}\xi^\alpha\eta^\beta}{[(g_{\alpha\beta}\xi^\alpha\xi^\beta)(g_{\alpha\beta}\eta^\alpha\eta^\beta)]^{1/2}} \tag{1.8}$$

does not change under the gauge transformation $\hat{g}_{\alpha\beta} \to f(x^\lambda)g_{\alpha\beta}$. Again if one has a Weyl geometry characterized by $g_{\alpha\beta}$ and ϕ_α with connections determined by (1.7) one may replace the geometric quantities by use of a scalar field f with

$$\hat{g}_{\alpha\beta} = f(x^\lambda)g_{\alpha\beta},\ \hat{\phi}_\alpha = \phi_\alpha + (1/2)(log(f)_{|\alpha};\ \hat{\Gamma}^\alpha{}_{\beta\gamma} = \Gamma^\alpha{}_{\beta\gamma} \tag{1.9}$$

without changing the intrinsic geometric properties of vector fields; the only change is that of local lengths of a vector via $\hat{\ell}^2 = f(x^\lambda)\ell^2$. Note that one can reduce

$\hat{\phi}_\alpha$ to the zero vector field if and only if ϕ_α is a gradient field, namely $F_{\alpha\beta} = \phi_{\alpha|\beta} - \phi_{\beta|\alpha} = 0$ (i.e. $\phi_\alpha = (1/2)\partial_a log(f) \equiv \partial_\beta\phi_\alpha = \partial_\alpha\phi_\beta$). In this case one has length preservation after transplantation around an arbitrary closed curve and the vanishing of $F_{\alpha\beta}$ guarantees a choice of metric in which the Weyl geometry becomes Riemannian; thus $F_{\alpha\beta}$ is an intrinsic geometric quantity for Weyl geometry; note $F_{\alpha\beta} = -F_{\beta\alpha}$ and

$$\{F_{\alpha\beta|\gamma}\} = 0;\ \{F_{\mu\nu|\lambda}\} = F_{\mu\nu|\lambda} + F_{\lambda\mu|\nu} + F_{\nu\lambda|\mu} \tag{1.10}$$

Similarly the concept of covariant differentiation depends only on the idea of vector transplantation. Indeed one can define covariant derivatives via

$$\xi^\alpha_{||\beta} = \xi^\alpha_{|\beta} - \Gamma^\alpha_{\ \beta\gamma}\xi^\gamma \tag{1.11}$$

In Riemannian geometry the curvature tensor is

$$\xi^\alpha_{||\beta|\gamma} - \xi^\alpha_{||\gamma|\beta} = R^\alpha_{\eta\beta\gamma}\xi^\eta;\ R^\alpha_{\ \beta\gamma\delta} = -\Gamma^\alpha_{\ \beta\gamma|\delta} + \Gamma^\alpha_{\ \beta\delta|\gamma} + \Gamma^\alpha_{\ \tau\delta}\Gamma^\tau_{\ \beta\gamma} - \Gamma^\alpha_{\ \tau\gamma}\Gamma^\tau_{\ \beta\delta} \tag{1.12}$$

Using (1.8) one then can express this in terms of $g_{\alpha\beta}$ and ϕ_α but this is complicated. Equations for $R_{\beta\delta} = R^\alpha_{\ \beta\alpha\delta}$ and $R = g^{\beta\delta}R_{\beta\delta}$ are however given in [**12**]. One notes that in Weyl geometry if a vector ξ^α is given, independent of the metric, then $\xi_\alpha = g_{\alpha\beta}\xi^\beta$ will depend on the metric and under a gauge transformation one has $\hat{\xi}_\alpha = f(x^\lambda)\xi_\alpha$. Hence the covariant form of a gauge invariant contravariant vector becomes gauge dependent and one says that a tensor is of weight n if, under a gauge transformation, $\hat{T}^{\alpha\cdots}_{\beta\cdots} = f(x^\lambda)^n T^{\alpha\cdots}_{\beta\cdots}$. Note ϕ_α plays a singular role in (1.9) and has no weight. Similarly $\sqrt{-\hat{g}} = f^2\sqrt{-g}$ (weight 2) and $F^{\alpha\beta} = g^{\alpha\mu}g^{\beta\nu}F_{\mu\nu}$ has weight -2 while $\mathfrak{F}^{\alpha\beta} = F^{\alpha\beta}\sqrt{-g}$ has weight 0 and is gauge invariant. Similarly $F_{\alpha\beta}F^{\alpha\beta}\sqrt{-g}$ is gauge invariant. Now for Weyl's theory of electromagnetism one wants to interpret ϕ_α as an EM potential and one has automatically the Maxwell equations

$$\{F_{\alpha\beta|\gamma}\} = 0;\ \mathfrak{F}^{\alpha\beta}_{|\beta} = \mathfrak{s}^\alpha \tag{1.13}$$

(the latter equation being gauge invariant source equations). These equations are gauge invariant as a natural consequence of the geometric interpretation of the EM field. For the interaction between the EM and gravitational fields one sets up some field equations as indicated in [**12**] and the interaction between the metric quantities and the EM fields is exhibited there (there is much more on EM theory later and see also Section 2.1.1).

REMARK 3.1.1. As indicated earlier in [**12**] $R^i_{\ jk}$ is defined with a minus sign compared with e.g. [**950, 1309**] for example. There is also a difference in definition of the Ricci tensor which is taken to be $G^{\beta\delta} = R^{\beta\delta} - (1/2)g^{\beta\delta}R$ in [**12**] with $R = R^\delta_\delta$ so that $G_{\mu\gamma} = g_{\mu\beta}g_{\gamma\delta}G^{\beta\delta} = R_{\mu\gamma} - (1/2)g_{\mu\gamma}R$ with $G^\gamma_\eta = R^\eta_\eta - 2R \Rightarrow G^\eta_\eta = -R$ (recall $n = 4$). In Ohanian-Ruffini [**950**] the Ricci tensor is simply $R_{\beta\mu} = R^\alpha_{\beta\mu\alpha}$ where $R^\alpha_{\beta\mu\nu}$ is the Riemann curvature tensor and $R = R^\eta_\eta$ again. This is similar to Willmore [**1309**] where the Ricci tensor is defined as $\rho_{j\ell} = R^i_{ji\ell}$. To clarify all this we note that

$$R_{\eta\gamma} = R^\alpha_{\ \eta\alpha\gamma} = g^{\alpha\beta}R_{\beta\eta\alpha\gamma} = -g^{\alpha\beta}R_{\beta\eta\gamma\alpha} = -R^\alpha_{\ \eta\gamma\alpha} \tag{1.14}$$

which reveals the minus sign difference. ■

2. SKETCH OF DEBROGLIE-BOHM-WEYL THEORY

From Chapters 1 and 2 we know something about Bohmian mechanics and the quantum potential and we go now to the papers [**1151, 1152, 1153, 1154, 1155, 1156, 1157, 1158, 1159, 1160, 1161, 1162, 1163, 1164, 1165, 1166, 1167**] by A. and F. Shojai and M. Golshani to begin the present discussion (cf. also [**8, 146, 147, 354, 901, 902, 1104, 1105, 1107, 1108, 1109, 1110, 1111**] for related work from the Iranian school and [**237, 270, 828, 962, 1115, 1116**] for linking of dBB theory with Weyl geometry). In nonrelativistic deBroglie-Bohm theory the quantum potential is $Q = -(\hbar^2/2m)(\nabla^2|\Psi|/|\Psi|)$. The particles trajectory can be derived from Newton's law of motion in which the quantum force $-\nabla Q$ is present in addition to the classical force $-\nabla V$. The enigmatic quantum behavior is attributed here to the quantum force or quantum potential (with Ψ determining a "pilot wave" which guides the particle motion). Setting $\Psi = \sqrt{\rho}exp[iS/\hbar]$ one has

$$\frac{\partial S}{\partial t} + \frac{|\nabla S|^2}{2m} + V + Q = 0;\ \frac{\partial \rho}{\partial t} + \nabla \cdot \left(\rho \frac{\nabla S}{m}\right) = 0 \tag{2.1}$$

The first equation in (2.1) is a Hamilton-Jacobi (HJ) equation which is identical to Newton's law and represents an energy condition $E = (|p|^2/2m) + V + Q$ (recall from HJ theory $-(\partial S/\partial t) = E(= H)$ and $\nabla S = p$. The second equation represents a continuity equation for a hypothetical ensemble related to the particle in question. For the relativistic extension one could simply try to generalize the relativistic energy equation $\eta_{\mu\nu}P^\mu P^\nu = m^2c^2$ to the form

$$\eta_{\mu\nu}P^\mu P^\nu = m^2c^2(1+\mathcal{Q}) = \mathcal{M}^2c^2;\ \mathcal{Q} = (\hbar^2/m^2c^2)(\Box|\Psi|/|\Psi|) \tag{2.2}$$

$$\mathcal{M}^2 = m^2\left(1 + \alpha\frac{\Box|\Psi|}{|\Psi|}\right);\ \alpha = \frac{\hbar^2}{m^2c^2} \tag{2.3}$$

This could be derived e.g. by setting $\Psi = \sqrt{\rho}exp(iS/\hbar)$ in the Klein-Gordon (KG) equation and separating the real and imaginary parts, leading to the relativistic HJ equation $\eta_{\mu\nu}\partial^\mu S\partial^\nu S = \mathfrak{M}^2c^2$ (as in (2.1) - note $P^\mu = -\partial^\mu S$) and the continuity equation is $\partial_\mu(\rho\partial^\mu S) = 0$. The problem of $\mathcal{M}^2$ not being positive definite here (i.e. tachyons) is serious however and in fact (2.2) is not the correct equation (see e.g. [**1158, 1160, 1163**]). One must use the covariant derivatives ∇_μ in place of ∂_μ and for spin zero in a curved background there results ($\mathcal{Q}$ as above)

$$\nabla_\mu(\rho\nabla^\mu S) = 0;\ g^{\mu\nu}\nabla_\mu S\nabla_\nu S = \mathfrak{M}^2c^2;\ \mathfrak{M}^2 = m^2e^{\mathcal{Q}} \tag{2.4}$$

To see this one must require that a correct relativistic equation of motion should not only be Poincaré invariant but also it should have the correct nonrelativistic limit. Thus for a relativistic particle of mass $\mathfrak{M}$ (which is a Lorentz invariant quantity) $\mathfrak{A} = \int d\lambda(1/2)\mathfrak{M}(r)(dr_\mu/d\lambda)(dr^\nu/d\lambda)$ is the action functional where λ is any scalar parameter parametrizing the path $r_\mu(\lambda)$ (it could e.g. be the proper time τ). Varying the path via $r_\mu \to r'_\mu = r_\mu + \epsilon_\mu$ one gets (cf. [**1158**])

$$\mathfrak{A} \to \mathfrak{A}' = \mathfrak{A} + \delta\mathfrak{A} = \mathfrak{A} + \int d\lambda\left[\mathfrak{M}\frac{dr_\mu}{d\lambda}\frac{d\epsilon^\mu}{d\lambda} + \frac{1}{2}\frac{dr_\mu}{d\,gl}\frac{dr^\mu}{d\lambda}\epsilon_\nu\partial^\nu\mathfrak{M}\right] \tag{2.5}$$

By least action the correct path satisfies $\delta\mathfrak{A} = 0$ with fixed boundaries so the equation of motion is

$$(2.6)\qquad (d/d\lambda)(\mathfrak{M}u_\mu) = (1/2)u_\nu u^\nu \partial_\mu \mathfrak{M};$$

$$\mathfrak{M}(du_\mu/d\lambda) = ((1/2)\eta_{\mu\nu}u_\alpha u^\alpha - u_\mu u_\nu)\partial^\nu \mathfrak{M}$$

where $u_\mu = dr_\mu/d\lambda$. Now look at the symmetries of the action functional via $\lambda \to \lambda+\delta$. The conserved current is then the Hamiltonian $\mathfrak{H} = -\mathfrak{L}+u_\mu(\partial\mathfrak{L}/\partial u_\mu) = (1/2)\mathfrak{M}u_\mu u^\mu = E$. This can be seen by setting $\delta\mathfrak{A} = 0$ where

$$(2.7)\qquad 0 = \delta\mathfrak{A} = \mathfrak{A}' - \mathfrak{A} = \int d\lambda \left[\frac{1}{2}u_\mu u^\mu u^\nu \partial_\nu \mathfrak{M} + \mathfrak{M}u_\mu \frac{du^\mu}{d\lambda}\right]\delta$$

which means that the integrand is zero, i.e. $(d/d\lambda)[(1/2)\mathfrak{M}u_\mu u^\mu] = 0$. Since the proper time is defined as $c^2d\tau^2 = dr_\mu dr^\mu$ this leads to $(d\tau/d\lambda) = \sqrt{(2E/\mathfrak{M}c^2)}$ and the equation of motion becomes

$$(2.8)\qquad \mathfrak{M}(dv_\mu/d\tau) = (1/2)(c^2\eta_{\mu\nu} - v_\mu v_\nu)\partial^\nu \mathfrak{M}$$

where $v_\mu = dr_\mu/d\tau$. The nonrelativistic limit can be derived by letting the particles velocity be ignorable with respect to light velocity. In this limit the proper time is identical to the time coordinate $\tau = t$ and the result is that the $\mu = 0$ component is satisfied identically via $(r \sim \vec{r})$

$$(2.9)\qquad \mathfrak{M}\frac{d^2r}{dt^2} = -\frac{1}{2}c^2\nabla\mathfrak{M} \Rightarrow m\left(\frac{d^2r}{dt^2}\right) = -\nabla\left[\frac{mc^2}{2}log\left(\frac{\mathfrak{M}}{\mu}\right)\right]$$

where μ is an arbitrary mass scale. In order to have the correct limit the term in parenthesis on the right side should be equal to the quantum potential so $(mc^2/2)log(\mathfrak{M}/\mu) = (\hbar^2/2m)(\nabla^2|\psi|/|\psi|)$ and hence

$$(2.10)\qquad \mathfrak{M} = \mu exp[-(\hbar^2/m^2c^2)(\nabla^2|\Psi|/|\Psi|)]$$

One infers that the relativistic quantum mass field is $\mathfrak{M} = \mu exp[(\hbar^2/2m)(\Box|\Psi|/|\Psi|)]$ (manifestly invariant) and setting $\mu = m$ we get (cf. also (2.12) below)

$$(2.11)\qquad \mathfrak{M} = m exp[(\hbar^2/m^2c^2)(\Box|\Psi|/|\Psi|)]$$

If one starts with the standard relativistic theory and goes to the nonrelativistic limit one does not get the correct nonrelativistic equations; this is a result of an improper decomposition of the wave function into its phase and norm in the KG equation (cf. also [**133**] for related procedures). One notes here also that (2.11) leads to a positive definite mass squared. Also from [**1158**] this can be extended to a many particle version and to a curved spacetime. However, for a particle in a curved background we will take (cf. [**1160**] which we follow for the rest of this section)

$$(2.12)\quad \nabla_\mu(\rho\nabla^\mu S) = 0;\ g^{\mu\nu}\nabla_\mu S\nabla_\nu S = \mathfrak{M}^2c^2;\ \mathfrak{M}^2 = m^2e^{\mathfrak{Q}};\ \mathfrak{Q} = \frac{\hbar^2}{m^2c^2}\frac{\Box_g|\Psi|}{|\Psi|}$$

((2.11) suggests that $\mathfrak{M}^2 = m^2exp(2\mathfrak{Q})$ but (2.12) is used for compatibility with the KG approach, etc., where $exp(\mathfrak{Q}) \sim 1+\mathfrak{Q}$ (see Noldus [**940**] and Chapter 8 for a derivation of (2.12) from first principles). Since, following deBroglie, the quantum

HJ equation (QHJE) in (2.12) can be written in the form $(m^2/\mathfrak{M}^2)g^{\mu\nu}\nabla_\mu S\nabla_\nu S = m^2c^2$, **the quantum effects are identical to a change of spacetime metric**

$$g_{\mu\nu} \to \tilde{g}_{\mu\nu} = (\mathfrak{M}^2/m^2)g_{\mu\nu} \tag{2.13}$$

which is a conformal transformation. The QHJE becomes then $\tilde{g}^{\mu\nu}\tilde{\nabla}_\mu S\tilde{\nabla}_\nu S = m^2c^2$ where $\tilde{\nabla}_\mu$ represents covariant differentiation with respect to the metric $\tilde{g}_{\mu\nu}$ and the continuity equation is then $\tilde{g}_{\mu\nu}\tilde{\nabla}_\mu(\rho\tilde{\nabla}_\nu S) = 0$. The important conclusion here is that the presence of the quantum potential is equivalent to a curved spacetime with its metric given by (2.13). This is a geometrization of the quantum aspects of matter and it seems that there is a dual aspect to the role of geometry in physics. The spacetime geometry sometimes looks like "gravity" and sometimes reveals quantum behavior. The curvature due to the quantum potential may have a large influence on the classical contribution to the curvature of spacetime. The particle trajectory can now be derived from the guidance relation via differentiation of (2.12) leading to the Newton equations of motion

$$\mathfrak{M}\frac{d^2x^\mu}{d\tau^2} + \mathfrak{M}\Gamma^\mu_{\nu\kappa}u^\nu u^\kappa = (c^2g^{\mu\nu} - u^\mu u^\nu)\nabla_\nu\mathfrak{M} \tag{2.14}$$

Using the conformal transformation above (2.14) reduces to the standard geodesic equation.

Now a general "canonical" relativistic system consisting of gravity and classical matter (no quantum effects) is determined by the action

$$\mathcal{A} = \frac{1}{2\kappa}\int d^4x\sqrt{-g}\mathcal{R} + \int d^4x\sqrt{-g}\frac{\hbar^2}{2m}\left(\frac{\rho}{\hbar^2}\mathcal{D}_\mu S\mathcal{D}^\mu S - \frac{m^2}{\hbar^2}\rho\right) \tag{2.15}$$

where $\kappa = 8\pi G$ and $c = 1$ for convenience. It was seen above that via deBroglie the introduction of a quantum potential is equivalent to introducing a conformal factor $\Omega^2 = \mathfrak{M}^2/m^2$ in the metric. Hence in order to introduce quantum effects of matter into the action (2.15) one uses this conformal transformation to get $(1 + Q \sim exp(Q))$

$$\mathfrak{A} = \frac{1}{2\kappa}\int d^4x\sqrt{-\bar{g}}(\bar{\mathcal{R}}\Omega^2 - 6\bar{\nabla}_\mu\Omega\bar{\nabla}^\mu\Omega) + \tag{2.16}$$

$$+\int d^4x\sqrt{-\bar{g}}\left(\frac{\rho}{m}\Omega^2\bar{\nabla}_\mu S\bar{\nabla}^\mu S - m\rho\Omega^4\right) + \int d^4x\sqrt{-\bar{g}}\lambda\left[\Omega^2 - \left(1 + \frac{\hbar^2}{m^2}\frac{\Box\sqrt{\rho}}{\sqrt{\rho}}\right)\right]$$

where a bar over any quantity means that it corresponds to the nonquantum regime. Here only the first two terms of the expansion of $\mathfrak{M}^2 = m^2 exp(\mathfrak{Q})$ in (2.12) have been used, namely $\mathfrak{M}^2 \sim m^2(1 + \mathfrak{Q})$. No physical change is involved in considering all the terms. λ is a Lagrange multiplier introduced to identify the conformal factor with its Bohmian value. One uses here $\bar{g}_{\mu\nu}$ to raise of lower indices and to evaluate the covariant derivatives; the physical metric (containing the quantum effects of matter) is $g_{\mu\nu} = \Omega^2\bar{g}_{\mu\nu}$. By variation of the action with respect to $\bar{g}_{\mu\nu}$, Ω, ρ, S, and λ one arrives at the following quantum equations of motion:

(1) The equation of motion for Ω

$$\bar{\mathcal{R}}\Omega + 6\Box\Omega + \frac{2\kappa}{m}\rho\Omega(\bar{\nabla}_\mu S\bar{\nabla}^\mu S - 2m^2\Omega^2) + 2\kappa\lambda\Omega = 0 \tag{2.17}$$

(2) The continuity equation for particles $\bar{\nabla}_\mu(\rho\Omega^2\bar{\nabla}^\mu S) = 0$

(3) The equations of motion for particles (here $a' \equiv \bar{a}$)

$$(\bar{\nabla}_\mu S\bar{\nabla}^\mu S - m^2\Omega^2)\Omega^2\sqrt{\rho} + \frac{\hbar^2}{2m}\left[\Box'\left(\frac{\lambda}{\sqrt{\rho}}\right) - \lambda\frac{\Box'\sqrt{\rho}}{\rho}\right] = 0 \tag{2.18}$$

(4) The modified Einstein equations for $\bar{g}_{\mu\nu}$

$$\Omega^2\left[\bar{\mathcal{R}}_{\mu\nu} - \frac{1}{2}\bar{g}_{\mu\nu}\bar{\mathcal{R}}\right] - [\bar{g}_{\mu\nu}\Box' - \bar{\nabla}_\mu\bar{\nabla}_\nu]\Omega^2 - 6\bar{\nabla}_\mu\Omega\bar{\nabla}_\nu\Omega + 3\bar{g}_{\mu\nu}\bar{\nabla}_\alpha\Omega\bar{\nabla}^\alpha\Omega + \tag{2.19}$$

$$+\frac{2\kappa}{m}\rho\Omega^2\bar{\nabla}_\mu S\bar{\nabla}_\nu S - \frac{\kappa}{m}\rho\Omega^2\bar{g}_{\mu\nu}\bar{\nabla}_\alpha S\bar{\nabla}^\alpha S + \kappa m\rho\Omega^4\bar{g}_{\mu\nu} +$$

$$+\frac{\kappa\hbar^2}{m^2}\left[\bar{\nabla}_\mu\sqrt{\rho}\bar{\nabla}_\nu\left(\frac{\lambda}{\sqrt{\rho}}\right) + \bar{\nabla}_\nu\sqrt{\rho}\bar{\nabla}_\mu\left(\frac{\lambda}{\sqrt{\rho}}\right)\right] - \frac{\kappa\hbar^2}{m^2}\bar{g}_{\mu\nu}\bar{\nabla}_\alpha\left[\lambda\frac{\bar{\nabla}^\alpha\sqrt{\rho}}{\sqrt{\rho}}\right] = 0$$

(5) The constraint equation $\Omega^2 = 1 + (\hbar^2/m^2)[(\Box\sqrt{\rho})/\sqrt{\rho}]$

Thus the back reaction effects of the quantum factor on the background metric are contained in these highly coupled equations (cf. also [**29**]). A simpler form of (2.17) can be obtained by taking the trace of (2.19) and using (2.17) which produces $\lambda = (\hbar^2/m^2)\bar{\nabla}_\mu[\lambda(\bar{\nabla}^\mu\sqrt{\rho})/\sqrt{\rho}]$. A solution of this via perturbation methods using the small parameter $\alpha = \hbar^2/m^2$ yields the trivial solution $\lambda = 0$ so the above equations reduce to

$$\bar{\nabla}_\mu(\rho\Omega^2\bar{\nabla}^\mu S) = 0;\ \bar{\nabla}_\mu S\bar{\nabla}^\mu S = m^2\Omega^2;\ \mathfrak{G}_{\mu\nu} = -\kappa\mathfrak{T}^{(m)}_{\mu\nu} - \kappa\mathfrak{T}^{(\Omega)}_{\mu\nu} \tag{2.20}$$

where $\mathfrak{T}^{(m)}_{\mu\nu}$ is the matter energy-momentum (EM) tensor and

$$\kappa\mathfrak{T}^{(\Omega)}_{\mu\nu} = \frac{[g_{\mu\nu}\Box - \nabla_\mu\nabla_\nu]\Omega^2}{\Omega^2} + 6\frac{\nabla_\mu\Omega\nabla_\nu\Omega}{\omega^2} - 2g_{\mu\nu}\frac{\nabla_\alpha\Omega\nabla^\alpha\Omega}{\Omega^2} \tag{2.21}$$

with $\Omega^2 = 1 + \alpha(\Box\sqrt{\rho}/\sqrt{\rho})$. Note that the second relation in (2.20) is the Bohmian equation of motion and written in terms of $g_{\mu\nu}$ it becomes $\nabla_\mu S\nabla^\mu S = m^2c^2$.

In the preceeding one has tacitly assumed that there is an ensemble of quantum particles so what about a single particle? One translates now the quantum potential into purely geometrical terms without reference to matter parameters so that the original form of the quantum potential can only be deduced after using the field equations. Thus the theory will work for a single particle or an ensemble and in this connection we make

REMARK 3.2.1. One notes that the use of $\psi\psi^*$ automatically suggests or involves an ensemble if it is to be interpreted as a probability density. Thus the idea that a particle has only a probability of being at or near x seems to mean that some paths take it there but others don't and this is consistent with Feynman's use of path integrals for example. This seems also to say that there is no such thing as a particle, only a collection of versions or cloud connected to the particle idea. Bohmian theory on the other hand for a fixed energy gives a one parameter family of trajectories associated to ψ (see here Section 2.2 and [**244**] for details). This is because the trajectory arises from a third order differential while fixing the solution ψ of the second order stationary Schrödinger equation involves only two "boundary" conditions. As was shown in [**244**] this automatically generates a Heisenberg inequality $\Delta x \Delta p \geq c\hbar$; i.e. the uncertainty is built in when using the wave function ψ and amazingly can be expressed by the operator theoretical framework of quantum mechanics. Thus a one parameter family of paths can be associated with the use of $\psi\psi^*$ and this generates the cloud or ensemble automatically associated with the use of ψ. In fact, based on Remark 2.2.2, one might conjecture that upon using a wave function discription of quantum particle motion, one opens the door to a cloud of particles, all of whose motions are incompletely governed by the SE, since one determining condition for particle motion is ignored. Thus automatically the quantum potential will give rise to a force acting on any such particular trajectory and the "ensemble" idea naturally applies to a cloud of identical particles (cf. also Theorem 1.2.1 and Corollary 1.2.1). ■

Now first ignore gravity and look at the geometrical properties of the conformal factor given via

$$g_{\mu\nu} = e^{4\Sigma}\eta_{\mu\nu};\; e^{4\Sigma} = \frac{\mathfrak{M}^2}{m^2} = exp\left(\alpha\frac{\Box_\eta\sqrt{\rho}}{\sqrt{\rho}}\right) = exp\left(\alpha\frac{\Box_\eta\sqrt{|\mathfrak{T}|}}{\sqrt{|\mathfrak{T}|}}\right) \tag{2.22}$$

where $\mathfrak{T}$ is the trace of the EM tensor and is substituted for ρ (true for dust). The Einstein tensor for this metric is

$$\mathfrak{G}_{\mu\nu} = 4g_{\mu\nu}\Box_\eta exp(-\Sigma) + 2exp(-2\Sigma)\partial_\mu\partial_\nu exp(2\Sigma) \tag{2.23}$$

Hence as an Ansatz one can suppose that in the presence of gravitational effects the field equation would have a form

$$\mathcal{R}_{\mu\nu} - \frac{1}{2}\mathcal{R}g_{\mu\nu} = \kappa\mathfrak{T}_{\mu\nu} + 4g_{\mu\nu}e^{\Sigma}\Box e^{-\Sigma} + 2e^{-2\Sigma}\nabla_\mu\nabla_\nu e^{2\Sigma} \tag{2.24}$$

This is written in a manner such that in the limit $\mathfrak{T}_{\mu\nu} \to 0$ one will obtain (2.22). Taking the trace of the last equation one gets $-\mathcal{R} = \kappa\mathfrak{T} - 12\Box\Sigma + 24(\nabla\Sigma)^2$ which has the iterative solution $\kappa\mathfrak{T} = -\mathcal{R} + 12\alpha\Box[(\Box\sqrt{\mathcal{R}})/\sqrt{\mathcal{R}}]$ leading to

$$\Sigma = \alpha[(\Box\sqrt{|\mathfrak{T}|}/\sqrt{|\mathfrak{T}|})] \simeq \alpha[(\Box\sqrt{|\mathcal{R}|})/\sqrt{|\mathcal{R}|})] \tag{2.25}$$

to first order in α.

One goes now to the field equations for a toy model. First from the above one sees that $\mathfrak{T}$ can be replaced by $\mathcal{R}$ in the expression for the quantum potential or for the conformal factor of the metric. This is important since the explicit reference to ensemble density is removed and the theory works for a single particle or an ensemble. So from (2.24) for a toy quantum gravity theory one assumes the following field equations

$$\mathfrak{G}_{\mu\nu} - \kappa\mathfrak{T}_{\mu\nu} - \mathfrak{Z}_{\mu\nu\alpha\beta}exp\left(\frac{\alpha}{2}\Phi\right)\nabla^\alpha\nabla^\beta exp\left(-\frac{\alpha}{2}\Phi\right) = 0 \tag{2.26}$$

where $\mathfrak{Z}_{\mu\nu\alpha\beta} = 2[g_{\mu\nu}g_{\alpha\beta} - g_{\mu\alpha}g_{\nu\beta}]$ and $\Phi = (\Box\sqrt{|\mathcal{R}|}/\sqrt{|\mathcal{R}|})$. The number 2 and the minus sign of the second term are chosen so that the energy equation derived later will be correct. Note that the trace of (2.26) is

$$\mathcal{R} + \kappa\mathfrak{T} + 6exp(\alpha\Phi/2)\Box exp(-\alpha\Phi/2) = 0 \tag{2.27}$$

and this represents the connection of the Ricci scalar curvature of space time and the trace of the matter EM tensor. If a perturbative solution is admitted one can expand in powers of α to find $\mathcal{R}^{(0)} = -\kappa\mathfrak{T}$ and $\mathcal{R}^{(1)} = -\kappa\mathfrak{T} - 6exp(\alpha\Phi^0/2)\Box exp(-\alpha\Phi^0/2)$ where $\Phi^{(0)} = \Box\sqrt{|\mathfrak{T}|}/\sqrt{|\mathfrak{T}|}$. The energy relation can be obtained by taking the four divergence of the field equations and since the divergence of the Einstein tensor is zero one obtains

$$\kappa\nabla^\nu\mathfrak{T}_{\mu\nu} = \alpha\mathcal{R}_{\mu\nu}\nabla^\nu\Phi - \frac{\alpha^2}{4}\nabla_\mu(\nabla\Phi)^2 + \frac{\alpha^2}{2}\nabla_\mu\Phi\Box\Phi \tag{2.28}$$

For a dust with $\mathfrak{T}_{\mu\nu} = \rho u_\mu u_\nu$ and u_μ the velocity field, the conservation of mass law is $\nabla^\nu(\rho\mathfrak{M}u_\nu) = 0$ so one gets to first order in α $\nabla_\mu\mathfrak{M}/\mathfrak{M} = -(\alpha/2)\nabla_\mu\Phi$ or $\mathfrak{M}^2 = m^2 exp(-\alpha\Phi)$ where m is an integration constant. This is the correct relation of mass and quantum potential.

In [**1160**] there is then some discussion about making the conformal factor dynamical via a general scalar tensor action (cf. also [**1154**]) and subsequently one makes both the conformal factor and the quantum potential into dynamical fields and creates a scalar tensor theory with two scalar fields. Thus start with a general action

$$\mathfrak{A} = \int d^4x\sqrt{-g}\left[\phi\mathcal{R} - \omega\frac{\nabla_\mu\phi\nabla^\mu\phi}{\phi} - \frac{\nabla_\mu Q\nabla^\mu Q}{\phi} + 2\Lambda\phi + \mathfrak{L}_m\right] \tag{2.29}$$

The cosmological constant generally has an interaction term with the scalar field and here one uses an ad hoc matter Lagrangian

$$\mathfrak{L}_m = \frac{\rho}{m}\phi^a\nabla_\mu S\nabla^\mu S - m\rho\phi^b - \Lambda(1+Q)^c + \alpha\rho(e^{\ell Q} - 1) \tag{2.30}$$

(only the first two terms $1 + Q$ from $exp(Q)$ are used for simplicity in the third term). Here a, b, c are constants to be fixed later and the last term is chosen (heuristically) in such a manner as to have an interaction between the quantum potential field and the ensemble density (via the equations of motion); further the interaction is chosen so that it vanishes in the classical limit but this is ad hoc. Variation of the above action yields

(1) The scalar fields equation of motion

$$\mathcal{R} + \frac{2\omega}{\phi}\Box\phi - \frac{\omega}{\phi^2}\nabla^\mu\phi\nabla_\mu\phi + 2\Lambda + \tag{2.31}$$

$$+\frac{1}{\phi^2}\nabla^\mu Q\nabla_\mu Q + \frac{a}{m}\rho\phi^{a-1}\nabla^\mu S\nabla_\mu S - mb\rho\phi^{b-1} = 0$$

(2) The quantum potential equations of motion

$$(\Box Q/\phi) - (\nabla_\mu Q\nabla^\mu\phi/\phi^2) - \Lambda c(1+Q)^{c-1} + \alpha\ell\rho exp(\ell Q) = 0 \tag{2.32}$$

(3) The generalized Einstein equations

$$\mathfrak{G}^{\mu\nu} - \Lambda g^{\mu\nu} = -\frac{1}{\phi}\mathfrak{T}^{\mu\nu} - \frac{1}{\phi}[\nabla^\mu\nabla^\nu - g^{\mu\nu}\Box]\phi + \frac{\omega}{\phi^2}\nabla^\mu\phi\nabla^\nu\phi - \tag{2.33}$$

$$-\frac{\omega}{2\phi^2}g^{\mu\nu}\nabla^\alpha\phi\nabla_\alpha\phi+\frac{1}{\phi^2}\nabla^\mu Q\nabla^\nu Q-\frac{1}{2\phi^2}g^{\mu\nu}\nabla^\alpha Q\nabla_\alpha Q$$

(4) The continuity equation $\nabla_\mu(\rho\phi^a\nabla^\mu S)=0$

(5) The quantum Hamilton Jacobi equation

(2.34) $$\nabla^\mu S\nabla_\mu S=m^2\phi^{b-a}-\alpha m\phi^{-a}(e^{\ell Q}-1)$$

In (2.31) the scalar curvature and the term $\nabla^\mu S\nabla_\mu S$ can be eliminated using (2.33) and (2.34); further on using the matter Lagrangian and the definition of the EM tensor one has

(2.35) $$(2\omega-3)\Box\phi=(a+1)\rho\alpha(e^{\ell Q}-1)-2\Lambda(1+Q)^c+2\Lambda\phi-\frac{2}{\phi}\nabla_\mu Q\nabla^\mu Q$$

(where $b=a+1$). Solving (2.32) and (2.35) with a perturbation expansion in α one finds

(2.36) $$Q=Q_0+\alpha Q_1+\cdots;\ \phi=1+\alpha Q_1+\cdots;\ \sqrt{\rho}=\sqrt{\rho_0}+\alpha\sqrt{\rho_1}+\cdots$$

where the conformal factor is chosen to be unity at zeroth order so that as $\alpha\to 0$ (2.34) goes to the classical HJ equation. Further since by (2.34) the quantum mass is $m^2\phi+\cdots$ the first order term in ϕ is chosen to be Q_1 (cf. (2.12)). Also we will see that $Q_1\sim\Box\sqrt{\rho}/\sqrt{\rho}$ plus corrections which is in accord with Q as a quantum potential field. In any case after some computation one obtains $a=2\omega k$, $b=a+1$, and $\ell=(1/4)(2\omega k+1)=(1/4)(a+1)=b/4$ with $Q_0=[1/c(2c-3)]\{[-(2\omega k+1)/2\Lambda]k\sqrt{\rho_0}-(2c^2-c+1)\}$ while ρ_0 can be determined (cf. [**1160**] for details). Thus heuristically the quantum potential can be regarded as a dynamical field and perturbatively one gets the correct dependence of quantum potential upon density, modulo some corrective terms.

One goes next to a number of examples and we only consider here the conformally flat solution (cf. also [**1156**]). Thus take $g_{\mu\nu}=exp(2\Sigma)\eta_{\mu\nu}$ where $\Sigma<<1$. One obtains from (2.24)

(2.37) $$\mathcal{R}_{\mu\nu}=\eta_{\mu\nu}\Box\Sigma+2\partial_\mu\partial_\nu\Sigma\Rightarrow\mathfrak{G}_{\mu\nu}=2\partial_\mu\partial_\nu\Sigma-2\eta_{\mu\nu}\Box\Sigma$$

One can solve this iteratively to get

(2.38) $$\mathcal{R}^{(0)}=-\kappa\mathfrak{T}\Rightarrow\Sigma^{(0)}=-\frac{\kappa}{6}\Box^{-1}\mathfrak{T};$$

$$\mathcal{R}^{(1)}=-\kappa\mathfrak{T}+3\alpha\Box\frac{\Box\sqrt{|\mathfrak{T}|}}{\sqrt{|\mathfrak{T}|}}\Rightarrow\Sigma^{(1)}=-\frac{\kappa}{6}\Box^{-1}\mathfrak{T}+\frac{\alpha}{2}\frac{\Box\sqrt{|\mathfrak{T}|}}{\sqrt{|\mathfrak{T}|}}$$

Consequently

(2.39) $$\Sigma=-\frac{\kappa}{6}\Box^{-1}\mathfrak{T}+\frac{\alpha}{2}\frac{\Box\sqrt{|\mathfrak{T}|}}{\sqrt{|\mathfrak{T}|}}+\cdots$$

The first term is pure gravity, the second pure quantum, and the remaining terms involve gravity-quantum interactions. Other impressive examples are given (cf. also [**1156**]).

One goes now to a generalized equivalence principle. The gravitational effects determine the causal structure of spacetime as long as quantum effects give its conformal structure. This does not mean that quantum effects have nothing to do with the causal structure; they can act on the causal structure through back reaction terms appearing in the metric field equations. The conformal factor of the metric is a function of the quantum potential and the mass of a relativistic particle is a field produced by quantum corrections to the classical mass. One has shown that the presence of the quantum potential is equivalent to a conformal mapping of the metric. Thus in different conformally related frames one feels different quantum masses and different curvatures. In particular there are two frames with one containing the quantum mass field and the classical metric while the other contains the classical mass and the quantum metric. In general frames both the spacetime metric and the mass field have quantum properties so one can state that different conformal frames are identical pictures of the gravitational and quantum phenomena. We feel different quantum forces in different conformal frames. The question then arises of whether the geometrization of quantum effects implies conformal invariance just as gravitational effects imply general coordinate invariance. One sees here that Weyl geometry provides additional degrees of freedom which can be identified with quantum effects and seems to create a unified geometric framework for understanding both gravitational and quantum forces. Some features here are: (i) Quantum effects appear independent of any preferred length scale. (ii) The quantum mass of a particle is a field. (iii) The gravitational constant is also a field depending on the matter distribution via the quantum potential (cf. [**1154, 1161**]). (iv) A local variation of matter field distribution changes the quantum potential acting on the geometry and alters it globally; the nonlocal character is forced by the quantum potential (cf. [**1155**]).

2.1. DIRAC-WEYL ACTION. Next (still following F. and A. Shojai [**1160**]) one goes to Weyl geometry based on the Weyl-Dirac action

$$\mathfrak{A} = \int d^4x\sqrt{-g}(F_{\mu\nu}F^{\mu\nu} - \beta^2\ {}^W\mathcal{R} + (\sigma+6)\beta_{;\mu}\beta^{;\mu} + \mathfrak{L}_{matter}) \tag{2.40}$$

Here $F_{\mu\nu}$ is the curl of the Weyl 4-vector ϕ_μ, σ is an arbitrary constant and β is a scalar field of weight -1. The symbol ";" represents a covariant derivative under general coordinate and conformal transformations (Weyl covariant derivative) defined as $X_{;\mu} = {}^W\nabla_\mu X - \mathcal{N}\phi_\mu X$ where $\mathcal{N}$ is the Weyl weight of X. The equations of motion are then

$$\mathfrak{G}^{\mu\nu} = -\frac{8\pi}{\beta^2}(\mathfrak{T}^{\mu\nu} + M^{\mu\nu}) + \frac{2}{\beta}(g^{\mu\nu\,W}\nabla^{\alpha\,W}\nabla_\alpha\beta - {}^W\nabla^{\mu\,W}\nabla^\nu\beta) + \tag{2.41}$$

$$+\frac{1}{\beta^2}(4\nabla^\mu\beta\nabla^\nu\beta - g^{\mu\nu}\nabla^\alpha\beta\nabla_\alpha\beta) + \frac{\sigma}{\beta^2}(\beta^{;\mu}\beta^{;\nu} - \frac{1}{2}g^{\mu\nu}\beta^{;\alpha}\beta_{;\alpha});$$

$${}^W\nabla_\mu F^{\mu\nu} = \frac{1}{2}\sigma(\beta^2\phi^\mu + \beta\nabla^\mu\beta) + 4\pi J^\mu;$$

$$\mathcal{R} = -(\sigma+6)\frac{{}^W\Box\beta}{\beta} + \sigma\phi_\alpha\phi^\alpha - \sigma\,{}^W\nabla^\alpha\phi_\alpha + \frac{\psi}{2\beta}$$

where

$$M^{\mu\nu} = (1/4\pi)[(1/4)g^{\mu\nu}F^{\alpha\beta}F_{\alpha\beta} - F^\mu_\alpha F^{\nu\alpha}] \tag{2.42}$$

and

$$8\pi\mathfrak{T}^{\mu\nu} = \frac{1}{\sqrt{-g}}\frac{\delta\sqrt{-g}\mathfrak{L}_{matter}}{\delta g_{\mu\nu}};\ 16\pi J^{\mu} = \frac{\delta\mathfrak{L}_{matter}}{\delta\phi_{\mu}};\ \psi = \frac{\delta\mathfrak{L}_{matter}}{\delta\beta} \tag{2.43}$$

For the equations of motion of matter and the trace of the EM tensor one uses invariance of the action under coordinate and gauge transformations, leading to

$$^W\nabla_\nu\mathfrak{T}^{\mu\nu} - \mathfrak{T}\frac{\nabla^\mu\beta}{\beta} = J_\alpha\phi^{\alpha\mu} - \left(\phi^\mu + \frac{\nabla^\mu\beta}{\beta}\right){}^W\nabla_\alpha J^\alpha; \tag{2.44}$$

$$16\pi\mathfrak{T} - 16\pi{}^W\nabla_\mu J^\mu - \beta\psi = 0$$

The first relation is a geometrical identity (Bianchi identity) and the second shows the mutual dependence of the field equations. Note that in the Weyl-Dirac theory the Weyl vector does not couple to spinors so ϕ_μ cannot be interpreted as the EM potential; the Weyl vector is used as part of the spacetime geometry and the auxillary field (gauge field) β represents the quantum mass field. The gravity fields $g_{\mu\nu}$ and ϕ_μ and the quantum mass field determine the spacetime geometry. Now one constructs a Bohmian quantum gravity which is conformally invariant in the framework of Weyl geometry. If the model has mass this must be a field (since mass has non-zero Weyl weight). The Weyl-Dirac action is a general Weyl invariant action as above and for simplicity now assume the matter Lagrangian does not depend on the Weyl vector so that $J_\mu = 0$. The equations of motion are then

$$\mathfrak{G}^{\mu\nu} = -\frac{8\pi}{\beta^2}(\mathfrak{T}^{\mu\nu} + M^{\mu\nu}) + \frac{2}{\beta}(g^{\mu\nu}{}^W\nabla^\alpha{}^W\nabla_\alpha\beta - {}^W\nabla^\mu{}^W\nabla^\nu\beta) + \tag{2.45}$$

$$+\frac{1}{\beta^2}(4\nabla^\mu\beta\nabla^\nu\beta - g^{\mu\nu}\nabla^\alpha\beta\nabla_\alpha\beta) + \frac{\sigma}{\beta^2}\left(\beta^{;\mu}\beta^{;\nu} - \frac{1}{2}g^{\mu\nu}\beta^{;\alpha}\beta_{;\alpha}\right);$$

$$^W\nabla_\nu F^{\mu\nu} = \frac{1}{2}\sigma(\beta^2\phi^\mu + \beta\nabla^\mu\beta);\ \mathcal{R} = -(\sigma+6)\frac{^W\Box\beta}{\beta} + \sigma\phi_\alpha\phi^\alpha - \sigma{}^W\nabla^\alpha\phi_\alpha + \frac{\psi}{2\beta}$$

The symmetry conditions are

$$^W\nabla_\nu\mathfrak{T}^{\mu\nu} - \mathfrak{T}(\nabla^\mu\beta/\beta) = 0;\ 16\pi\mathfrak{T} - \beta\psi = 0 \tag{2.46}$$

(recall $\mathfrak{T} = \mathfrak{T}^{\mu\nu}_{\mu\nu}$). One notes that from (2.45) results $^W\nabla_\mu(\beta^2\phi^\mu + \beta\nabla^\mu\beta) = 0$ so ϕ_μ is not independent of β. To see how this is related to the Bohmian quantum theory one introduces a quantum mass field and shows it is proportional to the Dirac field. Thus using (2.45) and (2.46) one has

$$\Box\beta + \frac{1}{6}\beta\mathcal{R} = \frac{4\pi}{3}\frac{\mathfrak{T}}{\beta} + \sigma\beta\phi_\alpha\phi^\alpha + 2(\sigma-6)\phi^\gamma\nabla_\gamma\beta + \frac{\sigma}{\beta}\nabla^\mu\beta\nabla_\mu\beta \tag{2.47}$$

This can be solved iteratively via

$$\beta^2 = (8\pi\mathfrak{T}/\mathcal{R}) - \{1/[(\mathcal{R}/6) - \sigma\phi_\alpha\phi^\alpha]\}\beta\Box\beta + \cdots \tag{2.48}$$

Now assuming $\mathfrak{T}^{\mu\nu} = \rho u^\mu u^\nu$ (dust with $\mathfrak{T} = \rho$) we multiply (2.46) by u_μ and sum to get

$$^W\nabla_\nu(\rho u^\nu) - \rho(u_\mu\nabla^\mu\beta/\beta) = 0 \tag{2.49}$$

Then put (2.46) into (2.49) which yields

$$u^\nu{}^W\nabla_\nu u^\mu = (1/\beta)(g^{\mu\nu} - u^\mu u^\nu)\nabla_\nu\beta \tag{2.50}$$

To see this write (assuming $g^{\mu\nu}\nabla_\nu\beta = \nabla^\mu\beta$)

$$ {}^W\nabla_\nu(\rho u^\mu u^\nu) = u^{\mu W}\nabla_\nu \rho u^\mu + \rho u^{\nu W}\nabla_\nu u^\mu \Rightarrow \tag{2.51}$$

$$\Rightarrow u^\mu\left(\frac{u_\mu \nabla^\mu\beta}{\beta}\right) + u^{\nu W}\nabla_\nu u^\mu - \frac{\nabla^\mu\beta}{\beta} = 0 \Rightarrow u^{\nu W}\nabla_\nu u^\mu = (1 - u^\mu u_\mu)\frac{\nabla^\mu\beta}{\beta} =$$

$$(g^{\mu\nu} - u^\mu u_\mu g^{\mu\nu})\frac{\nabla_\nu\beta}{\beta} = (g^{\mu\nu} - u^\mu u^\nu)\frac{\nabla_\nu\beta}{\beta}$$

which is (2.49). Then from (2.48)

$$\beta^{2(1)} = \frac{8\pi\mathfrak{T}}{\mathcal{R}};\ \beta^{2(2)} = \frac{8\pi\mathfrak{T}}{\mathcal{R}}\left(1 - \frac{1}{(\mathcal{R}/6) - \sigma\phi_\alpha\phi^\alpha}\frac{\Box\sqrt{\mathfrak{T}}}{\sqrt{\mathfrak{T}}}\right);\cdots \tag{2.52}$$

Comparing with (2.14) and (2.3) shows that we have the correct equations for the Bohmian theory provided one identifies

$$\beta \sim \mathfrak{M};\ \frac{8\pi\mathfrak{T}}{\mathcal{R}} \sim m^2;\ \frac{1}{\sigma\phi_\alpha\phi^\alpha - (\mathcal{R}/6)} \sim \alpha = \frac{\hbar^2}{m^2c^2} \tag{2.53}$$

Thus β is the Bohmian quantum mass field and the coupling constant α (which depends on $\hbar$) is also a field, related to geometrical properties of spacetime. One notes that the quantum effects and the length scale of the spacetime are related. To see this suppose one is in a gauge in which the Dirac field is constant; apply a gauge transformation to change this to a general spacetime dependent function, i.e.

$$\beta = \beta_0 \to \beta(x) = \beta_0 exp(-\Xi(x));\ \phi_\mu \to \phi_\mu + \partial_\mu\Xi \tag{2.54}$$

Thus the gauge in which the quantum mass is constant (and the quantum force is zero) and the gauge in which the quantum mass is spacetime dependent are related to one another via a scale change. In particular ϕ_μ in the two gauges differ by $-\nabla_\mu(\beta/\beta_0)$ and since ϕ_μ is a part of Weyl geometry and the Dirac field represents the quantum mass one concludes that the quantum effects are geometrized (cf. also (2.45) which shows that ϕ_μ is not independent of β so the Weyl vector is determined by the quantum mass and thus the geometrical aspects of the manifold are related to quantum effects).

2.2. REMARKS ON CONFORMAL GRAVITY. We mention here a series of papers by Arias, Bonal, Cardenas, Gonzalez, Leyva, Martin, and Quiros (cf. [**59, 60, 61, 164, 227, 228, 1052, 1053, 1054, 1055**]) which were discussed at some length in [**254**]. The material concerns in particular Brans-Dicke theory, conformal gravity, and deBroglie-Bohm-Weyl (dBBW) theory (cf. also [**26, 45, 1059, 1277, 1279**]). Questions about the physical significance of Riemannian geometry in relativity have been raised in the past (cf. [**180, 376**]) due to the arbitrariness in the metric tensor resulting from the indefiniteness in the choice of units of measure. In fact Brans-Dicke (BD) theory with a changing dimensionless gravitational coupling constant $Gm^2 \sim \phi^{-1}$ (with m the intertial mass of some elementary particle and ϕ the BD field - $\hbar = c = 1$ here) can be formulated in two different ways since either m or G could vary with position in spacetime. The choice $G \sim \phi^{-1}$ with $m = const.$ leads to the Jordan frame (JF) formalism based on the Lagrangian

$$L^{BD}[g,\phi] = \frac{\sqrt{-g}}{16\pi}\left(\phi R - \frac{\omega}{\phi}g^{nm}\nabla_n\phi\nabla_m\phi\right) + L_M[g] \tag{2.55}$$

where R is the curvature scalar, ω is the BD coupling constant, and $L_M[g]$ is the Lagrangian density for ordinary matter minimally coupled to the scalar field. On the other hand the choice $m \sim \phi^{-1/2}$ with G constant leads to the Einstein frame (EF) BD theory based on the Lagrangian

$$\hat{L}^{BD} = \frac{\sqrt{-g}}{16\pi}\left(\hat{R} - \left(\omega + \frac{3}{2}\right)\hat{g}^{nm}\hat{\nabla}_n\hat{\phi}\hat{\nabla}_m\hat{\phi}\right) + \hat{L}_M[\hat{g},\hat{\phi}] \tag{2.56}$$

where now in the EF metric $\hat{g}$ the ordinary matter is nonminimally coupled to the scalar field $\hat{\phi} \equiv log(\phi)$ through the Lagrangian density $\hat{L}_M[\hat{g},\hat{\phi}]$. Both JF and EF formulations of BD gravity are equivalent representations of the same physical situation since they both belong to the same conformal class (cf. [**180**]); in particular $L^{BD}_{EF} \equiv L^{BD}_{JF}$ via a rescaling of spacetime metric $g \to \hat{g} = \phi g$ or $\hat{g}_{ab} = \phi g_{ab}$ where ϕ is smooth and nonvanishing. This rescaling can be interpreted as a particular transformation of the physical units and any dimensionless number (e.g. Gm^2) is invariant; experimental observations are unchanged since spacetime coincidences are not affected. Hence both based formulations (one based on varying G and the other on varying m are indistinguishable) and one has physically equivalent representations of a same physical situation. The same line of reasoning can be applied if minimal and nonminimal coupling to matter are interchanged via

$$(\mathbf{A})\ L^{GR}[g,\phi] = \frac{\sqrt{-g}}{16\pi}\left(\phi R - \frac{\omega}{\phi}g^{nm}\nabla_n\phi\nabla_m\phi\right) + L_M[g,\phi]; \tag{2.57}$$

$$(\mathbf{B})\ \hat{L}^{GR} = \frac{\sqrt{-g}}{16\pi}\left(\hat{R} - \left(\omega + \frac{3}{2}\right)\hat{g}^{nm}\hat{\nabla}_n\hat{\phi}\hat{\nabla}_m\hat{\phi}\right) + \hat{L}_M[\hat{g}]$$

Both Lagrangians represent equivalent pictures of GR and **(B)** is simply GR with a scalar field as an additional source of gravity (EFGR) and its conformally equivalent Lagrangian **(A)** refers to Jordan frame GR (JFGR). The field equations derivable from Lagrangian **(B)** are

$$\hat{G}_{ab} = 8\pi\hat{T}_{ab} + \left(\omega + \frac{3}{2}\right)\left(\hat{\nabla}_a\hat{\phi}\hat{\nabla}_b\hat{\phi} - \frac{1}{2}\hat{g}_{ab}\hat{g}^{nm}\hat{\nabla}_n\hat{\phi}\hat{\nabla}_m\hat{\phi}\right); \tag{2.58}$$

$$\Box\hat{\phi} = 0;\ \hat{\nabla}_n\hat{T}^{na} = 0;\ \Box = \hat{g}^{nm}\hat{\nabla}_n\hat{\nabla}_m$$

where $\hat{G}_{ab} = \hat{R}_{ab} - (1/2)\hat{g}_{ab}\hat{R}$ and $\hat{T}_{ab} = (2/\sqrt{\hat{g}})(\partial/\partial\hat{g}^{ab})(\sqrt{-\hat{g}}\hat{L}_M)$. Some disadvantages for JFGR historically involve first that the BD scalar field is nonminimally coupled both to scalar curvature and to ordinary matter so the gravitational constant G varies as $G \sim \phi^{-1}$. At the same time the material test particles don't follow the geodesics of the geometry since they are acted on by both the metric field and the scalar field. In particular masses vary from point to point in spacetime so as to preserve a constant Gm^2 (so $m \sim \phi^{1/2}$). The most serious (but illusory) objection is linked with the formulation of the theory in unphysical variables so that the kinetic energy of the scalar field is not positive definite (cf. Faraoni [**451**]). However one shows in [**1055**] that the indefiniteness in the sign of the energy density in the Jordan frame is only apparent; in fact once the scalar field energy density is positive definite in the Einstein frame it is also in the Jordan frame.

One can show that GR with an extra scalar field and its conformal formulation (JFGR) are different but physically equivalent representations of the same theory. The claim is based on the argument that spacetime coincidences (coordinates) are not affected by a conformal rescaling of the spacetime metric (★) $\hat{g}_{ab} = \Omega^2 g_{ab}$ where Ω^2 is a smooth nonvanishing function on the manifold. Thus the experimental observations (measuements) being nothing but verifications of these coincidences are unchanged too by (★). This means that canonical GR and its conformal image are experimentally indistinguishable.

Another question regarding metric theories of spacetime is also clarified, namely the physical content of a given theory of spacetime should be contained in the invariants of the group of position dependent transformations of units and coordinate transformations (cf. [**180**]). All known metric theories of spacetime, including GR, BD, and scalar-tensor theories in general fufill the requirement of invariance under the group of coordinate transformations. It is also evident that any consistent formulation of a given effective theory of spacetime must be invariant also under the group of transformations of units of length, time, and mass. This aspect is treated below and one shows that the only consistent formulation of gravity (among those studied here) is the conformal representation of GR. One shows that in particular canonical GR and the Einstein frame formulation of BD theory are not invariant under transformations of length, time, and reciprocal mass. This leads to an argument for a Weyl geometry which we have developed (and will further develop) in other ways. In any event the moral here is that Weyl geometry implicitly contains the quantum effects of matter - it is already a quantum geometry! In particular a free falling test particle would not "feel" any special quantum force since the effect is built into the free fall.

3. THE SCHRÖDINGER EQUATION IN WEYL SPACE

We go now to Santamato [**1115**] and derive the SE from classical mechanics in Weyl space (i.e. from Weyl geometry - cf. also [**81, 82, 236, 237, 270, 261, 276, 659, 1116, 1299**]). The idea is to relate the quantum force (arising from the quantum potential) to geometrical properties of spacetime; the Klein-Gordon (KG) equation is also treated in this spirit in [**270, 1116**] and we discuss this later. In fact the general theme is picked up again in Chapter 8 and expanded in various ways. One wants to show how geometry acts as a guidance field for matter (as in general relativity). Initial positions are assumed random (as in the Madelung approach) and thus the theory is statistical and is really describing the motion of an ensemble. Thus assume that the particle motion is given by some random process $q^i(t,\omega)$ in a manifold M (where ω is the sample space tag) whose probability density $\rho(q,t)$ exists and is properly normalizable. Assume that the process $q^i(t,\omega)$ is the solution of differential equations

$$\dot{q}^i(t,\omega) = (dq^i/dt)(t,\omega) = v^i(q(t,\omega),t) \tag{3.1}$$

with random initial conditions $q^i(t_0,\omega) = q_0^i(\omega)$. Once the joint distribution of the random variables $q_0^i(\omega)$ is given the process $q^i(t,\omega)$ is uniquely determined by (3.1). One knows that in this situation $\partial_t\rho + \partial_i(\rho v^i) = 0$ (continuity equation) with initial Cauchy data $\rho(q,t) = \rho_0(q)$. The natural origin of v^i arises via a least action principle based on a Lagrangian $L(q,\dot{q},t)$ with

$$L^*(q,\dot{q},t) = L(q,\dot{q},t) - \Phi(q,\dot{q},t);\ \Phi = \frac{dS}{dt} = \partial_t S + \dot{q}^i\partial_i S \tag{3.2}$$

Then $v^i(q,t)$ arises by minimizing

$$I(t_0,t_1) = E[\int_{t_0}^{t_1} L^*(q(t,\omega),\dot{q}(t,\omega),t)dt] \tag{3.3}$$

where t_0, t_1 are arbitrary and E denotes the expectation (cf. [**234, 235, 908, 910, 926**] for stochastic ideas). The minimum is to be achieved over the class of all random motions $q^i(t,\omega)$ obeying (3.2) with arbitrarily varied velocity field $v^i(q,t)$ but having common initial values. One proves first ($H \sim p_i\dot{q}^i - L$)

$$\partial_t S + H(q,\nabla S,t) = 0;\ v^i(q,t) = \frac{\partial H}{\partial p_i}(q,\nabla S(q,t),t) \tag{3.4}$$

Thus the value of I in (3.3) along the random curve $q^i(t,q_0(\omega))$ is

$$I(t_1,t_0,\omega) = \int_{t_0}^{t_1} L^*(q(,q_0(\omega)),\dot{q}(t,q_0(\omega)),t)dt \tag{3.5}$$

Let $\mu(q_0)$ denote the joint probability density of the random variables $q_0^i(\omega)$ and then the expectation value of the random integral is

$$I(t_1,t_0) = E[I(t_1,t_0,\omega)] = \int_{\mathbf{R}^n}\int_{t_0}^{t_1} \mu(q_0)L^*(q(t,q_0),\dot{q}(t,q_0),t)d^n q_0 dt \tag{3.6}$$

Standard variational methods give then

$$\delta I = \int_{\mathbf{R}^n} d^n q_0\mu(0)\left[\frac{\partial L^*}{\partial \dot{q}^i}(q(t_1,q_0),\partial_t q(t_1,q_0),t)\delta q^i(t_1,q_0) - \right. \tag{3.7}$$

$$\left. - \int_{t_0}^{t_1} dt\left(\frac{\partial}{\partial t}\frac{\partial L^*}{\partial \dot{q}^i}(q(t,q_0),\partial_t q)t,q_0),t) - \frac{\partial L^*}{\partial q^i}(q(t,q_0),\partial_t q(t,q_0),t)\right)\delta q^i(t,q_0)\right]$$

where one uses the fact that $\mu(q_0)$ is independent of time and $\delta q^i(t_0,q_0) = 0$ (recall common initial data is assumed). Therefore

$$(\mathbf{A})\ (\partial L^*/\partial \dot{q}^i)(q(t,q_0),\partial_t q(t,q_0),t) = 0; \tag{3.8}$$

$$(\mathbf{B})\ \frac{\partial}{\partial t}\frac{\partial L^*}{\partial \dot{q}^i}(q(t,q_0),\partial_t q(t,q_0,t) - \frac{\partial L^*}{\partial q^i}(q(t,q_0),\partial_t q(t,q_0),t) = 0$$

are the necessary conditions for obtaining a minimum of I. Conditions **(B)** are the usual Euler-Lagrange (EL) equations whereas **(A)** is a consequence of the fact that in the most general case one must retain varied motions with $\delta q^i(t_1,q_0)$ different from zero at the final time t_1. Note that since L^* differs from L by a total time derivative one can safely replace L^* by L in **(B)** and putting (3.2) into **(A)** one obtains the classical equations

$$p_i = (\partial L/\partial \dot{q}^i)(q(t,q_0),\dot{q}(t,q_0),t) = \partial_i S(q(t,q_0),t) \tag{3.9}$$

It is known now that if $det[(\partial^2 L/\partial\dot{q}^i\partial\dot{q}^j] \neq 0$ then the second equation in (3.4) is a consequence of the gradient condition (3.9) and of the definition of the Hamiltonian function $H(q,p,t) = p_i\dot{q}^i - L$. Moreover **(B)** in (3.8) and (3.9) entrain the HJ equation in (3.4). In order to show that the average action integral (3.6) actually gives a minimum one needs $\delta^2 I > 0$ but this is not necessary for Lagrangians whose Hamiltonian H has the form

$$H_C(q,p,t) = \frac{1}{2m}g^{ik}(p_i - A_i)(p_k - A_k) + V \tag{3.10}$$

with arbitrary fields A_i and V (particle of mass m in an EM field A) which is the form for nonrelativistic applications; given positive definite g_{ik} such Hamiltonians involve sufficiency conditions $det[\partial^2 L/\partial\dot{q}^i\partial\dot{q}^k] = mg > 0$. Finally **(B)** in (3.8) with L^* replaced by L) shows that along particle trajectories the EL equations are satisfied, i.e. the particle undergoes a classical motion with probability one. Notice here that in (3.4) no explicit mention of generalized momenta is made; one is dealing with a random motion entirely based on position. Moreover the minimum principle (3.3) defines a 1-1 correspondence between solutions $S(q,t)$ in (3.4) and minimizing random motions $q^i(t,\omega)$. Provided v^i is given via (3.4) the particle undergoes a classical motion with probability one. Thus once the Lagrangian L or equivalently the Hamiltonian H is given, $\partial_t\rho + \partial_i(\rho v^i) = 0$ and (3.4) uniquely determine the stochastic process $q^i(t,\omega)$. Now suppose that some geometric structure is given on M so that the notion of scalar curvature $R(q,t)$ of M is meaningful. Then we assume (ad hoc) that the actual Lagrangian is

$$L(q,\dot{q},t) = L_C(q,\dot{q},t) + \gamma(\hbar^2/m)R(q,t) \tag{3.11}$$

where $\gamma = (1/6)(n-2)/(n-1)$ with $n = dim(M)$. Since both L_C and R are independent of $\hbar$ we have $L \to L_C$ as $\hbar \to 0$.

Now for a differential manifold with $ds^2 = g_{ik}(q)dq^i dq^k$ it is standard that in a transplantation $q^i \to q^i + \delta q^i$ one has $\delta A^i = \Gamma^i_{k\ell}A^\ell dq^k$ with $\Gamma^i_{k\ell}$ general affine connection coefficients on M (Riemannian structure is not assumed). In [**1115**] it is assumed that for $\ell = (g_{ik}A^iA^k)^{1/2}$ one has $\delta\ell = \ell\phi_k dq^k$ where the ϕ_k are covariant components of an arbitrary vector (Weyl geometry). Then the actual affine connections $\Gamma^i_{k\ell}$ can be found by comparing this with $\delta\ell^2 = \delta(g_{ik}A^iA^k)$ and using $\delta A^i = \Gamma^i_{k\ell}A^\ell dq^k$. A little linear algebra gives then

$$\Gamma^i_{k\ell} = -\left\{ \begin{matrix} i \\ k\ \ell \end{matrix} \right\} + g^{im}(g_{mk}\phi_\ell + g_{m\ell}\phi_k - g_{k\ell}\phi_m) \tag{3.12}$$

Thus we may prescribe the metric tensor g_{ik} and ϕ_i and determine via (3.12) the connection coefficients. Note that $\Gamma^i_{k\ell} = \Gamma^i_{\ell k}$ and for $\phi_i = 0$ one has Riemannian geometry. Covariant derivatives are defined via

$$A^k_{,i} = \partial_i A^k - \Gamma^k_{i\ell}A^\ell;\ A_{k,i} = \partial_i A_k + \Gamma^\ell_{ki}A_\ell \tag{3.13}$$

for covariant and contravariant vectors respectively (where $S_{,i} = \partial_i S$). Note Ricci's lemma no longer holds (i.e. $g_{ik,\ell} \neq 0$) so covariant differentiation and operations of raising or lowering indices do not commute. The curvature tensor $R^i_{k\ell m}$ in Weyl geometry is introduced via $A^i_{,k,\ell} - A^i_{,\ell,k} = F^i_{mk\ell}A^m$ from which arises the standard formula of Riemannian geometry

(3.14) $$R^i_{mk\ell} = -\partial_\ell \Gamma^i_{mk} + \partial_k \Gamma^i_{m\ell} + \Gamma^i_{n\ell}\Gamma^n_{mk} - \Gamma^i_{nk}\Gamma^n_{m\ell}$$

where (3.12) is used in place of the Christoffel symbols. The tensor $R^i_{mk\ell}$ obeys the same symmetry relations as the curvature tensor of Riemann geometry as well as the Bianchi identity. The Ricci symmetric tensor R_{ik} and the scalar curvature R are defined by the same formulas also, viz. $R_{ik} = R^\ell_{i\ell k}$ and $R = g^{ik}R_{ik}$. For completeness one derives here

(3.15) $$R = \dot{R} + (n-1)[(n-2)\phi_i\phi^i - 2(1/\sqrt{g})\partial_i(\sqrt{g}\phi^i)]$$

where $\dot{R}$ is the Riemannian curvature built by the Christoffel symbols. Thus from (3.12) one obtains

(3.16) $$g^{k\ell}\Gamma^i_{k\ell} = -g^{k\ell}\left\{ \begin{array}{c} i \\ k\ \ell \end{array} \right\} - (n-2)\phi^i;\ \Gamma^i_{k\ell} = -\left\{ \begin{array}{c} i \\ k\ \ell \end{array} \right\} + n\phi_k$$

Since the form of a scalar is independent of the coordinate system used one may compute R in a geodesic system where the Christoffel symbols and all $\partial_\ell g_{ik}$ vanish; then (3.12) reduces to $\Gamma^i_{k\ell} = \phi_k \kappa^i_\ell + \phi_\ell \delta^i_k - g_{k\ell}\phi^i$ and hence

(3.17) $$R = -g^{km}\partial_m \Gamma^i_{k\ell} + \partial_i(g^{k\ell}\Gamma^i_{k\ell}) + g^{\ell m}\Gamma^i_{n\ell}\Gamma^n_{mi} - g^{m\ell}\Gamma^i_{n\ell}\Gamma^n_{m\ell}$$

Further one has $g^{\ell m}\Gamma^i_{n\ell}\Gamma^n_{mi} = -(n-2)(\phi_k\phi^k)$ at the point in consideration. Putting all this in (3.17) one arrives at

(3.18) $$R = \dot{R} + (n-1)(n-2)(\phi_k\phi^k) - 2(n-1)\partial_k\phi^k$$

which becomes (3.15) in covariant form. Now the geometry is to be derived from physical principles so the ϕ_i cannot be arbitrary but must be obtained by the same averaged least action principle (3.3) giving the motion of the particle. The minimum in (3.3) is to be evaluated now with respect to the class of all Weyl geometries having arbitrarily varied gauge vectors but fixed metric tensor. Note that once (3.11) is inserted in (3.2) the only term in (3.3) containing the gauge vector is the curvature term. Then observing that $\gamma > 0$ when $n \geq 3$ the minimum principle (3.3) may be reduced to the simpler form $E[R(q(t,\omega),t)] = min$ where only the gauge vectors ϕ_i are varied. Using (3.15) this is easily done. First a little argument shows that $\hat{\rho}(q,t) = \rho(q,t)/\sqrt{g}$ transforms as a scalar in a coordinate change and this will be called the scalar probability density of the random motion of the particle (statistical determination of geometry). Starting from $\partial_t\rho + \partial_i(\rho v^i) = 0$ a manifestly covariant equation for $\hat{\rho}$ is found to be $\partial_t\hat{\rho} + (1/\sqrt{g})\partial_i(\sqrt{g}v^i\hat{\rho}) = 0$. Now return to the minimum problem $E[R(q(t,\omega),t)] = min$; from (3.15) and $\hat{\rho} = \rho/\sqrt{g}$ one obtains

(3.19) $$E[R(q(t,\omega),t)] = E[\dot{R}(q(t,\omega),t)] +$$
$$+(n-1)\int_M [(n-2)\phi_i\phi^i - 2(1/\sqrt{g})\partial_i(\sqrt{g}\phi^i)]\hat{\rho}(q,t)\sqrt{g}d^nq$$

Assuming fields go to 0 rapidly enough on ∂M and integrating by parts one gets then

(3.20) $$E[R] = E[\dot{R}] - \frac{n-1}{n-2}E[g^{ik}\partial_i(log(\hat{\rho})\partial_k(log(\hat{\rho})] +$$

$$+\frac{n-1}{n-2}E\{g^{ik}[(n-2)\phi_i+\partial_i(log(\hat{\rho})][(n-2)\phi_k+\partial_k(log(\hat{\rho})]\}$$

Since the first two terms on the right are independent of the gauge vector and g^{ik} is positive definite $E[R]$ will be a minimum when

$$\phi_i(q,t)=-[1/(n-2)]\partial_i[log(\hat{\rho})(q,t)] \tag{3.21}$$

This shows that the geometric properties of space are indeed affected by the presence of the particle and in turn the alteration of geometry acts on the particle through the quantum force $f_i=\gamma(\hbar^2/m)\partial_i R$ which according to (3.15) depends on the gauge vector and its derivatives. It is this peculiar feedback between the geometry of space and the motion of the particle which produces quantum effects.

In this spirit one goes now to a geometrical derivation of the SE. Thus inserting (3.21) into (3.16) one gets

$$R=\dot{R}+(1/2\gamma\sqrt{\hat{\rho}})[1/\sqrt{g})\partial_i(\sqrt{g}g^{ik}\partial_k\sqrt{\hat{\rho}})] \tag{3.22}$$

where the value $(n-2)/6(n-1)$ for γ is used. On the other hand the HJ equation (3.2) can be written as

$$\partial_t S+H_C(q,\nabla S,t)-\gamma(\hbar^2/m)R=0 \tag{3.23}$$

where (3.11) has been used. When (3.22) is introduced into (3.23) the HJ equation and the continuity equation $\partial_t\hat{\rho}+(1/\sqrt{g})(\sqrt{g}v^i\hat{\rho})=0$, with velocity field given by (3.4), form a set of two nonlinear PDE which are coupled by the curvature of space. Therefore self consistent random motions of the particle (i.e. random motions compatible with (3.17)) are obtained by solving (3.23) and the continuity equation simultaneously. For every pair of solutions $S(q,t,\hat{\rho}(q,t))$ one gets a possible random motion for the particle whose invariant probability density is $\hat{\rho}$. The present approach is so different from traditional QM that a proof of equivalence is needed and this is only done for Hamiltonians of the form (3.10) (which is not very restrictive). The HJ equation corresponding to (3.10) is

$$\partial_t S+\frac{1}{2m}g^{ik}(\partial_i S-A_i)(\partial_k S-A_k)+V-\gamma\frac{\hbar^2}{m}R=0 \tag{3.24}$$

with R given by (3.22). Moreover using (3.4) as well as (3.10) the continuity equation becomes

$$\partial_t\hat{\rho}+(1/m\sqrt{g})\partial_i[\hat{\rho}\sqrt{g}g^{ik}(\partial_k S-A_k)]=0 \tag{3.25}$$

Owing to (3.22),(3.24) and (3.25) form a set of two nonlinear PDE which must be solved for the unknown functions S and $\hat{\rho}$. Now a straightforward calculations shows that, setting

$$\psi(q,t)=\sqrt{\hat{\rho}(q,t)}exp](i/\hbar)S(q,t)], \tag{3.26}$$

the quantity ψ obeys a linear PDE (corrected from [**1115**])

$$i\hbar\partial_t\psi=\frac{1}{2m}\left\{\left[\frac{i\hbar\partial_i\sqrt{g}}{\sqrt{g}}+A_i\right]g^{ik}(i\hbar\partial_k+A_k)\right\}\psi+\left[V-\gamma\frac{\hbar^2}{m}\dot{R}\right]\psi \tag{3.27}$$

where only the Riemannian curvature $\dot{R}$ is present (any explicit reference to the gauge vector ϕ_i having disappeared). (3.27) is of course the SE in curvilinear coordinates whose invariance under point transformations is well known. Moreover (3.26) shows that $|\psi|^2 = \hat{\rho}(q,t)$ is the invariant probability density of finding the particle in the volume element $d^n q$ at time t. Then following Nelson's arguments that the SE together with the density formula contains QM the present theory is physically equivalent to traditional nonrelativistic QM. One sees also from (3.26) and (3.27) that the time independent SE is obtained via $S = S_0(q) - Et$ with constant E and $\hat{\rho}(q)$. In this case the scalar curvature of space becomes time independent; since starting data at t_0 is meaningless one replaces the continuity equation with a condition $\int_M \hat{\rho}(q)\sqrt{g}d^n q = 1$.

REMARK 3.3.1. We recall (cf. [**236**]) that in the nonrelativistic context the quantum potential has the form $Q = -(\hbar^2/2m)(\partial^2\sqrt{\rho}/\sqrt{\rho})$ ($\rho \sim \hat{\rho}$ here) and in more dimensions this corresponds to $Q = -(\hbar^2/2m)(\Delta\sqrt{\rho}/\sqrt{\rho})$. Here we have a SE involving $\psi = \sqrt{\rho}exp[(i/\hbar)S]$ with corresponding HJ equation (3.24) which corresponds to the flat space 1-D $S_t + (s')^2/2m + V + Q = 0$ with continuity equation $\partial_t\rho + \partial(\rho S'/m) = 0$ (take $A_k = 0$ here). The continuity equation in (3.25) corresponds to $\partial_t\rho + (1/m\sqrt{g})\partial_i[\rho\sqrt{g}g^{ik}(\partial_k S)] = 0$. For $A_k = 0$ (3.24) becomes

$$\partial_t S + (1/2m)g^{ik}\partial_i S\partial_k S + V - \gamma(\hbar^2/m)R = 0 \quad (3.28)$$

This leads to an identification $Q \sim -\gamma(\hbar^2/m)R$ where R is the Ricci scalar in the Weyl geometry (related to the Riemannian curvature built on standard Christoffel symbols via (3.15)). Here $\gamma = (1/6)[(n-2)/(n-2)]$ as above which for $n = 3$ becomes $\gamma = 1/12$; further the Weyl field $\phi_i = -\partial_i log(\rho)$. Consequently (see below).

PROPOSITION 3.1. For the SE (3.27) in Weyl space the quantum potential is $Q = -(\hbar^2/12m)R$ where R is the Weyl-Ricci scalar curvature. For Riemannian flat space $\dot{R} = 0$ this becomes via (3.22)

$$R = \frac{1}{2\gamma\sqrt{\rho}}\partial_i g^{ik}\partial_k\sqrt{\rho} \sim \frac{1}{2\gamma}\frac{\Delta\sqrt{\rho}}{\sqrt{\rho}} \Rightarrow Q = -\frac{\hbar^2}{2m}\frac{\Delta\sqrt{\rho}}{\sqrt{\rho}} \quad (3.29)$$

as it should and the SE (3.27) reduces to the standard SE in the form $i\hbar\partial_t\psi = -(\hbar^2/2m)\Delta\psi + V\psi$ ($A_k = 0$). ■

3.1. FISHER INFORMATION REVISITED. Via Remarks 1.1.4, 1.1.5, and 1.1.6 from Chapter 1 (based on [**508, 509, 594, 596, 598, 599, 1062, 1063**]) we recall the derivation of the SE in Theorem 1.1.1. Thus with some repetition recall first that the classical Fisher information associated with translations of a 1-D observable X with probability density $P(x)$ is

$$F_X = \int dx\, P(x)([log(P(x)]')^2 > 0 \quad (3.30)$$

One has a well known Cramer-Rao inequality $Var(X) \geq F_X^{-1}$ where $Var(X) \sim$ variance of X. A Fisher length for X is defined via $\delta X = F_X^{-1/2}$ and this quantifies the length scale over which $p(x)$ (or better $log(p(x))$) varies appreciably. Then the root mean square deviation ΔX satisfies $\Delta X \geq \delta X$. Let now P be the momentum observable conjugate to X, and P_{cl} a classical momentum observable corresponding to the state ψ given via $p_{cl}(x) = (\hbar/2i)[(\psi'/\psi)-(\bar{\psi}'/\bar{\psi})]$. One has then the identity $<p>_\psi=<p_{cl}>_\psi$ following via integration by parts. Now define the nonclassical momentum by $p_{nc} = p - p_{cl}$ and one shows then

$$\Delta X \Delta p \geq \delta X \Delta p \geq \delta X \Delta p_{nc} = \hbar/2 \tag{3.31}$$

Then consider a classical ensemble of n-dimensional particles of mass m moving under a potential V. The motion can be described via the HJ and continuity equations

$$\frac{\partial s}{\partial t} + \frac{1}{2m}|\nabla s|^2 + V = 0; \ \frac{\partial P}{\partial t} + \nabla \cdot \left[P\frac{\nabla s}{m}\right] = 0 \tag{3.32}$$

for the momentum potential s and the position probability density P (note that there is no quantum potential and this will be supplied by the information term). These equations follow from the variational principle $\delta L = 0$ with Lagrangian $L = \int dt\, d^n x\, P\left[(\partial s/\partial t) + (1/2m)|\nabla s|^2 + V\right]$. It is now assumed that the classical Lagrangian must be modified due to the existence of random momentum fluctuations. The nature of such fluctuations is immaterial and one can assume that the momentum associated with position x is given by $p = \nabla s + N$ where the fluctuation term N vanishes on average at each point x. Thus s changes to being an average momentum potential. It follows that the average kinetic energy $< |\nabla s|^2 > /2m$ appearing in the Lagrangian above should be replaced by $< |\nabla s + N|^2 > /2m$ giving rise to

$$L' = L + (2m)^{-1} \int dt < N \cdot N >= L + (2m)^{-1} \int dt (\Delta N)^2 \tag{3.33}$$

where $\Delta N =< N \cdot N >^{1/2}$ is a measure of the strength of the quantum fluctuations. The additional term is specified uniquely, up to a multiplicative constant, by the three assumptions given in Remark 1.1.4 This leads to the result that

$$(\Delta N)^2 = c \int d^n x\, P|\nabla log(P)|^2 \tag{3.34}$$

where c is a positive universal constant (cf. Hall [**594**]). Further for $\hbar = 2\sqrt{c}$ and $\psi = \sqrt{P}exp(is/\hbar)$ the equations of motion for p and s arising from $\delta L' = 0$ are $i\hbar\frac{\partial\psi}{\partial t} = -\frac{\hbar^2}{2m}\nabla^2\psi + V\psi$.

A second derivation is given in Remark 1.1.5. Thus let $P(y^i)$ be a probability density and $P(y^i + \Delta y^i)$ be the density resulting from a small change in the y^i. Calculate the cross entropy via

$$J(P(y^i + \Delta y^i) : P(y^i)) = \int P(y^i + \Delta y^i) log\frac{P(y^i + \Delta y^i)}{P(y^i)} d^n y \simeq \tag{3.35}$$

$$\simeq \left[\frac{1}{2}\int \frac{1}{P(y^i)}\frac{\partial P(y^i)}{\partial y^i}\frac{\partial P(y^i)}{\partial y^k)} d^n y\right] \Delta y^i \Delta y^k = I_{jk}\Delta y^i \Delta y^k$$

The I_{jk} are the elements of the Fisher information matrix. The most general expression has the form

$$I_{jk}(\theta^i) = \frac{1}{2}\int \frac{1}{P(x^i|\theta^i)}\frac{\partial P(x^i|\theta^i)}{\partial\theta^j}\frac{\partial P(x^i|\theta^i)}{\partial\theta^k}d^nx \quad (3.36)$$

where $P(x^i|\theta^i)$ is a probability distribution depending on parameters θ^i in addition to the x^i. For $P(x^i|\theta^i) = P(x^i + \theta^i)$ one recovers (3.35). If P is defined over an n-dimensional manifold with positive inverse metric g^{ik} one obtains a natural definition of the information associated with P via

$$I = g^{ik}I_{ik} = \frac{g^{ik}}{2}\int \frac{1}{P}\frac{\partial P}{\partial y^i}\frac{\partial P}{\partial y^k}d^ny \quad (3.37)$$

Now in the HJ formulation of classical mechanics the equation of motion takes the form

$$\frac{\partial S}{\partial t} + \frac{1}{2}g^{\mu\nu}\frac{\partial S}{\partial x^\mu}\frac{\partial S}{\partial x^\nu} + V = 0 \quad (3.38)$$

where $g^{\mu\nu} = diag(1/m, \cdots, 1/m)$. The velocity field u^μ is then $u^\mu = g^{\mu\nu}(\partial S/\partial x^\nu)$. When the exact coordinates are unknown one can describe the system by means of a probability density $P(t, x^\mu)$ with $\int Pd^nx = 1$ and

$$(\partial P/\partial t) + (\partial/\partial x^\mu)(Pg^{\mu\nu}(\partial S/\partial x^\nu) = 0 \quad (3.39)$$

These equations completely describe the motion and can be derived from the Lagrangian

$$L_{CL} = \int P\{(\partial S/\partial t) + (1/2)g^{\mu\nu}(\partial S/\partial x^\mu)(\partial S/\partial x^\nu) + V\}\, dtd^nx \quad (3.40)$$

using fixed endpoint variation in S and P. Quantization is obtained by adding a term proportional to the information I defined in (3.37). This leads to
(3.41)

$$L_{QM} = L_{CL} + \lambda I = \int P\left\{\frac{\partial S}{\partial t} + \frac{1}{2}g^{\mu\nu}\left[\frac{\partial S}{\partial x^\mu}\frac{\partial S}{\partial x^\nu} + \frac{\lambda}{P^2}\frac{\partial P}{\partial x^\mu}\frac{\partial P}{\partial x^\nu}\right] + V\right\} dtd^nx$$

Fixed endpoint variation in S leads again to (3.39) while variation in P leads to

$$\frac{\partial S}{\partial t} + \frac{1}{2}g^{\mu\nu}\left[\frac{\partial S}{\partial x^\mu}\frac{\partial S}{\partial x^\nu} + \lambda\left(\frac{1}{P^2}\frac{\partial P}{\partial x^\mu}\frac{\partial P}{\partial x^\nu} - \frac{2}{P}\frac{\partial^2 P}{\partial x^\mu \partial x^\nu}\right)\right] + V = 0 \quad (3.42)$$

These equations are equivalent to the SE if $\psi = \sqrt{P}exp(iS/\hbar)$ with $\lambda = (2\hbar)^2$ (recall also Remark 1.1.6 for connections to entropy). Now following ideas in [**270, 275, 941**] we note in (3.41) for $\phi_\mu \sim A_\mu = \partial_\mu log(P)$ (which arises in (3.21)) and $p_\mu = \partial_u S$, a complex velocity can be envisioned leading to (cf. also [?])

$$|p_\mu + i\sqrt{\lambda}A_\mu|^2 = p_\mu^2 + \lambda A_\mu^2 \sim g^{\mu\nu}\left(\frac{\partial S}{\partial x^\mu}\frac{\partial S}{\partial x^\nu} + \frac{\lambda}{P^2}\frac{\partial P}{\partial x^\mu}\frac{\partial P}{\partial x^\nu}\right) \quad (3.43)$$

Further I in (3.37) is exactly known from ϕ_μ so one has a direct connection between Fisher information and the Weyl field ϕ_μ, along with motivation for a complex velocity (cf. Sections 1.2 and 1.3).

REMARK 3.3.2. Comparing now with [**237**] and quantum geometry in the form $ds^2 = \sum(dp_j^2/p_j)$ on a space of probability distributions (to be discussed in Chapter 5) we can define (3.37) as a Fisher information metric in the present context. This should be positive definite in view of its relation to $(\Delta N)^2$ in (3.34) for example. Now for $\psi = Rexp(iS/\hbar)$ one has ($\rho \sim \hat{\rho}$ here)

$$(3.44) \qquad -\frac{\hbar^2}{2m}\frac{R''}{R} \equiv -\frac{\hbar^2}{2m}\frac{\partial^2\sqrt{\rho}}{\sqrt{\rho}} = -\frac{\hbar^2}{8m}\left[\frac{2\rho''}{\rho} - \left(\frac{\rho'}{\rho}\right)^2\right]$$

in 1-D while in more dimensions we have a form ($\rho \sim P$)

$$(3.45) \qquad Q \sim -2\hbar^2 g^{\mu\nu}\left[\frac{1}{P^2}\frac{\partial P}{\partial x^\mu}\frac{\partial P}{\partial x^\nu} - \frac{2}{P}\frac{\partial^2 P}{\partial x^\mu \partial x^\nu}\right]$$

as in (3.44) (arising from the Fisher metric I of (3.37) upon variation in P in the Lagrangian). It can also be related to an osmotic velocity field $u = D\nabla log(\rho)$ via $Q = (1/2)u^2 + D\nabla \cdot u$ connected to Brownian motion where D is a diffusion coefficient (cf. [**275, 508, 941**]). For $\phi_\mu = -\partial_\mu log(P)$ we have then $\mathbf{u} = -D\phi$ with $Q = D^2((1/2)(|\mathbf{u}|^2 - \nabla\cdot\phi)$, expressing Q directly in terms of the Weyl vector. This enforces the idea that QM is built into Weyl geometry! ■

3.2. THE KG EQUATION. The formulation above from Santamato [**1115**] was modified in [**1116**] to a derivation of the Klein-Gordon (KG) equation via an average action principle. The spacetime geometry was then obtained from the average action principle to obtain Weyl connections with a gauge field ϕ_μ (thus the geometry has a statistical origin). The Riemann scalar curvature $\dot{R}$ is then related to the Weyl scalar curvature R via an equation

$$(3.46) \qquad R = \dot{R} - 3[(1/2)g^{\mu\nu}\phi_\mu\phi_\nu + (1/\sqrt{-g})\partial_\mu(\sqrt{-g}g^{\mu\nu}\phi_\nu)]$$

Explicit reference to the underlying Weyl structure disappears in the resulting SE (as in (3.27)). The HJ equation in [**1116**] has this form (for $A_\mu = 0$ and $V = 0$) $g^{\mu\nu}\partial_\mu S\partial_\nu S = m^2 - (R/6)$ so in some sense (recall here $\hbar = c = 1$) $m^2 - (R/6) \sim \mathfrak{M}^2$ where $\mathfrak{M}^2 = m^2 exp(Q)$ and $Q = (\hbar^2/m^2c^2)(\Box\sqrt{\rho}/\sqrt{\rho}) \sim (\Box\sqrt{\rho}/m^2\sqrt{\rho})$ via Section 3.2 (for signature $(-,+,+,+)$ - recall here $g^{\mu\nu}\nabla_\mu S\nabla_\nu S = \mathfrak{M}^2c^2$). Thus for $exp(Q) \sim 1 + Q$ one has $m^2 - (R/6) \sim m^2(1+Q) \Rightarrow (R/6) \sim -Qm^2 \sim -(\Box\sqrt{\rho}/\sqrt{\rho})$. This agrees also with [**270**] where the whole matter is analyzed incisively (cf. also Remark 3.3.5). We recall also here from [**1054**] (cf. also [**254**]) that in the conformal geometry the particles do not follow geodesics of the conformal metric alone. We will sketch an elaboration of this now from [**1116**] (paper one). Thus summarizing [**1115**] and the second paper in [**1116**] one shows that traditional QM is equivalent (in some sense) to classical statistical mechanics in Weyl spaces. The following two points of view are taken to be equivalent

(1) **(A)** The spacetime is a Riemannian manifold and the statistical behavior of a spinless particle is described by the KG equation while probabilities combine according to Feynman quantum rules.
(2) **(B)** The spacetime is a generic affinely connected manifold whose actual geometric structure is determined by the matter content. The statistical behavior of a spinless particle is described by classical statistical mechanics and probabilities combine according to Laplace rules.

(3) In nonrelativistic applications the words spacetime, Riemannian, and KG are to be replaced by space, Euclidean, and SE.

We are skipping over the second paper in [**1116**] here and going to the first paper which treates matters in a gauge invariant manner. The moral seems to be (loosly) that quantum mechanics in Riemannian spacetime is the same as classical statistical mechanics in a Weyl space. In particular one wants to establish that traditional QM, based on wave equations and ad hoc probability calculus (as in (1) above) is merely a convenient mathematical construction to overcome the complications arising from a nontrivial spacetime geometric structure. Here one works from first principles and includes gauge invariance (i.e. invariance with respect to an arbitrary choice of the spacetime calibration). The spacetime is supposed to be a generic 4-dimensional differential manifold with torsion free connections $\Gamma^{\lambda}_{\mu\nu} = \Gamma^{\lambda}_{\nu\mu}$ and a metric tensor $g_{\mu\nu}$ with signature $(+,-,-,-)$ (one takes $\hbar = c = 1$ - which I deem unfortunate since the role and effect of such quantities is not revealed). Here the (restrictive) hypothesis of assuming a Weyl geometry from the beginning is released, both the particle motion and the spacetime geometric structure are derived from a single average action principle. A result of this approach is that the spacetime connections are forced to be integrable Weyl connections by the extremization principle.

The particle is supposed to undergo a motion in spacetime with deterministic trajectories and random initial conditions taken on an arbitrary spacelike 3-dimensional hypersurface; thus the theory describes a relativistic Gibbs ensemble of particles (cf. Remark 3.3.3). Both the particle motion and the spacetime connections can be obtained from the average stationary action principle

$$\delta\left[E\left(\int_{\tau_1}^{\tau_2} L(x(\tau,\dot{x}(\tau))d\tau\right)\right] = 0 \tag{3.47}$$

This action integral must be parameter invariant, coordinate invariant, and gauge invariant. All of these requirements are met if L is positively homogeneous of the first degree in $\dot{x}^{\mu} = dx^{\mu}/d\tau$ and transforms as a scalar of Weyl type $w(L) = 0$. The underlying probability measure must also be gauge invariant. A suitable Lagrangian is then

$$L(x,dx) = (m^2 - (R/6))^{1/2}ds + A_{\mu}dx^{\mu} \tag{3.48}$$

where $ds = (g_{\mu\nu}\dot{x}^{\mu}\dot{x}^{\nu})^{1/2}d\tau$ is the arc length and R is the space time scalar curvature; m is a parameterlike scalar field of Weyl type (or weight) $w(m) = -(1/2)$. The factor 6 is essentially arbitrary and has been chosen for future convenience. The vector field A_{μ} can be interpreted as a 4-potential due to an externally applied EM field and the curvature dependent factor in front of ds is an effective particle mass. This seems a bit ad hoc but some feeling for the nature of the Lagrangian can be obtained from Section 3.2 (cf. also [**81**]). The Lagrangian will be gauge invariant provided the A_{μ} have Weyl type $w(A_{\mu}) = 0$. Now one can split A_{μ} into its gradient and divergence free parts $A_{\mu} = \bar{A}_{\mu} - \partial_{\mu}S$, with both S and $\bar{A}_{\mu}$ having Weyl type zero, and with $\bar{A}_{\mu}$ interpreted as and EM 4-potential in the Lorentz gauge. Due to the nature of the action principle regarding fixed endpoints in variation one notes that the average action principle is not invariant under EM gauge transformations $A_\mu \to A_\mu + \partial_\mu S$; but one knows that QM is also not invariant under EM gauge transformations (cf. [**19**]) so there is no incompatability with QM here.

Now the set of all spacetime trajectories accessible to the particle (the particle path space) may be obtained from (3.47) by performing the variation with respect to the particle trajectory with fixed metric tensor, connections, and an underlying probability measure. Thus (cf. Remark 3.3.3) the solution is given by the so-called Carathéodory complete figure (cf. Rund [**1095**]) associated with the Lagrangian

$$\bar{L}(x, dx) = (m^2 - (R/6))^{1/2} ds + \bar{A}_\mu dx^\mu \tag{3.49}$$

(note this leads to the same equations as (3.48) since the Lagrangians differ by a total differential dS). The resulting complete figure is a geometric entity formed by a one parameter family of hypersurfaces $S(x) = const.$ where S satisfies the HJ equation

$$g^{\mu\nu}(\partial_\mu S - \bar{A}_\mu)(\partial_\nu S - \bar{A}_\nu) = m^2 - \frac{R}{6} \tag{3.50}$$

and by a congruence of curves intersecting this family given by

$$\frac{dx^\mu}{ds} = \frac{g^{\mu\nu}(\partial_\nu S - \bar{A}_\nu)}{[g^{\rho\sigma}(\partial_\rho S - \bar{A}_\rho)(\partial_\sigma S - \bar{A}_\sigma)]^{1/2}} \tag{3.51}$$

The congruence yields the actual particle path space and the underlying probability measure on the path space may be defined on an arbitrary 3-dimensional hypersurface intersecting all of the members of the congruence without tangencies (cf. [**591**]). The measure will be completely identified by its probability current density j^μ (see [**1116**] and Remark 3.3.3). Moreover, since the measure is independent of the arbitrary choice of the hypersurface, j^μ must be conserved, i.e. $\partial_\mu j^\mu = 0$ (see Remark 3.3.3). Since the trajectories are deterministically defined by (3.51), j^μ must be parallel to the particle 4-velocity (3.51), and hence

$$j^\mu = \rho\sqrt{-g}\, g^{\mu\nu}(\partial_\nu S - \bar{A}_\nu) \tag{3.52}$$

with some $\rho > 0$. Now gauge invariance of the underlying measure as well as of the complete figure requires that j^μ transforms as a vector density of Weyl type $w(j^\mu) = 0$ and S as a scalar of Weyl type $w(S) = 0$. From (3.52) one sees then that ρ transforms as a scalar of Weyl type $w(\rho) = -1$ and ρ is called the scalar probability density of the particle random motion.

The actual spacetime affine connections are obtained from (3.47) by performing the variation with respect to the fields $\Gamma^\lambda_{\mu\nu}$ for a fixed metric tensor, particle trajectory, and probability measure. It is expedient to tranform the average action principle to the form of a 4-volume integral

$$\delta\left[\int_\Omega d^4x[(m^2 - (R/6))(g_{\mu\nu}j^\mu j^\nu]^{1/2} + A_\mu j^\mu\right] = 0 \tag{3.53}$$

where Ω is the spacetime region occupied by the congruence (3.51) and j^μ is given by (3.52) (cf. [**1116**] and Remark 3.3.3 for proofs). Since the connection fields

$\Gamma^\lambda_{\mu\nu}$ are contained only in the curvature term R the variational problem (3.53) can be further reduced to

$$\delta\left[\int_\Omega \rho R\sqrt{-g}d^4x\right]=0 \tag{3.54}$$

(here the HJ equation (3.50) has been used). This states that the average space-time curvature must be stationary under a variation of the fields $\Gamma^\lambda_{\mu\nu}$ (principle of stationary average curvature). The extremal connections $\Gamma^\lambda_{\mu\nu}$ arising from (3.54) are derived in [**1116**] using standard field theory techniques and the result is

$$\Gamma^\lambda_{\mu\nu}=\left\{\begin{array}{c}\lambda\\ \mu\ \nu\end{array}\right\}+\frac{1}{2}(\phi_\mu\delta^\lambda_\nu+\phi_\nu\delta^\lambda_\mu-g_{\mu\nu}g^{\lambda\rho}\phi_\rho);\ \phi_\mu=\partial_\mu log(\rho) \tag{3.55}$$

This shows that the resulting connections are integrable Weyl connections with a gauge field ϕ_μ (cf. [**1115**], Section 3, and Section 3.1). The HJ equation (3.50) and the continuity equation $\partial_\mu j^\mu=0$ can be consolidated in a single complex equation for S, namely

$$e^{iS}g^{\mu\nu}(iD_\mu-\bar{A}_\mu)(iD_\nu-\bar{A}_\nu)e^{-iS}-(m^2-(R/6))=0;\ D_\mu\rho=0 \tag{3.56}$$

Here D_μ is (doubly covariant - i.e. gauge and coordinate invariant) Weyl derivative given by (cf. [**81**])

$$D_\mu T^\alpha_\beta=\partial_\mu T^\alpha_\beta+\Gamma^\alpha_{\mu\epsilon}T^\epsilon_\beta-\Gamma^\epsilon_{\mu\beta}T^\alpha_\epsilon+w(T)\phi_\mu T^\alpha_\beta \tag{3.57}$$

It is to be noted that the probability density (but not the rest mass) remains constant relative to D_μ. When written out (3.56) for a set of two coupled partial differential equations for ρ and S. To any solution corresponds a particular random motion of the particle.

Next one notes that (3.56) can be cast in the familiar KG form, i.e.

$$[(i/\sqrt{-g})\partial_\mu\sqrt{-g}-\bar{A}_\mu]g^{\mu\nu}(i\partial_\nu-\bar{A}_\nu)\psi-(m^2-(\dot{R}/6))\psi=0 \tag{3.58}$$

where $\psi=\sqrt{\rho}exp(-iS)$ and $\dot{R}$ is the Riemannian scalar curvature built out of $g_{\mu\nu}$ only. We have the (by now) familiar formula

$$R=\dot{R}-3[(1/2)g^{\mu\nu}\phi_\mu\phi_\nu+(1/\sqrt{-g})\partial_\mu(\sqrt{-g}g^{\mu\nu}\phi_\nu)] \tag{3.59}$$

According to point of view **(A)** above in the KG equation (3.58) any explicit reference to the underlying spacetime Weyl structure has disappeared; thus the Weyl structure is hidden in the KG theory. However we note that no physical meaning is attributed to ψ or to the KG equation. Rather the dynamical and statistical behavior of the particle, regarded as a classical particle, is determined by (3.56), which, although completely equivalent to the KG equation, is expressed in terms of quantities having a more direct physical interpretation.

REMARK 3.3.3. We extract here from the Appendices to paper 1 of Santamato [**1116**]. In Appendix A of [**254**] one shows (following [**1116**]) that the Carathéodory complete figure formed by the congruence (3.51) solves the variational problem (3.47). One needs the notion of the Gibbs ensemble in relativistic mechanics (cf. Hakim [**591**]). Roughly a relativistic Gibbs ensemble of particles may be assimilated to an incoherent globule of matter moving in spacetime. More precisely a relativistic Gibbs ensemble is given by (i) A congruence of timelike curves in spacetime (the path space of the particles) and (ii) A probability measure defined on this congruence (note a congruence of spacelike curves could also be envisioned but causality is affected - a physical intepretation is unclear although it could be related to a statistical formulation of virtual phenomena). The construction here goes as follows. Let K be a 3-parameter congruence of time like curves in spacetime be given via (♦) $x^\mu = x^\mu(\tau, u^k)$ where $k = 1, 2, 3$ and τ is an arbitrary parameter along each curve of the congruence. For simplicity assume that the congruence covers a region Ω of spacetime simply (i.e. one and only one curve of K passes through each point of Ω). Then one can regard (♦) as a change of coordinates from x^μ to y^μ where $y^0 = t,\ y^k = u^k$ (assume the Jacobian is nonzero in Ω). Consider then the action integral $L = \int_{\tau_1}^{\tau_2} L(x(\tau, u^k), \dot{x}(\tau, u^k)d\tau$ with L homogeneous of the first degree in the derivatives $\dot{x}^\mu = \partial x^\mu/\partial\tau$. Given a 1-1 correspondence between the u^k and members of the congruence K one may introduce a formula for the probability that the particle follows a sample path having parameters u^k in some 3-dimensional region B as $prob(B) = \int_{B\subset\mathbf{R}} \mu(u^k)du^1du^2du^3$ where $\mu(u^k)$ is some probability density defined on $\mathbf{R}^3$. Hence the average action integral in (3.47) may be written as

$$I = E\left[\int_{\tau_1}^{\tau_2} Ld\tau\right] = \int_{\mathbf{R}^3}\int_{\tau_1}^{\tau_2} \mu(u^k)L(x^\mu(\tau, u^k), \dot{x}^\mu(\tau, u^k)d\tau \prod du^i \tag{3.60}$$

The last term is a 4-dimensional volume integral over the zone between the hyperplanes $y^0 = \tau_1$ and $y^0 = \tau_2$ in the y coordinate. In the x coordinates these hyperplanes are mapped on two 3-dimensional hypersurfaces $\tau(x^\mu) = \tau_1$ and $\tau(x^\mu) = \tau_2$ where $\tau(x^\mu)$ is obtained by solving (♦) with respect to τ; since they are merely a result of the parametrization of K they can be regarded as essentially arbitrary. The integrand in (3.60) depends on the 4 unknown functions $x^\mu(y^\nu)$ and on their first derivatives $\partial x^\mu/\partial y^0$, and on the coordinates y^ν themselves. Therefore the variational problem $\delta I = 0$ is reduced to a standard variational problem whose solution will yield the functions $x^\mu(\tau, u^k)$, i.e. the actual congruence that renders the average action stationary.

Now the Lagrangian density in (3.60) is $\Lambda = \mu(u^k)L(x^\mu(\tau, u^k), x^\mu_{,\tau}(\tau, u^k)$ in which $x^\mu_{,\tau} = \dot{x}^\mu$ with τ and u^k are the independent variables. By standard methods the EL expressions are ($x^\mu_{,k} = \partial x^\mu/\partial u^k$)

$$E(\Lambda) = \frac{\partial}{\partial u^k}\left[\frac{\partial\Lambda}{\partial x^\mu_{,k}}\right] + \frac{\partial}{\partial\tau}\left[\frac{\partial\Lambda}{\partial x^\mu_{,\tau}}\right] - \frac{\partial\Lambda}{\partial x^\mu} \tag{3.61}$$

In this case howver $\partial\Lambda/\partial x^\mu_{,k} = 0$ and hence the fixed equations $E(\Lambda) = 0$ reduce to (note μ does not depend explicitly on τ)

$$\frac{\partial}{\partial\tau}\left[\mu\frac{\partial L}{\partial x^\mu_{,\tau}}\right] - \mu\frac{\partial L}{\partial x^\mu} = 0 \Rightarrow \frac{\partial}{\partial\tau}\left[\frac{\partial L}{\partial\dot{x}^\mu}\right] - \frac{\partial L}{\partial x^\mu} = 0 \tag{3.62}$$

and this coincides with the EL equations associated with the action integral above. This means that the actual congruence must be a congruence of extremals or equivalently that the particle obeys equations of motion (3.62) with probability one. Even if the congruence is extremal however we are left with nonvanishing surface terms in the variation of I, namely

(3.63)

$$\delta I = \int_{\mathbf{R}^3} \mu(u^k) \prod du^i \left[\frac{\partial L}{\partial \dot{x}^\mu}(\tau_2, u^k)\delta x^\mu(\tau_2, u^k) - \frac{\partial L}{\partial \dot{x}^\mu}(\tau_1, u^k)\delta x^\mu(\tau_1, u^k) \right] = 0$$

In (3.63) the quantities δx^μ at $\tau = \tau_2$ and $\tau = \tau_1$ are displacements between points P and $P + \delta P$ where the curves x^μ and $x^\mu + \delta x^\mu$ intersect the hypesurfaces $\tau = \tau_2$ and $\tau = \tau_1$ so $\delta x^\mu(\tau_1, u^k)$ and $\delta x^\mu(\tau_1, u^k)$ are tangential to the hypersurfaces. Since the hypersurfaces $\tau(x^\mu) = const.$ are essentially arbitrary so must be the displacements δx^μ and $\delta I = 0$ implies then $(\bullet)\ \partial L/\partial \dot{x}^\mu(\tau, u^k) = 0$. Finally relating L with the Lagrangian (3.48) and comparing with $\bar{L}$ as defined in (3.49) one has $\partial L/\partial \dot{x}^\mu = \partial \bar{L}/\partial \dot{x}^\mu - \partial_\mu S$ so $(\bullet)$ yields $\partial \bar{L}/\partial \dot{x}^\mu = \partial_\mu S$. Moreover L and $\bar{L}$, differing only by a total differential dS, lead to the same EL equations and hence one can replace L by $\bar{L}$ in (3.62). In conclusion the congruence that renders the average action stationary must be (i) A congruence of curves that are extremal with respect to Lagrangian $\bar{L}$ and (ii) A congruence satisfying the integrability conditions $\partial \bar{L}/\partial \dot{X}^\mu = \partial_\mu S$. However by standard HJ theory such a congruence is given by (3.51) provided $S(x^\mu)$ obeys the HJ equation associated with $\bar{L}$, namely (3.50).

In appendix B of the paper 1 in Santamato [**1116**] the current density j^μ is introduced and the equivalence between the average action (3.47) and the 4-volume integral (3.53) is proved. This provides a useful connection between ensemble averages and 4-volume integrals appearing in field theories. Here (3.60) is expressed in terms of the y coordinates (τ, u^k) and it can also be expressed in terms of the x coordinates. For this one introduces the current density j^μ associated with the relativistic Gibbs ensemble. The surface element normal to the hypersurface $\tau(u^k) = const.$ is given by $d\sigma_\mu = \pi_\mu du^1 du^2 du^3$ where π_μ are Jacobians

(3.64)
$$\pi_0 = \frac{\partial(x^1, x^2, x^3)}{\partial(u^1, u^2, u^3)};\ \pi_1 = \frac{\partial(x^0, x^2, x^3)}{\partial(u^1, u^2, u^3)}, \cdots$$

Then define the current density via $\mu = j^\mu \pi_\mu$ so that $prob(B)$ becomes

(3.65)
$$prob(B) = \int_{B \subset \mathbf{R}^3} \mu du^1 du^2 du^3 = \int_{B \subset \mathbf{R}^3} j^\mu d\sigma_\mu$$

The direction of j^μ is still not defined so one is free to choose the current direction parallel to the congruence K, i.e. $j^\mu = \lambda \dot{x}^\mu$. The independence of the underlying measure on the chosen hypersurface $\tau = const.$ is exprssed analytically by the fact that $\mu = \mu(u^1, u^2, u^3)$ does not depend on τ explicitly. Consequently $\partial_\mu j^\mu = 0$ since by the Gauss theorem

(3.66)
$$\int_{\tau(x^\mu)=\tau_2} j^\mu d\sigma_\mu - \int_{\tau(x^\mu)=\tau_1} j^\mu d\sigma_\mu = \int_\Omega \partial_\mu j^\mu d^4x = 0$$

where Ω is the strip between the essentially arbitrary hypersurfaces $\tau = \tau_1$ and $\tau = \tau_2$. The same result could be obtained by differentiating $\mu = j^\mu \pi_\mu$ and using properties of Jacobians. Passing to x coordinates (3.60) becomes

$$I = \int_\Omega \mu L J^{-1} d^4x; \; J = \frac{\partial(x^0, x^1, x^2, x^3)}{\partial(\tau, u^1, u^2, u^3)} \tag{3.67}$$

Note that by definition $J = (\partial x^\mu/\partial\tau)\pi_\mu$ so

$$I = \int_\Omega \mu[L(x^\mu, \dot{x}^\mu)/(\dot{x}^\mu \pi_\mu)]d^4x \tag{3.68}$$

Since L is homogeneous of the first degree in the $\dot{x}^\mu$ the term in square brackets in (3.68) is homogeneous of degree zero in the $\dot{x}^\mu$. Hence we can replace $\dot{x}^\mu$ with the current $j^\mu = \lambda\dot{x}^\mu$ without affecting the integral to obtain $I = \int_\Omega L(x^\mu j^\mu)d^4x$ where $\mu = j^\mu\pi_\mu$ has been used. Thus the average action I may be converted to a four volume integral of $L(x^\mu, j^\mu)$. When this formal substitution is made in (3.48), (3.53) is obtained. This substitution does not alter the functional dependence of the average action integral I on the connection fields $\Gamma^\lambda_{\mu\nu}$ so the variational problems (3.47) and (3.53) are equivalent as long as the variation is performed with respect to these fields.

In Appendix C one derives (3.55); since similar calculations have already been used earlier (and will recur again) we omit this here. ■

REMARK 3.3.4. The formula (3.59) goes back to Weyl [**1294**] and the connection of matter to geometry arises from (3.55). The time variable is treated in a special manner here related to a Gibbs ensemble and $\rho > 0$ is built into the theory. Thus problems of statistical transparancy as in Remark 2.3.3 will apparently not arise. ■

REMARK 3.3.5. As mentioned at the beginning of Section 3.2, in Castro [**270**] the Santamato theory is analyzed in depth from several points of view and a number of directions for further study are indicated (in Castro-Mahecha [**276**] the importance of a complex velocity is emphasized). There is also a related development for the Dirac equation using an approach related to [**504, 622, 623**], where both relativistic and nonrelativistic spin 1/2 particles can be classically treated using anticommuting Grassmanian variables. However we prefer to treat the Dirac equation in a different manner later (cf. also [**110**] and Section 2.1.1).■

4. SCALE RELATIVITY AND KG

In [**234**] and Section 1.2 we sketched a few developments in the theory of scale relativity. This is by no means the whole story and we want to give a taste of some further main ideas while deriving the KG equation in this context (cf. [**11, 287, 336, 337, 338, 941, 942, 943, 944, 945, 946, 947**]). A main idea here is that the Schrödinger, Klein-Gordon, and Dirac equations are all geodesic equations in the fractal framework. They have the form $D^2/ds^2 = 0$ where D/ds represents the appropriate covariant derivative. The complex nature of the SE and KG equation arises from a discrete time symmetry breaking based on nondifferentiability. For the Dirac equation further discrete symmetry breakings are needed on the spacetime variables in a biquaternionic context (cf. here Célérier-Nottale

[**287**]). First we go back to [**941, 942, 946**] and sketch some of the fundamentals of scale relativity. This is a very rich and beautiful theory extending in both spirit and generality the relativity theory of Einstein (cf. also Castro [**278**] for variations involving Clifford theory). The basic idea here is that (following Einstein) the laws of nature apply whatever the state of the system and hence the relevant variables can only be defined relative to other states. Standard scale laws of power-law type correspond to Galilean scale laws and from them one actually recovers quantum mechanics (QM) in a nondifferentiable space. The quantum behavior is a manifestation of the fractal geometry of spacetime. In particular (as indicated in Section 1.2) the quantum potential is a manifestation of fractality in the same way as the Newton potential is a manifestation of spacetime curvature. In this spirit one can also conjecture (cf. Nottale [**946**]) that this quantum potential may explain various dynamical effects presently attributed to dark matter (cf. also [**18**] and Chapter 4). Now for basics one deals with a continuous but nondifferentiable physics. It is known for example that the length of a continuous nondifferentiable curve is dependent on the resolution ϵ. One approach now involves smoothing a nondifferentiable function f via $f(x,\epsilon)=\int_{-\infty}^{\infty}\phi(x,y,\epsilon)f(y)dy$ where ϕ is smooth and say "centered" at x (we refer also to Remark 1.2.8 and [**11, 336, 337, 338**] for a more refined treatment of such matters). There will now arise differential equations involving $\partial f/\partial log(\epsilon)$ and $\partial^2 f/\partial x\partial log(\epsilon)$ for example and the $log(\epsilon)$ term arises as follows. Consider an infinitesimal dilatation $\epsilon\to\epsilon'=\epsilon(1+d\rho)$ with a curve length

$$\ell(\epsilon)\to\ell(\epsilon')=\ell(\epsilon+\epsilon d\rho)=\ell(\epsilon)+\epsilon\ell_\epsilon d\rho=(1+\tilde{D}d\rho)\ell(\epsilon) \tag{4.1}$$

Then $\tilde{D}=\epsilon\partial_\epsilon=\partial/\partial log(\epsilon)$ is a dilatation operator and in the spirit of renormalization (multiscale approach) one can assume $\partial\ell(x,\epsilon)/\partial log(\epsilon)=\beta(\ell)$ (where $\ell(x,\epsilon)$ refers to the curve defined by $f(x,\epsilon)$). Now for Galilean scale relativity consider $\partial\ell(x,\epsilon)/\partial log(\epsilon)=a+b\ell$ which has a solution

$$\ell(x,\epsilon)=\ell_0(x)\left[1+\zeta(x)\left(\frac{\lambda}{\epsilon}\right)^{-b}\right] \tag{4.2}$$

where $\lambda^{-b}\zeta(x)$ is an integration constant and $\ell_0=-a/b$. One can choose $\zeta(x)$ so that $<\zeta^2(x)>=1$ and for $a\neq 0$ there are two regimes (for $b<0$)

(1) $\epsilon<<\lambda\Rightarrow\zeta(x)(\lambda/\epsilon)^{-b}>>1$ and ℓ is given by a scale invariant fractal like power with dimension $D=1-b$, namely $\ell(x,\epsilon)=\ell_0(\lambda/\epsilon)^{-b}$.
(2) $\epsilon>>\lambda\Rightarrow\zeta(x)(\lambda/\epsilon)^{-b}<<1$ and ℓ is independent of scale.

Here $\epsilon=\lambda$ constitutes a transition point between fractal and nonfractal behavior. Only the special case $a=0$ yields unbroken scale invariance of $\ell=\ell_0(\lambda/\epsilon)^\delta$ ($\delta=-b$) and one has then $\tilde{D}\ell=b\ell$ so the scale dimension is an eigenvalue of $\tilde{D}$. Finally the case $b>0$ corresponds to the cosmological domain.

Now one looks for scale covariant laws and checks this for power laws $\phi=\phi_0(\lambda/\epsilon)^\delta$. Thus a scale transformation for $\delta(\epsilon')=\delta(\epsilon)$ will have the form

$$log\frac{\phi(\epsilon')}{\phi_0}=log\frac{\phi(\epsilon)}{\phi_0}+V\delta(\epsilon);\ V=log\frac{\epsilon}{\epsilon'} \tag{4.3}$$

In the same way that only velocity differences have a physical meaning in Galilean relativity here only V differences or scale differences have a physical meaning. Thus V is a "state of scale" just as velocity is a state of motion. In this spirit laws of linear transformation of fields in a scale transformation $\epsilon \to \epsilon'$ amount to finding $A, B, C, D(V)$ such that

$$(4.4) \quad log\frac{\phi(\epsilon')}{\phi_0} = A(V)log\frac{\phi(\epsilon)}{\phi_0} + B(V)\delta(\epsilon);\ \delta(\epsilon') = C(V)log\frac{\phi(\epsilon)}{\phi_0} + D(V)\delta(\epsilon)$$

Here $A = 1$, $B = V$, $C = 0$, $D = 1$ corresponds to the Galileo group. Note also $\epsilon \to \epsilon' \to \epsilon'' \Rightarrow V'' = V + V'$. Now for the analogue of Lorentz transformations there is a need to preserve the Galilean dilatation law for scales larger than the quantum classical transition. Note $V = log(\epsilon/\epsilon') \sim \epsilon/\epsilon' = exp(-V)$ and set $\rho = \epsilon'/\epsilon$ with $\rho' = \epsilon''/\epsilon'$ and $\rho'' = \epsilon''/\epsilon$; then $log\rho'' = \log\rho + \log\rho'$ and one is thinking here of $\rho : \epsilon \to \epsilon'$, $\rho' : \epsilon' \to \epsilon''$ and $\rho'' : \epsilon \to \epsilon''$ with compositions (the notation is meant to somehow correspond to (3.1)). Now recall the Einstein-Lorentz law $w = (u+v)/[1+(uv/c^2)]$ but one now has several regimes to consider. Following [**942, 946**] small scale symmetry is broken by mass via the emergence of $\lambda_c = \hbar/mc$ (Compton length) and $\lambda_{dB} = \hbar/m\nu$ (deBroglie length), while for extended objects $\lambda_{th} = \hbar/m < \nu^2 >^{1/2}$ (thermal deBroglie length) affects transitions. The transition scale in (4.2) is the Einstein-deBroglie scale (in rest frame $\lambda \sim \tau = \hbar/mc^2$) and in the cosmological realm the scale symmetry is broken by the emergence of static structure of typical size $\lambda_g = (1/3)(GM/ < \nu^2 >)$. The scale space consists of three domains (quantum, classical - scale independent, and cosmological). Another small scale transition factor appears in the Planck length scale $\lambda_P = (\hbar G/c^3)^{1/2}$ and at large scales the cosmological constant Λ comes into play. With this background the composition of dilatations is taken to be

$$(4.5) \qquad log\frac{\epsilon'}{\lambda} = \frac{log\rho + \log\frac{\epsilon}{\lambda}}{1 + \frac{log\rho log\frac{\epsilon}{\lambda}}{log^2(L/\lambda)}} = \frac{log\rho + \log\frac{\epsilon}{\lambda}}{1 + \frac{log\rho log(\epsilon/\lambda)}{C^2}}$$

where $L \sim \lambda_P$ near small scales and $L \sim \Lambda$ near large scales (note $\epsilon = L \Rightarrow \epsilon' = L$). Comparing with $w = (u+v)/(1+(uv/c^2))$ one thinks of $log(L/\lambda) = C \sim c$ (note here $log^2(a/b) = log^2(b/a)$ in comparing formulas in [**942, 946**]). Lengths now change via

$$(4.6) \qquad log\frac{\ell'}{\ell_0} = \frac{log(\ell/\ell_0) + \delta log\rho}{\sqrt{1 - \frac{\log^2\rho}{C^2}}}$$

and the scale variable δ (or djinn) is no longer constant but changes via

$$(4.7) \qquad \delta(\epsilon') = \frac{\delta(\epsilon) + \frac{log\rho log(\ell/\ell_0)}{C^2}}{\sqrt{1 - \frac{log^2\rho}{C^2}}}$$

where $\lambda \sim$ fractal-nonfractal transition scale.

We have derived the SE in Section 1.2 (cf. also [**234**]) and go now to the KG equation via scale relativity. The derivation in the first paper of [**287**] seems the most concise and we follow that at first (cf. also [**942**]). All of the elements of the approach for the SE remain valid in the motion relativistic case with the time replaced by the proper time s, as the curvilinear parameter along the geodesics. Consider a small increment dX^μ of a nondifferentiable four coordinate along one of the geodesics of the fractal spacetime. One can decompose this in terms of a large scale part $\overline{LS} < dX^\mu >= dx^\mu = v_\mu ds$ and a fluctuation $d\xi^\mu$ such that $\overline{LS} < d\xi^\mu >= 0$. One is led to write the displacement along a geodesic of fractal dimension $D = 2$ via

(4.8) $$dX^\mu_\pm = d_\pm x^\mu + d\xi^\mu_\pm = v^\mu_\pm ds + u^\mu_\pm \sqrt{2\mathcal{D}} ds^{1/2}$$

Here $u^\mu_\pm$ is a dimensionless fluctuation andd the length scale $2\mathcal{D}$ is introduced for dimensional purposes. The large scale forward and backward derivatives d/ds_+ and d/ds_- are defined via

(4.9) $$\frac{d}{ds_\pm} f(s) = lim_{s \to 0_\pm} \overline{LS} \left\langle \frac{f(s + \delta s) - f(s)}{\delta s} \right\rangle$$

Applied to x^μ one obtains the forward and backward large scale four velocities of the form

(4.10) $$(d/dx_+)x^\mu(s) = v^\mu_+;\ (d/ds_-)x^\mu = v^\mu_-$$

Combining yields

(4.11) $$\frac{d'}{ds} = \frac{1}{2}\left(\frac{d}{ds_+} + \frac{d}{ds_-}\right) - \frac{i}{2}\left(\frac{d}{ds_+} - \frac{d}{ds_-}\right);$$

$$\mathcal{V}^\mu = \frac{d'}{ds} x^\mu = V^\mu - iU^\mu = \frac{v^\mu_+ + v^\mu_-}{2} - i\frac{v^\mu_+ - v^\mu_-}{2}$$

For the fluctuations one has

(4.12) $$\overline{LS} < d\xi^\mu_\pm d\xi^\nu_\pm >= \mp 2\mathcal{D}\eta^{\mu\nu} ds$$

One chooses here $(+, -, -, -)$ for the Minkowski signature for $\eta^{\mu\nu}$ and there is a mild problem because the diffusion (Wiener) process makes sense only for positive definite metrics. Various solutions were given in [**394, 1145, 1334**] and they are all basically equivalent, amounting to the transformatin a Laplacian into a D'Alembertian. Thus the two forward and backward differentials of $f(x, s)$ should be written as

(4.13) $$(df/ds_\pm) = (\partial_s + v^\mu_\pm \partial_\mu \mp \mathcal{D}\partial^\mu\partial_\mu) f$$

One considers now only stationary functions f, not depending explicitly on the proper time s, so that the complex covariant derivative operator reduces to

(4.14) $$(d'/ds) = (\mathcal{V}^\mu + i\mathcal{D}\partial^\mu)\partial_\mu$$

Now assume that the large scale part of any mechanical system can be characterized by a complex action $\mathfrak{S}$ leading one to write

(4.15) $$\delta\mathfrak{S} = -mc\delta \int_a^b ds = 0;\ ds = \overline{LS} < \sqrt{dX^\nu dX_\nu} >$$

This leads to $\delta\mathfrak{S} = -mc \int_a^b \mathcal{V}_\nu d(\delta x^\nu)$ with $\delta x^\nu = \overline{LS} < dX^\nu >$. Integrating by parts yields

(4.16) $$\delta\mathfrak{S} = -[mc\delta x^\nu]_a^b + mc \int_a^b \delta x^\nu (d\mathcal{V}_\mu/ds) ds$$

To get the equations of motion one has to determine $\delta\mathfrak{S} = 0$ between the same two points, i.e. at the limits $(\delta x^\nu)_a = (\delta x^\nu)_b = 0$. From (3.15) one obtains then a differential geodesic equation $d\mathcal{V}/ds = 0$. One can also write the elementary variation of the action as a functional of the coordinates. So consider the point a as fixed so $(\delta x^\nu)_a = 0$ and consider b as variable. The only admissable solutions are those satisfying the equations of motion so the integral in (3.15) vanishes and writing $(\delta x^\nu)_b$ as δx^ν gives $\delta\mathfrak{S} = -mc\mathcal{V}_\nu\delta x^\nu$ (the minus sign comes from the choice of signature). The complex momentum is now

$$\mathcal{P}_\nu = mc\mathcal{V}_\nu = -\partial_\nu\mathfrak{S} \tag{4.17}$$

and the complex action completely characterizes the dynamical state of the particle. Hence introduce a wave function $\psi = exp(i\mathfrak{S}/\mathfrak{S}_0)$ and via (3.16) one gets

$$\mathcal{V}_\nu = (i\mathfrak{S}_0/mc)\partial_\nu log(\psi) \tag{4.18}$$

Now for the scale relativistic prescription replace the derivative in d/ds by its covariant expression d'/ds. Using (3.17) one transforms $d\mathcal{V}/ds = 0$ into

$$-\frac{\mathfrak{S}_0^2}{m^2c^2}\partial^\mu log(\psi)\partial_\mu\partial_\nu log(\psi) - \frac{\mathfrak{S}_0\mathcal{D}}{mc}\partial^\mu\partial_\mu\partial_\nu log(\psi) = 0 \tag{4.19}$$

The choice $\mathfrak{S}_0 = \hbar = 2mc\mathcal{D}$ allows a simplification of (3.18) when one uses the identity

$$\frac{1}{2}\left(\frac{\partial_\mu\partial^\mu\psi}{\psi}\right) = \left(\partial_\mu log(\psi) + \frac{1}{2}\partial_\mu\right)\partial^\mu\partial^\nu log(\psi) \tag{4.20}$$

Dividing by $\mathcal{D}^2$ one obtains the equation of motion for the free particle $\partial^\nu[\partial^\mu\partial_\mu\psi/\psi] = 0$. Therefore the KG equation (no electromagnetic field) is

$$\partial^\mu\partial_\mu\psi + (m^2c^2/\hbar^2)\psi = 0 \tag{4.21}$$

and this becomes an integral of motion of the free particle provided the integration constant is chosen in terms of a squared mass term $m^2c^2/\hbar^2$. Thus the quantum behavior described by this equation and the probabilistic interpretation given to ψ is reduced here to the description of a free fall in a fractal spacetime, in analogy with Einstein's general relativity. Moreover these equations are covariant since the relativistic quantum equation written in terms of d'/ds has the same form as the equation of a relativistic macroscopic and free particle using d/ds. One notes that the metric form of relativity, namely $V^\mu V_\mu = 1$ is not conserved in QM and it is shown in Pissondes [**1028**] that the free particle KG equation expressed in terms of $\mathcal{V}$ leads to a new equality

$$\mathcal{V}^\mu\mathcal{V}_\mu + 2i\mathcal{D}\partial^\mu\mathcal{V}_\mu = 1 \tag{4.22}$$

In the scale relativistic framework this expression defines the metric that is induced by the internal scale structures of the fractal spacetime. In the absence of an electromagnetic field $\mathcal{V}^\mu$ and $\mathfrak{S}$ are related by (3.16) which can be writen as $\mathcal{V}_\mu = -(1/mc)\partial_\mu\mathfrak{S}$ so (3.21) becomes

$$\partial^\mu\mathfrak{S}\partial_\mu\mathfrak{S} - 2imc\mathcal{D}\partial^\mu\partial_\mu\mathfrak{S} = m^2c^2 \tag{4.23}$$

which is the new form taken by the Hamilton-Jacobi equation.

REMARK 3.4.1. We go back to [**942, 1028**] now and repeat some of their steps in a perhaps more primitive but revealing form. Thus one omits the $\overline{LS}$ notation and uses $\lambda \sim 2\mathcal{D}$; equations (3.8) - (4.14) and (3.11) are the same and one writes now $\eth/ds$ for d'/ds. Then $\eth/ds = \mathcal{V}^\mu\partial_\mu + (i\lambda/2)\partial^\mu\partial_\mu$ plays the role of a scale covariant derivative and one simply takes the equation of motion of a free relativistic quantum particle to be given as $(\eth/ds)\mathcal{V}^\nu = 0$, which can be interpreted as the equations of free motion in a fractal spacetime or as geodesic equations. In fact now $(\eth/ds)\mathcal{V}^\nu = 0$ leads directly to the KG equation upon writing $\psi = exp(i\mathfrak{S}/mc\lambda)$ and $\mathfrak{P}^\mu = -\partial^\mu\mathfrak{S} = mc\mathcal{V}^\mu$ so that $i\mathfrak{S} = mc\lambda log(\psi)$ and $\mathcal{V}^\mu = i\lambda\partial^\mu log(\psi)$. Then

$$\left(\mathcal{V}^\mu\partial_\mu + \frac{i\lambda}{2}\partial^\mu\partial_\mu\right)\partial^\nu log(\psi) = 0 = i\lambda\left(\frac{\partial^\mu\psi}{\psi}\partial_\mu + \frac{1}{2}\partial^\mu\partial_\mu\right)\partial^\nu \log(\psi) \tag{4.24}$$

Now some identities are given in [**1028**] for aid in calculation here, namely

$$\frac{\partial^\mu\psi}{\psi}\partial_\mu\frac{\partial^\nu\psi}{\psi} = \frac{\partial^\mu\psi}{\psi}\partial^\nu\left(\frac{\partial_\mu\psi}{\psi}\right) = \tag{4.25}$$

$$= \frac{1}{2}\partial^\nu\left(\frac{\partial^\mu\psi}{\psi}\frac{\partial_\mu\psi}{\psi}\right);\ \partial_\mu\left(\frac{\partial^\mu\psi}{\psi}\right) + \frac{\partial^\mu\psi}{\psi}\frac{\partial_\mu\psi}{\psi} = \frac{\partial^\mu\partial_\mu\psi}{\psi}$$

The first term in the last equation of (3.23) is then $(1/2)[(\partial^\mu\psi/\psi)(\partial_\mu\psi/\psi)]$ and the second is

$$(1/2)\partial^\mu\partial_\mu\partial^\nu log(\psi) = (1/2)\partial^\mu\partial^\nu\partial_\mu log(\psi) = \tag{4.26}$$

$$= (1/2)\partial^\nu\partial^\mu\partial_\mu log(\psi) = (1/2)\partial^\nu\left(\frac{\partial^\mu\partial_\mu\psi}{\psi} - \frac{\partial^\mu\psi\partial_\mu\psi}{\psi^2}\right)$$

Combining we get $(1/2)\partial^\nu(\partial^\mu\partial_\mu\psi/\psi) = 0$ which integrates then to a KG equation

$$-(\hbar^2/m^2c^2)\partial^\mu\partial_\mu\psi = \psi \tag{4.27}$$

for suitable choice of integration constant (note $\hbar/mc$ is the Compton wave length).

Now in this context or above we refer back to Section 2.2 for example and write $Q = -(1/2m)(\Box R/R)$ (cf. Section 2.2 before Remark 2.2.1 and take $\hbar = c = 1$ for convenience here). Then recall $\psi = exp(i\mathfrak{S}/m\lambda)$ and $\mathfrak{P}_\mu = m\mathcal{V}_\mu = -\partial_\mu\mathfrak{S}$ with $i\mathfrak{S} = m\lambda log(\psi)$. Also $\mathcal{V}_\mu = -(1/m)\partial_\mu\mathfrak{S} = i\lambda\partial_\mu log(\psi)$ with $\psi = Rexp(iS/m\lambda)$ so $log(\psi) = i\mathfrak{S}/m\lambda = log(R) + iS/m\lambda$, leading to

$$\mathcal{V}_\mu = i\lambda[\partial_\mu log(R) + (i/m\lambda)\partial_\mu S] = -\frac{1}{m}\partial_\mu S + i\lambda\partial_\mu log(R) = V_\mu + iU_\mu \tag{4.28}$$

Then $\Box = \partial^\mu\partial_\mu$ and $U_\mu = \lambda\partial_\mu log(R)$ leads to

$$\partial^\mu U_\mu = \lambda\partial^\mu\partial_\mu log(R) = \lambda\Box log(R) \tag{4.29}$$

Further $\partial^\mu\partial_\nu log(R) = (\partial^\mu\partial_\nu R/R) - (R_\nu R_\mu/R^2)$ so

$$\Box log(R) = \partial^\mu\partial_\mu log(R) = (\Box R/R) - (\sum R_\mu^2/R^2) = \tag{4.30}$$

$$= (\Box R/R) - \sum(\partial_\mu R/R)^2 = (\Box R/R) - |U|^2$$

for $|U|^2 = \sum U_\mu^2$. Hence via $\lambda = 1/2m$ for example one has

$$Q = -(1/2m)(\Box R/R) = -\frac{1}{2m}\left[|U|^2 + \frac{1}{\lambda}\Box log(R)\right] = \tag{4.31}$$

$$= -\frac{1}{2m}\left[|U|^2 + \frac{1}{\lambda}\partial^\mu U_\mu\right] = -\frac{1}{2m}|U|^2 - \frac{1}{2}div(\vec{U})$$

(cf. Section 2.2). ∎

REMARK 3.4.2. The words fractal spacetime as used in the scale relativity methods of Nottalle et al for producing geodesic equations (SE or KG equation) are somewhat misleading in that essentially one is only looking at continuous nondifferentiable paths for example. Scaling as such is of course considered extensively at other times. It would be nice to create a fractal derivative based on scaling properties and H-dimension alone for example which would permit the powerful techniques of calculus to be used in a fractal context. There has been of course some work in this direction already in e.g. [**235, 322, 531, 581, 629, 637, 764, 947, 994, 1080**]. ∎

5. QUANTUM MEASUREMENT AND GEOMETRY

We consider here a paper J. Wheeler [**1299**], which is based in part on a famous paper of London [**828**] (reprinted in [**962**]). In [**828**] it was shown that the ratio of the Weyl scale factor to the Schrödinger wave function is constant if the proportionality constant between the Weyl potential and the EM potential is taken to be imaginary; this observation gave birth to modern gauge theories and the original Weyl theory was absorbed into QM with the original scale freedom becoming invariance under unitary gauge transformations (cf. also Section 3.5.1). Both the Weyl theory and the Schrödinger theory describe the evolution of a field in time and given the factor of i and the Kaluza-Klein framework used by London, those evolutions are the same. In the Weyl picture the field characterizes the length scales of fundamental matter, while in the Schrödinger picture it is the wave function corresponding to a fundamental particle. This analogy is pursued further in [**1299**] with a main theme being the equivalence between Weyl measurement and quantum measurement; a complete theory of measurement in a Weyl geometry is said to contain the crucial elements of quantization and analogies of the following sort are indicated.

(5.1)

Weyl – quantum correspondence	*Quantum mechanics*
Zero – Weyl – weight number	*Real eigenvalue*
Diffusion equation	*SE*
Weiner path integral	*Feynman path integral*
Weightful length field ψ_w	*Complex state function* ψ
Weyl conjugate ψ_{-w}	ψ^*
Probability $\psi_w\psi_{-w}$	*Probability* $|\psi|^2$
$\psi_w \to e^{w\phi}\psi_w$ (*conformal*)	$\psi \to e^{i\phi}\psi$ (*unitary*)

We will try to make sense out of this following [**1299**] (cf. also Audretsch [**81, 82**]). Begin with a real 4-dimensional manifold $(M, [g])$ where $[g]$ is a conformal equivalence class of Lorentz metrics. In addition to local coordinate transformations one has Weyl (conformal) transformations given via $T(x)' = exp[w(T)\Lambda(x)]T(x)$ where T is a tensor field and $w(T)$ is the Weyl weight (a real number). One takes a coordinate basis $E_\alpha = \partial/\partial x^\alpha$ and $E^\alpha = dx^\alpha$ in the tangent and cotangent space satisfying $w(E_\alpha) = w(E^\alpha) = 0$.

DEFINITION 5.1. One defines a torsion free derivative D via

- Linearity: $D(aT_1 + bT_2) = aDT_1 + bDT_2$ for real a, b
- Leibniz: $D(T_1T_2) = (DT_1)T_2 + T_1(DT_2)$
- Weyl covariant: $D(fT) = [df + w(f)Wf]T + fDT$ where W is a real 1-form (Weyl potential)
- Zero weight: $w(DT) = w(T)$

Under a Weyl transformation $W \to W' = W - d\Lambda$ and one has

(5.2) $$DT = D_\mu T^\alpha{}_\beta E^\mu \otimes E_\alpha \otimes E^\beta;\ D_\mu T^\alpha{}_\beta = \\ = \partial_\mu T^\alpha{}_\beta + T^\rho{}_\beta \Gamma^\alpha{}_{\rho\mu} - T^\alpha{}_\rho \Gamma^\rho{}_{\beta\mu} + w(T)W_\mu T^\alpha{}_\beta$$

There is no unique metric on the space; instead the metric is to be taken of the Weyl type $w(g) = 2$ so that under a Weyl transformation $g' = exp[2\Lambda(x)]g$. The principle fields of the theory are related by the requirement $Dg = 0$, or in components

(5.3) $$D_\mu g_{\alpha\beta} = 0 = \partial_\mu g_{\alpha\beta} - g_{\rho\beta}\Gamma^\rho{}_{\alpha\mu} - g_{\alpha\rho}\Gamma^\rho{}_{\beta\mu} + 2W_\mu g_{\alpha\beta}$$

This can be solved to give

(5.4) $$\Gamma^\alpha{}_{\beta\mu} = \left\{ \begin{array}{c} \alpha \\ \beta\ \mu \end{array} \right\} + (\delta^\alpha{}_\beta W_\mu + \delta^\alpha{}_\mu W_\beta - g_{\beta\mu} W^\alpha)$$

Vanishing torsion has been assumed in (5.4) so that the bracket expression is the usual Christoffel connection. The curvature tensor is then

(5.5) $$R^\alpha{}_{\beta\mu\nu} = \Gamma^\alpha{}_{\beta\nu,\mu} - \Gamma^\alpha{}_{\beta\mu,\nu} + \Gamma^\alpha{}_{\rho\mu}\Gamma^\rho{}_{\beta\nu} - \Gamma^\alpha{}_{\rho\nu}\Gamma^\rho{}_{\beta\mu}$$

Unlike the Riemannian curvature tensor the Weyl curvature has nonvanishing trace on the first pair of indices so that $(1/2)R^\alpha{}_{\alpha\mu\nu} = W_{\nu,\mu} - W_{\mu,\nu} = W_{\mu\nu}$ where $W_{\mu\mu}$ is the gauge invariant field strength of the Weyl potential. One says that two fields are Weyl conjugate if they have the same Lorenz transformation properties but opposite Weyl weights.

Now for a theory of measurement one first looks at zero weight fields. In this direction note that fields with nonvanishing Weyl weight will experience changes under parallel transport. For example the mass squared transported along a path with unit tangent vector $u^\mu = dx^\mu/d\tau$ satisfies

(5.6) $$0 = u^\mu D_\mu(m^2) = u^\mu \partial_\mu(m^2) + w(m^2)u^\mu W_\mu m^2$$

Integrating along the path of motion one finds a path dependence of the form $m^2 = m_0^2 exp[w(m^2)\int W_\mu u^\mu d\tau]$ where the line integral has been written in terms of the path parameter τ. Note this is analogous to $m^2 = m_0^2 exp(Q)$ in the

Shojai theory of Section 3.2 suggesting some relation to a quantum potential $Q \sim w(m^2)\int W_\mu u^\mu d\tau$. However at this point there is no quantum matter posited and no density ρ so a Weyl vector $W_\mu \sim \partial_\mu log(\rho)$ as in Remark 3.3.1 is untenable and no comparison to (3.28) can be undertaken. However this does show a geometrical dependence of mass in general and in the flat space of Remark 3.3.1 it is replaced by a quantum potential. Indeed this (Schouten-Haantjes) conformal mass thus depends on the Weyl vector and if two particles of identical mass are allowed to propagate freely (by parallel transport) along different paths and brought together there will be a mass difference

$$\Delta m^2 = m_0^2 e^{w(m^2)\oint W_\mu u^\mu d\tau} \equiv m_0^2 e^{w(m^2)\int_S W_{\mu\nu} dS^{\mu\nu}} \tag{5.7}$$

where $dS^{\mu\nu}$ is an element of any 2-surface S bounded by the closed curve defined by the two particles. Hence unless the surface integral of the Weyl field strength vanishes there will be a path dependence for masses and of any other field of nonzero weight. One postulates now **(I)** that all quantities of vanishing Weyl weight should be physically meaningful (observables) and **(II)** that all fields occur in conjugate pairs satisfying conjugate equations of motion. Assume that $M_\pm$ evolves by parallel transport along a path as above via

$$0 = u^\mu D_\mu D_\pm = u^\mu \bar{D}_\mu M_\pm \pm w(M) M_\pm W_\mu u^\mu \tag{5.8}$$

where $\bar{D}$ is a derivation using the full connection (5.4) and one sets $w(M) = w(M_+) > 0$ for convenience. One has also

$$M_\pm = \mathfrak{M} exp[\mp w(M)\int W_\mu u^\mu d\tau] \tag{5.9}$$

where $\mathfrak{M}$ is weightless with $u^\mu \bar{D}_\mu \mathfrak{M} = 0$. Now suppose one wants to measure some characteristic of M (i.e. of M_+ or M_-). M can be scaled by an arbitrary gauge function and one transports M along a path so that its covariant derivative in the dirction of motion vanishes. Then the change in size is specified by (5.9) but it is not clear that we can tell what path a particle has taken. In a Riemannian space there are geodesics determining the paths of classical matter but that is not true in a Weyl space (in this regard we refer to [**236**], Section 3.2, and to [**164, 1052, 1053, 1054, 1055**]).

In order to study the motion of M one begins with the observation that a Weyl geometry provides a probability $P_{AB}(M)$ of finding a value M at a point B for a system which is known to have had a value M_0 at point A. Finding $P_{AB}(M)$ is tantamount to finding the fraction of paths which the system may follow leading to any given value of M. Since there may be no special paths in a Weyl geometry one has to settle for moments of the distribution. To find the average value of magnitude of M denoted by $<M>$ one integrates (5.9) over all paths via

$$<M>= \int \mathcal{D}[x] M_0 exp[w(M)\int_A^B W_\mu u^\mu d\tau] \tag{5.10}$$

where the usual path integral normalization is included implicitly in $\mathcal{D}[x]$ (see e.g. [**463, 616, 1137**]) for path integrals). However this gives no information as to whether one should expect M to actualy reach B. In [**1299**] there is then a long discussion (and a detailed Appendix) involving path averages, probability, Wiener integrals, etc. plus a postulate **(III)** that the probability a system will undergo a given infinitesimal displacement x^μ is inversely proportional to the change in length such a displacement produces in the system. Now $d\ell = w(M)W_\mu dx^\mu = w(M)W_\mu u^\mu d\tau$ and a plausible (rigorous) argument is given then to represent the probability of the system reaching any spacetime point x from x_0 as

$$G(x_0;x) = \int \mathcal{D}[x]exp[w(M)\int_{x_0}^{x} W_\mu u^\mu d\tau] \tag{5.11}$$

(which bears an obvious resemblence to (5.10)). Comparison of (5.10) and (5.11) involves noting first that (5.11) is gauge dependent but the gauge dependence comes out of the path integral since it depends only on the end points. Thus

$$G'(x_0;x) = \int \mathcal{D}[x]exp[w(M)\int_{x_0}^{x}(W_\mu - \partial_\mu\phi)u^\mu d\tau] = \tag{5.12}$$

$$= e^{-w(M)[\phi(x)-\phi(x_0)]}\int \mathcal{D}[x]exp[w(M)\int_{x_0}^{x} W_\mu u^\mu d\tau]$$

This means that one can eleiminate the gauge factor by multiplying by the Weyl conjugate expression

$$\bar{G}'(x_0;x) = e^{w(M)[\phi(x)-\phi(x_0)]}\int \mathcal{D}[x]exp[-w(M)\int_{x_0}^{x} W_\mu u^\mu d\tau] \tag{5.13}$$

to give a meaningful gauge invariant probability $P(x_0,x) = \bar{G}(x_0;x)G(x_0;x)$ which is the probability of detecting the dilating system at x given its presence at x_0. It may be thought of as the joint probability of finding both M and $\bar{M}$ at x. Here one is dealing with a real path integral, unlike QM, and the phase invariance of a wave function $\psi' = exp(i\phi)\psi$ is replaced by conformal invariance $M' = exp(\phi)M$ (this is the same factor of i introduced by London in 1927). Since that time gauge transformations have appeared as phases and the wave interpretation has been maintained; now one maintains a real gauge transformation and changes the interpretation of physical phenomena (see [**1299**] for more discussion in this direction).

Now one shows the equivalence to QM of the nonrelativistic limit of (5.11) when the exponent in the path integral is identified with a multiple of the classical action, i.e. $\int W_\mu u^\mu d\tau = \lambda S = \lambda\int Ld\tau$. The integrands here may also be equated except for the possible addition of the total derivative of a function of τ. But such a derivative is already known to be both a gauge freedom of W_μ and a transformation of L that leaves the equations of motion unaltered. So the possible equivalent versions of L may be understood as gauge changes of the underlying geometry. This identification fixes the physical interpretation of W_μ up to the gauge choice and since $u^\mu = \dot{x}^\mu$ equating the integrands gives

$$\lambda P_\mu = \lambda(\partial L/\partial u^\mu) = W_\mu \tag{5.14}$$

so that W_μ is proportional to the generalized momentum P_μ conjugate to x^μ. Now Weyl had originally identified W_μ with the derivative of an EM potential $\partial_\mu U \sim A_\mu$ and the present approach suggests $W_\mu = \lambda(p_\mu + A_\mu)$ so that all energy provides a surce of expansion rather than just EM energy. This still allows gauge transformations of W_μ to be identified with gauge transformations of A_μ. Next one goes to the nonrelativistic limit of the path integral to find a differential equation for the amplitudes $G(x_0; x)$. It is convenient to explicitly separate the kinetic term $p_\mu u^\mu$ from $W_\mu u^\mu$ which will enable one to identify the path integral in (5.11) with a Wiener integral. Thus with full generality one writes $W_\mu = \lambda(p_\mu + \tilde{W}_\mu)$ where any gauge transformation is understood to apply to $\tilde{W}_\mu$. Now consider the nonrelativistic limit where the integral $\int p_\mu u^\mu d\tau \sim mc^2 \int d\tau$ so that $mc^2 \int d\tau \sim \int [mc^2 - (m/2)\mathbf{v}^2]dt$. To this order the path integral becomes (suppressing limits of integration)

$$G(x_0; x) = \int \mathcal{D}[x] e^{\lambda w(M) \int [(1/2)m\mathbf{v}^2 + \tilde{W}\cdot\mathbf{v} - \tilde{W}^0 - mc^2)]dt} \tag{5.15}$$

This is of the form

$$P(x_0; x) = \int \mathcal{D}[x] exp[-(1/2) \int ((\dot{\mathbf{q}} + \mathbf{w})^2 - \nabla\cdot\mathbf{w})dt] \tag{5.16}$$

where $P(x_0; x)$ is the propagator for the Fokker-Planck equation $\partial_t P = (1/2)\nabla^2 P + \nabla\cdot(\mathbf{w}P)$ provided one makes the identifications

$$\dot{\mathbf{q}} = \sqrt{-w(M)\lambda m}\mathbf{v};\ \nabla_x = \sqrt{-w(M)\lambda m}\nabla_q;\ \psi = Pe^{-2mc^2}; \tag{5.17}$$

$$\mathbf{w} = \sqrt{-w(M)\lambda/m}\tilde{\mathbf{W}};\ 2w(M)\lambda(mc^2 + \tilde{\mathbf{W}}^0) = \mathbf{w}^2 - \nabla\cdot\mathbf{w}$$

(cf. [**616, 908, 910, 926, 1138**]). Carrying out the substitutions and setting $\lambda\tilde{\mathbf{W}}_\mu = -U(\lambda\phi, \mathbf{A})$ one obtains $\psi(x) = \int \psi(x')G(x, x')dx'$ as a solution to

$$\frac{1}{w(M)\lambda}\partial_t\psi = -\frac{1}{2m[w(M)\lambda]^2}[\nabla + w(M)\lambda\mathbf{A}]^2 + (mc^2 + U\phi)\psi \tag{5.18}$$

with initial condition $\psi = \psi(x')$ (this should be checked to clarify the roles of U and ϕ). If one sets $\lambda = \hbar^{-1}$ and the time is allowed to become imaginary the SE minimally coupled to EM arises. Thus choose $\lambda = \hbar^{-1}$ but leave time alone since it is not needed; then (5.18) can be interpreted as a stochastic form of QM. Evidently the Weyl weight serves the function of i, changing sign appropriately for the conjugate field. The emergence of the Fokker-Planck equation indicates diffusion and this is discussed at length in [**234, 908, 910, 926, 1138**]. In addition the matter is discussed in [**1299**] from various points of view. In particular one takes $(1/\hbar)S = \int W_\mu u^\mu d\tau$ and observes that a classical limit of the Weyl geometry will exist whenever there is an extremum to the action (as in the Feynman path integral). Thus a classical limit of (5.11) occurs whenever $\Psi = exp[w(M)\int_{x_0}^x W_\mu u^\mu d\tau]$ is extremal. However there is a difference here involving Ψ as a length factor. One shows that $\delta\Psi = 0$ corresponds to a special case of the Weyl field since $\int_A^B d\tau(W_{\mu,\nu} - W_{\nu,\mu})u^\mu\delta x^\nu = 0$ arises via variation which means $W_{\mu\nu}u^\nu = 0$. Some calculation then shows that $W_\alpha = \xi\partial_\alpha\chi$ (up to a gauge transformation) for any appropriately normalized functions ξ, χ satisfying

$$(D_\mu\chi)u^\nu = (D_\mu\chi)v^\mu = (D_\mu\xi)u^\nu = (D_\mu\xi)v^\mu = 0; \tag{5.19}$$

$$(1/2)\epsilon^{\mu\nu\alpha\beta}W_{\alpha\beta} = u^\mu v^\nu - u^\nu v^\mu$$

with ϵ the Levi-Civita tensor (cf. [**377, 1299**]). Now $W_\alpha = \xi\partial_\alpha\chi$ is a rather remarkable relation; it represents a restricted form of W^α since it is easy to find a Weyl vector such that $W_{\mu\nu}u^\nu \sim W_{\mu 0} \neq 0$ for all nonspacelike u^ν. Since this formula arises for an arbitrary set of paths u^α it is clear that not all Weyl fields will have a classical limit. Thus as argued at the beginning the generic Weyl geometry lacks preferred paths and requires a path average. On the other hand if one chooses a gauge where $W_\alpha u^\alpha = 0$ (which is possible) then weightful bodies followed the preferred classical trajectories and experience no dilation. There is considerable discussion along these lines in [**1299**] which is omitted here; there is also interesting material on relations to general relativity. In particular it is pointed out that size changes associated with nonvanishing Weyl field strength are not necessarily classically observable. However the Weyl field itself must be present and consequently must be detectable. Finding the physical field that it corresponds to simply requires substituting the appropriate conjugate momentum for W_μ in the classical equation of motion $W_{\mu\nu}u^\nu = 0$. Since the only long range forces are gravity and EM and gravity is still accounted for by the Riemannian curvature, W_μ must be electromagnetic. The most general classical conjugate momentum is therefore that of a point particle with charge q moving in an EM field. Then in an arbitrary gauge

$$W_\mu = (1/\hbar)(p_\mu + qA_\mu + \partial_\mu\Lambda) \tag{5.20}$$

where $p_\mu = mu_\mu$ and $u_\mu u^\mu = -1$. Then

$$0 = W_{\mu\nu}u^\nu = (1/\hbar)(p_{\mu,\nu} - p_{\nu,\mu} + qA_{\mu,\nu} - qA_{\nu\mu})u^\nu \tag{5.21}$$

or (using $(u_\mu u^\mu)_{,\nu} = 0$) $dp^\mu/d\tau = qu_\nu F^{\mu\nu}$ which is the Lorenz force law (note that Planck's constant drops out). For the interpretation of W_μ itself one can combine the curl of (5.20) with

$$W_{\alpha\beta} = D_\alpha\chi D_\beta\xi - D_\beta\chi D_\alpha\xi = \partial_\alpha\chi\partial_\beta\xi - \partial_\beta\chi\partial_\alpha\xi \tag{5.22}$$

(cf. (5.19) and the surrounding discussion); this leads to

$$\partial_\alpha\chi\partial_\beta\xi - \partial_\beta\chi\partial_\alpha\xi = (1/\hbar)(p_{\alpha,\beta} - p_{\beta,\alpha} + qA_{\alpha,\beta} - qA_{\beta,\alpha}) \tag{5.23}$$

the time component of which gives again the Lorenz law. The spatial components can be solved for the magnetic field to give

$$\mathbf{B} = (\hbar/q)(\nabla\chi \times \nabla\xi) - (m/q)(\nabla \times \mathbf{v}) \tag{5.24}$$

The two fields χ and ξ on the right side of $\mathbf{B}$ are sufficient to guarantee the existence of any type of physical magnetic field. Conversely one can use (5.24) to solve for the Weyl field in terms of $\mathbf{B}$ and $\mathbf{v}$ (which of course depend on $\hbar$). One notes that for vanishing Weyl field (5.24) reduces to the London equation for superconductivity. This means that matter fields which conspire to produce a Riemannian geometry become superconducting.

5.1. MEASUREMENT ON A BICONFORMAL SPACE. We continue the theme of Section 3.5 with a more general perspective from Anderson-Wheeler [**42**] based on biconformal geometry (cf. Appendix E for some background material and see also [**42, 43, 136, 666, 759, 1295, 1284, 1285, 1299, 1300, 1301, 1302, 1303, 1304**]). We regard this approach via biconformal geometry as very interesting and will try to present it faithfully. The background material in Appendix E should be read first; results in J. Wheeler [**1304**] for example create a unified geometrical theory of gravity and electromagnetism based on biconformal geometry. One develops in [**42**] an interpretation for quantum behavior within the context of biconformal gauge theory based on the following postulates:

(1) A σ_C biconformal space provides the physical arena for quantum and classical physics.
(2) Quantities of vanishing conformal weight comprise the class of physically meaningful observables.
(3) The probability that a system will follow any given infinitesimal displacement is inversely proportional to the dilatation the displacement produces in the system.

From these assumptions follow the basic properties of classical and quantum mechanics. The symplectic structure of biconformal space is similar to classical phase space and also gives rise to Hamilton's equations, Hamilton's principal function, conjugate variables, fundamental Poisson brackets, and Liouville theory when postulate 3 is replaced by a postulate of extremal motion. We sketch this here (somewhat brutally) and refer to [**42**] for details, philosophy, and further references; the details for the biconformal geometry are spelled out in [**1302, 1304**]. Thus one wants a physical arena which contains 4-D spacetime in a straightforward manner but which is large enough and structured so as to contain both general relativity (GR) and quantum theory (QT) at the same time. One demands therefore invariance under global Lorentz transformations, translations, and scalings (see below) and the Lie group characterizing this is the conformal group $O(4,2)$ or its covering group $SU(2,2)$. In Appendix E the basic facts about Lorentz transformations $M^a_{\ b} = -M_{ba} = \eta_{ac}M^c_{\ b}$, translations P_a, special conformal transformations K^a, and dilatations D are exhibited in the context of conformal gauge theory $(a, b = 0, 1, 2, 3)$. One has two involutive automorphisms of the conformal algebra, first

$$\sigma_1 : (M^a_{\ b}, P_a, K^a, D) \to (M^a_{\ b}, -P_a, -K^a, D) \tag{5.25}$$

which identifies the residual local Lorentz and dilatation symmetry characteristic of biconformal gauging and this corresponds (resp. for the Poincaré Lie algebra or the Weyl algebra) to

$$\sigma_1 : (M^a_{\ b}, P_a) \to (M^a_{\ b}, -P_a) \ \textit{or}\ \sigma_1 : (M^a_{\ b}, P_a, D) \to (M^a_{\ b}, -P_a, D) \tag{5.26}$$

There is also a second involution for the conformal group, namely

$$\sigma_2 : (M^a_{\ b}, P_a, K^a, D) \to (M^a_{\ b}, K_a, P^a, -D) \tag{5.27}$$

Some representations of the conformal algebra, namely $su(2,2)$, are necessarily complex and σ_2 can be realized as complex conjugation. Specifically one thinks of a representation in which P_a and K^a are complex conjugates while $M^a_{\ b}$ is real and D is purely imaginary and such representations will be called σ_C representations. Biconformal spaces for which the connection 1-forms (and hence curvatures) have this property are then called σ_C spaces (see Appendix E for examples). This leads to postulate 1 above, namely the physical arena for QT and classical physics is a

σ_C biconformal space. Now biconformal gauging of the conformal group provides in particular a symplectic structure as follows. Gauging D introduces a single gauge 1-form ω (the Weyl vector) and the corresponding dilatational curvature 2-form is

$$\Omega = d\omega - 2\omega^a\omega_a \tag{5.28}$$

where ω^a, ω_a are 1-form gauge fields for the translation and special conformal transformations respectively, which span an 8-dimensional space as an orthonormal basis (note $\omega_a = \eta_{ab}\bar{\omega}^b$ for σ_C representations and products are wedge products). Now for all torsion free solutions to the biconformal field equations (i.e. $*d*d\omega_0^0 = J$, $\omega_a^0 = T_a + \cdot$, etc. - cf. Appendix E) the dilatational curvature takes the form (•) $\Omega = \kappa\omega^a\omega_a$ with κ constant, so the structure equation becomes (••) $d\omega = (\kappa+2)\omega^a\omega_a$. As a result $d\omega$ is closed and nondegenerate and hence symplectic (since ω^a, ω_a span the space). There is also a biconformal metric arising from the group invariant Killing metric $K_{\Sigma\Pi} = c^{\Lambda}_{\Delta\Sigma}c^{\Delta}_{\Lambda\Pi}$ where $c^{\Lambda}_{\Delta\Sigma}$ $(\Sigma, \Pi, \cdots = 1, 2, \cdots, 15)$ are the real structure constants from the Lie algebra. This metric has a nondegenerate projection to the 8-D subspace spanned by P_a, K^a and provides a natural pseudo-Riemannian metric on biconformal manifolds. The projection takes the form

$$K_{AB} = \begin{pmatrix} & \eta_{ab} \\ \eta_{ab} & \end{pmatrix} \quad (A, B = 0, 1, \cdots, 7) \tag{5.29}$$

One defines now conformal weights w of a definite weight field F via (♦) $D_\phi : F \to [exp(w\phi)]F$ where D_ϕ is dilatation by $exp(\phi)$ (cf. [**1299**] and Section 3.5). One assumes now postulate 2 and concludes that for a field with nontrivial Weyl weight to have physical meaning it must be possible to construct weightless scalars by combining it with other fields (easily done with conjugate fields); one notes that zero weight fields are self conjugate. The symplectic form $\Theta = \omega^a\omega_a$ defines a symplectic bracket via

$$\{f, g\} = \Theta^{MN}\frac{\partial f}{\partial u^M}\frac{\partial g}{\partial u^N} \tag{5.30}$$

where $u^M = (x^a, y^b)$. For real solutions f, g to the field equations f and g are conjugate if they satisfy $\{f, f\} = 1$, $\{f, f\} = \{g, g\} = 0$. However for σ_C representations ω is a pure imaginary 1-form since it is defined as the dual to the dilatation generator D which is pure imaginary. One sees then that

$$\overline{\omega^a\omega_a} = \bar{\omega}^a\bar{\omega}_a = \eta^{ab}\omega_b\eta_{ac}\omega^c = -\omega^a\omega_a \tag{5.31}$$

so the dilatational curvature and the symplectic form are imaginary (cf. also [**42, 710**]). Consequently, for use of a complex gauge vector with real gauge transformations, the fundamental brackets should take here the form

$$\{f, g\} = i; \ \{f, f\} = \{g, g\} = 0; \ w_f = -w_g \tag{5.32}$$

In an arbitrary biconformal space one sets either

$$\frac{1}{\hbar}S = \frac{1}{\hbar}\int Ld\lambda = \int\omega = \int(W_a dx^a + \bar{W}_a dy^a) \ or \tag{5.33}$$

$$\frac{i}{\hbar}S = \frac{i}{\hbar}\int Ld\lambda = \int\omega = \int(W_a dx^a + \bar{W}_a dy^a)$$

The second form holds in a σ_C representation for the conformal group. An arbitrary parameter λ is OK since the integral of the Weyl 1-form is independent of parametrization. This integral also governs measurable size change since under parallel transport the Minkowski length of a vector V^a changes by

$$\ell = \ell_0 exp \int \omega; \ \ell^2 = \eta_{ab} V^a V^b \tag{5.34}$$

(cf. Appendix E). This change occurs because $\eta_{ab} = (-1, 1, 1, 1)$ is not a natural structure for biconformal space. This is in contrast to the Killing metric K_{AB} where lengths are of zero conformal weight. In a σ_C representation the Weyl vector is imaginary so the measurable part of the change in ℓ is not a real dilatation - rather, it is a change of phase. Now for classical mechanics one uses a variation of postulate 3, namely: **The motion of a (classical) physical system is given by extrema of the integral of the Weyl vector**. Biconformal spaces are real symplectic manifolds so the Weyl vector can be chosen so that the symplectic form satisfies the Darboux theorem $\omega = W_a dx^a = -y_z dx^a$; for σ_C representations the Darboux equations still holds but now with

$$\omega = W_a dx^a = -iy_a dx^a \tag{5.35}$$

and the classical motion is independent of which form is chosen. Thus the symplectic form for the σ_C case is $\Theta = d\omega = -idy_a dx^a$ and one has (♦♦) $\{x^a, y_b\} = i\delta^a_b$. Thus from (♦♦) it follows that y_b is the conjugate variable to the position coordinate x^b and in mechanical units one may set $y_a = \alpha p_a$ with

$$i\alpha S = \int \omega = -i\alpha \int (p_0 dt + p_i dx^i) \tag{5.36}$$

(α can be any constant with appropriate dimensions). Now if one requires t as an invariant parameter (so $\delta t = 0$) one can vary the corresponding canonical bracket to find

$$0 = \delta\{t, p_0\} = \{\delta t, p_0\} + \{t, \delta p_0\} = \frac{\partial(\delta p_0)}{\partial p_0} \tag{5.37}$$

Thus δp_0 can depend only on the remaining coordinates so $\delta p_0 = -\delta H(y_i, x^j, t)$ and the existence of a Hamiltonian is a consequence of choosing time as a nonvaried parameter of the motion. Applying the postulate $\delta S = 0$ variation leads to

$$0 = i\alpha\delta S = -i\alpha \int (\delta p_0 dt + \delta p_i dx^i - dp_i \delta x^i) = \tag{5.38}$$

$$= -i\alpha \int \left(-\frac{\partial H}{\partial x^i} \delta x^i dt - \frac{\partial H}{\partial p_i} \delta p_i dt + \delta p_i dx^i - dp_i \delta x^i \right)$$

and this gives the standard Hamilton's equations

$$0 = -\frac{\partial H}{\partial p_i} dt + dx^i; \ 0 = -\frac{\partial H}{\partial x^i} dt - dp_i \tag{5.39}$$

(note i and α drop out of the equations).

In the presence of nonvanishing dilatational curvature one then considers a classical experiment to measure size (or phase) change along C_1, while a ruler measured by λ moves along C_2 (C_i are classical paths between two fixed points). Some argument (see [**42**]) leads to an unchanged ratio of lengths via

$$\frac{\ell}{\lambda} = \frac{\ell_0}{\lambda_0} exp \int_{C-1-C-2} \omega = \frac{\ell_0}{\lambda_0} exp \oint \omega = \frac{\ell_0}{\lambda_0} exp \int\int_S d\omega = \frac{\ell_0}{\lambda_0} \tag{5.40}$$

where S is any surface bounded by the closed curve $C_1 - C_2$ (cf. also Section 3.5). Thus no dilatations are observable along classical paths. This calculation also shows that the restriction of ω to classical paths is exact and proves the existence of Hamilton's principal function S with

$$\alpha S(x) = \int^x W_a dx^a = \int^x W_a \frac{dx^a}{dt} dt \tag{5.41}$$

There is further argument in [**42**] via gauge freedom to show that classical objects do not exhibit measurable length change (in the complex case the phase changes cannot be removed by gauge choice but they are unobservable). Relations between phase space and biconformal space are discussed and one arrives at QM.

From the above one knows that there is no measurable size change along classical pathes in a biconformal geometry but for systems evolving along other than extremal paths (where the Hamilton equations do not apply for example) there may be measurable dilatation. To deal with this one needs nonclassical motion and one goes to the basic postulate 3, namely that the probability a system will follow any given infinitesimal displacement is inversely proportional to the dilatation the displacement produces in the system. The properties of biconformal space determine the evolution of Minkowski lengths along arbitrary curves and the imaginary Weyl vector produces measurable phase changes in the same way as the wave function. Combining this with the classically probabilistic motion of postulate 3, together with the necessary use of a standard of length to comply with postulate 2, one concludes that the probability of a system at x_0^a arriving at the point x_1^a is given by

$$P(x_1^i) = \int \mathcal{D}[x_{C'}] exp\left(\int_{C'} \omega\right) \int \mathcal{D}[x_C] exp\left(-\int_C \omega\right) = \tag{5.42}$$

$$= \mathcal{P}(x_1^i)\mathcal{P}(-x_1^i) = \mathcal{P}(x_1^i)\bar{\mathcal{P}}(x_1^i)$$

where a path average over all paths connecting the two points is involved and $\bar{\mathcal{P}}(x)$ is simulaneously the probability amplitude of the conformally conjugate system reaching x_1^i. Here one considers ratios ℓ/λ as above and includes all possible ruler paths. These are standard Feynman path integrals which are known to lead to the Schrödinger equation (not Wiener integrals as in [**1299**]) and it is the requirement of a length standard that forces the product structure in (5.42). Note that the phase invariance of a wave function $\psi' = exp(i\phi)\psi$ is created by the σ_C conformal invariance $M' = exp(\lambda w)M$. The i in the Weyl vector is the crucial i noted by London in [**828**] (cf. [**1299**] and Section 3.5). Note also that the path integral in (5.42) and the biconformal paths depend generically on the spacetime and momentum variables so one can immediately generalize to the usual integrals of QM, namely

$$\mathcal{P}(x_1^i) = \int \mathcal{D}[x_C]\mathcal{D}[y_C]exp(\int C\omega) \tag{5.43}$$

Note also that the failure of the base space to break into space like and momentum like submanifolds indicates a fundamental coupling between position and momentum and suggests a connection to the Heisenberg uncertainty principle. The arguments in [**42**] have a somewhat heuristic flavor at times but are certainly plausible and do refine the techniques of [**1299**] (sketched in Section 3.5) in many ways. Given the success of biconformal geometry in unifying GR and EM it would seem only natural and just that QM could be encompassed as well in the same framework.

REMARK 3.5.1 We note from [**1303**] that when identifying biconformal coordinates (x^μ, y_ν) with phase space coordinates (x^μ, p_ν) one sets naturally $y_\nu = \beta p_\nu$. This β must account for a sign difference in $\eta^{\mu\nu}\beta p_\mu \beta p_\nu = -\eta^{\mu\nu}y_\mu y_\nu$ (cf. [**1303**]) so β is pure imaginary. Further to account for the different units of y_ν ($length^{-1}$) and p_ν (momentum) one chooses $y_\nu = (i/\hbar)p_\nu$ and this relation between the geometric variables of conformal gauge theory and the physical momentum variables is the source of complex quantities in QM. ∎

CHAPTER 4

GEOMETRY AND COSMOLOGY

This chapter and the next will cover a number of more or less related topics having to do with cosmology, the zero point field (ZPF), the aether and vacuum, quantum geometry, electromagnetic (EM) phenomena, and Dirac-Weyl geometry.

1. DIRAC WEYL GEOMETRY

A sketch of Dirac Weyl geometry following Dirac [**377**] was given in [**236**] in connection with deBroglie-Bohm theory in the spirit of the Iranian school (cf. [**8, 146, 147, 901, 902, 1104, 1105, 1106, 1107, 1108, 1109, 1110, 1111, 1151, 1152, 1153, 1154, 1155, 1156, 1157, 1158, 1159, 1160, 1161, 1162, 1163, 1164, 1165, 1166, 1167**]). We go now to [**668, 669, 670, 671, 672, 673, 1082**] for generalizations of the Dirac Weyl theory involved in discussing magnetic monopoles, dark matter, quintessence, matter creation, etc. We skip [**670**] where some notational problems seem to arise in the Lagrangian and go to Israelit [**669**] where in particular an integrable Weyl-Dirac theory is developed (the book [**668**] by Israelit is a lovely exposition but the work in [**669**] is somwhat newer). Note, as remarked in [**871**] (where twistors are used), the integrable Weyl-Dirac geometry is desirable in order that the natural frequency of an atom at a point should not depend on the whole world line of the atom. The first paper in [**669**] is designed to investigate the integrable Weyl-Dirac (Int-W-D) geometry and its ability to create massive matter. For example in this theory a spherically symmetric static geometric formation can be spatially confined and an exterior observer will recognize it as a massive entity. This may be either a fundamental particle or a cosmic black hole both confined by a Schwarzschild surface. We summarize again some basic features in order to establish notation, etc. Thus in the Weyl geometry one has a metric $g_{\mu\nu} = g_{\nu\mu}$ and a length connection vector w_μ along with an idea of Weyl gauge transformation (WGT)

$$g_{\mu\nu} \to \tilde{g}_{\mu\nu} = e^{2\lambda} g_{\mu\nu};\ g^{\mu\nu} \to \tilde{g}^{\mu\nu} = e^{-2\lambda} g^{\mu\nu} \quad (1.1)$$

where $\lambda(x^\mu)$ is an arbitrary differerentiable function (cf. also [**401, 402, 1015, 1131, 1132**] for Weyl geometry). One is interested in covariant quantities satisfying $\psi \to \tilde{\psi} = exp(n\lambda)\psi$ where the Weyl power n is described via $\pi(\psi) = n$, $\pi(g_{\mu\nu}) = 2$, and $\pi(g^{\mu\nu}) = -2$. If $n = 0$ the quantity ψ is said to be gauge invariant (in-invariant). Under parallel displacement one has length changes and for a vector

$$(i)\ dB^\mu = -B^\sigma \Gamma^\mu_{\sigma\nu} dx^\nu;\ (ii)\ B = (B^\mu B^\nu g_{\mu\nu})^{1/2};\ (iii)\ dB = Bw_\nu dx^\nu \quad (1.2)$$

(note $\pi(B) = 1$). In order to have agreement between (i) and (iii) one requires

(1.3) $$\Gamma^{\lambda}_{\mu\nu} = \left\{ \begin{array}{c} \lambda \\ \mu\ \nu \end{array} \right\} + g_{\mu\nu}w^{\lambda} - \delta^{\lambda}_{\nu}w_{\mu} - \delta^{\lambda}_{\mu}w_{\nu}$$

where $\left\{ \begin{array}{c} \lambda \\ \mu\ \nu \end{array} \right\}$ is the Christoffel symbol based on $g_{\mu\nu}$. In order for (iii) to hold in any gauge one must have the WGT $w_\mu \to \tilde{w}_\mu = w_\mu + \partial_\mu \lambda$ and if the vector B^μ is transported by parallel displacement around an infinitesimal closed parallelogram one finds

(1.4) $$\Delta B^{\lambda} = B^{\sigma}K^{\lambda}_{\sigma\mu\nu}dx^{\mu}\delta x^{\nu};\ \Delta B = BW_{\mu\nu}dx^{\mu}\delta x^{\nu}$$

where

(1.5) $$K^{\lambda}_{\sigma\mu\nu} = -\Gamma^{\lambda}{}_{\sigma\mu,\nu} + \Gamma^{\lambda}{}_{\sigma\nu,\mu} - \Gamma^{\alpha}{}_{\sigma\mu}\Gamma^{\lambda}_{\alpha\nu} + \Gamma^{\alpha}_{\sigma\nu}\Gamma^{\lambda}{}_{\alpha\mu}$$

is the curvature tensor formed from (1.3) and $W_{\mu\nu} = w_{\mu,\nu} - w_{\nu,\mu}$. Equations for the WGT $w_\mu \to \tilde{w}_\mu$ and the definition of $W_{\mu\nu}$ led Weyl to identify w_μ with the potential vector and $W_{\mu\nu}$ with the EM field strength; he used a variational principle $\delta I = 0$ with $I = \int L\sqrt{-g}d^4x$ with L built up from $K^{\lambda}_{\sigma\mu\nu}$ and $W_{\mu\nu}$. In order to have an action invariant under both coordinate transformations and WGT he was forced to use R^2 (R the Riemannian curvature scalar) and this led to the gravitational field.

Dirac revised this with a scalar field $\beta(x^\nu)$ which under WGT changes via $\beta \to \tilde{\beta} = e^{-\lambda}\beta$ (i.e. $\pi(\beta) = -1$). His in-invariant action integral is then ($f_{,\mu} \equiv \partial_\mu f$)

(1.6) $$I = \int [W^{\lambda\sigma}W_{\lambda\sigma} - \beta^2 R + \beta^2(k-6)w^{\sigma}w_{\sigma} + 2(k-6)\beta w^{\sigma}\beta_{,\sigma} +$$

$$+k\beta_{,\underline{\sigma}}\beta_{,\sigma} + 2\Lambda\beta^4 + L_M]\sqrt{-g}d^4x$$

Here k is a parameter, Λ is the cosmological constant, L_M is the Lagrangian density of matter, and an underlined index is to be raised with $g^{\mu\nu}$. Now according to (1.4) this is a nonintegrable geometry but there may be situations when geometric vector fields are ruled out by physical constraints (e.g. the FRW universe). In this case one can preserve the WD character of the spacetime by assuming that w_ν is the gradient of a scalar function w so that $w_\nu = w_{,\nu} = \partial_\nu w$. One has then $W_{\mu\nu} = 0$ and from (1.4) results $\Delta B = 0$ yielding an integrable spacetime (Int-W-D spacetime). To develop this begin with (1.6) but with w_ν given by $w_\nu = \partial_\nu w$ so the first term in (1.6) vanishes. The parameter k is not fixed and the dynamical variables are $g_{\mu\nu}$, w, and β. Further it is assumed that L_M depends on ($g_{\mu\nu}$, w, β). For convenience write

(1.7) $$b_\mu = (log(\beta))_{,\mu} = \beta_{,\mu}/\beta$$

and use a modified Weyl connection vector $W_\mu = w_\mu + b_\mu$ which is a gauge invariant gradient vector. Write also $k - 6 = 16\pi\kappa$ and varying w in (1.6) one gets a field equation

(1.8) $$2(\kappa\beta^2 W^{\nu})_{;\nu} = S$$

where the semicolon denotes covariant differentiation with the Christoffel symbols and S is the Weylian scalar charge given by $16\pi S = \delta L_M/\delta w$. Varying $g_{\mu\nu}$ one gets also

$$G^\nu_\mu = -8\pi\frac{T^\nu_\mu}{\beta^2} + 16\pi\kappa\left(W^\nu W_\mu - \frac{1}{2}\delta^\nu_\mu W^\sigma W_\sigma\right) + \tag{1.9}$$

$$+2(\delta^\nu_\mu b^\sigma_{;\sigma} - b^\nu_{;\mu}) + 2b^\nu b_\mu + \delta^\nu_\mu b^\sigma_\sigma - \delta^\nu_\mu \beta^2\Lambda$$

where G^ν_μ represents the Einstein tensor and the energy momentum density tensor of ordinary matter is

$$8\pi\sqrt{-g}T^{\mu\nu} = \delta(\sqrt{-g}L_M)/\delta g_{\mu\nu} \tag{1.10}$$

Finally the variation with respect to β gives an equation for the β field

$$R + k(b^\sigma_{;\sigma} + b^\sigma b_\sigma) = 16\pi\kappa(w^\sigma w_\sigma - w^\sigma_{;\sigma}) + 4\beta^2\Lambda + 8\pi\beta^{-1}B \tag{1.11}$$

Note in (1.11) R is the Riemannian curvature scalar and the Dirac charge B is a conjugate of the Dirac gauge function β, namely $16\pi B = \delta L_M/\delta\beta$.

By a simple procedure (cf. Dirac [**377**]) one can derive conservation laws; consider e.g. $I_M = \int L_M\sqrt{-g}d^4x$. This is an in-invariant so its variation due to coordinate transformation or WGT vanishes. Making use of $16\pi S = \delta L_M/\delta w$, (1.10), and $16\pi B = \delta L_M/\delta\beta$ one can write

$$\delta I_M = 8\pi\int(T^{\mu\nu}\delta g_{\mu\nu} + 2S\delta w + 2B\delta\beta)\sqrt{-g}d^4x \tag{1.12}$$

Via $x^\mu \to \tilde{x}^\mu = x^\mu + \eta^\mu$ for an arbitrary infinitesimal vector η^μ one can write

$$\delta g_{\mu\nu} = g_{\lambda\nu}\eta^\lambda_{;\mu} + g_{\mu\lambda}\eta^\lambda_{;\nu};\ \delta w = w_{,\nu}\eta^\nu;\ \delta\beta = \beta_{,\nu}\eta^\nu \tag{1.13}$$

Taking into account $x^\mu \to \tilde{x}^\mu$ we have $\delta I_M = 0$ and making use of (1.13) one gets from (1.12) the energy momentum relations

$$T^\lambda_{\mu;\lambda} - Sw_\mu - \beta B b_\mu = 0 \tag{1.14}$$

Further considering a WGT with infinitesimal $\lambda(x^\mu)$ one has from (1.12) the equation $S+T-\beta B = 0$ with $T = T^\sigma_\sigma$. One can contract (1.9) and make use of (1.8) and $S + T = \beta B$ giving again (1.11), so that (1.11) is a corollary rather than an independent equation and one is free to choose the gauge function β in accordance with the gauge covariant nature of the theory. Going back to the energy-momentum relations one inserts $S + T = \beta B$ into (1.14) to get $T^\lambda_{\mu;\lambda} - Tb_\mu = SW_\mu$. Now go back to the field equation (1.9) and introduce the energy momentum density tensor of the W_μ field

$$8\pi\Theta^{\mu\nu} = 16\pi\kappa\beta^2[(1/2)g^{\mu\nu}W^\lambda W_\lambda - W^\mu W^\nu] \tag{1.15}$$

Making use of (1.8) one can prove $\Theta^\lambda_{\mu;\nu} - \Theta b_\mu = -SW_\mu$ and using $T^\lambda_{\mu;\lambda} - TB_\mu = SW_\mu$ one has an equation for the joint energy momentum density

$$(T^\lambda_\mu + \Theta^\lambda_\mu)_{;\lambda} - (T + \Theta)b_\mu = 0 \tag{1.16}$$

One can derive now the equation of motion of a test particle (following [**1082**]). Consider matter consisting of identical particles with rest mass m and Weyl scalar charge q_s, being in the stage of a pressureless gas so that the energy momentum density tensor can be written in the form $T^{\mu\nu} = \rho U^\mu U^\nu$ where U^μ is the 4-velocity and the scalar mass density ρ is given by $\rho = m\rho_n$ with ρ_n the particle density. Taking into account the conservation of particle number one obtains from $T^\lambda_{\mu;\lambda} - Tb_\mu = SW_\mu$ the equation of motion

$$\frac{dU^\mu}{ds} + \left\{ \begin{matrix} \mu \\ \lambda\ \sigma \end{matrix} \right\} U^\lambda U^\sigma = \left(b_\lambda + \frac{q_s}{m} W_\lambda\right)(g^{\mu\lambda} - U^\mu U^\lambda) \tag{1.17}$$

In the Einstein gauge ($\beta = 1$) we are then left with

$$\frac{dU^\mu}{ds} + \left\{ \begin{matrix} \mu \\ \lambda\ \sigma \end{matrix} \right\} U^\lambda U^\sigma = \frac{q_s}{m} w_\lambda (g^{\mu\lambda} - U^\mu U^\lambda) \tag{1.18}$$

EXAMPLE 1.1. Following Israelit [**669**] one considers a static spherically symmetric situation with line element

$$ds^2 = e^\nu dt^2 - e^\lambda dr^2 - r^2(d\theta^2 + Sin^2(\theta) d\phi^2) \tag{1.19}$$

and all functions λ, ν, β, w, T^ν_μ, S, B depend only on r. One looks for local phenomena so $\Lambda = 0$. The field equations (1.9) can be written explicitly for $(\mu\nu) = (0,0)$, $(1,1)$, $(2,2)$, or $(3,3)$ to obtain

$$e^{-\lambda}\left(-\frac{\lambda'}{r} + \frac{1}{r^2}\right) - \frac{1}{r^2} = -\frac{8\pi T^0_0}{\beta^2} + \tag{1.20}$$

$$+2e^{-\lambda}\left(-\frac{(b')^2}{2} - b'' + \frac{\lambda' b'}{2} - \frac{2b'}{r}\right) + 8\pi\kappa e^{-\lambda}(W')^2;$$

$$e^{-\lambda}\left(\frac{\nu'}{r} + \frac{1}{r^2}\right) - \frac{1}{r^2} = -\frac{8\pi T^1_1}{\beta^2} - 2e^{-\lambda}\left(\frac{\nu' b'}{2} + \frac{2b'}{r} + \frac{3(b')^2}{2}\right) - 8\pi\kappa e^{-\lambda}(W')^2;$$

$$\frac{1}{4}\left(\nu'' + \frac{(\nu')^2}{2} + \frac{\nu' - \lambda'}{r} - \frac{\nu'\lambda'}{2}\right) =$$

$$= -\frac{4e^\lambda \pi T^2_2}{\beta^2} - \left(b'' + \frac{(\nu' - \lambda')b'}{2} + \frac{b'}{r} + \frac{(b')^2}{2}\right) + 4\pi\kappa e^{-\lambda}(W')^2$$

From (1.8) one has the equation for the W field

$$2\kappa\left[W'' + \left(2b' + \frac{\nu' - \lambda'}{2} + \frac{2}{r}\right) W'\right] = -\frac{e^\lambda S}{\beta^2} \tag{1.21}$$

The most intriguing situation is when ordinary matter is absent, so $T^\mu_\nu = 0$, and then from $S + T = \beta B$ and $T^\lambda_{\mu;\lambda} - Tb_\mu = SW_\mu$ one has $S = 0$ and $B = 0$. Take first the simple case when $W' = 0$ or $\kappa = 0$ so (1.21) is satisfied identically and (1.20) takes the simple form

$$e^{-\lambda}\left(-\frac{\lambda'}{r} + \frac{1}{r^2}\right) - \frac{1}{r^2} = 2e^{-\lambda}\left(-\frac{(b')^2}{2} - b'' + \frac{\lambda' b'}{2} - \frac{2b'}{r}\right); \tag{1.22}$$

$$e^{-\lambda}\left(\frac{\nu'}{r} + \frac{1}{r^2}\right) - \frac{1}{r^2} = -2e^{-\lambda}\left(\frac{\nu' b'}{2} + \frac{2b'}{r} + \frac{2(b')^2}{2}\right);$$

$$\frac{e^{-\lambda}}{2}\left(\nu'' + \frac{(\nu')^2}{2} + \frac{\nu' - \lambda'}{r} - \frac{\nu'\lambda'}{2}\right) = -2e^{-\lambda}\left(b'' + \frac{\nu' - \lambda'}{2} + \frac{b'}{r} + \frac{(b')^2}{2}\right)$$

Subtracting the first equation from the second one obtains $(1/r)(\lambda' + \nu') = 2b' - (\lambda' + \nu')b' - 2(b')^2$. The scalar $b = log(\beta)$ is still arbitrary so that one can impose a condition on it. Thus writing $b'' - (b')^2 = 0$ one can integrate to get $b(r) = log[1/(a - cr)]$ (curiously enough this is true) with a, c arbitrary constants which are taken to be positive. Using $b'' = (b')^2$ one obtains the equation $\lambda' + \nu' = 0$ and hence via $b = log[1/(a - cr)]$ there rsults from (1.22) a solution $exp(\nu) = exp(-\lambda) = (a - cr)^2/a^2$. Now go back to the Einstein equations $G^{\nu}_{\mu} = -8\pi T^{\nu}_{\mu}$; if one thinks of β as creating matter we can then calculate the matter density and pressure. From (1.22) the density is given by $8\pi\rho = -(3c^2/a^2) + (4c/ar)$ and the radial pressure is $P_r = -\rho$ (so there is tension rather than pressure). One notes that $P_r = -\rho$ has been used as the equation of state of prematter in cosmology (cf. [**672**]). Finally one can calculate the transverse pressure from (1.22) as $8\pi P_t = (3c^2/a^2) - (2c/ar)$ (which is anisotropic). Now suppose there is a spherically symmetric body filled with matter described by $8\pi\rho = -3c^2/a^2 + 4c/ar$, $P_r = -\rho$, and $8\pi P_t = (3c^2/a^2(-(2c/ar)$. Since the matter density can take only nonnegative values one has a limit on the size of the body $r_{boundary} \leq (4a/3c)$. Several models are possible; take e.g. a body with maximum radius $r_{bound} = 4a/3c$. One sees that on the boundary the density and radial pressure vanish so that this is an open model. Go back for a moment to the first equation in (1.3). It may be integrated, giving $exp(-\lambda) = exp(\nu) = 1 - (8\pi/r)\int_0^r \rho r^2 dr$. Assume that outside of the body the Einstein gauge holds, i.e. $\beta = 1$ ($b = 0$) for $r > (4a/3c)$ so that one is left with the ordinary Riemannian geometry and with the exterior Schwarzschld solution $exp(-\lambda) = exp(\nu) = 1 - (2m/r)$. Comparing and using the equations at hand one obtains

$$m = 4\pi \int_0^{4a/3c} \rho r^2 dr = (16a/27c) = (4/9) r_{bound} \tag{1.23}$$

Note that in the body (at $r_s = a/c$) there is a singularity of β and of the metric; however the physical quantities ρ, P_r, P_t are regular there (cf. Israelit-Rosen [**673**]). An external observer staying in the Riemannian spacetime will recognize the above entity, made of Weyl-Dirac geometry, as a body having mass (1.23) and radius $4a/3c$. ∎

EXAMPLE 1.2. Another example of matter creation via geometry is also given in [**669**] with a homogeneous and isotropic FRW universe and line element

$$ds^2 = dt^2 - R^2(t)\left[\frac{dr^2}{1 - \tilde{k}r^2} + r^2 d\Omega^2\right] \tag{1.24}$$

Here $R(t)$ is the cosmic scale factor, $\tilde{k} = 0, \pm 1$ stands for the spatial curvature parameter, and $d\Omega^2 = d\theta^2 + Sin^2(\theta)d\phi^2$ is the line element on the unit sphere. The universe is filled with ordinary cosmic matter in the state of a perfect fluid at rest and with the cosmic scalar fields β and w. One considers, for $\kappa \geq 0$, $T^0_0 = \rho(t)$, $T^1_1 = T^2_2 = T^3_3 = -P(t)$, $T = \rho - 3P$, $\beta = \beta(t)$, and $= w(t)$. Use also the Einstein gauge $\beta = 1 \sim b = 0$ and the Einstein-Friedmann equations (cf. [**669**]). After much calculation one looks at the expansion of the universe (with no ordinary matter) and the model provides a high rate of matter creation from an initial empty egg (i.e. geometry brings matter into being). Another model along similar lines looks at interaction between geometric fields and matter during radiation and dust dominated periods with a number of interesting results. In particular matter creation takes place in the radiation dominated universe and also for open and flat models in a dust dominated era while in a closed dust universe there is matter creation for awhile after which matter annihilation arises stimulated by the w field. ■

EXAMPLE 1.3. We go now to the second paper in Israelit [**668**] which builds up a singularity free cosmological model that originates from pure geometry. The Planckian state (characterized by $\rho_P = c^3/\hbar G = 3.83 \cdot 10^{65}\ cm^{-2}$, $R_I = (3/8\pi\rho_P)^{1/2} = 5.58 \cdot 10^{-34}\ cm$, and $T_I = 2.65 \cdot 10^{-180}\ K$) is preceeded by a pre-Planckian period. This starts from a primary empty spacetime entity, described by an integrable WD geometry. During the pre-Planckian period geometry creates cosmic matter and a the end of this creation process one has the Planckian cosmic egg filled with prematter. The prematter model of [**669**] will be updated according to present observational data and also modified by the introduction of a nonzero cosmological constant. Thus, reviewing a bit, we have $g_{\mu\nu}$, β, and w_μ as above with $w_\mu = \partial{-}\mu w$ to provide an integrable WD theory. There is an action (1.6) and one uses (1.7) and $W_\mu = w_\mu + b_\mu$. Also $k-6 = 16\pi\kappa$ (here σ is used in place of κ and later σ becomes $-\kappa^2$ so we will think of $\kappa \to -\kappa^2$ later on. As before one has (1.8), $16\pi S = \delta L_M/\delta w$, (1.9), (1.10), (1.11), $16\pi B = \delta L_M/\delta B$, $I_M = \int L_M\sqrt{-g}d^4x$, (1.14), $S+T = \beta B$, and $T^\lambda_{\mu;\lambda} - Tb_\mu = SW_\mu$. Recall also (1.11) is a corollary so that β and the Dirac charge B can be chosen arbitrarily. Further (1.15), (1.16), etc. will still apply after which one has an Int-W-D theory. One considers a homogeneous isotropic spatially closed ($\tilde{k} = 1$) universe described by the FRW line element of (1.24). For a universe filled with cosmic matter in the state of a perfect fluid at rest its EM tensor has nonvanishing components T^i_i ($i = 0, 1, 2, 3$) and in addition to any matter field there are two cosmic scalar fields $\beta(t)$ and $W(t)$ stemming from the geometric framework. Taking into account (1.24), one obtains from (1.9) the cosmological equations

$$\text{(1.25)} \qquad \frac{\dot{R}^2}{R^2} = \frac{8\pi T^0_0}{3\beta^2} - \frac{8\pi\sigma\dot{W}^2}{3} - \frac{2\dot{R}\dot{b}}{R} - \dot{b}^2 + \frac{\beta^2\Lambda}{3} - \frac{1}{R^2}$$

$$\frac{\ddot{R}}{R} = \frac{4\pi}{\beta^2}\left(T^1_1 - \frac{T^0_0}{3}\right) + \frac{16\pi\sigma\dot{W}^2}{3} - \ddot{b} - \frac{\dot{R}\dot{b}}{R} + \frac{\beta^2\Lambda}{3}$$

From (1.9) one also gets the trace equation

$$\text{(1.26)} \qquad G^\sigma_\sigma = -\frac{8\pi T}{\beta^2} - 16\pi\sigma\dot{W}^2 + 6\ddot{b} + 18\frac{\dot{R}\dot{b}}{R} + 6\dot{b}^2 - 4\beta\Lambda$$

Comparing (1.25) with the usual Einstein equations for cosmology one concludes that the observable density and pressure of the cosmic matter are $\rho = T^0_0/\beta^2$ and $P = -T^1_1/\beta^2$. Now we let $\sigma \to -\kappa^2$ and regard $\beta(t)$ as a function of $R(t)$ so that

$$\text{(1.27)} \qquad \beta = \beta(R),\ \dot{\beta} = \beta'\dot{R};\ \ddot{\beta} = \beta''\dot{R}^2 + \beta'\ddot{R}$$

Taking into account (1.27) one can rewrite (1.25) as

$$\frac{\dot{R}^2}{R^2}\left(1+\frac{\beta' R}{\beta}\right)^2 = \frac{8\pi}{3}(\rho+\kappa^2\dot{W}^2)+\frac{\beta^2\Lambda}{3}-\frac{1}{R^2}; \tag{1.28}$$

$$\frac{\ddot{R}}{R}\left(1+\frac{\beta' R}{\beta}\right)+\frac{\dot{R}^2}{R^2}\left(\frac{\beta'' R^2}{\beta}-\frac{(\beta')^2R^2}{\beta^2}+\frac{\beta' R}{\beta}\right) = -\frac{4\pi}{3}(3P+\rho+4\kappa^2\dot{W}^2)+\frac{\beta^2\Lambda}{3}$$

Further the equation (1.8) of the W field takes the form

$$\ddot{W}+\left(\frac{2\dot{\beta}}{\beta}+\frac{3\dot{R}}{R}\right) = -\frac{S}{2\kappa^2\beta^2} \equiv \partial_t(\dot{W}\beta^2R^3) = -\frac{SR^3}{2\kappa^2} \tag{1.29}$$

With (1.24) (and the density and pressure terms above) the EM relation for matter is now

$$\dot{\rho}+3(\dot{R}/R)(\rho+P)+(\dot{\beta}/\beta)(\rho+3P) = (S\dot{W}/\beta^2) \tag{1.30}$$

and $S+T=\beta B$ can be written as

$$S+(\rho-3P)\beta^2-B\beta = 0 \tag{1.31}$$

For the FRW line element (1.24) the β field equation (1.11) takes the form ($\sigma \sim -\kappa^2$)

$$R^\sigma_\sigma+\kappa\left[\frac{(\dot{b}R^3)_{,t}}{R^3}+\dot{b}^2\right]+16\pi\sigma\left[\frac{(\dot{w}R^3)_{,t}}{R^3}-\dot{w}^2\right]-4\beta^2\Lambda-8\pi\beta^{-1}B = 0 \tag{1.32}$$

By (1.29) and (1.31) this turns out to be identical with the trace equation (1.26) and B may be cancelled from the equations (i.e. β and B are arbitrary - recall that (1.11) is a corollary, etc.). Now introduce the energy density and pressure of the W field via

$$\rho_w = \Theta^0_0;\ P_w = -\Theta^1_1 = -\Theta^2_2 = -\Theta^3_3 \tag{1.33}$$

Making use of (1.24) one obtains $\rho_w = P_w = \kappa^2\beta^2\dot{W}^2$ leading to

$$(\rho+\kappa^2\dot{W}^2)_{,t}+\frac{\dot{\beta}}{\beta}(\rho+3P+4\kappa^2\dot{W}^2)+\frac{3\dot{R}}{R}(\rho+P+2\kappa^2\dot{W}^2) = 0 \tag{1.34}$$

Then introduce the reduced energy density and pressure $\bar{\rho}_w$ and $\bar{P}_w$ via

$$\bar{\rho}_w = \rho_w/\beta^2;\ \bar{P}_w = P_w/\beta^2;\ \bar{\rho}_w = \bar{P}_w = \kappa^2\dot{W}^2 \tag{1.35}$$

and write $\bar{\rho} = \rho+\bar{\rho}_w = \rho+\kappa^2\dot{W}^2$ and $\bar{P} = P+\bar{P}_w = P+\kappa^2\dot{W}^2$; then (1.28) becomes

$$\frac{\dot{R}^2}{R^2}\left(1+\frac{\beta' R}{\beta}\right)^2 = \frac{8\pi\bar{\rho}}{3}+\frac{\beta^2\Lambda}{3}-\frac{1}{R^2}; \tag{1.36}$$

$$\frac{\ddot{R}}{R}\left(1+\frac{\beta' R}{\beta}\right)+\frac{\dot{R}^2}{R^2}\left(\frac{\beta'' R^2}{\beta}-\frac{(\beta')^2R^2}{\beta^2}+\frac{\beta' R}{\beta}\right) = -\frac{4\pi}{3}(3\bar{P}+\bar{\rho})+\frac{\beta^2\Lambda}{3}$$

and the energy momentum relation (1.34) is

$$\bar{\rho}+\frac{\dot{\beta}}{\beta}(\bar{\rho}+3\bar{P})+3\frac{\dot{R}}{R}(\bar{\rho}+\bar{P}) = 0 \tag{1.37}$$

Now for the pre-Planckian period one looks for matter production by geometry and returns to (1.29) and (1.30). From (1.30) one sees that the W field can act as a creator of matter even if at the beginning moment no matter was present (surely someone has thought of a Higgs role for the W field ?). According to (1.29) the W field depends on the source function S. On the whole one adopts the singularity free cosmological model of Israelit-Rosen [**672**] with its initial prematter period but completed with a nonzero cosmological constant. Also some constants such as the Hubble constant, matter densities, etc. are updated. The initial Planckian egg is thus preceded by a pre-Planckian period originating from a primary geometric state (primary state) at $R_0 = 5.58\cdot 10^{-36}\ cm$ and lasting up to the initial Planckian egg at $R_I = 5.58\cdot 10^{-34}\ cm$. This spherically symmetric homogeneous and isotropic universe is described by the Int-W-D geometry (no cosmic matter) with nothing but geometry, including the W and β fields, at $R_0 = 5.58\cdot 10^{-36}\ cm,\ \rho_0 = 0,\ P_0 = 0,\ \beta_0 \neq 0$, and $W_0 \neq 0$. Assume this was quasistatic with $\dot{R}_0 = 0$ and via (1.27) also $\dot{\beta}_0 = 0$. Thus one obtains from (1.28) at the beginning moment

$$(1.38)\qquad (8\pi/3)\kappa^2\dot{W}_0^2 + (1/3)\beta_0^2\Lambda - (1/R_0^2) = 0$$

Since B may be chosen arbitrarily one can take a suitable function for the Weylian scalar charge S and then calculate the Dirac charge B according to (1.31). An "appropriate" choice is $S = S_0(\beta^2\dot{\beta}/R^3$ with S_0 constant (explanation omitted). Inserting this into (1.29) and integrating one gets $\dot{W} = -(S_0/6\kappa^2)(\beta/R^3)$ so $\ddot{W} = 0$ and one can rewrite (1.38) as

$$(1.39)\qquad (8\pi/3)(S_0^2/36\kappa^2)(\beta_0^2/R_0^6) + (1/3)\Lambda\beta_0^2 - (1/R_0^2) = 0$$

One will use this below to calculate the value of β_0 but first the scenario of the very early universe must be completed by an equation of state of cosmic matter during the pre-Planckian period. According to (1.30) and (1.34) the matter is created by the W field which has an energy momentum density tensor (1.14), etc. The components of this tensor are related by (1.33), etc., so that the pressure of this field is equal to its energy density. Thus one writes $P = \rho$ and then one can rewrite (1.30) as

$$(1.40)\qquad \dot{\rho} + 6(\dot{R}/R)\rho + 4(\dot{\beta}/\beta)\rho = S\dot{W}/\beta^2 \equiv \rho = (1/\beta^4R^6)\int S\dot{W}\beta^2R^6dt$$

The density of matter created by geometry is given now by the expression

$$(1.41)\qquad \rho = (S_0^2\beta_0^6/36\kappa^2\beta^4R^6)[1 - (\beta/\beta_0)^6]$$

Thus there was a zero matter density and pressure (with the assumptions above) at the beginning moment and for a rapidly decreasing $\beta(R)$ one can have $(\beta_I/\beta_0)^6 << 1$ where $\beta_I = \beta(R_I)$. According to this scenario at $R = R_I$ the matter density reaches its maximum $\rho_P = 3.83\cdot 10^{65}\ cm^{-2}$ there so that from (1.41), etc., one gets $\rho_P = S_0^2\beta_0^6/36\kappa^2\beta_I^4R_I^6$. From this one can calculate $S_0^2/36\kappa^2$ for a given gauge function $\beta(R)$. Now for a moment go back to the energy equation (1.40); it can be rewritten as

$$(1.42)\qquad \dot{\rho} + 6(\dot{R}/R)\rho + 4(\dot{\beta}/\beta)\rho = -(S_0^2/6\kappa^2)(\beta^2/R^6)(\dot{\beta}/\beta)$$

Then the term on the right side of (1.42), which describes matter creation by the W field can be compared with the third term on the left side which represents the existing amount of matter. Making use of (1.41) this gives

$$(1.43)\qquad (1/4\rho)(S_0^2\beta^2/6\kappa^2R^6) = [3\beta^6/2(\beta_0^6 - \beta^6)]$$

Further, comparing the matter energy with that of the W field in the equations (1.28) one gets

$$(1.44)\qquad (\kappa^2\dot{W}^2/\rho) = [\beta^6/(\beta_0^6 - \beta^6)]$$

Thus in the beginning when $\beta_0 - \beta$ is small $\dot{W}$ dominates the matter creation while for large R, when $\beta << \beta_0$, the matter creation term becomes negligible. ■

This is just a sampling of results in [**668, 669, 670, 671, 672, 673**]. Many other cosmological questions of great interest including dark matter, quintessence, etc. are also treated. There are in addition many fascinating papers speculating about the original universe from many points of view and we attempt no survey here. The approach here via Weyl-Dirac geometry seems however too lovely to ignore and it may provide further insight into questions of quantum fluctuations. The inroads into cosmology here are an inevitable consequence of the presence of Weyl-Dirac theory in dealing with quantum fluctuations and once wave functins and Bohmian ideas are introduced the quantum potential will automatically arise via β and w_μ.

2. REMARKS ON COSMOLOGY

We begin with some background information (cf. [**234, 235, 236, 237, 974, 1171, 1172, 1193**]. Thus recall that the deBroglie wave length is $\lambda = \hbar/p$ and the Compton wave length is $\Lambda = \hbar/mc$. The uncertainty principle states that $(\Delta x)(\Delta p) \geq \hbar$ and the diffusion coefficient for Brownian motion is proportional to $D = \hbar/m$. The fractal dimension of a quantum path is $d_f = 2$ at scales between λ and Λ but becomes $d_f = 1$ at scales smaller than Λ. Brownian motion characterizes the domain between Λ and λ (cf. [**1, 1193**]). Heuristically from $\Delta p = m(\Delta x/\Delta t)$ and $\Delta x\Delta p \geq \hbar$ we have $m(\Delta x/\Delta t) = \Delta p \geq \hbar/\Delta x$ or $(\Delta x)^2 \geq (\hbar/m)\Delta t$ which can be rephrased as $< x^2 >\sim Dt$ for $D = \hbar/m$. Further note that from $E \sim p^2/2m$ one has $\Delta E \sim (1/2m)(\Delta p)^2$ (working around $p_0 = 0$ say). Then from $\Delta p \geq \hbar/\Delta x$ and $(\Delta x)^2 \sim D\Delta t$ (with $\Delta x\Delta p \geq \hbar$) we obtain $(1/2m)(\Delta p)^2 \geq (\hbar^2/2mD\Delta t)$ from which $\Delta E\Delta t \sim \hbar/2$.

Now from Sidharth [**1172**] one looks at Weyl geometry and refers to the Lagrangian of Santamato [**1115**] of the form $L = L_C(q,\dot{q},t) + \gamma(\hbar^2/m)R(q,t)$ where R is the Ricci scalar curvature in the Weyl geometry (cf. Section 3.3). Then it turns out that the quantum potential Q has the form $Q = -\gamma(\hbar^2/m)R$ and the Q can be related to quantum fluctuations via Fisher information. In [**1172**] one replaces the Weyl vector ϕ (which measures length dilations) by a noncommutative (NC) geometry $ds^2 = (h_{\mu\nu} + \bar{h}_{\mu\nu})dx^\mu dx^\nu$ with a tensor density $\bar{h}_{\mu\nu}$ arising via the antisymmetric part. This corresponds then to $[dx^\mu, dx^\nu] \sim \ell^2 \neq 0$ and in a certain sense legitimizes the approach of [**1115**]. Moreover the NC geometry produces a multiply connected space in which a closed circuit cannot be shrunk to a point so for a circle C of diameter λ in e.g. a doubly connected space one will have ($V = (\hbar/m)\vec{\nabla}S$ and v is some average velocity)

$$\Gamma = \int_C m\vec{V}\cdot d\vec{r} = \hbar\int_C \vec{\nabla}S\cdot d\vec{r} = \hbar\oint dS = mv\pi\lambda = \pi\hbar \tag{2.1}$$

Consequently $\lambda = \hbar/mv$ and this shows an emergence of the deBroglie wavelength following from the NC geometry. We note also from [**236, 237, 1115**] that for $\psi = \sqrt{\rho}exp(iS/\hbar)$ the Weyl vector $\phi \sim -\nabla log(\rho)$ and ψ satisfies (for $A_k = 0$) an equation

$$i\hbar\psi_t = -\frac{\hbar^2}{2m}\left(\frac{1}{\sqrt{g}}\partial_i\sqrt{g}\right)g^{ik}\partial_k\psi + \left(V - \frac{\gamma\hbar^2}{m}\dot{R}\right)\psi \tag{2.2}$$

where $\dot{R}$ is the Riemannian curvature. Here $grad(f) \sim \partial_k f$ and Δf corresponds to taking the divergence of the associated contravariant vector $g^{ik}\partial_k f$, i.e. $\Delta f = div(grad(f)) = (1/\sqrt{g})\partial_i\sqrt{g}g^{ik}\partial_k f)$. Note also in forming a Lagrangian $\dot{x}^\alpha = 2g^{\alpha\beta}p_\beta$ or $p_\alpha = (1/2)g_{\alpha\beta}\dot{x}^\beta$ so that (cf. [**12**])

$$H = g^{\alpha\beta}p_\alpha p_\beta \mapsto L = \dot{x}^\alpha p_\alpha - g^{\alpha\beta}p_\alpha p_\beta = (1/4)g_{\alpha\beta}\dot{x}^\alpha\dot{x}^\beta \tag{2.3}$$

Thus one has a complete geometrization of quantum mechanics (QM) via (2.2) (recall also that the Ricci-Weyl curvature has the form

$$R = \dot{R} + (1/2\gamma\sqrt{\rho})[(1/\sqrt{g})\partial_i(\sqrt{g}g^{ik}\partial_k\sqrt{\rho})] \tag{2.4}$$

where $\gamma = 1/12$ here.

REMARK 4.2.1. We recall now that in the Nottale derivation of the Schrödinger equation SE) one has a complex velocity $V - iU$ due to fractal quantum paths (cf. [**234, 235, 236, 336, 941**]). Here $U = D(d/dx)(log(\rho))$ can be "conveniently" taken to be a constant α (cf. Cresson [**336**]) which would imply $log(\rho) = (\alpha/D)x = \beta x$ and $Q = -(\hbar^2/8m)\beta^2$ as pointed out in [**1171**] (recall $Q = -(\hbar^2/2m)(\partial^2\sqrt{\rho}/\sqrt{\rho})$ with $\sqrt{\rho} = exp(\beta x/2)$ here). Now one can apparently make a case for the Zitterbewegung or self interaction effects within a minimum cutoff Compton wavelength to generate inertial mass. If Q is inertial energy, say $Q = -\delta mc^2$, with $(\hbar^2/8m)\beta^2 = (\hbar^2/8m)(\alpha^2/D^2) = m\alpha^2/8 = \delta mc^2$ one arrives at $\alpha \sim (8\delta)c$ (omitting constants such as 8δ etc. in approximations involving large and small numbers). It is then argued that the stochastic-fractal formulation of Nottale leads to the emergence of spacetime coordinates (x, ict) and such matters are obviously intriguing. ■

There is a great deal of fascinating information available concerning various fundamental constants and large numbers in physics (we refer here to [**818, 941, 950, 1262, 1274**] for example and for more "adventurous" material to e.g. [**1110, 1128, 1173**] and the numerous papers of M. El Naschie in the journal Chaos, Solitons, and Fractals). Dirac had previously spoken eloquently about the importance of large numbers and their relations and in this spirit one is compelled to look as such matters. One seems at times to be simply playing with numbers (sometimes called numerology) but there are too many remarkable coincidences to be ignored and e.g. El Naschie's program of tying matters together via Cantor sets and the golden mean ϕ is in my opinion worth serious consideration. Sidharth's arguments about Cantorian $\mathcal{E}^{\infty}$ also lend some structural meaning and should be pursued further. In any event we gather here first a collection of numbers and ideas in no particular order following [**1171, 1172, 1173**].

(1) The average distance ℓ covered in N steps in a random walk is $\ell = R/\sqrt{N}$ where R is the dimension of the system. Such a relation with $R \sim 10^{28}$ cm (radius of the universe) and $N \sim 10^{80}$ (number of particles in the universe) gives $\ell \sim 10^{-12} \sim 10^{-13} = \ell_\pi$ which is the Compton wavelength of a typical elementary particle (the pion). The stipulation that $10^{-12} \sim 10^{-13}$ is of course reasonable but somewhat unsettling).

(2) The Planck scale is defined via $\ell_P = (\hbar G/c^3)^{1/2} \sim 10^{-33}$ cm and $t_P = (\hbar G/c^5)^{1/2} \sim 10^{-42}$ sec with $m_P = 10^{-5}$ gm the Planck mass. Here t_P is also the Compton time for the Planck mass ($\ell_P = ct_P$).

(3) $R \sim cT$ where T is the age of the universe so, from item 1, $T \sim \sqrt{N}\tau$ where $\tau = \ell_\pi/c$ is the Compton time for a pion. Further $M = Nm$ where M is the mass of the universe and m is the (typical) pion mass.

(4) The energy of fluctuations of the magnetic field $\vec{B}$ in a region of length ℓ is $B^2 \sim \hbar c/\ell^4$ so for $\ell \sim$ Compton wave length the resultant particle fluctuation energy in a volume $\sim \ell^3$ is $\ell^3 B^2 \sim \hbar c/\ell = mc^2$. Thus the entire energy of an elementary particle of mass m is generated by fluctuations alone.

(5) The fluctuation in particle number is of order $\sqrt{N}$ (cf. [**649**]) and a typical time interval of uncertainty is $\Delta t \sim \hbar/mc^2$ (via $\Delta E \Delta t \sim \hbar$). In the spirit of Prigogine Heisenberg uncertainty gives rise to production of energy over short intervals of time leading to a one way creation of particles. Thus $dN/dt \sim \sqrt{N}(mc^2/\hbar)$ leading to $T \sim (\hbar/2mc^2)\sqrt{N} = \sqrt{N}\tau$ as in item 3 ($\tau = \ell/c = (\hbar/mc)/c$ - a factor of 2 is included here).

(6) Now recall $R \sim GM/c^2$ where $Nm = M$ where m is the mass of a typical elementary particle. Then random walk considerations and fluctuations of order $\sqrt{N}$ from the ZPF give $R = \sqrt{N}\ell$. Going to [**1173**] we note first for $H = Gm^3c/\hbar^2$ ($H \sim$ Hubble constant) and $R = GmN/c^2$ one has, for G constant, $\dot{R} = (Gm/c^2)\dot{N} \sim (Gm/c^2)(mc^2\sqrt{N}/\hbar) = (Gm^3c/\hbar^2)(\hbar/mc)\sqrt{N} = HR$ as it should. In fact H is often defined as $\dot{R}/R$. In particular one can conclude from $\dot{R} = RH$ and H constant that $\ddot{R} = RH^2$.

Consider now [**941, 942, 1078**] where scale relations in micro and macro physics abound. Let us begin with $H = Gm^3c/\hbar^2$ or $m = (\hbar^2 H/cG)^{1/3}$ (m presumably refers to pion mass here). Then one notes that the cosmological constant Λ has dimension $1/L^2$ and this suggests that there should be a maximal scale length $L = 1/\sqrt{\Lambda}$. Next a version of Mach's principle is achieved by requiring that the gravitational energy of interaction of a body with the universe (described as a mass M at average distance R) should be equal to its self energy of inertial origin $E = mc^2$, namely

$$GmM/R = mc^2 \Rightarrow (GM/Rc^2) \sim 1 \qquad (2.5)$$

Now $2GM/c^2$ corresponds to the classical radius of a Schwartzschild black hole so (2.5) says that the universe is like a black hole. Next imagine $M = 4\pi\rho R^3/3$ with $\dot{R} = c = RH$ so from $2GM/c^2R = 1$ there results $(2G/c^2R)(4\pi\rho R^3/3) = 1$ which implies $(8\pi G\rho/3H^2) = \Omega = 1$ (space flatness condition - with a cosmological constant the Schwartzschild relation is $(2GM/c^2R) + (\Lambda R^2/3) = 1 \sim (8\pi G\rho + \Lambda c^2)/3H^2 = 1$ - cf. [**942, 946**]). Another formula arises by introducing the Planck mass as a natural unit and writing Newton's law with $Gm_P^2 = \hbar c$ (following from $R = \hbar/mc$ and $(Gm^2/R) = mc^2$ which implies $Gm^2 = \hbar c$ for $m = m_P$) in the form $F = \hbar c[(m/m_P)(m'/m_P)]/R^2$ (since $\hbar cmm'/m_P^2R^2 = Gmm'/R^2$); such a formula appears also in [**219, 547**]. Regarding the cosmological constant one notes first that the Planck length $\Lambda_P = (\hbar G/c^3)^{1/2}$ is the only length that can be envisioned with the three fundamental constants $\hbar$, G, c. Here the maximum scale length L should have the form $L = \Lambda_P K = 1/\sqrt{\Lambda}$. This gives a number $K \sim 10^{61}$. Now if the universe is a black hole, looking at a resolution scale $1/L$ in the Einstein model the maximal separation between points is πL so the effective mass should be characterized by $2GM/c^2\pi L = 1$. This leads to one of the classical large number coincidences $m/m_P = (\pi/2)K$ which for $K \sim 10^{61}$ gives a characteristic mass $\sim 10^{23}$ solar masses (which corresponds to 10^{11} galaxies of 10^{12} solar masses). To get $m/m_P = (\pi/2)K$ one uses an argument comparing lengths $\ell = Gm/ < v^2 >$ and $\lambda = \hbar/mv$ which if equivalent yields Planck mass with $v = c$ and $Gm_P^2 = \hbar c$ or $m_P^2 = \hbar c/G$. Then from $2GM/c62\pi L = 1$ one has

$$\frac{\pi}{2}K = \frac{GM}{c^2L}\frac{L}{\Lambda_P} = \frac{GM}{c^2\Lambda_P} = \frac{\hbar cM}{m_P^2c^2\Lambda_P} = \frac{\hbar M}{m_P^2c\Lambda_P} = \frac{M}{m_P} \tag{2.6}$$

Finally consider a characteristic minimal energy $E_{min} = \hbar c/L$ and, for the electron of purely electromagnetic (EM) origin, a scale r_0 is defined where $e^2/r_0 = m_ec^2$ (i.e. $r_0 = \alpha\lambda_C$ is the classical radius of the electron - note $r_0 = e^2/mc^2 = \lambda(\hbar/mc) = \alpha\lambda_C$ where λ_C is the Compton wavelength and $\alpha\hbar = e^2/c$ or $\alpha = e^2/\hbar c$ is the fine structure constan - sometimes written as $\alpha = e^2/4\pi\hbar c$ in suitable units). Then assume the gravitational self energy of the electron at scale r_0 equals the minimal energy E_{min}; this implies $Gm^2(r_0)/r_0 = \hbar c/L$ (here $m(r_0) \sim \alpha^{-1}m_e$ modulo a small scale dependence of α) and leads to $\alpha(m_P/m_e) = K^{1/3}$. To see this write $G = \hbar c/m_P^2$ as before and recll $m = \alpha^{-1}m_e$; then

$$\frac{Gm^2}{r_0} = \frac{\hbar c}{L} = \frac{\hbar c}{\Lambda_P K} \Rightarrow \frac{\hbar cm^2}{r_0m_P^2} = \frac{\hbar c}{\Lambda_P K} \Rightarrow K = \frac{r_0m_P^2}{m^2\Lambda_P} \tag{2.7}$$

But $\Lambda_P = \hbar/m_Pc$ and $r_0 = \hbar/mc$ so $K = m_P^2\hbar/m^3c\Lambda_P = m_P^4/m^3$ which means $K^{1/3} = m_P/m = \alpha(m_P/m_e)$. For completeness we note

$$\ell_P = \left(\frac{\hbar G}{c^3}\right)^{1/2} \sim 1.62 \cdot 10^{-33}\ cm;\ \lambda_P = \frac{\ell_P}{c} \sim 5.4 \cdot 10^{-44}\ sec; \tag{2.8}$$

$$m_P = \frac{\hbar}{\ell_P c} \sim 2.17 \cdot 10^{-5}\ gram$$

We jump ahead now to the more recent articles [**942, 943, 946, 1078**] and to the discussion in [**272, 273, 274**] (we also find it curious that Nottale's scale relativity has not been "blessed" with any establishment interest). We recall first a few basic facts following [**1078**]. The simplest form for a scale differential equation describing the dependence of a fractal coordinate X in terms of resolution ϵ is given by a first order, linear, renormalization group like equation $(\partial X(t,\epsilon)/\partial log(\epsilon)) = a - \delta X$ with solution $X(t,\epsilon) = x(t)[1 + \zeta(t)(\lambda/\epsilon)^{\delta}]$. This involves a fractal asymptotic behavior at small scales with fractal dimension $D = 1 + \delta$, which is broken at large scale beyond the transition scale λ (cf. [**941, 942, 946**] for more detail). By differentiating one obtains $dX = dx + d\xi$ where $d\xi$ is a scale dependent fractal part and dx is a scale independent classical part such that $dx = vdt$ and $d\xi = \eta\sqrt{2\mathcal{D}}(dt^2)^{1/2D}$ where $<\eta>=0$ and $<\eta^2>=1$ (one considers here only $D = 2$). Here each individual trajectory is assumed fractal and the test paricles can follow an infinity of possible trajectories. This leads one to a nondeterministic, fluid like description, in terms of $v = v(x(t),t)$. The reflection invariance $dt \to -dt$ is broken via nondifferentiablity leading to a two valued velocity vector (cf. [**234, 235, 336, 941**]) and one arrives at a complex time derivative $(d'/dt) = \partial_t + \mathcal{V}\cdot\nabla - i\mathcal{D}\Delta$. Setting $\psi = exp(i\mathfrak{S}/2m\mathcal{D})$ one obtains a geodesic equation in fractal space via $(d'/dt)\mathcal{V} = 0$ which becomes the free Schrödinger equation (SE) $\mathcal{D}^2\Delta\psi + i\mathcal{D}\partial_t\psi = 0$. Now in [**1078**] one is interested in macrophysical applications and considers a free particle in a curved space time whose spatial part is also fractal beyond some time or space transition. The equation of motion can be written (to first order approximation) by a free motion geodesic equation combining the relativistic covariant derivative (describing curvature) and the scale relativistic covariant derivative d'/dt (describing fracticality). We refer to [**236, 237, 1160**] for variations on this. Thus one considers in the Newtonian limit

$$(D/dt)\mathcal{V} = (d'/dt)\mathcal{V} + \nabla(\phi/m) = 0 \tag{2.9}$$

where ϕ is the Newton potential energy. In terms of ψ one obtains

$$\mathcal{D}^2\Delta\psi + i\mathcal{D}\partial_t\psi = (\phi/2m)\psi \tag{2.10}$$

Since the imaginary part of this equation is the equation of continuity (and thinking of the motion in terms of an infinite family of geodesics) $\rho = \psi\psi^\dagger$ can be interpreted as the probability density of the particle postions. For a Kepler potential (in the stationary case) one has then

$$2\mathcal{D}^2\Delta\psi + [(E/m) + (GM/r)]\psi = 0 \tag{2.11}$$

Via the equivalence principle (cf. Agnese-Festa [**15**]) this must be independent of the test particle mass while GM provides the natural length unit; hence $\mathcal{D} = (GM/2w)$ where w is a fundamental constant with the dimensions of velocity. The ratio $\alpha_g = w/c$ actually plays the role of a macroscopic gravitation coupling constant (cf. [**15**]). One shows in LaRoche-Nottale [**1078**] that the solutions of this gravitational SE are characterized by a universal quantization of velocities in terms of the constant $w = 144.7 \pm 0.5$ km/s (or its multiples or submultiples); the precise law of quantization depends on the potential. Depending on the scale either the classical or the fractal part dominates. Various situations are examined and we only indicate a few here. The evolution equations are the Schrödinger -

Newton equation and the classical Poisson equation (cf. [**855**])

$$\mathcal{D}^2\Delta\psi + i\mathcal{D}\frac{\partial\psi}{\partial t} - \frac{\phi}{2m}\psi = 0;\ \Delta\Phi = 4\pi G\rho\ \ (\phi = m\Phi) \tag{2.12}$$

Here Φ is the potential and $\phi = m\Phi$ the potential energy. Separating the real and imaginary parts one arives at

$$m\left(\frac{\partial}{\partial t} + V\cdot\nabla\right)V = -\nabla(\phi+Q);\ \frac{\partial P}{\partial t} + div(PV) = 0;\ Q = -2m\mathcal{D}^2\frac{\Delta\sqrt{P}}{\sqrt{P}} \tag{2.13}$$

In the situation where the particles are assumed to fill the "orbitals" the density of matter becomes proportional to the probability density, i.e. $\rho \propto P = \psi\psi^\dagger$ and the two equations combine to form a single Hartree equation for matter alone, namely

$$\Delta\left(\frac{\mathcal{D}^2\Delta\psi + i\mathcal{D}\partial_t\psi}{\psi}\right) - 2\pi G\rho_0|\psi|^2 = 0 \tag{2.14}$$

Another case arises when the number of bodies is small and they follow at random one among the possible trajectories so that $P = \psi\psi^\dagger$ is nothing else than a probability density while space remains essentially empty. It is suggested that this allows one to explain some effects that up to now have been attributed to dark matter (cf. [**361, 942, 946**] and below for more on this).

We mention in particular the case where $\Phi = -GM/r$ and the SE becomes $\mathcal{D}^2\Delta\psi + i\mathcal{D}\partial_t\psi + (GM/2r)\psi = 0$. One looks for solutions $\psi = \psi(r)exp(-iEt/2m\mathcal{D})$ and makes substitutions $\hbar/2m \sim \mathcal{D}$ and $e^2 \sim GMm$ where m is the test particle inertial mass. This yields an equation similar to the quantum hydrogen atom equation whose solution involves Laguerre polynomials $\psi(r) = \psi_{n\ell m}(r,\theta,\phi) = R_{n\ell}(r)Y_\ell^m(\theta,\phi)$ and the energy/mass ratio is quantized as

$$E_n/m = -(G^2M^2/8\mathcal{D}^2n^2) = -(1/2)(w_0^2/n^2) \tag{2.15}$$

while the natural length unit is the Bohr radius $a_0 = 4\mathcal{D}^2/GM = GM/w_0^2$. Consider now particles such as gas, dust, etc. in a highly chaotic and irreversible motion in a central Kepler potential; via the SE (2.11) there are solutions characterized by well defined and quantized values of conservative quantities such as energy etc. One therefore expects the particles to self-organize into "orbitals" and then to form planets etc. by accretion. Once so accreted one can recover classical elements such as eccentricity, semi-major axis, etc. We refer to [**942, 946, 1078**] for more discussion and details of this and many other examples.

One refers later in the paper [**1078**] to the dark matter problem and its connection to the formation of galaxies and large scale structure. Scale relativity allows one to suggest some solutions. Indeed the fractal geometry of a nondifferentiable space time solves the problem of formation on many scales and it also implies the appearance of a new scalar potential (as in (2.13) which manifests the fractality of space in the same way as Newton's potential manifests its curvature. It is suggested that this new potential (Q) may explain the anomalous dynamical effects without needing any missing mass. In this direction consider the case of the flat rotation curves of spiral galaxies. The formation of an isolated galaxy from a cosmological background of uniform density is obtaind in its first steps as the fundamental level solution $n = 0$ of the SE with an harmonic oscillator gravitational potential $(2\pi/3)G\rho r^2$ (for which some details are worked out in [**1078**]). Once the galaxy is formed let r_0 be its outer radius beyond which the amount of visible matter becomes small. The potential energy at this point is given via $\phi_0 = -(GMm/r_0) = -mv_0^2$ where $M \sim$ mass of the galaxy. Observational data says that the velocity in the exterior region of the galaxy keeps the constant value v_0 and from the virial theorem the potential energy is proportional to the kinetic energy so that it also keeps the constant value $\phi_0 = -GMm/r_0$. Therefore r_0 is the distance at which the rotation curve begins to be flat and v_0 is the corresponding constant velocity. In the standard approach this flat rotation curve is in contradiction with the visible matter alone from which one would expect to observe a variable Keplerian potential energy $\phi = -GMm/r$. This means that one observes an additional potential energy $Q_{obs} = -(GMm/r_0)[1-(r_0/r)]$. Now the regions exterior to the galaxy are described in the scale relativity approach by a SE with a Kepler potential enery $\phi = -GMm/r$ where M is still the sole visible mass, since we assume here no dark matter. The radial solution for the fundamental level is $\sqrt{P} = 2exp(-r/r_B)$ where $r_B = GM/w_0^2$ is the macroscopic Bohr radius of the galaxy. Now one computes the theoretically predicted new potential Q from (2.13) (using $\mathcal{D} = GM/2w_0$) to get

$$Q_{pred} = -2m\mathcal{D}^2\frac{\Delta\sqrt{P}}{\sqrt{P}} = -\frac{GMm}{2r_B}\left(1-\frac{2r_B}{r}\right) = -\frac{1}{2}w_0^2\left(1-\frac{2r_B}{r}\right) \tag{2.16}$$

Thus one obtains, without any added hypotheses, the observed form Q_{obs} of the new potential. Moreover the visible radius and the Bohr radius are now related via $r_0 = 2r_B$. The constant velocity v_0 of the flat rotation curve is also linked to the fundamental gravitational constant w_0 via $w_0 = \sqrt{2}v_0$. Observational data supporting all this is also given.

3. WDW EQUATION

We go now to the famous Wheeler-deWitt equation (which might also be thought of as an Einstein-Schrödinger equation). Some background about this is given in Appendix C of [**254**] along with an introduction to the Ashtekar variables (following the beautiful exposition of [**86**]). The approach here follows F. and A. Shojai [**1157, 1163**] which provides a Bohmian interpretation of quantum gravity and we cite also [**68, 69, 70, 71, 72, 75, 87, 148, 254, 255, 540, 258, 259, 379, 505, 642, 738, 749, 757, 853, 898, 1085, 1086, 1183, 1184, 1220, 1221, 1222, 1223**] for material on WDW and quantum gravity. Extracting liberally (and optimistically) now from [**1163**] (first paper) one writes the Lagrangian density for general relativity (GR) in the form $(16\pi G = 1)$

$$\mathfrak{L} = \sqrt{-g}\mathfrak{R} = \sqrt{q}N({}^3\mathfrak{R} + Tr(K^2)) \tag{3.1}$$

where ${}^3\mathfrak{R}$ is the 3-dimensional Ricci scalar, K_{ij} is the extrinsic curvature, and q_{ij} is the induced spatial metric. The canonical momentum of the 3-metric is given by

$$p^{ij} = \frac{\partial\mathfrak{L}}{\partial\dot{q}_{ij}} = \sqrt{q}(K^{ij} - q^{ij}Tr(K)) \tag{3.2}$$

The classial Hamiltonian is

$$H = \int d^3x \mathfrak{H};\ \mathfrak{H} = \sqrt{q}(NC + N^iC_i) \tag{3.3}$$

where the lapse and shift functions, N and N_i, are given via (cf. [**86**])

$$C = {}^3\mathfrak{R} + \frac{1}{q}\left(Tr(p^2) - \frac{1}{2}(Tr(p))^2\right) = -2G_{\mu\nu}n^\mu n^\nu; \tag{3.4}$$

$$C_i = -2^3\nabla^j\left(\frac{p_{ij}}{\sqrt{q}}\right) = -2G_{\mu i}n^\mu$$

Here n^μ is the normal vector to the spatial hypersurfaces given by $n^\mu = (1/N, -\vec{N}/N)$. Now in the Bohmian approach one must add the quantum potential to the Hamiltonian to get the correct equations of motion so $H \to H + Q$ via $\mathfrak{H} \to \mathfrak{Q}$ where

$$Q - \int d^3x\mathfrak{Q};\ \mathfrak{Q} = \hbar^2 NqG_{ijk\ell}\frac{1}{|\psi|}\frac{\delta^2|\psi|}{\delta q_{ij}\delta_{k\ell}} \tag{3.5}$$

Here $G_{ijk\ell}$ is the superspace metric and ψ is the wavefunction satisfying the WDW equation. This means that we must modify the classical constraints via

$$C \to C + \frac{\mathfrak{Q}}{\sqrt{q}N};\ C_i \to C_i \tag{3.6}$$

Now for the constraint algebra one uses the integrated forms of the constraints defined as

$$C(N) = \int d^3x\sqrt{q}NC;\ \tilde{C}(\vec{N}) \int d^3x\sqrt{q}N^iC_i \tag{3.7}$$

Then (cf. [**86, 1157**] and Appendix C for notation)

$$\{\tilde{C}(\vec{N}), \tilde{C}(\vec{N}')\} = \tilde{C}(\vec{N}\cdot\nabla\vec{N}' - \vec{N}'\cdot\nabla\vec{N}) \equiv \tilde{C}(N^i\vec{\nabla}N_i' - N'^i\vec{\nabla}N_i); \tag{3.8}$$

$$\{\tilde{C}(\vec{N}), C(N)\} = C(\vec{N}\cdot\vec{\nabla}N);\ \{C(N), C(N')\} \sim 0$$

The first 3-diffeomorphism subalgebra does not change with respect to the classical situation and the second, representing the fact that the Hamiltonian constraint is a scalar under the 3-diffeomorphism, is also the same as in the classical case. In the third the quantum potential changes the Hamiltonian constraint algebra dramatically giving a result weakly equal to zero (i.e. zero when the equations of motion are satisfied). Following [**1163**] we will give a number of calculations now regarding this Hamiltonian constraint. Thus first write the Poisson bracket explicitly as

$$\{C(N), C(N')\} = \int d^3z\sqrt{q(z)}\left(\frac{\delta C(N)}{\delta q_{ij}(z)}\frac{\delta C(N')}{\delta p^{ij}(z)} - \frac{\delta C(N)}{\delta p^{ij}(z)}\frac{\delta C(N')}{\delta q_{ij}(z)}\right) = \tag{3.9}$$

$$= \tilde{C}(N\vec{\nabla}N' - N'\vec{\nabla}N) + 2\int d^3zd^3x\sqrt{q(z)}G_{ijk\ell}(z)p^{k\ell}(z)\times$$

$$\times(-N(z)N'(x) + N(x)N'(z))\frac{\delta(Q/\sqrt{q}N))}{\delta q_{ij}(z)}$$

To simplify one differentiates the Bohmian HJ equation to get (cf. [**1157**])

$$(3.10)\quad \frac{1}{N}\frac{\delta}{\delta q_{ij}}\frac{Q}{\sqrt{q}} = \frac{3}{4\sqrt{q}}q_{k\ell}p^{ij}p^{k\ell}\delta(x-z) - \frac{\sqrt{q}}{2}q^{ij}({}^3\mathfrak{R}-2\Lambda)\delta(x-z) - \sqrt{q}\frac{\delta^3\mathfrak{R}}{\delta q_{ij}}$$

and use this in evaluation of the Poisson bracket giving the result indicated in (3.8). This calculation is given in the Appendix to F. and A. Shojai [**1163**] and we repeat it here for clarity.

REMARK 4.3.1. We follow here the Appendix to [**1163**] and to evaluate the integral (3.9), in view of (3.10), one needs to consider $\int d^3zF(q,p,N,N')(\delta\,{}^3\mathfrak{R}/\delta q_{ij})$. First look at the variation of the Ricci scalar with respect to the metric. Using the Palatini identity and dropping the superscript 3 one has

$$(3.11)\quad \delta\mathfrak{R}_{ij} = (1/2)q^{k\ell}(\nabla_j\nabla_i\delta q_{k\ell} - \nabla_k\nabla_j\delta q_{\ell i} - \nabla_k\nabla_i\delta q_{\ell j} + \nabla_k\nabla_\ell q_{ij})$$

Consequently

$$(3.12)\quad \frac{\delta\mathfrak{R}_{ij}(x)}{\delta q_{ab}(z)} = \frac{1}{2}\sqrt{q}q^{k\ell}\left(\delta_k^a\delta_\ell^b\nabla_j\nabla_i\frac{\delta(x-z)}{\sqrt{q}} - \delta_\ell^a\delta_i^b\nabla_k\nabla_j\frac{\delta(x-z)}{\sqrt{q}} - \right.$$
$$\left. -\delta_\ell^a\delta_j^b\nabla_k\nabla_i\frac{\delta(x-z)}{\sqrt{q}} + \delta_i^a\delta_j^b\nabla_k\nabla_\ell\frac{\delta(x-z)}{\sqrt{q}}\right)$$

and therefore

$$(3.13)\quad \frac{\delta\mathfrak{R}(x)}{\delta q_{ab}} = \frac{\delta(q^{ij}\mathfrak{R}_{ij})(x)}{\delta q_{ab}} = -\mathfrak{R}^{ab}\delta(x-z) + \sqrt{q}(q^{ab}\nabla^2 - \nabla^a\nabla^b)\frac{\delta(x-z)}{\sqrt{q}}$$

Using this identity in the equation of motion (3.10) the only nonvanishing terms in (3.9) are

(3.14)

$$\{C(N),C(N')\} \sim \tilde{C}(N\vec{\nabla}N' - N'\vec{\nabla}N) - 2\int d^3zd^3x\sqrt{q(x)q(z)}G_{ijk\ell}(z)p^{k\ell}(z)$$
$$\times(N(x)N'(z) - N(z)N'(x))\left(q^{ij}(x)\nabla_x^2\frac{\delta(x-z)}{\sqrt{q}} - \nabla_x^i\nabla_x^j\frac{\delta(x-z)}{\sqrt{q}}\right)$$

where $\sim$ means the equality is weak (i.e. modulo the equation of motion). Integrating by parts gives

$$(3.15)\quad \{C(N),C(N')\} \sim 2\int d^3x(\nabla_j(N\nabla_iN') - \nabla_j(N'\nabla_iN))p^{ij} +$$
$$+\int d^3x\sqrt{q}G_{ijk\ell}p^{k\ell}(N'\nabla^i\nabla^jN - N\nabla^i\nabla^jN') -$$
$$-\int d^3x\sqrt{q}G_{ijk\ell}p^{k\ell}q^{ij}(N'\nabla^2N - N\nabla^2N')$$

Hence there results

$$(3.16)\quad \{C(N),C(N')\} \sim 2\int d^3x(N\nabla_j\nabla_iN' - N'\nabla_j\nabla_iN)p^{ij} +$$
$$+\int d^3xp^{k\ell}(N'\nabla_k\nabla_\ell N - N\nabla_k\nabla_\ell N' + N'\nabla_\ell\nabla_kN - N\nabla_\ell\nabla_kN' -$$
$$-q_{k\ell}(N'\nabla^2N - N\nabla^2N')) + \int d^3xq_{k\ell}p^{k\ell}(N'\nabla^2N - N\nabla^2N') = 0$$

as desired. ■

One sees that the presence of the quantum potential means that the quantum algebra is the 3-diffeomorphism algebra times an Abelian subalgebra and the only difference with Markopoulou [**854**] is that this algebra is weakly closed. One sees that the algebra (3.8) is a clear projection of the general coordinate transformations to the spatial and temporal diffeomorphisms and in fact the equations of motion are invariant under such transformations (cf. also Pinto-Neto and Santini [**1022**]). In particular although the form of the quantum potential will depend on regularization and ordering, in the quantum constraint algebra the form of the quantum potential is not important; the algebra holds independently of the form of the quantum potential. Further it appears that the inclusion of matter terms will not change anything.

One goes next to the quantum Einstein equations (QEI). For the dynamical part consider the Hamiltonian equations

$$\dot{q}^{ij} = \{H, q^{ij}\};\ \dot{p}_{ij} = \{H, p_{ij}\} \tag{3.17}$$

which produce the quantum equations (note the square bracket [ ] means that one is to antisymmetrize over all permutations of the enclosed indices, multiplying each term in the sum by the sign (±1) of the permutation)

$$\dot{q}_{ij} = \frac{2}{\sqrt{q}} N\left(p_{ij} - \frac{1}{2}p^k_k q_{ij}\right) + 2\,{}^3\nabla_{[i}N_{j]} \tag{3.18}$$

$$\dot{p}^{ij} = -N\sqrt{q}\left({}^3\mathfrak{R}^{ij} - \frac{1}{2}\,{}^3\mathfrak{R}q^{ij}\right) + \frac{N}{2\sqrt{q}}q^{ij}\left(p^{ab}p_{ab} - \frac{1}{2}(p^a_a)^2\right) - \tag{3.19}$$
$$-\frac{2N}{\sqrt{q}}\left(p^{ia}p^j_a - \frac{1}{2}p^a_a p^{ij}\right) + \sqrt{q}(\nabla^i\nabla^j N - q^{ij}\,{}^3\nabla^a\,{}^3\nabla_a N) +$$
$$+\sqrt{q}\,{}^3\nabla_a\left(\frac{N^a}{\sqrt{q}}p^{ij}\right) - 2p^{a[i}\,{}^3\nabla_a N^{j]} - \sqrt{q}\frac{\delta Q}{\delta q_{ij}}$$

Combining these two equations one obtains after some calculation

$$\mathfrak{G}^{ij} = -\frac{1}{N}\frac{\delta\Omega}{\delta q_{ij}} \tag{3.20}$$

which means that the quantum force modifies the dynamical part of the Einstein equations. For the nondynamical parts one uses the constraint relations (3.4) to get

$$\mathfrak{G}^{00} = \frac{\Omega}{2N^3\sqrt{q}};\ \mathfrak{G}^{0i} = -\frac{\Omega}{2N^3\sqrt{q}}N^i \tag{3.21}$$

These last two equations can be written via

$$\mathfrak{G}^{0\mu} = \frac{\Omega}{2\sqrt{-g}}g^{0\mu} \tag{3.22}$$

and the nondynamical parts are also modified by the quantum potential.

One addresses next the possibility that for a reparametrization invariant theory the equations obtained by the Hamiltonian may differ from those given by the phase of the wavefunction and the guidance formula (in a Bohmian spirit). However it is seen that there is no difference. Indeed write the Bohmian HJ equation (cf. [**153, 643, 758**]) by decomposing the phase part of the WDW equation; this gives

$$G_{ijk\ell}\frac{\delta S}{\delta q_{ij}}\frac{\delta S}{\delta q_{k\ell}} - \sqrt{q}\,({}^3\mathfrak{R} - \mathfrak{Q}) = 0 \tag{3.23}$$

where S is the phase of the WDW wave function. In order to get the equation of motion one differentiates the HJ equation with respect to q_{ab} and uses the guidance formula $p^{k\ell} \equiv \sqrt{q}(K^{k\ell} - q^{k\ell}K) = \delta S/\delta q_{k\ell}$. After considerable calculation one arrives again at (3.20). Thus the evolution generated by the Hamiltonian is compatible with the guidance formula, i.e. the Poisson brackets of the Hamiltonian and the guidance relation ($\chi^{k\ell} = p^{k\ell} - \delta S/\delta q_{k\ell}$) are zero. This can be evaluated explicitly and equals zero weakly so consistency prevails.

Next one shows explicitly that these modified Einstein equations (MEI) are covariant under spatial and temporal diffeomorphisms. Consider first $t \to t' = f(t)$ with $\vec{x}$ unchanged; one has

$$q'_{ij} = q_{ij};\ N'_i = (df/dt)N_i;\ N' = (df/dt)N \tag{3.24}$$

Putting these in the MEI one sees that the right side transforms as a second rank tensor under time reparametrization. Similarly consider $\vec{x} \to \vec{x}' = \vec{g}(\vec{x})$ with t unchanged; one has

$$q'_{ij} = \frac{\partial x^\ell}{\partial x'^i}\frac{\partial x^m}{\partial x'^j}q_{\ell m};\ N'_i = \frac{\partial x^\ell}{\partial x'^i}N_\ell;\ N' = N \tag{3.25}$$

Again the right side of MEI is a second rank tensor under a spatial 3-diffeomorphism. Inclusion of matter field is straightforward via

$$\mathfrak{G}^{ij} = -\kappa\mathfrak{T}^{ij} - \frac{1}{N}\frac{\delta(\mathfrak{Q}_G + \mathfrak{Q}_m)}{\delta g_{ij}};\ \mathfrak{G}^{0\mu} = -\kappa\mathfrak{T}^{0\mu} + \frac{\mathfrak{Q}_G + \mathfrak{Q}_m}{2\sqrt{-g}}g^{0\mu} \tag{3.26}$$

where

$$\mathfrak{Q}_m = \hbar^2\frac{N\sqrt{q}}{2}\frac{1}{|\psi|}\frac{\delta^2|\psi|}{\delta\phi^2} \tag{3.27}$$

where ϕ is the matter field and, as before,

$$\mathfrak{Q}_G = \hbar^2 NqG_{ijk\ell}\frac{1}{|\psi|}\frac{\delta^2|\psi|}{\delta q_{ij}\delta q_{k\ell}} \tag{3.28}$$

$$Q_G = \int d^3x\mathfrak{Q}_G;\ Q_m = \int d^3x\mathfrak{Q}_m \tag{3.29}$$

Equations (3.26) are the Bohm-Einstein equations which are in fact the quantum version of the Einstein equations; regularization only affects the quantum potential but the QEI are the same. They are invariant under temporal and spatial diffeomorphisms and can be written as

$$\mathfrak{G}^{\mu\nu} = -\kappa\mathfrak{T}^{\mu\nu} + \mathfrak{S}^{\mu\nu} \tag{3.30}$$

$$(3.31)\quad \mathfrak{S}^{0\mu} = \frac{\mathfrak{Q}_G + \mathfrak{Q}_m}{2\sqrt{-g}} g^{0\mu} = \frac{\mathfrak{Q}}{2\sqrt{-g}} g^{0\mu};\ \mathfrak{S}^{ij} = -\frac{1}{N}\frac{\delta(\mathfrak{Q}_G + \mathfrak{Q}_m)}{\delta g_{ij}} = -\frac{1}{N}\frac{\delta\mathfrak{Q}}{\delta g_{ij}}$$

($\mathfrak{S}^{\mu,\nu}$ is the quantum correction tensor - under the temporal $\otimes$ spatial diffeomorphism subgroup which is peculiar to the ADM decomposition). Note that the QEI were derived for a Robertson-Walker metric in Vink [**1262**] but without symmetry considerations or more general metrics. One concludes with the conservation law via taking the divergence of (3.30) to get

$$(3.32)\quad \nabla_\mu \mathfrak{T}^{\mu\nu} = \frac{1}{\kappa}\nabla_\mu \mathfrak{S}^{\mu\nu}$$

REMARK 4.3.2. We refer to [**951**] for a discussion of the Lichnerowicz-York equation as a solution to the constraint equations of GR. ■

3.1. CONSTRAINTS IN ASHTEKAR VARIABLES. We go now to the third paper in F. and A. Shojai [**1163**] where the new variables (or Ashtekar variables) are employed and it is shown that the Poisson bracket of the Hamiltonian with itself changes with respect to its classical counterpart but is still weakly equal to zero (as above in Section 4.3). Caution is advised however since ill defined terms have not been regularized; for this one needs a background metric and the result must be independent of such a metric. Thus the dynamical variables are the self dual connection A_a^i and the canonical momenta are $\tilde{E}_i^a$ with constraints given via

$$(3.33)\quad G_i = \mathfrak{D}_a \tilde{E}_i^a;\ C_b = \tilde{E}_i^a F_{ab}^i;\ H = \epsilon_k^{ij}\tilde{E}_i^a \tilde{E}_j^b \mathfrak{F}_{ab}^k$$

where $\mathfrak{D}_a$ represents the self-dual covariant derivative and F_{ab}^k is the self-dual curvature (we refer to Appendix C for the new variables). Canonical quantization of these constraints can be achieved via changing $\tilde{E}_i^a \to -\hbar(\delta/\delta A_a^i)$ to get

$$(3.34)\quad \hbar\mathfrak{D}_a \frac{\delta\psi(A)}{\delta A_a^i} = 0;\ \hbar F_{ab}^i \frac{\delta\psi(A)}{\delta A_a^i};\ \hbar^2 \epsilon_k^{ij} F_{ab}^k \frac{\delta^2\psi(A)}{\delta A_a^i \delta A_b^j} = 0$$

In order to get the causal interpretation one puts a definition $\psi = Rexp(iS/\hbar)$ into these relations to obtain

$$(3.35)\quad (\mathbf{A})\ \mathfrak{D}_a \frac{\delta R(A)}{\delta A_a^i} = 0;\ (\mathbf{B})\ \mathfrak{D}_a \frac{\delta S(A)}{\delta A_a^i} = 0;$$

$$(\mathbf{C})\ F_{ab}^i \frac{\delta R(A)}{\delta A_a^i} = 0;\ (\mathbf{D})\ F_{ab}^i \frac{\delta S(A)}{\delta A_a^i} = 0;$$

$$(\mathbf{E})\ \epsilon_k^{ij} F_{ab}^k \frac{\delta}{\delta A_a^i}\left(R^2 \frac{\delta S(A)}{\delta A_b^j}\right) = 0;\ (\mathbf{F})\ -\epsilon_k^{ij} F_{ab}^k \frac{\delta S(A)}{\delta A_a^i}\frac{\delta S(A)}{\delta A_b^j} + Q = 0$$

where the quantum potential is defined as

$$(3.36)\quad Q = -\hbar^2 \epsilon_k^{ij} F_{ab}^k \frac{1}{R}\frac{\delta^2 R(A)}{\delta A_a^i \delta A_b^j}$$

Here **(E)** is the continuity equation while **(F)** is the quantum Einstein-Hamilton-Jacobi (EHJ) equation. The quantum trajectories would be achieved via the guidance relation

$$(3.37)\quad \tilde{E}_i^a = i\frac{\delta S(A)}{\delta A_a^i}$$

Now the constraint algebra in terms of smeared out Gauss, vector, and scalar constraints ($\mathcal{N}$ is N densitized via $\sqrt{h}$) is written out (cf. Nikolić [**933**]) and one defines $K^a = \tilde{E}^a_i \tilde{E}^{bi}(\mathcal{N}\partial_b \mathcal{M} - \mathcal{M}\partial_b \mathcal{N})$. The quantum trajectories can be obtained from the quantum Hamiltonian given by $H_Q = H + Q$ and the smeared out gauge and diffeomorphism constraints will not change; the Hamiltonian constraint becomes

$$\mathfrak{H}_Q(\mathcal{N}) = \frac{1}{2}\int d^3x \mathcal{N}\epsilon^{ij}_k \tilde{E}^a_i \tilde{E}^b_j \mathfrak{F}^k_{ab} + \mathfrak{Q}(\mathcal{N}) \tag{3.38}$$

where $\mathfrak{Q}(\mathcal{N}) = \int d^3x \mathcal{N}Q$ (the notation Q and $\mathfrak{Q}$ is switched here from previous use above). The first three constraint Poisson brackets will not change and the fourth is still valid because the quantum potential is a scalar density and one has

$$\{C(\vec{N}), \mathfrak{H}_Q(\mathcal{M})\} = \mathfrak{H}_Q(\mathfrak{L}_{\vec{N}}\mathcal{M}) \tag{3.39}$$

This applies also to the fifth bracket because

$$\{\mathfrak{G}(\Lambda_i), \mathfrak{H}_Q(\mathcal{N})\} = 0 \tag{3.40}$$

but the last bracket changes via

$$\{\mathfrak{H}_Q(\mathcal{N}), \mathfrak{H}_Q(\mathcal{M})\} = \{\mathfrak{H}(\mathcal{N}), \mathfrak{H}(\mathcal{M})\} + \{\mathfrak{Q}(\mathcal{N}), \mathfrak{H}(\mathcal{M})\} + \tag{3.41}$$
$$+\{\mathfrak{H}(\mathcal{N}), \mathfrak{Q}(\mathcal{M})\} + \{\mathfrak{Q}(\mathcal{N}), \mathfrak{Q}(\mathcal{M})\}$$

Here the last term is identically zero, since the quantum potential is a functional of the connection only. The sum of the second and third terms is

$$\{\mathfrak{Q}(\mathcal{N}), \mathfrak{H}(\mathcal{M})\} + \{\mathfrak{H}(\mathcal{N}), \mathfrak{Q}(\mathcal{M})\} \sim \tag{3.42}$$
$$\sim -\int d^3x \left(\mathcal{N}\epsilon^{ij}_k F^k_{ab}\tilde{E}^b_j \mathfrak{D}_c(\mathcal{M}\epsilon^{\ell m}_i \tilde{E}^a_\ell \tilde{E}^c_m) - \mathcal{M}\epsilon^{ij}_k F^k_{ab}\tilde{E}^b_j \mathfrak{D}_c(\mathcal{N}\epsilon^{\ell m}_i \tilde{E}^a_\ell \tilde{E}^c_m)\right)$$

A calculation then shows that the Poisson bracket of the quantum Hamiltonian with itself is given via

$$\{\mathfrak{H}_Q(\mathcal{N}), \mathfrak{H}_Q(\mathcal{M})\} \sim 0 \tag{3.43}$$

which is similar to the situation with the old variables (cf. Remark 4.3.1).

Now in order to obtain the quantum equations of motion via the Hamilton equations one has

$$\dot{A}^i_a = -i\epsilon^{ijk}\mathcal{N}\tilde{E}^b_j F_{abk} - N^b F^i_{ab};\ \dot{\tilde{E}} = i\epsilon^{jk}_i \mathfrak{D}_b(\mathcal{N}\tilde{E}^a_j \tilde{E}^b_k); \tag{3.44}$$
$$-2\mathfrak{D}_b(N^{[a}\tilde{E}^{b]}_i) + \frac{i}{2}\int d^3x \frac{\delta \mathfrak{Q}(\mathcal{N})}{\delta A^i_a(z)}$$

Further to recover the real quantum general relativity one must set the reality conditions, which are

$$\tilde{E}^a_i \tilde{E}^{bi}\ \textit{must be real}; \tag{3.45}$$
$$i\epsilon^{ijk}\tilde{E}^{(a}_i \mathfrak{D}_a(\tilde{E}^{b)}_k \tilde{E}^c_j) + \frac{i}{2}\int d^3x \frac{\delta \mathfrak{Q}}{\delta A^i_{(a}(z)}\tilde{E}^{b)}(x)\ \textit{must be real}$$

(note round brackets ( ) mean symmetrization with respect to the indices concerned). Thus formally one can construct a causal version of canonical quantum gravity using the Bohm-deBroglie interpretation of QM. All of the quantum behavior is encoded in the quantum potential. One has a well defined trajectory and no operators arise; the algebra action is in fact the Poisson bracket and only the Poisson bracket of the Hamiltonian with itself will change relative to the classical algebra by being weakly instead of strongly equal to zero. The result is similar to that obtained above with the old variables and one can give meaning to the idea of time generator for the Hamiltonian constraint. The equations of motion when finally written out should contain the quantum force. Regularization of ill defined terms remains and is promised in forthcoming papers of F. and A. Shojai.

The approach here in Sections 4.3 and 4.3.1 is quite elegant and we suspend any attempt at criticism here. Eventually one will have to reconcile this with results of Pinto-Neto et al [**1023, 1025**] for example (cf. Section 4.5). We remark also that in [**1167**] one makes a preliminary study of Bohmian ideas in loop quantum gravity using Ashtekar variables.

4. REMARKS ON REGULARIZATION

In Vink [**1262**] (second paper) for example one considers the classical and WDW description of a gravity-minisuperspace model (cf. also Kowalski-Glikman and Vink[**772**]). Thus consider a homogeneous and isotropic metric defined via

$$ds^2 = -N(t)^2dt^2 + a(t)^2d\Omega_3^2 \tag{4.1}$$

where $d\Omega_3^2$ is a standard metric on 3-space. The lapse function N and the scale factor a depend on a time parameter t. A minisuperspace model represented by a single homogeneous mode ϕ is defined by the Lagrangian

$$L = -a^3\left[\frac{1}{2N}\left(\frac{\dot a}{a}\right)^2 + NV_G(a)\right] + a^3\left[\frac{\dot\phi^2}{2N} - NV_M(\phi)\right] \tag{4.2}$$

One uses the Planck mass $m_P^2 = 3/4\pi G$ to scale all dimensional quantities so $a \equiv am_P$, $\phi \equiv \phi/m_P$, etc.; V_M is the potential for the scalar mode ϕ and the gravitational potential $V_G(a) = -(1/2)Ka^{-2} + (1/6)\Lambda$ may contain a cosmological constant Λ and a curvature constant $K = 1$, 0, or -1 for a spherical, planar, or hyperspherical 3-space. From the Lagragian one derives now the classical equations of motion by varying N, a, and ϕ respectively to obtain

$$\frac{1}{2}\left(\frac{\dot a}{a}\right)^2 - V_G(a) = \frac{1}{2}\dot\phi^2 + V_M(\phi); \tag{4.3}$$

$$\frac{1}{2}\left(\frac{\dot a}{a}\right)^2 + \frac{\ddot a}{a} - 3V_g(a) - a\partial_a V_G(a) + 3\left(\frac{1}{2}\dot\phi^2 - V_M(\phi)\right) = 0;$$

$$\ddot\phi + 3\left(\frac{\dot a}{a}\right)\dot\phi + \partial_\phi V_M(\phi) = 0$$

Here one has chosen the gauge $N = 1$ and t is identified with the classical time. However the time parameter is not directly observble, only the correlations $a(\phi)$ or $\phi(a)$ which follow from the solutions of (4.3). One could imagine an additional degree of freedom $\tau(t)$ to be used as a clock but this need not be done. One derives the WDW equation from (4.2) in the standard manner. After computing the classical Hamiltonian and replacing the canonical momenta $p_a = -a\dot{a}$ and $p_\phi = a^3\dot{\phi}$ by operators $p_\phi \to -i\partial_\phi$ and $p_a \to -i\partial_a$ the WDW Hamiltonian is

$$(4.4)\quad H_{WDW} = \left[\frac{1}{2}a^{-3}(a\partial_a)^2 + a^3V_G(a)\right] + \left[\frac{1}{2}a^{-3}\partial_\phi^2 + a^3V_M(\phi)\right] = H_G + H_M$$

The WDW equations is $H_{WDW}\psi = 0$ and there is an operator ordering ambiguity; the "Lagrangian" ordering has been chosen which makes H_{WDW} formaly selfadjoint in the inner product $(\psi,\psi) = \int a^2 da d\phi \psi^*(a,\phi)\phi(a,\phi)$. It is not clear that H_{WDW} can be used as a generator for time evolution since time has disappeared altogether in (4.4).

Now in the dBB treatment one has

$$(4.5)\qquad \dot{S} + \frac{1}{2m}(\partial_x S)^2 + V + Q = 0;\ \dot{R}^2 + \partial_x(R^2\partial_x S) = 0$$

where $Q = -\partial_x^2 R/2mR$. The trajectories are found by solving the autonomous system $\dot{x} = (1/m)\partial_x S$ and the measure R^2dx gives the probability for trajectories crossing the interval $(x, x+dx)$. Now treat the WDW equation as a SE which happens to be time independent and write it in the quantum potential form with $\psi = R(a,\phi)exp(iS(a,\phi))$ leading to

$$(4.6)\qquad \frac{1}{2}a^{-3}[-(a\partial_a S)^2 + (\partial_\phi S)^2] + a^3[V_G + V_M + Q_{GM}] = 0;$$

$$-a\partial_a(R^2 a\partial_a S) + \partial_\phi(R^2\partial_\phi) = 0$$

These equations come from the real and imaginary part of $H_{WDW}\psi = 0$ and the quantum potential is

$$(4.7)\qquad Q_{GM} = -\frac{1}{2}a^{-6}\left[-\frac{-(a\partial_a)^2R}{R} + \frac{\partial_\phi^2 R}{R}\right]$$

Trajectories $(a(t),\phi(t))$ are obtained from $S(a,\phi)$ by identifying $\partial_a S$ with the momentum p_a and $\partial_\phi S$ with p_ϕ; thus one uses the definition of the canonical momenta given before (4.4) to define trajectories parametrized by a time parameter t, via

$$(4.8)\qquad \dot{a} = -a^{-1}\partial_a S(a,\phi);\ \dot{\phi} = a^{-3/2}\partial_\phi S(a,\phi)$$

One notes that the probability measure $R^2(a,\phi)a^2 da d\phi$ is conserved in time t if (a,ϕ) are solutions of (4.7). This is a consequence of using the measure $a^2 da d\phi$ together with the Lagrangian factor ordering making H_{WDW} formally self adjoint. In using t as in (4.8) one should eliminate t after solving in order to determine $a(\phi)$ or $\phi(a)$; for trajectories where e.g $a(t)$ is 1-1 the scale factor can be used as a clock. The analogues of equations (4.3) are obtained by differentiating the first equation in (4.6) with respect to a and ϕ and then eliminating S, using

$$(4.9)\qquad -\partial_t(a\dot{a}) = \partial_a^2 S\dot{a} + \partial_\phi\partial_a S\dot{\phi};\ \partial_t(a^3\dot{\phi}) = \partial_a\partial_\phi\dot{a} + \partial_\phi^2 S\dot{\phi}$$

leading to

$$(4.10)\qquad \frac{1}{2}a\dot{a}^2 - V_G(a) - a^3\left(\frac{1}{2}\dot{\phi}^2 + V_M(\phi)\right] = Q_{GM}(a,\phi);$$

$$\frac{1}{2}\dot{a}^2 + a\ddot{a} - \partial_a V_G(a) + a^2\left[\frac{1}{2}\dot{\phi}^2 - V_M(\phi)\right] = \partial_a Q_{GM}(a,\phi);$$

$$\ddot{\phi} + 3\frac{\dot{a}}{a} + \partial_\phi V_M(\phi) = -a^{-3/2}\partial_\phi Q_{GM}(a,\phi)$$

This evidently generalizes (4.3). There is more in this paper which we omit here, namely a semiclassical development is given as well as and some exact solutions of WDW for toy models.

In Blalut and Kowalski-Glikhman [**153**] one looks at an equation

$$H\psi(x) = \left(\frac{1}{2}g^{ij}(x)\nabla_i\nabla_j - V(x)\right)\psi(x) = \left(\frac{1}{2}\Box - V(x)\right)\psi(x) = 0 \tag{4.11}$$

Assume $\psi = R(x)exp(iS(x)/\hbar)$ with R, S real leading to

$$H[S(x)] = \frac{1}{2}g_{ij}\frac{\partial S}{\partial x_i}\frac{\partial S}{\partial x^j} + V(x) = \frac{\hbar^2}{2R}\Box R;\ R\Box S + 2g^{ij}\frac{\partial S}{\partial x^i}\frac{\partial S}{\partial x^j} = 0 \tag{4.12}$$

Introduce time via

$$\frac{dx^i}{dt} = g^{ij}\frac{\delta H[S(x)]}{\delta(\partial S/\partial x^j)} \tag{4.13}$$

This defines the trajectory $x^i(t)$ in terms of the phase of the wave function S and one obtains

$$\frac{1}{2}g_{ij}\dot{x}^i\dot{x}^j + V(x) + Q = 0;\ Q = -\frac{\hbar^2}{2R}\Box R \tag{4.14}$$

Now define classical momenta

$$p_i = \frac{\delta H[S(x)]}{\delta(\partial S/\partial x^i)} = g_{ij}\dot{x}^j \tag{4.15}$$

and write (4.14) in the form

$$H = \frac{1}{2}g^{ij}p_ip_j + V(x) + Q = 0;\ \dot{p}_i = -\frac{\partial H}{\partial x^i};\ \dot{x}^i = \frac{\partial H}{\partial p_i} \tag{4.16}$$

This gives a method of identifying the time evolution corresponding to the wave function of a WDW type equation (e.g. the wave function of the universe). The following points are mentioned in [**153**]:

- (4.16) is equivalent to the classical equations of motion except for the quantum potential which therefore. in some sense, clarifies the effect of quantum matter on gravity.
- A semiclassical theory can be constructed from the above.
- The quantum potential provides a natural way to introduce time even when the Hamiltonian constraint (which doesn't involve time) is acting. In this connection note that the definition of time is not unique. One could use a more general expression

$$\dot{x}^j = N(x)\frac{\delta H[S(x)]}{\delta(\partial S/\partial x^j)} \tag{4.17}$$

  where N could be identified with the lapse function for the ADM formulation.

In Blaut and Kowalski-Glikhman [**153**] one also studies situations of the form

$$ds^2 = -N^2dt^2 + \sum_1^3 g_{ii}(\omega^i)^2 \tag{4.18}$$

where N is a lapse function and ω^i are appropriate 1-forms (some Bianchi types are classified) and we refer to [**153**] for details and [**105**] for criticism.

In [**154**] one starts with the classical constraints of Einstein's gravity, namely the diffeomorphism and Hamiltonian constraint in the form

$$\mathfrak{D}_a = \nabla_a \pi^{ab};\ \mathfrak{H} = \kappa^2 G_{abcd}\pi^{ab}\pi^{cd} - \frac{1}{\kappa^2}\sqrt{h}(R+2\Lambda) \tag{4.19}$$

Here π^{ab} are momenta associated with the 3-metric h_{ab} where

$$G_{abcd} = \frac{1}{2\sqrt{h}}(h_{ac}h_{bd} + h_{ad}h_{bc} - h_{ab}h_{cd} \tag{4.20}$$

is the WDW metric where R is the 3-dimensional scalar curvature, κ is the gravitational constant, and Λ the cosmological constant. The constraints satisfy the algebra

$$[\mathfrak{D},\mathfrak{D}] \sim \mathfrak{D};\ [\mathfrak{D},\mathfrak{H}] \sim \mathfrak{H};\ [\mathfrak{H},\mathfrak{H}] \sim \mathfrak{D} \tag{4.21}$$

The rules of quantization given by the metric representation of the canonical commutation relations are

$$[\pi^{ab}(x), h_{cd}(y)] = -i\delta^a_c\delta^b_d\delta(x,y);\ \pi^{ab}(x) = -\frac{\delta}{\delta h_{ab}(x)} \tag{4.22}$$

There are problems here of all types (cf. [**154**]) which we will not discuss but in [**771**] a class of exact solutions of the WDW equation was found via heat kernel regularization of the Hamiltonian with a suitable ordering and the question addressed here is the level of arbitrariness in this construction. One bases now such constructions on the principle that the algebra of constraints should be anomaly free, i.e. the algebra should be weakly identical with the classical one. One chooses a starting space of states to consist of integrals over compact 3-space of scalar densities like $\mathfrak{V} = \int_M \sqrt{h},\ \mathfrak{R} = \int_M \sqrt{h}R$, etc. so $\psi = \psi(\mathfrak{V},\mathfrak{R},\cdots)$. For the diffeomorphism constraint one takes the representation $\mathfrak{D}_a(x) = -i\nabla^x_b(\delta/\delta h_{ab}(x))$ where ∇^x_b means the covariant derivative acting at the point x. This constraint then annihilates all the states and the first commutator relation is satisfied. Further the second relation in (4.21) reduces to the formal relation

$$\mathfrak{D}(\mathfrak{H}\psi) \sim \mathfrak{H}\psi \tag{4.23}$$

Now for the construction of the WDW operator one makes a point split in the kinetic term of the form

$$G_{abcd}(x)\pi^{ab}(x)\pi^{cd}(x) \rightsquigarrow \int dx' K_{abcd}(x,x',t)\frac{\delta}{\delta h_{ab}(x)}\frac{\delta}{\delta h_{cd}(x')} \tag{4.24}$$

where $lim_{t\to 0^+}K_{abcd}(x,x',t) = \delta(x,x')$ and in particular one takes

$$K_{abcd}(x,x',t) = G_{abcd}(x')\Delta(x,x',t)(1+K(x,t)); \tag{4.25}$$

$$\Delta = \frac{exp(-(1/4t)h_{ab}(x-x')^a(x-x')^b)}{4\pi t^{3/2}}$$

with $K(x,t)$ analytic in t. Next to resolve the ordering ambiguity in $\mathfrak{H}$ one adds a new term $L_{ab}(x)(\delta/\delta h_{ab}(x))$ where L_{ab} is a tensor to be derived along with $K(x,t)$. Thus the WDW operator will have the form

(4.26)
$$\mathfrak{H}(x) = \kappa^2 \int dx' K_{abcd}(x,x',t) \frac{\delta}{\delta h_{ab}(x)} \frac{\delta}{\delta h_{cd}(x')} +$$
$$+ L_{ab}(x) \frac{\delta}{\delta h_{ab}(x)} + \frac{1}{\kappa^2}\sqrt{h}(R+2\Lambda)$$

Next one needs to define the action of operators on states which involves discussions of regularization and renormalization. Regularization is a trade off here of + and - powers of t for $\delta(0)$ type singularities and for renormalization one drops positive powers of t and replaces singular terms $t^{-k/2}$ by renormalization coefficients ρ^k (cf. [**154**] for more details and references). Then to interpret (4.23) for example one thinks of an operator acting on a state and the resulting state after renormalization can be acted upon by another operator. Thus (4.23) means

(4.27)
$$\mathfrak{D}(\mathfrak{H}\psi)_{ren} \sim (\mathfrak{H}\psi)_{ren}$$

Similarly for the Hamiltonian constraint

(4.28)
$$(\mathfrak{H}[N](\mathfrak{H}(M)\psi)_{ren})_{ren} - (\mathfrak{H}[M](\mathfrak{H}[N]\psi)_{ren})_{ren} = 0$$

(since ψ is diffeomorphism invariant); here one is using the smeared form of the WDW operator $\mathfrak{H}[M] = \int dx M(x)\mathfrak{H}(x)$. After some calculation one also concludes that for the states which are integrals of scalar densities there is no anomaly in the diffeomorphism - Hamiltonian commutator. As for the Hamiltonian - Hamiltonian commutator one claims that if $(\mathfrak{H}\psi)_{ren}$ contains terms which contain 4 or more derivatives of the metric like R^2, $R_{ab}R^{ab}$, etc. then (4.28) cannot be satisfied. This was checked for some low order terms but is otherwise open. The form of the wave function must also be examined and this leads to further conditions on the WDW operator, which is finally advanced in the form

(4.29)
$$\mathfrak{H}(x) = \kappa^2 \int dx' G_{agcd}(x')\Delta(x,x',t) \frac{\delta}{\delta h_{ab}(x)} \frac{\delta}{\delta h_{cd}(x')} +$$
$$+ \left(\frac{1}{\kappa}\alpha h_{ab} + \kappa\gamma\left(\frac{1}{4}h_{ab}R + R_{ab}\right)\right)(x)\frac{\delta}{\delta h_{ab}(x)} + \frac{1}{\kappa^2}\sqrt{h}(R+2\Lambda)$$

where α, γ are independent constants.

When one now works with a Bohmian approach where $\psi = exp(\Gamma)exp(i\Sigma)$ and one considers only the real part of the resulting equation, namely

(4.30)
$$-\kappa^2 G_{abcd}(x)\frac{\delta\Sigma}{\delta h_{ab}(x)}\frac{\delta\Sigma}{\delta h_{cd}(x)} + \frac{1}{\kappa}\sqrt{h(x)}(R(x)+2\Lambda) + \Re(L)_{ab}(x)\frac{\delta\Gamma}{\delta h_{ab}(x)} -$$
$$-\Im(L)_{ab}(x)\frac{\delta\Sigma}{\delta h_{ab}(x)} + e^{-\Gamma}\kappa^2\left(\frac{\delta^2 e^{\Gamma}}{\delta h^2}\right)_{ren}(x) = 0$$

Identifying now

$$p^{ab}(x) = \frac{\delta\Sigma}{\delta h_{ab}(x)} \tag{4.31}$$

we see that the first two terms in (4.30) are the same as the Hamiltonian constraint of classical relativity. The remaining terms are quantum corrections (and if $\hbar$ were inserted they would all be proportional to $\hbar^2$). The wave function is subject to the second set of equations, namely those enforcing the 3-dimensional diffeomorphism invariance, which read (for the imaginary part)

$$\nabla^a \frac{\delta\Sigma}{\delta h_{ab}(x)} = \nabla^a p_{ab} = 0 \tag{4.32}$$

Thus the theory is defined by two equations (4.30) (with the p^{ab} inserted) and (4.32). This leads then to the full set of ten equations governing the quantum gravity in the quantum potential approach, namely

$$0 = \mathfrak{H}^a = \nabla_a p^{ab}; \tag{4.33}$$

$$0 = \mathfrak{H}_\perp = -\kappa^2 G_{abcd}(x)p^{ab}p^{cd} + \frac{1}{\kappa^2}\sqrt{h(x)}(R(x) + 2\Lambda) +$$

$$+\Re(L)_{ab}(x)\frac{\delta\Gamma}{\delta h_{ab}(x)} - \Im(L)_{ab}(x)p^{ab} + \kappa^2 e^{-\Gamma}\left(\frac{\delta^2 e^\Gamma}{\delta h^2}\right)_{ren}(x);$$

$$\dot{h}_{ab}(x,t) = \{h_{ab}(x,t), \mathfrak{H}[N,\vec{N}]\};\ \dot{p}^{ab}(x,t) = \{p^{ab}(x,t), \mathfrak{H}[N,\vec{N}]\}$$

where (cf. Section 4.3)

$$\mathfrak{H}[N,\vec{N}] = \int d^3x (N(x)\mathfrak{H}_\perp(x) + N^a(x)\mathfrak{H}_a(x)) \tag{4.34}$$

This all shows in particular that when questions of regularization and renormalization are taken into account life becomes more complicated. Various examples are treated in [**153, 154, 770**] where the quantum potential approach works very well but others where time translation becomes a problem.

5. PILOT WAVE COSMOLOGY

We refer here to [**96, 97, 99, 102, 105, 106, 107, 1022, 1023, 1024, 1025, 1170**] (other references to be given as we go along). First from Shtanov [**1170**] a set of nonrelativistic spinless particles are described via spatial coordinates $x = (x_1, \cdots, x_n)$ and the wave function ψ satisfies the SE

$$i\hbar\frac{\partial\psi}{\partial t} = -\frac{\hbar^2}{2}\sum \frac{1}{m_n}\Delta_n\psi + V\psi \tag{5.1}$$

Putting in $\psi = Rexp(iS/\hbar)$ gives then

$$\frac{\partial S}{\partial t} + \sum_n \frac{1}{2m_n}(\nabla_n S)^2 + V + Q = 0;\ Q = -\sum_n \frac{\hbar^2}{2m_n}\frac{\Delta_n R}{R}; \tag{5.2}$$

$$\frac{\partial R^2}{\partial t} + \sum_n \frac{1}{m_n}\nabla_n(R^2\nabla_n S) = 0$$

In the pilot wave interpretation the evolution of coordinates is governed by the phase via $m_n\dot{x}_n = \nabla_n S$. In the relativistic theory of spin 1/2 one continues to describe particles by their spatial coordinates but the guidance conditions and the equations for the wave are now different, namely for the multispinor $\psi_{\alpha_1 \cdots \alpha_n}(x,t)$ the Dirac equation is $(\alpha \sim (\alpha_1 \cdots \alpha_n))$

(5.3) $$i\dot{\psi}_\alpha = \sum_n (H_D)\psi)_\alpha;\ H_D = -i\gamma^0\gamma^i\nabla_i + m\gamma^0$$

The guidance condition is

(5.4) $$\frac{dx_n^\mu}{dt} = \frac{\psi^\dagger(\gamma^0\gamma^\mu)_n\psi}{\psi^\dagger\psi}$$

Here the label n enumerates the arguments of the multispinor ψ and the γ_n^μ act on the corresponding spinor index α_n; ψ is chosen to be antisymmetric with respect to interchange of any pair of its arguments. For integer spin the formulation in which the role of configuration variables is played by coordinates seems to be impossible; instead one has to consider the field spatial configurations as fundamental configuration variables guided by the corresponding wave functionals. For example the wave functional $\chi[\phi(\mathbf{x}),t]$ for a scalar field ϕ will obey the standard SE for the case of curved spacetime with guidance equation as indicated (notational gaps are filled in below)

(5.5) $$\dot{\phi}(\mathbf{x},t) = \frac{\delta}{\delta\phi(\mathbf{x})} S[\phi(\mathbf{x}),t]\Big|_{\phi(\mathbf{x})=\phi(\mathbf{x},t)}$$

(here $S[\phi(\mathbf{x},t]$ is $\hbar$ times the phase of $\chi[\phi(\mathbf{x}),t])$. The classical limit in the dynamics here is achieved for those configuration variables for which the quantum potential becomes negligible and such variables evolve in accord with classical laws (cf. Holland [**637**]). One can emphasize that the temporal dynamics of the particle coordinates and bosonic field configurations completely determine the state, be it microscopic or macroscopic. The role of the wave function in all physical situations is also the same, namely to provide the guidance laws for configuration variables. In order that the probabilities of of different measurements coincide with those calculated in the standard approach it is necessary to assume that the configuration variables of the system in a pure quantume ensemble are distributed in accord with the rule $p(x) = |\psi(x)|^2$ (quantum equilibrium condition - cf. [**1252, 415**], Section 2.5, and references to Dürr, Goldstein, Holland, Valentini, Zanghi, et al for discussion).

Now for quantum gravity recall the ADM formulation for bosonic fields

(5.6) $$I = \int_M d^3\mathbf{x}dt(\pi^{ab}\dot{g}_{ab} + \pi_\Phi\dot{\Phi} - N\mathfrak{H} - N^a\mathfrak{H}_a)$$

(here Φ is the set of bosonic fields). The restraints of GR are

(5.7) $$\mathfrak{H} = \frac{1}{2\mu}\mathfrak{G}_{abcd}\pi^{ab}\pi^{cd} + \mu\sqrt{g}(2\Lambda - {}^3\mathfrak{R}) + \mathfrak{H}^\Phi \approx 0;\ \mathfrak{H}_a \equiv -2\nabla_b\pi_a^b + \mathfrak{H}_a^\Phi \approx 0$$

(only the gravitational parts of the constraints are explicitly written out). Here $\mu = (16\pi G)^{-1}$ with G the Newton constant and

(5.8) $$\mathfrak{G}_{abcd} = \frac{1}{\sqrt{g}}(g_{ac}g_{bd} + g_{ad}g_{bc} - g_{ab}g_{cd})$$

while $\nabla_a \sim$ covariant derivative relative to g_{ab} and ${}^3\mathfrak{R}$ is the scalar curvature for the metric g. The classical equations of motion for g are

$$\dot{g}_{ab} = \frac{N}{\mu}\mathfrak{G}_{abcd}\pi^{cd} + \nabla_a N_b + \nabla_b N_a \tag{5.9}$$

Recall that in the Schrödinger representation the GR quantum system will be described by the wave functional $\Psi[g_{ab}(\mathbf{x}, \Phi(\mathbf{x}), t]$ over a manifold Σ with coordinates $\mathbf{x}$ and the quantum constraint equations are $\hat{\mathfrak{H}}_\mu \Psi = 0$ ($\hat{\mathfrak{H}}$ refers to all the components mentioned above (operator ordering and regularization are not treated in [**1170**]). Putting in now $\Psi = Rexp(iS/\hbar)$ one arrives at

$$\frac{1}{2\mu}\delta S \circ \delta S + \mu\sqrt{g}(2\Lambda - {}^3\mathfrak{R}) - \frac{\hbar^2}{2\mu}\frac{\delta \circ \delta R}{R} + \frac{\mathfrak{R}(\Psi^\dagger \hat{\mathfrak{H}}^\Psi)}{R^2} = 0; \tag{5.10}$$

$$\delta \circ (R^2 \delta S) - \frac{2\mu}{\hbar}\Im(\Psi^\dagger \hat{\mathfrak{H}}^\Phi \Psi) = 0$$

(here $\delta \sim \delta/\delta g_{ab}(\mathbf{x})$ and $\circ$ means contraction with respect to Wheeler's supermetric (5.8)). Note that for $\hbar \to 0$ the first equation in (5.10) reduces to the classical Einstein-HJ equation. Via the general guidance rules the quantum evolution of the g_{ab} is now given by (5.9) with the substitution

$$\pi^{ab}(\mathbf{x}) \to \left.\frac{\delta S}{\delta g_{ab}(\mathbf{x})}\right|_{g_{ab}(\mathbf{x}) = g_{ab}(\mathbf{x},t)} \tag{5.11}$$

The Lagrange multipliers N and N^a in (5.9) remain undetermined and are to be specified arbitrarily. This is analogous to the classical situation where this arbitrariness reflects reparameterization freedom. Thus to get a solution g_{ab} and Φ depending on $(\mathbf{x}, t)$ one must first solve the constraint equation $\hat{\mathfrak{H}}_\mu \Psi = 0$, then specify the initial configuration (e.g. at $t = 0$) for g_{ab} and Φ, then specify arbitrarily $N(\mathbf{x}, t)$ and $N^a(\mathbf{x}, t)$, and then solve the guidance equations (5.9) and the analogous equations for Φ. The solution should then represent a 4-geometry foliated by spatial hypersurfaces $\Sigma(t)$ on which the 3-metric is $g_{ab}(\mathbf{x}, t)$, the lapse function is $N(\mathbf{x}, t)$, the shift vector is $N^a(\mathbf{x}, t)$ and the field configuration is $\Phi(\mathbf{x}, t)$. This would be lovely but unfortunately there are complications as indicated below from [**1023, 1025**]; such matters are partially anticipated in [**1170**] however and there is some discussion and calculation. The question of quantum randomness in pilot wave QM is picked up again in the first paper of [**1170**] along with a continuation of time considerations. We only remark here that time in (5.9) is just a universal label of succession for spatial field configurations; it is not an observable.

5.1. EUCLIDEAN QUANTUM GRAVITY. The discussion here revolves around [**1023**] and the first paper of [**1025**] (cf. also [**326, 327, 329, 269**]). We go to Pinto-Neto [**1023**] directly and refer to Pinto-Neto and Santini [**1025**] for some background calculations and philosophy. [**1023**] is a review paper of deBroglie-Bohm theory in quantum cosmology. Extracting liberally one can argue convincingly against the Copenhagen interpretation of quantum phenomena in cosmology, in particular because it imposes the existence of a classical domain (cf. [**960**]). Decoherence is discussed but this does not seem to be a complete answer to the measurement problem (cf. [**749, 532, 1023**]) and one can also argue against the many-worlds theory (cf. [**402, 637, 1023**]). Thence one goes to deBroglie-Bohm as in [**105, 1170, 1252, 1262**] etc. and the quantum potential enters in a natural manner as we have already seen. Let us follow the notation of [**1023**] here (with some repetition of other discussions) and write $H = \int d^3x(N\mathfrak{H}+N^j\mathfrak{H}_j)$ where (in standard notation) for GR with a scalar field ϕ

$$\mathfrak{H}_j = -2D_i\pi^i_j\pi_\phi\partial_j\phi;\ \mathfrak{H} = \kappa G_{ijk\ell}\pi^{ij}\pi^{k\ell} + \frac{1}{2}h^{-1/2}\pi^2_\phi + \tag{5.12}$$
$$+h^{1/2}\left[-\kappa^{-1}(R^{(3)} - 2\Lambda) + \frac{1}{2}h^{ij}\partial_i\phi\partial_j\phi + U(\phi)\right]$$

The canonical momentum is expressed via (we use π^{ij} instead of Π^{ij})

$$\pi^{ij} = -h^{1/2}(K^{ij} - h^{ij}K) = G^{ijk\ell}(\dot{h}_{k\ell} - D_kN_\ell - D_\ell N_k); \tag{5.13}$$
$$K_{ij} = -\frac{1}{2N}(\dot{h}_{ij} - D_iN_j - D_jN_i)$$

K is the extrinsic curvature of the 3-D hypersurface Σ in question with indices lowered and raised via the surface metric h_{ij} and its inverse). The canonical momentum of the scalar field is

$$\pi_\phi = \frac{h^{1/2}}{N}(\dot{\phi} - N^j\partial_j\phi) \tag{5.14}$$

As usual $R^{(3)}$ is the intrinsic curvature of the hypersurfaces and N, N_j are the standard Lagrange multipliers for the super-Hamiltonian constraint $\mathfrak{H} \approx 0$ and the super momentum constraint $\mathfrak{H}^i \approx 0$. These multipliers are present due to the invariance of GR under spacetime coordinate transformations. Recall also

$$G^{ijk\ell} = \frac{1}{2}h^{1/2}(h^{ik}h^{j\ell} + h^{i\ell}h^{jk} - 2h^{ij}h^{k\ell}); \tag{5.15}$$
$$G_{ijk\ell} = \frac{1}{2}h^{-1/2}(h_{ik}h_{j\ell} + h_{i\ell}h_{jk} - h_{ij}h_{k\ell})$$

(called the deWitt metric). Here D_i is the i-component of the covariant derivative on the hypersurface and $\kappa = 16\pi G/c^4$. The classical 4-metric

$$ds^2 = -(N^2 - N^iN_i)dt^2 + 2N_idx^idt + h_{ij}dx^idx^j \tag{5.16}$$

and the scalar field which are solutions of the Einstein equations can be obtained from the Hamilton equations

$$\dot{h}_{ij} = \{h_{ij}, H\};\ \dot{\pi}^{ij} = \{\pi^{ij}, H\};\ \dot{\phi} = \{\phi, H\};\ \dot{\pi}_\phi = \{\pi_\phi, H\} \tag{5.17}$$

for some choice of N, N^j provided initial conditions are compatible with the constraints $\mathfrak{H} \approx 0$ and $\mathfrak{H}_j \approx 0$. It is a feature of the Hamiltonian structure that the 4-metrics (5.6) constructed in this way with the same initial conditions describe the same 4-geometry for any choice of N and N^j. The algebra of constraints closes in the following form (cf. [**636**])

$$\{\mathfrak{H}(x), \mathfrak{H}(x')\} = \mathfrak{H}^i(x)\partial_i\delta^3(x,x') - \mathfrak{H}^i(x')\partial_i\delta(x',x); \tag{5.18}$$
$$\{\mathfrak{H}_i(x), \mathfrak{H}(x')\} = \mathfrak{H}(x)\partial_i\delta(x,x');$$
$$\{\mathfrak{H}_i(x), \mathfrak{H}_j(x')\} = \mathfrak{H}_i(x)\partial_j\delta^3(x,x') + \mathfrak{H}_j(x')\partial_i\delta^3(x,x')$$

One quantizes following Dirac to get $\hat{\mathfrak{H}}_i|\psi>=0$ and $\hat{\mathfrak{H}}|\psi>=0$ and in the metric and field representation the first equation is

$$-2h_{\ell i}D_j\frac{\delta\psi(h_{ij},\phi)}{\delta h_{\ell j}}+\frac{\delta\psi(h_{ij}\phi)}{\delta\phi}\partial_i\phi=0 \tag{5.19}$$

which implies that the wave functional ψ is invariant under space coordinate transformations. The second equation is the WDW equation which (in unregularized form) is

$$\left\{-\hbar^2\left[\kappa G_{ijk\ell}\frac{\delta}{\delta h_{ij}}\frac{\delta}{\delta h_{k\ell}}+\frac{1}{2}h^{-1/2}\frac{\delta^2}{\delta\phi^2}\right]+V\right\}\psi(h_{ij},\phi)=0; \tag{5.20}$$

$$V=h^{1/2}\left[-\kappa^{-1/2}(R^{(3)}-2\Lambda)+\frac{1}{2}h^{ij}\partial_i\phi\partial_j\phi+U(\phi)\right]$$

(V is the classical potential). This equation involves products of local operators at the same point and hence must be regulated (cf. also Section 4.4). After that one should find a factor ordering which makes the theory free of anomalies, in the sense that the commutator of the operator version of the constraints closes in the same way as their respective classical Poisson brackets (cf. [**642, 771, 833**]).

Consider now the dBB interpretation of (5.19)-(5.20) where $\psi=Aexp(iS/\hbar)$ with A, S functionals of h_{ij} and ϕ. One arrives at

(5.21)

$$-2h_{\ell i}D_j\frac{\delta S(h_{ij}\phi)}{\delta h_{\ell j}}+\frac{\delta S(h_{ij},\phi)}{\delta\phi}\partial_i\phi=0;\ -2h_{\ell i}D_j\frac{\delta A(h_{ij},\phi)}{\delta h_{\ell j}}+\frac{\delta A(h_{ij},\phi)}{\delta\phi}=0$$

upon writing $\psi=Aexp(iS/\hbar)$. These equations will depend on the factor ordering; however in any case one of the equations will have the form

$$\kappa G_{ijk\ell}\frac{\delta S}{\delta h_{ij}}\frac{\delta S}{\delta h_{k\ell}}+\frac{1}{2}h^{-1/2}\left(\frac{\delta S}{\delta\phi}\right)^2+V+Q=0 \tag{5.22}$$

where V is the classical potential. Contrary to the other terms in (5.22), which are already well defined, the precise form of Q depends on the regularization and factor ordering which are prescribed for the WDW equation. In the unregulated form of (5.20)

$$Q=-\frac{\hbar^2}{A}\left(\kappa G_{ijk\ell}\frac{\delta^2 A}{\delta h_{ij}\delta h_{k\ell}}+\frac{h^{-1/2}}{2}\frac{\delta^2 A}{\delta\phi^2}\right) \tag{5.23}$$

The other equation (in addition to (5.22)) is

$$\kappa G_{ijk\ell}\frac{\delta}{\delta h_{ij}}\left(A^2\frac{\delta S}{\delta h_{k\ell}}\right)+\frac{1}{2}h^{-1/2}\frac{\delta}{\delta\phi}\left(A^2\frac{\delta S}{\delta\phi}\right)=0 \tag{5.24}$$

Now consider the dBB interpretation. First (5.21) and (5.22), which are valid irrespective of any factor ordering, are like the HJ equations for GR supplemented by an extra term Q for (5.22) (which does depend on factor ordering etc.). One postulates that the 3-metric, the scalar field, and their canonical momenta always exist (independent of observation) and that the evolution of the 3-metric and scalar field can be obtained from the guidance relations

$$\pi^{ij} = \frac{\delta S(h_{ab}, \phi)}{\delta h_{ij}}; \; \pi_\phi = \frac{\delta S(h_{ij}, \phi)}{\delta \phi} \tag{5.25}$$

(cf. (5.13)-(5.14)). The evolution of these fields will be different from the classical one due to the presence of Q in (5.22). The only difference between the cases of the nonrelativistic particle and QFT in flat spacetime is the fact that (5.24) for canonical QG cannot be interpreted as a continuity equation for a probability density A^2 because of the hyperbolic nature of the deWitt metric $G_{ijk\ell}$. However even without a notion of probability density one can extract a lot of information from (5.22), whatever Q may be. First note that whatever the form of Q it must be a scalar density of weight one (via the HJ equation (5.22)). From this equation one can express Q via

$$Q = -\kappa G_{ijk\ell} \frac{\delta S}{\delta h_{ij}} \frac{\delta S}{\delta h_{k\ell}} - \frac{1}{2} h^{-1/2} \left(\frac{\delta S}{\delta \phi}\right)^2 - V \tag{5.26}$$

Since S is an invariant (via (5.21)) it follows that $\delta S/\delta h_{ij}$ and $\delta S/\delta \phi$ must be a second rank tensor density and a scalar density respectively, both of weight one. When their products are contracted with $G_{ijk\ell}$ and multiplied by $h^{-1/2}$ they form a scalar density of weight one. As V is also a scalar density of weight one then so must be Q. Further Q must depend only on h_{ij} and ϕ because it comes from the wave functional which depends only on these variabes. Of course it can be nonlocal but it cannot depend on the momenta.

A minisuperspace is the set of spacelike geometries where all but a set of $h^{ij}_{(n)}(t)$ and the corresponding momenta $\pi^{(n)}_{ij}(t)$ are put identically equal to zero (this violates the uncertainty principle but one hopes to retain suitable qualitative features - cf. Pinto-Neto [**1023**] for references). In the case of a minisuperspace of homogeneous models, one puts $\mathfrak{H}^i = 0$ and N_i in H can be set equal to zero without losing any of the Einstein equations. The Hamiltonian becomes $H_{GR} = N(t)\mathfrak{H}(p^\alpha(t), q_\alpha(t))$ where the q_α and p^α represent homogeneous degrees of freedom coming from $\pi^{ij}(x,t)$ and $h_{ij}(x,t)$. Equations (5.23)-(5.25) become then

$$\frac{1}{2} f_{\alpha\beta}(q_\mu) \frac{\partial S}{\partial q_\alpha} \frac{\partial S}{\partial q_\beta} + U(q_\mu) + Q(q_\mu) = 0; \tag{5.27}$$

$$Q(q_\mu) = -\frac{1}{R} f_{\alpha\beta} \frac{\partial^2 R}{\partial q_\alpha \partial q_\beta}; \; p^\alpha = \frac{\partial S}{\partial q_\alpha} = f^{\alpha\beta} \frac{1}{N} \frac{\partial q_\beta}{\partial t} = 0$$

where $f_{\alpha\beta}(q_\mu)$ and $U(q_\mu)$ are the minisuperspace particularizations of $G_{ijk\ell}$ and $-h^{1/2} R^{(3)}(h_{ij})$. The last equation is invariant under time reparametrization and hence even at the quantum level different choices of $N(t)$ yield the same spacetime geometry for a given nonclassical solution $q_\alpha(x,t)$.

After some discussion and computations involving the avoidance of singularities and quantum isotropization of the universe one goes now in [**1023**] to the general situation in full superspace. From the guidance equations (5.25) one obtains

$$\dot{h}_{ij} = 2NG_{ijk\ell}\frac{\delta S}{\delta h_{k\ell}} + D_iN_j + D_jN_i;\ \dot{\phi} = Nh^{-1/2}\frac{\delta S}{\delta \phi} + N^i\partial_i\phi \tag{5.28}$$

The question here is, given some initial 3-metric and scalar field, what kind of structure do we obtain upon integration in t? In particular does this structure form a 4-dimensional geometry with a scalar field for any choice of the lapse and shift functions? Classically all is well but with S a solution of the modified HJ equation containing Q there is no guarantee. One goes to the Hamiltonian picture because strong results have been obtained from that point of view historically (cf. Hojman-Kuchar-Teitelboim [**636**]). One constructs now a Hamiltonian formalism consistent with the guidance relations (5.25) which yields the Bohmian trajectories (5.28). Given this Hamiltonian one can then use results from the literature to obtain new results for the dBB picture of quantum geometrodynamics. Thus from (5.21)-(5.22) one can easily guess that the Hamiltonian which generates the Bohmian trajectories, once the guidance relations (5.25) are satisfied initially, should be

$$H_Q = \int d^3x\,[N(\mathfrak{H} + Q) + N^i\mathfrak{H}_i]\,;\ \mathfrak{H}_Q = \mathfrak{H} + Q \tag{5.29}$$

Here $\mathfrak{H}$ and $\mathfrak{H}_i$ are the usual GR quantities from (5.12) and in fact the guidance relations (5.25) are consistent with the constraints $\mathfrak{H}_Q \approx 0$ and $\mathfrak{H}_i \approx 0$ because S satisfies (5.21)-(5.22). Furthermore they are conserved by the Hamiltonian evolution given by (5.29). Then one can show that indeed (5.28) can be obtained from H_Q with the guidance relations (5.25) viewed as additional constraints (cf. [**1022, 1025**] for details). Thus one has a Hamiltonian H_Q which generates the Bohmian trajectories once the guidance relations (5.25) are imposed initially. Now one asks about the evolution of the fields driven by H_Q forms a 4-geometry as in classical geometrodynamics. First recall a result from [**1216**] which shows that if the 3-geometries and field configurations defined on hypersurfaces are evolved by some Hamiltonian with the form $\tilde{H} = \int d^3x(N\tilde{\mathfrak{H}} + N^i\tilde{\mathfrak{H}}_i)$ and if this motion can be viewed as the motion of a 3-dimensional cut in a 4-dimensional spacetime then the constraints $\tilde{\mathfrak{H}} \approx 0$ and $\tilde{\mathfrak{H}}_i \approx 0$ must satisfy the algebra

$$\{\tilde{\mathfrak{H}}(x), \tilde{\mathfrak{H}}(x')\} = -\epsilon[\tilde{\mathfrak{H}}^i(x)\partial_i\delta^3(x',x) - \tilde{\mathfrak{H}}(x')\partial_i\delta^3(x,x'); \tag{5.30}$$
$$\{\tilde{\mathfrak{H}}_i(x), \tilde{\mathfrak{H}}(x')\} = \tilde{\mathfrak{H}}(x)\partial_i\delta^3(x,x');$$
$$\{\tilde{\mathfrak{H}}_i(x), \tilde{\mathfrak{H}}_j(x')\} = \tilde{\mathfrak{H}}_i(x)\partial_j\delta^3(x,x') - \tilde{\mathfrak{H}}_j(x')\partial_i\delta^3(x,x')$$

The constant ϵ can be ± 1 (if the 4-geometry is Euclidean $(+1)$ or hyperbolic (-1)); these are the conditions for the existence of spacetime. For $\epsilon = -1$ this algebra is the same as (5.18) for GR, but the Hamiltonian (5.29) is different via Q in $\mathfrak{H}_Q$. The Poisson bracket $\{\mathfrak{H}_i(x), \mathfrak{H}_j(x')\}$ satisfies (5.30) because the $\mathfrak{H}_i$ of H_Q defined in (5.29) is the same as in GR. Also $\{\mathfrak{H}_i(x), \mathfrak{H}_Q(x')\}$ satisfies (5.30) because $\mathfrak{H}_i$ is the generator of spatial coordinate transformations and since $\mathfrak{H}_Q$ is a scalar density of weight one (recall Q is a scalar density of weight one) it must satisfy this Poisson bracket relation with $\mathfrak{H}_i$. What remains to be verified is whether the Poisson bracket $\{\mathfrak{H}_Q(x), \mathfrak{H}_Q(x')\}$ closes as in (5.30). For this one recalls a result of [**636**] where there is a general super Hamiltonian $\tilde{\mathfrak{H}}$ which satisfies (5.30), is a scalar density of weight one (whose geometrical degrees of freedom are given only by h_{ij} and its canonical momentum), and which contains only even powers and no non-local term in the momenta. From [**636**] this means that $\tilde{\mathfrak{H}}$ must have the form

$$\tilde{\mathfrak{H}} = \kappa G_{ijk\ell}\pi^{ij}\pi^{k\ell} + \frac{1}{2}h^{-1/2}\pi_{\phi}^2 + V_G; \tag{5.31}$$

$$V_G = -\epsilon h^{1/2}\left[-\kappa^{-1}(R^{(3)} - 2\Lambda) + \frac{1}{2}h^{ij}\partial_i\phi\partial_j\phi + U(\phi)\right]$$

Note that $\mathfrak{H}_Q$ satisfies the hypotheses since it is quadraic in the momenta and the quantum potential does not contain any nonlocal term in the momenta. Consequently there are three possible scenarios for the dBB quantum geometrodynamics, depending on the form of the quantum potential. First assume that quantum geometrodynamics is consistent (independent of the choice of lapse and shift functions) and forms a nondegenerate 4-geometry. Then $\{\mathfrak{H}_Q, \mathfrak{H}_Q\}$ must satisfy the first equation in (5.30) and consequently, combining with (5.30) for $\tilde{\mathfrak{H}}$, Q must be such that $V + Q = V_G$ with V given by (5.20) yielding

(5.32)

$$Q = -h^{1/2}\left[(\epsilon+1)\left(-\kappa^{-1}R^{(3)} + \frac{1}{2}h^{ij}\partial_i\phi\partial_j\phi\right) + \frac{2}{\kappa}(\epsilon\tilde{\Lambda} + \Lambda) + \epsilon\tilde{U}(\phi) + U(\phi)\right]$$

For this situation there are two possibilities, namely

(1) The spacetime is hyperbolic with $\epsilon = -1$ and

$$Q = -h^{1/2}\left[\frac{2}{\kappa}(\Lambda - \tilde{\Lambda}) - \tilde{U}(\phi) + U(\phi)\right] \tag{5.33}$$

Hence Q is like a classical potential; its effect is to renormalize the cosmological constant and the classical scalar potential, nothing more. The quantum geometrodynamics is indistinguishable from the classical one. It is not necessary to require $Q = 0$ since $V_G = V + Q$ may already describe the classical universe in which we live.

(2) The spacetime is Euclidean with $\epsilon = 1$ in which case

$$Q = -h^{1/2}\left[2\left(-\kappa^{-1}R^{(3)} + \frac{1}{2}h^{ij}\partial_i\phi\partial_j\phi\right) + \frac{2}{\kappa}(\tilde{\Lambda} + \Lambda) + \tilde{U}(\phi) + U(\phi)\right] \tag{5.34}$$

Now Q not only renormalizes the cosmological constant and the classical scalar field but also changes the signature of spacetime. The total potential $V_G = V + Q$ may describe some era of the early universe when it had Euclidean signature but not the present era when it is hyperbolic. The transition between these two phases must happen in a hypersurface where $Q = 0$ which is the classical limit. This result points in the direction of Gibbons-Hawking [**527**].

There remains the possiblity that the evolution is consistent but does not form a nondegenerate 4-geometry. In this case the Poisson bracket $\{\mathfrak{H}_Q, \mathfrak{H}_Q\}$ does not satisfy (5.30) but is weakly zero in some other manner. Consider for example

(1) For real solutions of the WDW equation, which is a real equation, the phase S is zero and from (5.22) one can see that $Q = -V$. Hence the quantum super Hamiltonian (5.29) will contain only the kinetic term and $\{\mathfrak{H}_Q, \mathfrak{H}_Q\} = 0$ (strong equality). This case is connected with the strong gravity limit of GR (cf. [**1023**] for references and further discussion).

(2) Any nonlocal quantum potential breaks spacetime and as an example consider $Q = \gamma V$ where γ is a function of the functional S (hence is nonlocal). Calculating one obtains (cf. [**1022, 1025**])

$$\{\mathfrak{H}_Q(x), \mathfrak{H}_Q(x')\} = (1+\gamma)[\mathfrak{H}^i(x)\partial_i\delta^3(x,x') - \mathfrak{H}^i(x')\partial_i\delta^3(x',x)] - \tag{5.35}$$

$$-\frac{d\gamma}{dS}V(x')[2\mathfrak{H}_Q(x) - 2\kappa G_{ijk\ell}(x)\pi^{ij}(x)\Phi^{k\ell}(x) - h^{-1/2}\pi_\Phi(x)\Phi_\phi(x) +$$

$$+\frac{d\gamma}{dS}V(x)[2\mathfrak{H}_Q(x') - 2\kappa G_{ijk\ell}(x')\pi^{ij}(x')\Phi^{k\ell}(x') - h^{-1/2}\pi_\phi(x')\Phi_\phi(x')] \approx 0$$

The last equation is weakly zero because it is a combination of the constraints and the guidance relations of Bohmian theory. This means that the Hamiltonian evolution with the quantum potential $Q = \gamma V$ is consistent only when restricted to the Bohmian trajectories. For other trajectories it is inconsistent.

In these examples one makes contact with the structure constants of the algebra characterizing the foam like pregeometry structure pointed out long ago by J.A. Wheeler. Another fact here of interest is that there are no inconsistent Bohmian trajectories (cf. [**1025**]). We call attention also to [**99**] where in particular one considers noncommutative geometry and cosmology in connection with Bohmian theory; the results are very interesting.

6. BOHM AND NONCOMMUTATIVE GEOMETRY

We extract here from Barbosa, Pinto-Neto, et al [**99, 100, 511**] with other references as we go along (cf. in particular [**286, 325, 398, 438, 750, 1024, 1025, 1026, 1184, 1191**]). First from [**99**] one refers to [**100**] where a new interpretation of the canonical commutation

$$[\hat{X}^\mu, \hat{X}^\nu] = i\theta^{\mu\nu} \tag{6.1}$$

was proposed. The idea was that it is possible to interpret the commutation relation as a property of the particle coordinate observables rather than of the spacetime coordinates and this enforced a reinterpretation of the meaning of the wave function in noncommutative QM (NCQM). In [**99**] one develops a Bohmian interpretation for NCQM amd forms a deterministic theory of hidden variables that exhibit canonical noncommutativity (6.1) between the particle position observables. There are several motivations for reconsideration of hidden variable theory (see e.g. [**641**]) and we begin with a Moyal star product defined via

$$(f * g) = \frac{1}{(2\pi)^n}\int d^m k d^n p e^{i(k_\mu + p_\mu)x^\mu - (1/2)k_\mu\theta^{\mu\nu}p_\nu} f(k)g(p) = \tag{6.2}$$

$$= e^{(1/2)\theta^{\mu\nu}(\partial/\partial\xi^\mu)(\partial/\partial\eta^\nu)} f(x+\xi)g(x+\eta)|_{\xi=\eta=0}$$

(cf. [**239**] for an extensive treatment of star products and some noncommutative geometry). The commutative coordinates x^i are called the Weyl symbols (WS) of position operators $\hat{X}^i$ and one will consider them as spacetime coordinates following [**100**]. One assumes here that $\theta^{0i} = 0$ and the Hilbert space of states of NCQM can be taken as in commutative QM with noncommutative SE now given by

$$i\hbar\frac{\partial\psi(x^i,t)}{\partial t} = -\frac{\hbar^2}{2m}\nabla^2\psi(x^i,t) + V(x^i) * \psi(x^i,t) = \tag{6.3}$$
$$= \frac{\hbar^2}{2m}\nabla^2\psi(x^i,t) + V\left(x^j + i\frac{\theta^{jk}}{2}\partial_k\right)\psi(x^i,t)$$

The operators

$$\hat{X}^j = x^j + \frac{i\theta^{jk}\partial_k}{2} \tag{6.4}$$

are the observables corresponding to the physical positions of the particles and x^i are the associated canonical coordinates. For intition one could think here in terms of a "half dipole" whose extent is proportional to the canonical momentum $\Delta x^i = \theta^{ij}p_j/2\hbar$; one of its endpoints carries the mass and is responsible for its interaction. The NCQM formulated with (6.3)-(6.4) can be considered as the uaual QM with a Hamiltonian not quadratic in momenta and "unusual" position operators. From this point of view the BNCQM below can be considered as an extension along the same lines. Any attempt to localize the particles must satisfy the uncertainty relations

$$\Delta X^i \Delta X^j \geq |\theta^{ij}|/2 \tag{6.5}$$

The expression for the definition of probability density $\rho(x^i,t) = |\psi(x^i,t)|^2$ has a meaning that differs from ordinary QM. Namely the quantity $\rho(x^i,t)d^3x$ must be interpreted as the probability that the system is found in a configuration such that the canonical coordinate of the particle is contained in a volume d^3x around the point x at time t. Given an arbitrary physical observable characterized by a Hermitian operator $\hat{A}(\hat{x}^i,\hat{p}^i)$ (which includes e.g. $\hat{A}(\hat{X}^i(\hat{x}^i,\hat{p}^i),\hat{p}^i)$) its expected value is

$$< \hat{A} >_t = \int d^3x\psi^*(x^i,t)\hat{A}(x^j,-i\hbar\partial_j)\psi(x^i,t) \tag{6.6}$$

A HJ formalism for NCQM is found by writing $\psi = Rexp(iS/\hbar)$, putting it in (6.3), and splitting real and imaginary terms; the real part is

$$\frac{\partial S}{\partial t} + \frac{(\nabla S)^2}{2m} + V + V_{nc} + Q_K + Q_I = 0 \tag{6.7}$$

Here the three new potential terms are defined as

$$V_{nc} = V\left(x^i - \frac{\theta^{ij}}{2\hbar}\partial_j S\right) - V(x^i); \tag{6.8}$$

$$Q_K = \Re\left(-\frac{\hbar^2}{2m}\frac{\nabla^2\psi}{\psi}\right) - \left(\frac{\hbar^2}{2m}(\nabla S)^2\right) = -\frac{\hbar^2}{2m}\frac{\nabla^2 R}{R};$$

$$Q_I = \Re\left(\frac{V[x^j + (i\theta^{jk}/2)\partial_k]\psi}{\psi}\right) - V\left(x^i - \frac{\theta^{ij}}{2\hbar}\partial_j S\right)$$

V_{nc} is the potential that accounts for the NC classical interactions while Q_K and Q_I account for the quantum effects. The NC contributions contained in the latter two can be split out by defining

$$Q_{nc} = Q_K + Q_I - Q_c;\ Q_c = -\frac{\hbar^2}{2m}\frac{\nabla^2 R_c}{R_c};\ R_c = \sqrt{\psi_c^*\psi_c} \tag{6.9}$$

Here ψ_c is the wave function obtained from the commutative SE containing the usual potential $V(x^i)$, i.e. the equation obtained by setting $\theta^{ij} = 0$ in (6.3) before solving. The imaginary part of the SE which yields the probability conservation law, is

$$\frac{\partial R^2}{\partial t} + \nabla\cdot\left(\frac{R^2\nabla S}{m}\right) + \Sigma_\theta = 0;\ \Sigma_\theta = -\frac{2R}{\hbar}\Im\left[e^{-iS/\hbar}V * (Re^{iS/\hbar})\right] \tag{6.10}$$

By integrating the first equation we obtain

$$\frac{d}{dt}\int R^2 d^3x = 0;\ \int \Sigma_\theta d^3x = 0 \tag{6.11}$$

when R^2 vanishes at infinity.

Now for an ontological theory of motion one follows the traditional methods (cf. Bohm, Hiley, Holland [**158, 637**]). Necessary conditions for the theory to be capable of reproducing the same statistical results as the standard interpretation of NCQM constrain the admissible form for the functions $X^i(t)$ which eliminates a certain arbitrariness in the constructions. First for the rules, with an arbitrary physical characterized by a Hermitian operator $\hat{A}(\hat{x}^i, \hat{p}^i)$ it is possible to associate a function $\mathfrak{A}(x^i, t)$, the local expectation value of $\hat{A}$ (cf. [**637**]), which when averated over the ensemble of density $\rho(x^i, t) = |\psi|^2$ gives the same expectation value obtained by the standard operator formalism. Thus it is natural to define the ensemble average via

$$< \hat{A} >_t = \int \rho(x^i, t)\mathfrak{A}(x^i, t)d^3x \tag{6.12}$$

For this to agree with (6.6) $\mathfrak{A}(x^i, t)$ must be defined as

$$\mathfrak{A}(x^i, t) = \frac{\Re\left[\psi^*(x^i, t)\hat{A}(x^j, -i\hbar\partial_j)\psi(x^i, t)\right]}{\psi^*(x^i, t)\psi(x^i, t)} = A(x^i, t) + \mathfrak{Q}_A(x^i, t) \tag{6.13}$$

where the real part is taken to account for the hemiticity of $\hat{A}(\hat{x}^i, \hat{p}^i)$ and $A(x^i, t) = A[x^i, p^i = \partial^i S(x^i, t)]$ while $\mathfrak{Q}_A$ is defined via

$$\mathfrak{Q}_A = \Re\left[\frac{\hat{A}(x^j, -i\hbar\partial_j)\psi(x^k, t)}{\psi(x^i, t)}\right] - A(x^i, t) \tag{6.14}$$

and this is the quantum potential that accompanies $A(x^i, t)$. From (6.13) one finds that the local expectation value of (6.4) is

$$X^i = x^i - \frac{\theta^{ij}}{2\hbar}\partial_j S(x^i, t) \tag{6.15}$$

Now to find the $X^i(t)$ the relevant information for particle motion can be extracted from the guiding wave $\psi(x^i,t)$ by first computing the associated canonical position tracks $x^i(t)$ and then evaluating (6.15) at $x^i = x^i(t)$. In order to find a good equation for the $x^i(t)$ it is interesting to consider the Heisenberg formulation and the equations of motion for the observables. Thus for the $\hat{x}^i$ one has

$$\frac{d\hat{x}^i_H}{dt} = \frac{1}{i\hbar}[\hat{x}^i_H, \hat{H}] = \frac{\hat{p}^i_H}{m} + \frac{\theta^{ij}}{2\hbar}\frac{\partial\hat{V}(\hat{X}^k_H)}{\partial\hat{X}^j_H} \tag{6.16}$$

By passing the right side of (6.16) to the Schrödinger picture one can define the velocity operators

$$\hat{v}^i = \frac{1}{\hbar}[\hat{x}^i, \hat{H}] = \frac{\hat{p}^i}{m} + \frac{\theta^{ij}}{2\hbar}\frac{\partial\hat{V}(\hat{X})}{\partial\hat{X}} \tag{6.17}$$

The differential equation for the canonical positions $x^i(t)$ is now found by identifying $dx^i(t)/dt$ with the local expectation value of $\hat{v}^i$, thus

$$\frac{dx^i(t)}{dt} = \left[\frac{\partial^i S(x^i,t)}{m} + \frac{\theta^{ij}}{2\hbar}\frac{\partial V(X^k)}{\partial X^j} + \frac{\mathfrak{Q}^i}{2}\right]\Bigg|_{x^i=x^i(t)} \tag{6.18}$$

where X^i is given in (6.15), $S(x^i, t0$ is the phase of ψ, and

$$\mathfrak{Q}^i = \Re\left(\frac{(\theta^{ij}/\hbar)[\partial\hat{V}(\hat{X}^k)/\partial\hat{X}^j]\psi(x^i,t)}{\psi(x^i,t)}\right) - \frac{\theta^{ij}}{\hbar}\frac{\partial V(X^k)}{\partial X^j} \tag{6.19}$$

The potentials $\mathfrak{Q}^i$ account for quantum effects coming from derivatives of order 2 and higher contained in $\partial\hat{V}(\hat{X}^i)/\partial\hat{X}^j$. Then once the $x^i(t)$ are known the particle trajectories are given via

$$X^i(t) = x^i(t) - \frac{\theta^{ij}}{2\hbar}\partial_j S(x^k(t),t) \tag{6.20}$$

One notes that the particles positions are not defined on nodal regions of ψ, where S is undefined, so the particles cannot run through such regions. Hence the vanishing of the wave function can be adopted as a boundary condition, implying that the particle does not run through such a region (see Barbosa and Pinto-Neto[**99**] for more discussion). The preceeding theory is now summarized in a formal structure as follows:

(1) The spacetime is commutative and has a pointwise manifold structure with canonical coordinates x^i. The observables correspondin to operators of position coordinates $\hat{X}^i$ of particles satisfy the commutation rules $[\hat{X}^k, \hat{X}^j] = i\theta^{kj}$. The position observables can be represented in the coordinate space as $\hat{X}^j = x^j + i\theta^{jk}\partial_k/2$ and the x^j are canonical coordinates associated with the particle.

(2) A quantum system is composed of a point particle and a wave ψ. The particle moves in spacetime under the guidance of the wave which satisfies the SE $i\hbar\partial_t\psi(x^i,t) = -(\hbar^2/2m)\nabla^2\psi + V(\hat{X}^i)\psi$ $(\psi = \psi(x^i,t))$.

(3) The particle moves with trajectory $X^i(t) = x^i(t) - (\theta^{ij}/2\hbar)\partial_j S(x^i(t), t)$ independently of observation, where S is the phase of ψ and the $x^i(t)$ describe the canonical position trajectories found by solving

$$\frac{dx^i(t)}{dt} = \left[\frac{\partial^i S(\vec{x}, t)}{m} + \frac{\theta^{ij}}{2\hbar}\frac{\partial V(X^k)}{\partial X^j} + \frac{\mathfrak{Q}^i}{2}\right]\Bigg|_{x^i = x^i(t)}$$

To find the path followed by a particle, one must specify its initial canonical position $x^i(0)$, solve the second equation, and then obtain the physical path from the first equation.

These three postulates constitute on their own a consisent theory of motion, and is intended to give a finer view of QM, namely a detailed description of the individual physical processes and to provide the same statistical predictions. In ordinary commutative Bohmian mechanics, in order to reproduce the statistics the additional requirement that $\rho(x^i, t_0) = |\psi(\vec{x}, t_0)|^2$ is imposed for some initial time t_0. Then ρ is said to be equivariant if this property persists under evolution $\dot{x}^i(t) = f^i(x^j, t)$; in such a case

$$\frac{\partial\rho}{\partial t} + \frac{\partial(\rho\dot{x}^i)}{\partial x^i} = 0 \tag{6.21}$$

In ordinary commutative QM the equivariance property is satisfied via $\dot{x}^i(t) = J^i/\rho$ which is a consequenc of the identification between $\dot{x}^i$ and the local expectation value of the $\hat{v}^i$. In the BNCQM proposed here the same identification is valid but it is not sufficient to guarantee equivariance in all cases. This is clear after computing the canonical probability current

$$J^i(x^i, t) = \Re[\psi^*\hat{v}\psi] = |\psi|^2\left[\frac{\partial^i S(\vec{x}, t)}{m} + \frac{\theta^{ij}}{2\hbar}\frac{\partial V(X^k)}{\partial X^j} + \frac{\mathfrak{Q}^i}{2}\right] = \rho\dot{x}^i \tag{6.22}$$

Then regrouping the terms in (6.10) so that the canonical probability flux (6.22) appears explicitly one obtains

$$\frac{\partial\rho}{\partial t} + \frac{\partial(\rho\dot{x}^i}{\partial x^i} - \frac{\partial}{\partial x^i}\left[\rho\left(\frac{\theta^{ij}}{2\hbar}\frac{\partial V(X^k)}{\partial X^j} + \frac{\mathfrak{Q}^i}{2}\right)\right] + \Sigma_\theta = 0 \tag{6.23}$$

For equivariance to occur an additional condition that the sum of the last two terms in the right side of (6.23) vanishes is required. When $V(X^i)$ is a linear or quadratic function, as in many applications, such a condition is trivially satisfied, and then $\rho(x^i, t) = |\psi(x^i, t)|^2$ as desired. The same may occur for other situations when other potentials are considered in #2 above and this is also discussed in [**99**].

In [**99**] (second paper) one looks at a Kantowski-Sachs (KS) universe (see e.g. [**286, 325, 511, 1191**]). Recall in the ADM formulation a line element is written in the form

$$ds^2 = (N_iN^i - N^2)dt^2 + 2N_i dx^i dt + h_{ij}dx^i dx^j \tag{6.24}$$

and the Hamiltonian of GR without matter is

$$H = \int d^3x(N\mathfrak{H} + N)i\mathfrak{H}^j);\ \mathfrak{H} = G_{ijk\ell}\pi^{ij}\pi^{k\ell} - h^{1/2}R^{(3)};\ \mathfrak{H}^j = 2D_i\pi^{ij} \tag{6.25}$$

Units are chosen so that $\hbar = c = 16\pi G = 1$. The momenta π_{ij} are canonically conjugate to h^{ij} and the deWitt metric $G_{ijk\ell}$ are given via

$$\pi_{ij} = -h^{1/2}(K_{ij} - h_{ij}K);\ G_{ijk\ell} = \frac{1}{2}h^{-1/2}(h_{ik}h_{j\ell} + h_{i\ell}h_{jk} - h_{ij}h_{k\ell}) \tag{6.26}$$

where $K_{ij} = -(\partial_t h_{ij} - D_i N_j - D_j N_i)/(2N)$ is the second fundamental form. The super Hamiltonian constraint $\mathfrak{H} \approx 0$ yields the WDW equation

$$\left(G^{ijk\ell}\frac{\delta}{\delta h^{ij}}\frac{\delta}{\delta h^{k\ell}} + h^{1/2}R^{(3)}\right)\psi[h^{ij}] = 0 \tag{6.27}$$

In the Bohmian approach now one has

$$\pi_{ij} = -h^{1/2}(K_{ij} - h_{ij}D) = \Re\left\{\frac{1}{\psi^*\psi}\left[\psi^*\left(-i\frac{\delta}{\delta h^{ij}}\right)\psi\right]\right\} = \frac{\delta S}{\delta h^{ij}} \tag{6.28}$$

If one puts $\psi = Aexp(iS)$ in (6.27) there results

$$G^{ijk\ell}\frac{\delta S}{\delta h^{ij}}\frac{\delta S}{\delta h^{k\ell}} - h^{1/2}R^{(3)} + Q = 0; \tag{6.29}$$

$$G^{ijk\ell}\frac{\delta S}{\delta h^{ij}}\left(A^2\frac{\delta S}{\delta h^{k\ell}}\right) = 0;\ Q = -\frac{1}{A}G^{ijk\ell}\frac{\delta^2 A}{\delta h^{ij}h^{k\ell}}$$

(one should really include $\hbar^2$ here in dealing with Q).

The Kantowski-Sachs universe is an important anisotropic model; the line element is

$$ds^2 = -Ndt^2 + X^2(t)dr^2 + Y^2(t)(d\theta^2 + Sin^2(\theta)d\phi^2) \tag{6.30}$$

In the Misner parametrization this becomes (cf. Garcia-Compean et al [**511**])

$$ds^2 = -N^2dt^2 + e^{2\sqrt{3}\beta}dr^2 + e^{-2\sqrt{3}\beta}e^{-2\sqrt{3}\Omega}(d\theta^2 + Sin^2(\theta)d\phi^2) \tag{6.31}$$

The Hamiltonian is then

$$H = N\mathfrak{H} = Nexp\left(\sqrt{3}\beta + 2\sqrt{3}\Omega\right)\left[-\frac{P_\Omega^2}{24} + \frac{P_\beta^2}{24} - 2exp(-2\sqrt{3}\Omega)\right] \tag{6.32}$$

One sets $\Theta = V^\alpha_{;\alpha}$ (volume expansion - $V^\alpha = \delta^\alpha_0/N$), and $\sigma^2 = \sigma^{\alpha\beta}\sigma_{\alpha\beta}/2$ (shear) where $\sigma_{\alpha\beta} = (h^\mu_\alpha h^\nu_\beta + h^\mu_\beta h^\nu_\alpha)V_{\mu;\nu}/2$. The semicolon denotes 4-dimensional covariant differentiation and $h^\mu_\alpha = \delta^\mu_\alpha + V^\mu V_\alpha$ is the projector orthogonal to the observer V^α (cf. Hawking-Elllis [**618**]). A characteristic length scale is defined via $\Theta = 3\dot{\ell}/(\ell N)$ and in the gauge $N = 24exp(-\sqrt{3}\beta - 2\sqrt{3}\Omega)$ one has

$$\Theta(t) = \frac{1}{N}\left(\frac{\dot{X}}{X} + 2\frac{\dot{Y}}{Y}\right) = -\frac{\sqrt{3}}{24}(\dot{\beta} + 2\dot{\Omega})e^{\sqrt{3}\beta + 2\sqrt{3}\Omega}; \tag{6.33}$$

$$\sigma(t) = \frac{1}{N\sqrt{3}}\left(\frac{\dot{X}}{X} - \frac{\dot{Y}}{Y}\right) = \frac{1}{24}(2\dot{\beta} + \dot{\Omega})e^{\sqrt{3}\beta + 2\sqrt{3}\Omega};$$

$$\ell^3(t) = X(t)Y^2(t) = e^{-\sqrt{3}\beta(t) - 2\sqrt{3}\Omega(t)}$$

In order to distinguish the role of the quantum and noncommutative effects in a NC quantum universe one starts now with a KS geometry in the commutative classical version. The Poisson brackets for the classical phase space variables are

$$\{\Omega, P_\Omega\} = 1 = \{\beta, P_\beta\};\ \{P_\Omega, P_\beta\} = 0 = \{\Omega, \beta\} \tag{6.34}$$

For the metric (6.31) the constraint $\mathfrak{H} \approx 0$ is reduced to

$$\mathfrak{H} = \xi h \approx 0;\ \xi = \frac{1}{24} e^{\sqrt{3}\beta + 2\sqrt{3}\Omega};\ h = -P_\Omega^2 + P_\beta^2 - 48e^{-2\sqrt{3}\Omega} \approx 0 \tag{6.35}$$

The classical equations of motion for the phase space variables Ω, P_Ω, β, and P_β are then

$$\dot{\Omega} = N\{\Omega, \mathfrak{H}\} = -2P_\Omega;\ \dot{\beta} = N\{\beta, \mathfrak{H}\} = 2P_\beta; \tag{6.36}$$

$$\dot{P}_\Omega = N\{P_\Omega, \mathfrak{H}\} = -96\sqrt{3}e^{-2\sqrt{3}\Omega};\ \dot{P}_\beta = N\{P_\beta, \mathfrak{H}\} = 0$$

Explicit formulas are found and exhibited in [**99**].

Now for a NC classical model one considers

$$\{\Omega, P_\Omega\} = 1;\ \{\beta, P_\beta\} = 1;\ \{P_\Omega, P_\beta\} = 0;\ \{\Omega, \beta\} = \theta \tag{6.37}$$

The equations of motion can be written as

$$\dot{\Omega} = -2P_\Omega;\ \dot{P}_\Omega = -96\sqrt{3}e^{-2\sqrt{3}\Omega};\ \dot{\beta} = 2P_\beta - 96\sqrt{3}\theta e^{-2\sqrt{3}\Omega};\ \dot{P}_\beta = 0 \tag{6.38}$$

The solutions for Ω and β are then

$$\Omega(t) = \frac{\sqrt{3}}{6} log\left\{\frac{48}{P_{\beta_0}^2} Cosh^2\left[2\sqrt{3}P_{\beta_0}(t - t_0)\right]\right\}; \tag{6.39}$$

$$\beta(t) = 2P_{\beta_0}(t - t_0) + \beta_0 - \theta P_{\beta_0} Tanh[2\sqrt{3}P_{\beta_0}(t - t_0]$$

Further calculations appear in [**99**].

For the commutative quantum model one works with the minisuperspace construction of homogeneous models and freezing out an infinite number of degrees of freedom. First an Ansatz of the form (6.31) is introduced and the spatial dependence of the metric is integrated out. The WDW equation is then reduced to a KG equation which for the KS universe has the form

$$\left[-\hat{P}_\Omega^2 + \hat{P}_\beta^2 - 48e^{-2\sqrt{3}\Omega}\right]\psi(\Omega, \beta) = 0 \tag{6.40}$$

where $\hat{P}_\Omega = -i\partial/\partial\Omega$ and $\hat{P}_\beta = -i\partial/\partial\beta$. A solution to (6.40) is then (cf. [**511**])

$$\psi_\nu(\Omega, \beta) = e^{i\nu\sqrt{3}\beta} K_{i\nu}\left(4e^{-\sqrt{3}\Omega}\right) \tag{6.41}$$

where $K_{i\nu}$ is a modified Bessel function and ν is a real constant. Once a quantum state of the universe is given as, e.g. a superposition of states

$$\psi(\Omega, \beta) = \sum_\nu C_\nu e^{i\nu\sqrt{3}\beta} K_{i\nu}\left(4e^{-\sqrt{3}\Omega}\right) = Re^{iS} \tag{6.42}$$

the evolution can be determined by integrating the guiding equation (6.28). In the minisuperspace approach the analogue of that equation is

$$(6.43)\qquad P_\Omega = -\frac{1}{2}\dot{\Omega} = \Re\left\{\frac{[\psi^*(-i\hbar\partial_\Omega)\psi]}{\psi^*\psi}\right\} = \frac{\partial S}{\partial\Omega};$$

$$P_\beta = \frac{1}{2}\dot{\beta} = \Re\left\{\frac{[\psi^*(-i\hbar\partial_\beta)\psi]}{\psi^*\psi}\right\} = \frac{\partial S}{\partial\beta}$$

As before one has fixed the gauge $N = 24\ell^3 = 24exp(-\sqrt{3}\beta - 2\sqrt{3}\Omega)$. Usually different choices of time yield different quantum theories but when one uses the Bohmian interpretation in minisuperspace models the situation is identical to that of the classical case (but not beyond minisuperspace - cf. [**1025**]), namely different choices yield the same theory (cf. [**105**]). Hence as long as $\ell^3(t)$ does not pass through zero (a singularity) the above choice for $N(t)$ is valid for the history of the universe. The minisuperspace analogue of the HJ equation in (6.29) is

$$(6.44)\qquad -\frac{1}{24}\left(\frac{\partial S}{\partial\Omega}\right)^2 + \frac{1}{24}\left(\frac{\partial S}{\partial\beta}\right)^2 - 2e^{-2\sqrt{3}\Omega} + \frac{1}{24R}\left(\frac{\partial^2 R}{\partial\Omega^2} - \frac{\partial^2 R}{\partial\beta^2}\right) = 0$$

Explicit calculations are then given in [**99**] with graphs and pictures.

Now for the NC quantum model one takes

$$(6.45)\qquad [\hat{\Omega}, \hat{\beta}] = i\theta$$

According to the Weyl quantization procedure (cf. Douglas-Nekrasov [**398**]) the realization of (6.45) in terms of commutative functions is made by the Moyal star product defined via (cf. [**239**])

$$(6.46)\qquad f(\Omega_c, \beta_c) * g(\Omega_c, \beta_c) = f(\Omega_c, \beta_c)e^{i(\theta/2)(\overleftarrow{\partial}_{\Omega_c}\overrightarrow{\partial}_{\beta_c} - \overleftarrow{\partial}_{\beta_c}\overrightarrow{\partial}_{gO_c})}g((\Omega_c, \beta_c)$$

The commutative coordinates Ω_c, β_c are called Weyl symbols of the operators $\hat{\Omega}$, $\hat{\beta}$. In order to compare evolutions with the same time parameter as above one again fixes the gauge $N = 24exp(-\sqrt{3}\beta - 2\sqrt{3}\Omega)$ and the WDW equation for the NC KS model is (cf. [**511**])

$$(6.47)\qquad \left[-P_{\Omega_c}^2 + P_{\beta_c}^2 - 48e^{-2\sqrt{3}\Omega_c}\right] * \psi(\Omega_c, \beta_c) = 0$$

which is the Moyal deformed version of (6.40). By using properties of the Moyal bracket (cf. [**239**]) one can write the potential term (denoted by V to include the general case) as

$$(6.48)\qquad V(\Omega_c, \beta_c) * \psi(\Omega_c, \beta_c) = V\left(\Omega_c + i\frac{\theta}{2}\partial_{\beta_c}, \beta_c - i\frac{\theta}{2}\partial_{\Omega_c}\right)\psi(\Omega_c, \beta_c) =$$

$$= V(\hat{\Omega}, \hat{\beta})\psi(\Omega_c, \beta_c);\ \hat{\Omega} = \hat{\Omega}_c - \frac{\theta}{2}\hat{P}_{\beta_c};\ \hat{\beta} = \hat{\beta}_c + \frac{\theta}{2}\hat{P}_{\Omega_c}$$

The WDW equation then reads as

$$(6.49)\qquad \left[-\hat{P}_{\Omega_c}^2 + \hat{P}_{\beta_c}^2 - 48e^{-2\sqrt{3}\hat{\Omega}_c + \sqrt{3}\theta\hat{P}_{\beta_c}}\right]\psi(\Omega_c, \beta_c)$$

Two consistent interpretations for the cosmology emerging from these equations are possible. One consists in considering the Weyl symbols Ω_c and β_c as the constituents of the physical metric, which makes things essentially commutative with a modified interaction. The second, as adopted in Barbosa [**100, 101**], involves the Weyl symbols being considered as auxiliary coordinates, and thereby one studies the evolution of a NC quantum universe.

Next for the Bohmian formulation of the NC minisuperspace one looks at

$$[\hat{x}^i, \hat{p}^j] = i\hbar\delta^{ij} \tag{6.50}$$

Note the operator formalism of QM is not a primary concept in Bohmian mechanics. Thus it is reasonable to expect that in Bohmian NC quantum cosmology it should be possible to describe the metric variables as well defined entities, although the operators $\hat{\Omega}$ and $\hat{\beta}$ satisfy (6.45). This is exactly what is proposed here. One wants to give an objective meaning to the wavefunction and the metric variables Ω and β and the wave function is obtained by solving (6.47). What is missing is the evolution law for Ω and β. To find this one associates a function $\mathfrak{A}(\Omega_c, \beta_c)$ to $\hat{A}(\hat{\Omega}_c, \hat{\beta}_c, \hat{P}_{\Omega_c}, \hat{P}_{\beta_c})$ according to the rule

$$\mathfrak{B}[\hat{A}] = \frac{\Re[\psi^*(\Omega_c, \beta_c)\hat{A}(\Omega_c, \beta_c, -i\hbar\partial_{\Omega_c}, -i\hbar\partial_{\beta_c})\psi(\Omega_c, \beta_c)]}{\psi^*(\Omega_c, \beta_c)\psi(\Omega_c, \beta_c)} = \mathfrak{A}(\Omega_c, \beta_c) \tag{6.51}$$

where the real part takes into account the hermiticity of $\hat{A}$ (the $\mathfrak{B}$ here refers to the idea of "beable"). Applying this to the operators $\hat{\Omega}$ and $\hat{\beta}$ one arrives at

$$\Omega(\Omega_c, \beta_c) = \mathfrak{B}[\hat{\Omega}] = \frac{\Re[\psi^*(\Omega_c, \beta_c)\hat{\Omega}(\Omega_c, -i\hbar\partial_{\beta_c})\psi(\Omega_c, \beta_c)]}{\psi^*(\Omega_c, \beta_c)\psi(\Omega_c, \beta_c)} = \Omega_c - \frac{\theta}{2}\partial_{\beta_c}S; \tag{6.52}$$

$$\beta(\Omega_c, \beta_c) = \mathfrak{B}[\hat{\beta}] = \frac{\Re[\psi^*(\Omega_c, \beta_c)\hat{\beta}(\beta_c, -i\hbar\partial_{\Omega_c})\psi(\Omega_c, \beta_c)]}{\psi^*\psi} = \beta_c + \frac{\theta}{2}\partial_{\Omega_c}S$$

Thus the relevant information for universe evolution can be extracted from the guiding wave $\psi(\Omega_c, \beta_c)$ by first computing the associated canonical position tracks $\Omega_c(t)$ and $\beta_c(t)$. Then one obtains $\Omega(t)$ and $\beta(t)$ by evaluating (6.52) at $\Omega_c = \Omega_c(t)$ and $\beta_c = \beta_c(t)$ (similar procedures are worked out in Barbosa and Pinto-Neto [**99**] for the NC classical situation). Differential equations for the canonical positions $\Omega_c(t)$ and $\beta_c(t)$ can be found by identifying $\dot{\Omega}_c(t)$ and $\dot{\beta}_c(t)$ with the beables associated with their time evolution and formulas are worked out in [**99**].

The combination of NC geometry and Bohmian type quantum physics is somewhat like a fusion of two apparently opposite ways of thinking, one fuzzy and the other refering ontologically to point particles. Every Hermitian operator can be associated with an ontological element and by averaging the beable $\mathfrak{B}[\hat{A}]$ over an ensemble of particles with probability density $\rho = |\psi|^2$ one gets the same result as computing the expectation value of the observable $\hat{A}$ in standard operational formalism. In the KS universe, where WDW is of KG type (with no notion of probability) the beable mapping is well defined, even in the NC case. In the commutative context this formulation leads to the Bohmian quantum gravity proposed by Holland in [**637**] in the minisuperspace approximation. The work here shows that noncommutativity can modify appreciably the universe evolution in the quantum context (qualitatively as well as quantitively).

We go next to Barbosa [**100**] and consider NC in the evolution of Friedman-Robertson-Walker (FRW) universes with a conformally coupled scalar field. First take the commutative situation and restrict attention to the case of constant positive curvature of the spatial sections. The action is then

$$S = \int d^4x\sqrt{-g}\left[-\frac{1}{2}\phi_{;\mu}\phi^{;\mu} + \frac{R}{16\pi G} - \frac{R\phi^2}{12}\right] \tag{6.53}$$

Units are chosen so that $\hbar = c = 1$ and $8\pi G = 3\ell_P^2$ where ℓ_P is the Planck length. For the FRW model with a homogeneous scalar field the following Ansatz of minisuperspace can be adopted

$$ds^2 = -N^2(t)dt^2 + a^2(t)\left[\frac{dr^2}{1-r^2} + r^2(d\theta^2 + Sin^2(\theta)d\phi^2)\right];\ \phi = \phi(t) \tag{6.54}$$

Rescaling the scalar field via $\chi = \phi a \ell_P/\sqrt{2}$ one obtains the minisuperspace action, Hamiltonian, and momenta

$$S = \int dt\left(Na - \frac{a\dot{a}^2}{N} + \frac{a\dot{\chi}^2}{N} - \frac{N\chi^2}{a}\right);\ P_a = -\frac{2a\dot{a}}{N}; \tag{6.55}$$

$$H = N\left[-\frac{P_a^2}{4a} + \frac{P_\chi^2}{4a} - a + \frac{\chi^2}{a}\right] = N\mathfrak{H};\ P_\chi = \frac{2a\dot{\chi}}{N}$$

For the classical phase space variables one knows $\{a,\chi\} = 0 = \{P_a, P_\chi\}$ and $\{a, P_a\} = 1 = \{\chi, P_\chi\}$ and the equations for the metric and matter field variables following from this and (6.54) are

$$\dot{a} = \{a, H\} = -\frac{NP_a}{2a};\ \dot{P}_a = \{P_a, H\} = 2N; \tag{6.56}$$

$$\dot{\chi} = \{\chi, H\} = \frac{NP_\chi}{2a};\ \dot{P}_\chi = \{P_\chi, H\} = -\frac{2N\chi}{a}$$

Now one adopts the conformal time gauge $N = \dot{a}$ and the general solution of (6.56) in this gauge is

$$a(t) = (A+C)Cos(t) + (B+D)Sin(t); \tag{6.57}$$

$$\chi(t) = (A-C)Cos(t) + (B-D)Sin(t)$$

where the constraint $\mathfrak{H} \approx 0$ imposes the relation $AC + BD = 0$.

Next look at a NC deformation by keeping the Hamiltonian with the same functional form as in (6.55) but now with NC variables

$$H = N\left[-\frac{P_{a_{nc}}^2}{4a_{nc}} + \frac{P_{\chi_{nc}}^2}{4a_{nc}} - a_{nc} + \frac{\chi_{nc}^2}{a_{nc}}\right] \tag{6.58}$$

where

$$\{a_{nc}, \chi_{nc}\} = \theta;\ \{a_{nc}, P_{a_{nc}}\} = 1 = \{\chi_{nc}, P_{\chi_{nc}}\};\ \{P_{a_{nc}}, P_{\chi_{nc}}\} = 0 \tag{6.59}$$

Now make the substitution

$$a_{nc} = a_c - \frac{\theta}{2}P_{\chi_c};\ \chi_{nc} = \chi_c + \frac{\theta}{2}P_{a_c};\ P_{a_c} = P_{a_{nc}},\ P_{\chi_c} = P_{\chi_{nc}} \tag{6.60}$$

Then the theory defined by (6.58)-(6.59) can be mapped to a theory where the metric and matter variables satisfy

$$\{a_c, \chi_c\} = 0 = \{P_{a_c}, P_{\chi_c}\};\ \{a_c, P_{a_c}\} = \{\chi_c, P_{\chi_c}\} = 1 \tag{6.61}$$

In the case where a_c, χ_c are taken as the preferred variables one has a commutative theory referred to as a theory realized in the C-frame (cf. [**511**]). When a_{nc} and χ_{nc} are used as constituents of the physical metric and matter field one refers to the NC frame (cf. [**99, 100, 101**]). Some work assumes the difference between the C and NC variables is negligible (cf. [**289**]) but as shown in [**101**] even in simple models the difference in behavior between these two types of variables can be appreciable; here one shows that the assumption of C or NC frame leads to dramatic differences in the analysis of universe history. Calculations are made for the classical situation in both sets of variables which we omit here.

Next comes the quantum version of the commutative universe model for the FRW universe with conformally coupled scalar field; this has been investigated on the basis of the WDW equation in e.g. Acacio de Barros, Pinto-Neto, Leal [**108**] using Bohmian trajectories but there was there a restriction to the regime of small scale parameters and the wavefunctions there were different from those used here. One writes then $P_a = -\partial/\partial a$ and $P_\chi = -i\partial/\partial\chi$ to get from (6.55) the WDW equation

$$\left[-\frac{\partial^2}{\partial a^2} + \frac{\partial^2}{\partial\chi^2} + 4(a^2 - \chi^2)\right]\psi(a,\chi) = 0 \tag{6.62}$$

One can separate variables as in [**108, 750**] but here one chooses a different route more suitable for application in the NC situation. Thus write $a = \xi Cosh(\eta)$ and $\chi = \xi Sinh(\eta)$ to get

$$\left[\left(\frac{\partial^2}{\partial\xi^2} + \frac{1}{\xi}\frac{\partial}{\partial\xi} - \frac{1}{\xi^2}\frac{\partial^2}{\partial\eta^2}\right) - 4\xi^2\right]\psi(\xi,\eta) = 0 \tag{6.63}$$

Putting in $\psi = R(\xi)exp(i\alpha\eta)$ one obtains

$$\frac{\partial^2 R}{\partial\xi^2} + \frac{1}{\xi}\frac{\partial R}{\partial\xi} + \left(\frac{\alpha^2}{\xi^2} - 4\xi^2\right)R = 0 \tag{6.64}$$

A solution is $R(\xi) = AK_{i\alpha/2}(\xi^2) + BI_{i\alpha/2}(\xi^2)$ where K_ν and I_ν are Bessel functions of the second kind, A, B are constants, and α is a real number. The solution of WDW is then

$$\psi(\xi,\eta) = AK_{i\alpha/2}(\xi^2)e^{i\alpha\eta} + BI_{i\alpha/2}(\xi^2)e^{i\alpha\eta} \tag{6.65}$$

Such wavefunctions appear in the study of quantum wormholes (cf. [**286**]) and in quantum cosmology for the KS universe (cf. [**511, 1174**]). One often discards the I_ν solution (not always wisely) and uses

$$\psi(\xi,\eta) = \sum_\alpha A_\alpha K_{i\alpha/2}(\xi^2)e^{i\alpha\eta} \tag{6.66}$$

as a solution.

Now for the Bohmian approach one recalls from quantum information theory that the wavefunction has a nonphysical character (cf. Peres-Terno [**1011**]) and Bohmian theory should be in accord with this in some way. In the present picture the object of attention is the primordial quantum universe, characterized in the minisuperspace formalism by the configuration variables a and χ. Having fixed the ontological objects one must determine how they evolve in time and this is done with the aid of the wave function. Thus an evolution law is ascribed to point particles via

$$\dot{x}^i = \Re\left\{\frac{1}{m}\frac{[\psi^*(-i\hbar\partial_i)\psi]}{\psi^*\psi}\right\} = \frac{\nabla S}{m};\ i\hbar\partial_\psi = -\frac{\hbar^2}{2m}\nabla^2\psi + V\psi \tag{6.67}$$

where ψ is the wavefunction of the universe and S comes from $\psi = Aexp(iS)$. All phenomena governed by nonrelativistic QM follow from the analysis of this dynamical system. The expectation value of a physical quantity associated with a Hermitian operator $\hat{A}(\hat{x}^i, \hat{p}^i)$ can be computed in the Bohmian formalism via

$$\mathfrak{B}(\hat{A}) = \Re\left\{\frac{[\psi^*\hat{A}(\hat{x}^i, -i\hbar\partial_i)\psi]}{\psi^*\psi}\right\} = A(x^i, t) \tag{6.68}$$

which represents the same quantity when seen from the Bohmian perspective (cf. [**637**]). In the context of nonrelativistic QM it can be shown from first principles that for an ensemble of particles obeying the evolution law (6.67) (first equation) the associated probability density is $\rho = |\psi|^2$ (cf. [**415**]). This is why computing the ensemble average of $A(x^i, t)$ via

$$\int d^3x\rho A(x^i, t) = \int d^3x\psi^*\hat{A}(\hat{x}^i, -i\hbar\partial_i)\psi = <\hat{A}>_t \tag{6.69}$$

leads to the same result as the standard operator formalism. Note that the law of motion in (6.67) can itself be obtained from (6.68) by associating $\dot{x}^i$ with the beable corresponding to the velocity operator, namely

$$\dot{x}^i = \mathfrak{B}(i[\hat{H}, \hat{x}^i]) = \nabla S/m \tag{6.70}$$

Again Bohmian mechanics does not give to probability a privileged role, but, as discussed in [**415**], probability is a derived concept arising from the laws of motion of point particles. In this sense the Bohmian approach is suited to an isolated system (such as the universe) but on the other hand there might be an ensemble of universes.

Now for quantum cosmology on the present context of a FRW universe with a conformally coupled scalar field. In the commutative case the resulting Bohmian minisuperspace formalism matches with the minisuperspace version of the Bohmian quantum gravity proposed in [**637**] and employed in [**108**]. From (6.70) one has in the gauge $N = a$

$$\dot{a} = \Re\left\{\frac{[\psi^*(-i\partial_a/2)\psi]}{\psi^*\psi}\right\} = -\frac{1}{2}\frac{\partial S}{\partial a};\ \dot{\chi} = \Re\left\{\frac{[\psi^*(-i\partial_\chi/2)\psi]}{\psi^*\psi}\right\} = \frac{1}{2}\frac{\partial S}{\partial\chi} \tag{6.71}$$

Changing into the (ξ, η) coordinates we obtain

$$\frac{d\xi}{dt} = -\frac{1}{2}\frac{\partial S(\xi,\eta)}{\partial\xi};\ \frac{d\eta}{dt} = \frac{1}{2\xi^2}\frac{\partial S(\xi,\eta)}{\partial\eta} \tag{6.72}$$

For a single Bessel function in (6.66) one has $\psi(\xi,\eta) = AK_{i\alpha/2}(\xi^2)exp(i\alpha\eta)$ where A is a constant. From $S = \alpha\eta$ the equations of motion in (6.72) reduce to

$$\frac{d\xi}{dt} = 0;\ \frac{d\eta}{dt} = \frac{\alpha}{2\xi^2} \Rightarrow \xi = \xi_0;\ \eta = \frac{\alpha}{2\xi_0^2}t + \eta_0 \tag{6.73}$$

leading to

$$a(t) = \xi_0 Cosh\left(\frac{\alpha}{2\xi_0^2} + \eta_0\right);\ \chi(t) = \xi_0 Sinh\left(\frac{\alpha}{2\xi_0^2} + \eta_0\right) \tag{6.74}$$

Quantum effects can therefore remove the cosmological singularity giving rise to bouncing universes. The case of a superposition of two Bessel functions of type $K_{i\nu}$ is also written out in part but the calculations become difficult.

For the NC quantum model one takes the quantum version of (6.59) as

$$[\hat{a},\hat{\chi}] = i\theta;\ [\hat{a},\hat{P}_a] = i;\ [\hat{\chi},\hat{P}_\chi] = i;\ [\hat{P}_a,\hat{P}_\chi] = 0 \tag{6.75}$$

These commutation relations can be realized in terms of commutative functions by making use of star products as in (6.46). The commutative coordinates a_c, χ_c are called Weyl symbols as before and a WDW equation is

$$\left[\hat{P}^2_{a_c} - \hat{P}^2_{\chi_c}\right]\psi(a_c,\chi_c) + 4(a_c^2 - \chi_c^2) * \psi(a_c,\chi_c) = 0 \tag{6.76}$$

(obtained by Moyal deforming (6.62)). The resulting equations are simply operator versions of (6.60).

For the NC Bohmian version one departs from the C-frame and uses the beable mapping (6.68) to ascribe an evolution law to the canonical variables. In time gauge $N = a_{nc}$ the Hamiltonian (6.58) reduces to

$$h = \left[-\frac{P^2_{a_{nc}}}{4} + \frac{P^2_{\chi_{nc}}}{4} - a^2_{nc} + \chi^2_{nc}\right] \tag{6.77}$$

One can therefore use h to generate time dependence and obtain the Bohmian equations as

$$\frac{da_c}{dt} = \mathfrak{B}(i[\hat{h},\hat{a}_c]) = -\frac{1}{2}(1-\theta^2)\frac{\partial S}{\partial a_c} + \theta\chi_c; \tag{6.78}$$

$$\frac{d\chi_c}{dt} = \mathfrak{B}(i[\hat{g},\hat{\chi}_c]) = \frac{1}{2}(1-\theta^2)\frac{\partial S}{\partial \chi_c} + \theta a_c$$

The connection between the C and NC frame variables is established b applying the "beable" mapping to the operator equations based on (6.60), namely $\hat{a} = \hat{a}_c - (\theta/2)\hat{P}_{\chi_c}$, etc.; that is by defining $a \equiv \mathfrak{B}(\hat{a})$ and $\chi = \mathfrak{B}(\hat{\chi})$. Once the trajectories are determined in the C frame one can find their counterparts in the NC frame by evaluating the variables a and χ along the C-frame trajectories

$$a(t) = \mathfrak{B}(\hat{a})|_{\substack{a_c=a_c(t)\\ \chi_c=\chi_c(t)}} = a_c(t) - \frac{\theta}{2}\partial_{\chi_c}S[a_c(t),\chi_c(t)]; \tag{6.79}$$

$$\chi(t) = \mathfrak{B}(\hat{\chi})|_{\substack{a_c=a_c(t)\\ \chi_c=\chi_c(t)}} = \chi_c(t) + \frac{\theta}{2}\partial_{a_c}S[a_c(t),\chi_c(t)]$$

Now for an application to NC quantum cosmology use $P_{a_c} = -i\partial_{a_c}$ and $P_{\chi_c} = -i\partial_{\chi_c}$ with NC WDW from (6.76) written as

$$(6.80)\quad \left[\beta\left(-\frac{\partial^2}{\partial a_c^2}+\frac{\partial^2}{\partial \chi_c^2}\right)+4(a_c^2-\chi_c^2)+4i\theta\left(\chi_c\frac{\partial}{\partial a_c}+a_c\frac{\partial}{\partial \chi_c}\right)\right]\psi(a_c,\chi_c)=0$$

where $\beta = 1-\theta^2$. Separation of variables works, after writing $a_c = \xi Cosh(\eta)$ and $\chi_c = \xi Sinh(\eta)$, allowing (6.80) to be rewritten as

$$(6.81)\quad \left[\beta\left(\frac{\partial^2}{\partial\xi^2}+\frac{1}{\xi}\frac{\partial}{\partial\xi}-\frac{1}{\xi^2}\frac{\partial^2}{\partial\eta^2}\right)-4i\theta\frac{\partial}{\partial\eta}-r\xi^2\right]\psi(\xi,\eta)=0$$

Using (6.78) one can write the Bohmian equations of motion as

$$(6.82)\quad \frac{d\xi}{dt}=-\frac{1}{2}(1-\theta^2)\frac{\partial S(\xi,\eta)}{\partial\xi};\ \frac{d\eta}{dt}=\frac{1}{2\xi^2}(1-\theta^2)\frac{\partial S(\xi,\eta)}{\partial\eta}+\theta$$

(6.79) can be written in the new set of coordinates as

$$(6.83)\quad a_{nc}(t)=a_c(t)+\frac{\theta}{2}Sinh(\eta)\partial_\xi S[\xi(t),\eta(t)]-\frac{\theta}{2}\xi^{-1}Cosh(\eta)\partial_\eta S[\xi(t),\eta(t)];$$

$$\chi_{nc}(t)=\chi_c(t)+\frac{\theta}{2}Cosh(\eta)\partial_\xi S[\xi(t),\eta(t)]-\frac{\theta}{2}\xi^{-1}Sinh(\eta)\partial_\eta S[\xi(t),\eta(t)]$$

The paper continues with extensive computations for a variety of situations and we hope to have captured the spirit of investigtion.

7. EXACT UNCERTAINTY AND GRAVITY

We go here to Hall [**592**] (cf. Remarks 1.1.4 and 1.1.5 along with Section 3.1) and [**234, 235, 237, 250**] for background (for the original sources see e.g.[**594, 596, 598, 599, 600, 1062, 1063, 1064**]). The theme here is that the exact uncertainty approach may be promoted to *the* fundamental element distinguishing quantum and classical mechanics. Nonclassical fluctuations are addded to the usual deterministic connection between the configuration and momentum properties of a physical system. Assuming that the uncertainty introduced to the momentum (i.e. the fluctuation strength) is fully determined by the uncertainty in the configuration via the configuration probability density one arrives at QM from CM. We remark that the quantum potential arises from variation of the Fisher metric with respect to the probability P and this is another significant feature relating the quantum potential to fluctuations and indirectly to the Bohmian formulation of QM (cf. [**234, 235, 237, 250**]). For a quick review of the particle situation let $H = (p^2/2m) + V(x)$ be the Hamiltonian for a spinless particle with SE $i\hbar\partial_t\psi = H(x, -i\hbar\nabla, t)\psi = -(\hbar^2/2m)\nabla^2\psi + V\psi$ and recall that the probability density P is specified as $|\psi|^2$. Thus in this canonical approach there is a lot of black magic while in the exact uncertainty approach one uses statistical concepts from the beginning and the wavefunction and SE are derived rather than postulated. Thus assume an ensemble picture from the beginning (due to uncertainty in the position) and assume that a fundamental position probability density P follows from an action principle involving

$$(7.1)\quad A=\int dt\left[-\tilde{H}+\int dxP\frac{\partial S}{\partial t}\right];\ \frac{\partial P}{\partial t}=\frac{\delta\tilde{H}}{\delta S};\ \frac{\partial S}{\partial t}=-\frac{\delta\tilde{H}}{\delta P}$$

(no ψ is assumed here). One shows that conservation of probability requires $\tilde{H}$ is invariant under $S \to S+c$ and if $\tilde{H}$ has no explicit time dependence then its value is a conserved quantity corresponding to energy. As an example here consider the classical ensemble Hamiltonian

$$\tilde{H}_c[P,S] = \int dxP\left[\frac{|\nabla S|^2}{2m} + V\right] \tag{7.2}$$

Then as above

$$\frac{\partial P}{\partial t} + \nabla\cdot\left[P\frac{\nabla S}{m}\right] = 0;\ \frac{\partial S}{\partial t} + \frac{|\nabla S|^2}{2m} + V = 0 \tag{7.3}$$

This formalism based on an action principle for the position probability density successfully describes the motion of ensembles of classical particles; moreover it is considerably more general (see e.g. [**600**]). In particular the essential difference between classical and quantum ensembles becomes a matter of form, being characterized by a simple difference in the forms of the emsemble Hamiltonians $\tilde{H}_c$ and $\tilde{H}_q$.

Thus assume the physical momentum is given via (♣) $p = \nabla S + f$ where the fluctuation field f vanishes almost everywhere on average. This is not dissimilar from e.g. Nelson's mechanics where a Brownian motion is attached, or the approach of scale relativity (cf. [**234, 235, 236, 237, 250**] for an extensive treatment of such matters). One will see that such fluctuations introduce indeterminism at the level of individual particles but *not* at the ensemble level. Write now an overline to denote averaging over the fluctuations at a given position so $\overline{f} = 0$ and $\overline{p} = \nabla S$ by assumption and the classical ensemble energy is replaced by

$$<E>= \int dxP[(2m)^{-1}\overline{|\nabla S + f|^2} + V] = \tag{7.4}$$

$$= \int dxP[(2m)^{-1}(|\nabla S|^2 + 2\overline{f}\cdot\nabla S + \overline{f\cdot f}) + V] = \tilde{H}_c + \int dxP\frac{\overline{f\cdot f}}{2m}$$

Now one asks whether this modified classical ensemble can be subsumed within the general formalism above and this is OK proved that $\overline{f\cdot f}$ is determined by some function of P, S and their derivatives, i.e.

$$\overline{f\cdot f} = \alpha(x,P,S,\nabla P,\nabla S,\cdots) \tag{7.5}$$

In this case one can define a modified ensemble Hamiltonian

$$\tilde{H}_q = \tilde{H}_c + \int dxP\frac{\alpha(x,p,S,\nabla P,\nabla S,\cdots)}{2m} \tag{7.6}$$

The aim of the exact uncertainty approach is to fix the form of α uniquely and this is done by requiring first three generally desirable principles to be satisfied (causality, invariance, and independence), plus an exact uncertainty principle, and given this the resulting equations of motion are equivalent to the SE for a quantum ensemble of particles. This is covered at length in [**234, 250, 254**] and in Hall-Reginatto [**596**] for example and slightly different versions are given here, following Hall, Kumar, and Reginatto [**599**] for bosonic fields, in order to make a connection with qauntum gravity. The requirements are: (i) The modified ensemble

Hamiltonian $\tilde{H}_q$ leads to causal equations of motion (so α cannot depend on second and higher derivatives of P and S) (ii) The respective fluctuation strengths for noninteracting uncorrelated ensembles are independent (thus $\overline{f_1 \cdot f_1}$ and $\overline{f_2 \cdot f_2}$ are independent of P_1 and P_2 respectively when $P(x) = P_1(x_2)P_2(x_2)$) (iii) The fluctuations transform correctly under linear canonical transformations (thus $f \to L^T f$ for any invertible linear coordinate transformation $x \to L^{-1}x$). The fourth assumption is: (iv) The strength of the momentum fluctuations $\alpha = \overline{f \cdot f}$ is determined solely by the uncertainty in position - hence α can only depend on x, P and derivatives. It is shown in the references above that these four principles lead to the unique form

$$\tilde{H}_q[P,S] = \tilde{H}_c[P,S] + C\int dx \frac{\nabla P \cdot \nabla P}{2mP} \tag{7.7}$$

where C is a positive universal constant (i.e. having the same value for all ensembles). Moreover if one sets $\hbar = 2\sqrt{C}$ and $\psi = \sqrt{P}exp(iS/\hbar)$ the SE results as above.

Now for gravitational situations we have an ADM metric

$$ds^2 = -(N^2 - h^{ij}N_iN_j)dt^2 + 2N_i dx^i dt + h_{ij}dx^i dx^j \tag{7.8}$$

where N and N_j refer to lapse and shifts with h_{ij} the spatial metric. Consider now the possibility that the configuration of the field is an inherently imprecise notion, hence requiring a probability functional $P[h_{ij}]$ for its description. Assume that the dynamics of the corresponding statistical ensemble are generated by an action principle $\delta A = 0$ where $A = \int dt[\tilde{H} + \int DhP(\partial S/\partial t)]$ analogous to (7.1). Here $\int Dh$ denotes a functional integral over configuration space and $\tilde{H}$ depends on $P[h_{ij}]$ and its conjugate functional $S[h_{ij}]$. The equations of motion are then

$$\frac{\partial P}{\partial t} = \frac{\Delta \tilde{H}}{\Delta S};\ \frac{\partial S}{\partial t} = -\frac{\Delta \tilde{H}}{\Delta P} \tag{7.9}$$

where $\Delta/\Delta F$ denotes the variational derivative with respect to the functional F (cf. [**599**] for details and Remark 5.1). A suitable "classical" ensemble Hamiltonian may be constructed from knowledge of the classical equations of motion for an individual field via (cf. [**599**])

$$\tilde{H}_c[P,S] = \int DhPH_0[h_{ij}, \delta S/\delta h_{ij}] \tag{7.10}$$

where

$$H_0[h_{ij}, \pi^{ij}] = \int dx \left[N\left(\frac{1}{2}G_{ijk\ell}\pi^{ij}\pi^{k\ell} + V(h_{ij})\right) - 2N_i\pi^{ij}_{\ \ |j}\right] \tag{7.11}$$

where V is the negative of twice the product of the 3-curvature scalar with $[det(h)]^{1/2}$ and $|j$ denotes the covariant 3-derivative (this is the single field Hamiltonian). For the ensemble Hamiltonian $\tilde{H}_c$ in (7.10) one has now

$$\frac{\partial P}{\partial t} + \int dx \frac{\delta}{\delta h_{ij}}(P\dot{h}_{ij}) = 0;\ \frac{\partial S}{\partial t} + H_0[h_{ij}, \delta S/\delta h_{ij}] = 0; \tag{7.12}$$

$$\dot{h}_{ij} = NG_{ijk\ell}\frac{\delta S}{\delta h_{k\ell}} - N_{i|j} - N_{j|i}$$

These equations of motion correspond to the conservation of probality with probability flow $\dot{h}_{ij}$ and the HJ equation for an individual gravitational field with configuration h_{ij} As is well known the lack of conjugate momenta for the lapse and shift components N and N_i places constraints on the classical equations of motion. In the ensemble formalism these constraints take the form (cf. [**599**])

$$\frac{\delta P}{\delta N}=\frac{\delta P}{\delta N_i}=\frac{\partial P}{\partial t}=0;\ \left.\left(\frac{\delta P}{\delta h_{ij}}\right)\right|_{|j}=0; \tag{7.13}$$

$$\frac{\delta S}{\delta N}=\frac{\delta S}{\delta N_i}=\frac{\partial S}{\partial t}=0;\ \left.\left(\frac{\delta S}{\delta h_{ij}}\right)\right|_{|j}=0$$

and this corresponds to invariance of the dynamics with respect to the choice of lapse and shift functions and initial time - and to the invariance of P and S under arbitrary spatial coordinate transformations. Applying these constraints to the above classical equations of motion yields for the Gaussian choice $N=1,\ N_i=0$ the reduced classical equations

$$\frac{\delta}{\delta h_{ij}}\left(PG_{ijk\ell}\frac{\delta S}{\delta h_{k\ell}}\right)=0;\ \frac{1}{2}G_{ijk\ell}\frac{\delta S}{\delta h_{ij}}\frac{\delta S}{\delta h_{k\ell}}+V=0 \tag{7.14}$$

Now the exact uncertainty approach can be adapted in a straightforward way to obtain a modified ensemble Hamiltonian that generates the quantum equations of motion. It is assumed first that the classical deterministic relation between the field configuration h_{ij} and its conjugate momentum density π^{ij} is relaxed to

$$\pi^{ij}=\frac{\delta S}{\delta h_{ij}}+f^{ij} \tag{7.15}$$

analogous to (♣) where here f^{ij} vanishes on average for all configurations. This adds a kinetic term to the average ensemble energy analogous to (7.4) with

$$\tilde{H}_q=<E>=\tilde{H}_x+\frac{1}{2}\int DhP\int dxNG_{ijk\ell}\overline{f^{ij}f^{k\ell}} \tag{7.16}$$

Note here that the term in (7.11) linear in the derivative of π^{ij} can be integrated by parts to give a term directly proportional to π^{ij}, which remains unchanged when the fluctuations are added and averaging is performed. Next one fixes the form of $\tilde{H}_q$ using the same principles of causality, independence, invariance and exact uncertainty used before (cf. [**599**] for details) leading to

$$\tilde{H}_q[P,S]=\tilde{H}_c[P,S]+\frac{C}{2}\int Dh\int dxNG_{ijk\ell}\frac{1}{P}\frac{\delta P}{\delta h_{ij}}\frac{\delta P}{\delta h_{k\ell}} \tag{7.17}$$

analogous to (7.7) where C is a positive constant with the same value for all fields. The corresponding modified equations of motion may be calculated via (7.9) and the constraints in (7.13) applied to obtain reduced equations analogous to (7.14). If one now *defines* $\hbar=2\sqrt{C}$ and $\Psi[h_{ij}]=\sqrt{P}exp(iS/\hbar)$ these reduced equations can be rewritten in the form (cf. [**599**])

$$\left[-\frac{\hbar^2}{2}\frac{\delta}{\delta h_{ij}}G_{ijk\ell}\frac{\delta}{\delta h_{k\ell}}+V\right]\Psi=0 \tag{7.18}$$

This is of course the WDW equation for quantum gravity. Note also that the constraints in (7.13) may be rewritten in terms of the wavefunctional Ψ as

$$\frac{\delta\Psi}{\delta N} = \frac{\delta\Psi}{\delta N_i} = \frac{\partial\Psi}{\partial t} = 0; \quad \left(\frac{\delta\Psi}{\delta h_{ij}}\right)\Big|_{|j} = 0 \tag{7.19}$$

An interesting aspect of the WDW equation in (7.19) is that it has been obtained with a particular operator ordering, namely the supermetric $G_{ijk\ell}$ is sandwiched between the two functional derivatives. This constrasts with the canonical approach which is unable to specify a unique ordering (cf. Hall [**592**] for references). It should be noted that different orderings can lead to different physical predictions and hence the exact uncertainty approach is able to remove ambiguity in this respect. An analogous removal of ambiguity is obtained for quantum particles having a position dependent mass (cf. [**599**]) with the exact uncertainty approach specifying, via (7.7), the unique sandwich ordering

$$i\hbar\frac{\partial\psi}{\partial t} = -\frac{\hbar^2}{2}\nabla\cdot\frac{1}{m}\nabla\psi + V\psi \tag{7.20}$$

for the SE.

Summarizing now it follows that physical ensembles are described by a probability density on configuration space (P), a corresponding conjugate quantity (S), and an ensemble Hamiltonian $\tilde{H}[P,S]$. The transition from classical ensembles to quantum ensembles then follows as a consequence of the addition of nonclassical fluctuations, under the assumption that the fluctuation uncertainty is fully determined by the configuration uncertainty. In contrast to the canonical approach the SE and WDW equations are derived, rather than postulated, and the probability connection $P = |\psi|^2$ is a simple consequence of the definition of ψ in terms of P and S. Planck's constant appears as a consequence of a derived universal scale for the nonclassical momentum fluctuations instead of being an unexplained constant in the canonical approach. A (nonserious) limitation appear in that the momentum of a classical ensemble must contribute quadratically to the ensemble energy for the exact uncertainty approach to go through whereas the canonical approach is indifferent to this. We refer to [**592, 599**] for more details and discussion.

REMARK 4.7.1. We extract here from the appendix to [**599**] for some notation and constructions involving functional derivatives. One considers functionals $F[f]$ with

$$\delta F = F[f+\delta f] - F[f] = \int dx \frac{\delta F}{\delta f_x}\delta f_x \tag{7.21}$$

Here $f \sim f(x)$ refers to fields with real or complex values. Thus the functional derivative is a field density $\delta F/\delta f$ having the value $\delta F/\delta f_x$ at position x. For curved spaces one would need a more elaborate notation, and volume element, etc. The choice $F[f] = f_{x'}$ in (7.21) yields $\delta f_{x'}/\delta f_x = \delta(x - x')$ and if the field depends on a parameter t then writing $\delta f_x = f_x(t+\delta t) - f_x(t)$ one arrives at (♠) $dF/dt = \partial_t F + \int dx(\delta F/\delta f_x)\partial_t f_x$ for the rate of change of F with respect t. Functional integrals correspond to integration of functionals over the vector space of physical fields (or equivalence classes thereof) and the only property one requires for the present discussion is the existence of a measure Df on this vector space which is translation invariant (i.e. $\int Df \equiv \int Df'$ for $f' = f + h$). This property implies e.g.

$$\int Df\frac{\delta F}{\delta f} = 0 \;\; if \;\; \int DfF[f] < \infty \tag{7.22}$$

This follows immediately via

$$0 = \int Df(F[f+\delta f] - F[f]) = \int dx\delta f_x\left(\int Df\frac{\delta F}{\delta f_x}\right) \tag{7.23}$$

In particular if $F[f]$ has a finite expectation value with respect to some probability density functional $P[f]$ then (7.22) gives an integration by parts formula

$$\int DfP(\delta F/\delta f) = -\int Df(\delta P/\delta f)F \tag{7.24}$$

Moreover again via (7.22) the total probability $\int DfP$ is conserved for any probability flow satisfying a continuity condition

$$\frac{\partial P}{\partial t} + \int dx\frac{\delta}{\delta f_x}[PV_x] = 0 \tag{7.25}$$

provided that the average flow rate $< V_x >$ is finite. Next consider a functional integral of the form $I[F] = \int Df\xi(F, \delta F/\delta f)$; then variation of $I[F]$ with respect to F gives to first order

$$\Delta I = I[F+\Delta F] - I[F] = \tag{7.26}$$

$$= \int Df\left\{(\partial\xi/\partial F)\Delta F + \int dx[\partial\xi/\partial(\delta F/\delta f_x)][\delta(\Delta F)/\delta f_x]\right\} =$$

$$= \int Df\left\{(\partial\xi/\partial F) - \int dx\frac{\delta}{\delta f_x}[\partial\xi/\partial(\delta F/\delta f_x)]\right\}\Delta F +$$

$$+ \int dx\int Df\frac{\delta}{\delta f_x}\{[\partial\xi/\partial(\delta F/\delta f_x)]\Delta F\}$$

One assumes here that Df is translation invariant and hence if the functional integral of the expression in curly brackets in the last term of (7.26) is finite the term will vanish, yielding the result $\Delta I = \int Df(\Delta I/\Delta F)\Delta F$ where $\Delta I/\Delta F = \partial_F\xi - \int dx(\delta/\delta f_x)[\partial\xi/\partial(\delta F/\delta f_x)]$. ■

REMARK 4.7.2. Regarding the general HJ formulation of classical field theory we go to Appendix B of [**599**]. Two classical fields f, g are canonically conjugate if there is a Hamiltonian functional $H[f, g, t]$ such that

$$\frac{\partial f}{\partial t} = \frac{\delta H}{\delta g};\;\; \frac{\partial g}{\partial t} = -\frac{\delta H}{\delta f} \tag{7.27}$$

These equations follow from the action principle $\delta A = 0$ with $A = \int dt[-H + \int dxg_x(\partial f_x/\partial t))]$. The rate of change of an arbitrary functional $G[f, g, t]$ follows from (7.27) and (♠) as

$$\frac{dG}{dt} = \frac{\partial G}{\partial t} + \int dx\left(\frac{\delta G}{\delta f_x}\frac{\delta H}{\delta g_x} - \frac{\delta G}{\delta g_x}\frac{\delta H}{\delta f_x}\right) = \frac{\partial G}{\partial t} + \{G, H\} \tag{7.28}$$

A canonical transformation maps f, g, H to f', g', H' such that the equations of motion for the latter retain the canonical form of (7.27). Equating the variations of the corresponding actions A and A' to zero it follows that physical trajectories must satisfy

$$(7.29) \qquad -H + \int dx g_x(\partial f_x/\partial t) = -H' + \int dx g'_x(\partial f'_x/\partial t) + (dF/dt)$$

for some generating functional F. Now any two of the fields f, g, f', g' determine the remaining two fields for a given canonical transformation; choosing f, g' as independent and defining the new generating functional $G[f, g', t] = F + \int dx f'_x g'_x$ gives then via (♠)

$$(7.30) \qquad H' = H + \frac{\partial G}{\partial t} + \int dx \left[\frac{\partial f_x}{\partial t} \left(\frac{\delta G}{\delta f_x} - g_x \right) + \frac{\partial g'_x}{\partial t} \left(\frac{\delta G}{\delta g'_x} - f'_x \right) \right]$$

The terms in round brackets therefore vanish identically yielding the generating realtions

$$(7.31) \qquad H' = H + \partial G/\partial t; \; g = \delta G/\delta f; \; f' = \delta G/\delta g'$$

A canonical transformation is thus completely specified by the associated generating function. To obtain the HJ formulation of the equations of motion consider a canonical transformation to fields f', g' which are time independent. From (7.27) one may choose the corresponding Hamiltonian $H' = 0$ without loss of generality and hence from (7.31) the momentum density and the associated generating functional S are specified by

$$(7.32) \qquad g = \frac{\delta S}{\delta f}; \; \frac{\partial S}{\partial t} + H[f, \delta S/\delta f, t] = 0$$

the latter being the desired HJ equation. Solving this equation for S is equivalent to solving (7.27) for f and g.

Note that along a physical trajectory one has $g' = constant$ and hence from (♠) and (7.32)

$$(7.33) \qquad \frac{dS}{dt} = \frac{\partial S}{\partial t} + \int dx \frac{\delta S}{\delta f_x} \frac{\partial f_x}{\partial t} = -H + \int dx g_x \frac{\partial f_x}{\partial t} = \frac{dA}{dt}$$

Thus the HJ functional S is equal to the action functional A up to an additive constant. This relation underlies the connection between the derivation of the HJ equation from a particular type of canonical transformation as above and the derivation from a particular type of variation of the action. The HJ formulation has the feature that once S is specified the momentum density is determined by the relation $g = \delta S/\delta f$, i.e. it is a functional of f. Thus unlike the Hamiltonian formulation of (7.27) an ensemble of fields is specified by a probability density functional $P[f]$, not by a phase space density functional $\rho[f, g]$. In either case the equation of motion for the probability density corresponds to the conservation of probability, i.e. to a continuity equation as in (7.25). For example in the Hamiltonian formulation the associated continuity equation for $\rho[f, g]$ is

$$(7.34) \qquad \frac{\partial \rho}{\partial t} + \int dx \{ (\delta/\delta f_x)[\rho(\partial f_x/\partial t)] + (\delta/\delta g_x)[\rho(\partial g_x/\partial t)] \} = 0$$

which reduces to the Liouville equation $\partial_t \rho = \{H, \rho\}$ via (7.27). Similarly in the HJ formulation the rate of change of f follows from (7.27) and (7.32) as

$$V_x[f] = \frac{\partial f_x}{\partial t} = \left(\frac{\delta H}{\delta g_x}\right)\bigg|_{g=\delta S/\delta f} \tag{7.35}$$

and hence the associated continuity equation for an ensemble of fields described by $P[f]$ follows as in (7.25) to be

$$\frac{\partial P}{\partial t} + \int dx \frac{\delta}{\delta f_x}\left[P\frac{\delta H}{\delta g_x}\bigg|_{g=\delta S/\delta f}\right] \tag{7.36}$$

Everything generalizes naturally for multicomponent fields. ∎

Given the background in Remarks 4.7.1 and 4.7.2 the development in [**599**] is worth sketching in connection with general bosonic field calculations. Thus one looks at the HJ formalism which provides a straightforward mechanism for adding momentum fluctuations to an ensemble of fields. First the equation of motion for an individual classical field is given by the HJ equation (•) $\partial_t S + H[f, \delta S/\delta f, t] = 0$ where $S[f]$ denotes the HJ functional. The momentum density associated with the field f is $g = \delta S/\delta f$ and hence S is called a momentum potential. Next the description of an ensemble of such fields further requires a probability density functional $P[f]$ whose equation of motion corresponds to conservation of probability, i.e. to the continuity equation (cf. (7.36))

$$\frac{\partial P}{\partial t} + \sum_a \int dx \frac{\delta}{\delta f_x^a}\left(P\frac{\delta H}{\delta g_x^a}\bigg|_{g=\delta S/\delta f}\right) = 0 \tag{7.37}$$

Equations (7.37) and (•) describe the motion of the ensemble completely via P and S and this can be written in the Hamiltonian form

$$\frac{\partial P}{\partial t} = \frac{\Delta\tilde{H}}{\Delta S};\ \frac{\partial S}{\partial t} = -\frac{\Delta\tilde{H}}{\Delta P};\ \tilde{H}[P,S,t] = < H > = \int DfPH[f,(\delta S/\delta f),t] \tag{7.38}$$

(cf. Remark 4.7.1). The functional integral $\tilde{H}$ in (7.38) will be referred to then as the ensemble Hamiltonian and in analogy to (7.27) P and S may be regarded as canonically conjugate functionals. Note from (7.38) that $\tilde{H}$ typically corresponds to the mean energy of the ensemble; moreover (7.38) follows from the action principle $\Delta\tilde{A} = 0$ with action $\tilde{A} = \int dt[-\tilde{H} + \int DfS(\partial P/\partial t)]$. In the following one specializes to ensembles for which the associated Hamiltonian is quadratic in the momentum field density, i.e. of the form

$$H[f, g, t] = \sum_{a,b}\int dx K_x^{ab}[f]g_x^a g_x^b + V[f] \tag{7.39}$$

Here $K_x^{ab} = K_x^{ba}$ is a kinetic factor coupling components of the momentum density and $V[f]$ is a potential energy functional. The corresponding ensemble Hamiltonian is given via (7.38) and one notes that cross terms of the form $g_x^a g_{x'}^b$ with $x \neq x'$ are not permitted in local field theories and hence are not considered here.

The approach of [**599**] to obtain a quantum ensemble of fields is now simply to add nonclassical fluctuations to the momentum density with the magnitude of the fluctuations derermined by the uncertainty in the field. This leads to equations of motion equivalent to those of a bosonic field with the advantage of a unique operator ordering for the associated SE. Note here also the analogy with adding a quantum potential to the HJ equation in the Bohmian theory. Thus suppose now that $\delta S/\delta f$ is in fact an average momentum density associated with the field f in the sense that the true momentum density is given by

$$g = \frac{\delta S}{\delta f} + N \tag{7.40}$$

where N is a fluctuation field that vanishes on the average for any given f. The meaning of S becomes then that of being an average momentum potential. No specific underlying model for N is assumed or necessary; one may in fact interpret the source of the fluctuations as the field uncertainty itself. The main effect of the fluctuation field is to remove any deterministic connection between f and g. Since the momentum fluctuations my conceivably depend on the field f the average over such fluctuations for a given quantity $A[f,N]$ will be denoted by $\overline{A}[f]$ and the average over fluctuations and the field by $< A >$. Thus $\overline{N} = 0$ by assumption and in general $< A >= \int DfP[f]\overline{A}[f]$. Assuming a quadratic dependence on momentum as in (7.39) it follows that when the fluctuations are significant the classical ensemble Hamiltonian $\tilde{H} =< H >$ in (7.38) should be replaced by

$$\tilde{H}' =< H[f,(\delta S/\delta f) + N, t] >= \tilde{H} + \sum_{a,b} \int Df \int dx P K_x^{ab} \overline{N_x^a N_x^b} = \tag{7.41}$$

$$= \sum_{a,b} \int Df \int dx P K_x^{ab} \overline{[(\delta S/\delta f_x^a) + N_x^a][(\delta S/\delta f_x^b) + N_x^b]} + < V >$$

Thus the momentum fluctuations lead to an additional nonclassical term in the ensemble Hamiltonian specified via the covariance matrix

$$[Cov_x(N)]^{ab} = \overline{N_x^a N_x^b} \tag{7.42}$$

This covariance matrix is uniquely determined (up to a multiplicative constant) by the following four assumptions:

(1) Causality: $\tilde{H}'$ is an ensemble Hamiltonian for the canonical conjugate functionals P and S which yield causal equations of motion. Thus no higher than first order functional derivatives can appear in the additional term in (7.41) which implies

$$Cov_x(N) = \alpha\left(P, \frac{\delta P}{\delta f_x}, S, \frac{\delta S}{\delta f_x}, f_x, t\right)$$

for some symmetric matrix function α. Note the fourth assumption below removes the possibility of dependence here on auxillary fields and functionals.

(2) Independence: If the ensemble has two independent noninteracting subensembles 1 and 2 with a factorisable probability density functional $P[f^1, f^2] =$

$P_1[f^1]P_2[f^2]$ then any dependence of the corresponding N^1, N^2 on P only enters via P_1, P_2 in the form

$Cov_x(N^1)|_{P_1P_2} = Cov_x(N^1)|_{P_1}$; $Cov_x(N^2)|_{P_1P_2} = Cov_x(N^2)|_{P_2}$

This implies that the ensemble Hamiltonian $\tilde{H}'$ in (7.41) is additive for independent noninteracting ensembles (as is the corresponding action $\tilde{A}'$).

(3) Invariance: The covariance matrix transforms correctly under linear canonical transformations of the field components. Thus $f \to \Lambda^{-1}f$, $g \to \Lambda^T g$ is a canonical transformation for any invertible matrix Λ preserving the quadratic form of H in (7.39) and leaving the momentum potential S invariant (since $\delta/\delta f \to \Lambda^T\delta/\delta f$) and thus from (7.40) $N \to \Lambda^T N$ so $Cov_x(N) \to \Lambda^T cov_x(N)\Lambda$ for $f \to \Lambda^{-1}f$ is required. For single component fields this reduces to a scaling relation for the variance of the fluctuations at each point x.

(4) Exact uncertainty: The uncertanty of the momentum density fluctuations, as characterized by the covariance matrix, is specified by the field uncertainty and hence by the probability density functional P; hence $Cov_x(N)$ cannot depend on S or explicitly on t.

One proves then in [**599**]

THEOREM 7.1. Under the above four assumptions one has

$$\overline{N_x^a N_x^b} = \frac{C}{P^2}\frac{\delta P}{\delta f_x^a}\frac{\delta P}{\delta f_x^b} \tag{7.43}$$

where C is a positive universal constant.

COROLLARY 7.1. The equations of motion corresponding to the ensemble Hamiltonian $\tilde{H}'$ can be expressed in the form

$$i\hbar\frac{\partial\Psi}{\partial t} = H\left[f, -i\hbar\frac{\delta}{\delta f}, t\right]\Psi = -\hbar^2\left(\sum_{a,b}\int dx \frac{\delta}{\delta f_x^a}K_x^{ab}[f]\frac{\delta}{\delta f_x^b}\right)\Psi + V[f]\Psi$$

where $\hbar = 2\sqrt{C}$ and $\Psi = \sqrt{P}exp(iS/\hbar)$.

One notes that the corollary specifies a unique operator ordering for the functional derivative operators. The proofs are given below following [**599**] and this is substantially different from (and stronger than) proofs of analogous theorems for quantum particles in [**596**].

PROOF OF THEOREM AND COROLLARY:

From the causality and exact uncertainty assumptions one has $Cov_x(N) = \alpha(P, (\delta P/\delta f_x), f_x)$. Next to avoid issues of regularisation it is convenient to consider a position dependent canonical transformation $f_x \to \Lambda_x^{-1}f_x$ such that $A[\Lambda] = exp[\int dx log(|det(\Lambda_x)|)] < \infty$. Then the probability density functional P and the measure Df transform as $P \to AP$ and $Df \to A^{-1}Df$ respectively (for conservation of probability) and the invariance assumption requires

$$\alpha(AP, A\Lambda_x^T u, \Lambda_x^{-1}w) \equiv \Lambda_x^T\alpha(P, u, w)\Lambda_x \tag{7.44}$$

where u^a, w^a denote respectively $\delta P/\delta f^a_x$, f^a_x for a given value of x. This must hold for A and Λ_x independently so choosing Λ_x to be the identity matrix at some point x, one has $\alpha(AP, Au, w) = \alpha(P, u, w)$ for all A, which implies that α can involve P only via the combination $v = u/P$. Consequently

$$\alpha(\Lambda^T v, \Lambda^{-1} w) = \Lambda^T \alpha(v, w)\Lambda \tag{7.45}$$

This equation is linear and invariant under multiplication of α by any function of the scalar $J = v^T w$. Moreover one checks that if σ and τ are solutions then so are $\sigma\tau^{-1}\sigma$ and $\tau\sigma^{-1}\tau$. Choosing the two independent solutions $\sigma = vv^T$ and $\tau = (ww^T)^{-1}$ it follows that the general solution has the form

$$\alpha(v, w) = \beta(J)vv^T + \gamma(J)(ww^T)^{-1} \tag{7.46}$$

for arbitrary functions β and γ. Now for $P = P_1P_2$ one has $v = (v_1, v_2)$ and $w = (w_1, w_2)$ so the independence assumption reduces to the requirements

$$\beta(J_1 + J_2)v_1v_1^T + \gamma(J_1 + J_2)(w_1w_1^T)^{-1} = \beta_1(J_1)v_1v_1^T + \gamma_1(J_1)(w_1w_1^T)^{-1}; \tag{7.47}$$
$$\beta(J_1 + J_2)v_2v_2^T + \gamma(J_1 + J_2)(w_2w_2^T)^{-1} = \beta_2(J_2)v_2v_2^T + \gamma_2(J_2)(w_2w_2^T)^{-1}$$

Thus $\beta = \beta_1 = \beta_2 = C$ and $\gamma = \gamma_1 = \gamma_2 = D$ for universal C and D yielding the general form

$$[cov_x(N)]^{ab} = C\frac{1}{P^2}\frac{\delta P}{\delta f^a_x}\frac{\delta P}{\delta f^b_x} + DW^{ab}_x[f] \tag{7.48}$$

where $W_x[f]$ denotes the inverse of the matrix with ab-coefficient $f^a_x f^b_x$. Since $W_x[f]$ is purely a functional of f it merely contributes a classical additive potential term to the ensemble Hamiltonian of (7.41); thus it has no nonclassical role and can be absorbed directly into the classical potential $< V >$. In fact for fields with more than one component this term is ill-defined and can discarded on physical grounds; consequently one can take $D = 0$ without loss of generality. Positivity of C follows from positivity of the covariance matrix and the theorem is proved.

For the corollary one notes first that the equations of motion corresponding to the ensemble Hamiltonian $\tilde{H}'$ follow via the theorem and (7.38) as first: The continuity equation (7.37) as before (since the additional term does not depend on S) and following (7.39) this takes the form

$$\frac{\partial P}{\partial t} + 2\sum_{a,b}\int dx \frac{\delta}{\delta f^a_x}\left(PK^{ab}_x\frac{\delta S}{\delta f^b_x}\right) = 0 \tag{7.49}$$

and second: The modified HJ equation

$$\frac{\partial S}{\partial t} = -\frac{\Delta\tilde{H}'}{\Delta P} = -H[f, (\delta S/\delta f), t] - \frac{\Delta(\tilde{H}' - \tilde{H})}{\Delta P} \tag{7.50}$$

Calculating the last term via (7.43) and Remark 5.1 this simplifies to

$$\frac{\partial S}{\partial t} + H[f, (\delta S/\delta f), t] - 4CP^{-1/2}\sum_{a,b}\int dx\left(K^{ab}_x\frac{\delta^2 P^{1/2}}{\delta f^a_x \delta f^b_x} + \frac{\delta K^{ab}_x}{\delta f^a_x}\frac{\delta P^{1/2}}{\delta f^b_x}\right) = 0 \tag{7.51}$$

Now writing $\Psi = P^{1/2}exp(iS/\hbar)$, multiplying each side of the equation in the corollary by Ψ^{-1}, and expanding, one obtains a complex equation for P and S.

The imaginary part is just the continuity equation (7.49) and the real part is the modified HJ equation (7.50) provided $C = \hbar^2/4$. ■

EXAMPLE 7.1. This formulation is applied to gravitational fields for example to obtain a version of the WDW equation as in (7.18). Thus write (with some repetition from before)

$$ds^2 = g_{\mu\nu}dx^\mu dx^\nu = -(N^2 - \mathbf{N}\cdot\mathbf{N})dt^2 + 2N_i dx^i dt + h_{ij}dx^i dx^j \tag{7.52}$$

as before and recall that the Einstein field equations follow from the Hamiltonian functional

$$H[h,\pi,N,\mathbf{N}) = \int dx N\mathfrak{H}_G[h,\pi] - 2\int dx N_i \pi^{ij}{}_{|j} \tag{7.53}$$

where $\pi = (\pi^{ij})$ is the momentum density conjugate to h and $|j$ denotes the covariant 3-derivative. Further

$$\mathfrak{H}_G = (1/2)G_{ijk\ell}[h]\pi^{ij}\pi^{k\ell} - 2\,{}^3R[h](det(h))^{1/2} \tag{7.54}$$

Here ${}^3R$ is the scalar curvature corresponding to h and

$$G_{ijk\ell}[h] = (h_{ik}h_{j\ell} + h_{i\ell}h_{jk} - h_{ij}h_{k\ell})(det(h))^{-1/2} \tag{7.55}$$

The Hamiltonian functional corresponds to the standard Lagrangian given via $L = \int dx(-det(g))^{1/2}R[g]$ where the momenta π^0 and π^i conjugate to N and N_i vanish identically. However the lack of dependence of H on π^0 and π^i is consistently maintained only if the rates of change of these momenta also vanish, i.e. (noting (7.27), only if the constraints

$$\frac{\delta H}{\delta N} = \mathfrak{H}_G = 0;\quad \frac{\delta H}{\delta N_i} = -2\pi^{ij}{}_{|j} = 0 \tag{7.56}$$

are satisfied. Thus the dynamics of the field are independent of N and N_i so that these functions may be fixed arbitrarily. Moreover these constraints immediately yield $H = 0$ in (7.53) and hence the system is static with no explicit time dependence. It follows that in the HJ formulation of the equations of motion the momentum potential S is independent of N, $\mathbf{N}$, and t. Noting that $\pi = \delta S/\delta h$ in this formulation (7.56) yields the corresponding constraints

$$\frac{\delta S}{\delta N} = \frac{\delta S}{\delta N_i} = \frac{\partial S}{\partial t} = 0;\quad \left(\frac{\delta S}{\delta h_{ij}}\right)\Big|_{|j} = 0 \tag{7.57}$$

As shown in [**599**] a given functional $F[h]$ is invariant under spatial coordinate transformations if and only if $(\delta F/\delta h_{ij})|_{|j} = 0$ and hence the fourth constraint in (7.57) is equivalent to the invariance of S under such transformations. This constraint implies moreover that the second term in (7.53) may be dropped from the Hamiltonian yielding the reduced Hamiltonian

$$H_G[h,\pi,N] = \int dx N\mathfrak{H}_G[h,\pi] \tag{7.58}$$

For an ensemble of classical gravitational fields the independence of the dynamics with respect to (N, $\mathbf{N}$, t) implies that members of the ensemble are distinguishable only by their corresponding 3-metric h. Moreover it is natural to impose the additional geometric requirement that the ensemble is invariant under spatial coordinate transformations. Consequently one has constraints

$$\frac{\delta P}{\delta N} = \frac{\delta P}{\delta N_i} = \frac{\partial P}{\partial t} = 0; \quad \left(\frac{\delta P}{\delta h_{ij}}\right)\Big|_{|j} = 0 \tag{7.59}$$

The first two constraints imply that ensemble averages only involve integration over h_{ij}.

Now in view of (7.54) the Hamiltonian H_G in (7.58) has the quadratic form of (7.39) so the exact uncertainty approach is applicable and leads immediately to the SE

$$i\hbar\frac{\partial\Psi}{\partial t} = \int dx N\mathfrak{H}_G[h, -i\hbar(\delta/\delta h)]\Psi \tag{7.60}$$

for a quantum ensemble of gravitational fields. One follows the guiding principle used before that all constraints imposed on the classical ensemble should be carried over to the corresponding constraints on the quantum ensemble. Thus from (7.57) and (7.59) one requires that P and S and hence Ψ in the equation of the Corollary above are independent of N, $\mathbf{N}$, and t as well as invariant under spatial transformations. Thus

$$\frac{\delta\Psi}{\delta N} = \frac{\delta\Psi}{\delta N_i} = \frac{\partial\Psi}{\partial t} = 0; \quad \left(\frac{\delta\Psi}{\delta h_{ij}}\right)\Big|_{|j} = 0 \tag{7.61}$$

Applying the first and third of these constraints to (7.60) gives then, via (7.54), the reduced SE

(7.62)

$$H_G[h, -i\hbar(\delta/\delta h)]\Psi = (-\hbar^2/2)\frac{\delta}{\delta h_{ij}}G_{ijk\ell}[h]\frac{\delta}{\delta h_{k\ell}}\Psi - 2\,{}^3R[h](det(h))^{1/2}\Psi = 0$$

which is again the WDW equation. ∎

CHAPTER 5

FLUCTUATIONS AND GEOMETRY

1. THE ZERO POINT FIELD

The zero point field (ZPF) arising from the quantum vacuum is still a contentious idea and we only make a few remarks following [**20, 175, 176, 178, 230, 315, 316, 317, 364, 387, 403, 586, 587, 588, 589, 590, 655, 656, 659, 820, 879, 880, 1004, 1005, 1046, 1047, 1048, 1060, 1076, 1091, 1093, 1092, 1253, 1267, 1268**] (see also [**167, 482, 483, 484, 1322**] for stochastic spacetime and gravity fluctuations). It is to be noted that a certain amount of research on ZPF is motivated (and funded) by the desire to extract energy from the vacuum for "space travel" and in this direction one is referred to publications of the CIPA (see http://www.calphysics.org/sci.html) where a number of papers discussed here originate or are referenced. In a sense the quest here seems also to be an effort to really understand the equation $E = mc^2$. There is of course some firm physical evidence for forces induced by quantum fluctuations via the Casimir effect for example and it is very stimulating to see so much speculation now in the literature about questions of mass, inertia, Zitterbewegung, and the vacuum. We extract here first from Haisch, Rueda, and Puthoff [**586**] (cf. also [**589**]). No attempt is made here to be complete; the quantum vacuum is a relatively hot topic and there are many unsolved matters (for survey material see e.g. [**882, 880, 1004, 1267, 1268**]). There are apparently at least two main views on the origin of the EM ZPF as embodied in QED and Stochastic electrodynamics (SED). QED is "standard" physics and the arguments go as follows. The Heisenberg uncertainty principle sets a fundamental limit on the precision wieh thich conjugate quantities are allowed to be determined. Thus $\Delta x\Delta p \geq \hbar/2$ and $\Delta E\Delta t \geq \hbar/2$. It is standard to work via harmonic oscillators (see below) and there are two non-classical results for a quantized harmonic oscillator. First the energy levels are discrete; one can add energy but only in units of $\hbar\nu$ where ν is a frequency. The second stems from the fact that if an oscillator were able to come completely to rest Δx would be zero and this would violate $\Delta x\Delta p \geq \hbar/2$. Hence there is a minimum energy of $\hbar\nu/2$ and the oscillator can only take on values $E = (n + 1/2)\hbar\nu$ which can never be zero. The argument is then made that the EM field is analogous to a mechanical harmonic oscillator since the fields $\vec{E}$ and $\vec{B}$ are modes of oscillating plane waves with minimum energy $\hbar\nu/2$. The density of modes between ν and $\nu + d\nu$ is given by the density of states function $N_\nu d\nu = (8\pi\nu^2/c^3)d\nu$. Each state has a minimum $\hbar\nu/2$ of energy and thus the ZPF spectral density function is

$$\rho(\nu)d\nu = (8\pi\nu^2/c^3)(\hbar\nu/2)d\nu \tag{1.1}$$

It is instructive to compare this with the blackbody radiation

$$\rho(\nu, T)d\nu = \frac{8\pi\nu^2}{c^3}\left(\frac{\hbar\nu}{e^{\hbar\nu/kT}-1} + \frac{\hbar\nu}{2}\right)d\nu \tag{1.2}$$

If one takes away all the thermal energy ($T \to 0$) what remains is the ZPF term. It is traditionally assumed in QM that the ZPF can be ignored or subtracted away. In SED is is assumed that the ZPF is as real as any other EM field and just came with the universe. In this spirit the Heisenberg uncertainty relations for example are not a result of quantum laws but a consequence of ZPF. Philosophically one could ask whether or not the universe is classical with ZPF (as envisioned at times by Planck, Einstein, and Nerst and later by Nelson et al) or has two sets of laws, quantum and classical. A recent discussion [**746**] of fluctuations is of interest here.

It was shown in the 1970's that a Planck like component of the ZPF will arise in a uniformly accelerated coordinate system having constant proper acceleration a with what amounts to an effective temperature $T_a = \hbar a/2\pi ck$ (cf. [**364**]). More precisely one says that an observer who accelerates in the conventional quantum vacuum of Minkowski space will perceive a bath of radiation, while an inertial observer of course perceives nothing. This is a quantum phenomenon and the temperature is negligible for most accelerations, becoming significant only in extremely large gravitational fields. Thus for the case of no true external thermal radiation ($T = 0$), but including this acceleration effect T_a, equation (1.1) becomes

$$\rho(\nu, T_a)d\nu = \frac{8\pi\nu^2}{c^3}\left[1 + \left(\frac{a}{2\pi c\nu}\right)^2\right]\left[\frac{\hbar\nu}{2} + \frac{\hbar\nu}{e^{\hbar\nu/kT_a}-1}\right]d\nu \tag{1.3}$$

(the pseudo-Planckian component at the end is generally very small).

There have been (at least) two approaches demonstrating how a reaction force proportional to acceleration ($\vec{f}_r = -m_{ZP}\vec{a}$) arises out of properties of the ZPF. One could be called HRP based on the first paper in [**586**] identified the Lorentz force arising from the stochastically averaged magnetic component of the ZPF, namely $< \vec{B}^{ZP} >$, as the basis of $\vec{f}_r$. The second, called RH after Rueda-Haisch [**1091**], considers only the relativistic transformations of the ZPF itself to an accelerated frame, leading to a nonzero stochastically averaged Poynting vector $(c/4\pi) < \vec{E}^{ZP} \times \vec{B}^{ZP} >$ which leads immediately to a nonzero EM ZPF momentum flux as viewed by an accelerating object. If the quarks and electrons in such an accelerating object scatter this asymmetric radiation an acceleration dependent reaction force $\vec{f}_r$ arises. In this context $\vec{f}_r$ is the space part of a relativistic four vector so that the resulting equation of motion is not simply the classical $\vec{f} = m\vec{a}$ but rather the properly relativistic $\mathfrak{F} = d\mathfrak{P}/d\tau$ (which becomes exactly to $\vec{f} = m\vec{a}$ for subrelativistic velocities). The expression for inertial mass in HRP for an individual particle is $m_{ZP} = \Gamma_z\hbar\omega_c^2/2\pi c^2$ where Γ_z represents a damping constant for Zitterbewegung oscillations (a free parameter) and ω_c represents an assumed cutoff frequency for the ZPF spectrum (another free parameter). The expression for inertial mass in RH for an object with volume V_0 is

$$m_i = m_{ZP} = \left(\frac{V_0}{c^2}\int \eta(\omega)\frac{\hbar\omega^3}{2\pi^2 c^3}d\omega\right) = \frac{V_0}{c^2}\int \eta(\omega)\rho_{ZP}d\omega \tag{1.4}$$

(note from (1.1) $\omega \sim 2\pi\nu$ and m_i here refers to inertial mass). Also from [**588**] recall that 4-momentum is defined as $\vec{P} = (E/c, \vec{p}) = (\gamma m_0 c, \gamma m_0 \vec{v})$ where $|\vec{P}| = m_0 c$ and $E = \gamma m_0 c^2$ (γ is a Lorentz contraction factor). Recall that the Compton frequency is given via $\hbar\nu_C = m_0 c^2$ for a particle of rest mass m_0 (note $\lambda_C = \hbar/m_0 c$ so $c = (\hbar/m_0 c)\nu_C = \lambda_C \nu_C \Rightarrow \nu_C = c/\lambda_C$ has dimension T^{-1}). Further $\lambda_B = \hbar/p$ for a particle of momentum p and here $p \sim m_0\gamma v$ so one expects

$$\lambda_B = \hbar/m_0\gamma v = m_0 c\lambda_C/m_0\gamma v = (c/\gamma v)\lambda_C \tag{1.5}$$

(cf. also [**776**]). In any event in [**1092**] one argues that what appears as inertial mass in a local frame corresonds to gravitational mass m_g and identifies m_i in (1.4) with m_g.

REMARK 5.1.1. We mention here a thoughtful analysis about the origin of inertial mass etc. in Ibison [**656**] (second paper), which refers to the material developed above in [**586, 587, 1091, 1092**]. It is worthwhile extracting in some detail as follows (cf. also [**589, 655, 879, 1004**]). The purpose of the paper is stated to be that of suggesting qualifications in some claims that the classical equilibrium spectrum of charged matter is that of the classically conceived ZFF (cf. here Ibison-Haisch [**655**] where one introduces an alternative classical ZPF with a different stochastic character which reproduces the statistics of QED). It is pointed out that a classical massless charge cannot acquire mass from nothing as a result of immersion in any EM field and therefore that the ZPF alone cannot provide a full explanation of intertial mass. Thus as background one mentions several works where classical representations of the ZPF have been used to derive a variety of quantum results, e.g. the van der Waals binding (cf. [**176**]), the Casimir effect (cf. [**879**]), the Davies-Unruh effect (cf. [**364**]), the ground state behavior of the QM harmonc oscillator (cf. [**655**]), and the blackbody spectrum (cf. [**176, 316**]). In its role as the originator of inertial mass the ZPF has been evisined as an external energizing influence for a classical particle whose mass is to be explained. In SED a free charged particle is deemed to obey the (relativistic version of the) Braffort-Marshall equation

$$m_0 a^\mu - m_0\tau_0\left[\frac{da^\mu}{d\tau} + \frac{a^\lambda a_\lambda}{c^2}u^\mu\right] = eF^{\mu\lambda}u_\lambda \tag{1.6}$$

where $\tau_0 = e^2/6\pi\epsilon_0 m_0 c^3$ and F is the field tensor of the ZPF interpreted classically (cf. [**655**]) for the correspondence between this and the vacuum state of the EM field - indicated also later in this paper). If F is the ZPF field tensor operator then (1.6) is a relativistic generalization of the Heisenberg equation of motion for the QM position operator of a free charged particle, properly taking into account the vacuum sate of the quantized EM field (cf. [**879**]). From the standpoint of QED, once coupling to the EM field is switched on and radiation reaction admitted, the action of the vacuum field is not an optional extra, but a necessary component of the fluctuation dissipation relation between atom and field. In the ZPF inertial mass studies the electrodynamics of the charge in its pre-mass condition has not received attention, presumably on the grounds that the ZPF energization will quickly render the particle massive so that the intermediate state of masslessness is on no import. However letting $m_0 \to 0$ one sees that $F^{\mu\lambda}u_\lambda \to 0$ which demands that $\mathbf{E}\cdot\mathbf{v} \to 0$ (the massless particle moves orthogonal to E - cf. [**656**], first paper). It is concluded that if a charge is initially massless, there is no means by which it can acquire inertial mass energy from an EM field, including the ZPF, and one must discount the possibility that the given ZPF can alone explain inertial mass of such a particle (this does not take into consideration however the possible effects of Dirac-Weyl geomety on mass creation - cf. Section 4.1). This is not to deny that mass may yet emerge from a process involving the ZPF or some other EM field, only that it cannot be the whole story. Note here that in earlier works [**589, 1091**] an internal structure is implied by the use of a mass-specific frequency dependent coupling constant between the charged particles and the ZPF so there is no contradiction to the arguments above, namely to the statement that the classical structureless charge particle cannot acquire mass solely as a result of immersion in the ZPF. One must also be concerned with a distinction between inertia as a reaction force and inertial mass as energy. For example one could ask whether the ZPF could be the cause of resistance to acceleration without it having to be the cause of mass-energy. To address this consider the geometric form of the mass action

$$I = -m_0 c^2 \int \gamma^{-1} dt \tag{1.7}$$

which simultaneously gives both the mass acceleration $f^\mu = m_0 a^\mu$ and the Noether conserved quantity under time translations $E = \gamma m_0 c^2 = m(v)c^2$ (i.e. the traditional mechanical energy (γ is the Lorenz factor). Thus the distinction between the two qualities of mass appears to be one of epistemology. In some work (cf. [**586**]) the ZPF has been envisioned as an external energizing influence for an explicitly declared local interal degree of freedom, intrinsic to the charged particle whose mass is to be energized. Upon immersion in the ZPF this (Planck) oscillator is energized and the energy so acquired is some of all of its observed inertial mass. Such a particle is not a structureless point in the usual classical sense and so does not suffer from an inability to acquire mass from the ZPF, provided the proposed components (sub-electron charges) already carry some inertia. There is much more interesting discussion which should be read. Note that one is excluding here such matters as the (conjectured) Higgs field as well as renormalization. ∎

One of the first objections typically raised against the existence of a real ZPF is that the mass equivalent of the energy embodied in (1.1) would generate an enormous spacetime curvature that would shrink the universe to microscopic size (apparently refuted however via the principle of equivalence - cf. [**586**]). One notes that all matter at the level of quarks and electrons is driven to oscillate (Zitterbewegung) by the ZPF and every oscillating charge will generate its own minute EM fields. Thus any particle wil experience the ZPF as modified ever so slightly by the fields of adjacent particles - but that might be gravitation as a kind of long range van der Waals force. A ZPF based theory of gravitation is however only exploratory at this point and there are disputes (see e.g. [**230, 317, 387, 1047**] - discussed in part later). In any event following [**588, 589, 1092**] the preceeding analysis leads to $\mathfrak{F} = d\mathfrak{P}/d\tau = (d/d\tau)(\gamma m_i c, p)$ for the relativistic force where $m_i \sim m_{ZP} \sim m_g$. More generally via [**1091**] (which improves [**586**]) plausible arguments are given to show that the EM quantum vacuum makes a contribution to the inertial mass m_i in the sense that at least part of the inertial force of opposition to acceleration, or inertia reaction force, springs from the EM quantum vacuum. Specifically, the properties of the EM vacuum as experienced in a Rindler constant acceleration frame were investigated, and the existence of an EM flux was discovered, called for convenience the Rindler flux (RF). The RF, and its relative, Unruh-Davies radiation, both stem from event horizon effects in accelerating reference frames. The force of radiation pressure produced by the RF proves to be proportional to the acceleration of the reference fame, which leads to the hypothesis that at least part of the inertia of an object should be due to the individual and collective interaction of its quarks and electrons with the RF. This is called the quantum vacuum inertia hypothesis (QVIH). This is consistent with general relativity (GR) and it answers a fundamental question left open within GR, namely, whether there is a physical mechanism that generates the reaction force known as weight when a specific nongeodesic motion is imposed on an object. Put another way, while geometrodynamics dictates the spacetime metric and thus specifies geodesics, whether there is an identifiable mechanism for enforcing the motion of freely falling bodies along geodesic trajectories. The QVIH provides such a mechanism since by assuming local Lorentz invariance (LLI) one can immediately show that the same RF arises due to curved spacetime geometry as for acceleration in flat spacetime. Thus the previously derived expression for the inertial mass contribution from the EM quantum vacuum field is exactly equal to the corresonding contribution to the gravitational mass m_g and the Newtonian weak equivalence principle $m_i = m_g$ ensues. One also adopts the assumption of space and time uniformity (uniformity assumption (UA) stating that the laws of physics are the same at any time or place within the universe. With such hypotheses one also derives in [**1092**] the Newtonian gravitational law $\mathbf{f} = -(GMm/r^2)\dot{\mathbf{r}}$ while disclaiming the success of such attempts in earlier work (cf. [**230, 317, 587, 1047**]). We mention also some quite appropriate comments about the conjectured Higgs field in [**589**] (cf. also [**590**]), the main point being perhaps that even if the Higgs exists it does not necessaarily explain inertial mass.

ZPF also plays the role of a Lorentz invariant EM component of a Dirac ether (cf. [**587**]). Thus Newton's equation of motion in special relativity can be written as $m(d^2x^\mu/d\tau^2) = F^\mu$ where four vectors are involved and F^μ represents a nongravitational force. In general relativity this becomes

$$m\left(\frac{d^2x^\mu}{d\tau^2} + \Gamma^\mu_{\nu\rho}\frac{dx^\nu}{d\tau}\frac{dx^\rho}{d\tau}\right) = F^\mu \tag{1.8}$$

The velocity $dx^\mu/d\tau$ is a time-like 4-vector and F^μ is a space-like 4-vector orthogonal to the velocity. If $F^\mu \neq 0$ in any coordinate frame it will be nonzero in all coordinate frames. The $\Gamma^\mu_{\nu\rho}$ represent the gravitational force and can be set equal to zero by a coordinate transformation but F^μ cannot be transformed away.

Turning the arguments around the absolute nature of nongravitational acceleration is demonstrated by the manifestation of a force that cannot be transformed away. Thus there is need for a special reference frame that is not perceptible on account of uniform motion but that is perceptible on account of non gravitational acceleration and one proposes that the Lorentz invariant ZPF plays this role. We refer to [**175, 1093**] for background here.

REMARK 5.1.2. We mention for heuristic purposes some developments following [**590, 929**] (cf. also [**662**] and Section 5.2 for possible comparison). One addresses the quantum theoretic prediction that the Zitterbewegung of particles occurs at the speed of light, that particles exihbit spin, and that pair creation can occur. Given motion at the speed of light the particles involved in Zitterbewegung would seem to be massless (and in this respect we refer to [**662**]). A suitable equation for motion of a massless charge can be derived as a massless limit of the Lorentz force equation $ma^\mu = (q/c)F^{\mu\nu}u_\nu$ where q is charge and m will be go to zero; $F^{\mu\nu}$ is the EM field tensor of impressed fields, including the ZPF. The acceleration a and velocity u are four vectors and in terms of the usual three space quanties $\mathbf{a}$ and $\mathbf{v}$ one can write

$$m\left[\gamma^2 a^j + \gamma^4 \mathbf{v}\cdot\mathbf{a}\frac{v^j}{c}\right] = \frac{q}{c}\gamma\left[\sum_1^3 F^{jk}v_k - F^{j0}c\right]; \tag{1.9}$$

$$m\gamma^4\frac{\mathbf{v}\cdot\mathbf{a}}{c} = \frac{q}{c}\gamma\left[\sum_1^3 F^{0k}v_k - F^{00}c\right]$$

where $\gamma = 1/\sqrt{1-(v^2/c^2)}$. Combining leads to

$$m\gamma a^j = \frac{q}{c}\left[\sum_1^3\left(F^{jk} - F^{0k}\frac{v^j}{c}\right)v_k - \left(F^{j0} - F^{00}\frac{v^j}{c}\right)c\right] \tag{1.10}$$

Now let $m \to 0$ and $\gamma \to \infty$ with $m\gamma$ remaining finite; thus one will have $lim_{m\to 0, v\to c} m\gamma = m_*$ and m_* has the dimensions of mass but is not mass. We need further $\mathbf{v} = c\mathbf{n}$ where $\mathbf{n}$ is a unit vector in the direction of the particle motion. Acceleration can therefore only be due to changes in the direction of the form $\mathbf{a} = c(d\mathbf{n}/dt)$ so that

$$m_* c\frac{dn^j}{dt} = \frac{q}{c}\left[\sum_1^3\left(F^{jk} - F^{0k}\frac{v^j}{c}\right)v_k - \left(F^{j0} - F^{00}\frac{v^j}{c}\right)c\right] \tag{1.11}$$

In terms of EM fields this is $(d\mathbf{n}/dt) = (q/m_*c)[\mathbf{n}\times\mathbf{B} - (\mathbf{n}\cdot\mathbf{E})\mathbf{n} + \mathbf{E}]$. Since the particle is moving at the speed of light it should see a universe Lorentz contracted to two transverse dimensions and can only be accelerated by forces from the side. When the impressed fields include the ZPF this motion can be regarded as Schrödinger's Zitterbewegung. When a field above the vacuum is applied the charge will be observed to drift in a preferred direction in its Zitterbewegung wandering. When viewed in a zoom picture one sees spin line orbital motion driven by the ZPF. Now equation (1.11) does not exhibit inertia which seems in violation of relativity. However the ZPF fields have the vacuum energy density spectrum of the form $\rho(\omega)d\omega = (\hbar\omega^3/2\pi^2c^3)d\omega$ where $\omega = 2\pi\nu$ (cf. (1.1)). The cubic frequency dependence endows the spectrum with Lorentz invariance and all inertial frames see an isotropic ZPF. A Lorentz transformation will cause a Doppler shift of each frequency component but an equal amount of energy is shifted into and out of each frequency bin. When there are no fields above the vacuum in an inertial frame an observer in that frame should expect to see a zero-mean random walk due to the isotropic ZPF. Thus in our example an observer in a frame comoving with the average motion of the charge just before the driving field is switched off should expect to see continued zero-mean Zitterbewegung in his frame whereas (1.11) produces zero-mean motion in whatever frame the calculation is performed. To have a consistent theory Lorentz covariance must be restored. Thus assume (1.11) holds in the spacetime of the particle and describes a null geodesic. The curvature is defined by the EM fields in the particles history and since the particle is massless and moving at the speed of light one cannot use proper time as the affine parameter of the geodesic (since proper time intervals vanish for null geodesics). However normal time serves as well for time parameter and (1.11) is replaced with

$$\frac{dp^\mu}{dt} + \frac{1}{m_*}\Gamma^\mu_{\nu\rho}p^\nu p^\rho = 0;\ p^\mu = m_* c n^\mu;\ n^\mu = (n^0, \mathbf{n}) \tag{1.12}$$

One equates the connection terms with the Lorentz force terms of $ma^\mu = (q/c)F^{\mu\nu}u_\nu$, i.e.

$$\Gamma^\mu_{\nu\rho}p^\nu p^\rho = -(q/c)F^\mu_\nu p^\nu \tag{1.13}$$

and these equations can be solved for the metric of the particle's spacetime (although not uniquely - there is apparently a class of metrics). Further constraints are needed to select a particular solution and in particular the geodesic should be a null curve as expected for a massless object. Using (1.13) the geodesic equation in (1.12) is claimed to be

$$\frac{dn^j}{dt} = \frac{q}{m_* c}\left[F^j_\nu n^\nu - F^o_\nu n^\nu \frac{n^j}{n^0}\right] + \frac{n^j}{n^0}\frac{dn^0}{dt};\ \frac{dm_*}{dt} = \frac{q}{c}F^0_\nu \frac{n^\nu}{n^0} - \frac{m_*}{n^0}\frac{dn^0}{dt} \tag{1.14}$$

In general n^0 does not retain a value of 1 but should change in a way that preserves the null curve property $g_{\mu\nu}p^\mu p^\nu = 0$. We note that the second equation in (1.14) is an equation for the parameter m_*, which is therefore not a constant but rather varies in response to applied forces. The effect is to introduce time dilation (or Doppler shifting) in the EM 4-vector analogous to the gravitational redshift of GR. Also note that inertia is not assumed here by the requirement that the particle travel on a null geodesic; the particle is not restricted but its motion is used to define the spacetime and metric that the particle sees. It is only after a transformation to Minkowski spacetime that inertial behavior appears. To find solutions of (1.13) consider an infinitesimal region around the charge and require $g_{\mu\nu} = \eta_{\mu\nu} + g_{\mu\nu,\rho}dx^\rho$ where $\eta_{\mu\nu}$ is the flat spacetime metric (signature $(-1,1,1,1)$). The Christoffel symbols are then calculated in terms of the derivatives $g_{\mu\nu,\rho}$ and substituted into (1.13). A simple local solution is written down with some interesting properties for which we refer to [**590, 929**].

1.1. REMARKS ON THE AETHER AND VACUUM. Regarding massless particles and Maxwell's equations we refer here to [**9, 23, 34, 184, 288, 421, 522, 590, 736, 899, 929, 1150, 1151, 1211**] and consider first Gersten [**522**]. Thus the Dirac equation is derived from the relativistic condition $(E^2-c^2\mathbf{p}^2-m^2c^4)I^4\psi=0$ where I^4 is a 4×4 unit matrix and ψ is a fourcomponent (bispinor) wave function. This can be decomposed via

$$(1.15)\quad \left[EI^4+\begin{pmatrix} mc^2I^2 & c\mathbf{p}\cdot\sigma \\ c\mathbf{p}\cdot\sigma & -mc^2I^2\end{pmatrix}\right]\times\left[EI^4-\begin{pmatrix} mc^2I^2 & c\mathbf{p}\cdot\sigma \\ c\mathbf{p}\cdot\sigma & -mc^2I^2\end{pmatrix}\right]\psi=0;$$

$$\sigma_x=\begin{pmatrix}0&1\\1&0\end{pmatrix};\ \sigma_y=\begin{pmatrix}0&-i\\i&0\end{pmatrix};\ \sigma_z=\begin{pmatrix}1&0\\0&-1\end{pmatrix}$$

and I^2 is a 2×2 unit matrix. The two component neutrino equation can be derived from the decomposition

$$(1.16)\qquad (E^2-c^2\mathbf{p}^2)I^2\psi=[EI^2-c\mathbf{p}\cdot\sigma][EI^2+c\mathbf{p}\cdot\sigma[\psi=0$$

where ψ is a two component spinor wavefunction. The photon equation can be derived from the decomposition
(1.17)

$$\left(\frac{E^2}{c^2}-\mathbf{p}^2\right)I^3=\left(\frac{E}{c}I^3-\mathbf{p}\cdot\mathbf{S}\right)\left(\frac{E}{c}I^3+\mathbf{p}\cdot\mathbf{S}\right)-\begin{pmatrix} p_x^2 & p_xp_y & p_xp_z\\ p_yp_x & p_y^2 & p_yp_x\\ p_xp_z & p_xp_y & p_z^2\end{pmatrix}=0;$$

$$S_x=\begin{pmatrix}0&0&0\\0&0&-i\\0&i&0\end{pmatrix};\ S_y=\begin{pmatrix}0&0&i\\0&0&0\\-i&0&0\end{pmatrix};\ S_z=\begin{pmatrix}0&-i&0\\i&0&0\\0&0&0\end{pmatrix}$$

and I^3 is a 3 unit matrix. Further

$$(1.18)\qquad [S_x,S_y]=iS_z,\ [S_z,S_x]=iS_y,\ [S_y,S_z]=iS_x,\ \mathbf{S}^2=2I^3$$

One notes that the last matrix in (1.17) (first line) can be written as $(p_x\,p_y\,p_z)^T\cdot(p_x\,p_y\,p_z)$. This leads to the photon equation in the form

$$(1.19)\quad \left(\frac{E^2}{c^2}-\mathbf{p}^2\right)\psi=\left(\frac{E}{c}I^3-\mathbf{p}\cdot\mathbf{S}\right)\left(\frac{E}{c}I^3+\mathbf{p}\cdot\mathbf{S}\right)\psi-\begin{pmatrix}p_x\\p_y\\p_z\end{pmatrix}(\mathbf{p}\cdot\psi)=0$$

where ψ is a 3-component column wave function. (1.19) will be satisfied if

$$(1.20)\qquad \left(\frac{E}{c}I^3+\mathbf{p}\cdot\mathbf{S}\right)\psi=0;\ \mathbf{p}\cdot\psi=0$$

For real energies and momenta conjugation leads to

$$(1.21)\qquad \left(\frac{E}{c}I^3-\mathbf{p}\cdot\mathbf{S}\right)\psi^*=0;\ \mathbf{p}\cdot\psi^*=0$$

The only difference here is that (1.20) is the negative helicity equation while (1.21) represents positive helicity. Now in (1.20) make the substitutions $E\sim i\hbar\partial_t$ and $\mathbf{p}\sim -i\hbar\nabla$ with $\psi=\mathbf{E}-i\mathbf{B}$ (∼ Riemann-Silberstein vector). Then $(\mathbf{p}\cdot\mathbf{S})\psi=\hbar\nabla\times\psi$ and this leads to

$$(1.22)\qquad \frac{i\hbar}{c}\partial_t\psi=-\hbar\nabla\times\psi;\ -i\hbar\nabla\cdot\psi=0$$

Cancelling $\hbar$ (!) one obtains then

$$\nabla\times(\mathbf{E}-i\mathbf{B})=-\frac{i}{c}\partial_t(\mathbf{E}-i\mathbf{B});\ \nabla\cdot(\mathbf{E}-i\mathbf{B})=0 \tag{1.23}$$

If the electric and magnetic fields are real one obtains the Maxwell equations

$$\nabla\times\mathbf{E}=-\frac{1}{c}\partial_t\mathbf{B};\ \nabla\times\mathbf{B}=\frac{1}{c}\partial_t\mathbf{E};\ \nabla\cdot\mathbf{E}=\nabla\cdot\mathbf{B}=0 \tag{1.24}$$

The Planck constant $\hbar$ does not appear since it cancelled out! In any event this shows the QM nature of the Maxwell equations and hence of EM fields.

REMARK 5.1.3. One notes here a similar formula arising in Abreau-Hott [**9**] going back to work of Majorana, Weinberg, et al (see [**9**] for references). The idea there involves the matrices S of (1.17) written as s^1, s^2, s^3 with

$$i\frac{\partial\mathbf{E}}{\partial t}=\frac{1}{i}(\mathbf{s}\cdot\nabla)i\mathbf{B};\ i\frac{\partial(i\mathbf{B})}{\partial t}=\frac{1}{i}(\mathbf{s}\cdot\nabla)\mathbf{E};\ \mathbf{E}=\begin{pmatrix}E_1\\E_2\\E_3\end{pmatrix};\ \mathbf{B}=\begin{pmatrix}B_1\\B_2\\B_3\end{pmatrix} \tag{1.25}$$

corresponding to the first two equations in (1.24). Then a fermion-like formulation is created via 6×6 matrices

$$\Gamma_0=\begin{pmatrix}I&0\\0&-I\end{pmatrix};\ \vec{\Gamma}=\begin{pmatrix}0&\mathbf{s}\\-\mathbf{s}&0\end{pmatrix};\ \Gamma_5=\begin{pmatrix}0&I\\I&0\end{pmatrix} \tag{1.26}$$

For $S=\begin{pmatrix}0&\mathbf{s}\\\mathbf{s}&0\end{pmatrix}$ (1.25) becomes

$$i\partial_t\psi=(1/i)(S\cdot\nabla)\psi\Rightarrow(\partial_t+S\cdot\nabla)\psi=0 \tag{1.27}$$

Equation (1.27) resembles the massless Dirac equation and one can consider ψ as a (quantum) wave function for the photon of type $\psi=(\mathbf{E}\ i\mathbf{B})^T$ with $\bar{\psi}=(\mathbf{E}^\dagger\ i\mathbf{B}^\dagger)$ where $\bar{\psi}=\psi^\dagger\Gamma_0$ is an analogue of Hermitian conjugate. Note $\psi^\dagger\psi=\mathbf{E}^2+\mathbf{B}^2$ (where $\mathbf{E}^\dagger=\mathbf{E}$ and $\mathbf{B}^\dagger=\mathbf{B}$) corresponds to the local mean number of photons. The construction of ψ thus mimics a Dirac spinor and one can write

$$i\Gamma_0\partial-t\psi=\frac{1}{i}(\Gamma_0 S\cdot\nabla)\psi=\frac{1}{i}(\vec{\Gamma}\cdot\nabla)\psi\Rightarrow(\Gamma_0\partial_t+\vec{\Gamma}\cdot\nabla)\psi=0 \tag{1.28}$$

or more compactly $\Gamma^\mu\partial_\mu\psi=0$ (although this is not manifestly covariant). One goes on to construct a Lagrangian with a duality transformation and we refer to [**9**] for more details. ■

REMARK 5.1.4. Going to Dvoeglazov [**421**] one finds some generalizations of [**522**] in the form

$$\nabla\times\mathbf{E}=-\frac{1}{c}\partial_t\mathbf{B}+\nabla\Im(\chi);\ \nabla\times\mathbf{B}=\frac{1}{c}\partial_t\mathbf{E}+\nabla\Re(\chi); \tag{1.29}$$

$$\nabla\cdot\mathbf{E}=-\frac{1}{c}\partial_t\Re(\chi);\ \nabla\cdot\mathbf{B}=\frac{1}{c}\partial_t\Im(\chi)$$

and some further analysis of spin situations.. If one assumes no monopoles it may be suggested that $\chi(x)$is a real field and its derivatives play the role of charge and current densities; there is also some flexibility here in interpretation. ■

1.2. A VERSION OF THE DIRAC AETHER. We go here to Carvahlo, Oliveira, and Rabaca [**266, 267**] (cf. also [**377, 669, 670, 958**]). Dirac adopted the idea of using spurious degrees of freedom associated to the gauge potential to describe the electron via a gauge condition $A^2 = k^2$. In fact in Dirac [**377**] this was introduced as a gauge fixing term in the Lagrangian $L = -(1/4)F^2 + (\lambda/2)(A^2 - k^2)$ leading to $\partial_\nu F^{\mu\nu} = J^\mu \equiv \lambda A^\mu$. Here the gauge condition doesn't intend to eliminate spurious degrees of freedom but rather acquires a physical meaning as the condition allowing the right description of the physics without having to introduce extra fields. From this Dirac argued that it would be possible to consider an aether provided that one interprets its four velocity v as a quantity subjected to uncertainty conditions. Admitting the aether velocity as defining a point in a hyperboloid with equation $v_0^2 - \vec{v}^2 = 1$ with $v_0 > 0$ it could be related to the gauge potential (satisfying $A^2 = k^2$) via $(1/k)A_\mu = v_\mu$. Then v (the aether velocity) would be the velocity with which an electric charge would flow if placed in the aether. The model in [**266**] is a continuation of [**958**] and it is implicit in the present formulation that there is an inertial frame (the aether frame) in which the aether is at rest; one writes $(1,0,0,0)$ $(\sim v^\mu_{aether})$ for the aether frame. Then an action $S = \int dx(-(1/4)F^2 + \sigma v_\alpha F^{\alpha\mu} A_\mu)$ is proposed with v being the aether's velocity relative to a generic observer (inertial or not). For inertial observers $v^\mu = \Lambda^\mu_\nu v^\nu_{aether} = \Lambda^\mu_0$ is still constant and one obtains from the equations of motion a current $J^\mu = -\sigma v^\mu \partial \cdot A + \sigma v_\nu \partial^\mu A^\nu$ that is understood as being induced in the aether by the presence of the EM field. Therefore it defines a polarization tensor $M_{\alpha\beta}$ from which one obtains the vectors of polarization and magnetization of the medium. In the aether reference frame this allows one to define the electric displacement vector as $\mathbf{D} = \mathbf{E} + \sigma\mathbf{A}$ while $\mathbf{H} = \mathbf{B}$; the resulting equations are similar in form with the macroscopic Maxwell equations in a medium, in agreement with [**731, 1259**] but with the difference that $\mathbf{D} \neq \epsilon\mathbf{E}$. Hence the aether cannot be thought of as an isotropic medium. Moreover in a generic refernce frame moving relative to the aether $\mathbf{D}$ and $\mathbf{H}$ will depend on v. There is more summary material in [**266**] but we proceed directly here to the equations.

For flat spacetime one takes for $\eta_{\mu\nu} = diag(+,-,-,-)$

$$x^\mu = (x^0, x^i) = (t, \mathbf{x});\ x_\mu = (x_0, x_i) = (t, -\mathbf{x});\ A_\mu = (A_0, A_i) = (\phi, -\mathbf{A}); \tag{1.30}$$

$$\partial^\mu = (\partial^0, \partial^i) = (\partial_t, -\nabla);\ \partial_\mu = (\partial_0, \partial_i) = (\partial_t, \nabla);\ A^\mu = (A^0, A^i) = (\phi, \mathbf{A});$$

$$F_{0i} = \mathbf{E}_i;\ F_{ij} = -\epsilon_{ijk}\mathbf{B}_k;\ F^{0i} = -F_{0i} = -\mathbf{E}_i;\ F^{ij} = F_{ij} = -\epsilon_{ijk}\mathbf{B}_k$$

Take now

$$S = \int dx(-(1/4)F^2 + \tilde{\mathbf{J}} \cdot \mathbf{A}); \tag{1.31}$$

$$\tilde{J}^\mu = \sigma v_\alpha F^{\alpha\mu};\ F_{\mu\nu} = \partial_\mu A_\nu - \partial_\nu A_\mu$$

The constant σ is associated to the aether conductivity and $\tilde{J}$ will be conserved (we drop the bold face here); $\tilde{J} \cdot A$ defines an interaction of the gauge field with itself. Then the equation of motion for A_μ is

$$\partial_\nu F^{\nu\mu} + \sigma v^\mu \partial \cdot A - \sigma v_\nu \partial^\mu A^\nu = 0 \tag{1.32}$$

and this assumes the form of the Maxwell equations in the presence of a source $\partial_\nu F^{\nu\mu} \equiv J^\mu$ provided that one identifies $J^\mu = -\sigma v^\mu \partial \cdot A + \sigma v_\nu \partial^\mu A^\nu$ with a conserved 4-current. This is the same as Dirac in which the term $j^\mu = \lambda A^\mu$ is interpreted as a 4-current but here this arises from the interaction term $\tilde{J} \cdot A$ instead of via a gauge fixing term $(1/2)\lambda(A^2 - k^2)$ in the action. Now taking the divergence of $\partial_\nu F^{\nu\mu} = J^\mu$ one gets

$$(1.33)\quad 0 = \partial_\mu J^\mu = \sigma v_\mu(\Box A^\mu - \partial^\mu \partial \cdot A) = \sigma v_\mu \partial_\nu F^{\nu\mu} = \sigma v_\mu J^\mu = \sigma^2(-v^2 \partial \cdot A + $$
$$+ v_\alpha v_\beta \partial^\alpha A^\beta);\ \partial \cdot A = (v_\alpha v_\beta / v^2)\partial^\alpha A^\beta$$

This constraint is a new feature of this model; its origin is independent of any local symmetry of the action.

For global gauge invariance one considers an invariance defined via the equation $A_\mu \to A'_\mu = A_\mu + \lambda v_\mu$ (where $\partial_\mu \lambda = 0$); this is associated to the Noether current

$$(1.34)\qquad \Theta^\mu = F^{\mu\nu} v_\nu - \sigma v^\mu A \cdot v + \sigma v^2 A^\mu \equiv \Theta^{\mu\nu} v_\nu$$

where $\Theta^{\mu\nu} = F^{\mu\nu} - \sigma v^\mu A^\nu + \sigma v^\nu A^\mu$. Later this $\Theta^{\mu\nu}$ will be interpreted as $H^{\mu\nu}$ which in the aether frame becomes $H^{\mu\nu} = (\mathbf{D}, \mathbf{H})$. In a system at rest relative to the aether one has $\Theta^\mu = (0, \mathbf{E} + \sigma \mathbf{A})$ and (1.33) implies $\nabla \cdot \mathbf{A} = 0$. Then the conservation equation for Θ^μ gives $\nabla \cdot E = 0$. In this model there is another conserved current

$$(1.35)\qquad \hat{J}^\mu = -\sigma v^\mu \partial \cdot A + \sigma v \cdot \partial A^\mu$$

where $\tilde{J} = \hat{J} - J$ so that conservation of $\tilde{J}$ will follow immediately. Equivalently one can think of $\tilde{J}^\nu$ as originating from the divergence of $\Theta^{\mu\nu}$, i.e. $\partial_\mu \Theta^{\mu\nu} = -\tilde{J}^\nu$. In the classical formulation of electrodynamics in conducting media the nonhomogeneous Maxwell equations are written covariantly as $\partial_\mu H^{\mu\nu} = -j^\nu_{ext}$ with $H^{\mu\nu}$ having $\mathbf{D}$ and $\mathbf{H}$ as its components. Here $\Theta^{\mu\nu}$ above generalizes $H^{\mu\nu}$ and $\partial_\mu \Theta^{\mu\nu} = -\tilde{J}^\nu$ corresponds to $\partial_\mu H = -j^\nu_{ext}$. This allows one to interpret $\tilde{J}^\mu$ as the corresponding 4-current in much the same way as Dirac interpreted $j_\mu = \lambda A_\mu$ as a 4-current in [**377**]. The interaction term $\tilde{J} \cdot A$ then parallels the same term in the usual electrodynamics. The global symmetry above is a new feature with no counterpart in the Maxwell formulation. It is also possible to add a mass term to S above that preserves this global symmetry; in fact the action

$$(1.36)\qquad S = \int dx(-(1/4)F^2 + \tilde{J} \cdot A - (1/2)\sigma^2 A_\mu (v^2 g^{\mu\nu} - v^\mu v^\nu) A_\nu)$$

is invariant. From this one obtains the equation

$$(1.37)\qquad \partial_\nu F^{\nu\mu} \equiv \bar{J}^\mu = -\sigma v^\mu \partial \cdot A + \sigma v_\nu \partial^\mu A^\nu + \sigma^2 (v^2 g^{\mu\nu} - v^\mu v^\nu) A_\nu \sim$$
$$\sim [g^{\mu\nu}(\Box - \sigma^2 v^2) = \partial^\mu \partial^\nu + \sigma(v^\mu \partial^\nu - v^\nu \partial^\mu) + \sigma^2 v^\mu v^\nu] A_\nu = 0$$

Here the conservation of the current $\bar{J}$ that follows from (1.37) doesn't produce any constraint on A; further the conserved current in (1.34) is associated to the global symmetry. Finally the local gauge invariance depends on a parameter $\theta(x)$ and has the usual form $A_\mu \to A'_\mu = A_\mu + \partial_\mu \theta$ and (1.33) adds some new features to the analysis. In fact let A' and A be two fields related by $A'_\mu = A_\mu + \partial_\mu \theta$. Since both field configurations should obey (1.33) one must have

$$\partial \cdot A' = (1/v^2)(v \cdot \partial)(v \cdot A') \iff \Box \theta = (v_\alpha v_\beta / v^2)\partial^\alpha \partial^\beta \theta \tag{1.38}$$

Now let $\partial \cdot A \neq 0$ and choose θ so that it ensures $\partial \cdot A' = 0$; one should then have θ satisfying $\Box\theta = -\partial \cdot A$. This last condition together with the constraints (1.33) and (1.38) give then $\partial^\alpha \partial^\beta \theta = -(1/2)(\partial^\alpha A^\beta + \partial^\beta A^\alpha)$ which represents a stronger restriction than that shown in $\Box\theta = -\partial \cdot A$. Equivalently one can obtain this stronger restriction directly from $0 = \partial \cdot A' = (v_\alpha v_\beta/v^2)\partial^\alpha(A^\beta + \partial^\beta \theta)$ using (1.33).

Now one considers this model in the aether reference frame where $v = (1,0,0,0)$ and it is supposed that the aether is a medium without any given density of charge or current. From $\partial_\nu F^{\nu\mu} = J^\mu$ one has

$$\nabla \cdot \mathbf{E} = -\sigma \nabla \cdot \mathbf{A};\ \nabla \times \mathbf{B} = \frac{\partial \mathbf{E}}{\partial t} + \sigma \frac{\partial \mathbf{A}}{\partial t} + \sigma \mathbf{E} \tag{1.39}$$

to which is added the homogeneous equations $\nabla \cdot \mathbf{B} = 0$ and $\nabla \times \mathbf{E} = -\partial_t \mathbf{B}$. Since $\nabla \cdot \mathbf{E} \neq 0$ we see that (1.39) introduces a new feature for the physical vacuum (cf. also [**809**]). Essentially a divergenceless equation for $\mathbf{E}$ signals that the vacuum is not merely an empty space but it is also capable of becoming electrically polarized. The presence of additional terms depending on the potential vector in (1.39) indicates the response of the medium to the presence of the fields $(\mathbf{E}, \mathbf{B})$, which resembles the phenomena of polarization and magnetization of a medium. Therefore one rewrites (1.39) as $\nabla \cdot \mathbf{D} = 0$ with $\mathbf{D} = \mathbf{E} + \sigma \mathbf{A} + \nabla \times \mathbf{K}$. At this point $\mathbf{K}$ is an arbitrary vector that can be thought of as playing the role of a gauge parameter. Now rewrite (1.39) with an assignment for $\mathbf{K}$ as

$$\nabla \times \mathbf{B} = \partial_t \mathbf{D} + \sigma \mathbf{E} - \partial_t \nabla \times \mathbf{K} \ \ \textit{with} \ \ \partial_t \nabla \times \mathbf{K} = \sigma \mathbf{E} \tag{1.40}$$

Then using $\nabla \cdot \mathbf{B} = 0$ and $\nabla \times \mathbf{E} = -\partial_t \mathbf{B}$ one obtains $\partial_t \nabla \times (\sigma \mathbf{A} + \nabla \times \mathbf{K}) = 0$. The vector $\mathbf{K}$ can be still further restricted so that $\sigma \mathbf{A} + \nabla \times \mathbf{K} = 0$ and this gives $\mathbf{E} = \mathbf{D}$ and $\mathbf{B} = \mathbf{H}$ with the Maxwell equations in free space, namely

$$\nabla \cdot \mathbf{E} = 0;\ \nabla \times \mathbf{B} = \partial_t \mathbf{E};\ \nabla \cdot \mathbf{B} = 0;\ \nabla \times \mathbf{E} = -\partial_t \mathbf{B} \tag{1.41}$$

Conditions (1.40) with $\sigma \mathbf{A} + \nabla \times \mathbf{K} = 0$ can be interpreted as originating from the imposition of the temporal gauge and together they imply $\nabla \mathbf{A}_0 = 0$ which is naturally satisfied by putting $\mathbf{A}_0 = 0$. It is possible to give another description for the present electrodynamics without using $\mathbf{K}$. Thus from (1.39) one can simply identify $\mathbf{D} = \mathbf{E} + \sigma \mathbf{A}$ and $\mathbf{B} = \mathbf{H}$, leading to

$$\nabla \cdot \mathbf{D} = 0;\ \nabla \times \mathbf{B} = \partial_t \mathbf{D} + \sigma \mathbf{E};\ \nabla \cdot \mathbf{B} = 0;\ \nabla \times \mathbf{E} = -\partial_t \mathbf{B} \tag{1.42}$$

and this coincides with the aether equations of [**731, 1259**]. In the identification $\mathbf{D} = \mathbf{E} + \sigma \mathbf{A}$ and $\mathbf{B} = \mathbf{H}$ the aether behaves like a medium that responds to the presence of the electric field by creating a polarization $\mathbf{P} = \sigma \mathbf{A}$. One also has a curreent $\mathbf{J} = \sigma \mathbf{E}$ which is in agreement with the supposition of the aether being a medium with conductivity σ. According to Schwinger's idea of a structureless vacuum an EM field disturbs the vacuum affecting its properties of homogeneity and isotropy. This is exactly the situation obtained in the present model where the presence of an EM field in a vacuum with conductivity σ produces a response of the medium ($\mathbf{D} \neq \epsilon\mathbf{E}$) that signals its nonisotropy.

Thus for a flat spacetime and a reference frame at rest relative to the aether one has $\mathbf{D} = \mathbf{E} + \sigma\mathbf{A}$ and $\mathbf{B} = \mathbf{H}$. In the case of a curved spacetime and a noninertial reference frame moving relative to the aether one will have a more complicated relation between H and F. Indeed in a medium that is at rest in any reference fame with a metric $g_{\alpha\beta}$ one knows from [**1266**] that the relation between H and F has the form

$$\sqrt{-g}H^{\alpha\beta} = \sqrt{-g}g^{\alpha\gamma}g^{\beta\kappa}S^{\mu}_{\gamma}S^{\nu}_{\kappa}F_{\mu\nu} \tag{1.43}$$

where S^{α}_{β} characterizes the EM properties of the medium. It is convenient to rewrite (1.43) as

$$\sqrt{-g}H^{\alpha\beta} = \sqrt{-g}g^{\alpha\gamma}g^{\beta\kappa}S^{\mu}_{\gamma\kappa}A_{\mu}; \quad S^{\mu}_{\gamma\kappa} = (S^{\nu}_{\gamma}S^{\mu}_{\kappa} - S^{\mu}_{\gamma}S^{\nu}_{\kappa})\partial_{\nu} \tag{1.44}$$

As an application of (1.43) it was shown in Volkov-Kiselev [**1266**] that for the vacuum (considered from an inertial reference frame) the tensor S^{α}_{β} has the form $S^{\alpha}_{\beta} = \delta^{\alpha}_{\beta}$ and the material equations become $\sqrt{-g}H^{\alpha\beta} = \sqrt{-g}g^{\alpha\mu}g^{\beta\nu}F_{\mu\nu}$ which reproduces the usual equations of free electrodynamics in a curved background (cf. Puthoff [**1048**]). Also in the case of a linear isotropic medium that is at rest in an inertial reference frame the tensor S^{α}_{β} is given by $S^0_0 = \epsilon\sqrt{\mu}$, $S^1_1 = S^2_2 = S^3_3 = 1/\sqrt{\mu}$ and one obtains the usual relations $\mathbf{D} = \epsilon\mathbf{E}$ and $\mathbf{H} = (1/\mu)\mathbf{B}$. In the present model in order to define $H^{\alpha\beta}$ for a generic reference frame in a curved background and t find a suitable material relation of the type (1.44) one should first follow the preceeding approach which allows one to define $H^{\mu\nu}$ directly from the equation of motion for A_{μ}. Explicitly take the equation of motion in a curved background to be

$$\partial_{\nu}(\sqrt{-g}F^{\mu\nu} - \sigma\sqrt{-g}v^{\nu}A^{\mu} + \sigma\sqrt{-g}v^{\mu}A^{\nu}) = J^{\mu} = -\sigma\sqrt{-g}v_{\nu}F^{\nu\mu} \tag{1.45}$$

Then one defines

$$\sqrt{-g}H^{\alpha\beta} = \sqrt{-g}(F^{\alpha\beta} - \sigma v^{\alpha}A^{\beta} + \sigma v^{\beta}A^{\alpha}) \tag{1.46}$$

This is equivaent to the introduction of an antisymmetric polarization tensor $M^{\alpha\beta}$ (cf. [**1266**]) of the form

$$\sqrt{-g}H^{\alpha\beta} = \sqrt{-g}(F^{\alpha\beta} + M^{\alpha\beta}) \iff \sqrt{-g}H^{\alpha\beta} = \sqrt{-g}g^{\alpha\nu}(F_{\nu\mu} + M_{\nu\mu}) \tag{1.47}$$

provided we identify

$$M^{\alpha\beta} = -\sigma v^{\alpha}A^{\beta} + \sigma v^{\beta}A^{\alpha} \iff M_{\alpha\beta} = -\sigma v_{\alpha}A_{\beta} + \sigma v_{\beta}A_{\alpha}; \quad F^{\mu\nu} = g^{\mu\alpha}g^{\nu\beta}F_{\alpha\beta},\ F_{\mu\nu} = \partial_{\mu}A_{\nu} - \partial_{\nu}A_{\mu};\ M^{\mu\nu} = g^{\mu\alpha}g^{\nu\beta}M_{\alpha\beta} \tag{1.48}$$

Finally in order to obtain the material equations one extends (1.44) by allowing $S^{\mu}_{\alpha\beta}$ to be a generic operator not restricted as in (1.44) but given by $S^{\mu}_{\alpha\beta} = \delta^{\mu}_{\beta}(\partial_{\alpha} - \sigma v_{\alpha}) - \delta^{\mu}_{\alpha}(\partial_{\beta} - \sigma v_{\beta})$. Here the tensor $S^{\mu}_{\alpha\beta}$ may contain not only EM properties of the medium (as in the case of (1.44)) but also information about the reference frame (implicit in the 4-velocity v_{α}). Adopting the convention $D_i = \sqrt{-g}H^{i0}$ and

$H_i = -(1/2)\epsilon_{ijk}\sqrt{-g}H^{jk}$ (cf. [**1266**]) and considering a flat spacetime one obtains (in the aether frame) the relations $\mathbf{D} = \mathbf{E} + \sigma\mathbf{A}$ and $\mathbf{B} = \mathbf{H}$.

1.3. MASSLESS PARTICLES. We review again some classical and quantum features of EM with some repetition of earlier material. First, following [**879**], recall the Maxwell equations

$$\nabla \cdot E = 0;\ \nabla \cdot B = 0;\ \nabla \times E = -\frac{1}{c}B_t;\ \nabla \times B = \frac{1}{c}E_t \tag{1.49}$$

(bold face is omitted). Recall now the vector potential A eners via $B = \nabla \times A$ and from the third equation in (1.49) $E = -(1/c)A_t - \nabla\phi$ where ϕ is the scalar potential. Hence $\nabla^2 A - (1/c^2)A_{tt} = 0$ in the Coulomb gauge defined via $\nabla \cdot A = 0$ and in the absence of sources $\phi = 0$. Separation of variables gives monochromatic solutions

$$A(x,t) = \alpha(t)A_0(x) + \alpha^*(t)A_0^*(x) = \alpha(0)e^{-i\omega t}A_0(x) + \alpha^*(0)e^{i\omega t}A_0^*(x) \tag{1.50}$$

where $\nabla^2 A_0(x) + k^2 A_0(x) = 0$ ($k = \omega/c$) and $\ddot{\alpha} = -\omega^2\alpha$. Consequently

$$E(x,t) = -\frac{1}{c}[\dot{\alpha}(t)A_0(r) + \dot{\alpha}^*(t)A_0^*(x); \tag{1.51}$$
$$B(x,t) = \alpha(t)\nabla \times A_0(x) + \alpha^*(t)\nabla \times A_0^*(x)$$

and a calculation gives the EM energy as

$$H_F = (1/8\pi)\int d^3x(E^2 + B^2) = (k^2/2\pi)|\alpha(t)|^2 \tag{1.52}$$

where A_0 is normalized via $\int d^3x|A_0(x)|^2 = 1$ Now defining

$$q(t) = \frac{i}{c\sqrt{4\pi}}[\alpha(t) - \alpha^*(t)];\ p(t) = \frac{k}{\sqrt{4\pi}}[\alpha(t) + \alpha^*(t)] \tag{1.53}$$

gives $H_F = (1/2)(p^2 + \omega^2 q^2)$ so the field mode of frequency ω is mathematically equivalent to a harmonic oscillator of frequency ω. Note q and p are canonically conjugate since $\dot{q} = p$ and $\dot{p} = -\omega^2 q$ corresponds to Hamiltonian equations with Hamiltonian H_F.

REMARK 5.1.5. We assume familiarity with harmonic oscillator calculations. For $H = (p^2/2m) + (1/2)m\omega^2 q^2$ one has $\dot{q} = (i\hbar)^{-1}[q,H] = p/m$ and $\dot{p} = (i\hbar)^{-1}[p,H] = -m\omega^2 q$. Then defining

$$a = \frac{1}{\sqrt{2m\hbar\omega}}(p - iq);\ a^\dagger = \frac{1}{\sqrt{2m\hbar\omega}}(p + im\omega q) \equiv \tag{1.54}$$
$$\equiv q = i\sqrt{\frac{\hbar}{2m\omega}}(a - a^\dagger);\ p = \sqrt{\frac{m\hbar\omega}{2}}(a + a^\dagger)$$

which yields $[q,p] = i\hbar$ and $[a,a^\dagger] = 1$ with $H = (1/2)\hbar\omega(aa^\dagger + a^\dagger a) = \hbar\omega(a^\dagger a + (1/2))$. For $N = a^\dagger a$ one has eigenkets $N|n> = n|n>$ and $a|n> = \sqrt{n}|n-1>$ with $a^\dagger|n> = \sqrt{n+1}|n+1>$; further $E_n = [n + (1/2)]\hbar\omega$. ■

One can now replace (1.50) by

$$A(x,t) = \left(\frac{2\pi\hbar c^2}{\omega}\right)^{1/2}[a(t)A_0(x) + a^\dagger(t)A_0^*(x)]; \tag{1.55}$$

$$E(x,t) = i(2\pi\hbar\omega)^{1/2}[a(t)A_0(x) - a^\dagger(t)A_0^*(x)];$$
$$B(x,t) = \left(\frac{2\pi\hbar c^2}{\omega}\right)^{1/2}[a(t)\nabla\times A_0(x) + a^\dagger(t)A_0^*(x)]$$

and H_F becomes $H_F = \hbar\omega[a^\dagger a+(1/2)]$. Now the vacuum state $|0>$ has no photons but has an energy $(1/2)\hbar\omega$ and QM thus predicts the ZPF. In all stationary states $|n>$ one has $<E(x,t)>=<B(x,t)>=0$ since $<n|a|n>=0$. This means that E and B fluctuate with zero mean in the state $|n>$ even though the state has a definite nonfluctuating energy $[n+(1/2)]\hbar\omega$. A computation also gives

(1.56) $$<E^2(x,t)[n+(1/2)]4\pi\hbar\omega|A_0(x)|^2 = 4\pi\hbar\omega|A_0(x)|^2 n + <E^2(x)>_0$$

The factor n is the number of photons and $|A_0(x)|^2$ gives the same spatial intensity as in the classical theory. Normal ordering is used of course in QM ($a^\dagger$ to the left of a) and this eliminates contributions of ZPF to various calculations. However one does not eliminate the ZPF by dropping its energy from the Hamiltonian. In the vacuum state E and B do not have definite values (they fluctuate about a zero mean value) and one arrives at energy densities etc. as before. An atom is often considered to be "dressed" by emmision and reabsorption of virutal photons from the vacuum which itself has an infinite energy. Some thermal aspects were mentioned before (cf. (1.2) for example) and this is enormously important (in this direction there is much information in [**879, 1247**]).

REMARK 5.1.6. We refer next to the first paper in Ibison [**656**] on massless classical electrodynamics for some interesting calculations (some details are omitted here). One considers a bare charge, free of self action, compensating forces, and radiation reaction (cf. [**656**] for a long discussion on all this). Use the convention $u^a v_a = u_0 v_0 - \mathbf{u}\cdot\mathbf{v}$ and Heavyside-Lorentz units with $c = 1$ in general. With the fields given the Euler equation for the (massless) lone particle degree of freedom is simply that the Lorentz force on the particle in question must vanish, i.e. $F^{\nu\mu}u_{\mu\ell} = 0$ where F is the EM field stregth tensor and the fields E, B are to be evaluated along the trajectory. In $3+1$ form, omitting particle labels, this is

(1.57) $$(dt(\lambda)/d\lambda)E(x(\lambda),t(\lambda)) + (dx(\lambda)/d\lambda)\times B(x(\lambda),t(\lambda)) = 0$$

In order for $F^{\nu\mu}u_{\mu\ell} = 0$ to have a solution the determinant of F must vanish which gives

(1.58) $$S(x(\lambda)) \equiv E(x(\lambda))\cdot B(x(\lambda)) = 0$$

This imposes a constraint on the fields along the trajectory which can be interpreted as the condition that the Lorentz force on a particle must vanish (recognized as the constraint on the fields such that there exist a frame in which the electric field is zero). Calculation leads to

(1.59) $$v(x,t) = \frac{dx/d\lambda}{dt/d\lambda} = \frac{E\times\nabla S - BS_t}{B\cdot\nabla S}$$

as the ordinary velocity of the trajectory passing through $(t(\lambda), x(\lambda))$. The right side is an arbitrary function of (x,t) decided by the fields and in general (1.59) will not admit a solution of the form $x = f(t)$ since the solution trajectory may be nonmonotonic in time. Various situations are discussed including the particles advanced and retarded fields. In particular the particle does not respond to force in the traditional sense of Newton's second law. Its motion is precisely that which causes it to feel no force. Yet its motion is uniquely prescribed by E and B, which decide the particle trajectory (given some initial condition) just as the Lorentz force determines the motion of a massive particle. The important difference is that traditionally the fields determine acceleration whereas here they determine the velocity. The massless particle discussed cannot be a relative of the neutrino and it does not seem to be a traditional classical object in need of quantization. ■

1.4. EINSTEIN AETHER WAVES. We go now to Jacobson-Mattingly [**686**] where the violation of Lorentz invariance by quantum gravity effects is examined (cf. also [**33, 64, 429, 430, 486, 620, 698, 866**]). In a nongravitational setting it suffices to specify fixed background fields violating Lorentz symmetry in order to formulate the Lorentz violating (LV) matter dynamics. However this would break general covariance, which is not an option, so one promotes the LV background fields to dynamical fields, governed by a generally covariant action. Virtually any configuration of matter fields breaks Lorentz invariance but here the LV fields contemplated are constrained dynamically or kinematically not to vanish, so that every relevant field configuration violates local Lorentz symmetry everywhere, even in the "vacuum". If the Lorentz violation preserves a 3-dimensional rotation subgroup then the background field must be a timelike vector and one considers here the case where the LV field is a unit timelike vector u^a which can be viewed as the minimal structure required to determine a local preferred rest frame. One opts to call this field the "aether" as it is ubiquitous and determines a local preferred rest frame. Kinetic terms in the action couple the ether directly to the spacetime metric in addition to any couplings that might be present between the aether and the matter fields. This system of the metric coupled to the aether will be referred to as Einstein-aether theory. In Gasperini [**515**] an essentially equivalent theory appears based on a tetrad formalism (cf. also [**686, 866**] for various special cases).

In the spirit of effective field theory consider a derivative expansion of the action for the metric g_{ab} and aether u^a. The most general action that is diffeomorphism invariant and quadratic in derivatives is

$$(1.60)\quad S = \frac{1}{16\pi G}\int d^4x\sqrt{-g}(-R+\mathfrak{L}_u-\lambda(j^a u_a-1));\ \mathfrak{L}_u = -K^{ab}{}_{mn}\nabla_a u^m\nabla_b u^n;$$

$$K^{ab}{}_{mn} = c_1 g^{ab}g_{mn} + c_2\delta^a_m\delta^b_n + c_3\delta^a_n\delta^b_m + c_4 u^a u^b g_{mn}$$

Here R is the Ricci scalar and λ is a Lagrange multiplier enforcing the unit constraint. The signature is $(+,-,-,-)$ and $c=1$ (cf. [**1274**] for other notation). The possible term $R_{ab}u^a u^b$ is proportional to the difference of the c_2 and c_3 terms via integration by parts and hence has been omitted. Also any matter coupling is omitted since one wants to concentrate on the metric-aether sector in vacuum. Varying the action with respect t u^a, g^{ab}, and λ yields the field equations

$$(1.61)\qquad \nabla_a J^a{}_m = c_4\dot{u}_a\nabla_m u^a = \lambda u_m;\ G_{ab} = T_{ab};\ g_{ab}u^a u^b = 1$$

Here one has $J^a_{\ m} = K^{ab}_{\ \ mn}\nabla_b u^n$ and $\dot{u}^m = u^a\nabla_a u^m$ and the aether stress tensor is

$$T_{ab} = \nabla_m(J_{(a}^{\ \ m}u_{b)} - J^m_{\ (a}u_{b)} - J_{(ab)}u^m) + \\ +c_1[(\nabla_m u_a)(\nabla^m u_b) - (\nabla_a u_m)(\nabla_b u^m)] + c_4\dot{u}_a\dot{u}_b + \\ +[u_n(\nabla_m J^{mn}) - c_4\dot{u}^2]u_a u_b - \frac{1}{2}g_{ab}\mathfrak{L}_u \tag{1.62}$$

Here the constraint has been used to eliminate the term that arises from varying $\sqrt{-g}$ in the constraint term in (3.1) and in the last line λ has been eliminated using the aether field equations.

Now the first step in finding the wave modes is to linearize the field equaitons about the flat Minkowski background η_{ab} and the constant unit vector $\underline{u}^a$. The expanded fields are then

$$g_{ab} = \eta_{ab} + \gamma_{ab}; \ u^a = \underline{u}^a + v^a \tag{1.63}$$

The Lagrange multiplier λ vanishes in the background so we use the same notation for its linearized version. Indices will be raised and lowered now with η_{ab} and one adopts Minkowski coordinates (x^0, x^i) aligned with $\underline{u}^a$; i.e. for which $\eta_{ab} = diag(1,-1,-1,-1)$ and $\underline{u}^a = (1,0,0,0)$. Keeping only the first order terms in v^a and γ_{ab} the field equations become

$$\partial_a J^{(1)a}_{\ \ \ \ m} = \lambda\underline{u}_m; \ G^{(1)}_{ab} = T^{(1)}_{ab}; \ v^0 + \frac{1}{2}\gamma_{00} = 0 \tag{1.64}$$

where the superscript (1) denotes the first order part. The linearized Einstein tensor is

$$G^{(1)}_{ab} = -\frac{1}{2}\Box\gamma_{ab} - \frac{1}{2}\gamma_{,ab} + \gamma_{m(a,b)}^{\ \ m} + \frac{1}{2}\eta_{ab}(\Box\gamma - \gamma_{mn,}^{\ \ \ \ mn}) \tag{1.65}$$

where γ_m^m is the trace, while the linearized aether stress tensor is

$$T^{(1)}_{ab} = \partial_m[J^{(1)m}_{(a}\underline{u}_{b)} - J^{(1)m}_{(a}\underline{u}_{b)} - J^{(1)}_{(ab)}\underline{u}^m] + [\underline{u}_n(\partial_m J^{(1)mn})]\underline{u}_a\underline{u}_b \tag{1.66}$$

In one imposes the linearized aether field equation (3.4) then the second and last terms of this expression for $T^{(1)}_{ab}$ cancel, yielding

$$T^{(1)}_{ab} = -\partial_0 J^{(1)}_{(ab)} + \partial_m J^{(1)m}_{(a}\underline{u}_{b)}; \\ J^{(1)}_{ab} = c_1\nabla_a u_b + c_2\eta_{ab}\nabla_m u^m + c_3\nabla_b u_a + c_4\underline{u}_a\nabla_0 u_b \tag{1.67}$$

where the covariant derivatives of u^a are expanded to linear order, i.e. replaced by

$$(\nabla_a u_b)^{(1)} = (v_b + (1/2)\gamma_{0b})_{,a} + (1/2)\gamma_{ab,0} - (1/2)\gamma_{a0,b} \tag{1.68}$$

Were it not for the aether background the linearized aether stress tensor (3.6) would vanish and the metric would drop out of the aether field equation, leaving all modes uncoupled.

Now diffeomorphism invariance of the action (3.1) implies that the field equations are tensorial and hence covariant under diffeomorphisms. The linearized equations inherit the linearized version of this symmetry and to find the independent physical wave modes one must fix the corresponding gauge symmetry. Thus an infinitesimal diffeomorphism generated by a vector field ξ^a transforms g_{ab} and u^a by

$$(1.69)\qquad \delta g_{ab} = \mathfrak{L}_\xi g_{ab} = \nabla_a\xi_b + \nabla_b\xi_a;\ \delta u^a = \mathfrak{L}_\xi u^a = \xi^m\nabla_m u^a - u^m\nabla_m\xi^a$$

ξ^a is itself first order in the perturbations so the linearized gauge transformations take the form

$$(1.70)\qquad \gamma'_{ab} = \gamma_{ab} + \partial_a\xi_b + \partial_b\xi_a;\ v'^a = v^a - \partial_0\xi^a$$

The usual choice of gauge in vacuum GR is the Lorentz gauge $\partial^a\bar{\gamma}_{ab} = 0$ where $\bar{\gamma}_{ab} = \gamma_{ab} - (1/2)\gamma\eta_{ab}$ but for various reasons this is inappropriate here (cf. [**686**]). Instead one imposes directly the four gauge conditions

$$(1.71)\qquad \gamma_{0i} = 0;\ v_{i,i} = 0$$

To see that this is accessible note that the gauge variations of γ_{0i} and $v_{i,i}$ are, according to (1.58)

$$(1.72)\qquad \delta\gamma_{0i} = \xi_{i,0} + \xi_{0,i};\ \delta v_{i,i} = -\xi_{i,i0}$$

Thus to achieve (1.59) one must choose ξ_0 and ξ_i to satisfy equations of the form

$$(1.73)\qquad (\mathbf{A})\ \xi_{i,0} + \xi_{0,i} = X_i\quad (\mathbf{B})\ \xi_{i,i0} = Y$$

Subtracting the second equation from the divergence of the first gives $\xi_{0,ii} = X_{i,i} - Y$ which determines ξ_0 up to constants of integration by solving a Poisson equation. Then ξ_i can be determined up to a time independent field by integrating **(A)** in (1.61) with respect to time. From these choices of ξ_0 and ξ_i **(A)** in (1.61) holds and the divergence of this gives **(B)** in (1.61). In the gauge (1.59) the tensors in the aether and spatial metric equations in (3.4) take the forms

$$(1.74)\qquad J_{ai,}{}^{a} = c_{14}(v_{i,00} - (1/2)\gamma_{00,i0}) - c_1 v_{i,kk} - (1/2)c_{13}\gamma_{ik,k0} - (1/2)c_2\gamma_{kk,0i};$$

$$G^{(1)}_{ij} = -(1/2)\Box\gamma_{ij} - (1/2)\gamma_{,ij} - \gamma_{k(i,j)k} - (1/2)\delta_{ij}(\Box\gamma - \gamma_{00,00} - \gamma_{k\ell,k\ell});$$

$$T^{(1)}_{ij} = -c_{13}(v_{(i,j)0} + (1/2)\gamma_{ij,00} - (1/2)c_2\delta_{ij}\gamma_{kk,00}$$

where e.g. $c_{14} = c_1 + c_4$ etc. Various wave modes are calculated and in particular there are a total of 5 modes, 2 with an unexcited aether which correspond to the usual GR modes, two "transverse" aether-metric modes, and a fifth trace aether-metric mode. We refer to [**686**] for details and discussion.

REMARK 5.2.1. The Lorentz violation theme and the idea of a preferred reference frame is presently of a certain general interest in connection with proposals of quantum gravity and we cite as before [**33, 64, 98, 429, 430, 486, 698, 866**]. We rephrase matters here as in Foster [**486**]. Thus in an effective field theory description the Lorentz symmetry breaking can be realized by a vector field that defines the preferred frame. In the flat spacetime of the standard model this field can be treated as non-dynamical background structure but in the context of GR diffeomorphism invariance (a symmetry distinct from local Lorentz invariance) can be preserved by elevating this field to a dynamical quantity. This leads to investigation of vector-tensor theories of gravity and one such model couples gravity to a vector field that is constrained to be everywhere timelike and of unit norm. The unit norm condition embodies the notion that the theory assigns no physical importance to the norm of the vector and this corresponds to the Einstein aether (AE) theory as in Eling-Jacobson [**429**]. In [**486**] one demonstrates the effect of a field redefinition on the conventional second order AE theory action. Thus take $g_{ab} \to g'_{ab} = A(g_{ab} - (1-B)u_a u_b)$ with $u^a \to (u')^a = (1/\sqrt{AB})u^a$ where g_{ab} is a Lorentzian metric and u^a is the aether field. The action is taken as the most general form which is generally covariant, second order in derivatives, and is consistent with the unit norm constraint. The redefinition preserves this form and the net effect is a rescaling of the action and a transformation of the coupling constants (generalizing the work of [**98**]). Thus start with $S = -(1/16\pi G)\int \sqrt{|g|}\mathfrak{L}$ where

$$\mathfrak{L} = R + c_1(\nabla_a u_b)(\nabla^a u^b) + c_2(\nabla_a u^a)(\nabla_b u^b) + \tag{1.75}$$
$$+ c_3(\nabla_a u^b)(\nabla_b u^a) + c_4(u^a\nabla_a u^c)(u^b\nabla_b u_c)$$

where R is the scalar curvature of g_{ab} with signature $(+,-,-,-)$ and the c_i are dimensionless constants. After the substitution indicated above one has

$$(g')^{ab} = \frac{1}{A}\left(g^{ab} - \left(1-\frac{1}{B}\right)u^a u^b\right);\ u'_a = \sqrt{AB}u_a \tag{1.76}$$

(note $(u')^a u'_a = 1$ is preserved). The net effect of A is a rescaling of the action by a factor of A (with the Lagrangian scaling as $1/A$ while the volume scales as A^2). There are many calculations (omitted here) and the constructions simplify the problem of characterizing solutions for a specific set of c_i by transforming that set into one in which one or more of the c_i vanish. If non-aether matter is included a metric redefinition not only changes the c_i but also modifies the matter action suggesting perhaps a universal metric to which the matter couples (cf. [**264, 698**]). ■

2. STOCHASTIC ELECTRODYNAMICS

From topics in Chapters 1,2,3, and 4 we are familiar with some stochastic aspects of QM. Further the ZPF has been seen to be related to quantum phenomena. The idea of stochastic electrodynamcs (SED) is essentially an attempt to establish SED as the foundation for QM. There has been some partial success in this direction but the methods break down when trying to deal with nonlinearities. The paper de la Pena-Cetto [**1005**] is a recent version of nonperturbative linear SED (LSED) which provides a speculative mechanism leading to the quantum behavior of field and matter based on 3 fundamental principles; it purports to explain for example why all systems described by it (and hence by QM ?) behave as if they consisted of a set of harmonic oscillators. We review here first some basic background issues in SED arising from de la Pena-Cetto [**1004**] (cf. also [**314, 684, 879**]) and then will sketch a few matters from [**1005**].

Thus start with the homogeneous Maxwell equations

$$\nabla\cdot\mathbf{D} = 0;\ -\frac{1}{c}\frac{\partial\mathbf{D}}{\partial t} + \nabla\times\mathbf{H} = 0;\ \nabla\cdot\mathbf{B} = 0;\ \frac{1}{c}\frac{\partial\mathbf{B}}{\partial t} + \nabla\times\mathbf{E} = 0 \tag{2.1}$$

One obtains then (we refer to [**1004**] for details and discussion)

$$\mathbf{B} = \nabla \times \mathbf{A};\ \mathbf{E} = -\nabla\Phi - \frac{1}{c}\frac{\partial \mathbf{A}}{\partial t};\ \nabla \times \left(\frac{1}{c}\frac{\partial \mathbf{A}}{\partial t} + \mathbf{E}\right) = 0 \tag{2.2}$$

Then the third and fourth Maxwell equations are satisfied identically and the first two in combination with $\mathbf{D} = \epsilon\mathbf{E}$ and $\mathbf{B} = \mu\mathbf{H}$ determine the evolution of the potentials $\mathbf{A}$ and Φ. In particular one has

$$\nabla^2\Phi + \frac{1}{c}\partial_t\nabla\cdot\mathbf{A} = 0;\ \nabla^2\underline{A} - \frac{1}{c^2}\frac{\partial^2\mathbf{A}}{\partial t^2} - \nabla\left(\nabla\cdot\mathbf{A} + \frac{1}{c}\frac{\partial\Phi}{\partial t}\right) = 0 \tag{2.3}$$

There is then room for gauge transformations

$$\mathbf{A} \to \mathbf{A}' = \mathbf{A} + \nabla\Lambda;\ \Phi \to \Phi' = \Phi - \frac{1}{c}\frac{\partial\Lambda}{\partial t} \tag{2.4}$$

Now choose the potentials to uncouple (2.3) via

$$\nabla\cdot\mathbf{A} + \frac{1}{c}\frac{\partial\Phi}{\partial t};\ \nabla^2\Phi - \frac{1}{c^2}\frac{\partial^2\Phi}{\partial t^2} = 0;\ \nabla^2\mathbf{A} - \frac{1}{c^2}\frac{\partial^2\mathbf{A}}{\partial t^2} = 0 \tag{2.5}$$

This set is equivalent in all respects to the Maxwell equations in vacuum (Lorentz gauge) and this arrangement can always be achieved via

$$\nabla^2\Lambda - \frac{1}{c^2}\frac{\partial^2\Lambda}{\partial t^2} = 0 \tag{2.6}$$

Another gauge selection is the Coulomb gauge defined via

$$\nabla\cdot\mathbf{A} = 0;\ \nabla^2\Phi = 0;\ \nabla^2\mathbf{A} - \frac{1}{c^2}\frac{\partial^2\mathbf{A}}{\partial t^2} = \frac{1}{c}\nabla\partial_t\Phi \tag{2.7}$$

One can take $\Phi = 0$ and the last equation reduces to the last equation of (2.5). In any case for any gauge one has

$$\nabla^2\mathbf{E} - \frac{1}{c^2}\frac{\partial^2\mathbf{E}}{\partial t^2} = 0;\ \Box^2\mathbf{A} = \Box^2\mathbf{B} = \Box^2\mathbf{E} = 0 \tag{2.8}$$

where $\Box^2 = (1/c^2)\partial_t^2 - \nabla^2$. One now goes into mode expansions with Fourier series and integrals and we simply list formulas of the form ($a_\alpha = a_\alpha(t) = a_{n\lambda}exp(-i\omega_n t) = a_{n\lambda}(t)$)

$$q_\alpha = i\sqrt{\frac{E_\alpha}{2\omega_\alpha^2}}(a_\alpha - a_\alpha^*);\ p_\alpha = \sqrt{\frac{E_\alpha}{2}}(a_\alpha + a_\alpha^*); \tag{2.9}$$

$$H = \sum H_\alpha = \sum E_\alpha a_\alpha^* a_\alpha = \sum \frac{1}{2}(p_\alpha^2 + \omega_\alpha^2 q_\alpha^2)$$

$$\mathbf{A} = \sum\sqrt{\frac{4\pi c^2}{\omega_n^2 V}}e_n^\lambda[p_{n\lambda}Cos(k\cdot x) + \omega_n q_{n\lambda}Sin(k\cdot x)]; \tag{2.10}$$

$$\mathbf{E} = \sum\sqrt{\frac{4\pi}{V}}e_n^\lambda[-p_{n\lambda}Sin(k\cdot x) + \omega_n q_{n\lambda}Cos(k\cdot x)];$$

$$\mathbf{B} = \sum\sqrt{\frac{4\pi c^2}{\omega_n^2 V}}(k\times e_n^\lambda)[-p_{n\lambda}Sin(k\cdot x) + \omega_n q_{n\lambda}Cos(k\cdot x)]$$

(we have written k for $\mathbf{k} \sim \mathbf{k_n}$, x for $\mathbf{x}$, e_n^λ for $\mathbf{e}_n^\lambda$, n for $\mathbf{n} \sim (n_1, n_2, n_3)$, etc. and one is thinking of a reference volume $V = L_1L_2L_3$ with sides L_i and $\omega_n = c|\mathbf{k_n}|$. Further $\mathbf{k_n}$ has components $k_i = (2\pi/L_i)$ and, setting $\hat{\mathbf{k}}_\mathbf{n} = \mathbf{k_n}/|\mathbf{k_n}|$, one has

$$(2.11)\quad \mathbf{k_n} \cdot e_n^\lambda = 0;\; e_n^\lambda \cdot e_n^{\lambda'} = \delta_{\lambda\lambda'};\; \hat{\mathbf{k}}_\mathbf{n} = e_n^1 \times e_n^2;\; e_n^1 = -\hat{\mathbf{k}}_n \times e_n^2;\; e_n^2 = \hat{\mathbf{k}}_\mathbf{n} \times e_n^1$$

Now the nonrelativistic Hamiltonian describing a charged particle in interaction with the radiation field is

$$(2.12)\qquad H = \frac{1}{2m}\left(\mathbf{p} - \frac{e}{c}\mathbf{A}\right)^2 + e\Phi + \frac{1}{8\pi}\int d^3x(\mathbf{E}^{\perp 2} + \mathbf{B}^2)$$

Here $\mathbf{E} = \mathbf{E}^\perp + \mathbf{E}^{||}$ and the contribution from the longitudinal part has been written as the Coulomb potential $e\Phi$; in the Coulomb gauge $\mathbf{E}^{||} = -\nabla\Phi$ so

(2.13)

$$\int d^3\mathbf{E}^{||2} = \int d^3x(\nabla\Phi)^2 = \int d^3x\nabla\cdot(\Phi\nabla\Phi) - \int d^3x\Phi\nabla^2\Phi = 4\pi\int d^3x\rho(x)\Phi$$

since $\nabla^2\Phi = -\nabla\cdot\mathbf{E}^{||} = -4\pi\rho(x)$ (Gauss law). From this one obtains the equations of motion for the particle ($x \sim \mathbf{x}$)

$$(2.14)\qquad m\dot{x} = \mathbf{p} - \frac{e}{c}\mathbf{A};\; \dot{\mathbf{p}} = \frac{e}{c}[\dot{x} \times \mathbf{B} + (\dot{x}\cdot\nabla)\mathbf{A}] - e\nabla\Phi$$

and also the following Maxwell equations in the presence of the charge and its current

(2.15)

$$\nabla\cdot\mathbf{D} = 4\pi\rho;\; \nabla\times\mathbf{H} = \frac{4\pi}{c}\mathbf{J} + \frac{1}{c}\frac{\partial\mathbf{D}}{\partial t};\; \rho = e\delta^3(x - x_p(t));\; \mathbf{J} = e\dot{x}_p\delta^3(x - x_p(t))$$

where $x_p(t)$ is the actual position of the particle. Combining these equations one obtains the Newton second law with the Lorentz force

$$(2.16)\qquad m\ddot{x} = e\mathbf{E} + \frac{e}{c}\dot{x}\times\mathbf{B}$$

Some simplification arises if one expresses matters via field modes $(q_\alpha(t), p_\alpha(t))$ and $(x(t), p(t))$ in the form ($p \sim \mathbf{p}$, $x \sim \mathbf{x}$, etc. with $e' = e\sqrt{4\pi/V}$)

$$(2.17)\quad m\dot{x} = p - \frac{e}{c}\mathbf{A};\; \dot{p} = F + e'\sum(\dot{x}\cdot e_\alpha)k\left(q_\alpha cos(k\cdot x) - \frac{p_\alpha}{\omega_\alpha}Sin(k\cdot x)\right);$$

$$\dot{q}_\alpha = p_\alpha - e'(\dot{x}\cdot e_\alpha)\frac{1}{\omega_\alpha}Cos(k\cdot x);\; \dot{p}_\alpha = -\omega_\alpha^2 q_\alpha + e'(\dot{x}\cdot e_\alpha)Sin(k\cdot x)$$

where $F = -\nabla(e\Phi)$ (without the Coulomb self-interaction). Write now

$$(2.18)\qquad c_\alpha(t) = \frac{1}{\sqrt{2E_\alpha}}(p_\alpha - i\omega_\alpha q_\alpha);\; c_\alpha^* = \frac{1}{\sqrt{2E_\alpha}}(p_\alpha + i\omega_\alpha q_\alpha)$$

Some calculation gives ($x' = x(t')$)

$$(2.19)\qquad c_\alpha(t) = a_\alpha(t) - ie\sqrt{\frac{2\pi}{E_\alpha V}}e^{-i\omega_\alpha t}\int_0^t(\dot{x}'\cdot e_\alpha)e^{-ik\cdot x' + i\omega_\alpha t'}dt'$$

Combining (2.17) and (2.19) one gets the equation of motion $m\ddot{x} = F + F_{Lor}^{free} + F_{self}$ where

(2.20) $$F_{Lor}^{free} = ie\sum\sqrt{\frac{2\pi E_\alpha}{V}}\left(e_\alpha + \frac{\dot{x}}{c}\times(\hat{k}\times e_\alpha)\right)\times(a_\alpha e^{-ik\cdot x} - a_\alpha^* e^{ik\cdot x});$$

$$F_{self} = -\frac{4\pi e^2}{V}\sum\left(e_\alpha + \frac{\dot{x}}{c}\times(\hat{k}\times e_\alpha)\right)\times\int dt'(\dot{x}'\cdot e_\alpha)Cos(\omega_\alpha(t'-t) - k(x'-x))$$

The self force is too complicated to handle here and some approximations are made (cf. [**1004**] for details) leading to an expression

(2.21) $$F_{self} = -\frac{4e^2}{3c^2}\left(\frac{1}{2}\dddot{x}(t) - \frac{\ddot{x}(t)}{\pi}\int_0^\infty d\omega\right) = m\tau\dddot{x}(t) - \delta m\ddot{x}(t)$$

where $\tau = (2e^2/3mc^3)$ and $\delta m = (4e^2/3\pi c^2)\int_0^\infty d\omega$. The self radiation has two effects now, within the present approximation (cf. also [**879**]). First there is a reaction force on the particle proportional to the time derivative of the acceleration (radiation reaction) and secondly there is a contribution to the term $m\ddot{x}(t)$ involving a total or dressed mass of the particle $m_T = m + \delta m$. This contribution is infinite for the point particle since $\int_0^\infty d\omega$ is divergent but a cure is to take a cutoff ω_c so that $\delta m = (2/\pi)m\tau\omega_c$. Even for huge ω_c the quantity $\delta m/m$ is smaller than 1 since τ is very small and one conjectures that in a more precise calculation the mass correction would be at most of order $\alpha = e^2/\hbar c = 1/137$ (fine structure constant). In any case one is led to the Abraham-Lorentz equation

(2.22) $$m_T\ddot{x} \equiv (m+\delta m)\ddot{x} = F + F_{Lor}^{free} + m_T\tau\dddot{x}$$

Note that the self field terms have this simple form only in free space. In any event such an equation is beset with problems; as an example one looks at a homogeneous time dependent force $F(t)$ and the equation $m\ddot{x} = F(t) + m\tau\dddot{x}$ or $a - \tau a = F(t)/m$ with solution

(2.23) $$a(t) = e^{t/\tau}\left(a(0) - \frac{1}{m\tau}\int_0^t e^{t'/\tau}F(t')dt'\right)$$

Problems with acausality arise and one comes to the conclusion that there seems to be no (classical or relativistic) equation of motion for a radiating particle in interaction with the radiation field that is free of conceptual difficulties.

A variation on the Abraham-Lorentz equation in the form

(2.24) $$m\ddot{x} = -m\omega_0^2 x + m\tau\dddot{x} + eE_x(x,t) + e\left(\frac{\dot{x}}{c}\times\mathbf{B}\right)$$

is the Braffort-Marshall equation; it is the analogue of the Langevin equation in Brownian motion (cf. here one refers back to the form $m\ddot{x} = F + F_{Lor}^{free} + F_{self}$ with $m \sim m_T$ and $\tau = 2e^2/3mc^3$). Upon approximating $\dddot{x}$ by$-\omega_0^2\dot{x}$ and linearizing one has something tractable but we do not discuss this here. There is also considerable discussion of harmonic oscillators and Fokker-Planck equations which we omit here. A Braffort-Marshall equation arises again in linear SED in the form

(2.25) $$m\ddot{x} = m\tau\dddot{x} + F(x) + eE(t)$$

(1-D suffices here and one observes on p.303 of [**1004**] that SED and QM are incompatible theories). This is subsequently modified to

$$m\ddot{x} = m\tau\dddot{x} + F(x) + e\sum \tilde{E}_k a_k^0 e^{-i\omega t} + c.c. \tag{2.26}$$

Here one has started from a standard Fourier representation of the ZPF in the form

$$\mathbf{E} = \sum \tilde{E}_k a_k e^{-i\omega_n t} + c.c.;\ a_k^0 \sim e^{i\phi_k};\ a_k \to a_k^0 \sim c \to \infty \tag{2.27}$$

with random phases ϕ_k uniformly distributed over $(0, 2\pi)$ (x here describes the response of the particle to the effective field). A lot of partial averaging has gone into this and we refer to [**1004**] for details and discussion. Even for simple examples the calculations are a kind of horror story!

Let us try to summarize now some of [**1005**]. One recalls that the central premise of SED is that the quantum behavior of the particle is a result of its interaction with the vacuum radiation field or ZPF. This field is assumed to pervade the space and is considered here to be in a stationary state with well defined stochastic properties. Its action on the particle is to impress upon it at every point a stochastic motion with an intensity characterized by Planck's constant which is a measure of the magnitude of fluctuations of the vacuum field. One begins with the Braffort-Marshall equation in the form

$$m\ddot{\mathbf{x}} = \mathbf{f}(\mathbf{x}) + m\tau\dddot{\mathbf{x}} + e\mathbf{E}(t) \tag{2.28}$$

where $\tau = 2e^2/3mc^3 \sim 10^{-23}$ for the electron. The term $e\mathbf{E}(t)$ stands for the electric force exerted by the ZPF on the particle (the magnetic term is omitted since one deals here with the nonrelativistic case). LSED is now to be based on three principles:

(1) Principle 1. **The system under study reaches an equilibrium state at which the rate of energy radiated by the particle equals the average rate of energy absorbed by it from the field.** To make this quantitative multiply (2.28) by $\dot{\mathbf{x}}$ to obtain

$$\left\langle \frac{dH}{dt} \right\rangle = -m\tau\left\langle \ddot{\mathbf{x}} \right\rangle + e\left\langle \dot{\mathbf{x}} \cdot \mathbf{E} \right\rangle \tag{2.29}$$

where H is the particle Hamiltonian including the Schott energy

$$H = \frac{1}{2}m\dot{\mathbf{x}}^2 + V(\mathbf{x}) - m\tau\dot{\mathbf{x}} \cdot \ddot{\mathbf{x}} \tag{2.30}$$

and V is the potential associated to the external force $\mathbf{f}$. The average is over the realizations of the background ZPF and when the system has reached the state of energetic equilibrium we have

$$\left\langle \frac{dH}{dt} \right\rangle = 0 \Rightarrow m\tau\left\langle \ddot{\mathbf{x}}^2 \right\rangle = e\left\langle \dot{\mathbf{x}} \cdot \mathbf{E} \right\rangle \tag{2.31}$$

When this equilibrium is reached (or nearly so) one says the system has reached the quantum regime. In LSED it is claimed that detailed energy balance will hold (i.e. for every frequency). One sees that at equilibrium the term $< \ddot{\mathbf{x}} >$ is determined by the ZPF and hence the acceleration itself should be determined by the field.

(2) Principle 2. **Once the quantum regime has been attained the vacuum field has gained control over the motion of the material part of the system.** To apply this principle consider the free particle with $m\ddot{x} = m\tau\dddot{x} + eE(t)$ and express the field via

$$E(t) = \sum \tilde{E}_\beta a_\beta e^{i\omega_\beta t} = \sum_{\omega_\beta > 0} (\tilde{E}_\beta^+ a_\beta e^{i\omega_\beta t} + \tilde{E}_\beta^- a_\beta^* e^{-i\omega_\beta t}) \tag{2.32}$$

The $a_\beta = a_\beta(\omega_\beta)$ are stochastic variables and the present approach leaves them momentarily unspecified. The amplitudes $\tilde{E}_\beta$ will be selected to assign to each mode of the field the mean energy $E_\beta = (1/2)\hbar\omega_\beta$ and this is the unique door through which Planck's constant enters the theory. Write now

$$E_\beta = \frac{1}{2} < p_\beta^2 + \omega_\beta^2 q_\beta^2 >; \; p_\beta = \sqrt{\frac{E_\beta}{2}}(a_\omega + a_\omega^*); \; i\omega_\beta q_\beta = \sqrt{\frac{E_\beta}{2}}(a_\omega - a_\omega^*) \tag{2.33}$$

The solution to $m\ddot{x} = m\tau\,[\dddot{x}] + eE(t)$ is then

$$x(t) = \sum \tilde{x}_\beta a_\beta e^{i\omega_\beta t} = -\sum \frac{e\tilde{E}_\beta a_\beta}{m\omega_\beta^2 + im\tau\omega_\beta^3} e^{i\omega_\beta t} \tag{2.34}$$

Here all quantities except the a_β are "sure" numbers but upon introduction of an external force $f(x)$ these parameters become in principle stochastic variables. Indeed from (2.28)

$$\sum \left(-m\omega_\beta^2 \tilde{x}_\beta - im\tau\omega_\beta^2 \tilde{x}_\beta + \frac{\tilde{f}_\beta}{a_\beta} \right) = e \sum \tilde{E}_\beta a_\beta e^{i\omega_\beta t} \tag{2.35}$$

For a generic force the Fourier coefficients $\tilde{f}_\beta$ will be a complicated function of $(\tilde{x}_\beta)$ and (a_β). Writing now

$$\tilde{x}_\beta = -\frac{e\tilde{E}_\beta}{m\omega_\beta^2 + im\tau\omega_\beta^3 + (\tilde{f}_\beta/\tilde{x}_\beta a_\beta)} \tag{2.36}$$

and putting this into (2.34) one gets

$$x(t) = -\sum \frac{e\tilde{E}_\beta a_\beta}{m\omega_\beta^2 + im\tau\omega_\beta^3 + (\tilde{f}_\beta/\tilde{x}_\beta a_\beta)} \tag{2.37}$$

The problem of determining $x(t)$ in general seems impossible. One tries now to simplify matters by looking for stable "orbits" and this leads to

(3) Principle 3. **There exist states of matter (quantum states) that are unspecific to (or basically independent of) the particular realization of the ZPF.** The ensuing calculations in [**1005**] seem to be somewhat mysterious and we refer to this paper for further discussion (cf. also [**588, 1038, 1144**]).

REMARK 5.2.1. We refer here to [**220, 465, 466, 467, 468**]) where, following Feynman, the idea is to introduce QM via the relativistic theory of free photons. The arguments are very physical and historically based which would make a welcome complement to the mainly mathematical features of the rest of this book and there is a lovely interplay of physical ideas. ■

3. PHOTONS AND EM

We replace here Section 5.3 of [**254**] by an expanded treatment taken from [**256**]. Let us begin with [**1236**] (which is also sketched in [**254**] in a somewhat different manner) and will provide a related description following [**256**]. We will also examine further various points of view concerning the massless Klein-Gordon (KG) equation, the SE, the Maxwell equations (ME), and the quantum vacuum. For background we mention here [**58, 85, 254, 314, 387, 441, 442, 489, 522, 584, 588, 617, 656, 655, 701, 748, 783, 879, 911, 1004, 1005, 1046, 1048, 1094, 1092, 1142, 1230**]. One takes massless photons as objects with energy E, momentum $\mathbf{P}$, and internal angular momentum (or spin) $\mathbf{S}$ with $E = c|\mathbf{P}|$ and $\mathbf{S}\times\mathbf{P} = 0$. It is presumed to have velocity c in the direction $\mathbf{k}$ and to spin in a plane perpendicular to $\mathbf{k}$, which is spanned by two vectors $\mathbf{e}$ and $\mathbf{b}$ where

$$\mathbf{k}\cdot\mathbf{e} = \mathbf{k}\cdot\mathbf{b} = 0;\ \mathbf{k}\times\mathbf{e} = \mathbf{b};\ \mathbf{k}\times\mathbf{b} = -\mathbf{e};\ \mathbf{e}| = |\mathbf{b}|;\ \mathbf{e}\cdot\mathbf{b} = 0 \tag{3.1}$$

One sets $\omega = e = b$ (frequency) and $E = \hbar\omega$ historically (with $|\mathbf{S}| = \pm\hbar$) while $\lambda = 2\pi c/\omega$ (which will eventually be identified with a wave length). The photon is considered as following a right of left handed helix generated by the tip of $\mathbf{e}$ where the plane of $\mathbf{e}$, $\mathbf{b}$ moves along the direction $\mathbf{k}$ with velocity c. These objects are exhibited via a photon tensor

$$f^{\mu\nu} = \begin{pmatrix} 0 & e_1 & e_2 & e_3 \\ -e_1 & 0 & b_3 & -b_2 \\ -e_2 & -b_3 & 0 & b_1 \\ -e_3 & b_2 & -b_1 & 0 \end{pmatrix} \tag{3.2}$$

which is <u>not</u> a field like the EM tensor $F^{\mu\nu}$ (see Section 5 for more comments on the tensor nature of $f^{\mu\nu}$). The dual tensor is

$$f^{*\mu\nu} = \frac{1}{2}\epsilon^{\mu\nu\sigma\rho}f_{\sigma\rho} = \begin{pmatrix} 0 & -b_1 & -b_2 & -b_3 \\ b_1 & 0 & e_3 & -e_2 \\ b_2 & -e_3 & 0 & e_1 \\ b_3 & e_2 & -e_1 & 0 \end{pmatrix} \tag{3.3}$$

with $f^{\mu\nu}f_{\mu\nu} = 2(e^2 - b^2) = 0$ and $f^{\mu\nu}f^*_{\mu\nu} = -4\mathbf{e}\cdot\mathbf{b} = 0$. One works here in a Hilbert space $H = H^S\otimes H^K$ with $\mathbf{S}\sim\mathbf{S}\otimes 1$, $\mathbf{P}\sim 1\otimes\mathbf{P}$, and $\mathbf{R}\sim 1\otimes\mathbf{R}$. Now spin is colinear with momentum (recall $\mathbf{S}\times\mathbf{P} = 0$) and the spin eigenstates $\chi_\pm$ correspond to helicities ± 1 satisfying

$$(\mathbf{k}\cdot\mathbf{S})\chi_\pm = \pm\hbar\chi_\pm \tag{3.4}$$

where $\mathbf{k}$ is a unit vector in the direction of $\mathbf{P}$. The spin operators will be expressed via

$$(3.5)\quad S_x = \hbar\begin{pmatrix} 0 & 0 & 0\\ 0 & 0 & -i\\ 0 & i & 0\end{pmatrix};\ S_y = \hbar\begin{pmatrix} 0 & 0 & i\\ 0 & 0 & 0\\ -i & 0 & 0\end{pmatrix};\ S_z = \hbar\begin{pmatrix} 0 & -i & 0\\ i & 0 & 0\\ 0 & 0 & 0\end{pmatrix}$$

with $(S_j)_{k\ell} = -i\hbar\epsilon_{jk\ell}$. One must distinguish here $\mathbf{k} \in H^K$ and $\mathbf{S} \in H^S$; the 2-dimensional spin space is orthogonal to $\mathbf{k}$ with $\mathbf{k}\cdot\mathbf{S} \sim \pm\hbar$ as indicated in (1.4). Now write $\psi_j \in H^S \otimes H^K$ with $j = 1, 2, 3$ denoting components in H^S and set $\mathbf{k} = \mathbf{p}/|\mathbf{p}|$. An operator leaving invariant a photon state $\chi^k_\pm\otimes\phi_p$ is $\mathbf{S}\cdot\mathbf{P} = \mathbf{k}\cdot\mathbf{S}\otimes|\mathbf{P}|$ where $|\mathbf{P}|\phi_p = |\mathbf{p}|\phi_p$ (with $E = c|\mathbf{p}| = c|p|$) and one has

$$(3.6)\qquad \mathbf{S}\cdot\mathbf{P}\chi^k_\pm \otimes \phi_p = \pm\frac{\hbar E}{c}\chi^k_\pm \otimes \phi_p$$

(the Hamiltonian is $H = (c/\hbar)\mathbf{S}\cdot\mathbf{P}$ and a minus sign should be interpreted as positive energy but negative helicity). Then the time evolution of a general photon state is

$$(3.7)\qquad i\hbar\partial_t\psi_j = (H)_{jk}\psi_k = \frac{c}{\hbar}(\mathbf{S}\cdot\mathbf{P})_{jk}\psi_k$$

(H is a 3×3 matrix in H^S whose components are operators in H^K). Putting $\mathbf{P} = -i\hbar\nabla$ one has a SE for the photon, namely

$$(3.8)\qquad \frac{i}{c}\partial_t\psi_j(t,\mathbf{r}) = -\epsilon_{jk\ell}\partial_\ell\psi_k(t,\mathbf{r})$$

(since $(c/\hbar)(-i\hbar\epsilon_{jk\ell})(-i\hbar\partial_\ell) = -c\hbar\epsilon_{jk\ell}\partial_\ell$). Note $\hbar$ has disappeared and although this is a QM equation it does not have a classical limit.

REMARK 5.3.1. It is pointed out in de la Torre-Daleo [**1237**] that there are conceptual errors in writing $\psi_j = E_j + iB_j$ and deriving the Maxwell equations via $(i/c)\partial_t(E_j + iB_j) = -\epsilon_{jk\ell}\partial_\ell(E_k + iB_k)$ in the form

$$(3.9)\qquad \frac{1}{c}\partial_t E_j = -\epsilon_{jk\ell}\partial_\ell B_k;\ \frac{1}{c}\partial_t B_j = \epsilon_{jk\ell}\partial_\ell E_k$$

(e.g. $(1/c)\partial_t E_1 = -\epsilon_{123}\partial_3 B_2 - \epsilon_{132}\partial_2 B_3 = \partial_2 B_3 - \partial_3 B_2$, etc. - note $\epsilon_{jk\ell} = -\epsilon_{j\ell k}$ in Tiwari [**1229**]). The equations are correct but the derivation is faulty since it identifies the 3-D space of states with the 3-D physical space! ■

REMARK 5.3.2. Using the momentum representation one can write as in [**1237**]

$$(3.10)\qquad \frac{\hbar}{c}\partial_t\psi_j(t,\mathbf{p}) = \epsilon_{jk\ell}p_\ell\psi_k(t,\mathbf{p})$$

but this is not $\vec{\psi}\times\vec{p}$ because the two vectors belong to different spaces. One can look also at stationary state solutions $\psi_j = exp[-(i/\hbar)Et)\Phi_{j,E}$ where $(\mathbf{k} = \mathbf{p}/|\mathbf{p}|)$

$$(3.11)\qquad (\mathbf{S}\cdot\mathbf{P})_{jk}\Phi_{k,E} = \frac{E\hbar}{c}\Phi_{j,E};\ \Phi_{j,E}(\mathbf{p}) = \chi^k_\pm\otimes\delta\left(|\mathbf{p}| - \frac{E}{c}\right)$$

For the corresponding position representation one would use $\chi_{\pm}^{k_0}\otimes\phi_{p_0}$ where

$$\phi_{p_0}(\mathbf{r})=\frac{1}{\sqrt{2\pi\hbar}^3}exp\left(\frac{i}{\hbar}\mathbf{p}_0\cdot\mathbf{r}\right) \tag{3.12}$$

(where $\mathbf{k}_0=\mathbf{p}_0/|\mathbf{p}_0|$. ■

3.1. THE EM FIELDS. In the last paper of [**1237**] it is shown how to construct the EM fields from knowledge of photons. First one defines

$$\epsilon_+=\frac{1}{\sqrt{2}}(\hat{\mathbf{e}}+i\hat{\mathbf{b}});\ \epsilon_-=\frac{1}{\sqrt{2}}(i\hat{\mathbf{e}}+\hat{\mathbf{b}}) \tag{3.13}$$

where $\mathbf{e}=\omega\hat{\mathbf{e}}$ and $\mathbf{b}=\omega\hat{\mathbf{b}}$. Then write

$$\mathbf{e}_+(t)=\left(\frac{\omega}{\sqrt{2}}\epsilon_+e^{-i\omega t}+c.c.\right);\ \mathbf{e}_-(t)=\left(\frac{\omega}{\sqrt{2}}\epsilon_-e^{-i\omega t}+c.c.\right) \tag{3.14}$$

One writes $\mathbf{e}_s(t)=[(\omega/\sqrt{2})\epsilon_s exp(-i\omega t)+c.c.)$ and $\mathbf{b}_s(t)=\mathbf{k}\times\mathbf{e}_s(t)$, uses momentum eigenfunctions as in Remark 5.3.2, and writes $\phi_{s,p}=\chi_s^k\otimes\phi_p$. For a state with n photons having helicity s_j and momentum p_j annihilation and creation operators are defined via

$$a_s^{\dagger}(p)\phi_{s_1p_1,\cdots,s_np_n}=\sqrt{n+1}\phi_{sp,s_1p_1,\cdots,s_np_n}; \tag{3.15}$$

$$a_s(p)\phi_{s_1p_1,\cdots,s_np_n}=\frac{1}{\sqrt{n}}\sum_1^n\delta_{s,s_i}\delta(\mathbf{p}-\mathbf{p}_i)\phi_{s_1p_1,\cdots,\widehat{s_ip_i},\cdots,s_np_n}$$

A vacuum state ϕ_0 with zero photons is defined via $a_s(p)\phi_0=0$ and n-photon states are built up via

$$\phi_{s_1p_1,\cdots,s_np_n}=\frac{1}{\sqrt{n!}}a_{s_1}^{\dagger}(p_1)\cdots a_{s_n}^{\dagger}(p_n)\phi_0 \tag{3.16}$$

A number operator is defined via $N_s(p)=a_s^{\dagger}(p)a_s(p)$ and $N=\sum_s\int d^3pN_s(p)$. The total energy, momentum, and spin of a system of photons (each with energy $E=c|p|=\hbar\omega$ and spin $\pm\hbar$) is then

$$H=\sum_s\int d^3p\hbar\omega N_s(p);\ \mathbf{P}=\sum_s\int d^3p\mathbf{p}N_s(p); \tag{3.17}$$

$$\mathbf{S}=\int d^3p\hbar\mathbf{k}(N_+(p)-N_-(p))$$

One defines then Hermitian operators

$$\mathbf{E}(\mathbf{r},t)=\frac{1}{2\pi\hbar}\sum_s\int d^3p\sqrt{\omega}\left(ia_s(p)\epsilon_se^{(i/\hbar)(\mathbf{p}\cdot\mathbf{r}-Et)}+h.c.\right); \tag{3.18}$$

$$\mathbf{B}(\mathbf{r},t)=\frac{1}{2\pi\hbar}\sum_s\int d^3p\sqrt{\omega}\left(ia_s(p)(\mathbf{k}\times\epsilon_s)e^{(i/\hbar)(\mathbf{p}\cdot\mathbf{r}-Et)}+h.c.\right);$$

$$\mathbf{A}(\mathbf{r},t)=\frac{c}{2\pi\hbar}\sum_s\int d^3p\frac{1}{\sqrt{\omega}}\left(a_s(p)\epsilon_se^{(i/\hbar)(\mathbf{p}\cdot\mathbf{r}-Et)}+h.c.\right)$$

Then

$$\mathbf{E}=-\frac{1}{c}\partial_t\mathbf{A};\ \mathbf{B}=\nabla\times\mathbf{A};\ H=\frac{1}{8\pi}\int d^3r(\mathbf{E}^2+\mathbf{B}^2); \tag{3.19}$$

$$\mathbf{P} = \frac{1}{8\pi c}\int d^3r(\mathbf{E}\times\mathbf{B} - \mathbf{B}\times\mathbf{E};\ \mathbf{S} = \frac{1}{8\pi c}\int d^3r(\mathbf{E}\times\mathbf{A} - \mathbf{A}\times\mathbf{E})$$

and one checks the Maxwell equations

$$-\nabla\times\mathbf{E} = \frac{1}{c}\partial_t\mathbf{B};\ \nabla\times\mathbf{B} = \frac{1}{c}\partial_t\mathbf{E};\ \nabla\cdot\mathbf{E} = \nabla\cdot\mathbf{B} = 0 \tag{3.20}$$

Thus photons are posited as the fundamental objects and they generate EM fields as a collective manifestation.

Next one defines the "singular" function (cf. [**1237**] for details)

$$D(\vec{\rho},\tau) = \frac{-1}{(2\pi\hbar)^3}\int d^3p e^{(i/\hbar)\mathbf{p}\cdot\vec{\rho}}\frac{Sin(\omega\tau)}{\omega} = \tag{3.21}$$

$$= \frac{-1}{8\pi^2 c\rho}[\delta(\rho - c\tau) - \delta(\rho + c\tau)]$$

Here $\rho = |\vec{\rho}|$ where $\vec{\rho}\sim\mathbf{r}_1 - \mathbf{r}_2$ and one can say that $D(\vec{\rho},\tau)$ has support on the light cone (cf. also [**855**]). This leads to

$$[E_i(\mathbf{r}_1,t_1),E_j(\mathbf{r}_2,t_2)] = -4\pi i\hbar c^2\left(\frac{\delta_{ij}}{c^2}\partial_{t_1}\partial_{t_2} + \partial_{r_1,i}\partial_{r_2,j}\right)D(\mathbf{r}_1-\mathbf{r}_2,t_1-t_2) \tag{3.22}$$

$$[B_i(\mathbf{r}_1,t_1),B_j(|bfr_2,t_2)] = -4\pi i\hbar c^2\left(\frac{\delta_{ij}}{c^2}\partial_{t_1}\partial_{t_2} + \partial_{r_1,i}\partial_{r_2,j}\right)D(\mathbf{r}_1 - \mathbf{r}_2,t_1 - t_2)$$

$$[E_i(\mathbf{r}_1,t_1),B_j(\mathbf{r}_2,t_2)] = 4\pi i\hbar c\epsilon_{ijk}\partial_{t_1}\partial_{r_1,k}D(\mathbf{r}_1 - \mathbf{r}_2,t_2 - t_1)$$

Note that the singular nature of D is really unacceptable in QM (e.g. because of the uncertainty principle) and one could conclude that the field strengths are not measurable quantities (cf. the first paper in [**1237**]). On the other hand field averages can be accepted in QM. This is one feature leading to the approach in [**1237**] based on the photon as fundamental. The EM fields are considered essentially as a classical macroscopic ideas and are not "basic". Such an argument might be extendable quite generally to cast suspicion on many results involving singular behavior or generalized solutions of partial differential equations (distributions). The "classical" theory might require e.g. averaging of dependent variables or some new physics (not necessarily QM) in order to retain any meaning.

One looks next at the expectation values of fields in the quantum state describing a system of photons. For the vacuum described via ϕ_0 one has

$$<\phi_0,E(\mathbf{r},t)\phi_0> = <\phi_0,\mathbf{B}(\mathbf{r},t)\phi_0> = 0 \tag{3.23}$$

as expected. However one can show that e.g.

$$<\phi_0,\mathbf{E}^2(\mathbf{r},t)\phi_0> = \frac{2}{(2\pi\hbar)^2}\int d^3p\,\omega \tag{3.24}$$

indicating that there are fluctuations of the electric field in vacuum. For a quantum state of n photons in the same state with fixed helicity and momentum one has (cf. [**1237**])

$$\phi = \phi_{n(s_1p_1)} = \frac{1}{\sqrt{n!}}(a^\dagger_{s_1}(\mathbf{p}_1))^n\phi_0;\ <\phi,\mathbf{E}(\mathbf{r},t)\phi> = <\phi,\mathbf{B}(\mathbf{r},t)\phi> = 0 \tag{3.25}$$

which is somewhat strange. However for an indefinite number of photons in a superposition of states $\psi = \sum_n C_n \phi_{n(s_1 p_1)}$ one has

$$(3.26) \qquad <\psi, \mathbf{E}(\mathbf{r},t)\psi> = \frac{\sqrt{\omega_1}}{2\pi\hbar}\left(i\sum_n C_n^* C_{n+1}\epsilon_{s_1} e^{(i/\hbar)(\mathbf{p}_1\cdot\mathbf{r}-E_1 t)} + c.c.\right)$$

One concludes here that the EM field of an indefinite number of photons all with the same helicity and momentum is a plane wave with circular polarization. The quantum state where all photons are in the same one photon state of fixed helicity and momentum apparently is a Bose-Einstein condensate.

REMARK 5.3.3. We extract here from Nair [**911**] for a few philosophical observations. The photon, as an elementary "particle" is unique; it is the only elementary particle of energy (cf. [**569, 814, 1290, 1291**] for other approaches to photons and light - this is discussed briefly in Chapter 8). A relativistic energy equation should be $E^2 = p^2c^2 + m_0c^2 = p^2c^2$ since the rest mass $m_0 = 0$. In the frame of the moving photon the photon's energy is stored as rotational (spin) energy where $E = \hbar\nu = \hbar c/\lambda$ with ν the frequency and λ the wave length. Hence the greater the energy the smaller the wave length and one expects to find a lower bound for the wavelength. For a "particle" the angular momentum is $L = mrw$ limited by $L = mrc$ and replacing L by the spin S one has (♠) $\hbar = mrc$ where m is a putative mass presumably "generated" by the spin (see here also [**1237**] for toy models with extended energy distributions). Assume the concept of Schwartzschild radius R is valid for the photon where for a black hole $R = 2Gm/c^2$ or (♣) $(R/m) = (2G/c^2)$. The right side of (♣) is a constant but for the photon the radius decreases as the "mass" increases; hence there is a unique value of radius and mass for which a photon can behave as a black hole. Combining (♣) with (♠) one finds (♦) $m = \sqrt{\hbar c/2G}$ for the Planck mass, which here is the maximum "pseudomass" permitted for the photon. This corresponds to a maximum energy of $mc^2 = (\sqrt{\hbar c/2G})c^2 = 8.61 \times 10^{22}$ MeV and the highest energy so far observed for a photon is apparently less than this. It is suggested that pair production or photon "splitting" will ensue at the energy limit. ■

3.2. MORE ON ZPF. This is a murky subject and essentially involves understanding the quantum vacuum, which seems very complicated. We gave some heuristic comments on ZPF in [**254**] (as in Section 5.1) based on [**85, 387, 588, 586, 589, 590, 656, 655, 879, 1004, 1005, 1046, 1048, 1094, 1092**], which upon hindsight seem woefully inadequate. Some of this is also summarized and enhanced in [**1094**] (first paper). We go here to the lovely collection of papers by J. Field (see e.g. [**465, 466, 467, 468, 469**]) for an aperçu of basic physical connections between QM, thermodynamics, and special relativity. This will serve as a complement to later Sections. We begin with [**465**] which in a sense follows the spirit of Feynman's QED where the fundamental concepts of QM are explained in terms of the interactions of photons and electrons. One recalls first the energy momentum vector $P = m(dX/d\tau) \sim ((E/c), p_x, p_y, p_z)$ with $X = (ct, x, y, z)$ and τ the proper time (time observed in the rest frame). If the inertial frame S' is moving with uniform velocity βc relative to the frame S along the common x, x'

axis with $0y$ parallel to $0y'$ then the 4-vectors as observed in S, S' are related by Lorentz transform (LT) equations

$$p'_x = \gamma(p_x - \beta p_t);\ p'_y = p_y;\ p'_z = p_z;\ p'_t = \gamma(p_t - \beta p_x); \tag{3.27}$$

$$\gamma = \frac{1}{\sqrt{1-\beta^2}};\ p_t = \frac{E}{c}$$

As $m \to 0$ P is still well defined so one has an energy momentum vector say $P_\gamma = [(E_\gamma/c), (E_\gamma/c)Cos(\phi), (E_\gamma/c)Sin(\phi), 0]$ for a photon of energy E_γ moving in the (x,y) plane in a direction making an angle ϕ with the x axis. A plane EM wave will be associated with a large number of photons in general and for such a collection, all with the same 4-vector P_γ, one finds from the LT equations an EM wave with

$$\nu' = \nu\gamma(1 - \beta Cos(\phi));\ E'_T = E_T\gamma(1 - \beta Cos(\phi)) \tag{3.28}$$

(here $\nu \sim$ frequency and $E_T \sim$ total energy). Using (3.1) the energies of the photons in the EM wave transform via $E'_\gamma = E_\gamma\gamma(1 - \beta Cos(\phi))$. If n_γ is the total number of photons then $E_T = n_\gamma E_\gamma$ which yields (3.2). Further one sees immediately that $E_\gamma/\nu = E'_\gamma/\nu' = constant$ and calling this constant $\hbar$ one finds that $E_\gamma = \hbar\nu$ which identifies $\hbar$ with Planck's constant. Consequently Planck's constant arises from consistency between the relativistic kinematics of photons, considered to be massless particles, and the relativistic Doppler effect for classical EM waves. Note also that using $\lambda = c/\nu$ and $E_\gamma = \hbar\nu = p_\gamma c$ one arrives at the deBroglie relation $p_\gamma = \hbar/\lambda = \hbar\nu/c$.

Now from the energy density of a plane EM wave, namely

$$\rho_W = \frac{\mathbf{E}^2 + \mathbf{B}^2}{8\pi} \tag{3.29}$$

the photon interpretation gives immediately Poynting's formula for the energy flow F per unit area per unit time, namely $F = c\rho_W$ as well as the formula for the radiation pressure P_{rad} of a plane wave at normal incidence on a perfect reflector, namely $P_{rad} = 2\rho_W$. Note the number of photons incident is F/E_γ per unit area per unit time so the total momentum transferred is then $p_\gamma(F/E_\gamma) = F/c = \rho_W$; but a perfect reflector will not absorb energy so an equal number of photons are re-emitted, yielding the factor of 2. Now consider a plane EM wave of wavelength λ moving in free space parallel to the positive x direction in the frame S, written as

$$E_y = E_0 e^\Phi;\ H_z = H_0 e^\Phi;\ \Phi = 2\pi i\frac{(x-ct)}{\lambda};\ E_0 = H_0 = A \tag{3.30}$$

The time averaged energy density per unit volume $\bar{\rho}_W$ is $\bar{\rho}_W = (E_0^2/8\pi) = (H_0^2/8\pi) = (A^2/8\pi)$. Assuming that the wave consists of a beam of photons of energy $\hbar\nu$ the average number density of photons $\bar{\rho}_\gamma$ in the wave is $\bar{\rho}_W/\hbar\nu$ so one gets $\bar{\rho}_\gamma = A^2/8\pi\hbar\nu$. This is the point where one now leaps across the chasm separating the classical and quantum worlds. First the use of a complex exponential to represent a classical EM wave is convenient but it is really the real or imaginary parts that come into play (Cosines and Sines); for QM the complex exponential is mandatory. Second one uses the definition of wavelength together with $E_\gamma = \hbar\nu$ and $p_\gamma = \hbar/\lambda$ to replace in the complex exponential the wave parameter λ by the particle parameters E_γ and p_γ. The parameter c is part of both descriptions (photons and EM waves) and this leads to the complex exponential describing photons in the form

$$(3.31)\qquad u_p = u_0 exp\left[\frac{2\pi i}{\hbar}(p_\gamma x - E_\gamma t)\right] = exp\left[\frac{2\pi i}{\hbar}(P\cdot X)\right];\ u_0 = \frac{A}{\sqrt{8\pi E_\gamma}}$$

In this situation $\bar{\rho}_\gamma = |u_p|^2 = u_0^2$ and for the case of very weak EM fields such that $\bar{\rho}_\gamma << 1$ it follows that $|u_p|^2 dV$ can be thought of as the probability that a photon is in the volume dV; for large numbers of photons or strong EM fields this probabilistic interpretation is not appropriate. Note also that one can write

$$(3.32)\qquad \mathcal{P}_x = -i\frac{\hbar}{2\pi}\frac{\partial}{\partial x};\ \mathcal{E} = i\frac{\hbar}{2\pi}\frac{\partial}{\partial t};\ \mathcal{P}_x u_p = p_\gamma u_p;\ \mathcal{E}u_p = E_\gamma u_p$$

Note also for f an arbitrary function on space-time

$$(3.33)\qquad \mathcal{P}_x(xf) = -\frac{i\hbar}{2\pi}\frac{\partial(xf)}{\partial x} = -\frac{i\hbar}{2\pi}f + x\mathcal{P}_x f$$

Repeated use of (3.5), (3.6), etc. and the relation $E_\gamma = p_\gamma c$ gives

$$(3.34)\qquad (c^2\mathcal{P}_x^2 - \mathcal{E}^2)u_p = \left(\frac{\hbar c}{2\pi}\right)\Box u_p = 0;\ \Box = \frac{1}{c^2}\frac{\partial^2}{\partial t^2} - \frac{\partial^2}{\partial x^2}$$

so u_p will satisfy the Maxwell-Lorentz equation $\Box u_P = 0$. If one uses instead of $E_\gamma = p_\gamma c$ the general energy momentum relation for massive particles $E^2 = p^2c^2 + m^2c^4$ one can arrive at the Klein-Gordon (KG) equation and expanding $E \simeq mc^2 + (p^2/2m) + \cdots$ the Schrödinger equation (SE) will result.

Consider next ($p_\gamma \to \hbar/\lambda$ and $E_\gamma \to \hbar c/\lambda$)

(3.35)
$$\chi = \frac{1}{2}(u_p + u_p^*) = \Re(u_p) = u_0 Cos\left[\frac{2\pi}{\hbar}(p_\gamma x - E_\gamma t)\right] \to u_0 Cos\left(\frac{2\pi(x-ct)}{\lambda}\right)$$

This equation is then a bridge back across the chasm from QM to the classical world (cf. (3.4)). Just as the quantum wave function is only meaningful in the limit of very low photon density so the function χ is meaningful only in the limit of high photon density. χ is not an eigenfunction of either E_γ or p_γ and is a real function. The time average of χ^2 is 1/2 the mean photon density $\bar{\rho}_\gamma$ and $\bar{\rho}_W = \hbar\nu\bar{\rho}_\gamma$. In a typical situation $\bar{\rho}_\gamma\Delta V$ is much larger than 1 and no probabilistic meaning can be attached to it.

We show next following Field [**465, 468, 469**] how to derive the Maxwell equations using only Coulomb's inverse square law, special relativity, and Hamilton's principle. Thus take two objects O_i of masses m_i and electric charges q_i with no external forces. The spatial distance separating them in the common center of mass frame is $\mathbf{x}_{12} = \mathbf{x}_1 - \mathbf{x}_2$. One constructs a most general Lorentz invariant Lagrangian in a nonrelativistic reference frame via ($x_i \sim \mathbf{x}_i$)

$$(3.36)\qquad L(x_1, u_1, x_2, u_2) = -\frac{m_1 u_1^2}{2} - \frac{m_2 u_2^2}{2} - \frac{j_1 \cdot j_2}{c^2\sqrt{-(x_1 - x_2)^2}}$$

where the $j_i = q_1 u_i$ are current 4-vectors; this is then put into the machinery of Hamilton's principle so that

$$\frac{d}{d\tau}\left(\frac{\partial L}{\partial u_i^\mu}\right) - \frac{\partial L}{\partial x_i^\mu} = 0;\ (i = 1, 2;\ \mu = 1, 2, 3, 4) \tag{3.37}$$

Since the Lagrangian is a Lorentz scalar this provides a description of the motion of the O_i in any inertial reference frame. Note that if one introduces a 4-vector potential $\mathbf{A}_2 = \mathbf{j}_2/cr_{12}$, $r_{12} = |\mathbf{r}_{12}|$ the standard Lorentz invariant Lagrangian, describing the motion of O_1 in the EM field created by O_2, namely $L(x_1, u_1) = -(m_1 u_1^2/2) - (1/c)q_1\mathbf{u}_1 \cdot \mathbf{A}_2$, is recovered (and similarly for motion of O_2 in the field of O_1). Now write $\partial_i = -\partial^1 \equiv (\partial/\partial x^i) \equiv \nabla_i$ and set $\mathbf{p} = m\mathbf{u}$ along with

$$E^i = \partial^i A^0 - \frac{1}{c}\frac{\partial A^i}{\partial t} = \partial^i A^0 - \partial^0 A^i;\ B^k = -\epsilon_{ijk}(\partial^i A^j - \partial^j A^i) = (\nabla \times \mathbf{A})^k \tag{3.38}$$

Some calculation gives then the 3-D Lorentz force equation and a relativistic Biot-Savart Law in the form

$$\frac{d\mathbf{p}}{dt} = q\left[\mathbf{E} + \frac{\mathbf{v}}{c} \times \mathbf{B}\right];\ \mathbf{B} = \frac{q_2\gamma_2(\mathbf{v}_2 \times \mathbf{r})}{cr^3} = \frac{\mathbf{j} \times \mathbf{r}}{cr^3}; \tag{3.39}$$

$$\mathbf{E} = \frac{j_2^0\mathbf{r}}{cr^3} - \frac{1}{c^2 r}\frac{d\mathbf{j}_2}{dt} - \frac{\mathbf{j}_2}{c^2}\frac{(\mathbf{r} \cdot \mathbf{v}_2)}{r^3}$$

where $\mathbf{r} = \mathbf{r}_{12}$. The Maxwell equations can be derived immediately from (1.59) along with the Faraday-Lenz law, Ampere's law, etc. (cf. [**465, 468, 469**] for details).

Now concerning the ZPF we collect some background information as follows.

(1) It seems well established that there is a unique Lorentz invariant spectral energy density in the EM vacuum of the form $\rho(\omega) = \rho_0(\omega) = \hbar\omega^2/2\pi^2 c^3$ (cf. Bacciagaluppi, Boyer, and Milonni [**85, 175, 879**]). An observer moving with constant velocity in the EM vacuum perceives no force.

(2) Following [**365, 1248**] an object undergoing uniform constant acceleration a in the vacuum perceives himself to be immersed in a thermal bath at temperature $T = \hbar a/2\pi kc$ ($k \sim$ Boltzman constant).

(3) One recalls also that there is a zero point energy $(1/2)\hbar\omega$ attached to a quantum harmonic oscillator. Also since there are $(\omega^2/2\pi^2 c^3)d\omega$ field nodes per unit volume in the frequency interval $[\omega, \omega + d\omega]$ one obtains the spectral density $\rho_0(\omega) = \hbar\omega^4/2\pi^2 c^3$ of Item 1 (cf. [**879**]).

(4) In [**85, 175**] one derives the Planck radiation law for the blackbody spectrum without the formalism of quantum theory. It is assumed only that (i) There is classical, homogeneous, and fluctuating EM EM radiation at absolute zero with Lorentz invariant spectrum. (ii) Classical EM theory holds for a dipole oscillator. (iii) A free particle in equilibrium with blackbody radiation has classical kinetic energy $(1/2)kT$ per degree of freedom. This leads then to the zero point energy density shown above and to Planck's formula

$$\rho(\omega, T) = \frac{\omega^2}{\pi^2 c^3}\left[\frac{\hbar\omega}{exp[(\hbar\omega/kT) - 1]} + \frac{1}{2}\hbar\omega\right] \tag{3.40}$$

If the zero point energy is ignored one obtains the Rayleigh-Jeans formula

$$\rho(\omega,T)=\left(\frac{\omega^2}{\pi^2c^3}\right)kT \tag{3.41}$$

Here (the quantum number) $\hbar$ arises in (1.61) as a linear factor in calculating the Lorentz invariant spectral density and can later be identified with Planck's constant (so the derivation is classical).

(5) Going again to [**85, 175**] one finds a lovely discussion involving entropy and energy fluctuations following and modifying Einstein's arguments. Thus one considers a cavity containing thermal radiation separated into large and small volumes V and $\mathcal{V}$. The energy $\mathcal{U}$ of EM radiantion in $\mathcal{V}$ between frequencies ω and $\omega+d\omega$ undergoes spontaneous fluctuations creating a change in the corresponding entropy. Let Σ (resp. $\mathfrak{S}$) be the entropy contributed between ω and $\omega+d\omega$ for V (resp. ($\mathcal{V}$). Then for ϵ the entropy fluctuation in $\mathcal{V}$

$$S(\epsilon)=\Sigma+\mathfrak{S}=\Sigma_0+\mathfrak{S}_0+(\partial_\epsilon\Sigma+\partial_\epsilon\mathfrak{S})\epsilon+\frac{1}{2}\left(\frac{\partial^2\Sigma}{\partial\epsilon^2}+\frac{\partial^2\mathfrak{S}}{\partial\epsilon^2}\right)\epsilon^2+\cdots \tag{3.42}$$

where $\Sigma_0,\ \mathfrak{S}_0$ signify equilibrium entropies where the fluctuation is zero. The first derivatives vanish at $\epsilon=0$ and if $V>>\mathcal{V}$ one finds $S(\epsilon)\simeq\Sigma_0+\mathfrak{S}_0+(1/2)(\partial^2\mathfrak{S}/\partial\mathcal{U}^2)\epsilon^2$. Now there is probabilistic entropy (♣) $S_{prob}=(S_{prob})_0+klog(W)$ (or $W=cexp(S_{prob}/k)$) where W is the number of microstates giving the same macrostate. There is also caloric entropy S_{cal} where $dS_{cal}=dQ/T$ for reversible processes. Then write

$$dW=cexp\left[\frac{S_{prob}}{k}\right]d\epsilon=\hat{c}exp\left[\frac{1}{2k}\frac{\partial^2S_{prob}}{\partial\mathcal{U}^2}\epsilon^2\right]d\epsilon \tag{3.43}$$

Some classical argument (cf. [**85, 175**]) involving $<\epsilon^2>=\int\epsilon^2dW\sim(\pi^2c^3/\omega^2)\rho^2d\omega$, $\mathcal{U}=\rho d\omega$, and $\partial^2\mathfrak{S}_{prob}/\partial\mathcal{U}^2=-k/<\epsilon^2>$ leads then to (♠) $\partial^2S_{prob}/\partial E^2=-(k/E^2)$ for average oscillator energy E. Note in fact directly from the definition (♣) one has $\partial S_{prob}/\partial E=k/E$ leading to (♠). Now Einstein assumed that $S_{prob}=S_{cal}$ in (♠) and produced $E=kT$ along with the Planck formula (cf. (1.61)) $E=\hbar\omega/[exp(\hbar\omega/kT)-1]$ (with the zero point term missing). Note here (using (4.14)) that the average energy of an oscillator is

$$<\epsilon>=\frac{\pi^2c^3}{\omega^2}\rho(\omega,T)=\frac{\hbar\omega}{exp(\hbar\omega/kT)-1}+\frac{1}{2}\hbar\omega=E \tag{3.44}$$

Now Boyer modifies Einstein's argument in a way which recovers the zero point term (and (4.18)). Indeed he writes $<\epsilon^2>=<\epsilon^2>_{ZPF}+<\epsilon^2>_{cal}$ and finds that

$$\frac{\partial^2S_{cal}}{\partial E^2}=-\frac{k}{E^2-(\hbar\omega/2)^2} \tag{3.45}$$

leading to (1.65).

3.3. MORE ON PHOTONS. We cite here [**135, 138, 151, 330, 514, 652, 653, 847, 1180**] for some interesting developments concerning the localization of photons and their structure. One knows of course that the methods of quantum field theory (QFT) work for a description of photon activity but we want to examine more direct connections to EM fields, Maxwell's equations, and wave-particle duality. First from [**1180**] one argues that a photon wave function can be introduced if one is willing to redefine in a physically meaningful manner what one wishes to mean by such a wave function. First one introduces a naive single photon wave function. Then one produces a second quantized many photon theory approached via many particle physics (which will correspond to the quantization of the free radiation field) and then recovers the naive single photon wave function by looking at the manifold of one photon states (cf. also [**151**]).

Now photons can be of positive or negative helicity and being massless one has $E = cp$ where $p = |\mathbf{p}|$. If one introduces probability amplitudes for photons of momentum $\mathbf{p}$ and helicity $\pm$, namely $\gamma_\pm(\mathbf{p}, t)$ which would be expected to satisfy a Schrödinger type equation (♦) $i\hbar\partial_t\gamma_\pm(\mathbf{p}, t) = cp\gamma_\pm(\mathbf{p}, t)$. Next for each $\mathbf{p}$ introduce two unit vectors $\hat{\mathbf{e}}_i(\hat{\mathbf{p}})$ where $\hat{\mathbf{p}} = \mathbf{p}/|\mathbf{p}|$ such that $\hat{\mathbf{e}}_1, \hat{\mathbf{p}}_2, \hat{\mathbf{p}}$ form a right handed triad (cf. here Section 1). Then define helicity vectors $\mathbf{e}_\pm(\hat{\mathbf{p}}) = \mp(1/\sqrt{2})[\hat{\mathbf{e}}_1 \pm i\hat{\mathbf{e}}_2]$ and write $\vec{\gamma}_\pm = \mathbf{e}_\pm\gamma_\pm$ with

$$\vec{\gamma}_+^* \cdot \vec{\gamma}_+ d\mathbf{p} = \gamma_+^*\gamma_+ d\mathbf{p} \tag{3.46}$$

for the probability of detecting a photon of positive helicity and momentum $\mathbf{p}$ between $\mathbf{p}$ and $\mathbf{p} + d\mathbf{p}$ (similarly for negative helicity). Note also that $\vec{\gamma}_+^* \cdot \vec{\gamma}_- = 0$. Then define Fourier transforms

$$\Phi_\pm(\mathbf{r}, t) = \int \frac{d\mathbf{p}}{(2\pi\hbar)^{3/2}}\gamma_\pm(\mathbf{p}, t)e^{i\mathbf{p}\cdot\mathbf{r}/\hbar} \tag{3.47}$$

One then checks that (♦) is satisfied if

$$i\hbar\partial_t\Phi_\pm(\mathbf{r}, t) = \pm c\hbar\nabla \times \Phi_\pm(\mathbf{r}, t) \tag{3.48}$$

From the assumption that one is dealing with a single photon there results

$$\int[\vec{\gamma}_+^* \cdot \vec{\gamma}_+ + \vec{\gamma}_-^* \cdot \vec{\gamma}_-]d\mathbf{p} = 1 \equiv \int[\Phi_+^* \cdot \Phi_+ + \Phi_-^* \cdot \Phi_-]d\mathbf{r} = 1 \tag{3.49}$$

The dynamical equations (♦) or (5.3) guarantee that if these equations (5.4) are true at one time then they are satisfied at all later times. In fact there results

$$\Phi_\pm(\mathbf{r}, t) = \int \frac{d\mathbf{p}}{(2\pi\hbar)^{3/2}}\vec{\gamma}_\pm(\mathbf{p}, 0)e^{-icpt/\hbar}e^{i\mathbf{p}\cdot\mathbf{r}/\hbar} \tag{3.50}$$

There is then a temptation to try and identify the $\Phi_\pm$ as position representation probability amplitudes for photons of positive or negative helicity or perhaps their sum $\Phi_+ + \Phi_-$ as a position representation of a photon. However photons are not localizable so this doesn't work. One way around (cf. Jauch-Piron [**693**]) is to show that an operator representating the number of photons in an arbitrary volume V can be defined but not as the integral over V of a photon density operator. Another approach (cf. Mandel [**848**]) is to determine an operator representing the number of photons in a volume V as the integral over V of a so called detection operator which (when the linear dimensions of V are large compared to the photon wavelengths) leads to a simple formula for the probability that n photons are present in V (cf. also [**330**] and above for coarse grained photon density and current density operators). Here one proceeds differently following [**1180**] and looks for a probability amplitude for the photon energy to be detected about $d\mathbf{r}$ of $\mathbf{r}$ in the form $\Psi^* \cdot \Psi d\mathbf{r}$ with normalizations in the sense that

$$\int \Psi^* \cdot \Psi d\mathbf{r} = \int cp[\vec{\gamma}_+^* \cdot \vec{\gamma}_+ + \vec{\gamma}_-^* \cdot \vec{\gamma}_-]d\mathbf{p} \tag{3.51}$$

To do this one sets $\Psi = \Psi_+ + \Psi_-$ with

$$\Psi_\pm(\mathbf{r},t) = \int \frac{\sqrt{cp}d\mathbf{p}}{(2\pi\hbar)^{3/2}}\vec{\gamma}_\pm(\mathbf{p},t)e^{i\mathbf{p}\cdot\mathbf{r}/\hbar} \tag{3.52}$$

One notes then that $i\hbar\partial_t\Psi_\pm = \pm c\hbar\nabla \times \Psi_\pm$ and one must satisfy (cf. (5.5)) an initial condition given by the $t = 0$ case of

$$\Psi_\pm(\mathbf{r},t) = \int \frac{\sqrt{cp}d\mathbf{p}}{(2\pi\hbar)^{3/2}}\vec{\gamma}_\pm(\mathbf{p},0)e^{-icpt/\hbar}e^{i\mathbf{p}\cdot\mathbf{r}/\hbar} \tag{3.53}$$

Note here that $\mathbf{p}$ (the usual photon momentum) and $\mathbf{r}$, the position associated with the photon energy, are not conjugate variables. One then builds up a QFT of of the free radiation field via many particle physics (not from a canonical formulation of the EM fields) and this is equivalent to standard canonical quantization. Moreover upon specializing to one photon the energy functions $\Psi_\pm$ above are recovered. Thus it is reasonable to describe the single photon energy distribution in a region $d\mathbf{r}$ about $\mathbf{r}$ via $\Psi^*(\mathbf{r},t)\cdot\Psi(\mathbf{r},t)d\mathbf{r}$. Further it is shown that in a spontaneous emission process the wave function $\Psi(\mathbf{r},t)$ generated is a causal field, propagating out from the emitting atom at the speed of light.

Now one goes to [**652, 653**] where the Bohr photon having a specific size and shape is discussed. This involves a circularly polarized photon being a monochromatic EM traveling wave confined within a circular ellipsoid of length equal to the wavelength (λ) and diameter λ/π propagating along the long axis of the ellipsoid. In this model the quantization of the photon's angular momentum (corresponding to spin $\hbar$) arises from an appropriately chosen of Maxwell's equations and the energy is quantized to be $\hbar\nu$. In a sense, not entirely clear (cf. [**252**]), one can think here of an ellipsoidal soliton arising from the imposition of causality upon the solution of the linear Maxwell equations where EM energy $\mathbf{E}^2 + \mathbf{H}^2$ integrated over the volume of the ellipsoid equals $\hbar\nu$ leading to an average intensity within the photon-soliton of $I_p = 4\pi\hbar c^2/\lambda^4$. The word wavicle is also used in [**588**]. For a wave traveling with the speed of light parallel to the z-axis the solution of Maxwell's equations can be any function of $z - ct$ and if monochromatic one has a term $S(z-ct) = exp[2\pi i(z-ct)/\lambda]$. Setting $x = rCos(\phi)$ and $y = rSin(\phi)$ in the already separated d'Alembert equation then leads to

$$\frac{1}{\Phi(\phi)}\frac{d^2\Phi}{d\phi^2} = m^2 = -\frac{1}{R(r)}\left[\frac{d^2R}{dr^2} + \frac{1}{r}\frac{dR}{dr}\right] \tag{3.54}$$

where m^2 is the real separation constant. The simple plane wave solutions with $m^2 = 0$ are rejected here since light is observed to travel along very narrow beams and for $m^2 = 1$ one has factors of r or $1/r$ with angular factors $exp(\pm i\phi)$. This corresponds to angular momentum $L_z = (\hbar/i)\partial_\phi$ leading to solutions

$$\psi(r,\phi,z-ct) = (\alpha r + \beta/r)(Ae^{i\phi} + Be^{-i\phi})e^{2\pi i(z-ct)/\lambda} \tag{3.55}$$

This yields then

$$E_z = H_z = 0;\ E_x = (\alpha r + \beta/r)\left[Ae^{i\phi} + Be^{-i\phi}\right]e^{2\pi i(z-ct)/\lambda} = \mu_0 c H_y; \tag{3.56}$$
$$E_y = i(\alpha r - \beta/r)\left[Ae^{i\phi} - Be^{-i\phi}\right]e^{2\pi i(z-ct)/\lambda} = -\mu_0 c H_x$$

Imposing causality leads to the result that if A or B is zero then the field must be contained within a circular ellipsoid of length λ and cross sectional diameter λ/π (cf. [**653**]). The amplitude is determined by integration of the energy $\mathbf{E}^2 + \mathbf{H}^2$ and the $1/r$ term is then discarded to preserve the ellipsoidal shape; there results $A^2 + B^2 = 1$ and $\alpha^2 = 120h\hbar c\pi^4 4/\epsilon_0\lambda^6$ (in suitable units). In addition one expects an evanescent wave decaying like $1/r$ (with $\alpha = 0$) described via

$$E_r = \frac{\beta}{r}[A+B] = \mu_0 c H_\phi;\ E_\phi = -i\frac{\beta}{r}[A-B] = -\mu_0 c H_r \tag{3.57}$$

where $\alpha r = \beta/r$ for $r = \lambda/2\pi$ and $\beta^2 = (\lambda/2\pi)^4 \times 120n\hbar c\pi^4/(\epsilon_0\lambda^6)$. The evanescent wave is believed to be responsible for diffraction and interference and some experimental material is sketched.

We sketch here from Cook [**330**] where it is shown that one can define the notions of photon density and photon current density with certain limits. As a trivial example think of geometrical optics where light is treated as an ensemble of point photons moving along definite trajectories with speed c. In the geometric limit the photon number density and current density are perfectly well defined as are the density and current density for any collection of point particles. As a second example one refers to [**848**] where Mandel defines an operator n_V representing the number of photons in V as the integral over V of the photon density $D_M(\mathbf{x}) = \mathbf{A}^\dagger(\mathbf{x})\cdot\mathbf{A}(\mathbf{x})$ where

$$\mathbf{A}(\mathbf{x}) = L^{-3/2}\sum_{\mathbf{k},\lambda}\vec{\epsilon}_{\mathbf{k},\lambda}a_{\mathbf{k},\lambda}e^{i(\mathbf{k}\cdot\mathbf{x}-\omega t)} \tag{3.58}$$

is the so-called detection operator (here $\vec{\epsilon}$, a, and $\omega = c\mathbf{k}$ are respectively the polarization unit vector, the annihilation operator and the frequency of a transverse photon of wave vector $\mathbf{k}$ and polarization λ $(= 1, 2)$ with L^3 the quantization volume). It is shown that when the linear dimensions of V are large compared to the photon wavelengths this definitiion of n_V yields a simple for the probability $p_V(n)$ that n photons are present in V. It was later shown by Amrein [**39**] that n_V agrees with that derived from the theory of [**693**] and this all has motivated the study in [**340**] that a coarse grained photon density operator can exist even though a fine grained or microscopic one may not.

Thus one derives a photon density $D(\mathbf{x})$ and a photon current density $\mathbf{C}(\mathbf{x})$ to satisfy ($\bullet$) $\partial_t D + \nabla\cdot\mathbf{C} = 0$ (conservation of photons ignoring absorption and emission). These will be defined in terms of vector field operators $\vec{\psi}(\mathbf{x})$ and $\vec{\phi}(\mathbf{x})$ which will be referred to as the photon field (cf. here also Section 2 again). For a volume V large compared to the photon wavelengths $D(\mathbf{x})$ will correctly predict the number statistics of photons in that volume while for a time interval $[t, t+T]$ long compared to λ/c $\mathbf{C}(\mathbf{x})$ correctly predicts the statistics of the number of photons that cross the surface S in time T. This was worked out in the first paper of [**330**] for a discrete situation and is redone in the second paper in a continuum context; we sketch this here for the free field case and, following [**330**], show that photon dynamics is a relativistically covariant theory. Thus write

$$\vec{\psi}(\mathbf{x},t) = \frac{1}{\sqrt{2(2\pi)^3}}\sum_{\lambda=1}^{2}\int d^3k\vec{\epsilon}_\lambda(\mathbf{k})a_\lambda(\mathbf{k})e^{i(\mathbf{k}\cdot\mathbf{x}-\omega t)}; \tag{3.59}$$

$$\vec{\phi}(\mathbf{x},t) = \frac{1}{\sqrt{2(2\pi)^3}}\sum_{\lambda}\int d^3k\left(\frac{\mathbf{k}}{k}\times\vec{\epsilon}_\lambda(\mathbf{k})\right)a_\lambda(\mathbf{k})e^{i(\mathbf{k}\cdot\mathbf{x}-\omega t)}$$

where $a_\lambda(\mathbf{k})$ is the annihilation operator and $\vec{\epsilon}_\lambda(\mathbf{k})$ the polarization vector of a transverse photon of wave vector $\mathbf{k}$ and polarization λ $(=1,2)$. Evidently we have the free field equations

$$\nabla\cdot\vec{\psi}=0;\ \nabla\cdot\vec{\phi}=0;\ \nabla\times\vec{\psi}+\frac{1}{c}\partial_t\vec{\phi}=0;\ \nabla\times\vec{\phi}-\frac{1}{c}\partial_t\vec{\psi}=0 \tag{3.60}$$

Then one defines

$$D=\vec{\psi}^\dagger\cdot\vec{\psi}+\vec{\phi}^\dagger\cdot\vec{\phi};\ \mathbf{C}=c(\vec{\psi}^\dagger\times\vec{\phi}-\vec{\phi}^\dagger\times\vec{\psi}) \tag{3.61}$$

Evidently (•) $\partial_t D+\nabla\cdot\mathbf{C}=0$ as required. Note that D is a positive definite operator whose integral over all space is the usual photon number operator

$$\int d^3xD(\mathbf{x} = \sum_{\lambda}\int d^3ka_\lambda^\dagger(\mathbf{k})a_\lambda(\mathbf{k}) \tag{3.62}$$

The interpretation of $\mathbf{C}$ as the photon current density is justified by (•) and by a calculation showing that the integral of the inward normal component of $\mathbf{C}$ over the surface of an ideal photon detector equals the counting rate of the detector (cf. [**330**] first paper). The operators for the number of photons in V and the number of photons crossing a given surface in the time interval $[t, t+T]$ are

$$n_V=\int_V d^3xD(\mathbf{x});\ n_T=\int_t^{t+T}dt'\int_S da\mathbf{n}\cdot\mathbf{C}(\mathbf{x},t) \tag{3.63}$$

($\mathbf{n}$ is the unit normal to S in the direction of interest). For a volume V large as described the probability that V contains m photons is (••) $p_V(m)=Tr[\rho : n_V^m exp(-n_V) :]/m!$ where ρ is the density operator of the radiation field and : : means normal ordering. Similarly for sufficently large T the probability that m photons cross the surface S in time T is (♦♦) $p_T(m)=Tr[\rho : n_T^m exp(-n_T) :]/m!$ where S is the sensitive surface of an ideal photon detector (with one unit quantum efficiency) and $p_T(m)$ is the photon count distribution measured by the detector. For these calculations see the first paper of Cook [**330**] and Mandel [**848**] (n_V and n_T are treated as number operators).

Now the transverse EM field operators $\mathbf{E}=\mathbf{E}^+ + \mathbf{E}^-$ and $\mathbf{B}=\mathbf{B}^+ + \mathbf{B}^-$ can be expressed via

(3.64)
$$\mathbf{E} = \frac{i}{2\pi}\sum_\lambda \int d^3k(\hbar\omega)^{1/2}\vec{\epsilon}(\mathbf{k})a_\lambda(\mathbf{k})e^{i(\mathbf{k}\cdot\mathbf{x}-\omega t)};$$

$$\mathbf{B} = \frac{i}{2\pi}\sum_\lambda \int d^3k(\hbar\omega)^{1/2}\left(\frac{\mathbf{k}}{k}\times\vec{\epsilon}(\mathbf{k})\right)a_\lambda(\mathbf{k})e^{i(\mathbf{k}\cdot\mathbf{x}-\omega t)}$$

and $\mathbf{E}^- = (\mathbf{E}^+)^\dagger$ with $\mathbf{B}^- = (\mathbf{B}^+)^\dagger$. One sees that the photon field vectors in (5.14) are obtained from $\mathbf{E}^+$ and $\mathbf{B}^+$ by multiplying the momentum components of $\mathbf{E}^+$ and $\mathbf{B}^+$ by $-i[4\pi\hbar\omega(\mathbf{k})]^{-1/2}$ which corresponds to a convolution in position space

(3.65)
$$\vec{\psi}(\mathbf{x},t) = \int d^3yg(\mathbf{x}-\mathbf{y})\mathbf{E}^+(\mathbf{y},t);\ \vec{\phi}(\mathbf{x},t) = \int d^3yg(\mathbf{x}-\mathbf{y})\mathbf{B}^+(\mathbf{y},t)$$

where

(3.66)
$$g(\mathbf{x}) = \frac{-i}{(2\pi)^3}\int d^3k(4\pi\hbar\omega)^{-1/2}e^{-i\mathbf{k}\cdot\mathbf{x}};\ \int d^3yg^{-1}(\mathbf{x}-\mathbf{y})g(\mathbf{y}-\mathbf{z} = \delta^3(\mathbf{x}-\mathbf{z});$$

$$g^{-1}(\mathbf{x}) = \frac{-i}{(2\pi)^3}\int d^3k(4\pi\hbar\omega)^{1/2}e^{-i\mathbf{k}}\cdot\mathbf{x}$$

One has also

(3.67)
$$\mathbf{E}^+ = \int d^3yg^{-1}(\mathbf{x}-\mathbf{y})\vec{\psi}(\mathbf{y},t);\ \mathbf{B}^+ = \int d^3yg^{-1}(\mathbf{x}-\mathbf{y})\vec{\phi}(\mathbf{y},t)$$

It is convenient now to express the photon field as a matrix

(3.68)
$$\psi_{\mu\nu} = \begin{pmatrix} 0 & -\psi_1 & -\psi_2 & -\psi_3 \\ \psi_1 & 0 & \phi_3 & -\phi_2 \\ \psi_2 & -\phi_3 & 0 & \phi_1 \\ \psi_3 & \phi_2 & -\phi_1 & 0 \end{pmatrix}$$

Note here the similarity to (1.2) (apparently de la Torre was unaware of Cook's work). An immediate relation to the EM field strength tensor is exhibited via

(3.69)
$$F^+_{\mu\nu} = \begin{pmatrix} 0 & -E_1^+ & -E_2^+ & -E_3^+ \\ E_1^+ & 0 & B_3^+ & -B_2^+ \\ E_2^+ & -B_3^+ & 0 & B_1^+ \\ E_3^+ & B_2^+ & -B_1^+ & 0 \end{pmatrix}$$

One is using coordinates $x^\mu = (ct, x, y, z)$ and metric $g^{00} = 1,\ g^{ii} = -1$ and will raise and lower indices with $g^{\alpha\beta}$or $g_{\alpha\beta}$ as in $\psi_{\mu\nu}$ were a tensor (it turns out to transform as a tensor under displacements and spatial rotations but not for boosts - cf. [**330**]). Now define

(3.70)
$$G(x) = \frac{-i}{(2\pi)^4}\int d^4k[4\pi\hbar\omega(k)]^{-1/2}e^{ikx}$$

where $x \sim x^\mu$, $k \sim k^\mu = (k^0, \mathbf{k})$, $kx \sim k^\mu x_\mu$, and $\omega(k) = c|\mathbf{k}|$. Clearly G has an inverse with $\int d^4yG^{-1}(x-y)G(y-z) = \delta^4(x-z)$ etc. Then one can write

(3.71)
$$\psi_{\mu\nu}(x) = \int d^4yG(x-y)F^+_{\mu\nu}(y);\ F^+_{\mu\nu}(x) = \int d^4yG^{-1}(x-y)\psi_{\mu\nu}(y)$$

To see that these are equivalent to (5.21) and (4.22) note that since $\omega(k) = c|\mathbf{k}|$ does not depend on k^0 the k^0 integrals can be evaluated immediately to give $G(x) = \delta(x^0)g(\mathbf{x})$ etc. Although (5.21) and (4.22) were derived from (5.14) and (5.20) for transverse photon and EM fields one assumes that they and hence (4.26) remain valid when the EM fields have a longitudinal component (for the free field this is of no concern). Now one shows that the photon field equations (4.22) are a direct consequence of the free field Maxwell equations

(3.72) $$\partial^\nu F^+_{\mu\nu} = 0;\ \partial_\alpha F^+_{\beta\gamma} + \partial_\beta F^+_{\gamma\alpha} + \partial_\gamma F^+_{\alpha\beta} = 0$$

The second of these equations is a general operator relation following from $F^+_{\mu\nu} = \partial_\nu A^+_\mu - \partial_\mu A^+_\nu$ while the first equation is valid in the sense that $\partial^\nu F^+_{\mu\nu}| >= 0$ for all physically admissable states $| >$ (which are defined as those satisfying the Gupta-Bleuler condition (★) $\partial^\mu A^+_\mu| >= 0$ - note the free field vector potential satisfies the wave equation $\partial^\nu\partial_\nu A^+_\mu = 0$). Now consider

(3.73) $$\frac{\partial\psi_{\mu\nu}(x)}{\partial x^\alpha} = \int d^4y \frac{\partial G(x-y)}{\partial x^\alpha} F^+_{\mu\nu}(y) =$$

$$= -\int d^4y \frac{\partial G(x-y)}{\partial y^\alpha} F^+_{\mu\nu}(y) = \int d^4y G(x-y) \frac{\partial F^+_{\mu\nu}(y)}{\partial y^\alpha}$$

Neglect of the integrated part is justified via $G(x) \to 0$ for $x \to \infty$ and $F^+_{\mu\nu}(x) \to 0$ at spatial infinity. From (4.27) one has then

(3.74) $$\partial^\nu \psi_{\mu\nu} = 0;\ \partial_\alpha\psi_{\beta\gamma} + \partial_\beta\psi_{\gamma\alpha} + \partial_\gamma\psi_{\alpha\beta} = 0$$

and these are equivalent to the original photon field equations (5.15); again one restricts to the subspace of physical states. Although these have the appearance of tensor equations they are not manifestly covariant since $\psi_{\mu\nu}$ is not a tensor (cf. Section 2 for comments in this direction). Nevertheless the equations are shown to be invariant under Lorentz transformations because the photon field $\psi_{\mu\nu}$ is defined in terms of the tensor $F^+_{\mu\nu}$ in the same way in each Lorentz frame (see here [**243**] for details).

Finally one considers the matrix of Hermitian operators

(3.75) $$N^{\alpha\beta} = \psi^{\dagger\alpha}_\lambda \psi^{\lambda\beta} + \psi^{\dagger\lambda\beta}\psi^\alpha_\lambda + \frac{1}{2} g^{\alpha\beta}\psi^{\dagger\mu\nu}\psi_{\mu\nu}$$

with is analogous to the EM energy momentum tensor. One checks easily that (♣♣) $\partial_\beta N^{\alpha\beta} = 0 \Rightarrow M^\alpha = \int d^3x N^{\alpha 0}$ is conserved as the photon field develops in time. In fact one can write

(3.76) $$N^{\alpha\beta} = \begin{pmatrix} D & C_1/c & C_2/c & C_3/c \\ C_1/c & S_{11}/c^2 & S_{12}/c^2 & C_{13}/c^2 \\ C_2/c & S_{21}/c^2 & S_{22}/c^2 & S_{23}/c^2 \\ C_3/c & S_{31}/c^2 & S_{32}/c^2 & S_{33}/c^2 \end{pmatrix}$$

Here (♠♠) $S_{ij} = c^2[D\delta_{ij} - (\psi^\dagger_i\psi_j + \psi^\dagger_j\psi_i) - (\phi^\dagger_i\phi_j + \phi^\dagger_j\phi_i)]$ is a 3×3 matrix analogous to the Maxwell stress tensor and (♣♣) now takes the form

(3.77) $$\partial_t D + \nabla\cdot\mathbf{C} = 0;\ \partial_t C_i + \frac{\partial S_{ij}}{\partial x^j} = 0;$$

$$M^0 = \int d^3x D(\mathbf{x}) = const.; \; cM^i = \int d^3x C_i(\mathbf{x} = const.$$

These equations express the local and global conservation of photons. One shows that $N^{\alpha\beta}$ does not transform as a tensor (except for coordinate displacements and spatial rotations) and in fact there is no general transformation law relating the components of $N^{\alpha\beta}$ in different Lorentz frames. Nevertheless (4.32) are covariant since the photon field equations are covariant and $N^{\alpha\beta}$ is constructed from the photon field in the same way in each frame. In particular the number of photons M^0 is independent of time in each Lorentz frame and considerable calculation also shows that M^0 is a scalar (using the condition (★)).

3.4. SOME SPECULATIONS ON THE AETHER. The aether has now been reviewed to a certain extent and in [**254**] some speculations were advanced concerning a possible geometry for the aether. These were based on work of [**2, 51, 245, 244, 238, 253, 250, 254, 445, 446, 662, 1257**] and we sketch here some variations and embellishments. First we note from (5.15) that the components ψ_i and ϕ_i satisfy the massless KG equation so for analysis of photons one needs 6 components (ψ_i, ϕ_i) each satisfying a massless KG equation. However the equations (5.15) are exactly the same as the Maxwell equations (2.7) so one could also imagine introducing a vector $\Psi = (-\mathbf{A}, \phi)$ with $A_{\mu\nu} = \Psi_{\mu,\nu} - \Psi_{\nu,\mu}$ to generate the photon equations for a free field with $\Box\Psi = 0$ (see e.g. [**1072**]). In this spirit then one would have a 4-vector Ψ satisfying the massless KG equation to serve as a generator of photon activity. In any event we will think of fields labeled ψ_i for $i = 0, 1, 2, 3$ as characterizing photon dynamics with each component satisfying the massless KG equation. Then we will apply the machinery of (x, ψ) duality of Faraggi-Matone and Vancea (see especially [**445, 1257**]) to express the coordinates x^μ in terms of the fields ψ_i arising from Ψ (which will be called aether fields); they are seen to be "potential" fields for the photon fields ψ_i, ϕ_i at the beginning of Section 3.

As background here we refer to a lovely paper [**662**] of P. Isaev where he makes conjectures, with supporting arguments, which arrive at a definition of the aether as a Bose-Einstein condensate of neutrino-antineutrino pairs of Cooper type (Bose-Einstein condensates of various types have been considered by others in this context - cf. [**326, 430, 686**]). The equation for the ψ-aether is then a solution of the massless Klein-Gordon (KG) equation (photon equation) $(\hbar^2\Delta - (\hbar^2/c^2)\partial_t^2)\psi = 0$. This ψ field heuristically acts as a carrier of waves (playground for waves) and one might say that special relativity (SR) is a way of including the influence of the aether on physical processes and consequently SR does not see the aether (cf. here also the idea of a Dirac aether in [**266, 267, 377, 958**], Einstein-aether theories as in [**430, 686**], and [**569, 814, 1290, 1291**] discussed briefly in Chapter 8). In the electromagnetic (EM) theory in [**662**] one looks at $\vec{\psi} = (\phi, \vec{A})$ with $\Box\psi_i = 0$ as the defining equation for a real ψ-aether, in terms of the potentials ϕ and $\vec{A}$ which therefore define the ψ-aether. EM waves are then considered as oscillations of the ψ aether and wave processes in the aether accompanying a moving particle determine wave properties of the particle. Interesting examples involving standing EM waves in a spherical resonator are attributed to oscillations of the ψ aether and references to superconductivity à la Volovik [**1269, 1267**] are indicated.

In [**445**] Faraggi and Matone develop a theory of $x-\psi$ duality, related to Seiberg-Witten theory in the string arena, which was expanded in various ways in [**2, 52, 51, 244, 238, 254, 864, 1257**]. Here one works from a stationary SE $[-(\hbar^2/2m)\Delta+V(x)]\psi=E\psi$, and, assuming for convenience one space dimension, the space variable x is determined by the wave function ψ from a prepotential $\mathfrak{F}$ via Legendre transformations. The theory suggests that x plays the role of a macroscopic variable for a statistical system with a scaling term involving $\hbar$. Thus define a prepotential $\mathfrak{F}_E(\psi)=\mathfrak{F}(\psi)$ such that the dual variable $\psi^D=\partial\mathfrak{F}/\partial\psi$ is a (linearly independent) solution of the same SE. Take V and E real so that $\bar{\psi}=\psi^D$ qualifies and write $\partial_x\mathfrak{F}=\psi^D\partial_x\psi=(1/2)[\partial_x(\psi\psi^D)+W)]$ where W is the Wronskian. This leads to $(\psi^D=\bar{\psi})$ the relation $\mathfrak{F}=(1/2)\psi\bar{\psi}+(W/2)x$ (setting the integration constant to zero). Consequently, scaling W to $-2i\sqrt{2m}/\hbar$ one obtains

$$\frac{i\sqrt{2m}}{\hbar}x=\frac{1}{2}\psi\frac{\partial\mathfrak{F}}{\partial\psi}-\mathfrak{F}\equiv\frac{i\sqrt{2m}}{\hbar}x=\psi^2\frac{\partial\mathfrak{F}}{\partial\psi^2}-\mathfrak{F} \tag{3.78}$$

which exhibits x as a Legendre transform of $\mathfrak{F}$ with respect to ψ^2. Duality of the Legendre transform then gives also

$$\mathfrak{F}=\phi\partial_\phi\left(\frac{i\sqrt{2m}\,x}{\hbar}\right)-\left(\frac{i\sqrt{2m}\,x}{\hbar}\right);\ \phi=\partial_{\psi^2}\mathfrak{F}=\frac{\bar{\psi}}{2\psi} \tag{3.79}$$

so that $\mathfrak{F}$ and $(i\sqrt{2m}\,x/\hbar)$ form a Legendre pair. In particular one has $\rho=|\psi|^2=\frac{2i\sqrt{2m}}{\hbar}x+2\mathfrak{F}$ which also relates $\mathfrak{F}$ and the probability density. In any event one sees that the wave function ψ specifically determines the location of the "particle" whose quantum evolution is described by ψ. We mention here also that the (stationary) SE can be replaced by a third order equation

$$4\mathfrak{F}'''+(V(x)-E)(\mathfrak{F}'-\psi\mathfrak{F}'')^3=0;\ \mathfrak{F}'\sim\frac{\partial\mathfrak{F}}{\partial\psi} \tag{3.80}$$

and a dual stationary SE has the form

$$\frac{\hbar^2}{2m}\frac{\partial^2 x}{\partial\psi^2}=\psi[E-V]\left(\frac{\partial x}{\partial\psi}\right)^3 \tag{3.81}$$

A noncommutative version of this is developed in the second paper of [**1262**].

We mention [**879, 1004, 1005**] for some material on the aether and the vacuum and refer to the bibliography for other references. We sketch first some material from [**2, 51, 445, 1257**] which extends the SE theory to the Klein-Gordon (KG) equation. Following Vancea [**1257**] take a spacetime manifold M with a metric field g and a scalar field ψ satisfying the KG equation. Locally one has cartesian coordinates x^α $(\alpha=0,1.\cdots,n-1)$ in which the metric is diagonal with $g_{\alpha\beta}(x)=\eta_{\alpha\beta}(x)$ and the KG equation has the form $(\Box_x+m^2)\psi(x)=0$ $(\Box_x\sim(\hbar^2/c^2)[(\partial_t^2/c^2)-\nabla^2])$. Defining prepotentials such that $\tilde{\psi}^{(\alpha)}=\partial\mathfrak{F}^{(\alpha)}[\psi^{(\alpha)}]/\partial\psi^{(\alpha)}$ where $\psi^{(\alpha)}$ and $\tilde{\psi}^{(\alpha)}$ are two linearly independent solutions of the KG equation depending on parameters x^α one has as above (with a different scaling factor)

$$\frac{\sqrt{2m}}{\hbar}x^\alpha = \frac{1}{2}\psi^{(\alpha)}\frac{\partial\mathfrak{F}^{(\alpha)}[\psi^{\alpha)}]}{\partial\psi^{(\alpha)}} - \mathfrak{F}^{(\alpha)};\ [\partial^\alpha\partial_\alpha - V^\alpha]\psi^\alpha = 0 \tag{3.82}$$

This is suggested in [**445**] and used in [**1257**]; the factor $\sqrt{2m}/\hbar$ is simply a scaling factor (possibly too stringent here) and it would be more productive to scale $x^0 \sim ct$ differently or in fact to scale all variables as indicated in [**254**] (cf. below for a general scaling). Locally $\mathfrak{F}^{(\alpha)}$ satisfies the third order equation

$$4\mathfrak{F}^{(\alpha)'''} + [V^{(\alpha)}(x^\alpha) + m^2](\psi^{(\alpha)}\mathfrak{F}^{(\alpha)''} - \mathfrak{F}^{(\alpha)'})^3 = 0 \tag{3.83}$$

where $' \sim \partial/\partial\phi^{(\alpha)}$ and a (quantum) potential V^α has the form

$$V^{(\alpha)}(x^\alpha) = \left[\frac{1}{\psi(x)}\sum_{\beta=0,\,\beta\neq\alpha}^{n-1}\partial^\beta\partial_\beta\psi(x)\right]\Bigg|_{x^\beta\neq\alpha\ fixed} \tag{3.84}$$

We go back to [**445**] now and derive equations for the KG equation with $m = 0$ from the beginning (rather than rescaling and then taking $m \to 0$). Further we proceed with more detail and show how a general scaling will involve insertion of some variable factors (cf. also [**240, 245**] for various scaling factors). Thus consider $(1/c^2)\psi_{tt} - \Delta\psi = 0$ with $x^0 = ct$ and write out explicitly $(i = 1, 2, 3)$

$$\frac{1}{c^2}\partial_t^2\psi^0 - V^0\psi^0 = 0;\ V^0 = \frac{\Delta\psi}{\psi} = \frac{(1/c^2)\psi_{tt}}{\psi}; \tag{3.85}$$

$$\partial_i^2\psi^i - V^i\psi^i = 0;\ V^i = \frac{\left(\frac{1}{c^2}\partial_t^2\psi - \sum_{j\neq i}\partial_j^2\psi\right)}{\psi}$$

Here V^i is thought of as $V^i(x^i)$ (where in fact $V^i = V^i(x^i, x^j, x^0)$ with $j \neq i$ and x^0, x^j are considered as parameters). Similarly $V^0 = V^0(x^0)$ $(\equiv V^0(x^0, x^i))$. Now e.g. for ψ^0 and $\tilde{\psi}^0$ linearly independent solutions of the first equation in (6.8) one has $\psi^0_{tt}\tilde{\psi}^0 = \psi^0\tilde{\psi}^0_{tt}$ which implies

$$W^0(t) = (\psi^0\tilde{\psi}^0_t(t) - \tilde{\psi}^0\psi^0_t)(t) = 2c\gamma(x^i) \tag{3.86}$$

Here, as specified above, $\tilde{\psi}^0 = \partial\mathfrak{F}^0/\partial\psi^0$, and

$$\partial_t\mathfrak{F}^0 = \mathfrak{F}^0_\psi\psi_t = \tilde{\psi}^0\psi_t \Rightarrow \tag{3.87}$$

$$\Rightarrow \frac{1}{2}\partial_t(\psi^0\tilde{\psi}^0) - \tilde{\psi}^0\psi^0_t = \frac{1}{2}(\psi^0\tilde{\psi}^0_t - \psi^0_t\tilde{\psi}^0) = \frac{1}{2}W^0 = c\gamma(x^i)$$

and consequently one can write

$$c\gamma(x^i)t = \frac{1}{2}\psi^0\frac{\partial\mathfrak{F}^0}{\partial\psi^0} - \mathfrak{F}^0 = \mathfrak{E}^0 \tag{3.88}$$

This leads to (for $\psi^0 \sim \phi$)

$$c\gamma(x^i) = \frac{\partial\mathfrak{E}^0}{\partial\phi}\frac{d\phi}{dt} = \left[\frac{1}{2}\left(\mathfrak{F}^0_\phi + \phi\frac{\partial^2\mathfrak{F}^0}{\partial\phi^2}\right) - \mathfrak{F}^0_\phi\right]\frac{d\phi}{dt} = \tag{3.89}$$

$$= \frac{1}{2}\left(\phi\frac{\partial^2\mathfrak{F}^0}{\partial\phi^2} - \mathfrak{F}^0_\phi\right)\frac{d\phi}{dt} = \frac{1}{2}E^0\frac{d\phi}{dt}$$

Similarly we write, using (3.85),

$$\tilde{\psi}^i = \frac{\partial \mathfrak{F}^i}{\partial \psi^i};\ W^i = \psi^i \partial_t \tilde{\psi}^i - \tilde{\psi}^i \partial_t \psi^i;\ \beta^i(x^0, x^j)x^i = \frac{1}{2}\psi^i \frac{\partial \mathfrak{F}^i}{\partial \psi^i} - \mathfrak{F}^i = \mathfrak{E}^i \tag{3.90}$$

Consequently ($\psi_i \equiv \psi^i$)

$$\gamma dx^0 = c\gamma dt = \frac{1}{2}E^0 d\psi^0;\ \beta^i dx^i = \frac{1}{2}E^i d\psi^i = \frac{1}{2}\left(\psi^i \frac{\partial^2 \mathfrak{F}^i}{\partial \psi_i^2} - \frac{\partial \mathfrak{F}^i}{\partial \psi^i}\right) d\psi^i \tag{3.91}$$

Since $\partial_t = \partial_\phi(d\phi/dt)$, etc. one can write then

$$\partial_t = \left(\frac{2c\gamma}{E^0}\right)\partial_\phi;\ \partial_i = \left(\frac{2\beta}{E^i}\right)\frac{\partial}{\partial \psi^i} \tag{3.92}$$

The extraneous variables are considered as parameters when concentrating on one x^i or x^0 and we note from (3.88) or (3.90) that x^0 or x^i can be considered as a function of $\phi = \psi^0$ or ψ^i and $\mathfrak{F}^i$ is a function of ψ^i (satisfying ordinary differential equations as in (3.83) - with $m = 0$). Here (6.14)-(6.15) represents an induced parametrization on the spaces $T_P(U)$ and $T_P^*(U)$ ($P \in U$ - local tangent and cotangent spaces). Now using the linearity of the metric tensor field one sees that the components of the metric in the $\{(\psi^\alpha, \mathfrak{F}^\alpha)\}$ parametrization are ($\beta^0 = c\gamma$)

$$G_{\alpha\sigma}(\psi) = \frac{E^\alpha E^\sigma}{4\beta^\alpha \beta^\sigma}\eta_{\alpha\sigma}(x) \tag{3.93}$$

(cf. [**1257**]). Now following [**1257**] let z^μ ($\mu = 0, 1, \cdots, n-1$) be a general coordinate system in U and write the coordinate transformation matrices via

$$A_\mu^\alpha = \frac{\partial x^\alpha}{\partial z^\mu};\ (A^{-1})_\alpha^\mu = \frac{\partial z^\mu}{\partial x^\alpha} \tag{3.94}$$

The metric then takes the form

$$g_{\mu\nu}(z) = \frac{4\beta^\alpha \beta^\sigma}{E^\alpha E^\sigma} A_\mu^\alpha A_\nu^\sigma G_{\alpha\sigma}(\psi) \tag{3.95}$$

The components of the metric connection can be computed via

$$\Gamma_{\mu\nu}^\rho = \frac{1}{2}g^{\rho\sigma}(z)\sum_{\mathcal{P}} \epsilon_{\mathcal{P}} \mathcal{P}\left[\frac{\partial g_{\sigma\nu}(z)}{\partial z^\mu}\right] \tag{3.96}$$

where $\mathcal{P}$ is a cyclic permutation of the ordered set of indices $\{\sigma\nu\mu\}$ and $\epsilon_{\mathcal{P}}$ is the signature of $\mathcal{P}$. Via the coordinate transformation (3.94) the function ψ^α depends on all the z^μ. The metric connection (3.96) can be expressed in the $\{\psi^\alpha, \mathfrak{F}^\alpha\}$ parametrization and in [**1262**] one computes also the components of the curvature tensor, the Ricci tensor, and the scalar curvature and gives an expression for the Einstein equations (we omit the details here). The same procedure apply to our formulas above which leads us to state heuristically (cf. [**256**])

THEOREM 3.1. The formulas (3.91), (3.92), (3.93), (3.94), (3.95) and (3.96), with their continuations, determine a geometry for a putative aether, expressed in terms of our so-called aether fields ψ_i.

REMARK 5.3.4. These matters are taken up again in [**51**] for a general curved spacetime and some sufficient constraints are isolated which make the theory work. Also in both papers a quantized version of the KG equation is also treated and the relevant $x - \psi$ duality is spelled out in operator form. We omit this also in remarking that the main feature here for our purposes is the fact that one can describe spacetime geometry (at least locally) in terms of (field) solutions of a KG equation and prepotentials (which are themselves functions of the fields). In other words the coordinates are programmed by fields and if the motion of some particle of mass m is involved then its coordinates are choreographed by the fields with a quantum potential eventually entering the picture via (3.84). In [**2**] a similar duality is worked out for the Dirac field and cartesian coordinates and to connect this with the aether idea one should examine such formulas for $m \to 0$.■

EXAMPLE 3.1. One knows that general solutions of the massless KG equation will have the form $\psi = \psi(\mathbf{a}\cdot\mathbf{x} - ct)$ with $|\mathbf{a}| = 1$. For example take $\psi = exp(\sum a_i x_i - ct)$ with $(1/c^2)\psi_{tt} = \psi$ and $\psi_{ii} = a_i^2\psi$. This leads to

(3.97) $$V^0 = 1;\ V^i = 1 - \sum_{j\neq i} a_j^2$$

Hence

(3.98) $$\frac{1}{c^2}\partial_t^2\psi^0 - \psi^0 = 0;\ \partial_i^2\psi^i - (1 - \sum_{j\neq i} a_j^2)\psi^i = 0$$

On the other hand if $\psi = f(\mathbf{a}\cdot\mathbf{x} - ct)$ one gets

(3.99) $$V^0 = \left(\frac{f''}{f}\right)(\mathbf{a}\cdot\mathbf{x} - ct);\ V^i = \left(1 - \sum_{j\neq i} a_j^2\right)\left(\frac{f''}{f}\right)(\mathbf{a}\cdot\mathbf{x} - ct)$$

Setting $f''/f = g(x^i, x^0)$ one has

(3.100) $$\partial_0^2\psi^0 - g(x^i, x^0)\psi^0 = 0;\ \partial_i^2\psi^i - \left(1 - \sum_{j\neq i} a_j^2\right) g(x^i, x^j, x^0)\psi^i$$

Here the x^i or $(x^j,\ x^0)$ are considered as parameters. ■

EXAMPLE 3.2. Consider a simple situation with two x^i variables and $x^0 = ct$ and take $a_1 = a_2 = 1/\sqrt{2}$. Then $V^0 = 1$ and $V^i = 1 - (1/2) = 1/2$ with

(3.101) $$\frac{1}{c^2}\partial_t^2\psi^0 = \psi^0;\ \frac{\partial^2\psi^i}{\partial(x^i)^2}\psi^i = \frac{1}{2}\psi^i$$

Hence we can take

(3.102) $$\psi^0 = A_0 e^{ct};\ \psi^i = A_i e^{(1/\sqrt{2})x^i};\ \tilde{\psi}^i = \tilde{A}_i e^{-(1/\sqrt{2})x^i};\ \tilde{\psi}^0 = \tilde{A}_0 e^{-ct}$$

Now $\psi^i\tilde{\psi}^i = \kappa_i$ for $i = 0, 1, 2$ so (recall $\beta^0 = \gamma$ and $x^0 = ct$)

(3.103) $$\beta^i x^i = \frac{1}{2}\kappa_i - \mathfrak{F}^i = \mathfrak{E}^i\ \ (i = 0, 1, 2)$$

and

$$\frac{1}{2}E^i = \frac{\partial}{\partial \psi^i}\left(\frac{1}{2}\kappa_i - \mathfrak{F}^i\right) = -\frac{\partial \mathfrak{F}^i}{\partial \psi^i} = -\tilde{\psi}^i \ (i = 0,1,2) \tag{3.104}$$

(the β^i here need not depend on other variables). Consequently one has

$$G_{\alpha\sigma}(\psi) = \frac{E^\alpha E^\sigma}{4\beta^\alpha \beta^\sigma}\eta_{\alpha\sigma}(x) = \frac{\tilde{\psi}^\alpha \tilde{\psi}^\sigma}{\beta^\alpha \beta^\sigma}\eta_{\alpha\sigma}(x) \tag{3.105}$$

and this exhibits in a simple example the manner in which the metric can depend on the fields. ■

EXAMPLE 3.3. We look now at the more complicated situation for $\psi = f(\mathbf{a}\cdot\mathbf{x} - ct)$ as in (3.99)-(3.100). Here $f''/f = g$ could be a fairly general function with argument $\mathbf{a}\cdot\mathbf{x} - ct$ and in the equations $\partial_i^2\psi^i = \alpha_i g \psi^i$ the function g_i is considered as a function of x^i with the other x^j as parameters. Let ψ^i and $\tilde{\psi}^i$ be two solutions ($i = 0,1,2,3$ say) and look at ($\psi_i \equiv \psi^i$)

$$\mathfrak{E}^i = \beta^i x^i = \frac{1}{2}\psi^i \frac{\partial \mathfrak{F}}{\partial \psi^i} - \mathfrak{F}^i;\ E^i = \psi^i \frac{\partial^2 \mathfrak{F}}{\partial \psi_i^2} - \frac{\partial \mathfrak{F}}{\partial \psi^i} \tag{3.106}$$

Recall $\tilde{\psi}^i = \partial\mathfrak{F}/\partial\psi^i$ and we can write, from Item 3 in Section 2, $\phi^i = \partial\mathfrak{F}/\partial(\psi^i)^2 = \tilde{\psi}^i/2\psi^i$ (although this will not be used here). In terms of the two fields ψ and $\tilde{\psi}^i$ one has

$$\mathfrak{E}^i = \frac{1}{2}\psi^i\tilde{\psi}^i - \mathfrak{F}^i = \beta^i x^i;\ \mathfrak{F}^i = \mathfrak{F}^i(\psi^i, \tilde{\psi}^i, x^i, \beta^i); \tag{3.107}$$

$$E^i = \psi^i \frac{\partial \tilde{\psi}^i}{\partial \psi^i} - \tilde{\psi}^i;\ \beta^i dx^i = \frac{1}{2}E^i d\psi^i$$

In particular E^i is expresed directly in terms of the fields ψ^i and $\tilde{\psi}^i$; no extraneous variables are explicit. Now ψ^i and $\tilde{\psi}^i$ are linearly independent solutions of $\partial_i^2\psi^i = \alpha_i g\psi$ but they are linked by a Wronskian $W_i = (\partial_x\psi^i)\tilde{\psi}^i - \psi^i(\partial_x\tilde{\psi}^i) = -2\beta^i$ where β^i does not depend on x^i (only perhaps on the other x^j). One can write now

$$\partial_x\left(\frac{\tilde{\psi}^i}{\psi^i}\right) = \frac{W_i}{\psi_i^2} \Rightarrow \tilde{\psi}^i = \psi^i \int^x \frac{W_i dx}{\psi_i^2} + c\psi^i \tag{3.108}$$

Formally this suggests

$$\frac{\partial \tilde{\psi}^i}{\partial \psi^i} = -2\beta^i \int^x \frac{dx}{\psi_i^2} + 4\beta^i\psi^i \int^x \frac{dx}{\psi_i^3} + c \tag{3.109}$$

from which follows

$$E^i = \psi^i\left[-2\beta^i \int^x \frac{dx^i}{\psi_i^2} + c + 4\beta^i\psi^i \int^x \frac{dx^i}{\psi_i^3} - \frac{2\beta^i}{\psi^i}\frac{dx^i}{d\psi^i}\right] + \tag{3.110}$$

$$+2\beta^i\psi^i \int^x \frac{dx^i}{d\psi^i} - c\psi^i = 4\beta^i\psi_i^2 \int^x \frac{dx^i}{\psi_i^3} - 2\beta^i\frac{E^i}{2\beta^i} \Rightarrow E^i = 2\beta^i\psi_i^2 \int^x \frac{dx^i}{\psi_i^3}$$

Thus E^i can be expressed entirely in terms of the field ψ^i. ■

One notes here that these arguments and results hold for any ψ^α, V^α as in (3.82)-(3.84) so we state heuristically (cf. [**256**])

THEOREM 3.2. The objects E^α used in constructing the geometry can be expressed in terms of fields ψ^α as in (3.110).

EXAMPLE 3.4. For an interesting connection of the ψ-aether with QM arises as follows. Consider a hydrogen atom with spherically symetric and time independent potential $V(\mathbf{r}) = V(r)$ where $r = |\mathbf{r}|$. The solution to the SE $-i\hbar\partial_t\psi = -(\hbar^2/2m)\nabla^2\psi + V(r)\psi$ is obtained by separation of variables $\psi = u(\mathbf{r})f(t)$ with $u(\mathbf{r}) = R(r)Y(\theta,\phi)$. This is a problem of two body interaction (a proton and an electron) and for stationary states with energy E one looks at $\psi(x,t) = Cexp(-iEt/\hbar)$ satisfying

$$\frac{1}{sin(\theta)}\frac{\partial}{\partial\theta}\left(sin(\theta)\frac{\partial Y}{\partial\theta} + \frac{1}{sin^2(\theta)}\frac{\partial^2 Y}{\partial\phi^2}\right) + \lambda Y = 0; \tag{3.111}$$

$$\frac{1}{r^2}\frac{d}{dr}\left(r^2\frac{dR}{dr}\right) + \left\{\frac{2\mu}{\hbar^2}[E - V(r)] - \frac{\lambda}{r^2}\right\}R = 0$$

Here μ is the reduced mass of the system (proton + electron), E is the energy level for the bound state $p + e$ $(E < 0)$, and $V(r) = e^2/r$ is the potential energy. (3.111) is solved by further separation of variables $Y = \Theta(\theta)\Phi(\phi)$ leading to

$$\frac{\partial^2\Phi}{\partial\phi^2} + \nu\Phi = 0;\ \frac{1}{sin(\theta)}\frac{d}{d\theta}\left(sin(\theta)\frac{\partial\Theta}{\partial\theta}\right) + \left(\lambda - \frac{\nu}{sin^2\theta}\right)\Theta = 0 \tag{3.112}$$

The solution for Φ is $\Phi_m(\phi) = (1/2\pi)exp(im\phi)$ with $\nu = m^2$ and physically admissible solutions for Θ (associated Legendre polynomials) require $\lambda = \ell(\ell+1)$ with $|m| \leq \ell$. For R one has

$$\frac{1}{r^2}\frac{d}{dr}\left(r^2\frac{dR}{dr}\right) + \frac{2\mu}{\hbar^2}\frac{e^2}{r}R(r) + \frac{2\mu}{\hbar^2}ER(r) - \frac{\ell(\ell+1)}{r^2}R = 0 \tag{3.113}$$

Now here $V(r) = (2\mu/\hbar^2)(e^2/r) = [\ell(\ell+1)/r^2]$ and the term involving e^2/r is responsible for the Coulomb interaction of a proton with an electron; however the second term $\ell(\ell+1)/r^2$ does not depend on any physical interaction (even though in [**1126**] it is said to be connected with angular momentum). Now putting the Coulomb interaction to zero the $\ell(\ell+1)/r^2$ term does not disappear and it makes no sense to attribute it to angular momentum. It is now claimed that in fact this term arises because of the ψ-aether and an argument based on standing waves in a spherical resonator is given. Thus following [**188**] one considers an associated Borgnis function $U(r,\theta,\phi)$, having definite connections to **E** and **H**, and when it satisfies

$$\frac{\partial^2 U}{\partial r^2} + \frac{1}{r^2 sin(\theta)}\left[\frac{\partial}{\partial\theta}sin(\theta)\frac{\partial U}{\partial\theta} + \frac{\partial}{\partial\phi}\frac{1}{sin(\theta)}\frac{\partial U}{\partial\phi}\right] + k^2 U = 0 \tag{3.114}$$

the Maxwell equations are also valid. Further U is connected by definite relations with **A** and ϕ, i.e. with the ψ-aether (presumably all this is spelled out in [**188**]). To solve (3.114) one writes $U = F_1(r)F_2(\theta,\phi)$ (following the notation of [**188**]) and there results

$$(\mathbf{A})\quad \frac{1}{sin(\theta)}\frac{\partial}{\partial\theta}sin(\theta)\frac{\partial F_2}{\partial\theta} + \frac{1}{sin^2\theta}\frac{\partial^2 F_2}{\partial\phi^2} + \gamma F_2 = 0 \tag{3.115}$$

$$(\mathbf{B})\ \ r^2\frac{\partial^2 F_1}{\partial r^2} + k^2r^2F_1 - \gamma F_1 = 0$$

One considers here EM waves harmonic in time and characterized either by the frequency $\nu = kc/2\pi$ or by the wave vector $k = 2\pi\nu/c$ with $[k] = 1/cm$. Now **(A)** in (3.115) is the same as (3.111) with spherical function solutions and regular solutions of **(B)** in (3.115) exist when $\gamma = n(n+1)$. Setting $F_1(r) = rf(r)$ one obtains then

$$\frac{d^2f}{dr^2} + \frac{2}{r}\frac{df}{dr} + \left[k^2 - \frac{n(n+1)}{r^2}\right]f(r) = 0 \tag{3.116}$$

A little calculation puts (3.113) into the form

$$\frac{d^2R}{dr^2} + \frac{2}{r}\frac{dR}{dr} + \left(\frac{2\mu E}{\hbar^2} + \frac{2\mu e^2}{\hbar^2 r} - \frac{\ell(\ell+1)}{r^2}\right)R = 0 \tag{3.117}$$

Setting $2\mu e^2/\hbar^2 r = 0$ and replacing E by $E = p^2/2\mu$ in (3.117) one obtains

$$\frac{d^2R}{dr^2} + \frac{2}{r}\frac{dR}{dr} + \left(k^2 - \frac{\ell(\ell+1)}{r^2}\right)R = 0 \tag{3.118}$$

where $2\mu^2p^2/2\mu\hbar^2 = k^2\hbar^2/\hbar^2 = k^2$ with k the wave vector). Now (3.116) and (3.118) are identical and are solved under the same boundary conditions (i.e. $f(r)$ should be finite as $r \to 0$ and when $r \to \infty$ one wants $f(r) \to 0$ on the boundary of a sphere). The corresponding solutions to (3.116) represent standing waves inside the sphere at values $n = 0, 1, \cdots$ with $m \leq n$. Since EM waves are nothing but oscillations of the ψ-aether the term $n(n+1)/r^2$ in (1.20) is responsible for standing waves of the ψ-aether in a sphere resonator. Thus (mathematically at least) one can say that the problem of finding the energy levels in a hydrogen atom via the SE is equivalent to the problem of finding natural EM oscillations in a spherical resonator. One recalls that one of the basic postulates of QM (quantization of orbits in a hydrogen atom à la Bohr with $mvr = n\hbar/2\pi$) is equivalent to determination of conditions for existence of standing waves of the ψ-aether in a spherical resonator. This suggests that QM may be equivalent to "mechanics" of the ψ-aether. One remarks that until now only a small part of the alleged ψ-aether properties have been observed, namely in superfluidity and superconductivity (see e.g. Volovik [**1267**]). It is suggested that one might well rethink a lot of physics in terms of the aether, rather than, for example, the standard model. In any event there is much further discussion in [**662**], related to real physical situations, and well worth reading. ∎

CHAPTER 6

INFORMATION AND ENTROPY

Information and entropy have already been discussed and we continue with further elaboration (see in particular [**10, 25, 91, 177, 216, 217, 218, 300, 440, 498, 509, 513, 565, 594, 602, 647, 692, 858, 861, 863, 922, 972, 995, 1006, 1018, 1019, 1138, 1197, 1203, 1206, 1205, 1280**]). As before we will again encounter relations to the quantum potential which serves as a persistent theme of development. There is an enormous literature on entropy and we try to select aspects which fit in with ideas of quantum diffusion and information theory.

1. THE DYNAMICS OF UNCERTAINTY

We begin with some topics from Garbaczewski [**509**] to which we refer for certain tutorial aspects. Given events A_j $(1 \leq j \leq N)$ with probabilities μ_j of occurance in some game of chance with N possible outcomes one calls $log(\mu_j)$ an uncertainty function for A_j. We write the natural logarithm as log and recall that e.g. $log_2(b) = log(b)/ln(2)$ (the information theoretic base is taken as 2 in some contexts). The quantity (Shannon entropy)

$$\mathfrak{S}(\mu) = -\sum_1^N \mu_j log_2(\mu_j) \tag{1.1}$$

stands for the measure of the mean uncertainty of the possible outcomes of the game and at the same time quantifies the mean information which is accessible from an experiment (i.e. actually playing the game). Thus if one identifies the A_i as labels for discrete states of a system (7.1) can be interpreted as a measure of uncertainty of the state before this state is chosen and the Shannon entropy is a measure of the degree of ignorance concerning which possibility (event A_j) may hold true in the set of all $A_i^{'s}$ with a given a priori probability distribution (μ_i). Note also that $0 \leq \mathfrak{S}(\mu) \leq log_2(N)$ (since certainty means one entry with probability 1 and maximum uncertainty occurs when all events are equally probable with $\mu_j = 1/N$). There is some discussion of the Boltzman law $\mathfrak{S} = k_B log(W) = -k_B log(P)$ $(P = 1/W)$ and its relation to Shannon entropy, coarse graining, and differential entropy defined as

$$\mathfrak{S}(\rho) = -\int \rho(x) log(\rho(x)) dx \tag{1.2}$$

(cf. Sections 1.1.6 and 1.1.8). One recalls also the vonNeumann entropy

$$\mathfrak{S}(\hat{\rho}) = -k_B Tr(\hat{\rho} log(\hat{\rho})) \tag{1.3}$$

where $\hat{\rho}$ is the density operator for a quantum state ($\hat{\rho}log(\rho)$ is defined via functional calculus for selfadjoint operators (cf. [**1206**]). For diagonal density operators with eigenvalues p_i this will coincide with the Shannon entropy $\sum p_i log(p_i)$. We go now directly to an extension of the discussion in Sections 1.1.6-1.1.8. It is known from Shannon [**1148**] that among all one dimensional distributions $\rho(x)$ with a finite mean, subject to the condition that the standard deviation is fixed at σ, it is the Gaussian with half width σ which sets a maximum of the differential entropy. Thus for the Gaussian with $\rho(x) = (1/\sigma\sqrt{2\pi})exp[-(x-x_0)^2/2\sigma^2]$ one has

$$\mathfrak{S}(\rho) \leq \frac{1}{2}log(2\pi e\sigma^2) \Rightarrow \frac{1}{\sqrt{2\pi e}}exp[\mathfrak{S}(\rho)] \leq \sigma \tag{1.4}$$

A result of this is that the major role of the differential entropy is to be a measure of localization in the configuration space (note that even for relatively large mean deviations $\sigma < 1/\sqrt{2\pi e} \simeq .26$ the differential entropy $\mathfrak{S}(\rho)$ is negative. Consider now a one parameter family of probability densities $\rho_\alpha(x)$ on $\mathbf{R}$ whose first (mean) and second moments (variance) are finite. Write $\int x\rho_\alpha(x)dx = f(\alpha)$ with $\int x^2\rho_\alpha dx < \infty$. Under suitable hypotheses (implying that $\partial\rho_\alpha/\partial\alpha$ is bounded by a function $G(x)$ which together with $xG(x)$ is integrable on $\mathbf{R}$) one obtains

$$\int (x-\alpha)^2\rho_\alpha(x)dx \cdot \int \left(\frac{\partial log(\rho_\alpha)}{\partial\alpha}\right)^2 \rho_\alpha dx \geq \left(\frac{df(\alpha)}{d\alpha}\right)^2 \tag{1.5}$$

which results from

$$\frac{df}{d\alpha} = \int [(x-\alpha)\rho_\alpha^{1/2}] \left[\frac{\partial(log(\rho_\alpha))}{\partial\alpha}\rho_\alpha^{1/2}\right] dx \tag{1.6}$$

and the Schwartz inequality. Assume now that the mean value of ρ_α actually is α and fix at σ^2 the value of the variance $<(x-\alpha)^2>=<x^2>-\alpha^2$. Then (7.5) takes the familiar form

$$\mathfrak{F}_\alpha = \int \frac{1}{\rho_\alpha}\left(\frac{\partial\rho_\alpha}{\partial\alpha}\right)^2 dx \geq \frac{1}{\sigma^2} \tag{1.7}$$

where the left side is the Fisher information for ρ_α. This says that the Fisher information is a more sensitive indicator of the wave packet localization than the entropy power in (7.4). Consider now $\rho_\alpha = \rho(x-\alpha)$ so $\mathfrak{F}_\alpha = \mathfrak{F}$ is no longer dependent on α and one can transform this to the QM form (up to a factor of D^2 where $D = \hbar/2m$ which we acknowledge here via the symbol $\sim$)

$$\frac{1}{2}\mathfrak{F} = \int \frac{1}{\rho}\left(\frac{\partial\rho}{\partial x}\right)^2 dx \sim \int \rho\cdot\frac{u^2}{2}dx \sim -<\tilde{Q}> \tag{1.8}$$

where $u = \nabla log(\rho)$ is the osmotic velocity field and the average $<\tilde{Q}>=\int\rho\cdot\tilde{Q}dx$ involves a "quantum potential" $\tilde{Q} = 2(\Delta\sqrt{\rho}/\sqrt{\rho})$ (cf. equations (6.1.13) - (6.1.16)). Consequently $-<\tilde{Q}>\geq(1/2\sigma^2)$ for all relevant probability densities with any finite mean (with variance fixed at σ^2). We continue in this section with the notation $\tilde{Q} = 2(\Delta\sqrt{\rho}/\sqrt{\rho})$ and note that $D^2\tilde{Q} = -(1/m)Q$ where Q is our standard quantum potential ($D = \hbar/2m$).

Next one defines the Kullback entropy $K(\theta,\theta')$ for a one parameter family of probability densities ρ_θ so that the distance between any two densities can be directly evaluated. Let $p_{\theta'}$ be the reference density and one writes

$$K(\theta,\theta') = K(\rho_\theta|\rho_{\theta'}) = \int \rho_\theta(x) log\frac{\rho_\theta(x)}{\rho_{\theta'}(x)}dx \tag{1.9}$$

(note this is positive and sometimes one refers to $\mathfrak{H}_c = -K$ as a conditional entropy). If one takes $\theta' = \theta + \Delta\theta$ with $\Delta\theta << 1$ then under a number of standard assumptions

$$K(\theta,\theta+\Delta\theta) \simeq \frac{1}{2}\mathfrak{F}_\theta \cdot (\Delta\theta)^2 \tag{1.10}$$

where $\mathfrak{F}_\theta$ denotes the Fisher information measure as in (7.7). More generally for a two parameter family $\theta \sim (\theta_1,\theta_2)$ of densities one has

$$K(\theta, gt+\Delta\theta) \simeq \frac{1}{2}\sum \mathfrak{F}_{ij}\Delta\theta_i\Delta\theta_j;\ \mathfrak{F}_{ij} = \int \rho_\theta \frac{\partial log(\rho_\theta)}{\partial\theta_i}\frac{\partial log(\rho_\theta)}{\partial\theta_j}dx \tag{1.11}$$

For Gaussian densities at fixed σ with $\theta = \alpha$ one has then $K(\alpha,\alpha+\Delta\alpha) \simeq (\Delta\alpha)^2/2\sigma^2$. Various related formulas are derived and in particular one relates the Shannon entropy for a coarse grained density ρ_B to the differential entropy of the density ρ leading to a formula $\mathfrak{S}(\rho_B) - \mathfrak{S}(\rho'_B) \simeq \mathfrak{S}(\rho) - \mathfrak{S}(\rho')$. One considers also spatial Markov diffusion processes in $\mathbf{R}$ with a diffusion coefficient D which drive space-time inhomogeneous probability density densities $\rho(x,t)$. For example a free Brownian motion characterized by $v = -u = -D\nabla log(\rho(x,t))$ and diffusion current $j = v\cdot\rho$ obeys the continuity equation $\partial_t\rho = -\nabla j$ which is equivalent to the heat equation. As in Sections 1.1.6-1.1.8 and 6.1 we have the important relations

$$\tilde{Q} = 2D^2\frac{\Delta\rho^{1/2}}{\rho^{1/2}} = \frac{1}{2}u^2 + D\nabla\cdot u;\ \partial_t v + (v\cdot\nabla)v = -\nabla\tilde{Q} \tag{1.12}$$

A straightforward generalization refers to a diffusive dynamics of a mass m in a conservative force field $F = -\nabla V$. The associated Smoluchowski diffusion with a forward drift $b(x) = F/m\beta$ is analyzed in terms of a Fokker-Planck (FP) equation $\partial_t\rho = D\Delta\rho - \nabla(b\cdot\rho)$ with initial data $\rho_0(x) = \rho(x,0)$. For standard Brownian motion in an external force field one has $D = k_BT/m\beta$ where $\beta \sim$ friction, T is temperature and k_B is the Boltzman constant. With suitable hypotheses one has the following compatibility equations in the form of hydrodynamical conservation laws

$$\partial_t\rho + \nabla(v\rho) = 0;\ (\partial_t + v\cdot\nabla)v = \nabla(\Omega - \tilde{Q}) \tag{1.13}$$

where $\Omega(x)$ is the volume potential for the process, namely

$$\Omega = \frac{1}{2}\left(\frac{F}{m\beta}\right)^2 + D\nabla\cdot\left(\frac{F}{m\beta}\right) \tag{1.14}$$

Here $v = b - u = (F/m\beta) - D(\nabla\rho/\rho)$ defines the current velocity of Brownian particles in an external force field. With a solution ρ of the FP equation one associates a differential entropy $\mathfrak{S}(t) = -\int \rho log(\rho)dx$ which is typically not conserved.

With boundary conditions on ρ, $v\rho$, and $b\rho$ involving vanishing at boundaries or at infinity one obtains

(1.15) $$\frac{d\mathfrak{S}}{dt} = \int \left[\rho(\nabla \cdot b) + D\frac{(\nabla\rho)^2}{\rho}\right] dx$$

One emphasizes that it is not obvious whether the differential entropy grows, decreases, or whatever. One can rewrite (7.15) in the forms

(1.16) $$D\dot{\mathfrak{S}} = D < \nabla \cdot b > + < u^2 >= D < \nabla \cdot v >;$$

$$D\dot{\mathfrak{S}} =< v^2 > - < b \cdot v >= - < v \cdot u >$$

where $< >$ denotes the mean value relative to ρ. For $b = F/m\beta$ and $j = v\rho$ this leads to a characteristic "power release" expression

(1.17) $$\frac{d\tilde{Q}}{dt} = \frac{1}{D}\int \frac{1}{m\beta}F \cdot jcx = \frac{1}{D} < b \cdot v >$$

Again $\dot{Q}$ can be positive (power removal) or negative (power absorption). In thermodynamic terms one deals here with the time rate at which the mechanical work per unit of mass is dissipated (removed from the reservoir) in the form of heat in the course of the Smoluchowski diffusion process - i.e. $k_BT\dot{\tilde{Q}} = \int F \cdot jdx$ where T is the temperature of the bath. For $b = 0$ (no external forces) one has $D\dot{\mathfrak{S}} = D^2 \int[(\nabla\rho)^2/\rho]dx = D^2\mathfrak{F} = -D^2 < \tilde{Q} >$ and one can also write

(1.18) $$\frac{d\mathfrak{S}}{dt} = \left(\frac{d\mathfrak{S}}{dt}\right)_{in} - \frac{d\tilde{Q}}{dt}$$

from (7.15) and (7.16) (here $(\dot{\mathfrak{S}})_{in} = (1/D) < v^2 >$.

One goes now to mean energy and the dynamics of Fisher information and considers $-\rho$ and s where $v = \nabla s$ as canonically conjugate fields; then one can use variational calculus to derive the continuity and FP equations together with the HJ type equations whose gradient gives the hydrodynamical conservation law

(1.19) $$\partial_t s + (1/2)(\nabla s)^2 - (\Omega - \tilde{Q}) = 0$$

Here the mean Lagrangian is

(1.20) $$\mathfrak{L} = -\int \rho \left[\partial_t s + \frac{1}{2}(\nabla s)^2 - \left(\frac{u^2}{2} + \Omega\right)\right] dx$$

The related Hamiltonian (mean energy of the diffusion process per unit of mass) is

(1.21) $$\mathfrak{H} = \int \rho \left[\frac{1}{2}(\nabla s)^2 - \left(\frac{u^2}{2} + \Omega\right)\right] dx = \frac{1}{2}(< v^2 > - < u^2 >) - < \Omega >$$

(note here $v = \nabla s$ satisfies $v = b - u$ with $u = D\nabla log(\rho)$ and we refer to Section 1.1 for clarification). One defines a thermodynamic force $F_{th} = v/D$ associated with the Smoluchowski diffusion with a corresponding potential $-\nabla\Psi = k_BTF_{th} = F - k_BT\nabla log(\rho)$ so in the absence of external forces $F_{th} = -\nabla log(\rho) = -(1/D)u$. The mean value of the thermodynamic force associates with the diffusion process an analogue of the Helmholz free energy $<\Psi>=<V>-T\mathfrak{S}_G$ where the dimensional version $\mathfrak{S}_G=k_B\mathfrak{S}$ of information entropy has been introduced (it is a configuration space analogue of the Gibbs entropy). Here the term $<V>$ plays the role of (mean) internal energy and assuming ρv vanishes at boundaries (or infinity) one obtains the time rate of change of Helmholz free energy at a constant temperature, namely

(1.22)
$$\frac{d}{dt}<\Psi>=-k_BT\dot{\tilde{Q}}-T\dot{\mathfrak{S}}_G\Rightarrow\frac{d}{dt}<\Psi>=-(k_BT)\left(\frac{d\mathfrak{S}}{dt}\right)_{in}=-(m\beta)<v^2>$$

Now one can evaluate an expectation value of (7.19) which implies an identity $\mathfrak{H}=-<\partial_t s>$. Then using $\Psi=V+k_BTlog(\rho)$ (with time independent V) one arrives at $\dot{\Psi}=(k_BT/\rho)\nabla(v\rho)$ and since $v\rho=0$ at integration boundaries we get $<\dot{\Psi}>=0$. Since $v=-(1/m\beta)\nabla\Psi$ define then $s(x,t)=(1/m\beta)\Psi(x,t)$ so that $<\partial_t s>=0$ and hence $\mathfrak{H}=0$ identically. This gives an interplay between the mean energy and the information entropy production rate in the form

$$\frac{D}{2}\left(\frac{d\mathfrak{S}}{dt}\right)_{in}-\frac{1}{2}<v^2>=\int\rho\left(\frac{u^2}{2}+\Omega\right)dx\geq 0 \tag{1.23}$$

Next recalling (7.7)-(7.8) and setting $\mathfrak{F}=D^2\mathfrak{F}_\alpha$ one obtains

$$\mathfrak{F}=<v^2>-2<\Omega>\geq 0 \tag{1.24}$$

where $(1/2)\mathfrak{F}=-<\tilde{Q}>$ holds for probability densities with finite mean and variance. One also derives the following formulas (under suitable hypotheses)

$$\partial_t(\rho v^2)=-\nabla\cdot[(\rho v^3)]-2\rho v\cdot\nabla(\tilde{Q}-\Omega); \tag{1.25}$$

$$\frac{d}{dt}<\Omega>=<v\cdot\nabla\Omega>;\ \frac{d}{dt}\mathfrak{F}=\frac{d}{dt}[<v^2>-2<\Omega>]=-2<v\cdot\nabla\tilde{Q}>$$

Then since $\nabla\tilde{Q}=\nabla P/\rho$ where $P=D^2\rho\Delta log(\rho)$ the previous equation takes the form $\dot{\mathfrak{F}}=-\int\rho v\nabla\tilde{Q}dx=-\int v\nabla P dx$ which is an analogue of the familiar expression for the power release $(dE/dt)=F\cdot v$ with $F=-\nabla V$ in classical mechanics.

Next in [**509**] there is a discussion of differential entropy dynamics in quantum theory. Assume one has an arbitrary continuous function $\mathcal{V}(x,t)$ with dimensions of energy and consider the SE in the form $i\partial_t\psi=-D\Delta\psi+(\mathcal{V}/2mD)\psi$. Using $\psi=\rho^{1/2}exp(is)$ with $v=\nabla s$ one arrives at the standard equations $\partial_t\rho=-\nabla(v\rho)$ and $\partial_t s+(1/2)(\nabla s)^2+(\Omega-\tilde{Q})=0$ where $\Omega=\mathcal{V}/m$ and $\tilde{Q}$ has the same form as in (7.12) (note a sign change of the $\Omega-\tilde{Q}$ term in comparison with (7.19)). These two equations generate a Markovian diffusion type process the probability density of which is propagated by a FP dynamics as before with drift $b=v-u$ (instead of $v=b-u$) where $u=D\nabla log(\rho)$ is an osmotic velocity field. Repeating the variational calculations one looks at (cf. (7.21))

$$\mathfrak{H}=\int\rho\left[\frac{1}{2}(\nabla s)^2+\left(\frac{u^2}{2}+\Omega\right)\right]dx \tag{1.26}$$

Then

$$\mathfrak{H} = (1/2)[< v^2 > + < u^2 >] + < \Omega > = - < \partial_t s > \tag{1.27}$$

For time independent $\mathcal{V}$ one has $\mathfrak{H} = - < \partial_t s > = \mathcal{E} = const.$ and the FP equation propagates a probability density $|\psi|^2 = \rho$ whose differential entropy $\mathfrak{S}$ may nontrivially evolve in time. Maintaining the previous derivations involving $(\dot{\mathfrak{S}})_{in}$ one arrives at

$$(\dot{\mathfrak{S}})_{in} = \frac{2}{D}\left[\mathcal{E} - \left(\frac{1}{2}\mathfrak{F} + < \Omega >\right)\right] \geq 0 \tag{1.28}$$

One recalls $(1/2)\mathfrak{F} = - < \tilde{Q} > > 0$ so $\mathcal{E} - < \Omega > \geq (1/2)\mathfrak{F} > 0$. Hence the localization measure $\mathfrak{F}$ has a definite upper bound and the pertinent wave packet cannot be localized too sharply. Note also that $\mathfrak{F} = 2(\mathcal{E} - < \Omega >) - < v^2 >$ in general evolves in time (here $\mathcal{E}$ is a constant and $\dot{\Omega} = 0$). Using the hydrodynamical conservation laws one sees that the dynamics of Fisher information follows the rules

$$\frac{d\mathfrak{F}}{dt} = 2 < v\nabla\tilde{Q} >; \ \frac{1}{2}\dot{\mathfrak{F}} = -\frac{d}{dt}\left[\frac{1}{2} < v^2 > + < \omega >\right] \tag{1.29}$$

However $\dot{\mathfrak{F}} = \int v\nabla P dx$ where $P = D^2\rho\Delta log(\rho)$ and one interprets $\dot{\mathfrak{F}}$ as the measure of power transfer - keeping intact an overall mean energy $\mathfrak{H} = \mathcal{E}$. We refer to [**509**] for much more discussion and examples. We have concentrated on topics where the quantum potential appears in some form.

1.1. INFORMATION DYNAMICS. We go here to J and X. Calmet et al [**216, 217, 218**] and consider the idea of introducing some kind of dynamics in a reasoning process (Fisher information can apparently be linked to semantics - cf. [**1198, 1271**]). In [**216, 217**] one looks at the Fisher metric defined by

$$g_{\mu\nu} = \int_X d^4x p_\theta(x)\left(\frac{1}{p_\theta(x)}\frac{\partial p_\theta(x)}{\partial\theta^\mu}\right)\left(\frac{1}{p_\theta(x)}\right)\left(\frac{\partial p_\theta(x)}{\partial\theta^\nu}\right) \tag{1.30}$$

and constructs a Riemannian geometry via

$$\Gamma^\sigma_{\lambda\nu} = \frac{1}{2}g^{\nu\sigma}\left(\frac{\partial g_{\mu\nu}}{\partial\theta^\lambda} + \frac{\partial g_{\lambda\nu}}{\partial\theta^\mu} - \frac{\partial g_{\mu\lambda}}{\partial\theta^\nu}\right); \tag{1.31}$$

$$R^\lambda_{\mu\nu\kappa} = \frac{\partial\Gamma^\lambda_{\mu\nu}}{\partial\theta^\kappa} - \frac{\partial\Gamma^\lambda_{\mu\kappa}}{\partial\theta^\nu} + \Gamma^\eta_{\mu\nu}\Gamma^\lambda_{\kappa\eta} - \Gamma^\eta_{\mu\kappa}\Gamma^\lambda_{\nu\eta}$$

Then the Ricci tensor is $R_{\mu\kappa} = R^\lambda_{\mu\lambda\kappa}$ and the curvature scalar is $R = g^{\mu\kappa}R_{\mu\kappa}$. The dynamics associated with this metric can then be described via functionals

$$J[g_{\mu\nu}] = -\frac{1}{16\pi}\int\sqrt{g(\theta)}R(\theta)d^4\theta \tag{1.32}$$

leading upon variation in $g_{\mu\nu}$ to equations

$$R^{\mu\nu}(\theta) - \frac{1}{2}g^{\mu\nu}(\theta)R(\theta) = 0 \tag{1.33}$$

Contracting with $g_{\mu\nu}$ gives then the Einstein equations $R^{\mu\nu}(\theta) = 0$ (since $R = 0$). J is also invariant under $\theta \to \theta + \epsilon(\theta)$ and variation here plus contraction leads to a contracted Bianchi identity. Constraints can be built in by adding terms $(1/2)\int\sqrt{g}T^{\mu\nu}g_{\mu\nu}d^4\theta$ to $J[g_{\mu\nu}]$. If one is fixed on a given probability distribution

$p(x)$ with variable θ^μ attached to give $p_\theta(x)$ then this could conceivably describe some gravitational metric based on quantum fluctuations for example. As examples a Euclidean metric is produced in 3-space via Gaussian $p(x)$ and complex Gaussians will give a Lorentz metric in 4-space. However it seems to be very restrictive to have a fixed $p(x)$ as the basis; it would be nice if one could vary the probability distribution in some more general manner and study the corresponding Fisher metrics (and this seems eminently doable with a Fisher metric over a space of probability distributions).

1.2. INFORMATION MEASURES FOR QM. We follow here Parwani [**995**] and derive the SE within an information theoretic framework somewhat different from the exact uncertainty principle of Hall and Reginatto (cf. Sections 1.1, 3.1, and 4.7). Begin with a SE for N particles in $d+1$ dimensions of the form $i\hbar\psi_t = [-(\hbar^2/2m)g_{ij}\partial_i\partial_j + V]\psi$ with $g_{ij} = \delta_{ij}/m_{[i]}$ where $i,j = 1,\cdots,dN$ and $[i]$ is the smallest integer $\geq i/d$. Use the Madelung transformation $\psi = \sqrt{\rho}exp(iS/\hbar)$ (cf. [**831**]) to get
(1.34)

$$\partial_t S + \frac{g_{ij}}{2}\partial_i S\partial_j S + V - \frac{\hbar^2}{8}g_{ij}\left(\frac{2\partial_i\partial_j\rho}{\rho} - \frac{\partial_i\rho\partial_j\rho}{\rho^2}\right) = 0;\ \partial_t\rho + g_{ij}\partial_i(\rho\partial_j S) = 0$$

These equations can be obtained from a variational principle, minimizing the action

$$\Phi = \int \rho\left[\partial_t S + \frac{g_{ij}}{2}\partial_i S\partial_j S + V\right]dx^{Nd}dt + \frac{\hbar^2}{8}I_F; \tag{1.35}$$

$$I_F = \int dx^{Nd}dt g_{ij}\rho(\partial_i log(\rho))(\partial_j log(\rho))$$

Here I_F resembles the Fisher information of [**471**] whose inverse sets a lower bound on the variance of the probabiliy distribution ρ via the Cramer-Rao inequaliity (see Section 1.1). (7.44) was used to derive the SE through a procedure analogous to the principle of maximum entropy in Reginatto-Lenguel [**1063, 1064**] (cf. also Section 1.1). However the method of [**1063**] does not explain a priori the form of information measure that should be used; i.e. why must the Fisher information be minimized rather than something else. The aim of [**995**] is to construct permissible information measures I. Thus the relevant action is

$$\mathfrak{A} = \int \rho\left[\partial_t S + \frac{g_{ij}}{2}\partial_i S\partial_j kS + V\right]dx^{Nd}dt + \lambda I \tag{1.36}$$

with λ a Lagrange multiplier. Varying this action will lead in general to a nonlinear SE

$$i\hbar\partial_t\psi = \left[-\frac{\hbar^2}{2}g_{ij}\partial_i\partial_j + V\right]\psi + F(\psi,\psi^\dagger)\psi \tag{1.37}$$

In order to have deformations of the linear theory that permit maximal preservation of the usual interpretation of the wave function one considers the following conditions:

(1) I should be real valued and positive definite for all $\rho = \psi^\dagger\psi$ and should be independent of V.

(2) I should be of the form $I = \int dx^{Nd} dt\rho H(\rho)$ where H is a function of $\rho(x,t)$ and its spatial derivatives. This will insure the weak superposition principle in the equations of motion.
(3) H should be invariant under scaling, i.e. $H(\lambda\rho) = H(\rho)$ which allows solutions of (7.46) to be renormalized, etc.
(4) H should be separable for the case of two independent subsystems for which the wave function factorizes, i.e. $H(\rho_1\rho_2) = H(\rho_1) + H(\rho_2)$.
(5) H should be Galilean invariant.
(6) The action should not contain derivatives beyond second order (Absence of higher order derivatives or AHD condition). This will insure that the multiplier λ, and hence Planck's constant, will be the only new parameter that is required in making the transition from classical to quantum mechanics.

The conditions 2-6 are already satisfied by the classical part of the action so it is quite minimalist to require them also of I. The homogeneity requirement 3 cannot be satisfied if H depends only on ρ; it must contain derivatives and the AHD and rotational invariance conditions imply then that $H = g_{ij}(U_1\partial_i U_2 \partial_j U_3 + V_1\partial_i\partial_j V_2)$ where the U_i, V_i are functions of ρ. One can write then

$$H = g_{ij}\left(\frac{\partial_i\rho\partial_j\rho}{\rho^2}[U_1U_2'U_3'\rho^2 + V_1V_2''\rho^2] + \frac{\partial_i\partial_j\rho}{\rho}[V_1V_2'\rho]\right) \tag{1.38}$$

where the prime denotes a derivative with respect to ρ. Scaling conditions plus positivity and universality then lead to

$$I = \int dtdx^{Nd}\rho g_{ij}\frac{\partial_i\rho\partial_j\rho}{\rho^2} \tag{1.39}$$

Consequently the unique solution of the conditions 1-6 is the Fisher information measure and one arrives at the linear SE since the Lagrange multiplier must then have the dimension of *action*^{2} thereby introducing the Planck constant. Note condition 4 was not used but it will be useful below. Further one notes that the AHD condition ensures that within the information theoretic approach the SE is the unique single parameter extension of the classical HJ equations. One also argues that a different choice of metric in the information term would in fact lead back to the original g_{ij} after a nonlinear gauge transformation; this suggests that a nonlinear SE is not automatically pathological. Further argument also shows that I should not depend on S. The main difference between this and the Hall-Reginatto method is to replace the exact uncertainly principle by condition 3.

1.3. PHASE TRANSITIONS. Referring to Janke, Johnson, and Kenna [**692**] the introduction of a metric onto the space of parameters in models in statistical mechanics gives an alternative perspective on their phase structure. In fact the scalar curvature $\mathcal{R}$ plays a central role where for a flat geometry $\mathcal{R} = 0$ (noninteracting system) while $\mathcal{R}$ diverges at the critical point of an interacting one. Thus models are characterized by certain sets of parameters and given a probability distribution $p(x|\theta)$ and a sample x_i the object is to estimate the parameter θ. This can be done by maximizing the so-called likelihood function $L(\theta) = \prod_1^n p(x_i|\theta)$ or

its logarithm. Thus one writes
(1.40)

$$log(L(\theta)) = \sum_{1}^{n} log(p(x_i|\theta));\ U(\theta) = \frac{d\,log(L(\theta))}{d\theta};\ Var[U(\theta)] = -\left[\frac{-d^2log(L(\theta))}{d\theta^2}\right]$$

The last term $Var[U(\theta))]$ is called the expected or Fisher information and we note that it is the same as (7.39) (see below) and in multidimensional form is expressed via

$$G_{ij}(\theta) = -E\left[\frac{\partial^2 log(p(x|\theta))}{\partial\theta_i\partial\theta_j}\right] = -\int p(x|\theta)\frac{\partial^2 log(p(x|\theta))}{\partial\theta_i\partial\theta_j}dx \tag{1.41}$$

In generic statistical-physics models one often has two parameters β (inverse temperature) and h (external field); in this case the Fisher-Rao metric is given by $G_{ij} = \partial_i\partial_j f$ where f is the reduced free energy per site and this leads to a scalar curvature

$$\mathcal{R} = -\frac{1}{2G^2}\begin{vmatrix} \partial_\beta^2 f & \partial_\beta\partial_h f & \partial_h^2 f \\ \partial_\beta^3 f & \partial_\beta^2\partial_h f & \partial_\beta\partial_h^2 f \\ \partial_\beta^2\partial_h f & \partial_\beta\partial_h^2 f & \partial_h^3 f \end{vmatrix} \tag{1.42}$$

where $G = det(G_{ij})$. In some sense $\mathcal{R}$ measures the complexity of the system since for $\mathcal{R} = 0$ the system is not interacting and (in all known systems) the curvature diverges at, and only at, a phase transition point. As an example under standard scaling assumptions one can anticipate the behavior of $\mathcal{R}$ near a second order critical point. Set $t = 1 - (\beta/\beta_c)$ and consider

$$f(\beta, h) = \lambda^{-1} f(t\lambda^{a_t}, h\lambda^{a_h}) = t^{1/a_t}\psi(ht^{-a_h/a_t});\ a_t = \frac{1}{\nu d};\ a_h = \frac{\beta\delta}{\nu d} \tag{1.43}$$

a_t, a_h are the scaling dimensions for the energy and spin operators and d is the space dimension. For the scalar curvature there results

$$\mathcal{R} = -\frac{1}{2G^2}\begin{vmatrix} t^{(1/a_t)-2} & 0 & t^{(1/a_t)-2(a_h/a_t)} \\ t^{(1/a_t)-3} & 0 & t^{(1/a_t)-2(a_h/a_t)-1} \\ 0 & t^{(1/a_t)-2(a_h/a_t)-1} & t^{(1/a_t)-3(a_h/a_t)} \end{vmatrix}; \tag{1.44}$$

$$G \sim t^{(2/a_t)+2(a_h/a_t)-2} \Rightarrow \mathcal{R} \sim \xi^d \sim |\beta - \beta_c|^{\alpha-2}$$

where hyperscaling ($\nu d = 2 - \alpha$) is assumed and ξ is the correlation length. We refer to [**692**] for more details, examples, and references.

1.4. FISHER INFORMATION AND HAMILTON EQUATIONS. Going to Pennini and Plastino [**1006**] one shows that the mathematical form of the Fisher information I for a Gibbs canonical probability distribution incorporates important features of the intrinsic structure of classical mechanics and has a universal form in terms of forces and accelerations (i.e. one that is valid for all Hamiltonians of the form $T+V$). First one has shown that the Fisher information measure provides a powerful variational principle, that of extreme information, which yields most of the canonical Lagrangians of theoretical physics. In addition I provides an interesting characterization of the "arrow of time", alternative to the one associated with the Boltzman entropy (cf. [**1030, 1031**]). Following Frieden and Soffer [**490, 493**] one considers a (θ, z) "scenario" in which we deal with a system specified by a physical parameter θ while z is a stochastic variable ($z \in \mathbf{R}^M$) and $f_\theta(z)$ is a probability density for z. One makes a measurement of z and has to infer θ, calling the resulting estimate $\tilde{\theta} = \tilde{\theta}(z)$. Estimation theory states that the best possible estimator $\tilde{\theta}(z)$, after a large number of samples, suffers a mean-square error e^2 from θ that obeys a relationship involving Fisher's I, namely $Ie^2 = 1$, where $I(\theta) = \int dz f_\theta(z)[\partial log(f_\theta(z))/\partial\theta]^2$ (only unbiased estimators with $<\tilde{\theta}>=\theta$ are in competition). The result here is that $Ie^2 \geq 1$ (Cramer-Rao bound). A case of great importance here concerns shift invariant distribution functions where the form does not change under θ displacements and one can write

$$I = \int dz f(z) \left(\frac{\partial log(f(z))}{\partial z}\right)^2 \tag{1.45}$$

If one is dealing with phase space where z is a M=2N dimensional vector with coordinates r and p then $I(z) = I(r) + I(p)$ (cf. [**1006**]). Now assume that one wishes to describe a classical system of N identical particles of mass m with Hamiltonian

$$\mathfrak{H} = \mathfrak{T} + \mathfrak{V} = \sum_1^N \frac{p_i^2}{2m} + \sum_1^N V(r_i) \tag{1.46}$$

This is a simple situation but the analysis is not limited to such systems. Assume also that the system is in equilibrium at temperature T so that in the canonical ensemble the probability density is

$$\rho(r,p) = \frac{e^{-\beta\mathfrak{H}(r,p)}}{Z};\ Z = \int \frac{d^{3N}rd^{3N}p}{N!h^{3N}} e^{-\beta\mathfrak{H}(r,p)} \tag{1.47}$$

(here for h an elementary cell in phase space one writes $d\tau = d^{3N}rd^{3N}p/(N!h^{3N}$, $\beta = 1/kT$ with k the Boltzman constant, and Z is the partition function). Then from Hamilton's equations $\partial_p\mathfrak{H} = \dot{r}$ and $\partial_r\mathfrak{H} = -\dot{p}$ there results

$$-kT\frac{\partial log(\rho(r,p))}{\partial p} = \dot{r};\ -kT\frac{\partial log(\rho(r,p))}{\partial r} = -\dot{p} \tag{1.48}$$

One can now write the Fisher information measure in the form
(1.49)

$$I_\tau = \int \frac{d^{3N}rd^{3N}p}{N!h^{3N}} \rho(r,p)\mathfrak{A}(r,p);\ \mathfrak{A} = a\left(\frac{\partial log(\rho(r,p))}{\partial p}\right)^2 + b\left(\frac{\partial log(\rho(r,p))}{\partial r}\right)^2$$

One needs two coefficients for dimensional balance (cf. [**1006**]). One notes that

$$\frac{\partial log(\rho(r,p))}{\partial p} = -\beta\frac{\partial\mathfrak{H}}{\partial p};\ \frac{\partial log(\rho(r,p))}{\partial r} = -\beta\frac{\partial\mathfrak{H}}{\partial r} \tag{1.50}$$

leading to the Fisher information in the form

$$(kT)^2 I_\tau = a\left\langle\left(\frac{\partial\mathfrak{H}}{\partial p}\right)^2\right\rangle + b\left\langle\left(\frac{\partial\mathfrak{H}}{\partial r}\right)^2\right\rangle \Rightarrow I_\tau = \beta^2[a<\dot{r}^2>+b<\dot{p}^2>] \tag{1.51}$$

This gives the universal Fisher form for any Hamiltonian of the form (1.46) and we refer to Liboff [**821**] for connections to kinetic theory. Many other interesting results on Fisher can be found in [**490, 491, 1006**].

1.5. UNCERTAINTY AND FLUCTUATIONS. We go first to Anderson and Halliwell [**46**] and recall the idea of a phase space distribution in the form (♣) $\mu(p,q) =< z|\rho|z >$ where ρ is the density matrix and $|z>$ denotes coherent states (cf. [**238, 1010**] for coherent states and Chapter 9 for more on these themes). The chosen measure of uncertainty here is the Shannon information

$$I = -\int \frac{dpdq}{2\pi\hbar}\mu(p,q)log(\mu(p,q)) \tag{1.52}$$

The uncertainty principle manifests itself via the inequality (♠) $I \geq 1$ with equality if and only if ρ is a coherent state (cf. [**823, 1295**]). In [**46**] one wants to generalize this to include the effects of thermal fluctuations in nonequilibrium systems and we sketch some of the ideas at least for equilibrium systems. There are in general three contributions to the uncertainty:

(1) The quantum mechanical uncertainty (quantum fluctuations) which is not dependent on the dynamics.
(2) The uncertainty due to spreading or reassembly of the wave packet. This is a dynamical effect and it may increase or decrease the uncertainty.
(3) The uncertainty due to the coupling to a thermal environment (diffusion and dissipation).

The time evolution I_t of I is studied for nonequilibrium systems and it is shown to generally settle down to monotone increase. I_t^{min} is a measure of the amount of quantum and thermal noise the system must suffer after a nonunitary evolution for time t (we do not deal with this here but refer to [**46**] for the nonequilibrium situation where the system decomposes into a distinguished system $\mathcal{S}$ plus the rest, referred to as the environment; the resulting time evolution of ρ is then nonunitary). In any event the lower bound I_t^{min} includes the effects of 1 and 3 but avoids 2.

One recalls the Shannon information (discussed earlier)

$$I(S) = -\sum_1^N p_i log(p_i);\ 0 \leq I(S) \leq log(N) \tag{1.53}$$

This is often referred to as entropy but here the word entropy is reserved for the vonNeumann entropy. In a similar manner, for continuous distributions (X a random variable with probability density $p(x)$ and $\int p(x)dx = 1$), the information of X is defined as

$$I(X) = -\int dxp(x)log(p(x)) \tag{1.54}$$

One emphasize that $p(x)$ here is a density (so it may be greater that 1 and $I(X)$ may be negative). However it retains its utility as a measure of uncertainty and e.g. for a Gaussian

(1.55) $$p(x) = \frac{1}{[2\pi(\Delta x)^2]^{1/2}} exp\left(-\frac{(x-x_0)^2}{2(\Delta x)^2}\right);\ I(X) = log\left(2\pi e(\Delta x)^2\right)^{1/2}$$

Thus $I(X)$ is unbounded from below and goes to $-\infty$ as $\Delta x \to 0$ and $p(x)$ goes to a delta function. $I(X)$ is also unbounded from above but if the variance is fixed then $I(X)$ is maximized by the Gaussian distribution (1.55). Hence one has

(1.56) $$I(X) \le log\left(2\pi e(\Delta x)^2\right)^{1/2}$$

The generalization to more than one variable is straightforward, e.g.

(1.57) $$I(X,Y) = -\int dxdyp(x,y)log(p(x,y)) \Rightarrow I(X,Y) \le I(X) + I(Y)$$

where e.g. $I(X) = \int dyp(x,y)$. It is useful to introduce QM phase space distributions of the form

(1.58)
$$\mu(p,q) = < z|\rho|z >;\ < x|z > = < x|p,q > = \left(\frac{1}{2\pi\sigma_q^2}\right)^{1/4} exp\left(-\frac{(x-q)^2}{4\sigma_q^2} + ipx\right)$$

Here $< x|z >$ is a coherent state with $\sigma_q\sigma_p = (1/2)\hbar$ and there is a normalization $\int(dpdq/2\pi\hbar)\mu(p,q) = 1$. One can also show that

(1.59) $$\mu(p,q) = 2\int dp'dq' exp\left(-\frac{(p-p')^2}{2\sigma_p^2} - \frac{(q-q')^2}{2\sigma_q^2}\right) W_\rho(p',q');$$

$$W_\rho(p,q) = \frac{1}{2\pi\hbar}\int d\xi e^{-(i/\hbar)p\xi}\rho(q+(1/2)\xi, q-(1/2)\xi)$$

(Wigner function - cf. [**238, 239**]). One is interested in the extent to which $\mu(p,q)$ is peaked about some region in phase space and the Shannon information is a natural measure of the extent to which a probability distribution is peaked. Thus one takes as a measure of uncertainty the information

(1.60) $$I(P,Q) = -\int \frac{dpdq}{2\pi\hbar}\mu(p,q)log(\mu(p,q))$$

One expects there to be a lower bound for I and it should be achieved on a coherent state and this was in fact proved (cf. [**823, 1295**]) in the form $I(P,Q) \ge 1$ with equality if and only if ρ is the density matrix of a coherent state $|z' >< z'|$. Further

(1.61) $$log\left(\frac{e}{\hbar}\Delta_\mu q\Delta_\mu p\right) \ge I(Q) + I(P) \ge I(P,Q)$$

The variances here have the form

(1.62) $$(\Delta_\mu q)^2 = (\Delta_\rho q)^2 + \sigma_q^2;\ (\Delta_\mu p)^2 = (\Delta_\rho p)^2 + \sigma_p^2$$

where Δ_ρ denotes the QM variance and hence

(1.63) $$\left((\Delta_\rho q)^2 + \sigma_q^2\right)\left((\Delta_\rho p)^2 + \sigma_p^2\right) \ge \hbar^2$$

Minimizing (1.63) over σ_q (and recalling that $\sigma_q\sigma_p = (1/2)\hbar$) one obtains the standard uncertainty relation $\Delta x\Delta p \geq (\hbar/2)$. Now suppose one has a genuinely mixed state so that

$$(1.64) \qquad \rho = \sum_n p_n|n><n|;\; p_n < 1;\; \mu(p,q) = \sum p_n|<z|n>|^2$$

The information of (1.64) will always satisfy $I(P,Q) \geq 1$ but this is a very low lower bound; indeed from the inequality

$$(1.65) \qquad -\left(\int dxf(x)g(x)\right) log\left(\int dyf(y)g(y)\right) \geq -\int dxg(x)g(x)log(x)$$

we have

(1.66)

$$I \geq -\int \frac{dpdq}{2\pi\hbar}\sum_n |<z|n>|^2 p_n log(p_n) = -\sum p_n log(p_n) = -Tr(\rho log(\rho)) \equiv S[\rho]$$

Thus I is bounded from below by the vonNeumann entropy $S[\rho]$ and this is a virtue of the chosen measure of uncertainty. One sees that I is a useful measure of both quantum and thermal fluctuations. It has a lower bound expressing the effect of quantum fluctuations which is connected to entropy and this in turn is a measure of thermal fluctuations.

Consider now the situation of thermal equilibrium. Let the density matrix be thermal, $\rho = Z^{-1}exp(-\beta H)$ where $Z = Tr(e^{-\beta H})$ is the partition function and $\beta = 1/kT$. Then

$$(1.67) \qquad <z|\rho|z> = \frac{1}{Z}\sum e^{-\beta E_n}|<z|n>|^2$$

where $|n>$ are energy eigenstates with eigenvalue E_n. For simplicity look at a harmonic oscillator for which

$$(1.68) \quad H = \frac{1}{2}\left(\frac{p^2}{M} + M\omega^2 q^2\right);\; |<z|n>|^2 = \frac{|z|^{2n}}{n!}e^{-|z|^2};\; E_n = \hbar\omega(n + (1/2))$$

Here $z = (1/2)[(q/\sigma_q) + i(p/\sigma_p)]$ where $\sigma_q\sigma_q = (1/2)\hbar$ and $\sigma_q = (\hbar/2M\omega)^{1/2}$ (cf. [**238, 760, 1010**] for coherent states). There results

$$(1.69) \qquad \mu(q,p) = <z|\rho|z> = \left(1 - e^{-\beta\hbar\omega}\right) exp\left(-(1 - e^{-\beta\hbar\omega})|z|^2\right)$$

The information (1.60) is then ($\bullet$) $I = 1 - log(1 - e^{-\beta\hbar\omega})$ which is exactly what one expects; as $T \to 0$ one has $\beta \to \infty$ and the uncertainty reduces to the Lieb-Wehrl result $I(P,Q) \geq 1$ expressing purely quantum fluctuations. For nonzero temperature however the uncertainty is larger tending to the value $-log(\beta\hbar\omega)$ as $T \to \infty$ which expresses purely thermal fluctuations. It is interesting to compare ($\bullet$) with the entropy $S = -Tr(\rho log(\rho))$. Here the partition function is $Z = [2Sinh((1/2)\beta\hbar\omega)]^{-1}$ and the entropy is then $S = -\beta(\partial_\beta(log(Z)) + log(Z)$ or

$$(1.70) \qquad S = -log[2Sinh((1/2)\beta\hbar\omega)] + (1/2)\beta\hbar\omega Coth[(1/2)\beta\hbar\omega]$$

For large T one has then $S \simeq -log(\beta\hbar\omega)$ coinciding with I but $S \to 0$ as $T \to 0$ while I goes to a nontrivial lower bound. Hence one sees that I is a useful measure of uncertainty in both the quantum and thermal regimes. We refer also to [**4**] where an information theoretic uncertainty relation including the effects of thermal fluctuations at thermal equilibrium has been derived using thermofield dynamics (cf. [**1247**]); their information theoretic measure is however different than that in [**46**]. On goes next to non-equilibrium systems and proves for linear systems that, for each t, I has a lower bound I_t^{min} over all possible initial states. It coincides with the Lieb-Wehrl bound in the absence of an environment and is related to the vonNeumann entropy in the long time limit. We refer to [**46**] for details.

2. A TOUCH OF CHAOS

For quantum chaos we refer to [**40, 120, 121, 232, 269, 350, 458, 579, 650, 660, 697, 698, 782, 887, 993, 1042, 1228**] and begin here with Partovi [**993**]. Chaos is quantitatively measured by the Lyapunov spectrum of characteristic exponents which represent the principal rates of orbit divergence in phase space, or alternatively by the Kolmogorov-Sinai (KS) invariant, which quantifies the rate of information production by the dynamical system. Chaos is conspicuously absent in finite quantum systems but the chaotic nature of a given classical Hamiltonian produces certain characteristic features in the dynamical behavior of its quantized version; these features are referred to as quantum chaos (cf. [**269, 579**]). They include short term instabilities and diffusive behavior versus dynamical localization and other effects. One is concerned here with an approach to the information dynamics of the quantum-classical transition based on the HJ formalism with the KS invariant playing a central role. The extension to the quantum domain is accomplished via the orbits introduced by Madelung and Bohm (cf. [**161, 831**]); these are natural extensions of the classical phase space flow to QM and provide the required bridge across the transition. One striking result is that the quantum KS invariant for a given Madelung-Bohm (MB) orbit is equal to the mean decay rate of the probability density along the orbit. Further one shows that the quantum KS invariant averaged over the ensemble of MB orbits equals the mean growth rate of configuration space information and a general and rigorous argument is given for the conjecture that the standard quantum-classical correspondence (or the classical limit) breaks down for classically chaotic Hamiltonians.

We give only a sketch of results here. Thus consider a classical system of N degrees of freedom described by canonical variables (q_i, p_i) with $1 \leq i \leq N$ and denote the Hamiltonian as $H(\mathbf{q}, \mathbf{p}, t)$ with Hamilton principal function $S(\mathbf{q}, t, \mathbf{p}_0)$ where $\mathbf{p}_0$ being the initial momenta. In matrix form Hamilton's equations are $\dot{\xi} = \mathcal{J}\nabla_\xi H(\xi, t)$ where ξ stands for the 2N dimensional phase space vector $(\mathbf{q}, \mathbf{p})$. Here $\mathcal{J}$ is a real antisymmetric matrix of order 2N with a $2 \otimes N$ block form $(0_N, I_N, -I_N, 0_N)$ which is a listing of blocks in the order $(11, 12, 21, 22)$. The tangent dynamics of the system is described by the $2N \times 2N$ nonsingular matrix $\mathcal{T}_{\mu\nu}(t, \xi_0) = \partial\xi_\mu(t, \xi_0)/\partial\xi_{0\nu}$ (the sensitivity matrix) where $\xi(t, \xi_0)$ is the trajectory starting from ξ_0 at time t_0. One can in fact write ($\tilde{S} \equiv S^T$ - matrix transpose)

$$\mathcal{T} = (S^{-1}_{\mathbf{p}_0\mathbf{q}}, -S_{\mathbf{p}_0\mathbf{q}}S_{\mathbf{p}_0\mathbf{p}_0}, S_{\mathbf{q}\mathbf{q}}S^{-1}_{\mathbf{p}_0\mathbf{q}}, \tilde{S}_{\mathbf{p}_0\mathbf{q}} - S_{\mathbf{q}\mathbf{q}}S^{-1}_{\mathbf{p}_0\mathbf{q}}S_{\mathbf{p}_0\mathbf{p}_0}) \tag{2.1}$$

where $(S_{\mathbf{p}_0\mathbf{q}})_{ij} = \partial^2 S/\partial_{q_j}\partial_{p_0 i}$. It is shown that one can write $\mathcal{T}$ in an upper triangular block form $\Gamma = \Omega(\Theta)\mathcal{T}$ where $\Omega(\Theta) = (cos(\Theta), -Sin(\Theta), Sin(\Theta), Cos(\Theta))$

and $-\sigma = Tan(\Theta) = -S_{\mathbf{qq}}$. Here Θ is a real symmetric matrix of order N while Ω is orthogonal and symplectic (symplectic phase matrix). The upper triangular form $(\Gamma_{11}, \Gamma_{12}, 0_N, \Gamma_{22})$ of Γ satisfies $\Gamma_{11}^{-1} = \tilde{\Gamma}_{22}$ and the upper half of the Lyapunov spectrum is obtained from the singular values of Γ_{11} (see [**993**]). In particular the Kolmogorov-Sinai (KS) entropy is given via

(2.2) $$k = lim_{t\to\infty} log[det(\Gamma_{11})]/t$$

For illustration consider the standard form $H = \mathbf{p}^2/2 + V(\mathbf{q}, t)$ with N-dimensional vectors $\mathbf{q}$, $\mathbf{p}$. Then

(2.3) $$k =< Tr(\sigma) >_{p.v.};\ < f >= lim_{t\to\infty}\frac{1}{t}\int_0^t f(t')dt'$$

where p.v. stipulates a principal value evaluation (σ will have simple pole behavior near singularities and the principal value contribution vanishes). Since $Tr(\sigma) = \nabla_q^2 S$ along the orbit (7.51) simply states that the KS invariant equals the time average of the Laplacian of the action along the orbit. Now the MB formalism associates a phase space flow with a quantum system via

(2.4) $$\psi = exp[iS(\mathbf{x}, t)/\hbar + R(\mathbf{x}, t)];\ \dot{\mathbf{q}}(t, \mathbf{q}_0, \mathbf{p}_0) = \mathbf{p} = \nabla S[\mathbf{q}, t]$$

It can be verified that the expectation value of any observable in the state ψ is given by its average over the ensemble of orbits thus defined (e.g. Ehrenfest's equations arise in this manner). The correspondence thus allows us to define the quantum KS invariant for a given orbit as

(2.5) $$\mathbf{k} =< \nabla^2 S >_{p.v.}$$

(the averaging process is with respect to the time along the MB orbit to which S is restricted). Now intuitively one would expect that orbits neighboring a hypothetical chaotic orbit in the ensemble diverge from it on the average thus causing the orbit density along the chaotic orbit to decrease with a mean rate related to $\mathbf{k}$. This is fully realized here as one sees by considering the equation of motion for $R(\mathbf{x}, t)$ as inherited from the SE, namely $\partial_t R + \nabla R \cdot \nabla S = -(1/2)\nabla^2 S$. The characteristic curves for this equation are the MB orbits so that it takes the following form along these orbits;

(2.6) $$\frac{dR}{dt} = -\frac{1}{2}\nabla^2 S \Rightarrow \mathbf{k} = -2\left\langle \frac{dR}{dt}\right\rangle = -\left\langle \frac{d\,log(|\psi|^2)}{dt}\right\rangle_{p.v.}$$

This says that the quantum KS invariant for a given orbit is the mean decay rate of the probability density along the orbit. Comparing this to a classical system where $k \neq 0$ while $\mathbf{k} = 0$ for the quantum version one sees that the classical limit cannot hold for chaotic Hamiltonians and since chaotic classical Hamiltonians are certainly more common than regular ones the idea of classical limit is not a reliable test for quantum systems. Finally let $\bar{\mathbf{k}}$ be the MB ensemble average, which is the same as the QM expectation value, leading to

(2.7) $$\bar{\mathbf{k}} = lim_{t\to\infty}\frac{1}{t}\int dq|\psi|^2 log(|\psi|^2)$$

which is an information entropy measure. The discussion here is very incomplete but should motivate further investigation and we refer to [**993**] for more detail (cf. also [**1280, 1281**] involving chaos, fractals, and entropy.

2.1. CHAOS AND THE QUANTUM POTENTIAL. The paper Parameter and diRienzo [**991**] offers an interesting perspective on the quantum potential. Thus consider a system of n particles with the SE

$$i\hbar\partial_t\psi = \left[\sum_1^n \left(\frac{-\hbar^2}{2m_i}\right)\nabla_i^2 + V\right]\psi;\ \nabla_i = \left(\frac{\partial}{\partial x_i}, \frac{\partial}{\partial y_i}, \frac{\partial}{\partial z_i}\right) \tag{2.8}$$

(here $\mathbf{x}_i = (x_i, y_i, z_i)$). Set $\psi = Rexp[i(S/\hbar)]$ and there results as usual

$$\partial_t S + \sum_1^n (\nabla_i S)^2 (2m_i)^{-1} + Q + V = 0;\ \partial_t R^2 + \sum_1^n \nabla_i \cdot \left(\frac{R^2\nabla_i S}{m_i}\right) = 0 \tag{2.9}$$

where $Q = \sum_1^n (\hbar^2/2m_i R)\nabla_i^2 R$. Now just as the causal form of the HJ equation contains the additional term Q so the causal form of Newton's second law contains Q as follows

$$\dot{P}_i = -\nabla_i V - \nabla_i Q;\ P = \sum_1^n P_i;\ \dot{P} = \frac{dP}{dt} = -\sum_1^n \nabla_i V - \sum_1^n \nabla_i Q \tag{2.10}$$

The author cites a number of curious and conflicting statements in the literature concerning the effect of the quantum potential on Bohmian trajectories, for clarification of which he observes that for an isolated system one has

$$-\sum \nabla_i V = 0;\ \dot{P} = 0;\ -\sum \nabla_i Q = -\sum F_i = 0 \tag{2.11}$$

Thus the sum of all the quantum forces is zero so $F_i = \sum_{j\neq i}^n (-F_j)$. Thus the net quantum force on a given particle is the result of all the other particles exerting force on this particle via the intermediary of the quantum potential. This then is his explanation for the guidance role of the wave function.

Next it is noted that removing Q from the HJ equation is equivalent to adding the term

$$\left(\frac{\hbar^2}{2m}\right) exp(iS/\hbar)\nabla^2 R = \left(\frac{\hbar^2}{2m}\right)|\psi|^{-1}\psi\nabla^2|\psi| = -Q\psi \tag{2.12}$$

to the SE so that the effective Hamiltonian becomes

$$H_{eff} = -\left(\frac{\hbar^2}{2m}\right)\nabla^2 + V + \left(\frac{\hbar^2}{2m}\right)|\psi|^{-1}\nabla^2|\psi| \tag{2.13}$$

Since $H_{eff} = H_{eff}(\psi)$ depends on ψ the superposition principle no longer applies. When $\phi \neq \psi$ we have

$$\int (\phi^* H_{eff}\psi - \psi H_{eff}\phi^*)d\tau = \left(\frac{\hbar^2}{2m}\right)\int \phi^*\psi[|\psi|^{-1}\nabla^2|\psi| - |\phi|^{-1}\nabla^2|\phi|] \neq 0 \tag{2.14}$$

so H_{eff} is not Hermitian. Hence the time development operator $exp[(i/\hbar)H_{eff}t]$ is not unitary and the time dependent SE is a nonunitary flow. Then since

$i\hbar\partial_t(\psi^*\psi) = \psi^* H_{eff}\psi - \psi H_{eff}\psi^*$ one has

$$(2.15)\quad \partial_t \int |\psi - \phi|^2 d\tau = \left(\frac{\hbar^2}{2m}\right) \int i[\psi^*\phi - \phi^*\psi][|\psi|^{-1}\nabla^2|\psi| - |\phi|^{-1}|\phi|]d\tau \neq 0$$

Consider then the case where two initial conditions for the time dependent SE differ only infinitesimally. As time progresses the two corresponding wave functions can become quite different, indicating the possibity of deterministic chaos, and this is a consequence of H_{eff} being a functional of the state upon which it is acting. If the term $(\hbar^2/2m)|\psi|^{-1}\nabla^2|\psi|$ is removed from (7.35) one is left with a Hermitian Hamiltonian and the normalization of $(\psi - \phi)$ is time independent, so there can be no deterministic chaos. Thus in particular Q acts as a constraining force preventing deterministic chaos (cf. also [**845**]).

REMARK 7.2.1. There are many different aspects of quantum chaos and the perspective of [**991**] just mentioned does not deal with everything covered in the references already cited (cf. also [**856, 1289, 1311, 1312**] for additional referencers). We are not expert enough to attempt any kind of in depth coverage but extract here briefly from a few papers. First from [**1311**] one notes that the dBB theory of quantum motion provides motion in deterministic orbits under the influence of the quantum potential. This quantum potential can be very intricate because it generates wave interferences and further numerical work has shown the presence of chaos and complex behavior of quantum trajectories in various systems (cf. [**992**]). In [**1311**] one indicates that movement of the zeros of the wave function (called vortices) implies chaos in the dynamics of quantum trajectories. These vortices result from wave function interferences and have no classical explanation. In systems without magnetic fields the bulk vorticity $\nabla \times \mathbf{v}$ in the probability fluid is determined by points where the phase S is singular (which can occur when the wave function vanishes). Due to singlevaluedness of the wave function the circulation $\Gamma = \int_C \dot{\mathbf{r}} d\mathbf{r} = (2\pi n/m)$ around a closed contour C encircling a vortex is quantized with n an integer and the velocity must diverge as one approaches a vortex. This leads to a universal mechanism producing chaotic behavior of quantum trajectories (cf. also [**992, 1312**]).

Next in Weinstein, Lloyd, and Tsallis [**1289**] one speaks of the edge of quantum chaos (the border between chaotic and non-chaotic regions) where the Lyapunov exponent goes to zero; it is then replaced by a generalized Lyapunov coefficient describing power-law rather than exponential divergence of classical trajectories. In [**1289**] one characterizes quantum chaos by comparing the evolution of an initially chosen state under the chaotic dynamics with the same state evolved under a perturbed dynamics (cf. [**1014**]). When the initial state is in a regular region of a mixed system (one with regular and chaotic regions) the overlap remains close to one; however when the initial state is in a chaotic zone the overlap decay is exponential. It is shown that at the edge of quantum chaos there is a region of polynomial overlap decay. Here the overlap is defined as $O(t) = | < \psi_u(t)|\psi_p(t) > |$ where $\psi_u$is the state evolved under the unperturbed system operator and ψ_p is the state evolved under the perturbed operator.

In various papers (e.g. [**232, 650, 651, 697, 698, 782**]) one characterizes quantum chaos via the quantum action. This is defined via

$$\tilde{S}[x] = \int dt \frac{\tilde{m}}{2}\dot{x}^2 - \tilde{V}(x) \tag{2.16}$$

for a given classical action

$$S[x] = \int dt \frac{m}{2}\dot{x}^2 - V(x) \tag{2.17}$$

so that the QM transition amplitude is

$$G(x_f, t_f; x_i, t_i) = \tilde{Z}exp\left[\frac{i}{\hbar}\tilde{\Sigma}\Big| + x_i, t_i{}^{x_f,t_f}\right]; \tag{2.18}$$

$$\tilde{\Sigma}\Big|_{x_i,t_i}^{x_f,t_f} = \tilde{S}[\tilde{x}_{cl}]\Big|_{x_i,t_i}^{x_f,t_f} = \int_{t_i}^{t_f} dt\frac{\tilde{m}}{2}\dot{\tilde{x}}_{cl}^2 - \tilde{V})\tilde{x}_{cl})\Big| + x_i{}^{x_f}$$

where $\tilde{x}_{cl}$ is the classical path corresponding to the action $\tilde{S}$. One requires here 2-point boundary conditions $\tilde{x}_{cl}(t = t_i) = x_i$ and $\tilde{x}_{cl}(t = t_f) = x_f$ and $\tilde{Z}$ stands for a dimensionful normalisation factor. The parameters of the quantum action (i.e. mass and potential) are independent of the boundary points but depend on the transition time $T = t_f - t_i$. A general existence proof is lacking but such quantum actions exist in many interesting cases. Then quantum chaos is defined as follows. Given a classical system with action S the corresponding quantum system displays quantum chaos if the corresponding quantum action $\tilde{S}$ in the asymptotic regime $T \to \infty$ generates a chaotic phase space. ∎

3. GENERALIZED THERMOSTATISTICS

We refer to [**46, 298, 491, 493, 922, 924, 1006, 1032, 1033**] for discussion of various entropies based on deformed exponential functions (generalizations of the Boltzman-Gibbs formalism for equilibrium statistical physics), the entropies of Beck-Cohen, Kaniadakis, Renyi, Tsallis, etc., maximum entropy ideas, escort density operators, and a host of other matters in generalized theormstatistics. We sketch here first a few ideas following the third paper in of Naudts [**922**]. Thus a model of thermostatistics is described by a density of states $\rho(E)$ and a probability distribution $p(E)$ and for a system in thermal equilibrium at temperature T one has

$$p(E) = \frac{1}{Z(T)}e^{-E/T};\ Z(T) = \int dE\rho(E)e^{-E/T} \tag{3.1}$$

(Boltzman's constant is set equal to one here). Thermal averages are defined via $< f >= \int dE\rho(E)p(E)f(E)$ (this is a simplified treatment with T not made explicit - i.e. $p(E) \sim p(E,T)$). A microscopic model of thermostatistics is specified via an energy functional $H(\gamma)$ over phase space Γ which is the set of all possible microstates. Using $\rho(E)dE = d\gamma$ one can write

$$< f >= \int_\Gamma d\gamma p(\gamma)f(\gamma);\ p(\gamma) = \frac{e^{-H(\gamma)/T)}}{Z(T)};\ Z(T) = \int_\Gamma d\gamma e^{-H(\gamma)/T)} \tag{3.2}$$

In the quantum case the integration is replaced by a trace to obtain

$$(3.3)\qquad <f>=\frac{1}{Z(T)}Tr\,exp(-H/T)f;\;\; Z(T)=Tr\,exp(-H/T)$$

In relevant examples of thermostatistics the density of states $\rho(E)$ increases as a power law $\rho(E)\sim E^{\alpha N}$ with N the number of particles and $\alpha>0$. There is an energy - entropy balance where the increase of density of states $\rho(E)$ compensates for the exponential decrease of probability density $p(E)$ with a maximum of $\rho(E)p(E)$ reached at some macroscopic energy far above the ground state energy. One can write $\rho(E)p(E)=(1/Z)exp(log(\rho(E))-E/T)$ with the argument of the exponential maximal when E satisfies

$$(3.4)\qquad \frac{1}{\rho(E)}\rho'(E)=\frac{1}{T}$$

where $\rho'(E)$ is the derivative $d\rho/dE$. If $\rho(E)\sim E^{\alpha N}$ then $E\sim\alpha NT$ follows which is the equipartition theorem. The form of the theory here indicaters that the actual form of the probability distribution is not very essential; alternative expressions for $p(E)$ are acceptable provided they satisfy the equipartition theorem and reproduce thermodynamics. One begins here by generalizing the equipartition result (3.4) and postulates the existence of an increasing positive function $\phi(x)$ defined for $x\geq 0$ such that (•) $(1/T)=-[p'(E)/\phi(p(E))]$ holds for all E and T. Then the equation for the maximum of $\rho(E)p(E)$ becomes

$$(3.5)\quad 0=\frac{d}{dE}[\rho(E)p(E)]=\rho'(E)p(E)-\frac{1}{T}\rho(E)\phi(p(E))\equiv\frac{\rho'(E)}{\rho(E)}=\frac{1}{T}\frac{\phi(p(E))}{p(E)}$$

The Boltzman-Gibbs case is recovered when $\phi(x)=x$. Now (•) fixes the form of the probability distribution $p(E)$; to see this introduce a function $log_\phi(x)$ via

$$(3.6)\qquad log_\phi(x)=\int_1^x\frac{1}{\phi(y)}dy$$

The inverse is $exp_\phi(x)$ and from the identity $1=exp'_\phi(log_\phi(x))log'_\phi(x)$ there results (♦) $\phi(x)=exp'_\phi(log_\phi(x))$. Hence (•) can be written as

$$(3.7)\qquad p'(E)=-\frac{1}{T}exp'_\phi[log_\phi(p(E))]\Rightarrow p(E)=exp_\phi(G_\phi(T)-(E/T))$$

The function $G_\phi(T)$ is the integration constant and it must be chosen so that $1=\int dE\rho(E)p(E)$ is satisfied. The formula (3.7) resembles the Boltzman-Gibbs distribution but the normalization constant appears inside the function $exp_\phi(x)$; for $\phi(x)=x$ one has then $G_\phi(T)=-log(Z(T))$.

In general it is difficult to determine $G_\phi(T)$ but an expression for its temperature derivative can be obtained via escort probabilities (cf. [**177, 1240**]). The general definition is

$$(3.8)\qquad P(E)=\frac{1}{Z(T)}\phi(p(E));\;\; Z(T)=\int dE\rho(E)\phi(p(E))$$

Then expectation values for $P(E)$ are denoted by

$$<f>_* = \int dE\rho(E)P(E)f(E) \tag{3.9}$$

Note $P(E) = p(E)$ in the Boltzman-Gibbs case $\phi(x) = x$. Now calculate using (♦) and (3.8) to get

$$\frac{d}{dT}p(E) = exp'_\phi(G_\phi(T) - (E/T))\left(\frac{d}{dT}G_\phi(T) + \frac{E}{T^2}\right) = \tag{3.10}$$

$$= Z(T)P(E)\left(\frac{d}{dT}G_\phi(T) + \frac{E}{T^2}\right)$$

from which follows (recall $\int dE\rho(E)p(E) = 1$)

$$0 = \int dE\rho(E)\frac{d}{dT}p(E) = Z(T)\frac{d}{dT}G_\phi(T) + \frac{1}{T^2}Z(T) < E >_* \Rightarrow \tag{3.11}$$

$$\Rightarrow \frac{d}{dT}G_\phi(T) = -\frac{1}{T^2} < E >_*$$

Note also that combining (3.10) and (3.11) one obtains

$$\frac{d}{dT}p(E) = \frac{1}{T^2}Z(T)P(E)(E- < E >_*) \tag{3.12}$$

One wants now to show that generalized thermodynamics is compatible with thermodynamics begins by establishing thermal stability. Internal energy $U(T)$ is defined via $U(T) =< E >$ with $p(E)$ given by (3.7), so using (3.12) one obtains (3.13)

$$\frac{d}{dT}U(T) = \int dE\rho(E)E\frac{d}{dT}p(E) = \int dE\rho(E)\frac{E}{T^2}Z(T)P(E)(E- < E >_*) =$$

$$= \frac{1}{T^2}Z(T)(< E^2 >_* - < E >_*^2) \geq 0$$

Hence average energy is an increasing function of T but thermal stability requires more so define ϕ entropy (relative to $\rho(E)dE$ via

$$S_\phi(p) = \int dE\rho(E)[(1 - p(E))F_\phi(0) - F_\phi(p(E))];\ F_\phi(x) = \int_1^x dylog_\phi(y) \tag{3.14}$$

One postulates that thermodynamic entropy $S(T)$ equals the value of the above entropy $S_\phi(p)$ with p given by (3.7). Then

$$\frac{d}{dT}S(T) = \int dE\rho(E)(-log_\phi(p(E)) - F_\phi(0))\frac{d}{dT}p(E) = \tag{3.15}$$

$$= \int dE\rho(E)\left(-G_\phi(T) + \frac{E}{T} - F_\phi(0)\right)\frac{d}{dT}p(E) = \frac{1}{T}\frac{d}{dT}U(T)$$

(recall that $p(E)$ is normalized to 1). This shows that temperature T satisfies the thermodynamic relation $(1/T) = dS/dU$ and since E is an increasing function of T one concludes that S is a concave function of U; this is called thermal stability. One can also introduce the Helmholz free energy $F(T)$ via the well known $F(T) = U(T) - TS(T)$ so from (3.15) it follows that

$$\frac{d}{d\beta}\beta F(T) = U(T)\ \ (\beta = 1/T) \tag{3.16}$$

Going back to (3.11) which is similar to (3.16) with $F(T)$ replaced by $TG_\phi(T)$ and with $U(T) =< E >$ replaced by $< E >_*$ the comparison shows that $TG_\phi(T)$ is the free energy associated with the escort probability distribution $P(E)$ up to a constant independent of T.

The most obvious generalization now involves $\phi(x) = x^q$ with $q > 0$ and this essentially produces the Tsallis entropy where one has

$$\text{(3.17)} \quad log_q(x) = \int_1^x dyy^{-q} = \frac{1}{1-q}(x^{1-q} - 1);\ exp_q(x) = [1 + (1-q)x]_+^{1/(1-q)}$$

The probability distribution (3.17) becomes
(3.18)

$$p(E) = [1 + (1-q)(G_q(T) - (E/T))]_+^{1/(1-q)} = \frac{1}{z_q(T)}[1 - (1-q)\beta_q^*(T)E]_+^{1/(1-q)}$$

$$z_q(T) = (1 + (1-q)G_q(T))^{1/(1-q)};\ \beta_q^*(T) = z_q(T)^{1-q}/T$$

A nice feature of the Tsallis theory is that the correspondence between $p(E)$ and the escort $P(E)$ leads to a dual structure $q \leftrightarrow 1/q$; indeed

$$\text{(3.19)} \qquad P(E) = \frac{1}{Z_q(T)}p(E)^q \Rightarrow p(E) = \frac{1}{Z_{1/q}(T)}P(E)^{1/q}$$

Moreover there is also a $q - 2 \leftrightarrow q$ duality; given $log_\phi(x)$ a new deformed $log_\psi(x)$ is obtained via

$$\text{(3.20)} \quad log_\psi(x) = (x-1)F_\phi(0) - xF_\phi(1/x);\ \frac{1}{\psi(x)} = F_\phi(0) - F_\phi(1/x) + \frac{1}{x}log_\phi(1/x)$$

and for $\phi = x^q$ one has $\psi = (2-q)x^{2-q}$. One notes also that the definition (3.14) of entropy $S_\phi(p)$ can be written as

$$\text{(3.21)} \qquad S_\phi(p) = \int dE\rho(E)p(E)log_\psi(1/p(E))$$

and with $\psi(x) = x^q$ we get the Tsallis entropy

$$\text{(3.22)} \qquad S_q(p) = \int dE\rho(E)\frac{1}{1-q}(p(E)^q - p(E))$$

3.1. NONEXTENSIVE STATISTICAL THERMODYNAMICS. We go here to Tsallis, Baldovini, Cerbino, and Pierobon [**1241**] for an lovely introduction and extract liberally. The Boltzman-Gibbs entropy is given via

$$\text{(3.23)} \qquad S_{BG} = -k\sum_1^W p_i log(p_i);\ \sum_1^W p_i = 1$$

Here p_i is the probability for the system to be in the ith microstate and k is the Boltzman constant k_B (taken now to be 1). If every microstate has the same probability $p_i = 1/W$ then $S_{BG} = klog(W)$. The entropy (3.23) can be shown to be nonnegative, concave, extensive, and stable (or experimentally robust). By extensive one means that if A and B are two independent systems (i.e. $p_{ij}^{A+B} = p_i^A p_j^B$) then

$$\text{(3.24)} \qquad S_{BG}(A+B) = S_{BG}(A) + S_{BG}(B)$$

One can still not derive this form of entropy (3.23) from first principles. There is also good reason to conclude that physical entropies different from (3.23) would be more appropriate for anomalous systems. In this spirit the Tsallis entropy was proposed in [**1242**] and the property thereby generalized is extensivity. One discusses motivations etc. in [**1116**] and in particular observes that the function

$$y = \frac{x^{1-q}-1}{1-q} = log_q(x) \tag{3.25}$$

satisfies

$$log_q(x_A x_B) = log_q(x_A) + log_q(X_B) + (1-q)(log_q(X_A))(log_q(X_B)) \tag{3.26}$$

Now rewrite (3.23) in the form ($k = 1$)

$$S_{BG} = -\sum_1^W p_i log(p_i) = \sum_1^W p_i log(1/p_i) = \left\langle log\frac{1}{p_i}\right\rangle \tag{3.27}$$

The quantity $log(1/p_i)$ is called surprise or unexpectedness and one thinks of a q-surprise $log_q(1/p_i)$ in defining

$$S_q = \left\langle log_q\frac{1}{p_i}\right\rangle = \sum_1^W p_i log_q(1/p_i) = \frac{1-\sum_1^W p_i^q}{q-1} \tag{3.28}$$

In the limit $q \to 1$ one gets $S_1 = S_{BG}$ and assuming equiprobability $p_i = 1/W$ one gets

$$S_q = \frac{W^{1-q}-1}{1-q} = log_q(W) \tag{3.29}$$

Consequently S_q is a genuine generalization of the BG entropy and the pseudo-additivity of the q-logarithm implies (restoring momentarily k)

$$\frac{S_q(A+B)}{k} = \frac{S_q(A)}{k} + \frac{S_q(B)}{k} + (1-q)\frac{S_q(A)}{k}\frac{S_q(B)}{k} \tag{3.30}$$

if A and B are two independent systems (i.e. $p_{ij}^{A+B} = p_i^A p_j^B$). Thus $q = 1$, $q < 1$, and $q > 1$ respectively correspond to the extensive, superextensive, and subextensive cases and the q-generalization of statistical mechanics is referred to as nonextensive statistical mechanics. (3.30) is true for independent A and B but if A and B are correlated in some way one can ask if extensivity would hold for some q. For example a system whose elements are correlated at at scales might correspond to $W(N) \sim N^\rho$ $\rho > 0$ with entropy

$$S_q(N) = log_q W(N) \sim \frac{N^{\rho(1-q)}-1}{1-q} \tag{3.31}$$

and extensivity is obtained if and only if $q = 1 - (1/\rho) < 1$ or $S_q(N) \propto N$. Shannon and Khinchin gave early similar sets of axioms for the form of the entropy functional, both leading to (3.23). These were generalized in [**5, 1034, 1035, 1120**] leading to the entropy

$$S(p_1, \cdots, p_W) = k\frac{1-\sum_1^W p_i^q}{q-1} \tag{3.32}$$

and it was shown that S_q is the only possible entropy extending the Boltzman-Gibbs entropy maintaining all the basic properties except extensivity for $q \neq 1$.

Some other properties are also discussed, e.g. bias, concavity, and stability. First note

$$S_{BG} = -\left[\frac{d}{dx}\sum_1^W p_i^x\right]_{x=1} \tag{3.33}$$

(x here is referred to as a bias). Similarly

$$S_q = -\left[D_q\sum_1^W p_i^x\right]_{x=1}\ ;\ D_q h(x) = \frac{h(qx)-h(x)}{qx-1} \tag{3.34}$$

(Jackson derivative) and this may open the door to quantum groups (see e.g. [**239**]). As for concavity consider for $p_i'' = \mu p_i + (1-\mu)p_i'$ $(0 < \mu < 1)$ concavity defined via

$$S(\{p_i''\}) \geq \mu S(\{p_i\}) + (1-\mu)S(\{p_i'\}) \tag{3.35}$$

It can be shown that S_q is concave for every $\{p_i\}$ and $q > 0$. This implies theormdynamic stability in the framework of statistical mechanics (i.e. stability of the system with regard to energetic perturbations). This means that the entropy functional is defined such that the stationary state (thermodynamic equilibrium) makes it extreme.

There are also other generalizations of the BG entropy and we mention the Renyi entropy

$$S_q^R = \frac{log\sum_1^W p_i^q}{1-q} = \frac{log[1+(1-q)S_q]}{1-q} \tag{3.36}$$

and an entropy due to Landsberg, Vedral, Rajagopal, Abe defined via

$$S_q^N = S_q^{LVRA} = \frac{1-\frac{1}{\sum_1^W p_i^q}}{1-q} = \frac{S_q}{1+(1-q)S_q} \tag{3.37}$$

These are however not concave nor experimentally robust and seem unsuited for thermodynamical purposes; on the other hand Renyi entropy seems useful for geometrically characterizing multifractals.

Various connections of S_q to thermodynamics are indicated in [**1241**] and we mention here first the Legendre structure. Thus for all values of q

$$\frac{1}{T} = \frac{\partial S_q}{\partial U_q};\ T = \frac{1}{k\beta};\ U_q = -\frac{\partial}{\partial\beta}log_q Z_q; \tag{3.38}$$

$$log_q Z_q = \frac{Z_q^{1-q}-1}{1-q} = \frac{\bar{Z}^{1-q}-1}{1-q} - \beta U_q;\ F_q = U_q - TS_q = -\frac{1}{\beta}log_q Z_q$$

Here $U_q \sim$ internal energy and $F_q \sim$ free energy and the specific heat is

$$C_q = T\frac{\partial S_q}{\partial T} = \frac{\partial U_q}{\partial T} = -T\frac{\partial^2 F_q}{\partial T^2} \tag{3.39}$$

Finally a list of other properties follows supporting the thesis that S_q is a correct road for generalizing the BG theory (see [**1241**] for details and references); we mention a few here via

(1) Boltzmann H-theorem (macroscopic time irreversibility) $q(dS_q/dt) \geq 0\ (\forall q)$

(2) Ehrenfest theorem: For an observable $\hat{O}$ and a Hamiltonian $\hat{H}$ one has $d < \hat{O} >_q /dt = (i/\hbar) < [\hat{H}, \hat{O} >_q\ \ (\forall q)$

(3) Pesin theorem (connection between sensitivity to initial conditions and the entropy production per unit time). Define the q-generalized Kolmogorov-Sinai entropy as

$$K_q = lim_{t\to\infty} lim_{W\to\infty} lim_{N\to\infty} \frac{< S_q > (t)}{t} \tag{3.40}$$

where N is the number of initial conditions, W is the number of windows in the partition (fine graining), and t is discrete time (cf. also [**800**]). The q-generalized Lyapunov coefficient λ_q can be defined via sensitivity to initial conditions

$$\xi = lim_{\Delta x(0)\to 0} \frac{\Delta x(t)}{\Delta x(0)} = e_q^{\lambda_q t} \tag{3.41}$$

(focusing on a 1-D system, basically $x(t+1) = g(x(t))$ with g nonlinear). It was proved in [**92**] that for unimodal maps $K_q = \lambda_q$ if $\lambda_q > 0$ and $K_q = 0$ otherwise. More explicitly $K_1 = \lambda_1$ if $\lambda_1 \geq 0$ (and $K_1 = 0$ if $\lambda_1 < 0$). But if $\lambda_1 = 0$ then there is a special value of q such that $K_q = \lambda_q$ if $\lambda_q \geq 0$ (and $K_q = 0$ if $\lambda_q < 0$).

We refer also to [**35, 36, 860, 1040**] for other results and approaches to thermodynamics, temperature, fluctuations, etc. in generalized thermostatistics and to [**803**] for relativistic nonextensive thermodynamics.

4. FISHER PHYSICS

The book [**490**] by Frieden purports (with notable success) to unify several subdisciplines of physics via Fisher information and this theme appears also in many papers, e.g. [**268, 298, 299, 491, 492, 493, 494, 892, 1006, 1030, 1031, 1032, 1033, 1034, 1035, 1246**]. We sketch some of this here and note in passing an interesting classical-quantum trajectory in [**495**] which differs from a Bohmian trajectory (cf. also [**117, 332, 481, 777, 852, 1074, 1310**]). First let us sketch some summary items from [**490**] and then provide some details. Thus in Chapter 12 of [**490**] Frieden lists (among other things) the following items:

(1) Writing $p = q^2$ in the standard formulas one can express the Fisher information as $I = 4\int dx(dq/dx)^2$ with q a real probability amplitude for fluctuations in measurement. Under suitable conditions (see below) the information I obeys an I-theorem $dI/dt \leq 0$. In the same spirit by which a positive increment in therodynamic time corresponds to an increase in Boltzman entropy there is a positive increment in Fisher time defined by a decrease in information I (the two times do not always agree). Let θ

be the measured phenomenon and define the Fisher temperature T_θ via

(4.1) $$\frac{1}{T_\theta} = -k_\theta \frac{\partial I}{\partial \theta} \quad (k_\theta = const.)$$

When θ is taken to be the sysem energy E then the Fisher temperature has analogous properties to the ordinary Boltzman temperature, in particular there is a perfect gas law $\bar{p}V = k_E T_E I$ where $\bar{p}$ is the pressure. The I theorem can be extended to a multiparameter, multicomponent scenario with

(4.2) $$I = 4\int dx \sum_n \nabla q_n \cdot \nabla q_n$$

(2) Any measurement of physical parameters initiates a transformation of Fisher information $J \to I$ connecting the phenomenon with the "intrinsic data". The phenomenological or "bound" information is denoted by J and the acquired information is I; J is ultimately identified by an invariance principle that characterizes the measured phenomenon. In any exchange of information one must $\delta J = \delta I$ (conservation law) and for $K = I - J$ one arrives at a variational principle (extreme physical information or EPI) $K = I - J = extremum$. Since $J \geq I$ always the EPI zero principle involves $I - \kappa J = 0$ $(0 \leq \kappa \leq 1)$. These equations follow (independently of the axiomatic approach taken and of the I-theorem) if there is e.g. a unitary transformation connecting the measurement space with a physically meaningful conjugate space. In this manner one arrives at the Lagrangian approach to physics, often using the Fourier transform to connect I and J. This seems a little mystical at first but many convincing examples are given involving the SE, wave equations, KG equation, Dirac equation, Maxwell equations, Einstein equations, WDW equation, etc.

There is much more summary material in [**490**] which we omit here. A certain amount of metaphysical thinking seems necessary and Frieden remarks that John Wheeler (cf. [**1296**]) anticipated a lot of this in his remarks that "All things physical are information-theoretic in origin and this is a participatory universe....Observer participancy gives rise to information and information gives rise to physics." Going now to [**490**] recall $I = \int dx[(p')^2/p] = 4\int dx(q')^2$ for $p = q^2$ and one derives the inequality $e^2 I \geq 1$ as follows. Look at estimators $\hat{\theta}$ satisfying

(4.3) $$< \hat{\theta}(y) - \theta >= \int dy[\hat{\theta}(y) - \theta]p(y|\theta) = 0$$

where $p(y|\theta)$ describes fluctuations in data values y. Hence

(4.4) $$\int dy(\hat{\theta} - \theta)\frac{\partial p}{\partial \theta} - \int dyp = 0$$

Use now $\partial_\theta p = p(\partial log(p)/\partial\theta)$ and normalization to get $\int dy(\hat{\theta}-\theta)(\partial log(p)/\partial\theta)p = 1$ which becomes

(4.5)

$$\int dy\left[\frac{\partial log(p)}{\partial\theta}\sqrt{p}\right][(\hat{\theta}-\theta)\sqrt{p}] = 1 \Rightarrow \left[\int dy\left(\frac{\partial log(p)}{\partial\theta}\right)^2 p\right]\left[\int dy(\hat{\theta}-\theta)^2 p\right] \geq 1$$

For $e^2 = \int dy(\hat{\theta}-\theta)^2 p$ this gives immediately $e^2 I \geq 1$. One notes that if $p(y|\theta) = p(y-\theta)$ then I is simply $I = \int dx(\partial log(p(x))/\partial x)^2 p(x)$ where $x \sim y-\theta$. We recall also the Shannon entropy as $H = -\int dx p(x) log(p(x))$ and the Kullback-Leibler entropy is defined as

$$G = -\int dx p(x) log\frac{p(x)}{r(x)} \tag{4.6}$$

where $r(x)$ is a reference probability distribution function (PDF). Consider now a discrete form of Fisher information

(4.7)

$$I = (\Delta x)^{-1}\sum_n \frac{[p(x_{n+1}) - p(x_n)]^2}{p(x_n)} = (\Delta x)^{-1}\sum_n p(x_n)\left[\frac{p(x_n+\Delta x)}{p(x_n)} - 1\right]^2$$

Here $p(x_n+\Delta x)/p(x_n)$is close to 1 for Δx small and one writes $[p(x_n+\Delta x)/p(x_n)] - 1 = \nu$. Then $log(1+\nu) \sim \nu - (\nu^2/2)$ or $\nu^2 = 2[\nu - log(1+\nu)]$. Hence I becomes

$$I = -2(\Delta x)^{-1}\sum_n p(x_n) log\frac{p(x_n+\Delta x)}{p(x_n)} + \tag{4.8}$$

$$2(\Delta x)^{-1}\sum_n p(x_n+\Delta x) - 2(\Delta x)^{-1}\sum_n p(x_n)$$

But each of the last two terms is $(\Delta x)^{-1}$ by normalization so they cancel leaving

$$I = -\frac{2}{\Delta x}\sum p(x_n) log\frac{p(x_n+\Delta x)}{p(x_n)} \to -\frac{2}{\Delta x}G[p(x), p(x+\Delta x)] \tag{4.9}$$

One notes (cf. [**490**]) that I results as a cross information between $p(x)$ and $p(x+\Delta x)$ for many different types of information measure, e.g. Renyi and Wooters information and in this sense serves as a kind of "mother" information. Next the I-theorem says that $dI/dt \leq 0$ and this can be seen as follows. Start with (4.9) in the form

$$I(t) = -2lim_{\Delta x\to 0}(\Delta x)^{-2}\int dx\, p log\frac{p_{\Delta_x}}{p};\ p_{\Delta_x} = p(x+\Delta x|t);\ p = p(x|t) \tag{4.10}$$

Under certain conditions (cf. [**490**]) p obeys a FK equation

$$\frac{\partial p}{\partial t} = -\frac{d}{dx}[D_1(x,t)p] + \frac{d^2}{dx^2}[D_2(x,t)p] \tag{4.11}$$

where D_1 is a drift function and D_2 a diffusion function. Then it is shown (cf. [**1030, 1073**]) that two PDF such as p and p_{Δ_x} that obey the FP equation have a cross entropy satisfying an H-theorem

$$G(t) = -\int dx\, p log\frac{p}{p_{\Delta_x}};\ \frac{dG(t)}{dt} \geq 0 \tag{4.12}$$

Hence I obeys an I theorem $dI/dt \leq 0$. We refer to [**490**] for more on temperature, pressure, and gas laws.

For multivariable situations one writes $I = 4\int dx \sum \nabla q_n \cdot \nabla q_n$ with $p_n = q_n^2$. An interesting notation here is

$$(4.13)\quad \psi_n = \frac{1}{\sqrt{N}}(q_{2n-1} + iq_{2n})\ (n = 1, \cdots, N/2);\ \sum_1^{N/2} \psi_n^* \psi_n = \frac{1}{N}\sum q_n^2 = p(x)$$

In such situations one finds for $I_n = 4\int dx \nabla q_n \cdot \nabla q_n$ and $I = \sum I_n$ (cf. [**490**])

$$(4.14)\qquad I_n = -\frac{2}{(\Delta x)^2} G_n[p_n(x|t), p_n(x+\Delta x|t)];\ \frac{\partial I_n}{\partial t} \leq 0; I(t) \to \ min.$$

Now one looks at minimization problems for I where $\delta I[\mathbf{q}(\mathbf{x}|t)] = 0$ and for anything meaningful to happen the physics has to be introduced via constraints and covariance (we refer to [**490**] for a more thorough discussion of these matters). Thus one is considering $K = I - J$ and the physics is introduced via J. One can write e.g. $I = \int dx \sum i_n(x)$ and $J = \int dx \sum j_n(x)$ where $i_n = 4\nabla q_n \cdot \nabla q_n$. In general now the functional form of J follows from a statement about invariance for the system. Examples of invariance are (i) unitary transformations such as that between the space and momentum space in QM (ii) gauge invariance as in EM or gravitational theory (iii) a continuity equation for the flow, usually involving sources. The answer $\mathbf{q}$ for EPI is completely dependent on the particular $J(\mathbf{q})$ for that problem and that in turn depends completely on the invariance principle that is used. If the invariance principle is not sufficiently strong in defining the system then one can expect the EPI output $\mathbf{q}$ to be only approximately correct. One has $I \leq J$ generally but $I = J$ for an optimally strong invariance principle. Note $\kappa = I/J$ measures the efficiency of the EPI in transferring Fisher information from the phenomenon (specfied by J) to the output (specified by I). Thus $\kappa < 1$ indicates that the answer $\mathbf{q}$ is only approximate. When the invariance principle is the statement of a unitary transformation between the measurement space and a conjugate coordinate space then the solution to the requirement $I - \kappa J = 0$ will simply be the reexpression of I in the conjugate space; when this holds then one can show that in fact $I = J$ (i.e. $\kappa = 1$). In this situation the out put $\mathbf{q}$ will be "correct", i.e. not explicitly incorrect due to ignored quantum effects for example. There are in fact nonquantum and nonunitary theories for which $\kappa = 1$ (or in fact any real number) and the nature of κ is not yet fully understood.

Let us call attention also to the information demon of Frieden and Soffer (cf. [**490, 493**]). For real Fisher coordinates x the EPI process amounts to carrying through a zero sum game betweem an observer (who wants to acquire maximal information) and an information demon (who wants to minimize the information transfer) with a limited resource of intrinsic information. The demon represents nature (and always wins or breaks even of course) and $K = I - J \leq 0$. Further since $\Delta I = K$ one has $\Delta I \leq 0$ while $\Delta t \geq 0$; hence the I-theorem follows.

We run through the EPI procedure here for the KG equation which illustrates many points. Define $x_1 = ix$, $x_2 = iy$, $x_3 = iz$, $x_4 = ct$ with $r = (x, y, z)$ and

$\mathbf{x} = (x_1, x_2, x_3, x_4)$ and use the ψ_n notation of (4.13). From $I = 4\int dx \sum \nabla q_n \cdot \nabla q_n$ we get

(4.15) $$I = 4Nc\sum_1^{N/2}\int\int drdt\left[-(\nabla\psi_n)^*\cdot\nabla\psi_n + \left(\frac{1}{c^2}\right)^2(\partial_t\psi_n)^*(\partial_t\psi_n)\right]$$

The invariance principle here involves a unitary Fourier transformation from x to μ in the form

(4.16)
$$(ir, ct) \to (i\mu/\hbar, E/c\hbar);\ \psi_n(r,t) = \frac{1}{(2\pi\hbar)^2}\int\int d\mu dE\phi_n(\mu, E)e^{-i(\mu\cdot r - Et)/\hbar}$$

One recalls

(4.17) $$\int\int drdt\psi_m^*\psi_n = \int\int d\mu dE\phi_m^*\phi_n$$

Differentiating in (4.16) one has $(\nabla\psi_n, \partial_t\psi_n) \to (-i\mu\phi_n/\hbar, iE\phi_n/\hbar)$ and via $\nabla\psi_n \to -i\mu\phi_n/\hbar$ one gets

(4.18) $$\int\int drdt(\nabla\psi_n)^*\cdot\nabla\psi_n = \frac{1}{\hbar^2}\int\int d\mu dE|\phi_n(\mu, E)|^2\mu^2;$$
$$\int\int drdt(\partial_t\psi_n)^*\partial_t\psi_n = \frac{1}{\hbar^2}\int\int d\mu dE|\phi_n(\mu, E)|^2E^2$$

Putting this in (4.15) gives

(4.19) $$I = \left(\frac{4Nc}{\hbar^2}\right)\sum_1^{N/2}\int\int d\mu dE|\phi_n(\mu, E)|^2\left(-\mu^2 + \frac{E^2}{c^2}\right) = J$$

This is the invariance principle for the given scenario. The same value of I can be expressed in the new space (μ, E) where it is called J and J is then the bound (physical) information. Now one has from (4.17)

(4.20) $$c\int\int drdt|\psi_n|^2 = c\int\int d\mu dE|\phi_n|^2\ \ (n = 1, \cdots, N/2$$

Summing over n and using $p = \sum_1^{N/2}\psi_n^*\psi = (1/N)\sum q_n^2$ with normalization gives

(4.21) $$1 = \int d\mu deP(\mu, E);\ P(\mu, E) = c\sum_1^{N/2}|\phi_n(\mu, E)|^2$$

so P is a PDF in the (μ, E) space. One obtains then

(4.22)
$$I = J = \frac{4N}{\hbar^2}\int\int d\mu dEP(\mu, E)\left(-\mu^2 + \frac{E^2}{c^2}\right);\ J = \left(\frac{4N}{\hbar^2}\right)\left\langle-\mu^2 + \frac{E^2}{c^2}\right\rangle$$

One must have J a universal constant here so $-\mu^2 + (E^2/c^2) = const. = A^2(m, c)$ where A is some function of the rest mass m and c (which are the only other parameters ($\hbar$ must also be a constant). By dimensional analysis $A = mc$ so $E^2 = c^2\mu^2 + m^2c^4$ which links mass, momentum, and energy. This defines coordinates μ and E as momentum and energy values. One has then $I = 4N(mc/\hbar)^2 = J$ and the intrinsic information I in the 4-position of a particle is proportional to the square of its intrinsic energy mc^2. Since J is a universal constant (see comments below), c is fixed, and given that $\hbar$ has been fixed, one concludes that the rest mass m is a universal constant. Since I measures the capacity of the observed phenomenon to provide information about (in this case) 4-length it follows that I should translate into a figure for the ultimate fluctuation (resolution) length that is intrinsic to QM. Here the information is $I = (4N/L^2)$ with $L = \hbar/mc$ the reduced Compton wavelength. If all N estimates have the same accuracy some argument then leads to $e_{min} = L$ and e_{min} corresponds to a minimal resolution length (i.e. ability to know). Finally putting things together one gets

$$J = \frac{4Nm^2c^3}{\hbar^2}\int\int d\mu dE\sum_1^{N/2}\phi_n^*\phi_n = \frac{4Nm^2c^3}{\hbar^2}\int\int drdt\sum_1^{N/2}\psi_n^*\psi \tag{4.23}$$

$$K = I - J = \tag{4.24}$$

$$= 4Nc\sum_1^{N/2}\int\int drdt\left[-(\nabla\psi_n)^*\cdot\nabla\psi_n + \left(\frac{1}{c^2}\right)\partial_t\psi_n^*\partial_t\psi_n - \frac{m^2c^2}{\hbar^2}\psi_n^*\psi_n\right]$$

There is much more material in [**490**] to enhance and refine the above ideas. There are certain subtle features as well. In 4-dimensions the Fourier transform is unitary and covariance is achieved in all variables (treating t separately as in $q_n(x|t)$ is not a covariant formalism). EPI treats all phenomena as being statistical in origin and every Euler-Lagrange (EL) equation determines a kind of QM for the particular phenomenon (think here of the q_n as fields). This includes classical electromagnetism for example where the vector potential A is considered as a kind of probability "amplitude" for photons. In 4-D the Lorentz transformation satisfies the requirement that Fisher information I is invariant under a change of reference frame and this property is transmitted to J and K. Thus invariance of accuracy (or of error estimation) under a change of reference frame leads to the Lorentz transformation and to the requirement of covariance. Historically the classical Lagrangian has often been a contrivance for getting the correct answers and a main idea in [**490**] is to present a systematic approach to deriving Lagrangians. The Lagrangian represents the physical information $k(\mathbf{x}) = \sum k_n(\mathbf{x})$, $k_n(\mathbf{x}) = i_n(\mathbf{x}) - j_n(\mathbf{x})$, and $\int k(\mathbf{x})$ is the total physical information K for the system. The solution to the variational problem for the Lagrangian can represent then (for real coordinates) the payoff in a mathematical information game (e.g. the KG equation is a payoff expression). We exhibit now a derivation of the SE from [**490**] to show robustness of the EPI scheme. The position of a particle of mass m is measured as a value $y = x + \theta$ where x is a random excursion whose probability amplitude law $q(x)$ is sought. Since the time t is being ignored here one is in effect looking for a stationary solution to the problem. Note the issue of covariance does not arise here and the time dependent SE is not treated since in particular it is not covariant; it can however be obtained from the KG equation as a nonrelativistic limit. Assume that the particle is moving in a conservative field of scalar potential $V(x)$ with total energy W conserved. One defines complex wave functions as before and can write

$$I = 4N \sum_{1}^{N/2} \int dx \left| \frac{d\psi_n(x)}{dx} \right|^2 \qquad (4.25)$$

A Fourier transform space is defined via $\psi_n(x) = (1/\sqrt{2\pi\hbar} \int d\mu \phi_n(\mu) exp(-i\mu x/\hbar)$ where $\mu \sim$ momentum. The unitary nature of this transformation guarantees the validity of the EPI variational procedure. One uses the Parseval theorem to get

$$I = \frac{4N}{\hbar^2} \int d\mu \mu^2 \sum_n |\phi_n(\mu)|^2 = J \qquad (4.26)$$

This corresponds to (4.19) and is the invariance principle for the given measurement problem. The x-coordinate expressions analogous to (4.20) and (4.21) show that the sum in (4.26) is actually an expectation $J = (4N/\hbar^2) < \mu^2 >$. Now use the specifically nonrelativistic approximation that the kinetic energy E_{kin} of the particle is $\mu^2/2m$ and then

$$J = \frac{8Nm}{\mu^2} < E_{kin} >= \frac{8Nm}{\hbar^2} < [W - V(x)] >= \qquad (4.27)$$

$$= \frac{8Nm}{\hbar^2} \int dx [W - V(x)] \sum |\psi_n(x)|^2$$

where the last expression is the PDF $p(x)$. This J is the bound information functional $J[q] = J(\psi)$ and $\kappa = 1$ here. This leads to a variational problem

$$K = N \sum_{1}^{N/2} \int dx \left[4 \left| \frac{d\psi_n(x)}{dx} \right|^2 - \frac{8m}{\hbar^2} [W - V(x)] |\psi_n(x)|^2 \right] = extremum \qquad (4.28)$$

The Euler-Lagrange equations are then $(\psi^*_{nx} = \partial \psi^*_n / \partial x)$

$$\frac{d}{dx} \left(\frac{\partial \mathfrak{L}}{\partial \psi^*_{nx}} \right) = \frac{\partial \mathfrak{L}}{\partial \psi^*_n}; \ \psi''_n(x) + \frac{2m}{\hbar^2} [W - V(x)] \psi_n(x) = 0 \qquad (4.29)$$

which is the stationary SE. Since the form of equation (4.29) is the same for each index value n the scenario admits $N = 2$ degrees of freedom $q_n(x)$ or one complex degree of freedom $\psi(x)$; hence the SE defines a single complex wave function. Since this derivation works with a real coordinate x the information transfer game is being played here and the payoff is the Schrödinger wave function.

REMARK 6.4.1. There are generalizations of EPI to nonextensive information measures in [**268, 299, 298**] (cf. also [**1006, 1007, 1030, 1032**]). ■

4.1. LEGENDRE THERMODYNAMICS. We go to the last paper in [**491**] by Frieden, A. and A.R. Plastino, and Soffer which provides a discussion of Fisher thermodynamics and the Legendre transformation. It is shown that the Legendre transform structure of classical thermodynamics can be replicated without change if one replaces the entropy S by the Fisher information I. This produces a thermodynamics capable of treating equilibrium and nonequilibrium situations in a traditional manner. We recall the Shannon information measure $S = -\sum P(i) log[P(i)]$; it is known that if one chooses the Boltzmann constant as the informational unit and identifies Shannon's entropy with the thermodynamic entropy then the whole of statistical mechanics can be elegantly reformulated without any reference to the idea of ensemble. The success of thermodynamics and statistical physics depends crucially on the Legendre structure and one shows now that such relationships all hold if one replaces S by the Fisher information measure. We recall that for $\int g(x,\theta)dx = 1$ one writes $I = \int dx g(x,\theta)[\partial_\theta g/g]^2$ and for shift invariant g one has $I = \int dx[(g')^2/g]$. There are two approaches to using Fisher information, EPI and minimum Fisher information (MFI), and both lead to the same results here. We write (shifting to a probability function f)

$$\int dx f(x,\theta) = 1;\ I[f] = \int dx F_{Fisher}(f);\ F_{Fisher}(f) = f(x)[f'/f]^2 \tag{4.30}$$

Assume that for M functions $A_i(x)$ the mean values $< A_i >$ are known where

$$< A_i > = \int dx A_i(x) f(x) \tag{4.31}$$

This represents information at some appropriate (fixed) time t. The analysis will use MFI (or EPI) to find the probability distribution $f_I = f_{MFI}$ that extremizes I subject to prior conditions $< A_i >$ and the result will be given via solutions of a stationary Schrödinger like equation. The Fisher based extremization problem has the form $(F(f) = F_{Fisher}(f))$

$$\delta_f \left[I(f) - \alpha < 1 > - \sum_1^M \lambda_i < A_i > \right] = 0 \equiv \tag{4.32}$$

$$\delta_f \left[\int dx \left(F(f) - \alpha f - \sum_1^M \lambda_i A_i f \right) \right] = 0$$

Variation leads to $((\alpha, \lambda_1, \cdots, \lambda_M)$ are Lagrange multipliers)

$$\int dx \delta f \left[(f)^{-2} \left(\frac{\partial f}{\partial x} \right)^2 + \frac{\partial}{\partial x} \left(\frac{2}{f} \frac{\partial f}{\partial x} \right) + \alpha + \sum_1^M \lambda_i A_i \right] = 0 \tag{4.33}$$

and on account of the arbitrariness of δf this yields

$$(f)^{-2}(f')^2 + \frac{\partial}{\partial x}(2/f)f') + \alpha + \sum_1^M \lambda_i A_i = 0 \tag{4.34}$$

The normalization condition on f makes α a function of the λ_i and we assume $f_I(x,\lambda)$ to be a solution of (4.34) where $\lambda \sim (\lambda_i)$. The extreme Fisher information is then

$$I = \int dx f_I^{-1} (\partial_x f_I)^2 \tag{4.35}$$

Now to find a general solution of (4.34) define $G(x) = \alpha + \sum_1^M \lambda_i A_i(x)$ and write (4.34) in the form

$$\left[\frac{\partial log(f_I)}{\partial x} \right]^2 + 2 \frac{\partial^2 log(f_I)}{\partial x^2} + G(x) = 0 \tag{4.36}$$

Make the identification $f_I = (\psi)^2$ now the introduce a new variable $v(x) = \partial log(\psi(x))/\partial x$. Then (4.36) becomes

$$v'(x) = -\left[\frac{G(x)}{4} + v^2(x)\right] \tag{4.37}$$

which is a Riccati equation. This leads to

$$u(x) = exp\left[\int^x dx[v(x)]\right] = exp\left[\int^x dx\frac{dlog(\psi)}{dx}\right] = \psi; \tag{4.38}$$

$$-\frac{1}{2}\psi''(x) - \frac{1}{8}\sum_1^M \lambda_i A_i(x)\psi(x) = \frac{\alpha}{8}\psi(x)$$

where the Lagrange multiplier $\alpha/8$ plays the role of an energy eigenvalue and the sum of the $\lambda_i A_i(x)$ is an effective potential function $U(x) = (1/8)\sum_1^M \lambda_i A_i(x)$. We note (in keeping with the Lagrangian spirit of EPI) that the Fisher information measure corresponds to the expectation value of the kinetic energy of the SE. Note also that (4.38) has multiple solutions and it is reasonable to suppose that the solution leading to the lowest I is the equilibrium one. Now standard thermodynamics uses derivatives of the entropy S with respect to λ_i and $< A_i >$ and we start from (4.35) and write after an integration by parts

$$\frac{\partial I}{\partial \lambda_i} = \int dx \frac{\partial f_I}{\partial \lambda_i}\left[-f_I^{-2}(f_I')^2 - \frac{\partial}{\partial x}\left(\frac{2}{f_I}f_I'\right)\right] \tag{4.39}$$

Comparing this to (4.34) one arrives at

$$\frac{\partial I}{\partial \lambda_i} = \int dx \frac{\partial f_I}{\partial \lambda_i}\left[\alpha + \sum_1^M \lambda_j A_j\right] \tag{4.40}$$

which on account of normalization yields

$$\frac{\partial I}{\partial \lambda_i} = \sum_1^M \lambda_j \frac{\partial}{\partial \lambda_i}\int dx f_I A_j(x) \equiv \frac{\partial I}{\partial \lambda_i} = \sum_1^N \lambda_j \frac{\partial}{\partial \lambda_i} < A_j > \tag{4.41}$$

This is a generalized Fisher-Euler theorem whose thermodynamic counterpart is the derivative of the entropy with respect to the mean values. One computes easily

$$\sum_i \frac{\partial I}{\partial \lambda_i}\frac{\partial \lambda_i}{\partial < A_j >} = \sum_i\sum_k \lambda_k \frac{\partial < A_k >}{\partial \lambda_i}\frac{\partial \lambda_i}{\partial < A_j >} \Rightarrow \frac{\partial I}{\partial < A_j >} = \lambda_j \tag{4.42}$$

as expected. The Lagrange multipliers and mean values are seen to be conjugate variables and one can also say that $f_I = f_I(\lambda_1, \cdots, \lambda_M)$.

Now as the density f_I formally depends on $M + 1$ Lagrange multipliers, normalization $\int dx f_I(x) = 1$ makes α a function of the λ_i and we write $\alpha = \alpha(\lambda_1, \cdots, \lambda_M)$. One can assume that the input information refers to the λ_i and not to the $< A_i >$. Introduce then a generalized thermodynamic potential (Legendre transform of I) as

$$\lambda_J(\lambda_1, \cdots, \lambda_M) = I(< A_1 >, \cdots, < A_M >) - \sum_1^M \lambda_i < A_i > \tag{4.43}$$

Then

$$(4.44)\qquad \frac{\partial\lambda_J}{\partial\lambda_i}=\sum_1^M\frac{\partial I}{\partial<A_j>}\frac{\partial<A_j>}{\partial\lambda_i}-\sum_1^M\lambda_j\frac{\partial<A_j>}{\partial\lambda_i}-<A_i)=-<A_i>$$

where (4.42) has been used. Thus the Legendre structure can be summed up in

$$(4.45)\qquad \lambda_J=I-\sum_1^M\lambda_i<A_i>;\ \frac{\partial\lambda_J}{\partial\lambda_i}=-<A_i>;\ \frac{\partial I}{\partial<A_i>}=\lambda_i;$$

$$\frac{\partial\lambda_i}{\partial<A_j>}=\frac{\partial\lambda_j}{\partial<A_i>}=\frac{\partial^2 I}{\partial<A_i>\partial<A_j>};$$

$$\frac{\partial<A_j>}{\partial\lambda_i}=\frac{\partial<A_i>}{\partial\lambda_j}=-\frac{\partial^2\lambda_J}{\partial\lambda_i\partial\lambda_j}$$

As a consequence one can recast (4.41) in the form

$$(4.46)\qquad \frac{\partial I}{\partial\lambda_i}=\sum_1^M\lambda_j\frac{\partial}{\partial\lambda_j}<A_i>$$

Thus the Legendre transform structure of thermodynamics is entirely translated into the Fisher context.

4.2. FIRST AND SECOND LAWS. We go here to Plastino and Curado [**1033**] where one shows the coimplication of the first and second laws of thermodynamics. Thus macroscopically in classical phenomenological thermodynamics the first and second laws can be regarded as independent statements. In statistical mechanics an underlying microscopic substratum is added that is able to explain thermodynamics itself. Of this substratum a microscopic probability distribution (PD) that controls the population of microstates is a basic ingredient. Changes that affect exclusively microstate population give rise to heat and how these changes are related to energy changes provides the essential content of the first law (cf. [**1067**]). In [**1033**] one shows that the PD establishes a link between the first and second laws according to the following scheme.

- Given: An entropic form (or an information measure) S, a mean energy U and a temperature T, and for any system described by a microscopic PD p_i a heat transfer process via $p_i\to p_i+dp_i$ then
- If the PD p_i maximizes S this entails $dU=TdS$ and alternatively
- If $dU=TdS$ then this predetermines a unique PD that maximizes S.

For the second law one wants to maximize entropy S with M appropriate constraints A_k which take values $A_k(i)$ at the microstate i; the constrains have the form

$$(4.47)\qquad <A_k>=\sum_i p_iA_k(i)\ \ (k=1,\cdots M)$$

The Boltzman constant is k_B and assume that $k=1$ in (4.47) corresponds to the energy E with $A_1(i)=\epsilon_i$ so that the above expression specializes to

$$(4.48)\qquad U=<A_1>=\sum p_i\epsilon_i$$

One should now maximize the "Lagrangian" Φ given by

$$\Phi = \frac{S}{k_B} - \alpha \sum_i p_i - \beta \sum_i p_i \epsilon_i - \sum_2^M \lambda_k \sum_i p_i A_k(p_i) \tag{4.49}$$

in order to obtain the actual distribution p_i from the equation $\delta_{p_i}\Phi = 0$. Since here one is interested just in the "heat" part the last term on the right of (4.49) will not be considered. It is argued that if p_i changes to $p_i + dp_i$ because of $\delta_{p_i}\Phi = 0$ one will have

$$0 = \frac{dS}{k_B} - \beta dU \tag{4.50}$$

(note $\sum_i \delta p_i = 0$ via normalization). Since $\beta = 1/k_B T$ we get $dU = TdS$ so MaxEnt implies the first law.

The central goal here is to go the other way so assume one has a rather general information measure of the form

$$S = k \sum_i p_i f(p_i) \tag{4.51}$$

where $k \sim k_B$. The sum runs over a set of quantum numbers denoted by i (characterizing levels of energy ϵ_i) that specify an appropriate basis in Hilbert space, $\mathcal{P} = \{p_i\}$ is an (as yet unknown) probability distribution with $\sum p_i = constant$, and f is an arbitrary smooth function of the p_i. Assume further that mean values of quantities A that take the value A_i with probability p_i are evaluated via

$$< A >= \sum_i A_i g(p_i) \tag{4.52}$$

In particular the mean energy U is given by $U = \sum_i \epsilon_i g(p_i)$. Assume now that the set $\mathcal{P}$ changes in the fashion

$$p_i \to p_i + dp_i; \; \sum dp_i = 0 \tag{4.53}$$

(the last via $\sum p_i = constant$. This in turn generates corresponding changes dS and dU and one is thinking here of level population changes, i.e. heat. To insure the first law one assumes $(\bullet)$ $dU - TdS = 0$ and as a consequence of $(\bullet)$ a little algebra gives (up to first order in the dp_i the condition

$$\epsilon_i g'(p_i) - kT[f(p_i) + p_i f'(p_i)] = 0 \tag{4.54}$$

This equation is now examined for several situations. First look at Shannon entropy with

$$f(p_i) = -log(p_i); \; g(p_i) = p_i \tag{4.55}$$

In this situation (4.54) becomes

$$-\epsilon_i = kT[log(p_i) + 1] \Rightarrow p_i = \frac{1}{e} exp(-\epsilon_i/kT) \tag{4.56}$$

After normalization this is the canonical Boltzmann distribution and this is the only distribution that guarantees obedience to the first law for Shannon's information measure. A posteriori this distribution maximizes entropy as well with U as a constraint which establishes a link with the second law. Several other measures are considered, in particular the Tsallis measure, and we refer to [**1033**] for details. In summary if one assumes entropy is maximum one immediately derives the first law and if you assume the first law and an information measure this predetermines a probability distribution that maximizes entropy.

REMARK 6.4.2. There is currently a great interest in acoustic wave phenomena, sound and vortices, acoustic spacetime, acoustic black holes, etc. A prime source of material involves superfluid physics à la Volovik [**1267, 1268**] and Bose-Einstein condensates (see e.g. [**49, 50, 107, 102, 125, 144, 471, 472, 1201, 1202**]). We had originally written out material from [**1201, 1202**] in preparation for sketching some material from Volovik [**1267**]. This seems to be too big a project but we do include some material involving Bose-Einstein condensates in later Chapters. ∎

REMARK 6.4.3. We also note for comparison and analogy some relations between Legendre duals in mechanics, thermodynamics, and (x,ψ) duality. Thus (cf. [**249, 806**]) one has in mechanics $p\dot{x} - L = H$ via $L = (/2)m\dot{x}^2 - V$ and $H = (p^2/2m) + V$ with $p = \partial L/\partial\dot{x}$ and $\dot{x} = \partial H/\partial p$. In thermodynamics one has a Helmholtz free energy F with $F = U - TS$ for energy U, entropy S, and temperature T. Set $\mathcal{F} = -F$ to obtain $\mathcal{F} = T(\partial\mathcal{F}/\partial T) - U$ and $U = S\partial_S U - \mathcal{F}$ (where $\partial_T\mathcal{F} = S$ and $\partial_S U = T$). Now put this in a table where we write the (x,ψ) duality in the form $\chi = \psi^2(\partial\mathfrak{F}/\partial\psi^2) - \mathfrak{F}$ with $\mathfrak{F} = \phi(\partial\chi/\partial\phi) - \chi$ (for $\chi = (i\sqrt{2m}/\hbar)x$ and $\phi = (\partial\mathfrak{F}/\partial\psi^2)$). This leads to a table

Mechanics	*Thermodynamics*	(x,ψ) *duality*
$\dot{x}$, p, L, H	T, S, $\mathcal{F}$, U	ψ^2, ϕ, $\mathfrak{F}$, χ
$p\dot{x} - H = L$	$TS - U = \mathcal{F}$	$\psi^2\phi - \mathfrak{F} = \chi$
$L = \dot{x}\frac{\partial H}{\partial p} - H$	$\mathcal{F} = S\frac{\partial U}{\partial S} - U$	$\mathfrak{F} = \phi\frac{\partial\chi}{\partial\phi} - \chi$
$H = p\frac{\partial L}{\partial\dot{x}} - L$	$U = T\frac{\partial\mathcal{F}}{\partial T} - \mathcal{F}$	$\chi = \psi^2\frac{\partial\mathfrak{F}}{\partial\psi^2} - \mathfrak{F}$

One says that e.g. $(\mathfrak{F},\chi)$ or $(\mathcal{F},U)$ or (L,H) form a Legendre dual pair and in the first situation one refers to (x,ψ) duality. One sees in particular that $\mathfrak{F} = \psi^2\phi - \chi$ where $\psi^2\phi \sim \dot{x}p$ in mechanics. Note that $\phi = \partial\mathfrak{F}/\partial\psi^2 = (1/2\psi)(\partial\mathfrak{F}/\partial\psi) = \bar{\psi}/2\psi$ with $\psi^2\phi = (1/2)\psi\bar{\psi} = (1/2)|\psi|^2$. In any event $\chi = -i(\sqrt{2m}/\hbar)x$ and we will see below how the physics can be expressed via ψ, $\partial/\partial\psi$, $d\psi$ etc. without mentioning x. This allows one to think of the coordinate x as an emergent entity and we like to think of $x - \psi$ duality in this spirit. ∎

REMARK 6.4.4. In [**445**] Faraggi and Matone develop a theory of $x - \psi$ duality, related to Seiberg-Witten theory in the string arena, which was expanded in various ways in [**2, 52, 133, 245, 241, 238, 313, 864, 1002, 1257**]. Here one works from a stationary SE $[-(\hbar^2/2m)\Delta + V(x)]\psi = E\psi$, and, assuming for convenience one space dimension, the space variable x is determined by the wave function ψ from a prepotential $\mathfrak{F}$ via Legendre transformations. The theory suggests that x plays the role of a macroscopic variable for a statistical system with a scaling term $\hbar$. Thus define a prepotential $\mathfrak{F}_E(\psi) = \mathfrak{F}(\psi)$ such that the dual variable $\psi^D = \partial\mathfrak{F}/\partial\psi$ is a (linearly independent) solution of the same SE. Take V and E real so that $\bar{\psi} = \psi^D$ qualifies and write $\partial_x\mathfrak{F} = \psi^D\partial_x\psi = (1/2)[\partial_x(\psi\psi^D) +$

$W)]$ where W is the Wronskian. This leads to $(\psi^D = \bar{\psi})$ the relation $\mathfrak{F} = (1/2)\psi\bar{\psi} + (W/2)x$ (setting the integration constant to zero). Consequently, scaling W to $-2i\sqrt{2m}/\hbar$ one obtains

$$\frac{i\sqrt{2m}}{\hbar}x = \frac{1}{2}\psi\frac{\partial\mathfrak{F}}{\partial\psi} - \mathfrak{F} \equiv \frac{i\sqrt{2m}}{\hbar}x = \psi^2\frac{\partial\mathfrak{F}}{\partial\psi^2} - \mathfrak{F} \tag{4.57}$$

which exhibits x as a Legendre transform of $\mathfrak{F}$ with respect to ψ^2. Duality of the Legendre transform then gives also

$$\mathfrak{F} = \phi\partial_\phi\left(\frac{i\sqrt{2m}\,x}{\hbar}\right) - \left(\frac{i\sqrt{2m}\,x}{\hbar}\right);\ \phi = \partial_{\psi^2}\mathfrak{F} = \frac{\bar{\psi}}{2\psi} \tag{4.58}$$

so that $\mathfrak{F}$ and $(i\sqrt{2m}\,x/\hbar)$ form a Legendre pair. In particular one has $\rho = |\psi|^2 = \frac{2i\sqrt{2m}}{\hbar}x + 2\mathfrak{F}$ which also relates x and the probability density (but indirectly since the x term really only cancels the imaginary part of $2\mathfrak{F}$). In any event one sees that the wave function ψ specifically determines the exact location of the "particle" whose quantum evolution is described by ψ. We mention here also that the (stationary) SE can be replaced by a third order equation

$$4\mathfrak{F}''' + (V(x) - E)(\mathfrak{F}' - \psi\mathfrak{F}'')^3 = 0;\ \mathfrak{F}' \sim \frac{\partial\mathfrak{F}}{\partial\psi} \tag{4.59}$$

and a dual stationary SE has the form

$$\frac{\hbar^2}{2m}\frac{\partial^2 x}{\partial\psi^2} = \psi[E - V]\left(\frac{\partial x}{\partial\psi}\right)^3 \tag{4.60}$$

A noncommutative version of this is developed in the second paper of Vancea [**1257**]. ■

5. ENTROPY AND SPACETIME

In Padmanabhan [**976**] one takes an entropy functional ($u^a = \bar{x}^a - x^a$ is a perturbation)

$$S = \frac{1}{8\pi}\int d^4x\sqrt{g}\left[M^{abcd}\nabla_a u_b \nabla_c u_d + N_{ab}u^a u^b\right] \tag{5.1}$$

Extremizing with respect to u_b leads to $(N_{ab}u^a u^b = N^{ab}u_a u_b)$

$$\nabla_a\left(M^{abcd}\nabla_c\right)u_d = N^{bd}u_d \tag{5.2}$$

Note $\int d^4x\sqrt{-g}f\nabla_a u_b = -\int d^4x\sqrt{-g}u_b\nabla_a f$ since via [**12**] one can write $\delta\sqrt{-g} = -(1/2)\sqrt{-g}g_{\mu\nu}\delta g^{\mu\nu}$ and $\nabla_a g^{\mu\nu} = 0$. Choosing M and N such that (5.2) (for all u_d) implies the Einstein equations entails

$$M^{abcd} = g^{ad}g^{bc} - g^{ab}g^{cd};\ N_{ab} = 8\pi\left(T_{ab} - \frac{1}{2}g_{ab}T\right) \tag{5.3}$$

Consequently S becomes (cf. also Chapter 10)

$$S = \frac{1}{8\pi}\int d^4x\sqrt{-g}\left[(\nabla_a u^b)(\nabla_b u^a) - (\nabla_b u^b)^2 + N_{ab}u^a u^b\right] = \tag{5.4}$$

$$= \frac{1}{8\pi}\int d^4x\sqrt{-g}\left[Tr(J^2) - (Tr(J))^2 + 8\pi\left(T_{ab} - \frac{1}{2}g_{ab}T\right)u^a u^b\right]$$

where $J_a^b = \nabla_a u^b$. Note here

(5.5) $$\int d^4x\sqrt{-g}g^{ad}g^{bc}\nabla_a u_b \nabla_c u_d = \int d^4x\sqrt{-g}(\nabla^a u^b)(\nabla^c u^d)$$

and also

(5.6) $$\nabla_a\left(M^{abcd}\nabla_c\right)u_d = \nabla_a\left[g^{ad}g^{bc} - g^{ab}g^{cd}\right]\nabla_c u_d =$$

$$= \nabla_a g^{ad}g^{bc}\nabla_c u_d - \nabla_a g^{ab}g^{cd}\nabla_c u_d = \nabla_a\nabla^b u^a - \nabla^b\nabla_c u^c \sim (\nabla_a\nabla^b - \nabla^b\nabla_a)u^a$$

Further (as in (1.6))

(5.7) $$M^{abcd}\nabla_a u_b \nabla_c u_d = g^{ad}g^{bc}\nabla_a u_b\nabla_c u_d - g^{ab}g^{cd}\nabla_a u_b \nabla_c u_d =$$

$$= \nabla^d u_b \nabla^b u_d - \nabla_a u^a \nabla_c u^c$$

which confirms (5.4). We record also from [**950**] that

(5.8) $$(\nabla_\mu\nabla_\nu - \nabla_\nu\nabla_\mu)\alpha(w) = R(\alpha, \partial_\mu, \partial_\nu, w)$$

which identifies $\nabla_\mu\nabla_\nu - \nabla_\nu\nabla_\mu$ with $R_{\mu\nu}$ and allows us to imagine (5.6) as $R_a^b u^a$ with Einstein equations

(5.9) $$R_a^b u^a = N_a^b u^a \;(= N^{bc}g_{ca}g^{ca}u_c)$$

for example, which is of course equivalent to $R_{ab} = N_{ab}$ (cf. also [**977**]). Note also $G_{ab} = R_{ab} - (1/2)Rg_{ab} = kT_{ab}$ implies that $R^\mu_\nu - (1/2)R\delta^\mu_\nu = kT^\mu_\nu$ which upon contraction gives $R = -kT$ (since $\delta^\mu_\mu = 4$) and hence $R_{ab} = k(T_{ab} - (1/2)Tg_{ab})$.

For completeness we sketch here a derivation of the Einstein equations from an action principle (cf. [**12, 312, 884, 1288**]). The Einstein-Hilbert action is $A = \int_\Omega[\mathfrak{L}_G + \mathfrak{L}_M]d^4x$ where $\mathfrak{L}_G = (1/2\chi)\sqrt{-g}\,{}^4R$ ($\chi = 8\pi$ and ${}^4R$ is the Ricci scalar). Following Ciufolini-Wheeler [**312**] we list a few useful facts first (generally we will write if necessary $g_{ab}T^{cb} = T^c_{\cdot a}$ and $g_{ab}T^{bc} = T_a^{\cdot c}$).

1. (1) $\nabla_\gamma g^{\alpha\beta} = 0$ (by definitions of covariant derivative and Christoffel symbols).
2. (2) $\delta\sqrt{-g} = (1/2)\sqrt{-g}g^{\alpha\beta}\delta g_{\alpha\beta}$ and $(\delta g_{\alpha\beta})g^{\alpha\beta} = -(\delta g^{\alpha\beta})g_{\alpha\beta}$ (see e.g. Wald [**1274**] for the calculation).
3. (3) For a vector field v^a one has $\nabla_a v^a = \partial_a(\sqrt{-g}v^a)(1/\sqrt{-g})$ and $\nabla_\beta T^{\alpha\beta} = \partial_\beta(\sqrt{-g}T^{\alpha\beta})(1/\sqrt{-g}) + \Gamma^\alpha_{\sigma\beta}T^{\sigma\beta}$ (from $\Gamma^\sigma_{\sigma\alpha} = (1/2)(\partial_\alpha g_{\mu\nu})g^{\mu\nu}$ and $\partial_\alpha(log(\sqrt{-g})$ $\Gamma^\sigma_{\sigma\alpha})$.
4. (4) For two metrics g, g^* one shows that $\delta\Gamma^\alpha_{\beta\gamma} = \Gamma^{*\alpha}_{\beta\gamma} - \Gamma^\alpha_{\beta\gamma}$ is a tensor.
5. (5) $\delta R_{\alpha\beta} = \nabla_\sigma(\delta\Gamma^\sigma_{\alpha\beta} - \nabla_\beta(\delta\Gamma^\sigma_{\alpha\sigma})$ (see [**236**] for the calculations).
6. (6) Recall also Stokes theorem $\int_\Omega \nabla_\sigma v^\sigma\sqrt{-g}d^4x = \int_\Omega \partial_\sigma(v^\sigma\sqrt{-g})d^4x = \int_{\partial\Omega}\sqrt{-g}v^\sigma d^3\Sigma_\sigma$.

Now requiring a stationary action for arbitrary δg^{ab} (with certain derivatives of the g^{ab} fixed on the boundary of Ω one obtains ($\mathfrak{L}_M$ is the matter Lagrangian)

(5.10) $$\delta I = \frac{1}{2\chi}\int_\Omega\left(R_{\alpha\beta} - \frac{1}{2}g_{\alpha\beta}R\right)\sqrt{-g}\delta g^{\alpha\beta}d^4x +$$

$$+\frac{1}{2\chi}\int_\Omega g^{\alpha\beta}\sqrt{-g}\delta R_{\alpha\beta}d^4x + \int_\Omega \frac{\delta\mathfrak{L}_M}{\delta g^{\alpha\beta}}\delta g^{\alpha\beta}d^4x = 0$$

The second term can be written

(5.11) $$\frac{1}{2\chi}\int_\Omega g^{\alpha\beta}\sqrt{-g}\delta R_{\alpha\beta}d^4x = \frac{1}{2\chi}\int g^{\alpha\beta}\sqrt{-g}[\nabla_\sigma(\delta\Gamma^\sigma_{\alpha\beta}) - \nabla_\beta(\delta\Gamma^\sigma_{\alpha\sigma}]d^4x =$$

$$= \frac{1}{2\chi}\int_\Omega \sqrt{-g}[\nabla_\sigma(g^{\alpha\beta}\delta\Gamma^\sigma_{\alpha\beta}) - \nabla_\beta(g^{\alpha\beta}\delta\Gamma^\sigma_{\alpha\sigma})]d^4x =$$

$$= \frac{1}{2\chi}\int_\Omega \partial_\sigma[(\sqrt{-g}g^{\alpha\beta}\delta\Gamma^\sigma_{\alpha\beta}) - (\sqrt{-g}g^{\alpha\sigma}\delta\Gamma^\rho_{\alpha\rho})]d^4x$$

where $\delta\Gamma^\alpha_{\beta\gamma} = (1/2)[\nabla_\gamma(\delta g_{\beta\sigma}) + \nabla_\beta(\delta g_{\sigma\gamma}) - \nabla_\sigma(\delta g_{\gamma\beta})]$. This can be transformed into an integral over the boundary $\partial\Omega$ where it vanishes if certain derivatives of $g_{\alpha\beta}$ are fixed on the boundary. In fact the integral over the boundary $\partial\Omega = \sum S_i$ can be written as $\sum_i(\epsilon_I/2\chi)\int_{S_i}\gamma_{\alpha\beta}\delta\tilde{N}^{\alpha\beta}d^3x$ where $\epsilon_i = \mathbf{n}_i\cdot\mathbf{n}_i = \pm 1$ ($\mathbf{n}_i$ normal to S_i) and $\gamma_{\alpha\beta} = g_{\alpha\beta} - \epsilon_i\mathbf{n}_\alpha\cdot\mathbf{n}_\beta$ is the 3-metric on the hypersurface S_i (cf. [?]). Further

(5.12) $$\tilde{N}^{\alpha\beta} = \sqrt{|\gamma|}(K\gamma^{\alpha\beta} - K^{\alpha\beta}) = -\frac{1}{2}g\gamma^{\alpha\mu}\gamma^{\beta\nu}\mathcal{L}_{\mathbf{n}}(g^{-1}\gamma_{\mu\nu})$$

where $K_{\alpha\beta} = -(1/2)\mathcal{L}_{\mathbf{n}}\gamma_{\alpha\beta}$ is the extrinsic curvature of each S_i and $\mathcal{L}_{\mathbf{n}}$ is the Lie derivative. Consequently if the quantities $\tilde{N}^{\alpha\beta}$ are fixed on the boundary for an arbitrary $\delta g_{\alpha\beta}$ one gets from the first and last equations in (5.10) the Einstein field equations

(5.13) $$G_{\alpha\beta} = R_{\alpha\beta} - \frac{1}{2}Rg_{\alpha\beta} = \chi T_{\alpha\beta};\ T_{\alpha\beta} = -2\frac{\delta\mathfrak{L}_M}{\delta g^{ab}} + \mathfrak{L}_M g_{\alpha\beta}$$

We note here that

(5.14) $$\delta\int\mathfrak{L}_m\sqrt{-g}d^4x = \int\frac{\delta\mathfrak{L}_m}{\delta g^{ab}}\sqrt{-g}d^4x + \int\mathfrak{L}_m\delta(\sqrt{-g})d^4x =$$

$$= \int\frac{\delta\mathfrak{L}_m}{\delta g^{ab}}\sqrt{-g}d^4x - \frac{1}{2}\int\mathfrak{L}_m g_{ab}(\delta g^{ab})\sqrt{-g}d^4x$$

A factor of 2 then arises from the 2χ in (5.10).

REMARK 6.5.1. Let us rephrase some of this following Wald [**1272**] for clarity. Thus e.g. think of functionals $F(\psi)$ with $\psi = \psi_\lambda$ a one parameter family and set $\delta\psi = (d\psi_\lambda/d\lambda)|_{\lambda=0}$. For $F(\psi)$ one writes then $dF/d\lambda = \int\phi\delta\psi$ and sets $\phi = (\delta F/\delta\psi)|_{\psi_0}$. Then (assuming all functional derivatives are symmetric with no loss of generality) one has for $\mathfrak{L}_G = \sqrt{-g}R$ and $S_G = \int\mathfrak{L}_G d^4x$

(5.15) $$\frac{d\mathfrak{L}_G}{d\lambda} = \sqrt{-g}(\delta R_{ab})g^{ab} + \sqrt{-g}R_{ab}\delta g^{ab} + R\delta(\sqrt{-g})$$

But $g^{ab}\delta R_{ab} = \nabla^a v_a$ for $v_a = \nabla^b(\delta g_{ab}) - g^{cd}\nabla_a(\delta g_{cd})$. Further one has an equation $\delta\sqrt{-g} = -(1/2)\sqrt{-g}g_{ab}\delta g^{ab}$ and consequently

(5.16)
$$\frac{dS_G}{d\lambda} = \int\frac{d\mathfrak{L}_G}{d\lambda}d^4x = \int\nabla^a v_a\sqrt{-g}d^4x + \int\left(R_{ab} - \frac{1}{2}Rg_{ab}\right)(\delta g^{ab})\sqrt{-g}d^4x$$

Discarding the first term as a boundary integral we get the first term in (5.10).■

REMARK 6.5.2. From [**976**] we see that the entropy in S in (5.1) reduces to a 4-divergence when the Einstein equations are satisfied "on shell" making S a surface term

$$S = \frac{1}{8\pi}\int_V d^4x\sqrt{-g}\nabla_i(u^b\nabla_b u^i - u^i\nabla_b u^b) = \tag{5.17}$$

$$= \frac{1}{8\pi}\int_{\partial V} d^3x\sqrt{h}n_i(v^b\nabla_b u^i - u^i\nabla_b u^b)$$

Thus the entropy of a bulk region V of spacetime resides in its boundary ∂V when the Einstein equations are satisfied. In varying (5.1) to obtain (5.2) one keeps the surface contribution to be a constant. Thus in a semiclassical limit when the Einstein equations hold to the lowest order the entropy is contributed only by the boundary term and the system is holographic. ∎

6. REMARKS ON WDW

We gather now some information about the derivation of Einstein's equations from an action principle and also discuss the Hamiltonian theory involving the Einstein-Hamilton-Jacobi (EHJ) equation and the WDW equation (there is much more on WDW at various places in this book). One recalls from Gerlach [**521**] that the EHJ equation

$$^3R + \frac{1}{h}\left(\frac{1}{2}h_{ij}h_{k\ell} - h_{ik}h_{j\ell}\right)\left(\frac{\delta S}{\delta h_{ij}}\right)\left(\frac{\delta S}{\delta h_{k\ell}}\right) = 0 \tag{6.1}$$

(h_{ij} corresponds to the metric of a spatial hypersurface) plus a principle of constructive interference of deBroglie waves leads to the entire set of 10 Einstein equations. The idea of Tomonaga's multi fingered time is used here (cf. also [**254, 932**]).

Now there are a number of derivations of the WDW equations with connections to Bohmain dynamics and the quantum potential in [**254, 599, 592, 1023, 1022, 1025, 1168, 1169, 1157, 1164, 1163**] and we will go directly to [**599, 592**] after a few comments. First let us recall the deWitt metric for which we refer to [**111, 254, 374, 496, 506, 533, 534, 599, 592, 605, 749, 751, 752, 884, 1023, 1025, 1085, 1168, 1169, 1157, 1164, 1163, 1204, 1288, 1297**] (other references are given later). Various formulas arise for the WDW which involve a deWitt metric (or supermetric)

$$G^{\alpha}_{abcd} = \frac{1}{\sqrt{h}}(h_{ac}h_{bd} + h_{ad}h_{bc} - 2\alpha h_{ab}h_{cd}); \tag{6.2}$$

$$G^{abcd}_{\beta} = \frac{\sqrt{h}}{2}(h^{ac}h^{bd} + h^{ad}h^{bc} - 2\beta h^{ab}h^{cd})$$

where $\alpha+\beta = 3\alpha\beta$. For general relativity (GR) one takes $\beta = 1$ and $\alpha = 1/2$ (see e.g. [**533, 534, 749**] for this form of the metric). Here the WDW equation for GR is ($c = 1$)

$$\mathfrak{H}\psi[h_{ab},\phi] = [-16\pi G\hbar^2 G_{abcd}\frac{\delta^2}{\delta h_{ab}\delta h_{cd}} - \frac{\sqrt{h}}{16\pi G}(R - 2\Lambda) + \mathfrak{H}_m]\psi = 0 \tag{6.3}$$

where h_{ab} is a 3-metric, R the 3-D Ricci scalar, Λ the cosmological constant, $G_{abcd} = G^{1/2}_{abcd}$ the deWitt metric, and $\mathfrak{H}_m$ is the Hamiltonian density for non-gravitational fields. The integrated form of (1.3) is

$$\int d^3x\, N\mathfrak{H}\psi = \mathfrak{H}^N\psi = \left(\mathfrak{H}^N_G + \mathfrak{H}^N_m\right)\psi = 0 \tag{6.4}$$

Writing $\psi = exp\left(\frac{i(MS_0+S_1+M^{-1}S_2+\cdots}{\hbar}\right)$ for $M = (32\pi G)^{-1}$ leads to a power series in M with second term

$$\mathfrak{H}_x = \frac{1}{2}G_{abcd}\frac{\delta S_0}{\delta h_{ab}}\frac{\delta S_0}{\delta h_{cd}} - 2\sqrt{h}(R - 2\Lambda) = 0 \tag{6.5}$$

which is the Hamilton-Jacobi (HJ) equation for the gravitational field and we refer to [**533, 534, 749**] for more details.

It will be important to see here how the quantum potential arises and we go to [**1023, 1022**] with a metric

$$ds^2 = -(N^2 - N^iN_i)dt^2 + 2N_idx^idt + h_{ij}dx^idx^j \tag{6.6}$$

(classical ADM situation - cf. [**16, 85, 855**]) and Hamiltonian

$$H = \int d^3x(N\mathfrak{H} + N^j\mathfrak{H}_j);\ \mathfrak{H}_j = -2D_i\pi^i_j + \pi_\phi\partial_j\phi; \tag{6.7}$$

$$\mathfrak{H} = \kappa G_{ijk\ell}\pi^{ij}\pi^{k\ell} + \frac{1}{2}h^{-1/2}\pi^2_\phi + h^{1/2}\left[-\kappa^{-1}(^3R - 2\Lambda) + \frac{1}{2}h^{ij}\partial_i\phi\partial_j\phi + U(\phi)\right]$$

where $\kappa = 16\pi G/c^4$, D_k is the covariant derivative, and

$$\pi^{ij} = -h^{1/2}\left(K^{ij} - h^{ij}K\right) = G^{ijk\ell}[\dot{h}_{k\ell} - D_kN_\ell - D_\ell N_k]; \tag{6.8}$$

$$K_{ij} = -\frac{1}{2N}(\dot{h}_{ij} - D_iN_j - D_jN_i)$$

Thus K_{ij} is the extrinsic curvature of the hypersurface and (♣) $\pi_\phi = (h^{1/2}/N)(\dot{\phi} - N^i\partial_j\phi)$ where ϕ is a matter field. The classical 4-metric above and the scalar field which are solutions of the Einstein equations can be obtained from the Hamiltonian equations of motion

$$\dot{h}_{ij} = \{h_{ij}, H\};\ \dot{\pi}^{ij} = \{\pi^{ij}, H\};\ \dot{\phi} = \{\phi, H\};\ \dot{\pi}_\phi = \{\pi_\phi, H\} \tag{6.9}$$

for some choice of N and N^j, given suitable initial conditions compatible with the constraints (♠) $\mathfrak{H} \approx 0$ and $\mathfrak{H}_j \approx 0$ (in standard terminology). There is a standard constraint algebra involving Poisson brackets of the $\mathfrak{H}_i$ and $\mathfrak{H}$ (see e.g. Pinto-Neto [**1023**]) and for quantization the constraints become conditions on the possible states of the quantum system yielding equations (•) $\hat{\mathfrak{H}}_i|\psi >= 0$ and $\hat{\mathfrak{H}}|\psi >= 0$ leading to (♦) $-2h_{ij}D_j[\delta\psi/\delta h_{ij}] + [\delta\psi/\delta\phi]\partial_i\phi = 0$ and the WDW equation

$$\left\{-\hbar^2\left[\kappa G_{ijk\ell}\frac{\delta}{\delta h_{ij}}\frac{\delta}{\delta h_{k\ell}} + \frac{1}{2}h^{-1/2}\frac{\delta^2}{\delta\phi^2}\right] + V\right\}\psi(h_{ij}, \phi) = 0; \tag{6.10}$$

$$V = h^{1/2}\left[-\kappa^{-1}(^3R - 2\Lambda) + \frac{1}{2}h^{ij}\partial_i\partial_j\phi + U(\phi)\right]$$

This involves products of local operators at the same space point so regularization is indicated (we omit details).

Now for the Bohmian point of view one writes $\psi = Aexp(iS/\hbar)$ where A and S are functionals of h_{ij} and ϕ leading to two equations indicating that A and S are invariant under general space coordinate transformations, namely

$$(6.11)\qquad -2h_{ij}D_j\frac{\delta S}{\delta h_{ij}} + \frac{\delta S}{\delta\phi}\partial_i\phi = 0;\ -2h_{ij}D_j\frac{\delta A}{\delta h_{ij}} + \frac{\delta A}{\delta\phi}\partial_i\phi = 0$$

These could depend on factor ordering but in any event one will have e.g. the form

$$(6.12)\qquad \kappa G_{ijk\ell}\frac{\delta S}{\delta h_{ij}}\frac{\delta S}{\delta h_{k\ell}} + \frac{1}{2}h^{-1/2}\left(\frac{\delta S}{\delta\phi}\right)^2 + V + Q = 0;$$

$$Q = -\frac{\hbar^2}{A}\left(\kappa G_{ijk\ell}\frac{\delta^2 A}{\delta h_{ij}\delta h_{k\ell}} + \frac{h^{-1/2}}{2}\frac{\delta^2 A}{\delta\phi^2}\right)$$

where the unregularied Q above depends on the regularization and factor ordering prescribed for the WDW equation. In addition to (6.12) one has

$$(6.13)\qquad \kappa G_{ijk\ell}\frac{\delta}{\delta h_{ij}}\left(A^2\frac{\delta S}{\delta h_{k\ell}}\right) + \frac{h^{-1/2}}{2}\frac{\delta}{\delta\phi}\left(A^2\frac{\delta S}{\delta\phi}\right) = 0$$

One can stipulate that the 3-metric of spacelike hypersurfaces, the scalar field, and their canonical momenta always exist and the metric and scalar field can be determined via guidance relations

$$(6.14)\qquad \pi^{ij} = \frac{\delta S}{\delta h_{ij}};\ \pi_\phi = \frac{\delta S}{\delta\phi}$$

with π^{ij} and π_ϕ given via (6.8) etc. Note that one cannot interpret (6.13) as a continuity equation for a probability density due to the hyperbolic nature of the deWitt metric. Note also that whatever may be the form of Q it must be a scalar density of weight one; indeed from (1.12)

$$(6.15)\qquad Q = -\kappa G_{ijk\ell}\frac{\delta S}{\delta h_{ij}}\frac{\delta S}{\delta h_{k\ell}} - \frac{h^{-1/2}}{2}\left(\frac{\delta S}{\delta\phi}\right)^2 - V$$

and we refer to [**1023**] for the arguments. In addition note that Q can depend only on h_{ij} and ϕ.

7. EXACT UNCERTAINTY AND WDW

We go now to [**254, 599, 592, 1065, 1062, 1063, 1064**] and show how the WDW equation can be derived from a so called exact uncertainty principle of Hall and Reginatto (cf. also Chapter 8). The idea here is that uncertainty can be promoted to be the fundamental element distinguishing quantum and classical mechanics. In this approach nonclassical fluctuations are added to the deterministic connection between position and momentum (via the uncertainty principle) one essentially generates the quantum potential. In [**599, 592**] this is applied to gravity and a WDW equation is derived and originally this approach was used to generate the Schrödinger equation (SE) (see [**1065, 1062, 1063, 1064**] for further clarification). Thus take a metric as in (1.6) and think of the metric h_{ij} as being imprecise with a probability distribution $P[h_{ij}]$. Take a single field classical Hamiltonian of the form

$$H_0[h_{ij},\pi^{ij}] = \int dx \left[N\left(\frac{1}{2}G_{ijk\ell}\pi^{ij}\pi^{k\ell} + V(h_{ij})\right) - 2N_i\nabla_j\pi^{ij}\right] \tag{7.1}$$

(here $D_j \sim \nabla_j$ is the covariant derivative). As an ensemble Hamiltonian one takes now

$$\tilde{H}_c[P,S] = \int Dh\, P H_0[h_{ij},(\delta S/\delta h_{ij})] \tag{7.2}$$

leading to equations of motion

$$\partial_t P + \int dx \frac{\delta}{\delta h_{ij}}(P\dot{h}_{ij}) = 0;\ \partial_t S + H_0[h_{ij},(\delta S/\delta h_{ij})] = 0; \tag{7.3}$$

$$\dot{h}_{ij} = NG_{ijk\ell}\frac{\delta S}{\delta h_{k\ell}} - \nabla_j N^i - \nabla_i N_j$$

The lack of conjugate momenta for the lapse and shift components N and N_i places constraints on the classical equations of motion which in the ensemble formalism take the form

$$\frac{\delta P}{\delta N} = \frac{\delta P}{\delta N_i} = \frac{\partial P}{\partial t} = 0;\ \nabla_j\left(\frac{\delta P}{\delta h_{ij}}\right) = 0; \tag{7.4}$$

$$\frac{\delta S}{\delta N} = \frac{\delta S}{\delta N_i} = \frac{\partial S}{\partial t} = 0;\ \nabla_j\left(\frac{\delta S}{\delta h_{ij}}\right) = 0$$

This corresponds to invariance of the dynamics with respect to N, N_i, and the initial time; also to invariance of P and S under arbitrary spatial coordinate transformations. Applying these constraints to the above classical equations for the "Gaussian" choice $N = 1$ and $N_i = 0$ yields

$$\frac{\delta}{\delta h_{ij}}\left(PG_{ijk\ell}\frac{\delta S}{\delta h_{k\ell}}\right) = 0;\ \frac{1}{2}G_{ijk\ell}\frac{\delta S}{\delta h_{ij}}\frac{\delta S}{\delta h_{k\ell}} + V = 0;\ V \sim c\sqrt{h}(2\Lambda - {}^3R) \tag{7.5}$$

Now the exact uncertainty approach involves writing (★) $\pi^{ij} = (\delta S/\delta h_{ij}) + f^{ij}$ where f^{ij} vanishes on average for all configurations. This adds a kinetic term to the average ensemble energy leading to

$$\tilde{H}_q = \langle E \rangle = \tilde{H}_c + \frac{1}{2}\int Dh\, P \int dx\, NG_{ijk\ell}\overline{f^{ij}f^{k\ell}} \tag{7.6}$$

Note here that the term in (7.1) which is linear in the derivative of π^{ij} can be integrated by parts giving a term directly proportional to π^{ij} which remains unchanged when the fluctuations are added and averaged. Now using some general properties of causality, independence, invariance, and exact uncertainty (cf. [**234, 598, 592**]) one arrives at

$$\tilde{H}_q[P,S] = \tilde{H}_c[P,S] + \frac{c}{2}\int Dh \int dx\, NG_{ijk\ell}\frac{1}{P}\frac{\delta P}{\delta h_{ij}}\frac{\delta P}{\delta h_{k\ell}} \tag{7.7}$$

where C is a positive universal constant. Now if one defines $\hbar = 2\sqrt{c}$ and $\psi[h_{ij}] = \sqrt{P}exp(iS/\hbar)$ then, calculating as above, we obtain a WDW equation for quantum geometry in the form

$$\left[-\frac{\hbar^2}{2}\frac{\delta}{\delta h_{ij}}G_{ijk\ell}\frac{\delta}{\delta h_{k\ell}} + V\right]\psi = 0 \tag{7.8}$$

with a Q term $-(\hbar^2/2P)G_{ijk\ell}(\delta^2 P/\delta h_{ij}\delta h_{k\ell})$ added in the Hamiltonian equation (cf. (6.12)). Note further

$$\frac{\delta\psi}{\delta N} = \frac{\delta\psi}{\delta N_i} = \frac{\partial\psi}{\partial t} = 0;\ \nabla_j\left(\frac{\delta\psi}{\delta h_{ij}}\right) = 0 \tag{7.9}$$

An important feature of this WDW equation is that it is obtained with a particular operator ordering. Indeed $G_{ijk\ell}$ is sandwiched between the two functional derivatives and thus ambiguity is removed in this respect. One recalls that the same thing happens with the SE which is derived in the form (••) $i\hbar\partial_t\psi = -(\hbar^2/2)\nabla\cdot(1/m)\nabla\psi + V\psi$.

Now in the theory of Schrödinger equations there is a strong connection between terms of the form

$$I = \frac{1}{2}g^{ik}\int\frac{1}{P}\frac{\partial P}{\partial y^i}\frac{\partial P}{\partial y^k}d^n y \tag{7.10}$$

and concepts of Fisher information, entropy, and quantum potential (see [**234**] for an extensive development). Classical Fisher information is known to be connected to various forms of entropy via formulas like (cf. [**234, 538**])

$$\frac{\partial\mathfrak{S}}{\partial t} = \frac{\hbar}{2m}\mathfrak{F} = \frac{\hbar}{2m}\int\frac{(\nabla\rho)^2}{\rho} = \frac{4}{\hbar}\int\rho Q;\ Q = -\frac{\hbar^2}{2m}\frac{\delta\sqrt{\rho}}{\sqrt{\rho}} \tag{7.11}$$

Here $\mathfrak{S}\sim -\int\rho log(\rho)$ is a so-called differential entropy and ρ here corresponds to P or A in the notation of this paper (note P and A refer to 3-space quantities); $\mathfrak{F}$ is a Fisher information measure. There are relations between differential entropy and Shannon-Boltzman entropy for example and we refer to [**254, 521**] for details. We remark also that Olavo in [**955**] derives Schrödinger equations using entropy ideas where the entropy in [**955**] is of Shannon-Boltzman type $\mathcal{S} = k_B log(W) = -k_B log(P)$ where $P = 1/W$ is the probability of a microstate occurance (cf. Chapter 7 for more on this). One deals with momentum fluctuations $\overline{(\delta p)^2}$ and assumes $\overline{(\delta p)^2(\delta x)^2} = \hbar^2/4$. There results $\overline{(\delta p)^2}\sim -(\hbar^2/4)\partial^2 log(\rho)$ where ρ is a probability density. Then $\mathcal{S}_{equilib}\sim k_B log(\rho)$ implies $\overline{(\delta p)^2}\sim -(\hbar^2/4k_B)\partial^2\mathcal{S}_{equilib}$. Note also here that (calculating in 1-D for convenience) $\partial^2 log(\rho) = (\rho''/\rho) - (\rho'/\rho)^2$ and

$$Q = -\frac{\hbar^2}{2m}\frac{\partial^2\sqrt{\rho}}{\sqrt{\rho}} = \frac{\hbar^2}{8m}\left[2\frac{\rho''}{\rho} - \left(\frac{\rho'}{\rho}\right)^2\right] = \frac{\hbar^2}{8m}\left[2\partial^2\mathcal{S}_{equilib} + (\partial\mathcal{S}_{equilib})^2\right] \tag{7.12}$$

The theme here is to relate entropy, the quantum potential, and geometry in the relativistic context. One can think of entropy or of quantum fluctuations as generating quantum behavior (often via a quantum potential) and we want to connect these matters to the Einstein equations in a Bohmian spirit. Most of this is already done and sketched in [**254**] for example and we will make it more explicit in various forms as we go along.

REMARK 6.7.1. We call attentiion here to [**292, 291**] (cf. also [**830, 837, 1323**]) where Fisher information is related to uncertainty relations and a differential Shannon entropy is introduced. ■

REMARK 6.7.2. We refer to Chapters 3,4,7 and 8 for more on WDW from various points of view (cf. also [**254**] and references there). ■

8. FISHER INFORMATION AND ENTROPY

We will connect up here various ideas of entropy and Fisher information (following [**254, 490**]). First recall N_{ab} in Section 6.5 corresponds to $T_{ab}-(1/2)g_{ab}T$ and one can imagine this arising from a matter Lagrangian $\mathfrak{L}_m$ as in (1.10)-(1.13) where $G_{ab}=R_{ab}-(1/2)Rg_{ab}-\chi T_{ab}=\chi(-2(\delta\mathfrak{L}/\delta g^{ab})+\mathfrak{L}_m g_{ab})$. We recall that $R=-\chi T$ so $R_{ab}=\chi(T_{ab}-(1/2)g_{ab}T)$ and note that

$$T=g^{ab}\mathfrak{L}_m g_{ab}-2g^{ab}\frac{\delta\mathfrak{L}_m}{\delta g^{ab}}=\mathfrak{L}_m-2g^{ab}\frac{\delta\mathfrak{L}_m}{\delta g^{ab}} \tag{8.1}$$

Hence

$$R_{ab}=\chi\left(\mathfrak{L}_m g_{ab}-2\frac{\delta\mathfrak{L}_m}{\delta g^{ab}}-\frac{1}{2}\left[\mathfrak{L}_m-2g^{ab}\frac{\delta\mathfrak{L}_m}{\delta g^{ab}}\right]\right)= \tag{8.2}$$

$$=\chi\left(\frac{1}{2}g_{ab}\mathfrak{L}_m-\frac{\delta\mathfrak{L}_m}{\delta g^{ab}}\right)$$

and in the situation of Section 6.5 we have $R_{ab}u^b\sim N_{ab}u^b$. It is clear however from the preceeding that one does not need a matter potential in order to discuss the quantum potential in general spaces.

One goes to the deWitt 6-dimensional "superspace" with metric $G_{ijk\ell}$ (cf. deWitt [**374**]) (here $G_{ijk\ell}=(1/2\sqrt{h})(h_{ik}h_{j\ell}+h_{i\ell}h_{jk}-h_{ij}h_{k\ell})$ following [**374**] - cf. also (2.2) which differs by a factor of 2). The Fisher information will have a general form

$$I=4\int d\mathbf{x}\int Dg\sum_{ijk\ell}G_{ijk\ell}\frac{\partial\psi^*}{\partial h_{ij}}\frac{\partial\psi}{\partial h_{k\ell}} \tag{8.3}$$

where $dg\sim\prod dg_{ij}$ (cf. [**217, 216, 490, 493, 1006**]). To motivate and clarify this one thinks of a probability density function $f(y|\theta)$ used in estimating a parameter θ based on imperfect observations $y=\theta+x$ ($x\sim$ noise). Assume unbiased estimates, namely (♦♦) $<\hat{\theta}(y)-\theta>=0=\int dy[\hat{\theta}-\theta]p(y|\theta)$ where $p(y|\theta)$ is the probability for y in the presense of one parameter value θ. Differentiate (♦♦) to get $\int dy(\hat{\theta}-\theta)\partial_\theta p-\int dyp=0$ and via $\int p=1$ and $\partial_\theta=p\partial_\theta log(p)$ one arrives at

$$\int dy(\hat{\theta}-\theta)\partial_\theta log(p)=1=\int dy[\sqrt{p}\partial_\theta log(p)][(\hat{\theta}-\theta)\sqrt{p}] \tag{8.4}$$

The Schwartz inequality gives then

$$\int dy(\partial_\theta log(p))^2 p \int dy(\hat{\theta}-\theta)^2 p \geq 1 \tag{8.5}$$

(Cramer-Rao inequality) which links the mean square estimate e^2 (second factor) to the Fisher information I (first factor). In [**493**] one writes $p = q^2$ so (♠♠) $I \sim 4\int dx(q')^2$ and various quadratic Lagrangians in physics are considered, e.g. (1) $(1/2)m(\dot{q})^2 - V$, (2) $-\nabla\psi\cdot\nabla\psi^* + \cdots$, (3) $-(\hbar^2/2m)\nabla\psi\cdot\nabla\psi^* + \cdots$, (4) $\sum g_{mn}(q(\tau))\partial_\tau q_m \partial_\tau q_n$, etc. A principle of extreme physical information (EPI) is ennunciated (in a game theoretic context) and, setting e.g. $x_1 = ix$, $x_2 = iy$, $x_3 = iz$, and $x_4 = ct$ with $(x_1, x_2, x_3) \sim \mathbf{r}$, one posits modes $q_n = q_n(\mathbf{r}, t)$ with $\psi_n = q_{2n-1} + iq_{2n}$ $(n = 1, \cdots, N/2)$). Then take (♣♣) $\sum_1^{N/2} \psi_n^*\psi_n = \sum q_n^2 = p(\mathbf{r}, t)$ with $I \sim 4\sum_1^{N/2}\int d\tau \nabla q_n \cdot \nabla q_n$ (cf. (♠♠)). Then physical content is introduced via Fourier transform momentum-energy variables $(i\mathbf{r}, ct) \leftrightarrow [(i\mu/\hbar), (E/ct)]$ with $\psi_n \leftrightarrow \phi_n$ so that $(\nabla\psi_n, \partial_t\psi_n) \leftrightarrow [(-i\mu\phi_n/\hbar), (iE\phi_n/\hbar)]$ (only $E = mc^2$ will be assumed physically below). I is regarded as information obtained by an observer and this is to be balanced by the physical payoff J by a "demon" expressed in physical terms. The net information change $\Delta I = I - J$ should be zero (as in zero sum game) and EPI specifies that $I = J$ which means here $(\sum_1^{N'/2} \phi_n^*\phi_n = P(\mu, E))$

$$I = 4c\sum_1^{N/2}\int\int d\tau dt\left[-(\nabla\psi_n)^*\cdot\nabla\psi_n + \left(\frac{1}{c^2}\right)\left(\frac{\partial\psi_n}{\partial t}\right)^*\left(\frac{\partial\psi_n}{\partial t}\right)\right] = I = \tag{8.6}$$

$$= J = \frac{4c}{\hbar^2}\int\int d\mu dE\, P(\mu, E)(-\mu^2 + (E^2/c^2) = \frac{4c}{\hbar^2}\left\langle -\mu^2 + \frac{E^2}{c^2}\right\rangle$$

Some argument then gives $-\mu^2 + (E^2/c^2) = m^2c^2$ and minimizing ΔI $(\Delta I = 0)$ leads to the Klein-Gordon equation. This approach seems a little silly but it is also cute; it does in any case sort of motivate the use of (5.3) as a Fisher information.

REMARK 6.8.1. The entropy in Section 6.5 is of course contrived via perturbations in displacement and their derivatives (elastic deformation) and is not designed for quantization (cf. however [**955**]). We note also the apparent denial of an entropy functional for gravity without sources in [**301**]. However in [**235**] an entropy is introduced via a fluid stress energy tensor. The theme of [**976**] does not seem to be threatened; the Einstein equations arise as a consistency condition indicating that spacetime structure (as defined by the Einstein equations!) is robust under fluctuations. The quantum potential in Section 2 $(Q = Q_G + Q_M)$ arises via the Hamiltonian context when one looks at a complex wave function $\psi = Aexp(iS/\hbar)$ with A and S functionals of h_{ij} and a matter field. The interesting fact here is that Q_G automatically arises once a complex (quantum) solution is sought (cf. (2.12)). The same feature arises above where the introduction of a Bohmian context corresponds to the entrance of quantum theory and the quantum potential automatically appears. No matter potential is needed here and thus it seems that space time automatically contains a quantum aspect which emerges when one looks at a Hamiltonian formulation with a complex wave function (implicitly introducing a probability). The exact uncertainty approach above introduces perturbations or fluctuations in momentum based on fluctuations in h_{ij} and exhibits the associated quantum potential. The perturbations here are quite general in an explicit way and essentially generate the amplitude of the wave function. The form (3.7) in terms of Fisher information automatically gives the fluctuations an entropic character (cf. [**955**] where one derives the SE on entropy ideas - this is developed further at various places in this book. ■

REMARK 6.8.2. Going to Reginatto [**1065**] one considers

$$H = \frac{1}{2}G_{ijk\ell}\frac{\delta kS}{\delta h_{ij}}\frac{\delta S}{\delta h_{k\ell}} - \sqrt{h}R = 0;\ H_i = -2D_j\left(h_{ik}\frac{\delta S}{\delta h_{kj}}\right) = 0 \tag{8.7}$$

where D_j is the covariant derivative and (♦♦♦) $G_{ijk\ell} = (1/\sqrt{h})(h_{ik}h_{j\ell} + h_{i\ell}h_{jk} - h_{ij}h_{k\ell})$ with $G = 1/16\pi$ (this differs by a factor of 2 from a previous $G_{ijk\ell}$). As a consequence of the constraint $H = 0$, (8.7), and $H_i = 0$, S must satisfy various constraints (including $\partial_t S = 0$ - cf. [**147**]) and this is all subsumed in the invariance of the HJ functional S under spatial coordinate transformations. Hence one can keep the Hamiltonian constraint, ignore the momentum constraints, and reqire that S be invariant under the gauge group of spatial coordinate transformations. Now to define ensembles for gravitational fields one needs a measure Dh and a probability functional $P[h_{ij}]$ and this is discussed in some detail in [**1065**] following [**607, 903**] and at other places in this book. One is led to an ensemble Hamiltonian and derived equations

$$\tilde{H}_c = \int d^3x \int Dh\, PH;\ \partial_t P = \frac{\Delta\tilde{H}_c}{\Delta S};\ \partial_t S = -\frac{\Delta\tilde{H}_c}{\Delta P} \tag{8.8}$$

With $\partial_t S = \partial_t P = 0$ the equations take the form

$$H = 0;\ \int d^3x \frac{\delta}{\delta h_{ij}}\left(PG_{ijk\ell}\frac{\delta S}{\delta h_{k\ell}}\right) = 0 \tag{8.9}$$

The latter equation corresponds to a continuity equation and in this spirit some argument shows that it implies the standard rate equation

$$\partial_t h_{ij} = NG_{ijk\ell}\frac{\delta S}{\delta h_{k\ell}} + D_i N_j + D_j N_i \tag{8.10}$$

(as follows from the ADM formalism with N the lapse function and N_j the shift vector). Now writing $\pi^{k\ell} = (\delta S/\delta h_{k\ell}) + f^{k\ell}$ with $\overline{f^{k\ell}} = 0$ one obtains an ensemble Hamiltonian (3.6) and an equation (3.7) as before. Again putting $\hbar = 2\sqrt{c}$ and $\psi[h_{k\ell}] = \sqrt{P}exp(iS/\hbar)$ leads to the WDW equation (cf. (3.8))

$$\left[-\frac{\hbar^2}{2}\frac{\delta}{\delta h_{ij}}G_{ijk\ell}\frac{\delta}{\delta h_{k\ell}} - \sqrt{h}R\right]\psi = 0 \tag{8.11}$$

The procedures here suggest also replacing (8.10) by

$$\partial_t h_{ij} = NG_{ijk\ell}\left(\frac{\delta S}{\delta h_{k\ell}} + f^{k\ell}\right) + D_i N_j + D_j N_i \tag{8.12}$$

Since the field momenta are subject to fluctuations so must be the extrinsic curvature $K_{ij} = (1/2)G_{ijk\ell}(\delta S/\delta h_{k\ell})$ yielding then

$$K_{ij} = \frac{1}{2}G_{ijk\ell}\left(\frac{\delta S}{\delta h_{k\ell}} + f^{k\ell}\right) \tag{8.13}$$

for the extrinsic curvature. ■

9. STATISTICAL GEOMETRODYNAMICS

We sketch now some material from Caticha [**284, 285**] supporting the idea of a statistical geometrodynamics (SGD). Here one builds a model of SGD based on (i) Positing that the geometry of space is of statistical origin and is explained in terms of the distinguishability Fisher-Rao (FR) metric and (ii) Assuming the dynamics of the geometry is derived solely from principles of inference. There is no external time but an intrinsic one à la [**89**]. A scale factor $\sigma(x)$ is required to assign a Riemannian geometry and it is conjectured that it can be chosen so that the evolving geometry of space sweeps out a 4-D spacetime. The procedure defines only a conformal geometry but that is entirely appropriate d'après York [**1327**]. One uses the FR metric in two ways, one to distinguish neighboring points and the other to distinguish successive states. Consider then a "cloud" of dust with coordinate values y^i ($i = 1, 2, 3$) and estimates x^i with $p(y|x)dy$ the probability that the particle labeled x^i should have been labeled y^i (the FR metric encodes the use of probability distributions - instead of structureless points). One writes

$$\frac{p(y|x+dx) - p(y|x)}{p(y|x)} = \frac{\partial log[p(y|x)]}{\partial x^i}dx^i \tag{9.1}$$

$$d\lambda^2 = \int d^4yp(y|x)\frac{\partial log[p(y|x)]}{\partial x^i}\frac{\partial log[p(y|x)]}{\partial x^j}dx^i dx^j = \gamma_{ij}dx^i dx^j \tag{9.2}$$

and $d\lambda^2 = 0 \iff dx^i = 0$. The FR metric γ_{ij} is the only local Riemannian metric reflecting the underlying statistical nature of the manifold of distributions $p(y|x)$ and a scale factor σ giving a metric $g_{ij}(x) = \sigma(x)\gamma_{ij}(x)$ is needed for a Riemannian metric (cf. [**284, 285**]). Also the metric $d\lambda^2$ is related to the entropy of $p(y|x+dx)$ relative to $p(y|x)$, namely

$$S[p(y|x+dx)|p(y|x)] = -\int d^3yp(y|x+dx)log\frac{p(y|x+dx)}{p(y|x)} = -\frac{1}{2}d\lambda^2 \tag{9.3}$$

and maximizing the relative entropy S is equivalent to minimizing $d\lambda^2$. One thinks of $d\lambda$ as a spatial distance in specifying that the reason that particles at x and $x + dx$ are considered close is because they are difficult to distinguish. To assign an explicit $p(y|x)$ one assumes the relevant information is given via $< y^i >= x^i$ and the covariance matrix $< (y^i - x^i)(y^j - x^j) >= C^{ij}(x)$; this leads to

$$p(y|x) = \frac{C^{1/2}}{(2\pi)^{3/2}}exp\left[-\frac{1}{2}C_{ij}(y^i - x^i)(y^j - x^j)\right] \tag{9.4}$$

where $C^{ik}C_{kj} = \delta^i_j$ and $C = det(c_{ij})$. Subsequently to each x one associates a probability distribution

$$p(y|x,\gamma) = \frac{\gamma^{1/2}(x)}{(2\pi)^{3/2}} exp\left[-\frac{1}{2}\gamma_{ij}(x)(y^i - x^i)(y^j - x^j)\right] \tag{9.5}$$

where $\gamma_{ij}(x) = C_{ij}(x)$ (extreme curvature situations are avoided here). One deals with a conformal geometry described via γ_{ij} and a scale factor $\sigma(x)$ will be needed to compare uncertainties at different points; the choice of σ should then be based on making motion "simple".

Thus define a macrostate via

$$P[y|\gamma] = \prod_x p(y(x)|x,\gamma_{ij}(x)) = \tag{9.6}$$

$$= \left[\prod_x \frac{\gamma^{1/2}(x)}{(2\pi)^{3/2}}\right] exp\left[-\frac{1}{2}\sum_x \gamma_{ij}(x)(y^i - x^i)(y^j - x^j)\right]$$

Once a dust particle in an earlier state γ is identified with the label x one assumes that this particle can be assigned the same label x as it evolves into the later state $\gamma + \Delta\gamma$ (equilocal comoving coordinates). Then the change between $P[y|\gamma + \Delta\gamma]$ and $P[y|\gamma]$ is denoted by $\Delta\ell$ and is measured via their relative entropy (this is a form of Kullback-Leibler entropy - cf. [**254**])

$$S[\gamma + \Delta\gamma|\gamma] = -\int\left(\prod_x dy(x)\right) P[y|\gamma + \Delta\gamma] log\frac{P[y|\gamma + \Delta\gamma]}{P[y|\gamma]} = -\frac{1}{2}\Delta\ell^2 \tag{9.7}$$

Since $P[y|\gamma]$ and $P[y|\gamma + \Delta\gamma]$ are products one can write

$$S[\gamma + \Delta\gamma, \gamma] = \sum_x S[\gamma(x) + \Delta\gamma(x), \gamma(x)] = \tag{9.8}$$

$$= -\frac{1}{2}\sum_x \Delta\ell^2(x);\ \ \Delta\ell^2(x) = g^{ijk\ell}\Delta\gamma_{ij}(x)\Delta\gamma_{k\ell}(x)$$

where, using (5.5)

$$g^{ijk\ell} = \int d^3y p(y|x,\gamma)\frac{\partial log[p(y|x,\gamma)]}{\partial\gamma_{ij}}\frac{\partial log[p(y|x,\gamma)]}{\partial\gamma_{k\ell}} = \tag{9.9}$$

$$= \frac{1}{4}\left(\gamma^{ik}\gamma^{ji} + \gamma^{i\ell}\gamma^{jk}\right)$$

Then $\Delta L^2 = \sum_x \Delta\ell^2(x)$ can be written as an integral if we note that the density of distinguishable distributions is $\gamma^{1/2}$. Thus the number of distinguishable distributions, or distinguishable points in the interval dx is $dx\gamma^{1/2}$ ($dx \sim d^3x$) and one has

$$\Delta L^2 = \int dx\gamma^{1/2}\Delta\ell^2 = \int dx\gamma^{1/2}g^{ijk\ell}\Delta\gamma_{ij}\Delta\gamma_{k\ell} \tag{9.10}$$

Thus the effective number of distinguishable points in the interval dx is finite (due to the intrinsic fuzziness of space). Now to describe the change $\Delta\gamma_{ij}(x)$ one introduces an arbitrary time parameter t along a trajectory

$$\Delta\gamma_{ij} = \gamma_{ij}(t + \Delta t, x) - \gamma_{ij}(t,x) = \partial_t\gamma_{ij}\Delta t \tag{9.11}$$

Thus $\partial_t \gamma_{ij}$ is the "velocity" of the metric and (5.10) becomes

$$\Delta L^2 = \int dx \gamma^{1/2} g^{ijk\ell} \partial_t \gamma_{ij} \partial_t \gamma_{k\ell} \Delta t^2 \tag{9.12}$$

Now go to an arbitrary coordinate frame where equilocal points at t and $t+\Delta t$ have coordinates x^i and $\tilde{x}^i = x^i - \beta^i(x)\Delta t$. Then the metric at $t+\Delta t$ transforms into $\tilde{\gamma}_{ij}$ with

$$\gamma_{ij}(t+\Delta t, x) = \tilde{\gamma}_{ij}(t+\Delta t, x) - (\nabla_i \beta_j + \nabla_j \beta_i)\Delta t \tag{9.13}$$

where $\nabla_i \beta_j = \partial_i \beta_j - \Gamma^k_{ij}\beta_k$ is the covariant derivative associated to the metric γ_{ij}. In the new frame, setting $\tilde{\gamma}_{ij}(t+\Delta t, x) - \gamma_{ij}(t,x) = \Delta\gamma_{ij}$ one has

$$\Delta_\beta \gamma_{ij} = \Delta\gamma_{ij} - (\nabla_i \beta_j + \nabla_j \beta_i)\Delta t \sim \Delta_\beta \gamma_{ij} = \dot{\gamma}_{ij}\Delta t \tag{9.14}$$

$$\dot{\gamma}_{ij} = \partial_t \gamma_{ij} - \nabla_i \beta_j - \nabla_j \beta_i$$

leading to

$$\Delta_\beta L^2 = \int dx \gamma^{1/2} g^{ijk\ell} \dot{\gamma}_{ij} \dot{\gamma}_{k\ell} \Delta t^2 \tag{9.15}$$

Next one addresses the problem of specifying the best matching criterion, i.e. what choice of β^i provides the best equilocality match. This is treated as a problem in inference and asks for minimum $\Delta_\beta L^2$ over β. Hence one gets

$$\delta(\Delta_\beta L^2) = 2\int dx \gamma^{1/2} g^{ijk\ell} \dot{\gamma}_{ij} \dot{\gamma}_{k\ell} \Delta t^2 = 0 \Rightarrow \tag{9.16}$$

$$\Rightarrow \nabla_\ell (2 g^{ijk\ell} \dot{\gamma}_{ij}) = 0 \equiv \nabla_\ell \dot{\gamma}^{k\ell} = 0$$

(using (5.9) and $\dot{\gamma}^{k\ell} = \partial_t \gamma^{k\ell} + \nabla^k \beta^\ell + \nabla^\ell \beta^k$). These equations determine the shifts β^i giving the best matching and equilocality for the geometry γ_{ij} and alternatively they could be considered as constraints on the allowed change $\Delta\gamma_{ij} = \partial_t \gamma_{ij}\Delta t$ for given shifts β^i. In describing a putative entropic dynamics one assumes now e.g. continuous trajectories with each factor in $P[y|\gamma]$ evolving continuously through intermediate states labeled via $\omega(x) = \omega\zeta(x)$ where $\zeta(x)$ is a fixed positive function and $0 < \omega < \infty$ is a variable parameter (some kind of many fingered time à la Schwinger, Tomonaga, Wheeler, et al). It is suggested that they dynamics be determined by an action

$$J = \int_{t_i}^{t_f} dt \int dx \gamma^{1/2} [g^{ijk\ell} \dot{\gamma}_{ij} \dot{\gamma}_{k\ell}]^{1/2} \tag{9.17}$$

The similarities to "standard" geometrodynamics are striking.

9.1. INFORMATION DYNAMICS. We go here to J. and X. Calmet [**217, 216**] and consider the idea of introducing some kind of dynamics in a reasoning process. One looks at the Fisher metric defined by

$$g_{\mu\nu} = \int_X d^4 x p_\theta(x) \left(\frac{1}{p_\theta(x)} \frac{\partial p_\theta(x)}{\partial \theta^\mu}\right) \left(\frac{1}{p_\theta(x)}\right) \left(\frac{\partial p_\theta(x)}{\partial \theta^\nu}\right) \tag{9.18}$$

and constructs a Riemannian geometry via

$$\Gamma^{\sigma}_{\lambda\nu} = \frac{1}{2}g^{\nu\sigma}\left(\frac{\partial g_{\mu\nu}}{\partial\theta^{\lambda}} + \frac{\partial g_{\lambda\nu}}{\partial\theta^{\mu}} - \frac{\partial g_{\mu\lambda}}{\partial\theta^{\nu}}\right); \tag{9.19}$$

$$R^{\lambda}_{\mu\nu\kappa} = \frac{\partial\Gamma^{\lambda}_{\mu\nu}}{\partial\theta^{\kappa}} - \frac{\partial\Gamma^{\lambda}_{\mu\kappa}}{\partial\theta^{\nu}} + \Gamma^{\eta}_{\mu\nu}\Gamma^{\lambda}_{\kappa\eta} - \Gamma^{\eta}_{\mu\kappa}\Gamma^{\lambda}_{\nu\eta}$$

Then the Ricci tensor is $R_{\mu\kappa} = R^{\lambda}_{\mu\lambda\kappa}$ and the curvature scalar is $R = g^{\mu\kappa}R_{\mu\kappa}$. The dynamics associated with this metric can then be described via functionals

$$J[g_{\mu\nu}] = -\frac{1}{16\pi}\int\sqrt{g(\theta)}R(\theta)d^4\theta \tag{9.20}$$

leading upon variation in $g_{\mu\nu}$ to equations

$$R^{\mu\nu}(\theta) - \frac{1}{2}g^{\mu\nu}(\theta)R(\theta) = 0 \tag{9.21}$$

Contracting with $g_{\mu\nu}$ gives then the Einstein equations $R^{\mu\nu}(\theta) = 0$ (since $R = 0$). J is also invariant under $\theta \to \theta + \epsilon(\theta)$ and variation here plus contraction leads to a contracted Bianchi identity. Constraints can be built in by adding terms $(1/2)\int\sqrt{g}T^{\mu\nu}g_{\mu\nu}d^4\theta$ to $J[g_{\mu\nu}]$. If one is fixed on a given probability distribution $p(x)$ with variable θ^{μ} attached to give $p_{\theta}(x)$ then this could conceivably describe some gravitational metric based on quantum fluctuations for example. As examples a Euclidean metric is produced in 3-space via Gaussian $p(x)$ and complex Gaussians will give a Lorentz metric in 4-space.

CHAPTER 7

GEOMETRY AND THE QUANTUM POTENTIAL

The old Chapter 7 in [**254**] consisted partly of a summary and survey of Chapters 1-6 and we will not repeat this here. Rather we include some other material from the old Chapter 7 plus extracts from some recent papers.

1. QUANTUM GEOMETRY

First we sketch the relevant symbolism for geometrical QM from Ashtekar-Schilling [**68**] without much philosophy; the philosophy is eloquently phrased there and in [**186, 307, 747, 881**] for example (cf. also Bengtsson-Zyczkowski [**123**] for a more complete discussion). Thus let H be the Hilbert space of QM and write it as a real Hilbert space with a complex structure J. The Hermitian inner product is then $<\phi,\psi>=(1/2\hbar)G(\phi,\psi)+(i/2\hbar)\Omega(\phi,\psi)$ (note $G(\phi,\psi)=2\hbar\Re(\phi,\psi)$ is the natural Fubini-Study (FS) metric - cf. [**307**]). Here G is a positive definite real inner product and Ω is a symplectic form (both strongly nondegenerate). Moreover $<\phi,J\psi>=i<\phi,\psi>$ and $G(\phi,\psi)=\Omega(\phi,J\psi)$. Thus the triple (J,G,Ω) equips H with the structure of a Kähler space. Now, from Willmore [**1309**], on a real vector space V with complex structure J a Hermitian form satisfies $h(JX,JY)=h(X,Y)$. Then V becomes a complex vector space via $(a+ib)X=aX+bJX$. A Riemannian metric g on a manifold M is Hermitian if $g(X,Y)=g(JX,JY)$ for X,Y vector fields on M. Let ∇_X be he Levi-Civita connection for g (i.e. parallel transport preserves inner products and the torsion is zero. A manifold M with J as above is called almost complex. A complex manifold is a paracompact Hausdorff space with complex analytic patch transformation functions. An almost complex M with Kähler metric (i.e. $\nabla_X J=0$) is called an almost Kähler manifold and if in addition the Nijenhuis tensor vanishes it is a Kähler manifold (cf. (1.1) below). Here the defining equations for the Levi-Civita connection and the Nijenhuis tensor are

$$\Gamma^k_{ij}=\frac{1}{2}g^{hk}[\partial_i g_{jk}+\partial_j g_{ik}-\partial_k g_{ji}];\tag{1.1}$$

$$N(X,Y)=[JX,JY]-[X,Y]-J[X,JY]-J[JX,Y]$$

Further discussion can be found in [**1309**]. Material on the Fubini-Study metric will be provided later. Next by use of the canonical identification of the tangent space (at any point of H) with H itself, Ω is naturally extended to a strongly nondegenerate, closed, differential 2-form on H, denoted also by Ω. The inverse of Ω may be used to define Poisson brackets and Hamiltonian vector fields. Now in QM the observables may be viewed as vector fields, since linear operators associate a vector to each element of the Hilbert space. Moreover the Schrödinger equation, written here as $\dot{\psi} = -(1/\hbar)J\hat{H}\psi$, motivates one to associate to each quantum observable $\hat{F}$ the vector field $Y_{\hat{F}}(\psi) = -(1/\hbar)J\hat{F}\psi$. The Schrödinger vector field is defined so that the time evolution of the system corresponds to the flow along the Schrödinger vector field and one can show that the vector field $Y_{\hat{F}}$, being the generator of a one parameter family of unitary mappings on H, preserves both the metric G and the symplectic form Ω. Hence is is locally, and indeed globally, Hamiltonian. In fact the function which generates this Hamiltonian vector field is simply the expectation value of $\hat{F}$. To see this write $F : H \to \mathbf{R}$ via $F(\psi) =<\psi, \hat{F}\psi>=<\hat{F}>= (1/2\hbar)G(\psi, \hat{F}\psi)$. Then if η is any tangent vector at ψ

$$(1.2)\qquad (dF)(\eta) = \frac{d}{dt} <\psi + t\eta, \hat{F}(\psi + t\eta)> |_{t=0} =<\psi, \hat{F}\eta> + <\eta, \hat{F}\psi>=$$

$$= \frac{1}{\hbar}G(\hat{F}\psi, \eta) = \Omega(Y_{\hat{F}}, \eta) = (i_{Y_{\hat{F}}}\Omega)(\eta)$$

where one uses the selfadjointness of $\hat{F}$ and the definition of $Y_{\hat{F}}$ (recall the Hamiltonian vector field X_f generated by f satisfies the equation $i_{X_f}\Omega = df$ and the Poisson bracket is defined via $\{f, g\} = \Omega(X_f, X_g)$). Thus the time evolution of any quantum mechanical system may be written in terms of Hamilton's equation of classical mechanics; the Hamiltonian function is simply the expectation value of the Hamiltonian operator. Consequently Schrödinger's equation is simply Hamilton's equation in disguise. For Poisson brackets we have

$$(1.3)\qquad \{F, K\}_{\Omega} = \Omega(X_F, X_K) = \left\langle \frac{1}{i\hbar}[\hat{F}, \hat{K}] \right\rangle$$

where the right side involves the quantum Lie bracket. Note this is not Dirac's correspondence principle since the Poisson bracket here is the quantum one determined by the imaginary part of the Hermitian inner product.

Now look at the role played by G. It enables one to define a real inner product $G(X_F, X_K)$ between any two Hamiltonian vector fields and one expects that this inner product is related to the Jordan product. Indeed

$$(1.4)\qquad \{F, K\}_{+} = \frac{\hbar}{2}G(X_F, X_K) = \left\langle \frac{1}{2}[\hat{F}, \hat{K}]_{+} \right\rangle$$

Since the classical phase space is generally not equipped with a Riemannian metric the Riemann product does not have a classical analogue; however it does have a physical interpretation. One notes that the uncertainty of the observable $\hat{F}$ at a state with unit norm is $(\Delta\hat{F})^2 =<\hat{F}^2> - <\hat{F}^2>= \{F, F\}_{+} - F^2$. Hence the uncertainty involves the Riemann bracket in a simple manner. In fact Heisenberg's uncertainty relation has a nice form as seen via

$$(1.5)\qquad (\Delta\hat{F})^2(\Delta\hat{K})^2 \geq \left\langle \frac{1}{2i}[\hat{F}, \hat{K}] \right\rangle^2 + \left\langle \frac{1}{2}[\hat{F}_{\perp}, \hat{K}_{\perp}]_{+} \right\rangle^2$$

where $\hat{F}_\perp$ is the nonlinear operator defined by $\hat{F}_\perp(\psi) = \hat{F}(\psi) - F(\psi)$. Thus $\hat{F}_\perp(\psi)$ is orthogonal to ψ if $\|\psi\| = 1$. Using this one can write (4.5) in the form

$$(1.6) \qquad (\Delta\hat{F})^2(\Delta\hat{K})^2 \geq \left(\frac{\hbar}{2}\{F,K\}_\Omega\right)^2 + (\{F,K\}_+ - FK)^2$$

The last expression in (1.6) can be interpreted as the quantum covariance of $\hat{F}$ and $\hat{K}$.

The discussion in [**68**] continues in this spirit and is eminently worth reading; however we digress here for a more "hands on" approach following [**237, 307, 308, 309, 310, 311**]. Assume H is separable with a complete orthonormal system $\{u_n\}$ and for any $\psi \in H$ denote by $[\psi]$ the ray generated by ψ while $\eta_n = (u_n|\psi)$. Define for $k \in \mathbf{N}$

$$(1.7) \qquad U_k = \{[\psi] \in P(H);\ \eta_k \neq 0\};\ \phi_k : U_k \to \ell^2(\mathbf{C}) :$$
$$\phi_k([\psi]) = \left(\frac{\eta_1}{\eta_k}, \cdots, \frac{\eta_{k-1}}{\eta_k}, \frac{\eta_{k+1}}{\eta_k}, \cdots\right)$$

where $\ell^2(\mathbf{C})$ denotes square summable functions. Evidently $P(H) = \cup_k U_k$ and $\phi_k \circ \phi_j^{-1}$ is biholomorphic. It is easily shown that the structure is independent of the choice of complete orthonormal system. The coordinates for $[\psi]$ relative to the chart (U_k, ϕ_k) are $\{z_n^k\}$ given via $z_n^k = (\eta_n/\eta_k)$ for $n < k$ and $z_n^k = (\eta_{n+1}/\eta_k)$ for $n \geq k$. To convert this to a real manifold one can use $z_n^k = (1/\sqrt{2})(x_n^k + iy_n^k)$ with

$$(1.8) \qquad \frac{\partial}{\partial z_n^k} = \frac{1}{\sqrt{2}}\left(\frac{\partial}{\partial x_n^k} + i\frac{\partial}{\partial y_n^k}\right);\ \frac{\partial}{\partial \bar{z}_n^k} = \frac{1}{\sqrt{2}}\left(\frac{\partial}{\partial x_n^k} - i\frac{\partial}{\partial y_n^k}\right)$$

etc. Instead of nondegeneracy as a criterion for a symplectic form inducing a bundle isomorphism between TM and T^*M one assumes here that a symplectic form on M is a closed 2-form which induces at each point $p \in M$ a toplinear isomorphism between the tangent and cotangent spaces at p. For $P(H)$ one can do more than simply exhibit such a natural symplectic form; in fact one shows that $P(H)$ is a Kähler manifold (meaning that the fundamental 2-form is closed). Thus one can choose a Hermitian metric $\mathfrak{G} = \sum g_{mn}^k dz_m^k \otimes d\bar{z}_n^k$ with

$$(1.9) \qquad g_{mn}^k = (1 + \sum_i z_i^k \bar{z}_i^k)^{-1}\delta_{mn} - (1 + \sum_1 z_i^k \bar{z}_i^k)^{-2}\bar{z}_m^k z_n^k$$

relative to the chart $U_k, \phi_k)$. The fundamental 2-form of the metric $\mathfrak{G}$ is $\omega = i\sum_{m,n} g_{mn}^k dz_m^k \wedge d\bar{z}_n^k$ and to show that this is closed note that $\omega = i\partial\bar{\partial}f$ where locally $f = log(1 + \sum z_i^k \bar{z}_i^k)$ (the local Kähler function). Note here that $\partial + \bar{\partial} = d$ and $d^2 = 0$ implies $\partial^2 = \bar{\partial}^2 = 0$ so $d\omega = 0$ and thus $P(H)$ is a K manifold.

Now on $P(H)$ the observables will be represented via a class of real smooth functions on $P(H)$ (projective Hilbert space) called Kählerian functions. Consider a real smooth Banach manifold M with tangent space TM, and cotangent space T^*M. We remark that the extension of standard differential geometry to the infinite dimensional situation of Banach manifolds etc. is essentially routine modulo some functional analysis; there are a few surprises and some interesting technical machinery but we omit all this here. One should also use bundle terminology at various places but we will not be pedantic about this. One hopes here to simply give a clear picture of what is happening. Thus e.g. $L(T_x^*M, T_xM)$ denotes bounded linear operators $T_x^*M \to T_xM$ and $L_n(T_xM, \mathbf{R})$ denotes bounded n-linear forms on T_xM. An almost complex structure is provided by a smooth section J of $L(TM)$ = vector bundle of bounded linear operators with fibres $L(T_xM)$ such that $J^2 = -1$. Such a J is called integrable if its torsion is zero, i.e. $N(X,Y) = 0$ with N as in (1.1). An almost Kähler (K) manifold is a triple (M, J, g) where M is a real smooth Hilbert manifold, J is an almost complex structure, and g is a K metric, i.e. a Riemannian metric such that

- g is invariant; i.e. $g_x(J_xX_x, J_xY_x) = g_x(X_x, Y_x)$.
- The fundamental two form of the metric is closed; i.e.

$$\omega_x(X_x, Y_x) = g_x(J_xX_x, Y_x) \tag{1.10}$$

is closed (which means $d\omega = 0$).

Note that an almost K manifold is canonically symplectic and if J is integrable one says that M is a K manifold. Now fix an almost K manifold (M, J, g). The form ω and the K metric g induce two top-linear isomorphisms I_x and G_x between T_x^*M and T_xM via $\omega_x(I_xa_x, X_x) =< a_x, X_x >$ and $g_x(G_xa_z, X_x) =< a_x, X_x >$. Denoting the smooth sections by I, G one checks that $G = J \circ I$.

Definition 7.1.1. For $f, h \in C^\infty(M, \mathbf{R})$ the Poisson and Riemann brackets are defined via $\{f, h\} =< df, Idh >$ and $((f, h)) =< df, Gdh >$. From the above one can reformulate this as

(1.11)

$$\{f,h\} = \omega(Idf, Idh) = \omega(Gdf, Gdh); \;\; ((f,h)) = g(Gdf, Gdh) = g(Idf, Idh)$$

Definition 7.1.2. For $f, h \in C^\infty(M, \mathbf{C})$ the K bracket is $< f, h >= ((f,h)) + i\{f,h\}$ and one defines products $f \circ_\nu h = (1/2)\nu((f,h)) + fh$ (ν will be determined to be $\hbar$) and $f *_\nu h = (1/2)\nu < f, h > + fh$. One observes also that

$$f *_\nu h = f \circ_\nu h + (i/2)\nu\{f,h\}; \;\; f \circ_\nu h = (1/2)(f *_\nu h + h *_\nu f); \tag{1.12}$$
$$\{f,h\} = (1/i\nu)(f *_\nu h - h *_\nu f)$$

Definition 7.1.3. For $f \in C^\infty(M, \mathbf{R})$ let $X = Idf$; then f is called Kählerian (K) if $L_X g = 0$ where L_X is the Lie derivative along X (recall $L_X f = Xf$, $L_X Y = [X, Y]$, $L_X(\omega(Y)) = (L_X\omega)(Y) + \omega(L_X(Y)), \cdots$). More generally if $f \in C^\infty(M, \mathbf{C})$ one says that f is K if $\Re f$ and $\Im f$ are K; the set of K functions is denoted by $K(M, \mathbf{R})$ or $K(M, \mathbf{C})$.

REMARK 7.1.1. In the language of symplectic manifolds $X = df$ is the Hamiltonian vector field corresponding to f and the condition $L_X g = 0$ means that the integral flow of X, or the Hamiltonian flow of f, preserves the metric g. From this follows also $L_X J = 0$ (since J is uniquely determined by ω and g via (1.10)). Therefore if f is K the Hamiltonian flow of f preserves the whole K structure. Note also that $K(M, \mathbf{R})$ (resp. $K(M, \mathbf{C})$) is a Lie subalgebra of $C^\infty(M, \mathbf{R})$ (resp. $C^\infty(M, \mathbf{C})$). ■

Now $P(H)$ is the set of one dimensional subspaces or rays of H; for every $x \in H/\{0\}$, $[x]$ is the ray through x. If H is the Hilbert space of a Schrödinger quantum system then H represents the pure states of the system and $P(H)$ can be regarded as the state manifold (when provided with the differentiable structure below). One defines the K structure as follows. On $P(H)$ one has an atlas $\{(V_h, b_h, C_h)\}$ where $h \in H$ with $\|h\| = 1$. Here (V_h, b_h, C_h) is the chart with domain V_h and local model the complex Hilbert space C_h where

$$V_h = \{[x] \in P(H);\ (h|x) \neq 0\};\ C_h = [h]^{\perp}; \tag{1.13}$$
$$b_h : V_h \to C_h;\ [x] \to b_h([x]) = \frac{x}{(h|x)} - h$$

This produces a analytic manifold structure on $P(H)$. As a real manifold one uses an atlas $\{(V_h, R \circ b_h, RC_h)\}$ where e.g. RC_h is the realification of C_h (the real Hilbert space with $\mathbf{R}$ instead of $\mathbf{C}$ as scalar field) and $R : C_h \to RC_h;\ v \to Rv$ is the canonical bijection (note $Rv \neq \Re v$). Now consider the form of the K metric relative to a chart $(V_h, R \circ b_h, RC_h)$ where the metric g is a smooth section of $L_2(TP(H), \mathbf{R})$ with local expression $g^h : RC_h \to L_2(RC_h, \mathbf{R});\ Rz \mapsto g^h_{Rz}$ where

$$g^h_{Rz}(Rv, Rw) = 2\nu \Re\left(\frac{(v|w)}{1 + \|z\|^2} - \frac{(v|z)(z|w)}{(1 + \|z\|^2)^2}\right) \tag{1.14}$$

The fundamental form ω is a section of $L_2(TP(H), \mathbf{R})$, i.e. one can write $\omega^h : RC_h \to L_2(RC_h, \mathbf{R});\ Rz \to \omega^h_{Rz}$, given via

$$\omega^h_{Rz}(Rv, Rw) = 2\nu \Im\left(\frac{(v|w)}{1 + \|z\|^2} - \frac{(v|z)(z|w)}{(1 + \|z\|^2)^2}\right) \tag{1.15}$$

Then using e.g. (1.14) for the Fubini-Study (FS) metric in $P(H)$ consider a Schrödinger Hilbert space with dynamics determined via $\mathbf{R} \times P(H) \to P(H) : (t, [x]) \mapsto [exp(-(i/\hbar)tH)x]$ where H is a (typically unbounded) self adjoint operator in H. One thinks then of Kähler isomorphisms of $P(H)$ (i.e. smooth diffeomorphisms $\Phi : P(H) \to P(H)$ with the properties $\Phi^* J = J$ and $\Phi^* g = g$). If U is any unitary operator on H the map $[x] \mapsto [Ux]$ is a K isomorphism of $P(H)$. Conversely (cf. Cirelli et al [**309**]) any K isomorphism of $P(H)$ is induced by a unitary operator U (unique up to phase factor). Further for every self adjoint operator A in H (possibly unbounded) the family of maps $(\Phi_t)_{t \in \mathbf{R}}$ given via $\Phi_t : [x] \to [exp(-itA)x]$ is a continuous one parameter group of K isomorphisms of $P(H)$ and vice versa (every K isomorphism of $P(H)$ is induced by a self adjoint operator where boundedness of A corresponds to smoothness of the Φ_t). Thus in the present framework the dynamics of QM is described by a continuous one parameter group of K isomorphisms, which automatically are symplectic isomorphisms (for the structure defined by the fundamental form) and one has a Hamiltonian system. Next ideally one can suppose that every self adjoint operator represents an observable and these will be shown to be in $1-1$ correspondence with the real K functions.

Definition 7.1.4. Let A be a bounded linear operator on H and denote by $< A >$ the mean value function of A defined via $< A >: P(H) \to \mathbf{C},\ [x] \mapsto < A >_{[x]} = (x|Ax)/\|x\|^2$. The square dispersion is defined via $\Delta^2 A : P(H) \to$

$\mathbf{C}$, $[x] \mapsto \Delta^2_{[x]}A =< (A- < A >_{[x]})^2 >_{[x]}$.

These maps in Definition 2.4 are smooth and if A is self adjoint $< A >$ is real, $\Delta^2 A$ is nonnegative, and one can define $\Delta A = \sqrt{\Delta^2 A}$. To obtain local expressions one writes $< A >^h$: $C_h \to \mathbf{R}$ and $(d < A >)^h : C_h \to (C_h)^*$ via $< A >^h (R) = (z+h)|A(z+h))/(1+\|z\|^2)$ and

(1.16)

$$< (d < A >)^h_{Rz}|Rv >= 2\Re\left(\frac{A(z+h)}{1+\|z\|^2} - \frac{(h|A(z+h))}{1+\|z\|^2}h - \frac{(A(z+h)|z+h)}{(1+\|z\|^2)^2}z\middle| Rv\right)$$

Further the local expressions $X^h : RC_h \to RC_h$ and $Y^h : RC_h \to RC_h$ of the vector fields $X = Id < A >$ and $Y = Gd < A >$ are

$$X^h(Rz) = (1/\nu)R(i(h|A(z+h))(z+h) - iA(z+h)); \tag{1.17}$$
$$Y^h(Rz) = (1/\nu)R(-(h|A(z+h))(z+h) + A(z+h))$$

One proves then (cf. [**309, 594**]) that the flow of the vector field $X = Id < A >$ is complete and is given via $\Phi_t([x]) = [exp(-i(t/\nu)A)x]$. This leads to the statement that if f is a complex valued function on $P(H)$ then f is Kählerian if and only if there is a bounded operator A such that $f =< A >$. From the above it is clear that one should take $\nu = \hbar$ for QM if we want to have $< \mathfrak{H} >$ represent Hamiltonian flow ($\mathfrak{H} \sim$ a Hamiltonian operator) and this gives a geometrical interpretation of Planck's constant. The following formulas are obtained for the Poisson and Riemann brackets

(1.18)

$$\{< A >, < B >\}^h(Rz) =$$
$$= \frac{(z+h|(1/i\nu)(AB-BA)(z+h))}{1+\|z\|^2}; \; ((< A >, < B >))^h(Rz) =$$
$$= \frac{1}{\nu}\frac{(z+h|(AB+BA)(z+h))}{1+\|z\|^2} - \frac{2}{\nu}\frac{(z+h|A(z+h))}{1+\|z\|^2}\frac{(z+h|B(z+h))}{1+\|z\|^2}$$

This leads to the results

(1) $\{< A >, < B >\} =< (1/i\nu)[A, B] >$
(2) $((< A >, < B >)) = (1/\nu) < AB + BA > -(2/\nu) < A >< B >$; $((< A >< A >)) = (2/\nu)\Delta^2 A$
(3) $<< A >, < B >>= (2/\nu)(< AB > - < A >< B >)$
(4) $< A > \circ_\nu < B >= (1/2) < AB + BA >$
(5) $< A > *_\nu < B >=< AB >$

REMARK 7.1.2. One notes that (setting $\nu = \hbar$) item 1 gives the relation between Poisson brackets and commutators in QM. Further the Riemann bracket is the operation needed to compute the dispersion of observables. In particular putting $\nu = \hbar$ in item 2 one sees that for every observable $f \in K(P(H), \mathbf{R})$ and every state $[x] \in P(H)$ the results of a large number of measurements of f in the state $[x]$ are distributed with standard deviation $\sqrt{(\hbar/2)((f,f))([x])}$ around the mean value $f([x])$. This explains the role of the Riemann structure in QM, namely it is the structure needed for the probabilistic description of QM. Moreover the $\circ_\nu$ product corresponds to the Jordan product between operators (cf. item 5) and item 4 tells us that the $*_\nu$ product corresponds to the operator product. This allows one to formulate a functional representation for the algebra $L(H)$. Thus put $\|f\|_\nu = \sqrt{sup_{[x]}(\bar{f} *_\nu f)([x])}$. Equipped with this norm $K(P(H), \mathbf{C})$ becomes a W^* algebra and the map of W^* algebras between $K(P(H), \mathbf{C}$ and $L(H)$ is an isomorphism. This makes it possible to develop a general functional representation theory for C^* algebras generalizing the classical spectral representation for commutative C^* algebras. The K manifold $P(H)$is replaced by a topological fibre bundle in which every fibre is a K manifold isomorphic to a projective space. In particular a nonzero vector $x \in H$ is an eigenvector of A if and only if $d_{[x]} < A >= 0$ or equivalently if and only if $[x]$ is a fixed point for the vector field $Id < A >$ (in which case the corresponding eigenvalue is $< A >_{[x]}$). ■

1.1. PROBABILITY ASPECTS. We go here to [**40, 68, 69, 182, 186, 237, 307, 310, 322, 331, 500, 525, 526, 594, 596, 830, 840, 893, 985, 1018, 1019, 1062, 1229, 1320**] and refer also to Section 3.1 and Remark 3.3.2. First from [**182, 1320**] one defines a (Riemann) metric (statistical distance) on the space of probability distributions $\mathcal{P}$ of the form

$$ds^2_{PD} = \sum (dp_j^2/p_j) = \sum p_j(dlog(p_j))^2 \tag{1.19}$$

Here one thinks of the central limit theorem and a distance between probability distributions distinguished via a Gaussian $exp[-(N/2)(\tilde{p}_j - p_j)^2/p_j]$ for two nearby distributions (involving N samples with probabilities p_j, $\tilde{p}_j$). This can be generalized to quantum mechanical pure states via (note $\psi \sim \sqrt{p}exp(i\phi)$ in a generic manner)

(1.20)
$$|\psi >= \sum \sqrt{p_j}e^{i\phi_j}|j >; \ |\tilde{\psi} >= |\psi > + |d\psi >= \sum \sqrt{p_j + dp_j}e^{i(\phi_j + d\phi_j)}|j >$$

Normalization requires $\Re(< \psi|d\psi >) = -1/2 < d\psi|d\psi >$ and measurements described by the one dimensional projectors $|j >< j|$ can distinguish $|\psi >$ and $|\tilde{\psi} >$ according to the metric (1.19). The maximum (for optimal disatinguishability) is given by the Hilbert space angle $cos^{-1}(| < \tilde{\psi}|\psi > |)$ and the corresponding line element ($PS \sim$ pure state)

$$\frac{1}{4}ds^2_{PS} = [cos^{-1}(| < \tilde{\psi}|\psi > |)]^2 \sim 1 - | < \tilde{\psi}|\psi > |^2 =< d\psi_\perp|d\psi_\perp >\sim \tag{1.21}$$

$$\sim \frac{1}{4}\sum \frac{dp_j^2}{p_j} + \left[\sum p_j d\phi_j^2 - (\sum p_j d\phi_j)^2\right]$$

(called the Fubini-Study (FS) metric) is the natural metric on the manifold of Hilbert space rays. Here

$$|d\psi_\perp >= |d\psi > - |\psi >< \psi|d\psi > \tag{1.22}$$

is the projection of $|d\psi >$ orthogonal to $|\psi >$. Note that if $cos^{-1}(| < \tilde{\psi}|\psi > | = \theta$ then $cos(\theta) = | < \tilde{\psi}|\psi > |$ and $cos^2(\theta) = | < \tilde{\psi}|\psi > |^2 = 1 - Sin^2(\theta) \sim 1 - \theta^2$ for small θ. Hence $\theta^2 \sim 1 - cos^2(\theta) = 1 - | < \tilde{\psi}|\psi > |^2$. The term in square brackets (the variance of phase changes) is nonnegative and an appropriate choice of basis makes it zero. In Braustein-Caves [**182**] one then goes on to discuss distance formulas in terms of density operators and Fisher information but we omit this here. Generally as in Wooters [**1320**] one observes that the angle in Hilbert space is the only Riemannian metric on the set of rays which is invariant uder unitary transformations. In any event $ds^2 = \sum(dp_i^2/p_i)$, $\sum p_i = 1$ is referred to as the Fisher metric (cf. Moroianu [**893**]). Note in terms of $dp_i = \tilde{p}_i - p_i$ one can write $d\sqrt{p} = (1/2)dp/\sqrt{p}$ with $(d\sqrt{p})^2 = (1/4)(dp^2/p)$ and think of $\sum(d\sqrt{p_i})$ as a metric. Alternatively from $cos^{-1}(| < \tilde{\psi}|\psi > |$ one obtains $ds_{12} = cos^{-1}(\sum \sqrt{p_{1i}}\sqrt{p_{2i}})$ as a distance in $\mathcal{P}$. Note from (4.21) that $ds_{12}^2 = 4cos^{-1}| < \psi_1|\psi_2 > | \sim 4(1 - |(\psi_1|\psi_2)|^2 \equiv 4(< d\psi|d\psi > - < d\psi|\psi >< \psi|d\psi >)$ begins to look like a FS metric before passing to projective coordinates. In this direction we observe from [**893**] that the FS metric can be expressed also via

$$\partial\bar{\partial}log(|z|^2) = \phi = \frac{1}{|z|^2}\sum dz_i \wedge d\bar{z}_i - \frac{1}{|z|^4}\left(\sum \bar{z}_i dz_i\right) \wedge \left(\sum z_i d\bar{z}_i\right) \tag{1.23}$$

so for $v \sim \sum v_i\partial_i + \bar{v}_i\bar{\partial}_i$ and $w \sim \sum w_i\partial_i + \bar{w}_i\bar{\partial}_i$ and $|z|^2 = 1$ one has $\phi(v,w) = (v|w) - (v|z)(z|w)$ (cf. (3.4)).

REMARK 7.1.3. We refer now to Section 3.1 and Remark 3.3.2 for connections between quantum geometry and the quantum potential via Fisher information and probability. ■

1.2. GEOMETRIC PHASES. We go now to Dandaloff [**353**] for some remarks on geometric phase and the quantum potential. One refers back here to geometric phases of Berry [**108**] and Levy-Leblond [**817**] for example where the latter shows that when a quanton propagates through a tube, within which it is confined by impenetrable walls, it acquires a phase when it comes out of the tube. Thus consider a tube with square section of side a and length L. Before entering the tube the quanton's wave function is $\phi = exp(ipx/\hbar)$ where p is the initial momentum. In the tube the wave function has the form

$$\psi = Sin\left(n_x\pi\frac{x}{a}\right) Sin\left(n_y\pi\frac{y}{a}\right) exp(ip'x/\hbar) \tag{1.24}$$

with appropriate transverse boundary conditions. After entering the tube the energy E of the quanton is unchanged but satisfies

$$E = \frac{(p')^2}{2m}(n_x^2 + n_y^2)\frac{\pi^2}{2ma} \tag{1.25}$$

For the simplest case $n_x = n_y = 1$ it was found that after the quanton left the tube there was an additional phase

$$\Delta\Phi = \frac{\pi^2\hbar^2}{pa^2}L \tag{1.26}$$

Subsequently Kastner [**733**] related this to the quantum potential that arises in the tube. Thus let the wave function in the tube be $Rexp[(iS/\hbar)+(ipx/\hbar)]$ in polar form. The eventual changes in the phase of the wave function, due to the tube, are now concentrated in S. In order to single out the influence of the tube on the wave function write $\psi_1 = \psi exp(ipx/\hbar)$ and the quantum potential corresponding to ψ is then

$$Q = -\frac{\hbar^2}{2m}\frac{\Delta R}{R} = \frac{\pi^2\hbar^2}{ma^2} \tag{1.27}$$

Now turn to the laws of parallel transport where for the Berry phase the law of parallel transport for the wave function is (cf. [**1174**])

$$\Im < \psi|\dot{\psi} >= 0 \tag{1.28}$$

For the Levy-Leblond phase the law of parallel transport is given by

$$\Im < \psi|\dot{\psi} >= -\frac{1}{\hbar}Q|\psi|^2 \tag{1.29}$$

In this approach the wave function acquires an additional phase after the quanton has left the tube in the form

$$\psi(t + \Delta t) = exp(-iQ\Delta t/\hbar)\psi(t) \tag{1.30}$$

which after expansion in Δt leads to the law of parallel transport in (1.29). Indeed

$$Q\Delta t = \frac{\pi^2\hbar^2}{ma^2}\Delta t = \frac{\pi^2\hbar^2}{ma^2}\frac{mL}{p} = \frac{\pi^2\hbar^2}{pa^2}L = \Delta\Phi \tag{1.31}$$

If we use the polar form for the wave function (1.29) gives $(\partial S/\partial t) = -Q$ and this means that this new law of parallel transport eliminates the quantum potential from the quantum HJ equation. The whole quantum information is now carried by the phase of the wave function. One can see that the nature of this phase is quite different from Berry's phase; it is related to the presence or not of constraints in the system (in this case the tube).

Now consider a quite different type of constrained system where again a new geometric phase will arise. Look at a quantum particle constrained to move on a circle. The wave function has the form $\psi \sim sin(ns/\rho_0)$ where ρ_0 is the radius of the circle and s is the arc length with origin at a tangent point. Then the wave function will have a node at this tangent point. For a circle the value $n = 1/2$ is also allowed (cf. [**632, 769**]) and the corresponding quantum potential for $n = 1/2$ is

$$Q = \frac{h^2}{8m\rho_0^2} \tag{1.32}$$

which is exactly equal to the constant E_0 appearing in the Hamiltonian for a particle on a circle with radius ρ_0 following the Dirac quantization procedure for constrained systems (cf. [**1125**]). The phase which a quanton would acquire traveling along the circle is then

$$Q\frac{2\pi\rho_0 m}{p} = \frac{\pi\hbar^2}{4\rho_0 p} \tag{1.33}$$

Note that if the circle becomes very small then the geometrical phase can not get bigger than $\sim (\hbar/m)$. This limit is imposed by the Heisenberg uncertainty relation $\rho_0 p \sim \hbar$. This is not the case for the Levy-Leblond phase which can get very large provided $L >> a$.

2. HYDRODYNAMICS AND GEOMETRY

Some hydrodynamical aspects of the SE have already been discussed and we return to that now following Delphenich [**366**]. Here one wants to limit the role of statistics and measurement to unveil some geometric features of the so called Madelung approach. Thus, with some repetition, consider a SE $(\hbar/i)\psi_t + H(x,(\hbar/i)\nabla)\psi = 0$ with $\psi = Rexp(iS/\hbar)$ to arrive at

$$\frac{\partial S}{\partial t} + H(x,\nabla S) - \frac{\hbar^2}{2m}\frac{\Delta R}{R} = 0;\ \frac{\partial P}{\partial t} + \frac{\partial}{\partial x^i}(P\dot{x}^i) = 0;\ \dot{x}^i = \left[\frac{\partial H}{\partial p_i}\right]_{p=\nabla S} \tag{2.1}$$

(where $P = R^2$), and Madelung equations of the form (cf. (1.5))

$$\frac{\partial S}{\partial t} + H(x,\nabla S) - \frac{\hbar^2}{2m}\frac{\Delta\sqrt{\rho}}{\sqrt{\rho}} = 0;\ \frac{\partial \rho}{\partial t} + \frac{\partial}{\partial x^i}(\rho\dot{x}^i) = 0 \tag{2.2}$$

where, in a continuum picture, $\rho = mP$ is the mass density of an extended particle whose shape is dictated by P. Setting $v^i = \dot{x}^i$ one has then an Euler equation of the form

$$\frac{\partial}{\partial t}(\rho v^i) + \frac{\partial}{\partial x^k}(\rho v^i v^k) = -\frac{\rho}{m}\frac{\partial V}{\partial x^i} + \frac{\partial}{\partial x^k}\tau^{ik} \tag{2.3}$$

Following Takabayashi [**1214**] one has expressed the quantum force term here as the divergence of a symmetric "quantum stress" tensor

$$\tau_{ij} = \left(\frac{\hbar}{2m}\right)^2 \rho\frac{\partial^2(log(\rho))}{\partial x^i\partial x^j};\ p_i = \frac{\hbar^2}{2m^2}\rho\frac{\partial^2\sigma}{\partial(x^i)^2} \tag{2.4}$$

where p_i denotes diagonal elements or principal stresses expressed in normal coordinates (with $\rho = exp(2\sigma)$). The stress p_i is tension like (resp. pressure like) if $p_i > 0$ (resp. $p_i < 0$) and the mean pressure is

$$\bar{p} = -\frac{1}{3}Tr(\tau_{ij}) = -\frac{\hbar^2}{6m^2}\rho\Delta\sigma \tag{2.5}$$

In classical hydrodynamics negative pressures are often associated with cavitation which involves the formation of topological defects in the form of bubbles. For an ideal fluid one would need $\tau_{ij} = -\bar{p}\delta_{oj}$ and this occurs if and only if the mass density is Gaussian $\sigma \propto -x^i x_i$ in which case $\bar{p} \propto (\hbar^2/2m)\rho$. Generally the stress tensor will not be isotropic, and not an ideal fluid; moreover if one had a viscous fluid one would expect τ_{ij} to be coupled to the rate of deformation tensor (derived from $D\mathbf{v}$). Since this does not occur one does not call this form of matter a fluid but rather a Madelung continuum, corresponding to something like an inviscid fluid which also supports shear stresses, whereas the Gaussian wave packet of QM corresponds to an ideal compressible irrotational fluid medium.

If now one adds time as the zeroth coordinate and extends the velocity vector by $v^0 = 1$ then, defining the energy momentum tensor as

$$\mathfrak{T}^{\mu\nu} = \rho\left[v^\mu v^\nu - \left(\frac{\hbar}{2m}\right)^2\frac{\partial^2(log(\rho))}{\partial x^\mu\partial x^\nu}\right] \tag{2.6}$$

then the Euler and continuity equations can be combined in the form $\partial\mathfrak{T}^{\mu\nu}/\partial x^{\mu} = -(\rho/m)\partial^{\nu}V$; this is somewhat misleading since it is based on a nonrelativistic approach but it leads now to the relativistic theory. First start with the KG equation

$$\left[-\hbar^2\eta^{\mu\nu}\frac{\partial^2}{\partial x^{\mu}\partial x^{\nu}} + m_0^2c^2\right]\psi = 0 \tag{2.7}$$

For $\psi = Rexp(iS/\hbar)$ one gets now

$$\eta^{\mu\nu}\frac{\partial S}{\partial x^{\mu}}\frac{\partial S}{\partial x^{\nu}} + m_0^2c^2 - \hbar^2\frac{\Box R}{R} = 0;\ \eta^{\mu\nu}\frac{\partial}{\partial x^{\mu}}\left(P\frac{\partial S}{\partial x^{\nu}}\right) = 0 \tag{2.8}$$

Define now the 4-velocity, rest mass energy, energy momentum, and stress tensor via

$$u_{\mu} = \frac{1}{m_0}\frac{\partial S}{\partial x^{\mu}};\ \rho = m_0 P;\ p_{\mu} = \rho u_{\mu}; \tag{2.9}$$

$$\tau_{\mu\nu} = \left(\frac{\hbar}{2m_0}\right)^2\frac{\partial^2(log(\rho))}{\partial x^{\mu}\partial x^{\nu}};\ \mathfrak{T}_{\mu\nu} = \rho[u_{\mu}u_{\nu} + \tau_{\mu\nu}]$$

to arrive at relativistic equations for the medium described by ρ and u_{μ} in the form (♣) $\partial^{\mu}\mathfrak{T}_{\mu\nu} = 0$ and $\partial^{\mu}p_{\mu} = 0$ (the second equation is an incompressibility equation and this does not contradict the nonrelativistic compressibility of the medium since in relativity incompressibility in fluid media is equivalant to an infinite speed of light corresponding to rigidity in solid media).

One then erects an elegant mathematical framework involving spacetime foliations related to the complex character of ψ (see also the second paper in Delphenich [**366**] on foliated cobordism, etc.). This is lovely but rather too abstract for the style of this book so we will not try to reproduce it here; we can however skip to some calculations involving the geometric origin of the quantum potential. Thus consider the consequences of choosing a scale of unit norm via the function $\sqrt{\rho}$. One takes a conformally related metric $\bar{g} = \Omega^2 g$ on a manifold M (where $\Omega^2 > 0$) and, writing $\Omega = exp(\sigma)$, one obtains the following formulas for the Levi-Civita connection, Ricci curvature, and scalar curvature

$$\bar{\Gamma}^i_{jk} = \Gamma^i_{jk} = \delta^i_j\partial_k\sigma + \delta^i_k\partial_j\sigma - g_{jk}g^{i\ell}\partial_{\ell}\sigma; \tag{2.10}$$

$$\bar{R}_{ij} = R_{ij} - (n-2)\sigma_{ij} - [\Delta\sigma + (n-2)(\partial^k\sigma\partial_k\sigma)]g_{ij}$$

$$\bar{R} = e^{-2\sigma}[R - 2(n-1)\Delta\sigma - (n-1)(n-2)(\partial^i\sigma\partial_i\sigma)]$$

where $\sigma_{ij} = \partial_i\partial_j\sigma - \partial_j\sigma\partial_i\sigma$. Now if the constant m is replaced by the function ρ then one must contend with the derivatives $\partial_i\rho$ and for $\rho = exp(2\sigma)$ the Minkowski metric will be deformed from $g = \eta$, $\Gamma^i_{jk} = 0$, $R_{ij} = R = 0$ to

$$\bar{\Gamma}^i_{jk} = \delta^i_j\partial_k\sigma + \delta^i_k\partial_j\sigma - \eta_{jk}\eta^{i\ell}\partial_{\ell}\sigma;\ \bar{R}_{ij} = -2\sigma_{ij} - [\Box\sigma + 2(\partial^k\sigma\partial_k\sigma)]\eta_{ij} \tag{2.11}$$

$$\bar{R} = -6e^{-2\sigma}[\Box\sigma + (\partial^i\sigma\partial_i\sigma]$$

Putting Ω back into the equation for scalar curvature one obtains

$$\bar{R} = -\frac{6}{\Omega^2}\left(\frac{\Box\Omega}{\Omega}\right);\ \frac{\Box\sqrt{\rho}}{\sqrt{\rho}} = \frac{\Box\Omega}{\Omega} = -\frac{1}{6}\bar{R}\Omega^2 = -\frac{1}{6}\rho\bar{R} \tag{2.12}$$

This identifies the quantum potential as a mass density times a scalar curvature and resembles some results obtained earlier from [**1115, 1116**] for example (cf. Section 3.3 and 3.3.2). One has also

$$\partial^i \bar{R}_{ij} = \partial^i \left(\frac{1}{2} g_{ij} \bar{R}\right) = \frac{1}{2} \partial^j \bar{R} \Rightarrow \partial^i \bar{R}_{ij} = -3\partial^i \tau_{ij} \tag{2.13}$$

so the Takabayashi stress tensor differs from the Ricci curvature only by a term of vanishing divergence. Hence there is no loss of generality in using the Ricci curvature of $g = \rho\eta$ as the stress tensor since both define the same force field; this means in particular that one is dealing with principal curvatures instead of principal stresses. To extend all this to a more general Lorentz manifold one notes that under a conformal change of spacetime metric to the energy metric the Einstein tensor becomes

$$\bar{G}_{\mu\nu} = \bar{R}_{\mu\nu} - \frac{1}{2}\bar{g}_{\mu\nu}\bar{R} = G_{\mu\nu} - 2\sigma_{\mu\nu} + [2\Box\sigma + \partial^\lambda \sigma \partial_\lambda \sigma] g_{\mu\nu} \tag{2.14}$$

Thus it seems correct to assume that this implies a quantum correction to the Einstein equation

$$G_{\mu\nu} + \left(\Box\sigma + \frac{1}{2}\partial^\lambda \sigma \partial_\lambda \sigma\right) g_{\mu\nu} = 8\pi G T_{\mu\nu} + 2\sigma_{\mu\nu} \tag{2.15}$$

REMARK 7.2.1. There is also a discussion of hydrodynamic features of QM, EM, and BM in Holland [**640**]. ■

3. REMARKS ON TRAJECTORIES

There have been a number of papers written involving microstates and Bohmian mechanics (cf. [**168, 169, 170, 238, 241, 244, 250, 382, 383, 384, 385, 386, 446, 447, 448, 478, 479, 480, 481, 699**]) and we sketch here some features of the Bouda-Djama method following [**386**]. There are some disagreements regarding quantum trajectories, discussed in [**168, 480**], which we will not deal with here. Generally we have followed Faraggi-Matone [**446**] in our previous discussion and microstates were not explicitly considered (beyond mentioning the third order equation and the comments in Remark 2.2.2). Thus, referring to Djama [**386**] for philosophy, one begins with the SE $-(\hbar^2/2m)\Delta\psi + V\psi = i\hbar\psi_t$ where $\psi = Rexp(iS/\hbar)$ in 3-D and arrives at the standard

$$\frac{1}{2m}(\nabla S)^2 - \frac{\hbar^2}{2m}\frac{\Delta R}{R} + V = -S_t;\ \nabla\cdot\left(R^2\frac{\nabla S}{m}\right) + V = -\partial(R^2) \tag{3.1}$$

witeh $Q = -(\hbar^2/2m)(\Delta R/R)$. Then one sets

$$\mathbf{j} = \frac{\hbar}{2mi}(\psi^*\nabla\psi - \psi\nabla\psi^*) = R^2\frac{\nabla S}{m} \Rightarrow \nabla\cdot\mathbf{j} + \partial_t R^2 = 0 \tag{3.2}$$

and $\rho = |\psi|^2 = R^2$ as usual. The velocity $\mathbf{v}$ is taken as $\mathbf{v} = \mathbf{j}/\rho = \nabla S/m$ here in the spirit of Bohm (and Dürr, Goldstein, Zanghi, et al). Working in 1-D with $S = S_0(x, E) - Et$ one recovers the stationary HJ equation of Section 2.2 for example and there is some discussion about the situation $S_0 = constant$ referring to Floyd and Farragi-Matone. Explicit calculations for microstates are considered and comparisons are indicated. The EP of Faraggi-Matone is then discussed as in

Section 2.2 and the quantum mass field $m_Q = m(1 - \partial_E Q)$ is introduced. This leads to the third order differential equation for $\dot{x}$ (where $P = \partial_x S_0 = m_Q \dot{x}$)

$$\text{(3.3)}\quad \frac{m_Q^2}{2m} + V(x) - E + \frac{\hbar^2}{4m}\left(\frac{m_Q''}{m_Q} - \frac{3}{2}\frac{(m_Q')^2}{m_Q^2} - \frac{m_Q'}{m_Q}\frac{\ddot{x}}{\dot{x}^2} + \frac{\dddot{x}}{\dot{x}^3} - \frac{5}{2}\frac{\ddot{x}^2}{\dot{x}^4}\right) = 0$$

It is observed correctly that (3.3) is a difficult equation to manipulate, requiring a priori a solution of the QSHJE.

Now one proposes a Lagrangian which depends on x, $\dot{x}$ and the set of hidden variables Γ which is connected to constants of integration from an equation like (3.3). This approach was developed in order to avoid dealing with the Jacobi type formula $t - t_0 = \partial S_0/\partial E$ which the authors felt should be restricted to HJ equations of first order. Then one looks for a quantum Lagrangian L_q such that $(d/dt)(\partial L_q/\partial \dot{x}) - \partial_x L_q = 0$ and writes

$$\text{(3.4)}\quad L_q(x,\dot{x},\Gamma) = \frac{m}{2}\dot{x}^2 f(x,\Gamma) - V(x);\ \frac{\partial L_q}{\partial \dot{x}} = m\dot{x}f(x,\Gamma);\ \frac{\partial L_q}{\partial x} = \frac{m}{2}\dot{x}^2 f_x - V_x$$

This leads to

$$\text{(3.5)}\quad mf(x,\Gamma)\ddot{x} + \frac{m\dot{x}^2}{2}f_x + V_x = 0$$

Then set $H_q = (\partial_x L_q)\dot{x} - L_q$ and $P = \partial L_q/\partial\dot{x} = m\dot{x}f$ so

$$\text{(3.6)}\quad H_q = \frac{m\dot{x}^2}{2}f(x,\Gamma) + V(x) = \frac{P^2}{2mf} + V(x)$$

Working with the stationary situation $S = S_0(x,\Gamma) - Et$ some calculation gives then

$$\text{(3.7)}\quad \frac{1}{2mf}S_x^2 + V = -S_t \Rightarrow \frac{1}{2mf}(\partial_x S_0)^2 + V(x) - E = 0$$

Now referring to the general equation (2.18) in Chapter 2 (extracted from Faraggi-Matone [**446**]) one writes here $w = \tilde{\theta}/\tilde{\phi} \sim \psi^D/\psi \in \mathbf{R}$ with $(\alpha \sim \omega)$ so that (cf. [**244, 384**])

$$\text{(3.8)}\quad e^{2iS_0/\hbar} = e^{i\omega}\frac{(\tilde{\theta}/\tilde{\phi}) + i\bar{\ell}}{(\tilde{g}t/\tilde{\phi}) - i\ell} \rightsquigarrow S_0 = \hbar Tan^{-1}\frac{\theta + \mu\phi}{\nu\theta + \phi}$$

(cf. [**169**] for details). For the QSHJE the basic equation is (2.2.17) which we repeat as

$$\text{(3.9)}\quad \frac{1}{2m}(S_0')^2 + \mathfrak{W} + Q = 0;\ \mathfrak{W} = -\frac{\hbar^2}{4m}\left\{e^{2iS_0/\hbar}, x\right\} \sim V - E;\ Q = \frac{\hbar^2}{4m}\{S_0, x\}$$

There is a "quantum" transformation $x \to \hat{x}$ described in [**446, 447**] with the QSHJE arising then from a conformal modification of the CSHJE. Thus note (•) $\{x, S_0\} = -(S_0')^{-2}\{S_0, x\}$ and define $\mathfrak{U}(S_0) = \{x, S_0\}/2 = -(1/2)(S_0')^{-2}\{S_0, x\}$. This gives a conformal rescaling $\frac{1}{2m}(S_0')^2[1 - \hbar^2\mathfrak{U}] + V - E = 0$ since

(3.10)

$$\frac{1}{2m}(S_0')^2[1 - \hbar^2\mathfrak{U}] = \frac{1}{2m}(S_0')^2[1 - \frac{\hbar^2}{2}\{x, S_0\}] = \frac{1}{2m}(S_0')^2[1 + \frac{\hbar^2}{2}(S_0')^{-2}\{S_0, x\}] =$$

$$= \frac{1}{2m}(S_0')^2 + \frac{\hbar^2}{4m}\{S_0, x\} = \frac{1}{2m}(S_0')^2 + Q \Rightarrow Q = -\frac{\hbar^2}{2m}(S_0')^2\mathfrak{U}$$

which agrees with $Q = (\hbar^2/4m)\{S_0, x\}$ using (•). Then from (3.10)

$$\left(\frac{\partial x}{\partial \hat{x}}\right)^2 = 1 = \hbar^2\mathfrak{U}(S_0) = 1 + 2m(S_0')^2 Q \Rightarrow \hat{x} = \int^x \frac{dx}{\sqrt{1 + 2m(\partial S_0)^{-2}Q}} \quad (3.11)$$

Similarly using the QSHJE (2.2.17) we have

$$\left(\frac{\partial x}{\partial \hat{x}}\right)^2 = (S_0')^{-2}[(S_0')^2 + 2mQ] = [(S_0')^2 - 2m\mathfrak{W}](S_0')^{-2} \Rightarrow \quad (3.12)$$

$$\Rightarrow \hat{x} = \int^x \frac{S_0' dx}{\sqrt{2m(E-V)}}$$

This all follows from [**446, 447**] and is used in [**386**]. Now from (3.7), (3.9), (3.10), and the QSHJE one can write (correcting a sign in [**386**])

$$f(x, \Gamma) = \left[1 + \frac{\hbar^2}{2}(S_0')^{-2}\{S_0, x\}\right]^{-1} \Rightarrow f = \frac{(S_0')^2}{2m(E-V)} \quad (3.13)$$

and via (3.8) $\Gamma = \Gamma(E, \mu, \nu)$ with $f = f(x, E, \mu, \nu)$. Putting this in (3.7) gives then

$$E = \frac{m\dot{x}^2}{2}\frac{(S_0')^2}{2m(E-V)} + V \Rightarrow \dot{x}S_0' = 2(E-V) \quad (3.14)$$

Note that this equation also follows from (3.5), namely $m\dot{x}f = \partial L_q/\partial \dot{x}$, and integration (cf. [**386**]). Now for the appropriate third order trajectory equation in this framework, one finds from (3.14) and the QSHJE

$$(E-V)^4 - \frac{m\dot{x}^2}{2}(E-V)^3 + \frac{\hbar^2}{8}\left[\frac{3}{2}\left(\frac{\ddot{x}}{\dot{x}}\right)^2 - \frac{\dddot{x}}{\dot{x}}\right](E-V)^2 - \quad (3.15)$$

$$-\frac{\hbar^2}{8}\left[\dot{x}^2\frac{d^2V}{dx^2} + \ddot{x}\frac{dV}{dx}\right](E-V) - \frac{3\hbar^2}{16}\left[\dot{x}\frac{dV}{dx}\right]^2 = 0$$

(cf. [**168, 386**]). This is somewhat simpler to solve that (3.3) since it is independent of the SE and the QSHJE. We refer now to [**168, 169, 170, 382, 383, 384, 385, 386**] for more in this direction.

4. GEOMETRY AND THE QUANTUM POTENTIAL

In [**254**] (and in Chapters 1-6 of this book) we surveyed many feature of the so called quantum potential (QP) (cf. also [**257, 258, 259, 237, 234, 235, 244, 255, 250**]). Some matters were treated more thoroughly than others and we want to discuss here following [**261**] certain geometrical aspects in more detail (note the references to [**254**] can often be interpreted as references to Chapters 1-6 of the present book). To set the state we recall the Schrödinger equation (SE) in 1-D of the form (**4A**) $-(\hbar^2/2m)\psi'' + V\psi = i\hbar\psi_t$ so that for $\psi = Rexp(iS/\hbar)$ one has

$$S_t + \frac{1}{2m}S_x^2 + V + Q = 0;\ Q = -\frac{\hbar^2 R''}{2mR};\ \partial_t(R^2) + \frac{1}{m}(R^2 S_x)_x = 0 \quad (4.1)$$

Here Q is the quantum potential (QP) and one can argue that Bohmian mechanics is simply classical symplectic mechanics using the Hamiltonian (**4B**) $H_q = H_c + Q = (1/2m)S_x^2 + V + Q$ from the Hamiltonian-Jacobi (HJ) equation (4.1) (cf. here [**85, 109, 520, 598**]). One can write $P = R^2 = |\psi|^2$ (a probability density) with $\rho = mP$ a mass density and obtain a hydrodynamical version of (4.1). Note in particular (**4C**) $Q = -(\hbar^2/2m)(\partial^2\sqrt{\rho}/\sqrt{\rho})$ and using $p = S_x = m\dot{q} = mv$ one obtains

$$mv_t + mvv_x + \partial V + \partial Q = 0;\ \ \rho_t + (\rho\dot{q})_x = 0 \tag{4.2}$$

leading to

$$\partial_t(\rho v) + \partial(\rho v^2) + \frac{\rho}{m}\partial V + \frac{\rho}{m}\partial Q = 0 \tag{4.3}$$

which has the flavor of an Euler equation (cf. [**234, 239, 387**]). There is however a missing pressure term from the hydrodynamical theory (cf. [**234, 818, 1028**]) and looking at (4.2) one could imagine a pressure term supplied in the form (**4D**) $\partial Q = (1/R^2)\partial\mathfrak{P}$ (where $\mathfrak{P}$ denotes pressure). This suggests a hydrodynamical interpretation for Q, namely, going to 3-D for example, (**4E**) $\nabla\mathfrak{P} = R^2\nabla Q$ (cf. [**239**]). This will all be discussed in detail below and we make first a few background remarks about the QP.

REMARK 7.4.1. In [**257**] we considered given a function $Q \in L^\infty(\Omega)$ (for Ω a bounded domain) and looked for $R \in H_0^1(\Omega)$ satisfying $Q = -(\hbar^2/2m)(\Delta R/R) \equiv \Delta R + (2m/\hbar^2)QR = 0$. We showed that if $Q < 0$ ($\beta = (2m/\hbar^2)$) then there is a unique solution and if 0 is not in the countable spectrum of $\Delta R + \beta QR$ then $\Delta R + \beta QR = 0$ has a unique solution for any $Q \in L^\infty$. The corresponding HJ equation (**4F**) $\partial_t S + (1/2m)(\nabla S)^2 + Q + V = 0$ and the continuity equation (**4G**) $\partial_t R^2 + (1/m)\nabla(R^2\nabla S) = 0$ must then be solved to obtain some sort of generalized quantum theory. Here a priori V must be assumed unknown and there are then two equations for two unknowns S and V, namely (in 1-D for simplicity)

$$S_t + \frac{1}{2m}S_x^2 + Q + V = 0;\ \ \partial_t R^2 + \frac{1}{m}(R^2 S_x)_x = 0 \tag{4.4}$$

the solution of which would yield a SE based on Q (see here Remark 7.7.1 for more detail in this regard). ■

EXAMPLE 4.1. Now $(1/2m)p^2 + V = E$ (classical Hamiltonian - recall $p \sim S_x$) so we could perhaps treat E as an unknown here and try to solve

$$S_t + Q + E = 0;\ \ \partial_t R^2 + \frac{1}{m}(R^2 S_x)_x = 0 \tag{4.5}$$

Consider first $R^2 S_x = -\int^x m\partial_t R^2 dx + f(t)$ from which

$$2RR_tS_x + R^2S_{xt} = -\int^x m\partial_t^2 R^2 dx + f' \Rightarrow (Q_x + E_x)R^2 = \tag{4.6}$$

$$= -R^2 S_{xt} = \frac{2R_t}{R}\left(-\int^x m\partial_t R^2 dx + f\right) + \int^x m\partial_t^2 R^2 dx - f'$$

Hence

$$R^2S_x = -\int^x m\partial_t R^2 dx + f(t);\ R^2E_x = -Q_xR^2+ \tag{4.7}$$
$$+\int^x m\partial_t^2 R^2 dx - f' + \frac{2R_t}{R}\left(-\int^x m\partial_t R^2 dx + f\right)$$

giving S_x and E_x modulo an arbitrary differentiable function $f(t)$. Note also

$$R^2E_x = R^2V_x + \frac{R^2}{2m}(S_x^2)_x = R^2V_x + \frac{R^2}{m}S_xS_{xx} \tag{4.8}$$

and S_{xx} can be determined via $(1/m)(R^2S_x)_x = -\partial_t R^2$. Hence R^2V_x can be determined from R^2E_x. Note here

$$\frac{R^2}{m}S_xS_{xx} = -\partial_t R^2 - \frac{2}{m}RR_xf + 2RR_x\int^x \partial_t R^2 dx \tag{4.9}$$

(see Section 7.10 for more details). ∎

We mention also two examples from [**163, 257**]

EXAMPLE 4.2. For a free particle in 1-D there are possibilities such as $\psi_1 = Aexp[i[px-(p^2t/2m))/\hbar]$ and $\psi_2 = Aexp[-i(px+(p^2t/2m))/\hbar]$ in which case $Q = 0$ for both functions but for $\psi = (1/\sqrt{2})(\psi_1+\psi_2)$ there results $Q = p^2/2m$ ($p \sim \hbar k$ here). Hence $Q = 0$ depends on the wave function and cannot be said to represent a classical limit. Further we note that $S = \hbar kx - (\hbar^2k^2/2m)t$ in ψ_1 with $S_t = -\hbar^2k^2/2m \sim -E$, $S_x = \hbar k$, and $R = 1 \notin H_0^1$. For $\psi = (1/\sqrt{2})(\psi_1+\psi_2)$ on the other hand

$$R = \sqrt{2}ACos(kx) \notin H_0^1;\ \frac{R''}{R} = -k^2;\ Q = \frac{k^2\hbar^2}{2m};\ S = -\frac{k^2\hbar^2 t}{2m}; \tag{4.10}$$
$$S_t = -\frac{k^2\hbar^2}{2m} \sim -E;\ S_x = 0$$

Thus the same SE can arise from different Q (which is generally obvious of course) and S varies with Q. ∎

EXAMPLE 4.3. For $V = m\omega^2x^2/2$ and a stationary SE one has solutions of the form $\psi_n(x) = c_nH_n(\xi x)exp(-\xi^2x^2/2)$ where $\xi = (m\omega\hbar)^{1/2}$, $c_n = (\xi/\sqrt{\pi}2^n n!)$, and H_n is a Hermite function. One computes that $Q = \hbar\omega[n+(1/2)] - (1/2)m\omega^2x^2$ and hence $\hbar \to 0$ does not imply $Q \to 0$ and moreover $Q = 0$ corresponds to $x = \pm\sqrt{(2\hbar/m\omega)[n+(1/2)]}$ so not all systems in quantum mechanics have a classical limit. This example corresponds to $\Omega = \mathbf{R}$ and $\psi_n \in H_0^1$ is satisfied. ∎

5. REMARKS ON WEYL GEOMETRY

Now we recall how in various situations the QP is proportional to a Weyl-Ricci curvature R_w for example (cf. [**254, 237, 270, 271, 276, 1115, 1116**]) and this can be interpreted in terms of a statistical geometry for example (cf. also [**81, 82**]) and Chapter 3. In general (see e.g. [**163, 254, 257**]) one knows that each wave function $\psi = Rexp(iS/\hbar)$ for a given SE produces a different QP as in (1.1) (which in higher dimensions has the form $-(\hbar^2/2m)(\Delta R/R)$). Thus for $Q \sim R_w$ to make sense we have to think of a given R or $R^2 = P$ (or $\rho \sim mP$) as generating a (Weyl) geometry as in Chapter 3 (cf. also [**254, 237, 270, 271, 276**]); we will repeat here briefly some of the material in Chapter 3 with a view towards emphasizing the role of Q (cf. also Remark 7.10.4). This is in accord with having a Weyl vector (**5A**) $\phi_i \sim -\partial_i log(\hat{\rho})$ (where $\hat{\rho} = \rho/\sqrt{g}$ for a Riemannian metric g). Thus following Santamato [**1115, 1116**] one assumes that the motion of the particle is given by some random process $q^i(t,\omega)$ in a manifold M (ω is the random process label) with a probability density $\rho(q,t)$ and satisfying a deterministic equation (**5B**) $\dot{q}^i(t,\omega) = (dq^i/dt)(t,\omega) = v^i(q(t,\omega),t)$ with random initial conditions $q^i(t_0,\omega) = q_0^i(\omega)$. The probability density will satisfy (**5C**) $\partial_t\rho + \partial_i(\rho v^i) = 0$ with initial data $\rho_0(q)$. Let $L(q,\dot{q},t)$ be some Lagrangian for the particle and define an equivalent Lagrangian via

$$L^*(q,\dot{q},t) = L(q,\dot{q},t) - \partial_t S + q^i\partial_i S \tag{5.1}$$

for some function S. The velocity field $v^i(q,t)$ yielding a classical motion with probability one can be found by minimizing the action functional

$$I(t_0,t_1) = E\left[\int_{t_0}^{t_1} L^*(q(t,\omega),\dot{q}(t,\omega),t)dt\right] \tag{5.2}$$

This leads to (**5D**) $\partial_t S + H(q,\nabla S,t) = 0$ and $p_i = (\partial L/\partial\dot{q}^i) = \partial_t S$ where $H \sim p_i\dot{q}^i - L$ with $v^i(q,t) = (\partial H/\partial p_i)(q,\nabla S(q,t),t)$. Now suppose that some geometric structure is given on M via $ds^2 = g_{ij}dq^idq^j$ so that a scalar curvature $\mathcal{R}(q,t)$ is meaningful and write the acutal Lagrangian as (**5E**) $L = L_C + \gamma(\hbar^2/m)\mathcal{R}(q,t)$ where γ will turn out to have the form $\gamma = (1/8)[(n-2)/(n-1)] = 1/16$ for $n = 3$. Assume that in a transplantation $q^i \to q^i + dq^i$ the length of a vector $\ell = (g_{ik}A^iA^k)^{1/2}$ varies according to the law (**5F**) $\delta\ell = \ell\phi_k dq^k$ where the ϕ_k are covariant components of an arbitrary vector of M (this characterizes a Weyl geometry). One imagines that physics determines geometry so that the ϕ_k must be determined from some averaged least action principle yielding the motion of the particle; in particular the minimum now in (5.2) is to be evaluated with respect to the class of all Weyl geometries with fixed metric tensor. Since the only term containing the gauge vector $\vec{\phi} = (\phi_k)$ is the curvature term one requires $E[\mathcal{R}(q(t,\omega)t] = minimum$ ($\gamma > 0$ for $n \geq 3$). This minimization yields

$$\mathcal{R} = \dot{\mathcal{R}} + (n-1)\left[(n-2)\phi_i\phi^i - 2\left(\frac{1}{\sqrt{g}}\partial_i(\sqrt{g}\phi^i)\right)\right] \tag{5.3}$$

where $\phi^i = g^{ik}\phi_k$ and $\dot{\mathcal{R}}$ is the Riemannian curvature based on the metric. Note here that a Weyl geometry is assumed as the proper background for the motion. One shows that the quantity $\hat{\rho}(q,t) = \rho(q,t)/\sqrt{g}$ transforms as a scalar under coordinate changes and a covariant equation of the form (**5G**) $\partial_t\hat{\rho} + (1/\sqrt{g})\partial_i(\sqrt{g}v^i\hat{\rho}) = 0$ ensues (g_{ik} is assumed time independent). Some calculation gives then a minimum when (**5H**) $\phi_i(q,t) = -[1/(n-2)]\partial_i log(\hat{\rho})]$. This shows that the transplantation properties of space are determined by the presence of matter and in turn this change in geometry acts on the particle via a "quantum" force $f_i = \gamma(\hbar^2/m)\partial_i\mathcal{R}$ depending on the gauge vector $\vec{\phi}$. Putting this $\vec{\phi}$ in (5.3) yields

$$R_w = \mathcal{R} = \dot{\mathcal{R}} + \frac{1}{2\gamma\sqrt{\hat{\rho}}}\left[\frac{1}{\sqrt{g}}\partial_i(\sqrt{g}g^{ik}\partial_k\sqrt{\hat{\rho}}\right] \tag{5.4}$$

along with a (HJ) equation

$$\partial_t S + H_C(q, \nabla S, t) - \gamma\left(\frac{\hbar^2}{m}\right)\mathcal{R} = 0 \tag{5.5}$$

and for certain Hamiltonians of the form (**5I**) $H_C = (1/2m)g^{ik}(p_i - A_i)(p_k - A_k) + V$ with arbitrary fields A_k and V it is shown that the function $\psi = \sqrt{\hat{\rho}}exp[(i/\hbar)S(q,t)]$ satisfies a SE (omitting the A_i)

$$i\hbar\partial_t\psi = -\frac{\hbar^2}{2m}\frac{1}{\sqrt{g}}\left[\partial_i\left(\sqrt{g}g^{ik}\partial_k\right)\right]\psi + \left[V - \gamma\left(\frac{\hbar^2}{m}\right)\dot{\mathcal{R}}\right]\psi \tag{5.6}$$

This Hamiltonian is characteristic of a particle in an EM field and all Hamiltonians arising in nonrelativistic applications may be reduced to the above form with corresponding HJ equation

$$\partial_t S = \frac{1}{2m}g^{ik}\partial_i S\partial_k S + V - \gamma\frac{\hbar^2}{m}\mathcal{R} = 0 \tag{5.7}$$

(note there are mistakes in the SE in [**1115, 1116**] and in the improperly corrected form of [**254**]).

REMARK 7.5.1. Note that indices are lowered or raised via use of g_{ij} or its inverse g^{ij}. The most complete sources of notation for differential calculus on Riemannian manifolds seem to be [**9, 1275**]. It is seen that $\hbar$ arises only via (5.5) and for $\dot{\mathcal{R}} = 0$ there is no $\hbar$ in the HJ equation. If $\mathcal{R} = 0$ the quantum force is zero and "quantum mechanics" involves no $\hbar$; $\mathcal{R} = 0$ (with $\dot{\mathcal{R}} = 0$) involves (5.10) below giving $Q = 0$. ■

Now given (5.7), and comparing to (4.1) for example, we see that (**5J**) $Q \sim -\gamma(\hbar^2/m)\mathcal{R}$ with $\mathcal{R}$ given by (5.4) and $\gamma = 1/16$ for $n = 3$. Thus

$$Q \sim -\frac{\hbar^2}{16m}\left[\dot{\mathcal{R}} + \frac{8}{\sqrt{\hat{\rho}g}}\partial_i(\sqrt{g}g^{ik}\partial_k\sqrt{\hat{\rho}})\right] \tag{5.8}$$

and the SE (5.6) contains only $\dot{\mathcal{R}}$. Further from (**5H**) we have for the Weyl vector $\phi_i = -\partial_i log(\hat{\rho}) = -\partial_i\hat{\rho}/\hat{\rho}$ and there is an expression for $\mathcal{R}$ in the form (5.3) leading to

$$Q \sim -\frac{\hbar^2}{16m}\left[\dot{\mathcal{R}} + 2\left\{\phi_i\phi^i - \frac{2}{\sqrt{g}}\partial_i(\sqrt{g}\phi^i)\right\}\right] \tag{5.9}$$

showing how Q depends directly on the Weyl vector. When $\dot{\mathcal{R}} = 0$ (flat space) one sees that the SE is classical and (**5K**) $Q = -(\hbar^2/8m)[\phi_i\phi^i - (1/\sqrt{g})\partial_i(\sqrt{g}\phi^i)]$. Note that when $g = 1$ (so $\dot{\mathcal{R}} = 0$ automatically) and $\hat{\rho} = \rho$ we have then

$$\phi_k\phi^k - 2\partial_k\phi^k \sim -\left(\frac{|\nabla\rho|^2}{\rho^2} - \frac{2\Delta\rho}{\rho}\right) = 4\frac{\Delta\sqrt{\rho}}{\sqrt{\rho}} \tag{5.10}$$

which means (**5L**) $Q = -(\hbar^2/2m)(\Delta\sqrt{\rho}/\sqrt{\rho})$ as in the desired (**4C**).

REMARK 7.5.2. Thus starting with a manifold M with metric g_{ij} and random initial conditions as indicated for a particle of mass m, the resulting classical statistical dynamics based on a probability distribution P with $\rho = mP$ can be properly phrased in a Weyl geometry in which the particle undergoes classical motion with probability one. The assumed Weyl geometry as well as the particle motion is determined via $\hat{\rho}(\rho, g)$ which says that given a different P there will be a different ρ and $\hat{\rho}$ (since g is fixed). Hence writing $\psi = \sqrt{\hat{\rho}} exp(iS/\hbar)$ one expects a different quantum potential and a different Weyl geometry. The SE will however remain unchanged and this may be a solution to the apparent problems illustrated in Section 1 about different quantum potentials being attached to the same SE. Another point of view could be that for m fixed each P (or equivalently ρ) determines a P-dependent motion via it Weyl geometry and each such motion can be described by a P-dependent wave function. The choice of $\hbar$ is arbitrary; here it arises via ψ and any $\hbar$ will do. The identification with Planck's constant has to come from other considerations. ∎

We go now to [**1115, 1116**] and sketch an interesting role of Weyl geometry in the Klein-Gordon (KG) equation (cf. also [**254, 237, 250, 270, 271, 276**] for discussion of this approach). The idea is to start from first principles, extended to gauge invariance relative to an arbitrary choice of spacetime calibration. Weyl geometry is not assumed but derived with the particle motion from a single average action principle. Thus assume a generic 4-D manifold with torsion free connection $\Gamma^{\lambda}_{\mu\nu} = \Gamma^{\lambda}_{\nu\mu}$ and a metric tensor g with signature $(+,-,-,-)$; $\hbar = c = 1$ is taken for convenience (although this loses important information in the equations). The analysis will produce an integrable Weyl geometry with weights $w(g_{\mu\nu}) = 1$ and $w(\Gamma^{\lambda}_{\mu\nu}) = 0$ (cf. [**234, 655**] for Weyl geometry and Weyl-Dirac theory). One takes random initial conditions on a spacelike 3-D hypersurface and produces both particle motion and spacetime geometry via an average stationary action principle (**5M**) $\delta\left[E\int_{\tau_1}^{\tau_2} L(x(\tau), \dot{x}(\tau))d\tau\right] = 0$ where τ is an arbitrary parameter along the particle trajectory. Given L positively homogeneous of first degree in $\dot{x}^{\mu} = dx^{\mu}/d\tau$ and transforming as a scalar of weight $w(L) = 0$ as well as a gauge invariant probability measure it follows that the action integral will be parameter invariant, coordinate invariant, and gauge invariant. A suitable Lagrangian is (**5N**) $L(x, dx) = (m^2 - (1/6)\mathcal{R})^{1/2} ds + A_{\mu}dx^{\mu}$ where $ds = (g_{\mu\nu}\dot{x}^{\mu}\dot{x}^{\nu})^{1/2} d\tau$ and $w(m) = -1/2$ (m = rest mass corresponds to a scalar Weyl field with no equation needed and the factor (1/6) in L is for convenience later). One writes (**5O**) $A_{\mu} = \bar{A}_{\mu} - \partial_{\mu}S$ where $\bar{A}_{\mu} \sim$ EM 4-potential in Lorentz gauge and $w(S) = w(\bar{A}_{\mu}) = 0$.

Omitting here the considerable details of calculation (which are given in Santamato [**1115, 1116**] and sketched in [**254, 270, 271**]) we recall that one can work with a modified Lagrangian (**5P**) $\bar{L}(x, dx) = (m^2 - (1/6)\mathcal{R})^{1/2} + \bar{A}_{\mu}dx^{\mu}$. Variational methods lead to a 1-parameter family of hypersurfaces $S(x) = constant$ satisfying the HJ equation

$$g^{\mu\nu}(\partial_{\mu}S - \bar{A}_{\mu})(\partial_{\nu}S - \bar{A}_{\nu}) = m^2 - (1/6)\mathcal{R} \tag{5.11}$$

and a congruence of curves intersecting this family given via

$$\frac{dx^{\mu}}{ds} = \frac{g^{\mu\nu}(\partial_{\nu}S - \bar{A}_{\nu})}{[g^{\rho\sigma}(\partial_{\rho} - \bar{A}_{\rho})(\partial_{\sigma}S - \bar{A}_{\sigma})]^{1/2}} \tag{5.12}$$

The probability measure is determined by its probability current density j^μ where $\partial_\mu j^\mu = 0$ and (**5Q**) $j^\mu = \rho(\sqrt{-g}g^{\mu\nu}(\partial_\nu S - \bar{A}_\nu)$. Gauge invariance implies $w(j^\mu) = 0 = w(S)$ and $w(\rho) = -1$ so ρ is the scalar probability density of the particle random motion. To find the connection the variational principle for (**5M**) is rephrased as

$$(5.13) \qquad \delta\left[\int_\Omega d^4x[(m^2 - (1/6)\mathcal{R})(g_{\mu\nu}j^\mu j^\nu)]^{1/2} + A_\mu j^\mu\right] = 0$$

Since the $\Gamma^\lambda_{\mu\nu}$ arise only in $\mathcal{R}$ this reduces to (**5R**) $\delta[\int_\Omega \rho\mathcal{R}\sqrt{-g}d^4x] = 0$ where (2.11) has been used. This leads to

$$(5.14) \qquad \Gamma^\lambda_{\mu\nu} = \left\{\begin{array}{c}\lambda\\ \mu\nu\end{array}\right\} + \frac{1}{2}(\phi_\mu\delta^\lambda_\nu + \phi_\nu\delta^\lambda_\mu - g_{\mu\nu}g^{\lambda\rho}\phi_\rho);\ \phi_\mu = \partial_\mu log(\rho)$$

and shows that the connections are integrable Weyl connections with a gauge field ϕ_μ ((**5A**) suggests here perhaps $\phi_i = -(1/2)\partial_i log(\rho)$). The HJ equation (4.21) and $\partial_\mu j^\mu = 0$ can be combined into a single equation for $S(x)$, namely

$$(5.15) \qquad e^{iS}g^{\mu\nu}(iD_\mu - \bar{A}_\mu)(iD_\nu - \bar{A}_\nu)e^{-iS} - (m^2 - (1/6)\mathcal{R}) = 0$$

with $D_\mu\rho = 0$ where (cf. [**254, 234**])

$$(5.16) \qquad D_\mu T^\alpha_\beta = \partial_\mu T^\alpha_\beta + \Gamma^\alpha_{\mu\epsilon}T^\epsilon_\beta - \Gamma^\epsilon_{\mu\beta}T^\alpha_\epsilon + w(T)\phi_\mu T^\alpha_\beta$$

(D_μ is called the double-covariant Weyl derivative and one notes that it is ρ and not m, as in [**1**], which behaves as a constant under D_μ). Then to any solution (ρ, S) of these equations corresponds a particular random motion for the particle. One notes that (4.25)-(4.26)) can be written in a familiar KG form

$$(5.17) \qquad \left(\frac{i}{\sqrt{-g}}\partial_\mu\sqrt{-g} - \bar{A}_\mu\right)g^{\mu\nu}(i\partial_\nu - \bar{A}_\nu)\psi - (m^2 - (1/6)\dot{\mathcal{R}})\psi = 0$$

where $\psi = \sqrt{\rho}exp(-iS)$ and $\dot{\mathcal{R}}$ is the Riemannian scalar curvature. We have also from [**1115, 1116**]

$$(5.18) \qquad \mathcal{R} = \dot{\mathcal{R}} - 3\left[\frac{1}{2}g^{\mu\nu}\phi_\mu\phi_\nu + \frac{1}{\sqrt{-g}}\partial_\mu\sqrt{-g}g^{\mu\nu}\phi_\nu\right] = \dot{\mathcal{R}} + \mathcal{R}_w$$

in keeping also with [**270**].

REMARK 7.5.3. Note here $g^{\mu\nu}\phi_\nu = \phi^\mu$ so (5.18) gives for the last term (**2S**) $\mathcal{R}_w = -3[(1/2)\phi_\mu\phi^\mu + (1/\sqrt{-g})\partial_\mu(\sqrt{-g}\phi^\mu)]$ whereas (5.3) suggests here ($\bigstar$) $-3[2\phi_\mu\phi^\mu - (2/\sqrt{-g})\partial_\mu(\sqrt{-g}\phi^\mu)]$ which is similar [**1115, 1116**] in having a minus sign in the middle; we remark that a change $\phi_\mu \to -2\phi_\mu$ would produce some agreement and will stay with (5.18) or equivalently (**5S**) due to calculations in Remark 7.5.5. ∎

REMARK 7.5.4. We add here a few standard formulas involving derivatives; thus

$$(5.19) \qquad \nabla_\mu\lambda^\nu = \partial_\mu\lambda^\nu + \Gamma^\mu_{\rho\nu}\lambda^\rho;\ \nabla_\mu\lambda^\mu = \partial_\mu\lambda^\mu + \Gamma^\mu_{\rho\mu}\lambda^\rho\ (divergence);$$

$$\nabla_\mu\lambda_\nu = \partial_\mu - \Gamma^\rho_{\nu\mu}\lambda_\rho;\ \Gamma^\mu_{\rho\mu} = \partial_\rho log(\sqrt{g});\ \nabla_m\lambda^m = \frac{1}{\sqrt{g}}\partial_m(\sqrt{g}\lambda^m)$$

Also from (5.17)

(5.20) $$\Box \sim \frac{1}{\sqrt{-g}}\partial_\mu(\sqrt{-g}g^{\mu\nu}\partial_\nu) = \nabla_\mu g^{\mu\nu}\partial_\nu = \nabla_\mu\nabla^\mu$$

since $\nabla^\mu \sim \partial^\mu$ acting on functions (one could use $|g|$ instead of $\pm g$). ■

REMARK 7.5.5. We see that for $\bar{A}_\mu = 0$ the HJ equation (5.11) has the form (**5T**) $\partial_\mu S\partial^\mu S = m^2 - (1/6)\mathcal{R}$ and mention that it is shown in Castro [**270, 271**] that the 1/6 factor is essential if one wants a linear KG equation. We want now to identify $\mathcal{R}_w$ with a multiple of Q which should have a form like (**5U**) $Q \propto (1/\sqrt{\rho})\nabla^\mu\nabla_\mu(\sqrt{\rho})$. A crude calculation suggests

(5.21) $$\partial_\mu\partial^\mu\sqrt{\rho} = \partial_\mu\left[\frac{1}{2}\rho^{-1/2}\partial^\mu\rho\right] = \frac{1}{2}\left[-\frac{1}{2}\rho^{-3/2}\partial_\mu\rho\partial^\mu\rho + \rho^{-1/2}\partial_\mu\partial^\mu\rho\right] \Rightarrow$$
$$\Rightarrow \frac{\partial_\mu\partial^\mu\sqrt{\rho}}{\sqrt{\rho}} = \frac{1}{2}\left[-\frac{1}{2}\frac{\partial_\mu\rho\partial^\mu\rho}{\rho^2} + \frac{\partial_\mu\partial^\mu\rho}{\rho}\right]$$

and it is easy to check (cf. [**950**]) that $\nabla_m(fg^m) = (\nabla_m f)g^m + f(\nabla_m g^m)$. Hence $\nabla_m\nabla^m\sqrt{\rho}$ can be written out as in (5.21) to get

(5.22) $$\frac{\Box(\sqrt{\rho})}{\sqrt{\rho}} = \frac{1}{2}\left[-\frac{1}{2}\frac{\nabla_\mu\rho\nabla^\mu\rho}{\rho^2} + \frac{\Box(\rho)}{\rho}\right]$$

and hence from (5.18)

(5.23) $$\mathcal{R}_w = -3\left[\frac{1}{2}\frac{\nabla_\mu\rho\nabla^\mu\rho}{\rho^2} + \nabla_\mu\left(\frac{\nabla^\mu\rho}{\rho}\right)\right] = -3\left[\frac{1}{2}\frac{\nabla_\mu\rho\nabla^\mu\rho}{\rho^2} + \right.$$
$$\left. + \frac{\nabla_\mu\nabla^\mu\rho}{\rho} - \frac{\nabla_\mu\rho\nabla^\mu\rho}{\rho^2}\right] = -6\frac{\Box(\sqrt{\rho})}{\sqrt{\rho}}$$

The formula for Q is then (**5V**) $Q = -[\Box(\sqrt{\rho})/\sqrt{\rho}] = (1/6)\mathcal{R}_w$. We remark that in various contexts formulas for Q arise with multipliers $1/m^2$, $\hbar^2/2m$, etc. (cf. [**254**] and remarks below). ■

6. EMERGENCE OF Q IN GEOMETRY

In [**254**] and Chapters 1-6 we have indicated a number of contexts where Q arises in geometrical situations involving KG type equations and we review this here (cf. also [**250**]). We list a number of occasions (while omitting others).

(1) We omit any details for the Faraggi-Matone (FM) approach (see [**254, 446, 447, 448, 449**]) although it involves a whole philosophy (of considerable importance). Thus for $\eta^{\mu\nu} = diag(-1,1,1,1)$ and $q = (ct, q_1, q_2, q_3)$ one has

(6.1) $$\frac{1}{2m}\eta^{\mu\nu}\partial_\mu S^{cl}\partial_\nu S^{cl} + \mathfrak{W}'_{rel} = 0;$$

with $\mathfrak{W}'_{rel} = \frac{1}{2mc^2}[m^2c^4 - V^2(q) - 2cV(q)\partial_0 S^{cl}]$ where V is some potential which we could take to be zero. The quantum version attaches Q to (3.1) to get ($S^{cl} \to S$)

(6.2) $$\frac{1}{2m}(\partial S)^2 + \mathfrak{W}_{rel} + Q = 0;\ \mathfrak{W}_{rel} = \frac{1}{2mc^2}[m^2c^4 - V^2 - 2cV\partial_0 S]$$

This involves then

(6.3) $$\mathfrak{W}_{rel} = \left(\frac{\hbar^2}{2m}\right)\frac{\Box(Re^{iS/\hbar})}{Re^{iS/\hbar}};\ Q = -\frac{\hbar^2}{2m}\frac{\Box R}{R};\ \partial\cdot(R^2\partial S) = 0$$

where one uses $\partial \sim \nabla$ when $g_{\mu\nu} = \eta_{\mu\nu}$.

(2) One can derive the SE, the KG equation, and the Dirac equation using methods of scale relativity (cf. [**250, 254, 287, 336, 941, 942, 1028**]); here e.g. quantum paths are considered to be continuous nondifferentiable curves with left and right derivatives at any point. Using a "diffusion" coefficient $D = \hbar/2m$ as in the Nelson theory (cf. [**250, 254, 926**]) one defines "average" velocities $V = (1/2)[d_+x(t) + d_-x(t)]$ and $U = (1/2)[d_+x(t) - d_-x(t)]$. Then e.g. there is a SE $i\hbar\psi_t = -(\hbar^2/2m)\Delta\psi + \mathfrak{U}\psi$ with quantum potential $Q = -(m/2)U^2 - (\hbar/2)\partial U$ where $U = (\hbar/m)(\partial\sqrt{\rho}/\sqrt{\rho})$. The ideas should be extendible to a KG equation where $Q \sim (\hbar^2/m^2c^2)(\Box_g|\psi|/|\psi|)$ (see below).

(3) One can construct directly a KG theory following Nikolić [**930, 931**] in the form $(\partial_0^2 - \nabla^2 + m^2)\phi = 0$ where $\eta_{\mu\nu} = (1,-1,-1,-1)$. If $\psi = \phi^+$ with $\psi^* = \phi^-$ correspond to positive and negative frequency parts of $\phi = \phi^+ + \phi^-$ the particle current is $j_\mu = i\psi^*\overleftrightarrow{\partial}_\mu\psi$ and $N = \int d^3x j_0$ is the particle number. Trajectories have the form $d\mathbf{x}/dt = \mathbf{j}(t,\mathbf{x})/j_0(t,\mathbf{x})$ for $t = x_0$ and for $c = \hbar = 1$ one arrives at

(6.4) $$\partial^\mu(R^2\partial_\mu S) = 0;\ \frac{(\partial^\mu S)(\partial_\mu S)}{2m} - \frac{m}{2} + Q = 0;\ Q = -\frac{1}{2m}\frac{\partial^\mu\partial_\mu R}{R}$$

(4) A covariant field theoretic version is also given in [**931**] using deDonder-Weyl theory (cf. also [**250, 254**]). One works with a real scalar field $\phi(x)$ and defines (**6A**) $\mathfrak{A} = \int d^4x\mathfrak{L}$; $\mathfrak{L} = (1/2)(\partial^\mu\phi)(\partial_\mu\phi) - V(\phi)$ with (**6B**) $\pi^\mu = \partial\mathfrak{L}/\partial(\partial_\mu\phi) = \partial^\mu\phi$, $\partial_\mu\phi = \partial\mathfrak{H}/\partial\pi^\mu$, and $\partial_\mu\pi^\mu = -\partial\mathfrak{H}/\partial\phi$. One takes a preferred foliation of spacetime with R^μ normal to the leaf Σ and writes (**6C**) $\mathfrak{R}([\phi],\Sigma) = \int_\Sigma d\Sigma_\mu R^\mu$ with $\mathfrak{S}([\phi],\Sigma) = \int_\Sigma d\Sigma_\mu S^\mu$ and $\Psi = \mathfrak{R}exp(i\mathfrak{S}/\hbar)$. A covariant version of Bohmian mechanics ensues with

(6.5) $$\frac{1}{2}\frac{dS_\mu}{d\phi}\frac{dS^\mu}{d\phi} + V + Q + \partial_\mu S^\mu = 0;\ \frac{dR^\mu}{d\phi}\frac{dS^\mu}{d\phi} + J + \partial_\mu R^\mu = 0$$

(6.6) $$Q = -\frac{\hbar^2}{2\mathfrak{R}}\frac{\delta^2\mathfrak{R}}{\delta_\Sigma\phi^2};\ J = \frac{\mathfrak{R}}{2}\frac{\delta^2\mathfrak{S}}{\delta_\Sigma\phi^2}$$

The nature of this approach as a covariant version of the Bohmian hidden variable theory is spelled out in Nikolić [**931**]. This is a significant extension of earlier classical field theoretic approaches and another lovely extension is described by Nikolic in [**932**] involving a covariant many fingered time Bohmian interpretation of quantum field theory (QFT).

We preface the next set of examples with discussion of a formula $\mathfrak{M}^2 = m^2exp(\mathfrak{Q}_{rel})$ used in F. and A. Shojai [**1160**] in an important manner and produced also in Noldus [**940**]. This formula differs from the result $\mathfrak{M} = mexp(\mathfrak{Q}_{rel})$ of [**1158**] (which was abandoned in [**1111**]) and in order to clarify this we write out in more detail the approach of [**940**]. Thus one is dealing with a Bohmian theory and for a Klein-Gordon (KG) equation a wave function $\psi = Rexp(iS/\hbar)$ this leads to

$$\partial_\mu(R^2\partial^\mu S) = 0;\ \partial_\mu S\partial^\mu S = \mathfrak{M}^2c^2\ (\sim m^2c^2(1+\mathfrak{Q}_{rel})) \tag{6.7}$$

where $\mathfrak{Q}_{rel} = (\hbar^2/m^2c^2)(\partial_\mu\partial^\mu R/R)$ and (temporarily now) $\partial_\mu\partial^\mu \sim \Box = (1/c^2)\partial_t^2 - \Delta$ where $\eta_{\alpha\beta} = diag(1,-1,-1,-1)$. Now $\mathfrak{M} = 1 + \mathfrak{Q}_{rel}$ is only an approximation (leading e.g. to tachyon problems) and a better formula for $\mathfrak{M}$ can be found as follows. Thus (**6D**) $(dx^\mu(\tau)/d\tau) = (1/m)\partial^\mu S$ and differentiating one obtains

$$\partial_\tau\partial^\mu S = \partial_\nu\partial^\mu S\frac{dx^\mu}{d\tau} = \partial_\nu\mathfrak{M}\frac{dx^\nu}{d\tau}\frac{dx^\mu}{d\tau} + \mathfrak{M}\frac{d^2x^\mu}{d\tau^2} \tag{6.8}$$

But via the formula (valid for $g_{ab} = \eta_{ab}$ constant

$$\partial_b(\partial_a S\partial^a S) = (\partial_b\partial_a S)(\partial^a S) + (\partial_a S)(\partial_b\partial^a S); \tag{6.9}$$
$$\partial_a\partial_b\partial^a S = \partial_a S\eta^{ac}\partial_c\partial_b S = \partial^c S\partial_c\partial_b S$$

one has (**6E**) $\partial_\nu(\partial_\mu S\partial^\mu S) = 2(\partial^\mu S)(\partial_\mu\partial_\nu S)$ and therefore

$$\partial_\nu(\partial_\mu S\partial^\mu S) = \partial_\nu(\mathfrak{M}^2c^2) = 2\mathfrak{M}\partial_\nu\mathfrak{M}c^2 = 2(\partial^\nu S)(\partial_\mu\partial_\nu S) = \tag{6.10}$$
$$= 2\mathfrak{M}\frac{dx^\mu}{d\tau}(\partial_\mu\partial_\nu S)$$

Hence (**6F**) $\partial_\nu\mathfrak{M}c^2 = (\partial_\mu\partial_\nu S)(dx^\mu/d\tau)$ which implies

$$\eta^{\alpha\nu}c^2\partial_\nu\mathfrak{M} = \eta^{\alpha\nu}\partial_\mu\partial_\nu S(dx^\mu/d\tau) = \partial_\mu\partial^\alpha S(dx^\mu/d\tau) \tag{6.11}$$

Consequently (8.2) becomes

$$\eta^{\alpha\nu}c^2\partial_\nu\mathfrak{M} = \partial_\nu\mathfrak{M}\frac{dx^\nu}{d\tau}\frac{dx^\alpha}{d\tau} + \mathfrak{M}\frac{d^2x^\alpha}{d\tau^2} \equiv \tag{6.12}$$
$$\equiv \mathfrak{M}\frac{d^2x^\alpha}{d\tau^2} = \left(c^2\eta^{\alpha\nu} - \frac{dx^\nu}{d\tau}\frac{dx^\alpha}{d\tau}\right)\partial_\nu\mathfrak{M}$$

and this is equation (9) of [**940**]. For $|\dot{x}^\alpha| << c$ one obtains then $\mathfrak{M}\ddot{x}^\alpha \sim c^2\partial^\alpha\mathfrak{M} \sim -c^2\partial_\alpha\mathfrak{M}$ and comparing with the nonrelativistic equation $m\ddot{x}^\alpha = -\partial_\alpha Q_{cl}$ implies $\mathfrak{M} \sim mexp(\mathfrak{Q}_{cl}/mc^2)$ and suggests that $\mathfrak{M} \sim mexp(\mathfrak{Q}_{rel}/2)$ (recall $\mathfrak{Q}_{cl} = -(\hbar^2/2m)(\nabla^2|\psi|/|\psi|)$).

Now one observes that the quantum effects will affect the geometry and in fact are equivalent to a change of spacetime metric

$$g_{\mu\nu} \to \tilde{g}_{\mu\nu} = (\mathfrak{M}^2/m^2)g_{\mu\nu} \tag{6.13}$$

(conformal transformation). The QHJE becomes $\tilde{g}^{\mu\nu}\tilde{\nabla}_\mu S\tilde{\nabla}_\nu S = m^2c^2$ where $\tilde{\nabla}_\mu$ represents covariant differentiation with respect to the metric $\tilde{g}_{\mu\nu}$ and the continuity equation is then $\tilde{g}_{\mu\nu}\tilde{\nabla}_\mu(\rho\tilde{\nabla}_\nu S) = 0$. The important conclusion here is that the presence of the quantum potential is equivalent to a curved spacetime with its metric given by (6.13). This is a geometrization of the quantum aspects of matter and it seems that there is a dual aspect to the role of geometry in physics. The spacetime geometry sometimes looks like "gravity" and sometimes reveals quantum behavior. The curvature due to the quantum potential may have a large influence on the classical contribution to the curvature of spacetime. The particle trajectory can now be derived from the guidance relation via differentiation as in (**6D**) again, leading to the Newton equations of motion

$$\mathfrak{M}\frac{d^2x^\mu}{d\tau^2} + \mathfrak{M}\Gamma^\mu_{\nu\kappa}u^\nu u^\kappa = (c^2 g^{\mu\nu} - u^\mu u^\nu)\nabla_\nu \mathfrak{M} \tag{6.14}$$

Using the conformal transformation above this reduces to the standard geodesic equation.

We extract now from [**254, 1151, 1152, 1153, 1154, 1155, 1156, 1157, 1159, 1160, 1161, 1162, 1163, 1164, 1165, 1166, 1168**] with emphasis on the survey article [**1160**]. This material (partially summarized in [**1160**]) seems worthy of further emphasis in considering the ideas of "quantum gravity". Thus a general "canonical" relativistic system consisting of gravity and classical matter (no quantum effects) is determined by the action

$$\mathcal{A} = \frac{1}{2\kappa}\int d^4x\sqrt{-g}\mathcal{R} + \int d^4x\sqrt{-g}\frac{\hbar^2}{2m}\left(\frac{\rho}{\hbar^2}\mathcal{D}_\mu \mathcal{S}\mathcal{D}^\mu \mathcal{S} - \frac{m^2}{\hbar^2}\rho\right) \tag{6.15}$$

where $\kappa = 8\pi G$ and $c = 1$ for convenience and $\mathcal{D}_\mu$ is the covariant derivative based on $g_{\mu\nu}$ ($\mathcal{D}_\mu \sim \nabla_\mu$). It was seen above that via deBroglie the introduction of a quantum potential is equivalent to introducing a conformal factor $\Omega^2 = \mathfrak{M}^2/m^2$ in the metric. Hence in order to introduce quantum effects of matter into the action (6.15) one uses this conformal transformation to get ($1 + Q \sim exp(Q)$ and $Q \sim (\hbar^2/c^2m^2)(\Box(\sqrt{\rho})/\sqrt{\rho})$ with $c = 1$ here)

$$\begin{aligned}\mathfrak{A} = \frac{1}{2\kappa}\int & d^4x\sqrt{-\bar{g}}(\bar{\mathcal{R}}\Omega^2 - 6\bar{\nabla}_\mu\Omega\bar{\nabla}^\mu\Omega) + \\ & + \int d^4x\sqrt{-\bar{g}}\left(\frac{\rho}{m}\Omega^2\bar{\nabla}_\mu S\bar{\nabla}^\mu S - m\rho\Omega^4\right) + \\ & + \int d^4x\sqrt{-\bar{g}}\lambda\left[\Omega^2 - \left(1 + \frac{\hbar^2}{m^2}\frac{\bar{\Box}\sqrt{\rho}}{\sqrt{\rho}}\right)\right]\end{aligned} \tag{6.16}$$

where a bar over any quantity means that it corresponds to the nonquantum regime. Here only the first two terms of the expansion of $\mathfrak{M}^2 = m^2 exp(\mathfrak{Q})$ have been used, namely $\mathfrak{M}^2 \sim m^2(1 + \mathfrak{Q})$. λ is a Lagrange multiplier introduced to identify the conformal factor with its Bohmian value. One uses here $\bar{g}_{\mu\nu}$ to raise or lower indices and to evaluate the covariant derivatives; the physical metric (containing the quantum effects of matter) is $g_{\mu\nu} = \Omega^2\bar{g}_{\mu\nu}$. By variation of the action with respect to $\bar{g}_{\mu\nu}$, Ω, ρ, S, and λ one arrives at quantum equations of motion, including quantum Einstein equations (cf. [**254, 1160**]). There is a generalized equivalence principle. The gravitational effects determine the causal structure of spacetime as long as quantum effects give its conformal structure. This does not mean that quantum effects have nothing to do with the causal structure; they can act on the causal structure through back reaction terms appearing in the metric field equations. The conformal factor of the metric is a function of the quantum potential and the mass of a relativistic particle is a field produced by quantum corrections to the classical mass. One has shown that the presence of the quantum potential is equivalent to a conformal mapping of the metric. Thus in different conformally related frames one "feels" different quantum masses and different curvatures. In particular there are two frames with one containing the quantum mass field and the classical metric while the other contains the classical mass and the quantum metric. In general frames both the spacetime metric and the mass field have quantum properties so one can state that different conformal frames are identical pictures of the gravitational and quantum phenomena. One "feels" different quantum forces in different conformal frames. The question then arises of whether the geometrization of quantum effects implies conformal invariance just as gravitational effects imply general coordinate invariance. One sees here that Weyl geometry provides additional degrees of freedom which can be identified with quantum effects and seems to create a unified geometric framework for understanding both gravitational and quantum forces. Some features here are: (i) Quantum effects appear independent of any preferred length scale. (ii) The quantum mass of a particle is a field. (iii) The gravitational constant is also a field depending on the matter distribution via the quantum potential. (iv) A local variation of matter field distribution changes the quantum potential acting on the geometry and alters it globally; the nonlocal character is forced by the quantum potential (cf. [**254, 1160, 1154**]).

Next (still following [**1160**]) one goes to Weyl geometry based on the Weyl-Dirac action

$$\mathfrak{A} = \int d^4x\sqrt{-g}(F_{\mu\nu}F^{\mu\nu} - \beta^2\ {}^W\mathcal{R} + (\sigma+6)\beta_{;\mu}\beta^{;\mu} + \mathfrak{L}_{matter}) \tag{6.17}$$

Here $F_{\mu\nu}$ is the curl of the Weyl 4-vector ϕ_μ, σ is an arbitrary constant and β is a scalar field of weight -1. The symbol ";" represents a covariant derivative under general coordinate and conformal transformations (Weyl covariant derivative) defined as $X_{;\mu} = {}^W\nabla_\mu X - \mathcal{N}\phi_\mu X$ where $\mathcal{N}$ is the Weyl weight of X. The equations of motion are then given in [**234, 1108**]. There is then agreement with the Bohmian theory provided one identifies

$$\beta \sim \mathfrak{M};\ \ \frac{8\pi\mathfrak{T}}{\mathcal{R}} \sim m^2;\ \ \frac{1}{\sigma\phi_\alpha\phi^\alpha - (\mathcal{R}/6)} \sim \alpha = \frac{\hbar^2}{m^2c^2} \tag{6.18}$$

Thus β is the Bohmian quantum mass field and the coupling constant α (which depends on $\hbar$) is also a field, related to geometrical properties of spacetime. One notes that the quantum effects and the length scale of the spacetime are related. To see this suppose one is in a gauge in which the Dirac field is constant; apply a gauge transformation to change this to a general spacetime dependent function, i.e.

$$\beta = \beta_0 \to \beta(x) = \beta_0 exp(-\Xi(x));\ \ \phi_\mu \to \phi_\mu + \partial_\mu\Xi \tag{6.19}$$

Thus the gauge in which the quantum mass is constant (and the quantum force is zero) and the gauge in which the quantum mass is spacetime dependent are related to one another via a scale change. In particular ϕ_μ in the two gauges differ by $-\nabla_\mu(\beta/\beta_0)$ and since ϕ_μ is a part of Weyl geometry and the Dirac field represents the quantum mass one concludes that the quantum effects are geometrized which shows that ϕ_μ is not independent of β so the Weyl vector is determined by the quantum mass and thus the geometrical aspects of the manifold are related to quantum effects).

In [**1160, 1109**] (cf. also [**254**]) one can write down a scalar tensor theory where the conformal factor and the quantum potential are both dynamical fields but first we deal with (6.16). For the relativistic situation one will have e.g. $\mathfrak{Q} = (\hbar^2/m^2c^2)(\Box_g|\psi|/|\psi|)$ where $\Box_g|\psi| \sim \nabla_\alpha\nabla^\alpha|\psi| = g^{\alpha\beta}\nabla_\beta\nabla_\alpha|\psi|$ and the HJ equation is $\nabla_\mu S\nabla^\mu S = \mathfrak{M}^2c^2$ where $\mathfrak{M}^2 = m^2exp(\mathfrak{Q})$. Equivalently $\tilde{g}^{\mu\nu}\tilde{\nabla}_\mu S\tilde{\nabla}_\nu S = m^2c^2$ where $g_{\mu\nu} = (\mathfrak{M}/m)^2\tilde{g}_{\mu\nu}$ and $\tilde{\nabla}_\mu$ is the covariant derivative with respect to $\tilde{g}_{\mu\nu}$. The corresponding geodesic equation is given via (6.14). We write $\Omega^2 = (\mathfrak{M}/m)^2$ and this leads to (6.16) based on the fundamental action (6.15). Recall here $exp(\mathfrak{Q}) \sim m^2(1+\mathfrak{Q})$ has been used for $\mathfrak{M}$ in the last term in (6.16). We recall also the fundamental equations determined by varying the action (6.16) with respect to $\bar{g}_{\mu\nu}$, Ω, ρ, S, and λ are (cf. [**254, 1160**])

(1) The equation of motion for Ω

$$\bar{\mathcal{R}}\Omega + 6\,\bar{\Box}\,\Omega + \frac{2\kappa}{m}\rho\Omega(\bar{\nabla}_\mu S\bar{\nabla}^\mu S - 2m^2\Omega^2) + 2\kappa\lambda\Omega = 0 \tag{6.20}$$

(2) The continuity equation for particles $\bar{\nabla}_\mu(\rho\Omega^2\bar{\nabla}^\mu S) = 0$
(3) The equations of motion for particles

$$(\bar{\nabla}_\mu S\bar{\nabla}^\mu S - m^2\Omega^2)\Omega^2\sqrt{\rho} + \frac{\hbar^2}{2m}\left[\bar{\Box}\left(\frac{\lambda}{\sqrt{\rho}}\right) - \lambda\frac{\bar{\Box}\,\sqrt{\rho}}{\rho}\right] = 0 \tag{6.21}$$

(4) The modified Einstein equations for $\bar{g}_{\mu\nu}$

$$\Omega^2\left[\bar{\mathcal{R}}_{\mu\nu} - \frac{1}{2}\bar{g}_{\mu\nu}\bar{\mathcal{R}}\right] - [\bar{g}_{\mu\nu}\,\bar{\Box} - \bar{\nabla}_\mu\bar{\nabla}_\nu]\Omega^2 - 6\bar{\nabla}_\mu\Omega\bar{\nabla}_\nu\Omega + 3\bar{g}_{\mu\nu}\bar{\nabla}_\alpha\Omega\bar{\nabla}^\alpha\Omega + \tag{6.22}$$

$$+\frac{2\kappa}{m}\rho\Omega^2\bar{\nabla}_\mu S\bar{\nabla}_\nu S - \frac{\kappa}{m}\rho\Omega^2\bar{g}_{\mu\nu}\bar{\nabla}_\alpha S\bar{\nabla}^\alpha S + \kappa m\rho\Omega^4\bar{g}_{\mu\nu} +$$

$$+\frac{\kappa\hbar^2}{m^2}\left[\bar{\nabla}_\mu\sqrt{\rho}\bar{\nabla}_\nu\left(\frac{\lambda}{\sqrt{\rho}}\right) + \bar{\nabla}_\nu\sqrt{\rho}\bar{\nabla}_\mu\left(\frac{\lambda}{\sqrt{\rho}}\right)\right] - \frac{\kappa\hbar^2}{m^2}\bar{g}_{\mu\nu}\bar{\nabla}_\alpha\left[\lambda\frac{\bar{\nabla}^\alpha\sqrt{\rho}}{\sqrt{\rho}}\right] = 0$$

(5) The constraint equation $\Omega^2 = 1 + (\hbar^2/m^2)[(\bar{\Box}\,\sqrt{\rho})/\sqrt{\rho}]$

Thus the back reaction effects of the quantum factor on the background metric are contained in these highly coupled equations. A simpler form of (6.20) can be obtained by taking the trace of (6.21) and using (6.20) which produces $\lambda = (\hbar^2/m^2)\bar{\nabla}_\mu[\lambda(\bar{\nabla}^\mu\sqrt{\rho})/\sqrt{\rho}]$. A solution of this via perturbation methods using the small parameter $\alpha = \hbar^2/m^2$ yields the trivial solution $\lambda = 0$ so the above equations reduce to

$$\bar{\nabla}_\mu(\rho\Omega^2\bar{\nabla}^\mu S) = 0;\ \bar{\nabla}_\mu S\bar{\nabla}^\mu S = m^2\Omega^2;\ \mathfrak{G}_{\mu\nu} = -\kappa\mathfrak{T}^{(m)}_{\mu\nu} - \kappa\mathfrak{T}^{(\Omega)}_{\mu\nu} \tag{6.23}$$

where $\mathfrak{T}^{(m)}_{\mu\nu}$ is the matter energy-momentum (EM) tensor and

$$\kappa\mathfrak{T}^{(\Omega)}_{\mu\nu} = \frac{[g_{\mu\nu}\Box - \nabla_\mu\nabla_\nu]\Omega^2}{\Omega^2} + 6\frac{\nabla_\mu\Omega\nabla_\nu\Omega}{\omega^2} - 2g_{\mu\nu}\frac{\nabla_\alpha\Omega\nabla^\alpha\Omega}{\Omega^2} \tag{6.24}$$

with $\Omega^2 = 1+\alpha(\Box\sqrt{\rho}/\sqrt{\rho})$. Note that the second relation in (6.23) is the Bohmian equation of motion and written in terms of $g_{\mu\nu}$ it becomes $\nabla_\mu S \nabla^\mu S = m^2c^2$. Many examples with a lot of expansion is to be found in [**254, 1160**] and references there.

6.1. OTHER GEOMETRIC ASPECTS. The quantum potential arises in many geometrical and cosmological situations and we mention a few of these here.

(1) We have written about the Wheeler-deWitt (WDW) equation and the QP in [**254, 258, 259, 255**] at some length and in [**254**] have discussed the QP in related geometric situations following [**1023, 1022, 1025, 1026, 1119, 1156, 1163**] in particular. For background information on WDW we refer to Kiefer [**749**] (cf. also Chapters 6 and 8). One thinks of an ADM situation with (**6G**) $ds^2 = -(N^2 - h^{ij}H_iN_j)dt^2 + 2N_idx^idt + h_{ij}dx^idx^j$ and the deWitt metric (**6H**) $G_{ijk\ell} = (1/\sqrt{h})h_{ik}h_{j\ell} + h_{i\ell}h_{jk} - h_{ij}h_{k\ell})$. Given a wave function $\psi = \sqrt{P}exp(iS/\hbar)$ where P can be thought of in terms of momentum fluctuations $(1/P)(\delta P/\delta h_{ij})$ one finds a quantum potential

$$Q = -\frac{\hbar^2}{2}P^{-1/2}\frac{\delta}{\delta h_{ij}}\left(G_{ijk\ell}\frac{\delta P^{1/2}}{\delta h_{k\ell}}\right) \tag{6.25}$$

This is related to an intimate connection between Q and Fisher information based on techniques of Hall and Reginatto (cf. [**595, 596, 1062, 1063, 1065**]. The WDW equation is

$$\left[-\frac{\hbar^2}{2}\frac{\delta}{\delta h_{ij}}G_{ijk\ell}\frac{\delta}{\delta h_{k\ell}} + V\right]\psi = 0; \tag{6.26}$$

and there is a lovely relation

$$\int \mathfrak{D}hPQ = -\int \mathfrak{D}h\frac{\delta P^{1/2}}{\delta h_{ij}}G_{ijk\ell}\frac{\delta P^{1/2}}{\delta h_{k\ell}} \tag{6.27}$$

where the last term is Fisher information (cf. [**254, 255, 490, 595, 596, 1062, 1063, 1065**]).

(2) In F. and A. Shojai [**1157**] one uses again the attractive sandwich ordering and considers WDW in the form

$$\left[h^{-q}\frac{\delta}{\delta h_{ij}}h^qG_{ijk\ell}\frac{\delta}{\delta h_{k\ell}} + \sqrt{h}^{(3)}\mathcal{R} + \right. \tag{6.28}$$
$$\left. + \frac{1}{2\sqrt{h}}\frac{\delta^2}{\delta\phi^2} - \frac{1}{2}\sqrt{h}h^{ij}\partial_i\phi\partial_j\phi - \frac{1}{2}\sqrt{h}V(\phi)\right]\psi = 0$$

with momentum constraint (**6I**) $i[2\nabla_j(\delta/\delta h_{ij}) - h^{ij}\partial_j\phi(\delta/\delta\phi)]\psi = 0$ where ϕ is a matter field, q is an ordering parameter, and $h = det(h_{ij})$. Putting this in "polar" form $\psi = \sqrt{\rho}exp(iS/\hbar)$ leads to

$$G_{ijk\ell}\frac{\delta S}{\delta h_{ij}}\frac{\delta S}{\delta h_{k\ell}} + \frac{1}{2\sqrt{h}}\left(\frac{\delta S}{\delta\phi}\right)^2 - \sqrt{h}({}^{(3)}\mathcal{R} - \mathfrak{Q}_G) + \tag{6.29}$$
$$+\frac{\sqrt{h}}{2}h^{ij}\partial_i\phi\partial_j\phi + \frac{\sqrt{h}}{2}(V(\phi) - \mathfrak{Q}_M) = 0$$

where the gravity and matter quantum potentials are given via

(6.30) $$\mathfrak{Q}_G = -\frac{1}{\sqrt{\rho h}}\left(G_{ijk\ell}\frac{\delta^2\sqrt{\rho}}{\delta h_{ij}\delta h_{k\ell}} + h^{-q}\frac{\delta h^q G_{ijk\ell}}{\delta h_{ij}}\frac{\delta\sqrt{\rho}}{\delta h_{k\ell}}\right);\ \mathfrak{Q}_M = -\frac{1}{h\sqrt{\rho}}\frac{\delta^2\sqrt{\rho}}{\delta\phi^2}$$

There is a continuity equation

(6.31) $$\frac{\delta}{\delta h_{ij}}\left[2h^q G_{ijk\ell}\frac{\delta S}{\delta h_{k\ell}}\rho\right] + \frac{\delta}{\delta\phi}\left[\frac{h^q}{\sqrt{h}}\frac{\delta S}{\delta\phi}\rho\right] = 0$$

and the momentum constraint leads to equations (**6J**) $2\nabla_j(\delta\sqrt{\rho}/\delta h_{ij}) - h^{ij}\partial_j\phi(\delta\sqrt{\rho}/\delta\phi) = 0$ and $2\nabla_j(\delta S/\delta h_{ij}) - h^{ij}\partial_j\phi(\delta S/\delta\rho) = 0$ while the Bohmian "guidance" equations are

(6.32) $$\frac{\delta S}{\delta h_{ij}} = \pi^{k\ell} = \sqrt{h}(K^{k\ell} - h^{k\ell}K);\ \frac{\delta S}{\delta\phi} = \pi_\phi = \frac{\sqrt{h}}{N^\perp}\dot{\phi} - \sqrt{h}\frac{N^i}{N^\perp}\partial_i\phi$$

where K^{ij} is the extrinsic curvature. Since in the WDW equation the wavefunction is in the ground state with zero energy the stability condition of the metric and matter field is (**6K**) $h^{ij}\partial_i\phi\partial_j\phi + V(\phi) - 2^3\mathcal{R} + \mathfrak{Q}_M + 2\mathfrak{Q}_G = 0$ which is a pure quantum solution (this follows from (4.5) by setting all functional derivatives of S to be zero). In F. and A. Shojai [**1157**] these equations are examined perturbatively and we refer to [**1164, 1160, 1161**] for discussion of the constraint algebra and related matters.

(3) In F. and A. Shojai [**1164**] one studies the constraint algebra and equations of motion based on a Lagrangian (★) $\mathfrak{L} = \sqrt{-g}\mathcal{R} = \sqrt{h}N({}^{(3)}\mathcal{R} + Tr(K^2) - (TrK)^2)$ where ${}^{(3)}\mathcal{R}$ is the 3-D Ricci scalar, K_{ij} the extrinsic curvature, and h the induced spatial metric. The canonical momentum of the 3-metric is given via (**6M**) $P^{IJ} = \partial\mathfrak{L}/\partial\dot{h}_{ij}) = \sqrt{h}(K^{ij} - h^{ij}TrK)$ and the classical Hamiltonian is (**6N**) $H = \int d^3x\mathfrak{H}$ with $mfH = \sqrt{h}(NC + N^iC_i)$. Here one has

(6.33) $$C = -{}^{(3)}\mathcal{R} + \frac{1}{h}\left(Tr(p^2) - \frac{1}{2}(Trp)^2)\right) = -2G_{\mu\nu}n^{mu}n^\nu;$$

$$C_i = -2^{(3)}\nabla^j\left(\frac{p_{ij}}{\sqrt{h}}\right) = -2G_{\mu i}n^\mu$$

where n^μ is normal given via $n^\mu = (1/N, -\mathbf{N}/N)$. To get the quantum version one takes $H \to H + Q$ ($\mathfrak{H} \to \mathfrak{H} + \mathfrak{Q}$ (where $Q = \int d^3x\mathfrak{Q}$) and

(6.34) $$\mathfrak{Q} = \hbar^2 NhG_{ijk\ell}\frac{1}{|\psi|}\frac{\delta^2|\psi|}{\delta h_{ij}\delta h_{k\ell}}$$

The classical constraints are then modified via $C \to C + (\mathfrak{Q}/\sqrt{h}N)$ and $C_i \to C_i$. We disregard the constraint algebra here and go some formulas for quantum Einstein equations. First there is an HJ equation

(6.35) $$G_{ijk\ell}\frac{\delta S}{\delta h_{ij}}\frac{\delta S}{\delta h_{k\ell}} - \sqrt{h}({}^{(3)}\mathcal{R} - \mathfrak{Q}) = 0$$

where S is the phase of the wave function and this leads to Bohm-Einstein equations

(6.36)
$$\mathcal{G}^{ij} = -\kappa \mathcal{T}^{ij} - \frac{1}{N}\frac{\delta(\mathcal{Q}_G + \mathcal{Q}_m}{\delta g_{ij}}; \ \mathcal{G}^{0\mu} = -\kappa \mathcal{T}^{0\mu} + \frac{\mathcal{Q}_G + \mathcal{Q}_m}{2\sqrt{-g}} g^{0\mu};$$
$$\mathcal{Q}_m = \hbar^2 \frac{N\sqrt{H}}{2}\frac{\delta^2|psi|}{\delta\phi^2}; \ \mathcal{Q}_G = \hbar^2 NhG_{ijk\ell}\frac{1}{|\psi|}\frac{\delta^2|\psi|}{\delta h_{ij}\delta h_{k\ell}}$$

These are the quantum version of the Einstein equations and since regularization here only affects the quantum potential (cf. [**1164**]) for any regularization the quantum Einstein equations are the same and one can write (**6O**) $\mathcal{G}^{\mu\nu} = -\kappa \mathcal{T}^{\mu\nu} + \mathfrak{S}^{mu\nu}$ with

(6.37)
$$\mathfrak{S}^{0\mu} = -\frac{\mathcal{Q}_G + \mathcal{Q}_m}{2\sqrt{-g}} g^{0\mu} = \frac{\mathcal{Q}}{2\sqrt{-g}} g^{0\mu}; \ \mathfrak{S}^{ij} = -\frac{1}{N}\frac{\delta Q}{\delta g_{ij}}$$

(4) There are also developments of Bohmian theory and quantum geometrodynamics in [**96, 105, 153, 326, 770, 1022, 1025, 1026, 1119, 1170**] (cf. [**254**] for some survey and more references). In [**1022, 1025, 1026**] for example one writes the WDW equation in the form

(6.38)
$$\left\{-\hbar^2\left[\kappa G_{ijk\ell}\frac{\delta}{\delta h_{ij}}\frac{\delta}{\delta h_{k\ell}} + \frac{1}{2}h^{-1/2}\frac{\delta^2}{\delta\phi^2}\right] + \right\}\psi(h_{ij},\phi) = 0;$$
$$V = h^{1/2}\left[-\kappa^{-1}(\mathcal{R}^{(3)} - 2\Lambda) + \frac{1}{2}h^{ij}\partial_i\phi\partial_j\phi + U(\phi)\right]$$

(questions of factor ordering and regularization are ignored here) with a constraint (**6P**) $-2h_{ij}\nabla_j(\delta\psi/\delta h_{ij}) + (\delta\psi)/\delta\phi)\partial_i\phi = 0$. Writing now $\psi = Rexp(iS/\hbar)$ (**6P**) leads to

(6.39)
$$-2h_{ij}\nabla_j(\delta S/\delta h_{ij}) + (\delta S/\delta\phi)\partial_i\phi = 0; \ -2h_{ij}\nabla_j(\delta R/\delta h_{ij}) + (\delta R/\delta\phi)\partial_i\phi = 0$$

and (4.14) yields

(6.40)
$$\kappa G_{ijk\ell}\frac{\delta S}{\delta h_{ij}}\frac{\delta S}{\delta h_{k\ell}} + \frac{1}{2}h^{-1/2}\left(\frac{\delta S}{\delta\phi}\right)^2 + V + Q = 0;$$
$$Q = -\frac{\hbar^2}{R}\left(\kappa G_{ijk\ell}\frac{\delta^2 R}{\delta h_{ij}h_{j\ell}} + \frac{h^{-1/2}}{2}\frac{\delta^2 R}{\delta\phi^2}\right);$$
$$\kappa G_{ijk\ell}\frac{\delta}{\delta h_{ij}}\left(R^2\frac{\delta S}{\delta h_{k\ell}}\right) + \frac{1}{2}\frac{\delta}{\delta\phi}\left(R^2\frac{\delta S}{\delta\phi}\right) = 0$$

(5) In F. and A. Shojai [**1168**] one picks up again the approach of (2) to find a pure quantum state leading to a static Einstein universe whose classical counterpart is flat spacetime. For WDW one uses a form of (4.4) (with $16\pi G = 1$ and $\mathcal{R}$ the 3-curvature scalar), namely

(6.41)
$$\hbar^2 h^{-q}\frac{\delta}{\delta h_{ij}}\left(h^q G_{ijk\ell}\frac{\delta\psi}{\delta h_{k\ell}}\right) + \sqrt{h}\mathcal{R}\psi + \frac{1}{\sqrt{h}}\mathcal{T}^{00}\left(\frac{-i\hbar\delta}{\delta\phi_a},\phi_a\right)\psi = 0;$$

with 3-diffeomorphism constraint (**6Q**) $2\nabla_j(\delta/\delta h_{ij})\psi - \mathcal{T}^{i0}(\delta/\delta\phi_a,\phi_a)\pi = 0$. $\mathcal{T}^{\mu\nu}$ is the energy momentum tensor of matter fields ϕ_a in which the matter is quantized by replacing its conjugate momenta by $-i\hbar\delta/\delta\phi_a$. For the causal interpretation one sets again $\psi = Rexp(iS/\hbar)$ to obtain

$$G_{ijk\ell}\frac{\delta S}{\delta h_{ij}}\frac{\delta S}{\delta h_{k\ell}} - \sqrt{h}(\mathcal{R} - Q_G) + \frac{1}{\sqrt{h}}(\mathcal{T}^{00}(\delta S/\delta\phi_a, \phi_a) + Q_M) = 0; \tag{6.42}$$

$$\frac{\delta}{\delta h_{ij}}\left(2h^q G_{ijk\ell}\frac{\delta S}{\delta h_{k\ell}}R^2\right) + \sum \frac{\delta}{\delta\phi_a}\left(h^{q-(1/2)}\frac{\delta S}{\delta\phi_a}R^2\right) = 0 \tag{6.43}$$

$$Q_G = -\frac{\hbar^2}{\sqrt{h}R}\left(h^{-q}\frac{\delta}{\delta h_{ij}}h^q G_{ijk\ell}\frac{\delta R}{\delta h_{k\ell}}\right);\ Q_M - \frac{\hbar^2}{hR}\sum\frac{\delta^2 R}{\delta\phi_a^2} \tag{6.44}$$

$$2\nabla_j\frac{\delta R}{\delta h_{ij}} - \mathcal{T}^{i0}(\delta R/\delta\phi_a, \phi_a) = 0;\ 2\nabla_j\frac{\delta S}{\delta h_{ij}} - \mathcal{T}^{i0}(\delta S/\delta\phi_a, \phi_a) = 0 \tag{6.45}$$

One notes that all terms containing the second functional derivative are ill defined and can be regulated via $(\delta/\delta h_{ij}(x))(\delta/\delta h_{ij}(x) \to \int d^3x\sqrt{h}U(x-x')(\delta/\delta h_{ij}(x))(\delta/\delta h_{ij}(x'))$ where U is the regulator. Finally the guidance equations are (**6R**) $\pi^{k\ell} = \sqrt{h}(K^{k\ell} - Kh^{k\ell}) = \delta S/\delta h_{k\ell}$ and $\pi_{\phi_a} = \delta S/\delta\phi_a$ where $K_{ij} = (1/2N)(\dot{h}_{ij} - \nabla_i N_j - \nabla_j N_i)$ is the extrinsic curvature. Using the quantum Hamilton-Jacobi-Einstein (HJE) equation one can define a limit, called the pure quantum limit, where the total quantum potential is of the same order as the total classical potential and they can cancel each other. In this case one has (**6S**) $\delta S/\delta h_{ij} = \delta S/\delta\phi_a = 0$ and the continuity is satisfied identically. The resulting trajectory is not similar to any classical solution and the quantum HJE equation for a pure quantum state is an equation for spatial dependence of the metric and matter fields in terms of the norm of the wave function. Explicit calculations are given for some special situations.

REMARK 7.6.1 Note that for $\eta_{ab} \sim (1, -1, -1, -1)$ and $\hbar = c = 1$ one has $\partial_0^2 - \nabla^2 \sim \Box$ and (**6T**) $(\nabla S)^2 = m^2[1 + (\Box R/m^2 R)]$. This agrees with

$$(\nabla S)^2 = \mathfrak{M}^2 c^2(1 + Q);\ Q = \frac{\hbar^2}{m^2c^2}\frac{\Box R}{R} \tag{6.46}$$

For the FM theory with $\eta_{ab} \sim (-1, 1, 1, 1)$ one has (**6U**) $(1/2m)(\nabla S)^2 + (mc^2/2) - (\hbar^2/2m)(\Box R/R) = 0$ from (3.2). But $\Box R \to -\Box R$ and $(\nabla S)^2 \sim \eta^{ab}\nabla_b S\nabla_a S \to -(\nabla S)^2$ for $\eta_{ab} \to -\eta_{ab}$. Hence in the $\eta_{ab} = (1, -1, -1, -1)$ notation one obtains $(\nabla S)^2 = m^2c^2[1 + (\hbar^2/m^2c^2)(\Box R/R)]$ as in (4.22). ■

REMARK 7.6.2. There is a lot of motivation here for using the quantum potential as a generator of quantum gravity (cf. Budiyono [**207**]) and also for considering the conformal factor $\mathfrak{M}^2/m^2$ as a generator of Ricci flow (cf. [**339, 560, 742, 743, 744, 1009**] and Chapter 10). ■

REMARK 7.6.3. One finds fascinating connections between Bohmian theory and phase space mechanics in [**201, 199, 551, 552, 554, 915, 916, 917, 1188**]. In Brown-Hiley [**201**] one argues that if the quantum potential (QP) reflects the quantum aspects of a sysem it should be possible to to identify such aspects within the QP and in particular one shows how the balance between localisation and dispersion energies suggests a link between the QP and the Heisenberg uncertainty principle. Recall first that from the SE (**6V**) $i\hbar\partial_t\psi = [-(\hbar^2/2m)\nabla^2 + V]\psi$ there follows

$$(6.47)\qquad \partial_t S + \frac{(\nabla S)^2}{2m} - \frac{\hbar^2}{2m}\frac{\nabla^2 R}{r} + V = 0;\ \partial_t\rho + \nabla\cdot\left(\rho\frac{p}{m}\right) = 0$$

where $\psi = Rexp(iS/\hbar)$, $\rho = |\psi|^2$, and $p = \nabla S$. The QP is manifestly of the form $Q = -(\hbar^2/2m)(\nabla^2 R/R)$ and one writes $F = -\nabla(Q+V)$ and $v = j/\rho = p/m$. Now consider a more general derivation of the QP by writing the SE in the form (**6W**) $i\hbar\partial_t\psi = (T(\hat{p}) + V(\hat{x}))\psi$. Setting again $\psi = Rexp(iS/\hbar)$ one obtains

$$(6.48)\qquad \partial_t S + \Re\left(\frac{T\psi}{\psi}\right) + V(x) = 0;\ \partial_t\rho - \frac{2\rho}{\hbar}\Im\left(\frac{T\psi}{\psi}\right) = 0$$

Correspondingly in the momentum space with $\hat{x} = i\hbar\nabla_p$ and $\hat{p} = p$ the real and imaginary parts of the SE are

$$(6.49)\qquad \partial_t S + T(p) + \Re\left(\frac{V\psi}{\psi}\right) = 0;\ \partial_t\rho - \frac{2\rho}{\hbar}\Im\left(\frac{V\psi}{\psi}\right) = 0$$

Then expanding exponentials one writes

$$(6.50)\qquad \Re\left(\frac{\psi^* T\psi}{\rho}\right) = \Re\left(\frac{R[1-(iS/\hbar)-\cdots)T(\hat{p})R(1+(iS/\hbar)-\cdots]}{\rho}\right)$$

(note the formal equivalence $T\psi/\psi = \psi^* T\psi/\rho$). If now $T(\hat{p})$ is a general but analytic function of $\hat{p}$ one can expand in a power series in $\hat{p} = -i\hbar\nabla$ and the kinetic term may be separated into the sum of two parts

$$(6.51)\qquad \Re\left(\frac{T\psi}{\psi}\right) = T_h(x) + T_0(x);\ T_0(x) = T(\nabla S)$$

where $T_h(x)$ is an expansion in even positive powers of $\hbar$ and $T_0(x)$ is independent of $\hbar$ and identifies $p = \nabla S$. The same line of argument allows the potential term of the HJ equation in (2.3) to be separated as

$$(6.52)\qquad \Re\left(\frac{V\psi}{\psi}\right) = V_h(p) + V_0(p);\ V_0(p) = V(-\nabla_p S)$$

where V_h is an expansion in even positive powers of $\hbar$ and $V_0(p)$ is independent of $\hbar$ and identifies $x = -\nabla_p S$. We pursue this further in [**287**]. ■

7. THE QUANTUM POTENTIAL AND GEOMETRY

We begin with F. and A. Shojai [**1160, 1161**] and recall some features of Weyl geometry (some of which are indicated already in previous sections). We remember first that vectors change in length and direction under translation via (**7A**) $\delta\ell = \phi_\mu\delta x^\mu\ell$ so $\ell = \ell_0 exp(\int\phi_\mu dx^\mu)$ where ϕ_μ is the Weyl vector. Equivalently (**7B**) $g_{\mu\nu} \to exp(2\int\phi_\mu\delta x^\mu)g_{\mu\nu}$ which is a conformal transformation. Recall also that the metric is a Weyl covariant object of weight 2 and the Weyl connection is given via

$$(7.1)\qquad \Gamma^\mu_{\nu\lambda} = \left\{ \begin{matrix} \mu \\ \nu\lambda \end{matrix} \right\} + g_{\nu\lambda}\phi^\mu - \delta^\mu_\nu\phi_\lambda - \delta^\mu_\lambda\phi_\nu$$

A gauge transformation (**7C**) $\phi_\mu \to \phi'_\mu = \phi_\mu + \partial_\mu\Lambda$ transforms $g_{\mu\nu} \to g'_{\mu\nu} = exp(2\Lambda)g_{\mu\nu}$ with $\delta\ell \to \delta\ell' = \delta\ell + (\partial_\mu\Lambda)dx^\mu\ell$. We have seen how quantum effects are geometrized via the Dirac field β and gauge transformations. Let us now make more explicit some direct relations between the quantum potential and geometric ideas via the Weyl vector. We recall that for the SE, $Q = -(m/2)\mathbf{u}^2 - (\hbar/2)\partial\mathbf{u}$ where $\mathbf{u}$ is an osmotic velocity (see also [**254, 528**]). Similarly for the KG equation one has $Q = (\hbar^2/m^2c^2)(\Box_g|\psi|/|\psi|$ for example as in Remark 7.6.1 and we will also consider an appropriate osmotic velocity for this situation.

Consider now the situation $Q = 0$ for the SE which can be expressed in several forms.

(1) Defining the osmotic velocity as $\mathbf{u} = D\nabla log(\rho)$ with $D = \hbar/2m$ from [**254, 508, 509**] (cf. also [**941, 942**]) one has then (**5A**) $(m/2)\mathbf{u}^2 + (\hbar/2)\nabla\mathbf{u} = 0$.
(2) Another form is directly (for $g = 1$) $\Delta\sqrt{\rho} = 0$.
(3) There is a general form for (**7C**) $\phi_i = -\partial_i log(\hat{\rho})$ with $\hat{\rho} = \rho/\sqrt{g}$, namely

$$\dot{\mathcal{R}} + 2\left[\phi_i\phi^i - \frac{2}{\sqrt{g}}\partial_i(\sqrt{g}\phi^i)\right] = 0 \tag{7.2}$$

When $\dot{\mathcal{R}} = 0$ with $\sqrt{g} = 1$ and $\hat{\rho} = \rho$ this becomes (**7D**) $\phi_i\phi^i - 2\partial_i\phi^i = 0$. Note $\phi^i \sim g^{ik}\phi_k = -g^{ik}\partial_k log(\hat{\rho}) = -\partial^k log(\hat{\rho})$.
(4) From (2.8) another form of (5.2) above is

$$\dot{\mathcal{R}} + \frac{8}{\sqrt{\rho}}\partial_i(\sqrt{g}g^{ik}\partial_k\sqrt{\hat{\rho}}) = 0 \tag{7.3}$$

(5) In view of [**254, 255, 339, 595, 596, 1062, 1063, 1065**] one can say that the fundamental quantum fluctuation or perturbation in momentum has the form (**7E**) $\delta p \sim c(\nabla\rho/\rho)$ and this means (**7F**) $\delta p \sim \hat{c}\mathbf{u}$ or equivalently $\delta p \sim \tilde{c}\vec{\phi}$. We can assume that in a Weyl space situation an osmotic velocity $\mathbf{u} = Dlog(\hat{\rho})$ is meaningful. The "obligatory" nature of $\delta p \sim c(\nabla\rho/\rho)$ is made even more striking in the developments in [**608, 897**]. One shows there in particular that a classical momentum can be written as

$$\hat{p}_{cl} = \hat{p} + \left(\frac{i\hbar}{2}\right)\left(\frac{\nabla\rho}{\rho}\right) \Rightarrow \hat{p}_{cl} = -i\hbar\left(\nabla - \frac{1}{2}\frac{\nabla(\psi^*\psi)}{\psi^*\psi}\right) \tag{7.4}$$

It should now be possible to extract some analytic and geometric features of the situation $Q = 0$.

EXAMPLE 7.1. We think of $\psi = \sqrt{\rho}exp(iS/\hbar)$ with $\sqrt{\rho} = R$. Take #2 first and look for solutions of $\Delta R = 0$ in a finite region Ω with $R \in H_0^1(\Omega)$ (Sobolev space) for example (see [**251, 437**] for techniques and results in PDE). For a QM situation $R = 0$ on $\partial\Omega$ and $H_0^1(\Omega)$ is the natural setting with $R \in L^2(\Omega)$. However by Green's theorem $\int_\Omega R\Delta R dV = -\int_\Omega |\nabla R|^2 dS = 0$ which implies $\nabla R = R = 0$. this is consistent with the Example 4.2 where ψ involves plane waves and L^2 solutions are meaningless. ■

EXAMPLE 7.2. Consider next a situation $(m/2)|\mathbf{u}|^2 + (\hbar/2)\nabla\mathbf{u} = 0$ or in 1-D $\partial u + cu^2 = 0$ with $c > 0$. Then $u'/u^2 = -c \Rightarrow u = (\hat{c} + cx)^{-1}$ and setting $u = D\rho'/\rho$ yields $R^2 = \rho = k(\hat{c} + cx)^d$. This is not reasonable for $R = 0$ outside of a finite Ω. ∎

EXAMPLE 7.3. Consider $\phi_i\phi^i - 2\partial\phi^i = 0$ or equivalently (in 1-D for convenience) $\phi'/\phi^2 = 1/2$ leading to $-(1/\phi) = (1/2)x + c$ and problems similar to those in the second of the above examples. ∎

We can however think of ρ, $\vec{\phi}$, or $\mathbf{u}$ as functions of Q so for each admisible Q there will be in principle some well determined R, modulo spectral conditions as in Remark 7.4.1.

REMARK 7.7.1. In Remark 7.4.1 we saw that determining R from Q involved solving $\Delta R + \beta QR = 0$ $(\beta > 0)$ in say $H_0^1(\Omega)$. If $Q \leq 0$ this yields a unique solution while if 0 is not in the spectrum of $\Delta + \beta Q$ then (**5G**) $\Delta R + \beta QR = 0$ has a unique solution for say $Q \in L^\infty(\Omega)$. We also saw that modulo solvability of (1.4) one would obtain a "generalized" quantum theory based on Q. We can improve the statement of this in Remark 7.4.1 by saying that, given solutions V and S of (1.4) (via a solution R of (**7G**)), in converting this to a SE one eliminates Q from the picture entirely. For R unique $S(x,t)$ is determined up to a function $f(t)$ and a function $g(x)$ arising from

$$S = -\int^t (Q+V)dt - \frac{1}{2R^4}\left[f(t) - \int^x \partial_t R^2 dx\right]^2 + g(x) \tag{7.5}$$

However V_x is known via (1.4) in terms of S_{xt} which depends only on Q (via R), f, and f', hence only in terms of one function $f(t)$. If then $V = V(x)$ it may actually be almost determined and Q, instead of determining only one trajectory based on R and S, actually could lead to the SE itself (modulo f) for $\psi = Rexp(iS/\hbar)$; if it were to be the case that $V = V(x)$ does not use $f(t)$ this means that Q alone would determine a "generalized" quantum theory via the SE! This could eliminate some of the ambiguity connected with the idea of using Q as a quantization. It would be worthwhile checking the equations to find such situations (see below). We note also that if Q contains t it is transmitted to R as a parameter in solving the elliptic equation; if Q is independent of t then of course so is R and this could conceivably simplify matters in determining $V = V(x)$. ∎

We check this last idea in more detail now. Assume Q is a function of x alone, $Q = Q(x)$, and let it determine a unique $R \in H_0^1(\Omega)$ (normalized so that $\int_\Omega R^2 dx = 1$). Then look at

$$S_t + \frac{1}{2m}S_x^2 + Q + V = 0;\ \partial_t R^2 + \frac{1}{m}(R^2 S_x)_x = 0 \tag{7.6}$$

The second equation becomes $(R^2S_x)_x = 0$ which implies (**7H**) $R^2 S_x = f(t)$ for some "arbitrary" $f(t)$. Then (**7I**) $S_t + (1/2m)(f^2/R^4) + Q + V = 0$ and we can eliminate S from (**7H**) and (**7I**) via

$$R^2 S_{xt} = f_t;\ S_{xt} - \frac{2f^2 R_x}{mR^5} + Q_x + V_x = 0 \Rightarrow \frac{2f^2}{m}\frac{R_x}{R^3} - f_t = R^2(Q_x + V_x) \tag{7.7}$$

This determines V_x in terms of $Q(x)$ and $f(t)$ so we ask whether $V = V(x)$ can occur (no t dependence). In such a case the t derivatives of the last equation in (5.6) are zero yielding (**7J**) $f_{tt} = (2R_x/mR^3)\partial_t f^2$. This means

$$\frac{f_{tt}}{\partial_t f^2} = F(t) = \frac{2R_x}{mR^3} = \mathfrak{F}(x) \tag{7.8}$$

Consequently $F(t) = \mathfrak{F}(x) = c$ and (**7K**) $f_{tt} = (c/2)\partial_t f^2$ while $(R_x/R^3) = (cm/2)$ leading to

$$f_t = \frac{c}{2}f^2 + \hat{c};\ R^2 = \frac{1}{\tilde{c} - cmx} \tag{7.9}$$

Thus $x > (\tilde{c}/cm)$ but $R \notin H_0^1(\Omega)$ for any Ω. This seem to preclude $V = V(x)$ (or perhaps $Q = Q(x)$). Hence from (5.6) one has at least (cf. also below).

PROPOSITION 7.1. Given $Q = Q(x)$ determining a unique $R(x) \in H_0^1(\Omega)$ it follows that V_x is determined up to an "arbitrary" function $f(t)$ via

$$V_x = \frac{1}{R^2}\left[\frac{2f^2R_x}{mR^3} - f_t\right] - Q_x \tag{7.10}$$

This situation precludes V being a function of x alone. ■

REMARK 7.7.2. Note if $\int_\Omega R^2(x,t)dx = r^2(t)$ then to get a proper normalization one would take $\mathcal{R}(x,t) = (1/r(t))R(x,t)$ and note that Q computed on $\mathcal{R}$ is equal via (**7L**) $\mathcal{Q} = -(\hbar^2/2m)(\mathcal{R}_{xx}/\mathcal{R}) = -(\hbar^2/2m)(R_{xx}/R)$. Note also that r is determined by Q via R. We still think of $\psi \sim Rexp(iS/\hbar)$ so (**7H**) becomes (**7M**) $RS_x^2 = -m\int^x \partial_t R^2 dx + f(t) = A(f,Q)$ (since Q determines R). This leads to (**5N**) $2RR_tS_x + R^2S_{xt} = \partial_t A$ while (**7O**) $S_{xt} + (1/m)S_xS_{xx} + Q_x + V_x = 0$. Now we eliminate S_{xx} and S_{xt} to get V_x in terms of Q and f. First from (**7N**) one has

$$S_{xt} = \frac{1}{R^2}\left\{\partial_t A - 2\frac{R_t}{R}A\right\} \tag{7.11}$$

while (**7P**) $R^2S_{xx} + 2RR_x = \partial_x A \Rightarrow S_{xx} = (1/R^2)(\partial_x A - 2RR_x)$. Hence one arrives at

$$\frac{1}{R^2}\left[\partial_t A - \frac{2AR_t}{R}\right] = -Q_x - V_x - \frac{1}{m}\frac{A}{R^4}(\partial_x A - 2RR_x) \tag{7.12}$$

and we can state

PROPOSITION 7.2. Defining $A(f,Q) = f(t) = m\int^x \partial_t R^2 dx$ with f "arbitrary" one can determine V_x via (7.12) as $V_x(Q,f)$. Hence $V = \int^x V_x dx + h(t)$ for h "arbitrary" provides a potential $V(Q,f,h)$ and the associated SE is determined completely by V. If choices $f = h = 0$ are "natural" one can say that Q determines a natural SE and a corresponding "generalized" quantum theory. ■

REMARK 7.7.3. Consider the stationary case (cf. [**254**]) (**7Q**) $(1/2m)S_x^2 + Q + V - E = 0$ with $(R^2S_x)_x = 0$ where $R = R(x)$ is say uniquely determined via $Q = Q(x)$ (note however that both R and Q must contain E as a parameter). Then

$$S_x = \frac{c}{R^2};\ \frac{1}{2m}\left(\frac{c^2}{R^4}\right) + Q + V - E = 0 \tag{7.13}$$

This means (**7R**) $1 = \partial_E Q - (c^2/mR^5)\partial_E R$ (since V does not depend on E) and hence

$$\frac{2c^2 R_E}{mR^5} + 1 = \frac{\hbar^2 R'' R_E}{2mR^2} - \frac{\hbar^2 R''_E}{2mR} \tag{7.14}$$

Viewed in terms of $\rho = R^2$ this mean that $\rho = \rho(E, x)$ and the corresponding Weyl geometry based on $\vec{\phi} = -\nabla log(\rho)$ will depend on E (as will Q of course). We refer here also to the quantum mass idea of Floyd, namely $m_Q = m(1 - \partial_E Q)$ for stationary situations (this is sketched in [**307**] for example and we refer to [**491**] for more details). Some further ideas about this are sketched below. In particular one knows that $Q \sim -(\hbar^2/2m)\mathcal{R}$ where $\mathcal{R}$ is the Ricci-Weyl curvature with $\hbar$ essentially put in by hand to conform to the wave function idea and operator QM. We see that the geometry of the space in which a trajectory transpires is thereby determined by E (not surprisingly) which seems to say that the probability distribution ρ is the basic unknown here (and in the time dependent situation). Once one has a probability distribution one can posit a wave function and insert $\hbar$. In fact (given V) the two equations $(1/2m)(c^2/R^2) + Q + V - E = 0$ and $Q = -(\hbar^2/2m)(R''/R)$ determine $R = R(E, x)$ directly ($\hbar$ being gratuitously inserted). ∎

8. OLAVO THEORY

We go here to Olavo [**954, 955**] and sketch some matters dealing with uncertainty and the SE (cf. also [**163, 238**]). In the first paper of [**950**] an axiomatic formulation for quantum mechanics (QM) is given. Consider ensembles described by probability density functions in phase space described as $F(x, p, t)$; assume

(1) Newtonian particle mechanics is valid for particles in the ensemble.
(2) For an isolated system $dF(x, p, t)/dt = 0$.
(3) The Wigner-Moyal infinitesimal transformation is defined via

$$Z_Q(x, \delta x/2, t) = \int F(x, p, t) exp\left(\frac{ip\delta x}{\ell}\right) dp \tag{8.1}$$

where ℓ is a parameter which will necessarily be equal to $\hbar$.

From these axioms one can derive nonrelativistic QM as follows. First using (1) and (2) one has

$$\frac{dF}{dt} = \partial_t F + \dot{x} F_x + \dot{p} F_p = 0;\ \dot{x} = \frac{p}{m};\ \dot{p} = f = -V_x \tag{8.2}$$

Multiplying by the exponential in (6.1) and integrating one arrives at

$$-\partial_t Z_Q + \frac{i\hbar}{m}\frac{\partial^2 Z_Q}{\partial x \partial(\delta x)} - \frac{i}{\hbar}\delta V(x) Z_Q = 0 \tag{8.3}$$

where the infinitesimal nature of δx is used to write

$$\partial_x V \delta x = \delta V(x) = V\left(x + \frac{\delta x}{2}\right) - V\left(x - \frac{\delta x}{2}\right); \tag{8.4}$$

and one knows that (★) $F(x,p,t)exp\left(\frac{ip\delta x}{\ell}\right)^{p=\infty}_{p=-\infty} = 0$ by the nature of probability distributions. Writing $y = x + (\delta x/2)$ and $y' = x - (\delta x/2)$ one can rewrite (6.3) as

$$\left\{\frac{\hbar^2}{2m}[\partial_y^2 - \partial_{y'}^2] - [V(y) - V(y')]\right\} Z_Q(y,y',t) = -i\hbar\partial_t Z_Q(yy',t) \tag{8.5}$$

This is called a Schrödinger equation (SE) for the characteristic function Z_Q and is valid for all values of y, y' as long as they are infinitesimally close.

Now suppose one can write (this is a fundamental assumption)

$$Z_Q(y,y',t) = \psi^*(y',t)\psi(y,t);\ \ \psi(y,t) = R(t,y)e^{iS(y,t)/\hbar} \tag{8.6}$$

Then expanding Z_Q one obtains

$$Z_Q(y,y',t) = \left\{R(x,t)^2 - \left(\frac{\delta x}{2}\right)^2\left[(\partial_x R)^2 - R\frac{\partial^2 R}{\partial x^2}\right]\right\} exp\left(\frac{i}{\hbar}(\delta x)S_x\right) \tag{8.7}$$

Putting this in (8.3) leads to

$$P_t + \partial_x\left(\frac{PS_x}{m}\right) = 0;\ \ \frac{i\delta x}{\hbar}\partial_x\left[\frac{S_x^2}{2m} + V + S_t - \frac{\hbar^2}{2mR}\frac{\partial^2 R}{\partial x^2}\right] = 0 \tag{8.8}$$

where $P(x,t) = R^2 = lim_{\delta x\to 0}Z_Q(x + (\delta x/2), x - (\delta x/2), t)$ is the probability distribution in configuration space. Equation (8.8b) can be rewritten as

$$\frac{1}{2m}S_x^2 + V + S_t + Q = f(t);\ \ Q = -\frac{\hbar^2}{2mR}\partial_x^2 R \tag{8.9}$$

One can eliminate $f(t)$ by redefining $S(x,t)$ as $S'(x,t) = S(x,t) + \int_0^t f(t')dt'$; since this is only a new definition of the energy reference level one can simply take $f(t) = 0$ and obtain as well (**8A**) $(\hbar^2/2m)\psi_{xx} - V(x)\psi = -i\hbar\psi_t$. Thus if one can write Z_Q as the product (8.6) then ψ satisfies the SE (**8A**) and (8.8) holds.

Now define operators on Z_Q via primes (distinguished from operators acting on the probability amplitude) so that $\hat{p}' = -i\hbar(\partial/\partial(\delta x))$ and $\hat{x}' = x$; this is based on the fact that

$$\bar{p} = \lim_{\delta x\to 0} -i\hbar\frac{\partial}{\partial(\delta x)}\int F(x,p,t)exp\left(i\frac{p\delta x}{\hbar}\right)dxdp \tag{8.10}$$

and $\bar{x} = lim_{\delta x\to 0}\int xF(x,p,t)exp[ip\delta x/\hbar]dxdp$. Thus the result of position and momentum operators acting on Z_Q represents a mean value calculation for the ensemble components. Observe now via (8.7) that

$$\bar{p} = lim_{\delta x\to 0}\left[-i\hbar\frac{\partial}{\partial(\delta x)}\int F(x,p,t)exp\left(i\frac{p\delta x}{\hbar}\right)dxdp\right] = \int R^2(x)S_x dx \tag{8.11}$$

One can rewrite (8.11) in the form

$$\hat{p} = lim_{\delta x\to 0}\left\{-i\hbar\int\frac{\partial}{\partial(\delta x)}\left[\psi^*\left(x - \frac{\delta x}{2}, t\right)\psi\left(x + \frac{\delta x}{2}, t\right)\right]dx\right\} \tag{8.12}$$

Again via (8.7) this leads to (**8B**) $\hat{p} = \int\psi^* x, t)(-i\hbar\partial_x)\psi(x,t)dx$ with similar calculations for the position operator. Consequently (**8C**) $\hat{p}\psi = -i\hbar\partial_x\psi$ and $\hat{x}\psi = x\psi$. Moreover some calculation shows that $[\hat{x},\hat{p}] = i\hbar$ and consequently

$\Delta x\Delta p \geq \hbar/2$ (cf. [**950**] for details).

Next we go to [**954, 955**] to sketch another derivation of the SE from more physically grounded axioms and then the derivations are connected. Thus begin with the Liouville equation for $F(x,p,t)$

$$\partial_t F + \frac{p}{m}\partial_x F - \partial_x V \partial_p F = 0 \tag{8.13}$$

Integrating in p and using the definitions (**8D**) $\int F dp = \rho(x,t)$ and $\int pF dp = p(x,t)\rho(x,t)$ one obtains

$$\partial_t \rho + \partial_x \left[\frac{p(x,t)}{m}\rho(x,t)\right] = 0 \tag{8.14}$$

Then multiply the Liouville equation by p and integrate to get (**8E**) $\int p^2 F(x,p,t)dp = M_2(x,t)$ with

$$\partial_t[\rho(x,t)p(x,t)] + \frac{1}{m}\partial_x M_2 + (\partial_x V)\rho(x,t) = 0 \tag{8.15}$$

Putting (8.14) into (8.15) one gets after some calculation

$$\frac{1}{m}\partial_x[M_2(x,t) - p^2(x,t)\rho(x,t)] + \rho(x,t)\left[\partial_t p(x,t) + \partial_x\left(\frac{p^2(x,t)}{2m}\right) + V_x\right] = 0 \tag{8.16}$$

One can write then

$$M_2 - p^2(x,t)\rho(x,t) = \int [p^2 - p^2(x,t)]F(x,p,t)dp = \int [p - p(x,t)]^2 F(x,p,t)dp \tag{8.17}$$

and set (**8F**) $\overline{(\delta p)^2}\rho(x,t) = \int [p - p(x,t)]^2 F(x,p,t)dp$. Then one wants to find a functional expression for $\overline{(\delta p)^2}$ and we note here that this expression is related to the momentum fluctuations used in [**254, 595, 596, 1062, 1063, 1065**]. Thus consider the entropy of an isolated system in the form (**8G**) $\mathfrak{S}(x,t) = k_B log(\Omega(x,t))$ where k_B is the Boltzman constant and $\Omega(x,t)$ represents the system accessible states when the position x varies between x and $x+\delta x$. The equal a priori probability postulate then says that (**8H**) $\rho(x,t) \propto \Omega(x,t) = exp(\mathfrak{S}(x,t)/k_B)$ (cf. [**1076**]). One can now consider entropy defined on configuration space and write (**8I**) $\mathfrak{S} = \mathfrak{S}_{eq} + (1/2)(\partial^2\mathfrak{S}_{eq}/\partial x^2)(\delta x)^2$ where $\mathfrak{S}_{eq}(x)$ is the statistical equilibrium configuration entropy; here one has used the fact that the entropy must be a maximum giving $(\partial\mathfrak{S}_{eq}/\partial x) = 0$ for $\delta x = 0$. One could also have divided the system into N cells of dimension δx and written (each i refers to one specific cell)

$$\mathfrak{S}_i = (\mathfrak{S}_{eq})_i + (\partial_x \mathfrak{S}_{eq})\delta x_i + \frac{1}{2}\left(\frac{\partial^2\mathfrak{S}_{eq}}{\partial x^2}\right)(\delta x_i)^2; \tag{8.18}$$

Using known properties of entropy one can write ($\sum_1^N \delta x_i = 0$ and the system is adiabatically isolated)

$$\mathfrak{S} = \sum_1^N \mathfrak{S}_i = \mathfrak{S}_{eq} + \frac{1}{2}\left(\frac{\partial^2\mathfrak{S}}{\partial x^2}\right)\sum_1^N (\delta x_i)^2 \tag{8.19}$$

Now use (**8I**) to get

(8.20) $$\tilde{\rho}(x,\delta x,t) = \rho_{eq}(x,t)exp\left(-\frac{1}{2k_B}\left|\frac{\partial^2\mathfrak{S}_{eq}(x,t)}{\partial x^2}\right|(\delta x)^2\right)$$

(recall that the second derivative of the entropy is negative near an equilibrium point).

Thus at each point x the probability distribution with respect to small displacements $\delta x(x,t)$ is Gaussian and is related with the probability of having a fluctuation $\Delta\rho$ owing to a fluctuation δx. (6.20) guarantees that the system will tend to return to its equilibrium distribution represented by ρ_{eq} and using (6.20) the mean quadratic displacements related with the fluctuations are given via

(8.21) $$\overline{(\delta x)^2} = \frac{\int_{-\infty}^{\infty}(\delta x)^2 exp(-\gamma(\delta x)^2)d(\delta x)}{\int_{-\infty}^{\infty} exp(-\gamma(\delta x)^2 d(\delta x)} = \frac{1}{2\gamma}$$

where (**8J**) $(1/2\gamma) = k_B|\partial^2\mathfrak{S}_{eq}(x,t)/\partial x^2|^{-1}$. Note here that x is a constant so this is correct; it is δx which is variable in the integration. A priori there is no relation between the displacement and momentum fluctuations but in the statistical equilibrium situation one can impose the restriction that (**8K**) $\overline{(\delta p)^2}\cdot\overline{(\delta x)^2} = \hbar^2/4$ (compare here with the exact uncertainty principle of Hall and Reginatto in [**254, 594, 596, 1062**]). Using (8.21) and (**8K**) gives then

(8.22) $$\overline{(\delta p)^2} = -\frac{\hbar^2}{4}\frac{\partial^2 log(\rho(x,t))}{\partial x^2} \Rightarrow \overline{(\delta p)^2}\rho(x,t) = -\frac{\hbar^2}{4}\rho(x,t)\frac{\partial^2 log(\rho(x,t))}{\partial x^2}$$

Putting this in (8.16) and setting $\rho = R^2$ and $p = S_x$ one arrives at

(8.23) $$R^2\partial_x\left[S_t + \frac{1}{2m}S_x^2 + V + Q\right] = 0$$

which together with (8.14) is equivalent to the SE as before, namely

(8.24) $$-\frac{\hbar^2}{2m}\psi_{xx} + V\psi = i\hbar\psi_t;\ \psi = Rexp(iS/\hbar)$$

Thus (6.14) and (6.16) have the same content as the SE and one notes that $p \sim S_x$ is also an assumption (pilot wave condition).

To connect this with Z_Q and the previous derivation of the SE we recall (8.1) and note that $\rho(x,t) = lim_{\delta x\to 0}Z_Q(x,\delta x,t)$ with (**8L**) $\int p^2F(x,p,t)dp = lim_{\delta x\to 0}[-\hbar^2(\partial^2 Z_Q(x,\delta x,t)/\partial(\delta x)^2)]$. Then the right side of (8.16) can be written as

(8.25) $$\int [p - p(x,t)]^2 F(x,p,t)dp = \lim_{\delta x\to 0}\left[-\hbar^2\frac{\partial^2 Z_Q(x,\delta x,t)}{\partial(\delta x)^2} + \frac{\hbar^2}{Z_Q}\left(\frac{\partial Z_Q}{\partial(\delta x}\right)^2\right]$$

This can be rearranged as

(8.26) $$\overline{(\delta p)^2}\rho(x,t) = -\hbar^2 lim_{\delta x\to 0}Z_Q(x,\delta x,t)\frac{\partial^2 log(Z_Q)}{\partial(\delta x)^2}$$

It remains now to give the explicit appearance of this expression and show that it is equivalent to (8.22). Note that Z_Q can be written as

$$Z_Q = \int F(x,p,t)dp + \frac{i\delta x}{\hbar}\int pFdp - \frac{(\delta x)^2}{2\hbar^2}\int p^2 F dp + o((\delta x)^3) \tag{8.27}$$

and this is equivalent (using (**8D**) and (**8E**)) to

$$Z_Q = \rho(x,t) + \frac{i\delta x}{\hbar}p(x,t)\rho(x,t) - \frac{(\delta x)^2}{2\hbar^2}M_2(x,t) + o((\delta x)^3) \tag{8.28}$$

The left side has to be written as $Z_Q = \psi^*(x-(\delta x/2))\psi(x+(\delta x/2))$ and using $\psi = Rexp(iS/\hbar)$ one finds (up to second order in the infinitesimal parameter)

$$Z_Q = \left\{R^2 + \left(\frac{\delta x}{2}\right)^2 \left[RR_{xx} - (R_x)^2\right]\right\} exp\left(\frac{i}{\hbar}S_x\delta x\right) \tag{8.29}$$

Explicitly this is

$$Z_Q = R^2 + \frac{i\delta x}{\hbar}R^2 S_x + \frac{(\delta x)^2}{2}\left[\frac{1}{4}R^2\frac{\partial^2 log(R^2}{\partial x^2} - \frac{R^2}{\hbar^2}(S_x)^2\right] \tag{8.30}$$

Comparison with (8.28) gives (**8M**) $\rho(x,t) = R^2(x,t)$ and $p(x,t) = S_x(x,t)$ and (**8N**) $M_2(x,t) = -(\hbar^2/4)\rho(x,t)\partial^2 log(\rho(x,t)) + p(x,t)^2\rho(x,t)$. Using (8.28) one can write then

$$Z_Q = \rho(x,t)\left[1 + \frac{i\delta x}{\hbar}p(x,t) + \frac{(\delta x)^2}{2}\left[\frac{1}{4}\partial_x^2 log(\rho(x,t)) - \frac{p^2(x,t)}{\hbar^2}\right)\right] \tag{8.31}$$

which means that

$$lim_{\delta x\to 0}\frac{\partial^2}{\partial(\delta x)^2}log(Z_Q(x,\delta x,t) = \frac{1}{4}\partial_x^2 log(\rho(x,t)) \tag{8.32}$$

where one has expanded the logarithm in (8.31) up to the second order in δx. This last result gives then, using (8.26),

$$\overline{(\delta p)^2}\rho(x,t) = -\frac{\hbar^2}{4}\rho(x,t)\partial_x^2 log(\rho(x,t)) \tag{8.33}$$

which is equivalent to (8.22) as desired (this comes directly from (**8N**) and (8.17)). Another way of comparing the two approaches is to simply substitute (8.28) in the equation satisfied by the characteristic function which is (via (8.3)) (**6O**) $- i\hbar\partial_t Z_Q - (\hbar^2/m)(\partial^2 Z_Q/\partial x\partial(\delta x)) + \delta x(V_x)Z_Q = 0$. Taking the real and imaginary parts gives then

$$\partial_t\rho(x,t) + \partial_x\left[\frac{p(x,t)}{m}\rho(x,t)\right] = 0; \tag{8.34}$$

$$\partial_t[\rho(x,t)p(x,t)] + \frac{1}{m}\partial_x M_2(x,t) + V_x\rho(x,t) = 0$$

which are (8.14) and (8.15). Thus the restriction (**8K**) is equivalent to postulating the adequacy of the Wigner-Moyal infinitesimal transformation together with the restriction of writing $Z_Q = \psi^*(x-(\delta x/2),t)\psi(x+(\delta x/2),t)$.

9. THE UNCERTAINTY PRINCIPLE

We go now to [**185, 254, 292, 528, 595, 596, 597, 830, 941, 995, 807, 1062, 1063, 1065, 1117, 1181**] for some results involving fluctuations and the uncertainty principle. In particular Brenig [**185**] exhibits some fascinating relations between the Heisenberg uncertainty principle and the exact uncertainty principle of Hall-Reginatto and we sketch some of this here. The title of [**185**] is a question "Is QM based on an uncertainty principle" and we indicate how this is answered in [**185**] in terms of the quantum potential (QP). One modifies the classical mechanical definition of momentum uncertainty in order to satisfy certain transformation rules. This involves adding a new term to the classical quadratic momentum uncertainty which has to be proportional to the inverse of a measure of the quadratic position uncertainty. Then one imposes the Hall-Reginatto conditions of causality and additivity of kinetic energy which leads to a complete specification of the functional dependence of the new term requiring it to be essentially the QP (often associated with the idea of quantization - modulo some arguments at times). An observer is now characterized by parameters denoting the statistical position and momentum uncertainties Δx and Δp of its instruments. For example Δx^2 could be the trace of the covariance matrix associated to a given position probability density $\rho(x)$ or as the inverse of the trace of the Fisher matrix associated to ρ. The main postulate here is that under dilatations of space coordinates the parameters Δx and Δp must transform in such a manner that the relation (**9A**) $\Delta x\Delta p \geq (\hbar/2$ is kept invariant. The transformations allowed here are

$$\Delta x'^2 = e^{-\alpha}\Delta x^2;\ \Delta p'^2 = e^{-\alpha}\Delta p^2 + \frac{\hbar^2}{4}(e^{\alpha} - e^{-\alpha})\frac{1}{\Delta x^2} \tag{9.1}$$

where α is real and one sees that such transformations generate a group. Multiplying the terms together in (9.1) gives

$$\Delta x'^2\Delta p'^2 = e^{-2\alpha}\Delta x^2\Delta p^2 + \frac{\hbar^2}{4}(1 - e^{-2\alpha}) \tag{9.2}$$

One notes that for $\alpha \to \infty$ one has $\Delta x'^2\Delta p'^2 \to (\hbar^2/4)$, if $\Delta x^2\Delta p^2 = \hbar^2/4$ then it remains so, and for $\alpha \to -\infty$ one has $\Delta x'^2\Delta p'^2 \to \infty$ for any value of $\Delta x^2\Delta p^2 \geq (\hbar^2/4)$. Uniqueness of such transformations has not been established. There is some analogy here to special relativity and this is discussed in [**185**]. In any case one shows now that the stipulations (9.1) impose a radical modification of the laws of dynamics that corresponds precisely to that required in the passage from classical to quantum mechanics. Thus consider $\rho(x)$ and $S(x)$ as basic variables (fields) and specify that the time evolution of any functional via (**9B**) $\mathfrak{A} = \int d^3xF(x,\rho,\nabla\rho,\nabla^2\rho,\cdots,S,\nabla S,\nabla^2 S,\cdots)$ is given by (**9C**) $\partial_t\mathfrak{A} = \{\mathfrak{A},\mathfrak{H}_{cl}\}$ where $\mathfrak{H}_{cl} = \int d^3x(\rho|\nabla S|^2/2m)$ and

$$\{\mathfrak{A},\mathfrak{B}\} = \int d^3x\left[\frac{\delta\mathfrak{A}}{\delta\rho(x)}\frac{\delta\mathfrak{B}}{\delta S(x)} - \frac{\delta\mathfrak{B}}{\delta\rho(x)}\frac{\delta\mathfrak{A}}{\delta S(x)}\right] \tag{9.3}$$

The Poisson bracket provides a Lie algebra structure. Applying (**9C**) to ρ and S yields

$$(9.4)\qquad \partial_t\rho = -\nabla\cdot\left(\rho\frac{\nabla S}{m}\right);\ \partial_t S = -\frac{|\nabla S|^2}{2m}$$

(note $\nabla S \sim p$ which here is the classical momentum). Now consider the group of space dilatations $x \to exp(-\alpha/2)x$ and its action on ρ and S via (**9D**) $\rho'(x) = exp(3\alpha/2)\rho(exp(\alpha/2)x)$ and $S'(x) = exp(-\alpha)S(exp(\alpha/2)x)$ (α real). Such transformations preserve the normalization of ρ and keep the dynamical equations (9.3) invariant. Assume now that the average momentum of the particle is vanishing (i.e. a comoving frame is chosen which does not reduce the generality of the results). Then the classical definition of the scalar quadratic momentum uncertainty is given via (**9E**) $\Delta p_{cl}^2 = \int d^3x\rho|\nabla S|^2 = 2m\mathfrak{H}_{cl}$ and under (**9D**) this becomes (**9F**) $\Delta\rho_{cl}^{'2} = exp(-\alpha)\Delta p_{cl}^2$. Also any definition of the scalar quadratic position uncertainty measuring the dispersion Δx^2 of $\rho(x)$ transforms as (**9G**) $\Delta x^{'2} = exp(-\alpha)\Delta x^2$. The requirement of (9.1) to be fundamental now requires one to modify (**9E**) in order to get a quantity whose variance satisfies (9.1). Some argument shows that adding a supplementary term proportional to $\hbar^2$ is needed in order to have Δp^2 transform reasonably under (**9D**). This is accomplished via

$$(9.5)\qquad \Delta p_q^2 = \int d^3x\rho(x)|\nabla S(x)|^2 + \hbar^2\mathfrak{Q}$$

where $\mathfrak{Q}$ is to be determined. Applying (**9D**) to (9.4) leads to (**9H**) $\Delta p_q^{'2} = e^{-\alpha}\Delta p_{cl}^2 + \hbar^2(\mathfrak{Q}' - e^{-\alpha}\mathfrak{Q})$. Adding and subtracting $exp(\alpha)\hbar^2\mathfrak{Q}$ yields and equation (**9I**) $\Delta p_q^{'2} = exp(-\alpha)\Delta p_q^2 + \hbar^2(\mathfrak{Q}' - exp(-\alpha)\mathfrak{Q})$. Identifying this with (9.1) requires (**9J**) $\mathfrak{Q}' - exp(-\alpha)\mathfrak{Q} = (1/4\Delta x^2)(exp(\alpha) - exp(-\alpha))$ which, using (7.1) becomes (**9K**) $\mathfrak{Q}' - (1/4\Delta x^{'2}) = exp(-\alpha)[\mathfrak{Q} - (1/4\Delta x^2)]$. There are an infinity of solutions but the form indicates a relation between $\mathfrak{Q}$ and Δx^2 that is scale independent, namely (**9L**) $\mathfrak{Q} = (1/4\Delta x^2)$. This is the only solution for which the relation between Δp_q^2 and Δx^2 is independent of the scale α. Thus this is the needed term to obtain a definition of Δp_q^2 compatible with (9.1). Since Δx^2 depends only on $\rho(x)$ we see that $\mathfrak{Q}$ is a functional of the form (**9B**) that does not depend on S. For the precise form of Δx^2 one can now refer to the Hall-Reginatto results (cf. [**254**] for a survey and a discussion of relations to Fisher information) which work with entirely different arguments in discovering the additive term $\hbar^2\mathfrak{Q}$. At this point their additional requirements of causality and additivity can be invoked point to obtain $\mathfrak{H}_q = \Delta p_q^2/2m$ and $\mathfrak{Q} = \beta\int d^3x|\nabla\rho(x)^{1/2}|^2$ (with $\beta = 1$) and finally

$$(9.6)\qquad \mathfrak{H}_q = \int d^3x\left[\frac{\rho(x)|\nabla S|^2}{2m} + \frac{\hbar^2}{2m}|\nabla\rho(x)^{1/2}|^2\right]$$

The HJ equation $\partial_t S = -(|\nabla S|^2/2m) + (\hbar^2/2m)(\nabla^2\rho(x)^{1/2}/\rho^{1/2}(x)$ is given via (**9C**) along with the continuity equation in (9.3) leading to the standard SE for $\psi = \rho^{1/2}exp(iS/\hbar)$.

One looks next at dilatations under (**9D**) to get

$$\mathfrak{H}'_q = Cosh(\alpha)\int d^3x\left[\frac{\rho|\nabla S|^2}{2m} + \frac{\hbar^2}{2m}|\nabla\rho^{1/2}|^2\right] - \tag{9.7}$$
$$-Sinh(\alpha)\int d^3x\left[\frac{\rho|\nabla S|^2}{2m} - \frac{\hbar^2}{2m}|\nabla\rho^{1/2}|^2\right]$$

where (**9M**) $\mathfrak{H}'_q[\rho, S] = \mathfrak{H}_q[\rho', S']$. One writes then

$$\mathfrak{K}_q = \int d^3x\left[\frac{\rho|\nabla S|^2}{2m} - \frac{\hbar^2}{2m}|\nabla\rho^{1/2}|^2\right] \tag{9.8}$$

In more compact notation now write (**9N**) $\mathfrak{H}'_q = Cosh(\alpha)\mathfrak{H}_q - Sinh(\alpha)\mathfrak{K}_q$ with $\mathfrak{K}'_q = -Sinh(\alpha)\mathfrak{H}_q + Cosh(\alpha)\mathfrak{K}_q$ so that under (**9D**) $(\mathfrak{H}_q, \mathfrak{K}_q)$ transforms as a 2-D Minkowski vector under a Lorentz like transformation. This corresponds to (**9O**) $t' = Cosh(\alpha)t + Sinh(\alpha)\tau$ and $\tau' = Sinh(\alpha)t + Cosh(\alpha)\tau$ and one can write (**9P**) $\partial_t\mathfrak{A} = \{\mathfrak{A}, \mathfrak{H}_q\}$ with $\partial_\tau\mathfrak{A} = \{\mathfrak{A}, \mathfrak{K}_q\}$; further this transforms via (**9D**) to (**9Q**) $\partial_{t'}\mathfrak{A}' = \{\mathfrak{A}', \mathfrak{H}'_q\}$ and $\partial_{\tau'}\mathfrak{A}' = \{\mathfrak{A}', \mathfrak{K}'_q\}$. Thus this is all covariant under (**9D**). The SE is a particular case of (**9P**) where

$$\mathfrak{A} = \psi = \rho^{1/2}e^{iS/\hbar};\ i\hbar\partial_t\psi = -\frac{\hbar^2}{2m}\nabla^2\psi \tag{9.9}$$

and one obtains also

$$i\hbar\partial_\tau\psi = -\frac{\hbar^2}{2m}\nabla^2\psi + \frac{\hbar^2}{m}\frac{\nabla^2|\psi|}{|\psi|} \tag{9.10}$$

This is an interesting equation to be discussed later (see e.g. [**78, 389, 572, 573, 624, 1208, 1261, 1287**]). Thus (9.8) and (9.9) are covariant under space dilatations with

$$i\hbar\partial_{t'}\psi' = -\frac{\hbar^2}{2m}\nabla^2\psi';\ i\hbar\partial_{\tau'}\psi' = -\frac{\hbar^2}{2m}\nabla^2\psi' + \frac{\hbar^2}{m}\psi'\frac{\nabla^2|\psi'|}{|\psi'|} \tag{9.11}$$

where (cf. (**9D**))

$$\psi' = e^{3\alpha/4}[\psi(e^{\alpha/2}x)]^{(1/2)(1+exp(-\alpha))}[\psi^*(e^{\alpha/2}x)]^{(1/2)(1-exp(-\alpha))} \tag{9.12}$$

Note that when the transformation of the SE under dilatations is considered ψ transforms as the square root of a density $\rho^{1/2}$ (cf. also [**201, 199, 551, 552, 554, 627**]). Finally one notes that defining $\mathfrak{s} = \int d^3x\rho(x)S(x)$ (ensemble average of the quantum phase up to a factor of $\hbar$) one has

$$\{\mathfrak{s}, \mathfrak{H}_q\} = \int d^3x\left[\frac{\rho|\nabla S|^2|}{2m} - \frac{\hbar^2}{2m}|\nabla\rho^{1/2}|^2\right] = \mathfrak{K}_q \tag{9.13}$$
$$\{\mathfrak{s}, \mathfrak{K}_q\} = \int d^3x\left[\frac{\rho|\nabla S|^2}{2m} + \frac{\hbar^2}{2m}|\nabla\rho^{1/2}|^2\right] = \mathfrak{H}_q$$

This leads to a general statement $\mathfrak{A}' = \mathfrak{A} + \delta\alpha\{\mathfrak{A}, \mathfrak{s}\}$

REMARK 7.9.1. Let us gather together some of the ideas connecting [**185, 595, 596, 597, 1062, 1063, 1065**]. Thus in [**185**] (for 1-D here)

$$\Delta x^2 \sim \frac{1}{\int[(\rho')2/\rho]dx} = \frac{1}{F} \tag{9.14}$$

and the particle mean square deviation (or variance) is σ_x^2 with $\sigma_x^2 \geq \Delta x^2$. To check this one can follow [**490**] and write for suitable estimators $\hat{\theta}$ of θ (note $\rho_\theta = \partial_\theta \rho = \rho \partial_\theta log(\rho)$ and $\hat{\theta} = \hat{\theta}(y)$)

$$(9.15)\quad 0 =< \hat{\theta} - \theta >= \int dy \rho(y|\theta)[\hat{\theta}(y) - \theta] \Rightarrow 0 = \int dy \rho_\theta[\hat{\theta} - \theta] - \int \rho dy \Rightarrow$$

$$\Rightarrow 1 = \int dy \rho_\theta[\hat{\theta} - \theta] = \int dy \sqrt{\rho}[\hat{\theta} - \theta]\sqrt{\rho}\partial_\theta log(\rho) \leq \int \rho|\hat{\theta} - \theta|^2 \int \rho|\partial_\theta log(\rho)|^2$$

This says that $1 \leq \sigma_x^2 F$ with σ_x^2 the classical variance and $F \sim$ Fisher information. Now recall that it is natural to think of momentum fluctuations in the form $\delta p \sim \rho'/\rho = \partial_x log(\rho) \sim cu$ where u is an osmotic velocity. Then look at the exact uncertainty principle of [**595, 596, 597**] where one writes $\delta x \Delta p_{nc} = \hbar/2$ (recall $p_{nc} \sim \nabla S + \delta p$ with $\delta p \sim \hat{c}\partial log(\rho)$. In [**950**] on the other hand there is a crucial condition (**9K**) $\overline{(\delta p)^2 (\delta x)^2} = \hbar^2/4$ which is imposed in the context of a statistical equilibrium situation involving Boltzman type entropy. Here $\overline{(\delta p)^2}$ is determined via ((**9K**) and (9.21) in the form (**9R**) $\overline{(\delta x)^2} = 1/2\gamma$ where $2\gamma = (1/k_B)\partial^2 \mathfrak{S}_{eq}/\partial x^2$ leading to (9.22) where (**9S**) $\overline{(\delta p)^2} \sim -(\hbar^2/4)\partial_x^2 log(\rho)$. Since $\partial^2 log(\rho) \sim \partial(\rho'/\rho) = (\rho''/\rho - (\rho'/\rho)^2$ one sees that for $\rho' = \rho = 0$ outside of a compact Ω (**9T**) $\int \rho \partial^2 log(\rho) = -\int (\rho')^2/\rho = -\int \rho(\rho'/\rho)^2$ and such quantum perturbations to $\nabla S)^2$ would involve the same action as perturbations $(\rho'/\rho)^2$ (corresponding to $\int \rho Q$ where $Q \sim (\partial^2 \sqrt{\rho})/\sqrt{\rho} = (1/2)[(\rho''/\rho) - (\rho'^2/2\rho^2]$. ∎

We go next to Santamato [**1117**] and sketch some of the results. In "geometric quantum mechanics" (GQM) geometry is not prescribed but is determined by the space matter content. Further in [**1117**] one assumes that the affine connections are responsible for quantum phenomena. Given that GQM deals with a Gibbs ensemble of particles, rather than a single particle, one can treat it as a classical theory. Wave equations and the operational prescriptions of standard QM are devoid of physical meaning and are regarded as clever devices to overcome the difficulties inherent in a nontrivial geometric structure (see [**254, 1115, 1116**] and above for more on this). The object in [**1117**] is to show how the Heisenberg uncertainty principle arises in the context of GQM. Note that Planck's constant is conspicuously absent in the treatment. Thus in GQM one assumes that in a parallel displacement $d\mathbf{r}$ the length $\ell = (\mathbf{A} \cdot \mathbf{A})^{1/2}$ of a vector $\mathbf{A}$ changes by an amount (•) $\delta\ell = \ell(\vec{\phi} \cdot d\mathbf{r})$ where $\vec{\phi}$ is the Weyl gauge vector. In nonrelativistic mechanics the spatial metric is Euclidean so the dot means the standard scalar product and we have a Weyl space which generally will have a nonzero scalar curvature $\mathcal{R}$ even if its metric has zero curvature. In GQM the law (•) and hence the Weyl-Ricci curvature is determined by the motion of the particle itself and in turn it acts as a guidance field for the particle (this is perhaps another way to look at Bohmian mechanics). Following [**1117**] (cf. also above) the length transference law may be obtained from a minimum average curvature principle (••) $E[\mathcal{R}(\mathbf{r}(t), t)] = min$ where E denotes the ensemble expectation value. When the space is flat ($\dot{R} = 0$) one has (cf. [**254, 1115, 1116**]) (♦) $\mathcal{R} = 2|\vec{\phi}|^2 - 4\nabla \cdot \vec{\phi}$. Now let ρ be the probability density of the particle position (properly normalized and vanishing at infinity) and note that

(9.16) $$E[\nabla\cdot\vec{\phi}]=\int_{\mathbf{R}^3}\rho\nabla\cdot\vec{\phi}d^3\mathbf{r}=-\int_{\mathbf{R}^3}\rho\vec{\phi}\cdot\nabla(log(\rho)d^3\mathbf{r}=-E[\vec{\phi}\cdot\nabla log(\rho)]$$

Consequently

(9.17) $$E[\mathcal{R}]=2E[|\vec{\phi}|^2+2\vec{\phi}\cdot\nabla log(\rho)]=2E[(\vec{\phi}+\nabla log(\rho)^2]-2E[(\nabla log(\phi)^2]$$

The minimizing gauge vector ϕ is found from this to be (♦♦) $\vec{\phi}=-\nabla log(\rho)$ and the minimum average scalar curvature is (★) $E[\mathcal{R}]=-2E[|\vec{\phi}|^2]$. Note that the minimal average curvature is negative and moreover from (♦♦) one obtains
(9.18)
$$E[\vec{\phi}(\mathbf{r},t),t)]=\int_{\mathbf{R}^3}\rho(\mathbf{r},t)\vec{\phi}(\mathbf{r},t)d^3\mathbf{r}=-\int_{\mathbf{R}^3}\rho\nabla log(\rho)d^3\mathbf{r}=-\oint_S\rho\mathbf{n}dS=0$$

where S is a closed surface at ∞ enclosing $\mathbf{R}^3$ (where $\rho=0$). One notes also that the mean square deviation for $|\vec{\phi}|$ is not zero in general. In fact from (★) and (7.18) one finds the mean square deviation via

(9.19) $$\Delta|\vec{\phi}|=\{E[|\vec{\phi}|^2]-E^2[\vec{\phi}]\}^{1/2}]=2^{-1/2}[E(-\mathcal{R}]^{1/2}$$

Now let Δq be the root mean square deviation of the particle position so that (7.18) and the Schwarz inequality give

(9.20) $$\Delta q\Delta|\vec{\phi}|\geq Cov\{(\mathbf{r}-E[\mathbf{r}])\cdot(\vec{\phi}-E[\vec{\phi}])\}=E[\mathbf{r}\cdot\vec{\phi}]$$

where Cov denotes covariance. The average on the right side of (7.20) can be computed using (♦♦) and one gets (using (9.19))

(9.21) $$\Delta q\Delta|\vec{\phi}|^2\geq\int_{\mathbf{R}^3}\rho\nabla\cdot\mathbf{r}d^3\mathbf{r}=3\Rightarrow\Delta q(E[-\mathcal{R}])^{1/2}\geq\frac{3}{\sqrt{2}}$$

This is the fundamental relation between the particle root mean square deviation and the average space curvature. The space may be flat on average only if the particle is completely delocalized - particle localization forces the space to be curved. One notes that (7.21) is purely geometrical and Planck's constant is not involved. However (9.21) does imply the Heisenberg uncertainty principle provided the prescriptions of standard QM are used. Thus for $\psi=\sqrt{\rho}exp(iS/\hbar)$ with a particle of mass m and probability density $\rho=|\psi|^2$ one has trajectories (★★) $d\mathbf{r}/dt=\nabla S(\mathbf{r},t)/m$ where

(9.22) $$S_t+\frac{|\nabla S|^2}{2m}+V-\frac{\hbar^2}{16m}\mathcal{R}=0$$

(cf. above and [**1115, 1116**]). The averages obtained using the operator methods of standard QM do not coincide in general with the averages made on the classical ensemble. Thus using < , > for QM and $E[\ ]$ for ensemble averages one gets (using (7.22))

(9.23) $$<\mathbf{r}>=E[\mathbf{r}];\ <\mathbf{p}>=E[\nabla S];\ <\hat{p}^2>=E[|\nabla S|^2]+\frac{\hbar^2}{8}E[-\mathcal{R}]$$

where $-i\hbar\nabla \sim \hat{p}$. Note here also that defining $\Delta p = [< \mathbf{p} - < \mathbf{p} >^2]^{1/2}$ it follows immediately via (7.23) that

$$(9.24) \qquad (\Delta p)^2 = E[(\nabla S - E[|\nabla S|)^2] + \frac{\hbar^2}{8}E[-\mathcal{R}] \Rightarrow \Delta p \geq \frac{\hbar}{2\sqrt{2}}(E[-\mathcal{R}])^{1/2}$$

leading to the Heisenberg type inequality (**7U**) $\Delta p \delta Q \geq (3/2)\hbar$ via (7.21).

REMARK 7.9.2. The relation $\Delta x \Delta p \geq (\hbar/2)$ can be traced in an analogous manner using $\Delta x \Delta \phi_x \geq E[x\phi_x]$ in place of (7.20). ■

Finally one notes that in (9.24) there is a random motion part and a curvature part and it is the curvature part that forces Heisenberg's position momentum uncertainty relation. However this contribution is a mere consequence of the operator formulation of QM and, unlike (7.21), it has no physical meaning in a physically consistent approach to GQM. In fact, GQM being a classical theory, the particle momentum should be defined as $p = \nabla S$ and the mean square deviation should be given by the first term on the right in (9.24).

REMARK 7.9.3. In reading over the second paper in Smolin [**1181**] one is struck by similarities to the approach of Nottale (see [**254, 287, 941**]). The QP term arises basically because the quantum paths are not differentiable (à la Feynman they have fractal dimension 2) and as indicated in [**234**] this corresponds to expressing the QP in terms of an osmotic velocity **u**. In the Nottale approach **u** only appears because of "jagged" paths and in the presence of smooth (i.e. differentiable) paths the QP is zero and there is no SE. If there are nonsmooth paths then the fact that $\mathbf{u} = D\nabla log(\rho)$ (with $\vec{\phi} = -\nabla log(\rho)$) suggests that **u** is a curvature phenomena but that is misleading - it is really a diffusion phenomena and for D=0 **u** will not appear. In more detail, following Nottale, one can write

$$(9.25) \qquad \rho_t + div(\rho b_+) = D\Delta\rho; \;\; \rho_t + div(\rho b_-) = -D\Delta\rho$$

But $b_\pm$ represents e.g. right and left derivatives at a given point so smooth paths give $b_+ = b_-$ and hence $\mathbf{u} \sim (1/2)(b_+ - b_-) = 0$ with $D = 0$ necessarily via $D = -D$. Hence smooth paths imply no Wiener process and no SE. One can still have curvature via a nonzero Weyl vector $\phi \sim -\nabla log(\rho)$. The SE and QM arise via an assumption of diffusion and nonsmooth paths which is equivalent to introducing a nonzero QP. This does not answer the question "Why QM?" but does seem to provide evidence that (**9U**) is the fundamental GQM idea and the Heisenberg inequality arises only after introducing diffusion with a particular diffusion coefficient $\hbar/2m$ (cf. [**364**]), or introducing a Schrödinger wave function containing $\hbar$, or perhaps after introducing all the miracle machinery of Hilbert space QM, etc. This is all consistent with comments made in earlier Remarks. The factor $\hbar$ is "gratuitous" and is introduced in QM either via the wave function idea $\psi = Rexp(iS/\hbar)$ or via a diffusion coefficient $D = \hbar/2m$. ■

10. GEOMETRY AND QUANTUM MATTER

We have seen already a number of striking relations between geometry and quantum matter some of which we repeat here via equations (see [**254, 668, 669, 670, 671, 672, 673, 674, 675**] for much more on this). Thus

(1) From (2.9) we have

$$Q \sim -\frac{\hbar^2}{16m}\left[\dot{\mathcal{R}} + 2\left\{\phi_i\phi^i - \frac{2}{\sqrt{g}}\partial_i(\sqrt{g}\phi^i)\right\}\right] \tag{10.1}$$

(2) Generally

$$\phi_k\phi^k - 2\partial_k\phi^k \sim -\left(\frac{|\nabla\rho|^2}{\rho^2} - \frac{2\Delta\rho}{\rho}\right) = 4\frac{\Delta\sqrt{\rho}}{\sqrt{\rho}} \tag{10.2}$$

(3) For n=4

$$\mathcal{R} = \dot{\mathcal{R}} - 3\left[\frac{1}{2}g^{\mu\nu}\phi_\mu\phi_\nu + \frac{1}{\sqrt{-g}}\partial_\mu\sqrt{-g}g^{\mu\nu}\phi_\nu\right] = \mathcal{R} + \mathcal{R}_w \tag{10.3}$$

(4) For n=3

$$\mathcal{R} = \dot{\mathcal{R}} + 2\left[\phi_i\phi^i - 2\left(\frac{1}{\sqrt{g}}\partial_i(\sqrt{g}\phi^i)\right)\right] \tag{10.4}$$

(5) Further one knows that

$$\mathfrak{M}^2 = m^2 e^Q;\ Q = \frac{\hbar^2}{m^2c^2}\frac{\Box\sqrt{\rho}}{\sqrt{\rho}} = \frac{\hbar^2}{m^2c^2}\frac{\nabla^2|\psi|}{|\psi|} \tag{10.5}$$

(6) The following equations relate the Weyl vector $\vec{\phi}$ and the Dirac field β

$$\beta \sim \mathfrak{M};\ \beta_0 \to \beta = \beta_0 e^{-\Xi};\ \phi_\mu \to \phi_\mu + \partial_\mu\Xi \tag{10.6}$$

(7) Further $Q \sim -(m^2/2)\mathbf{u}^2 - (\hbar/2)\partial\mathbf{u}$ ($\mathbf{u} \sim$ osmotic velocity) with

$$\mathbf{u} = D\nabla log(\rho) = \frac{\hbar}{m}\frac{\nabla\sqrt{\rho}}{\sqrt{\rho}};\ \vec{\phi} = -\nabla log(\rho); \tag{10.7}$$

Thus in particular the quantum mass $m_Q \sim \mathfrak{M}$ is given via Q, or via ρ, or via $\vec{\phi}$ so $\vec{\phi}$ determines $\mathfrak{M}$ and conversely $\mathcal{M}$ determines ρ via Q. Recall that $\Delta R + \beta QR = 0$ has unique solutions in say $H_0^1(\Omega)$ where $\beta = -(2m/\hbar^2)$. In fact $\Delta R + \beta QR = \lambda R$ has a unique solution for $\lambda \neq \Sigma$ where Σ is a countable set $\Sigma \subset \mathbf{R}$ (cf. [**240**]) so if $0 \neq \Sigma$ then Q determines R uniquely. In any event ρ itself determines $\mathfrak{M}$ and $\vec{\phi}$ (as well as $\mathbf{u}$). Since ρ, or $\vec{\phi}$, or β determine the Weyl Dirac geometry (the Riemannian metric is assumed fixed here) one seems to have already an excellent theory for quantum perturbations of a classical Riemannian geometry. Is this not an important desideratum of quantum gravity theory? To ask for quantum perturbations that maintain a Riemannian structure seems perhaps excessive (cf. Castro-Granik [**273**] (although conformal equivalence is established via $\mathfrak{M}^2/m^2$) and we are not trying to answer the questions of basic structure (for this see e.g. [**114, 115, 283, 878, 1085, 1338**]).

10.1. REMARKS ON MANY WORLDS. We refer back to Remark 7.5.3, Remark 7.5.4, and earlier Sections. Consider a 3-D Riemannian manifold M with metric g_{ij} and Riemann curvature $\dot{\mathcal{R}}$. Let m be a mass and consider an ensemble of particles of mass m distributed via a probability density P which generates a mass density $\rho = mP(x,t)$ (see also Remark 7.10.4). Write $\hat{\rho} = \rho/\sqrt{g}$ and observe that a natural classical background for quantum evolution in M is provided via a Weyl geometry on M based on a Weyl vector $\vec{\phi} = -\nabla log(\hat{\rho})$. Then the SE for $\psi = Rexp(iS/\hbar)$, namely

$$i\hbar\psi_t - \frac{\hbar^2}{2m}\frac{1}{\sqrt{g}}[\partial_i(\sqrt{g}g^{ik}\partial_k)]\psi + [V - \frac{\hbar^2}{16m}\dot{\mathcal{R}}]\psi \tag{10.8}$$

corresponds to classical evolution

$$\partial_t S + \frac{1}{2m}g^{ik}\partial_i S\partial_k S + V - \frac{\hbar^2}{2m}\mathcal{R} = 0; \tag{10.9}$$

$$\partial_t\hat{\rho} + \frac{1}{m\sqrt{g}}\partial_i(\sqrt{g}\partial_i S\hat{\rho}) = 0$$

where

$$\mathcal{R} = \dot{\mathcal{R}} + \frac{8}{\sqrt{g\hat{\rho}}}\partial_i(\sqrt{g}g^{ik}\partial_k\sqrt{\hat{\rho}})]; \; Q \sim -\frac{\hbar^2}{16m}\mathcal{R} \tag{10.10}$$

Here $\mathcal{R} = \mathcal{R}_w$ is the Ricci-Weyl curvature.

REMARK 7.10.1. Thus assume M is Euclidean with $\dot{\mathcal{R}} = 0$ for simplicity so $g \sim 1$ and $\hat{\rho} = \rho$. The Weyl vector is then $\vec{\phi} = -\nabla log(\rho)$ and

$$\mathcal{R} = \frac{8\Delta\sqrt{\rho}}{\sqrt{\rho}}; \; Q \sim -\frac{\hbar^2}{2m}\frac{\Delta\sqrt{\rho}}{\sqrt{\rho}} \tag{10.11}$$

This suggests a kind of many worlds scenario for quantum motion. Thus given a probability distribution $P(x,t)$ with $\rho = mP$ as above one has wave functions $\psi = Rexp(iS/\hbar)$ where $R \sim \sqrt{\rho}$ (and some normalization $\int R^2 dx = \int \rho dx = 1$ for example is imposed). This seems to say that each wave function $Rexp(iS/\hbar)$ generates a family of time dependent Weyl geometries evolving via (9.2). The choice of $R \sim \sqrt{\rho}$ determines the geometry and we have an infinite number of Weyl space scenarios, each created by a choice of probability distribution $P(x,t)$. Thus a quantum particle is associated to an infinite number of time dependent Weyl space paths, each determined by some particular wave function $\psi = Rexp(iS/\hbar)$ where $R = |\psi|$ can be prescribed via a suitably normalized probability distribution $P(x,t)$. In view of (9.4) each ρ as above is associated to a quantum potential Q and if the association $R \leftrightarrow Q$ holds (as in Remark 7.7.1) then it could be said that Q determines the Weyl space. If $P(x,t) = P(x)$ is independent of time the Weyl space is fixed. ∎

REMARK 7.10.2. One can rephrase the above as follows. Take M as in Remark 7.10.1 with $\dot{\mathcal{R}} = 0$ and posit a SE in the form **(10A)** $i\hbar\psi_t = -(\hbar^2/2m)\Delta\psi + V\psi$. Every wave function $\psi = Rexp(iS/\hbar)$ gives rise to a probability distribution $R^2 = |\psi|^2 = \rho$ and thence to a time dependent set of Weyl geometries via

$\vec{\phi} = \nabla log(\rho)$ within which R and S evolve classically via (9.2) which we rewrite here as

$$\text{(10.12)} \qquad \partial_t S + \frac{1}{2m}S_x^2 + V + Q = 0;\ \partial_t\rho + \frac{1}{m}(\rho S_x)_x = 0;\ Q = -\frac{\hbar^2}{2m}\frac{\Delta\sqrt{\rho}}{\sqrt{\rho}}$$

Thus every wave function picks a world (family of Weyl spaces) describing the evolution of R and S. Recalling that $Q \sim -(\hbar^2/16m)\mathcal{R}_w$ one could also say that the quantum potential (defined via ρ) determines the family of Weyl spaces and indicates a quantization of the classical evolution (9.5) (with $Q = 0$). Here $\hbar$ and m arise via the SE and one could consider the original situation $\rho = mP = R^2$ here. If one expects $\int R^2 dx = 1$, which means $m \int P dx = 1$, it seems necessary to take $P = P(x)$, with a fixed Weyl space (otherwise one might have to deal with an equation $P\partial_t log(m) + P_t + P_x\dot{x} = 0$). For $P = P(x)$ the SE corresponds to an infinite number of classical systems each determined by a wave function of the SE with $S = S(\rho, V)$. It seems well motivated here to develop further and pursue a theme relating "normal" matter to Riemannian curvature and quantum matter to Weyl curvature, with some interaction (see also Remark 7.10.4). ∎

REMARK 7.10.3. It is not at all clear whether or not all this has any relation to the many worlds interpretation (MWI) (cf. [**200, 198, 626, 1100, 1207, 1227, 1261, 1275, 1307**]) and we will not belabor the idea. In many respects Bohmian mechanics as well as the consistent (or decoherent) histories approach (although different) seem to give a deeper insight into quantum processes than does the standard Copenhagen framework. Both are based on trajectories but represent a given history by different operators. We refer here to [**204, 517, 564, 604, 613, 614, 628**] for more on this. The important relation between quantum mechanics (QM) and mathematical statistics elucidated in [**157**] will be discussed below and this is relevant to any theory of wave functions. ∎

Let us now examine the situation from Section 7.4 in a more coherent manner, based on the Ricci-Weyl curvature $\mathcal{R} = \mathcal{R}_w \sim Q$ as the basic ingredient. Thus let Ξ be a class of quantum potentials, $Q = Q(x,t)$, for which $\Delta R + \beta Q R = 0$ ($\beta > 0$) has a unique solutiion $R \in H_0^1(\Omega)$ (where eventually $\beta \sim 2m/\hbar^2$). The t dependence in R arises directly from Q in this context. We specify now a Ricci-Weyl curvature (**10B**) $\mathcal{R}_w(x,t) = -(16m/\hbar^2)Q(x,t) = -8\beta Q(x,t)$ and think of $R(x,t) = R \sim \sqrt{\rho}$. We assume in the background a SE with potential V and via (10.5) one has e.g.

$$\text{(10.13)} \qquad \frac{1}{m}(\rho S_x)(x,t) = -\int_0^x \partial_t\rho dx + F(t)$$

where $(1/m)(\rho S_x)(0,t) = F(t)$ and $\rho(0,t)$ is known. If we are dealing with particle motion where $S_x \sim p$ then $S_x(0,t) \sim p(0,t)$ and we see also that S_t is known from (10.5) up to a factor $p(0,t)$. Hence given $\mathcal{R}_w \sim -8\beta Q$ based on a SE with potential V we can determine S_x and S_t up to a factor $p(0,t) = S_x(0,t)$ leading to

THEOREM 10.1. Given a SE based on a potential V in a region $\Omega \subset \mathbf{R}^3$ let $\mathcal{R}_w \sim -8\beta Q$ be given with $Q \in \Xi$ ($\beta \sim 2m/\hbar^2$). Then $\mathcal{R}_w$ can be associated with a wave function $\psi = Rexp(iS/\hbar)$ and a Weyl geometry based on a Weyl vector $\vec{\phi} = -\nabla log(\rho)$ where $R = \sqrt{\rho}$ is completely determined with S_x and S_t known up to a factor $S_x(0,t) \sim p(0,t)$. Thus for $Q \in \Xi$ the quantum potential, or equivalently the Weyl-Ricci curvature determine uniquely a Weyl space path describing the evolution of R and S.

Suppose for a given $\mathcal{R}_w \sim -8\beta Q$ and its corresponding $R(x,t)$ one had different $S(x,t)$, e.g. S and $\hat{S}$ with $S_x \sim F(t)$ and $\hat{S}_x \sim \hat{F}(t)$ so that from (10.5) and (10.6)

$$\frac{1}{m}\rho(\hat{S}_x \mp S_x) = \hat{F} \mp F = G_{\mp}(t) \Rightarrow \hat{S}_x \mp S_x = \frac{mG_{\mp}(t)}{\rho(x,t)} \tag{10.14}$$

$$\hat{S}_t - S_t = -\frac{1}{2m}(\hat{S}_x^2 - S_x^2) = -\frac{1}{2m}(\hat{S}_x - S_x)(\hat{S}_x + S_x) = -\frac{m}{2\rho^2}(G_-G_+) \tag{10.15}$$

COROLLARY 10.1. Given a SE based on a potential V, a wave function ψ determines a unique Weyl space path and Weyl-Ricci curvature $\mathcal{R}_w$ with corresponding evolution of R and S. On the other hand a given Weyl-Ricci curvatures $\mathcal{R}_w \sim -8\beta Q$ with $Q \in \Xi$ determines a unique Weyl space path for motion of R, S with R known completely and S determined modulo (10.7)-(10.8).

COROLLARY 10.2. As stated in Remark 7.10.2 a given SE is associated to a collection of Weyl spaces (determined via wave functions) in which R and S evolve classically. The Weyl space path is completely characterized via $Q \in \Xi$ (i.e. by Weyl-Ricci curvatures).

REMARK 7.10.4. The approach $\rho = mP(x,t)$ is intrinsic in Section 7.5 and [**1115, 1116**]. However one can argue as in Rylov [**1097**] that there is some incompatibility in claiming that a wave function can represent a single particle. The formula $\rho = mP$ suggests an ensemble interpretation and P is essentially arbitrary; fixing $P = P(x)$ with $\int Pdx = 1$, with a fixed Weyl space, seems then to determine an ensemble of particles corresponding to different $S = S(\rho, V)$ (or to different ψ). One can associate a mass m to each such wave function but one is indeed dealing with an ensemble so there seems to be no contradiction (cf. also [?]). ■

10.2. QUANTUM INFORMATICS. There is a fascinating series of papers indicated in Bogdanov [**157**] concerning **quantum informatics** and we only sketch here a few ideas. Let $\psi(x) \in L^2(\mathbf{R})$ be given and consider

$$\psi(x) = \frac{1}{\sqrt{2\pi}}\int \tilde{\psi}(p)e^{ipx}dp;\ \tilde{\psi}(p) = \frac{1}{\sqrt{2\pi}}\int \psi(x)e^{-ipx}dx \tag{10.16}$$

where (naturally) $\int |\psi|^2dx = \int |\tilde{\psi}|^2dp$. One writes (**10C**) $P(x) = |\psi(x)|^2$ and $\tilde{P}(p) = |\tilde{\psi}(p)|^2$ with normalization (**10D**) $\int P(x)dx = \int \tilde{P}(p)dp = 1$. The coordinate and momentum probability distributions $P(x)$ and $\tilde{P}(p)$ are called mutually complementary statistical distributions and one notes that e.g. information about the phase S of ψ is lost. For an experimental extracting of information it is not sufficient to use only one fixed representation (Bohr's complementarity principle). There is no distribution $P(x,p)$ corresponding to the $P(x)$ and $\tilde{P}(p)$ distributions since this would violate the Heisenberg uncertainty principle. One shows in [**157**] that classical statistics is incomplete here while quantum statistics is complete via the introduction of ψ. Thus write

$$(10.17)\qquad P(x) = \psi^*\psi = \frac{1}{2\pi}\int dpdp_1\tilde{\psi}^*(p)\tilde{\psi}(p_1)e^{-ix(p-p_1)} =$$

$$\frac{1}{2\pi}\int dudp\tilde{\psi}^*(p)\tilde{\psi}(p-u)e^{-ixu} = \frac{1}{2\pi}\int f(u)e^{-ixu}du$$

where (**10E**) $f(u) = \int dp\tilde{\psi}^*(p)\tilde{\psi}(p-u) = \int dp\tilde{\psi}^*(p+u)\tilde{\psi}(p)$. Thus one has a "mean value" formula (**10F**) $P(x) = (1/2\pi)\int f(u)exp(-ixu)du$ with (**10G**) $f(u) = M(exp(iux)) = \int P(x)exp(ixu)dx$. Similarly (**10H**) $\tilde{f}(t) = \int \tilde{P}(p)exp(ipt)dp = M(exp(ipt))$ for (**10I**) $\tilde{f}(t) = \int dx\psi^*(x-t)\psi(x)$. This leads to the assertion that in order for the function $f(u)$ to be a characteristic function (as above) it is necessary and sufficient that it be represented as a convolution as in (**10E**) with $\tilde{\psi}$ satisfying $\int dp|\tilde{\psi}(p)|^2 = 1$. To see the necessity let $f(u)$ be a characteristic function so by (**10F**) it defines a density $P(x)$. Let $\psi(x) = \sqrt{P(x)}exp(iS(x))$ for arbitrary real S (e.g. $S = 0$); this amounts to completing a classical statistical distribution to a quantum state. Then $\tilde{\psi}(p)$ defined via (9.9) provides the decomposition (**10E**) (thus $f(u) \to \tilde{\psi}(p)$). For sufficiency let $f(u)$ be represented via (**10E**) via $\tilde{\psi}(p)$ (normalized as in (**10C**)). Then $f(u)$ will be a characteristic function for $P(x)$ defined via (**10F**) (and normalized). Thus $\tilde{\psi}(p) \to f(u) \to P(x)$. We note that even in classical statistics the equation (**10E**) implicitly reveals the existence of a momentum space and the corresponding wave function $\tilde{\psi}(p)$ and suggests the paucity of classical notions of probability. Of course from (**10E**) one cannot derive a unique wave function $\tilde{\psi}(p)$ and (**10C**) does not yield $\psi(x)$ unambiguously. Hence one classical probability distribution may be described by a number of quantum objects; in order for the statistical theory to be "complete" it has to be expanded in a manner to obtain a quantum state vector ψ (e.g. via introduction of a phase multiplier).

One notes that moments of a random variable can be calculated via means of the characteristic function. Thus $f(u) = \int P(x)exp(ixu)dx$ yields (**10J**) $f^{(k)}(0) = i^kM(x^k)$ $(k = 0, 1, 2, \cdots)$. Simple calculations then lead to representations $\hat{x} \sim i\partial_p$ in momentum space and $\hat{p} = -i\partial_x$ in coordinate space. Consequently e.g. $\hat{p}\hat{x} - \hat{x}\hat{p} = -i$ is invariant under change of representation space. We recall also the standard Cauchy-Schwartz-Bunyakowski inequality (**10K**) $| < \phi|\psi > |^2 \leq < \phi|\phi >< \psi|\psi >$ and one introduces a "fidelity" F via (**10L**) $F = (| < \phi|\psi > |^2/ < \phi|\phi >< \psi|\psi >)$ with $0 \leq F \leq 1$. The Heisenberg uncertainty relation can be derived via consideration of

$$(10.18)\quad F(\xi) = < \psi|(-i\xi\hat{p}+\hat{x})(i\xi\hat{p}+\hat{x})\psi > = \xi^2M(\hat{p}^2) - i\xi M(\hat{p}\hat{x}-\hat{x}\hat{p}) + M(\hat{x}^2) \geq 0$$

Setting e.g. (**10M**) $D_p = M(\hat{p}^2) - (M(\hat{p}))^2$ this leads to (**10N**) $D_xD_p \geq 1/4$ with equality only when $(i\xi\hat{p} + \hat{x})|\psi >= 0$ for some ξ (which means Gaussian states). A more general result is the Robertson-Schrödinger uncertainty relation (cf. [**157, 1077**]). Thus let z_1 and z_1 be two observables (centered so $(M(z_1) =$

$M(z_2) = 0$) and consider

$$F(\xi) =< \psi|(\xi exp(-i\phi)z_2 + z_1)(\xi exp(i\phi)z_2 + z_1)|\psi >\geq 0 \tag{10.19}$$

Here ξ and ϕ are real and one defines the covariance as (**10O**) $cov(z_1, z_2) = (1/2) < \psi|z_1 z_2 + z_2 z_1|\psi >$. Let $z_1 z_2 - z_2 z_2 = iC$ where C is Hermitian so that (**10P**) $M(C) = -i < \psi|z_1 z_2 - z_2 z_1|\psi >$ leading to

$$F(\xi) = \xi^2 M(z_2^2) + \xi[2cov(z_1, z_2)cos(\phi) - M(C)Sin(\phi)] + M(z_1^2) \tag{10.20}$$

Set $\rho^2 = 4(cov(z_1, z_2))^2 + (M(C))^2$ and choose an angle β so that

$$2cov(z_1, z_2) = \rho Cos(\beta) \text{ and } M(C) = \rho Sin(\beta) \Rightarrow \tag{10.21}$$

$$\Rightarrow F(\xi) = \xi^2 M(z_2^2) + \xi\rho Cos(\phi + \beta) + M(z_1^2) \geq 0$$

Now choose ϕ so that $Cos(\phi + \beta) = 1$ which yields

$$M(z_1^2)M(z_2^2) = D(z_1)D(z_2) \geq \frac{\rho^2}{4} = \left((cov(z_1, z_2))^2 + \frac{(M(C))^2}{4}\right) \tag{10.22}$$

Then define a correlation coefficient (**10Q**) $r = [cov(z_1, z_2)/\sqrt{D(z_1)D(z_2)}]$ leading to

$$D(z_1)D(z_2) \geq \frac{(M(C))^2 K^2}{4}; \; K = \frac{1}{\sqrt{1 - r^2}} \tag{10.23}$$

Here K is analogous to the Schmidt number (cf. [**157**], papers 6 and 7 and [**425, 797, 1043**] for details). When $z_1 = x$ and $z_2 = p$, C is unitary, and one obtains (**10R**) $D(x)D(p) \geq (K^2/4)$ leading to $\Delta x \Delta p \geq (K/2)$. We note that since $\hat{x}$ and $\hat{p}$ do not commute their quantum covariance can not be estimated by their sampling as for a classical covariance; for the corresponding estimate one needs the wave function $\psi(x) = \sqrt{\rho(x)}exp(iS(x))$ which leads then to

$$cov(x, p) = \frac{1}{2} < \psi|xp + px|\psi >= \int x \frac{\partial S(x)}{\partial x} \rho(x) dx \tag{10.24}$$

where $\partial_x S = p$ is the momentum. Fisher information and the Cramer-Rao inequality are also discussed in [**157**]. In summarizing one postulates in [**157**]

(1) The principle object of quantum informatics is a quantum system. The evolution of the quantum system is described via probability amplitudes which construct state vectors in Hilbert space.
(2) The state vectors can be defined in different equivalent representations and are thus connected by unitary transformations which describe the time evolution of the quantum system.
(3) Measurements made in different unitary interconnected basis representations generate a set of mutually complementary statistical distributions. For a fixed basis the square of the absolute value of the probability amplitude defines the probability of the quantum system detection in a corresponding basis state.
(4) The space for a composite system state is produced by the tensor product of the states of individual systems.

A conclusion reached here is that QM is a root statistical model, based not on the actual probabilities but on their square roots; this is connected to the flow of half-densities under Bohmian flow in phase space as in [**551, 552, 554, 627**].

CHAPTER 8

ON NONLINEAR SCHRÖDINGER EQUATIONS

1. GENERAL REMARKS

Nonlinear Schrödinger equations (NLSE) have been considered in various contexts by various authors and we mention here only [**78, 139, 254, 275, 282, 360, 389, 391, 538, 540, 767, 1134, 1135**]. A general discussion arises in [**389, 391, 390, 538, 539, 540**] and we sketch first some of these results (there are many more references as we go along). One shows in particular that a whole family of NLSE are physically equivalent to the linear Schrödinger equation (SE) - cf. [**389, 391, 390**]. Thus let ψ be a quantum mechanical wave function with

$$\rho = \bar{\psi}\psi;\ \vec{\jmath} = \frac{\hbar}{2mi}[\bar{\psi}\nabla\psi - (\nabla\bar{\psi})\psi] \tag{1.1}$$

Then the NLSE considered have the form

$$i\hbar\psi_t = H_0\psi + i\mathfrak{I}(\psi)\psi + \mathfrak{R}(\psi)\psi;\ H_0\psi = -\frac{\hbar^2}{2m}\nabla^2\psi + V\psi \tag{1.2}$$

Here $\mathfrak{I}(\psi)$ and $\mathfrak{R}(\psi)$ are real valued functions given via

$$\mathfrak{R}(\psi) = \hbar D'\sum_1^5 c_j R_j(\psi);\ \mathfrak{I}(\psi) = \frac{1}{2}\hbar D R_2(\psi) \tag{1.3}$$

$$R_1 = \frac{\nabla\cdot\hat{\jmath}}{\rho};\ R_2 = \frac{\nabla^2\rho}{\rho};\ R_3 = \frac{\hat{\jmath}^2}{\rho^2};\ R_4 = \frac{\hat{\jmath}\cdot\nabla\rho}{\rho^2};\ R_5 = \frac{(\nabla\rho)^2}{\rho^2} \tag{1.4}$$

with (♣) $\hat{\jmath} = (m/\hbar)\vec{\jmath} = (1/2i)[\bar{\psi}\nabla\psi - (\nabla\bar{\psi})\psi]$ while D and D' are real numbers with the dimensions of diffusion coefficients. Further the usual continuity equation for ρ and $\vec{\jmath}$ is replaced by a Fokker-Planck type equation (♠) $\partial_t\rho = -\nabla\cdot\vec{\jmath} + D\nabla^2\rho$ (this goes back to Schuch [**1134, 1135**] for example). When one calculates the time rate of change of the expectation values $< -i\hbar\nabla >= \int\bar{\psi}[-i\hbar\nabla\psi(\mathbf{x},t)]d\mathbf{x}$ and $< i\hbar\partial_t >= \int\bar{\psi}[i\hbar\partial_t\psi]d\mathbf{x}$ there arise extra dissipative terms and there is a subfamily characterized by the conditions $D'c_1 = D = -D'c_4$ and $c_2 + 2c_5 = c_3 = 0$ for which the dissipative terms in $(d/dt) < -i\hbar\nabla >$ are zero (i.e. for which Ehrenfest's theorem formally holds). Equations in this subfamily are linearizable by means of nonlinear transformations. A larger subfamily of the equations are Galileian invariant (when $c_1 + c_4 = c_3 = 0$) so in particular the family obeying Ehrenfest's theorem is Galileian invariant. In Davidson [**360**] one considers nonlinear transformations of the form

$$\psi \to \psi' = N(\psi) = |\psi| exp[i(\gamma log(|\psi|) + \Lambda arg(\psi)] \tag{1.5}$$

where γ and Λ are real numbers with $\Lambda \neq 0$. If ψ solves an equation in the Ehrenfest class then ψ' satisfies the linear SE

$$i\frac{\hbar}{\Lambda_E}\partial_t\psi' = -\frac{\hbar^2}{2\Lambda_E^2 m}\nabla^2\psi' + V(\mathbf{x},t)\psi; \tag{1.6}$$

under the conditions

$$\gamma = \gamma_E = -\frac{2mD}{\hbar}\left(1 - \frac{4m}{\hbar}D'c_2 - \frac{4m^2D^2}{\hbar^2}\right)^{-1/2}; \tag{1.7}$$

$$\Lambda = \Lambda_E = \left(1 - \frac{4m}{\hbar}D'c_2 - \frac{4m^2D^2}{\hbar^2}\right)^{-1/2}; \quad \frac{4m}{\hbar}D'c_2 < 1 - \frac{4m^2D^2}{\hbar^2}$$

Note that the wave function ψ' in (6.5) involves an argument involving integer multiples of 2π but this can always be arranged and one has a concrete example of a nonlinear gauge transformation (in fact a "group" $\mathfrak{N}$ is involved so that distinct NLSE related by elements of $\mathfrak{N}$ should be regarded as belonging to equivalence classes predicting exactly the same physics - see here Czachor [**347**] for a 3-parameter extension to a group).

Let us rewrite (6.3)-(6.5) in terms of R and S where $\psi = Rexp(iS/\hbar)(\sim Re^{\sim})$. Thus working in one space dimension for convenience (with $\partial_x f = \partial f = f'$) one has $\psi_t = R_t e^{\sim} + R(iS_t/\hbar)e^{\sim}$, $\psi_x = R_x e^{\sim} + R(iS_x/\hbar)e^{\sim}$, and $\psi_{xx} = [R_{xx} + 2R_x(iS_x/\hbar) + R(iS_{xx}/\hbar) - R(S_x^2/\hbar^2)]e^{\sim}$ leading to

$$\hat{\jmath} = \frac{m}{\hbar}\vec{\jmath} = \frac{1}{2i}[\bar{\psi}\partial_x\psi - (\partial_x\bar{\psi})\psi] = \frac{R^2S_x}{\hbar}; \ \rho = R^2; \tag{1.8}$$

$$\nabla\cdot\hat{\jmath} \sim \partial_x\hat{\jmath} = \frac{1}{\hbar}(2RR_xS_x + R^2S_{xx}) \tag{1.9}$$

$$\partial\rho = \partial R^2 = 2RR_x; \ \partial^2\rho = 2R_x^2 + 2RR_{xx} \tag{1.10}$$

Hence

$$R_1 = \frac{\nabla\cdot\hat{\jmath}}{\rho} \sim \frac{1}{\hbar}\left[S_x\left(\frac{R_x}{R}\right) + S_{xx}\right]; \tag{1.11}$$

$$R_2 = \frac{\nabla^2\rho}{\rho} \sim 2\left(\frac{R_x}{R}\right)^2 + 2\left(\frac{R_{xx}}{R}\right); \ R_3 = \frac{\hat{\jmath}^2}{\rho^2} \sim \frac{S_x^2}{\hbar^2} \tag{1.12}$$

$$R_4 = \frac{\hat{\jmath}\cdot\nabla\rho}{\rho^2} \sim \frac{S_xR^2}{\hbar}\frac{2RR_x}{R^4} = 2\frac{R_x}{R}\frac{S_x}{\hbar} \tag{1.13}$$

$$R_5 = \frac{(\nabla\rho)^2}{\rho^2} \sim \frac{4R^2R_x^2}{R^4} = 4\left(\frac{R_x}{R}\right)^2$$

Note that all of the terms are homogeneous (i.e. remain the same for $\psi \to \lambda\psi$). We recall also that the quantum potential for a linear SE $i\hbar\psi_t = -(\hbar^2/2m)\nabla^2\psi + V\psi$

is $Q = -(\hbar^2/2m)(R''/R)$ and note that $\partial log(R) = R'/R = Z$ with $\partial^2 log(R) = (R''/R) - (R'/R)^2 = (R''/R) - (\partial log(R))^2$. Consequently

$$(1.14)\qquad Q = -\frac{\hbar^2}{2m}[\partial Z + Z^2] \equiv Z' + Z^2 = -\alpha Q;\ \alpha = \frac{2m}{\hbar^2};\ \frac{R''}{R} = -\alpha Q$$

Note that this is a Ricatti equation for Z in terms of Q and the standard linearization $Z = \phi'/\phi$ yields $\phi'' + \alpha Q\phi = 0$ corresponding to $\phi = R$. This equation $(\Delta\phi + \alpha Q\phi = 0)$ was considered briefly in [**254, 257**] and for a given suitable Q one can "usually" produce unique solutions in say $H_0^1(\mathbf{R}^3)$. However in general the correspondence $SE \leftrightarrow Q$ is not $1-1$ (depending in particular on ψ) and this feature is of possible interest for the Doebner-Goldin (DG) type NLSE. In any event the terms in (1.12)-(1.13) can all be written as functions of $Z = \partial log(R)$, S, and Q as follows:

$$(1.15)\qquad R_1 = \frac{1}{\hbar}[S'Z + S''];\ R_2 = 2Z^2 - 2\alpha Q;\ R_3 = \frac{S_x^2}{\hbar^2};$$

$$R_4 = \frac{2S'Z}{\hbar};\ R_5 = 4Z^2$$

Now write out the NLSE in terms of real and imaginary parts via

$$(1.16)\qquad i\hbar R_t - RS_t = -\frac{\hbar^2}{2m}\left[R_{xx} - \frac{RS_x^2}{\hbar} + \frac{2iS_xR_x}{\hbar} + \frac{iS_{xx}R}{\hbar}\right] + VR +$$

$$+\frac{i\hbar}{2}DR_2(\psi)R + \hbar D'\sum_1^5 c_jR_j(\psi)R$$

This leads to

$$(1.17)\qquad -RS_t = -\frac{\hbar^2}{2m}\left[R_{xx} - \frac{RS_x^2}{\hbar^2}\right] + VR + \hbar D'R\left[\frac{c_1}{\hbar}(S'' + S'Z) + \right.$$

$$\left. + 2c_2(Z^2 - \alpha Q) + c_3\frac{S_x^2}{\hbar^2} + c_4\frac{2S_x}{\hbar}Z + 4c_5Z^2\right]$$

$$(1.18)\qquad \hbar R_t = -\frac{\hbar^2}{2m}\left[\frac{2S_xR_x}{\hbar} + \frac{S_{xx}R}{\hbar}\right] + (\hbar D/2)(2Z^2 - 2\alpha Q)R$$

and consequently

$$(1.19)\qquad \partial_t log(R) = -\frac{1}{2m}(2S'Z + S'') + D(Z^2 - \alpha Q)$$

$$(1.20)\qquad -S_t = -\frac{\hbar^2}{2m}\left[\frac{R''}{R} - \frac{S_x^2}{\hbar^2}\right] + V + \hbar D'\left[\frac{c_1}{\hbar}(S'' + S'Z) + \right.$$

$$\left. + 2c_2(Z^2 - 2\alpha Q) + \frac{c_3S_x^2}{\hbar^2} + \frac{2c_4S_xZ}{\hbar} + 4c_5Z^2\right] =$$

$$= -\frac{\hbar^2}{2m}\left[-\alpha Q - \frac{S_x^2}{\hbar^2}\right] + V + D'[c_1(S'' + S'Z) + 2\hbar D'c_2(Z^2 - \alpha Q) +$$

$$+\frac{c_3D'S_x^2}{\hbar} + 2D'c_4S_xZ + 4\hbar D'c_5Z^2$$

Differentiating (1.19 in x yields then (using $Z' = -Z^2 - \alpha Q$)

$$\hbar\partial_t Z = -\frac{\hbar}{2m}(2S''Z + S''' + 2S'Z') + \hbar D(2Z(-Z^2 - \alpha Q) - \alpha Q') \tag{1.21}$$

Now we want to see first what (if anything) is added to the HJ equation, i.e. writing $V_N = V + \mathfrak{R}(\psi)$ and (1.20) as $S_t = -(\hbar^2/2m)S_x^2 + V + Q + \mathfrak{V}$, what is the difference between $\mathfrak{R}(\psi)$ and $\mathfrak{V}$. Thus consider

$$-S_t = -\frac{\hbar^2}{2m}\left[-\alpha Q - \frac{S_x^2}{\hbar^2}\right] + V + \mathfrak{R}(\psi) \Rightarrow V_N = V + \mathfrak{R}(\psi) \tag{1.22}$$

Consequently no further "quantization" of the problem arises; (1.20) is simply $S_t = -(1/2m)S_x^2 + Q + V + \mathfrak{R}(\psi)$. The $\mathfrak{J}(\psi)$ term on the other hand only contributes to the R equation in (1.19) which corresponds to the Fokker-Planck equation (♠). Indeed (♠) can be written as

$$\partial_t R^2 = -\nabla\cdot\vec{\jmath} + D\nabla^2 R^2 \sim -\partial(\hbar/m)\hat{\jmath} + D\nabla^2 R^2 = 2RR_t = \tag{1.23}$$

$$= -(1/m)(2RR_xS_x + R^2 S_{xx}) + 2D(R_x^2 + RR') \equiv$$

$$\equiv \frac{R_t}{R} = \frac{1}{2m}(2S_xZ + S_{xx}) + D\left[\left(\frac{R'}{R}\right)^2 + \frac{R''}{R}\right] \equiv$$

$$\equiv \partial_t log(R) = \frac{1}{2m}(2S_xZ + S_{xx}) + D(Z^2 - \alpha Q)$$

in agreement with (1.19).

It seems possibly interesting that Q appears in (1.23) and in the potential $\mathfrak{R}(\psi) = D'\sum_1^5 c_j R_j(\psi)$ where

$$\mathfrak{R}(\psi) = D'[c_1(S'' + S'Z) + 2\hbar c_2(Z^2 - \alpha Q) + c_3(S_x^2/\hbar) + 2c_4 S'Z + 4\hbar c_5 Z^2] \tag{1.24}$$

For the Ehrenfest family one has $D'c_1 = D = -D'c_4$ and $c_2 + 2c_5 = 0 = c_3$ so

$$\mathfrak{R}(\psi) = D(S'' + S'Z) + 2\hbar(c_2 + 2c_5)D'Z^2 + 2\hbar D'c_2(-\alpha Q) - 2DS'Z = \tag{1.25}$$

$$= D(S'' - S'Z) - 2\hbar c_2 \alpha Q D'$$

$$\mathfrak{J}(\psi) = (\hbar D/2)R_2 = \hbar D(Z^2 - \alpha Q) \tag{1.26}$$

When $Q = 0$ one has

$$\mathfrak{R}(\psi) = D(S'' - S'Z);\ \mathfrak{J}(\psi) = \hbar D Z^2 \tag{1.27}$$

Then (♠) holds and the NLSE has an additional term

$$D(S'' - S'Z) = \frac{D}{\rho^2}(\rho\partial_x\hat{\jmath} - \hat{\jmath}\partial_x\rho) \sim \frac{D}{\rho^2}[\rho\nabla\cdot\hat{\jmath} - \hat{\jmath}\cdot\nabla\rho] \tag{1.28}$$

We note also (checking the formula in [**366**]) that

$$\frac{\nabla^2\psi}{\psi} = i\frac{\nabla\cdot\hat{\jmath}}{\rho} + \frac{1}{2}\frac{\nabla^2\rho}{\rho} - \frac{\hat{\jmath}^2}{\rho^2} - \frac{1}{4}\frac{(\nabla\rho)^2}{\rho^2} = \tag{1.29}$$

$$= iR_1 + \frac{1}{2}R_2 - R_3 - \frac{1}{4}R_5$$

1.1. GAUGE TRANSFORMATIONS. We go here to Doebner, Goldin, and Natterman [**389, 390, 539**] (especially Goldin [**539**]). Utilizing (1.24) and (1.29) one considers first $i\hbar\partial_t\psi = H_0\psi + (i/2)\hbar D R_2[\psi]\psi + \hbar D'\sum_1^5 c_j R_j[\psi]\psi$ with

$$H_0\psi = \frac{1}{2m}\left[-i\hbar\nabla - \frac{e}{c}\mathbf{A}(x,t)\right]^2\psi + [V + e\Phi(x,t)]\psi \tag{1.30}$$

(here boldface denotes vectors and $x \sim \mathbf{x}$ in more than one space dimension). Using the expansion $\nabla^2\psi/\psi = iR_1[\psi] + (1/2)R_2[\psi] - R_3[\psi] - (1/4)R_5[\psi]$ (cf. (1.29)) one adds some additional terms to get

$$i\frac{\dot{\psi}}{\psi} = i\left[\sum_1^2 \nu_j R_j[\psi] + \frac{\nabla\cdot(\mathfrak{A}\rho)}{\rho}\right] + \tag{1.31}$$

$$\left[\sum_1^5 \mu_j R_j[\psi] + U(x,t) + \frac{\nabla\cdot(\mathfrak{A}_1\rho)}{\rho} + \frac{\mathfrak{A}_2\cdot\hat{\jmath}}{\rho} + \alpha_1 log(\rho) + \alpha_2 S\right]$$

$$\nu_1 = -\frac{\hbar}{2m};\ \nu_2 = \frac{D}{2};\ \mathfrak{A} = \frac{e}{2mc}\mathbf{A};\ \mu_1 = D'c_1;\ \mu_2 = -\frac{\hbar}{4m} + D'c_2; \tag{1.32}$$

$$\mu_3 = \frac{\hbar}{2m} + D'c_3;\ \mu_4 = D'c_4;\ \mu_5 = \frac{\hbar}{8m} + D'c_5;\ \alpha_1 = \alpha_2 = 0;$$

$$U = \frac{1}{\hbar}[V + e\Phi] + \frac{e^2}{2m\hbar c^2}\mathbf{A}^2;\ \mathfrak{A}_1 = 0;\ \mathfrak{A}_2 = -\frac{e}{mc}\mathbf{A}$$

where ν_j $(j = 1,2)$, μ_j $(j = 1,\cdots,5)$, and α_j $(j = 1,2)$ are continuously differentiable real valued functions of t. In fact in order to investigate the appropriate family of nonlinear gauge transformations one wants to complexify all this and deal with

$$i\frac{\dot{\psi}}{\psi} = i\left[\sum_1^5 \nu_j R_j[\psi] + \mathfrak{T}(x,t) + \frac{\nabla\cdot(\mathfrak{A}\rho)}{\rho} + \frac{\mathfrak{D}\cdot\hat{\jmath}}{\rho} + \delta_1 log(\rho) + \delta_2 S\right] + \tag{1.33}$$

$$+\left[\sum_1^5 \mu_j R_j[\psi] + U + \frac{\nabla\cdot(\mathfrak{A}_1\rho)}{\rho} + \frac{\mathfrak{A}_2\cdot\hat{\jmath}}{\rho} + \alpha_1 log(\rho) + \alpha_2 S\right]$$

where $\mathfrak{T}$ is a new external scalar field and $\mathfrak{D}$ a new external vector field. This extension of (1.38) to (1.33) is in fact quite natural and includes heat equations and some soliton type equations (cf. [**539**] for references). Now write $\psi = Rexp[iS]$ with real R and S so that $\rho = R^2$ and $\mathbf{j} = (\hbar/m)R^2\nabla S$ (note $\hat{\jmath} \sim \hat{\mathbf{j}}$). Under the usual unitary gauge transformations of QM $R' = R$ and $S' = S + \theta(\mathbf{x},t)$ so $\rho' = \rho$ while $\mathbf{j}' = \mathbf{j} + (\hbar/m)R^2\nabla\theta$. If one begins with a linear SE with no vector potential then $\psi' = R'exp[iS'] = exp[i\theta(x,t)]\psi$ will satisfy $i\hbar\partial_t\psi' = (\hbar^2/2m)[-i\nabla - \nabla\theta]^2\psi' + [V - \hbar\dot{\theta}]\psi'$. When one begins with H_0 as indicated one automatically generates gauge transformed potentials $\mathbf{A}' = \mathbf{A} + (\hbar c/e)\nabla\theta$ and $\Phi' = \Phi - (\hbar/e)\dot{\theta}$ so that a gauge invariant current can be defined via $\mathbf{J} = \mathbf{j} - (e/mc)\rho\mathbf{A}$ with $\partial_t\rho = -\nabla\mathbf{J}$; the physical fields $\mathbf{B} = \nabla\times\mathbf{A}$ and $\mathbf{E} = -\nabla\Phi - (1/c)\partial_t\mathbf{A}$ are also gauge invariant. This sets the stage for the DG nonlinear transformations (♦) $R' = R$ $S' = \Lambda S +$

$\gamma log(R) + \theta$ where in general γ and Λ are differentiable real valued functions of t with $\Lambda \neq 0$ and θ is a differentiable real valued function of (x, t). Now under (♦)

$$\rho' = \bar{\psi}'\psi' = \rho; \ \mathbf{J}' = \frac{1}{2i}[\bar{\psi}'\nabla\psi - (\nabla\bar{\psi}')\psi'] = \Lambda\hat{J} + \frac{\gamma}{2}\nabla\rho + \rho\nabla\theta \quad (1.34)$$

This equation has many nice properties, e.g. first, it is local and respects a certain separation condition for many particle product wave functions, and second, if ψ satisfies a SE of type (1.31) then ψ' transformed via (♦) satisfies another equation in the family with transformed coefficients (the coefficients and fields are given in [**539**]). Further there are gauge invariant parameters

$$\tau_1 = \nu_2 - \frac{1}{2}\mu_1; \ \tau_2 = \nu_1\mu_2 - \nu_2\mu_1; \ \tau_3 = \frac{\mu_3}{\nu_1}; \beta_2 = \alpha_2 - \frac{\dot{\nu}_1}{\nu_1}; \quad (1.35)$$

$$\tau_4 = \mu_4 - \mu_1\frac{\mu_3}{\nu_1}; \ \tau_5 = \nu_1\mu_5 - \nu_2\mu_4 + \nu_2^2\frac{\mu_3}{\nu_1}, \ \beta_1 = \nu_1\alpha_1 - \nu_2\alpha_2 + \nu_2\frac{\dot{\nu}_1}{\nu_1} - \dot{\nu}_2$$

(cf. [**390, 539**] for some discussion of these parameters. Following [**541**] one also writes down in [**539**] the gauge invariant fields and equations of motion corresponding to (1.31).

Now for the enlarged gauge group one writes $T = log(R)$ so that $log(\psi) = T + iS$ and considers the transformations

$$\begin{pmatrix} S' \\ T' \end{pmatrix} = \begin{pmatrix} \Lambda & \gamma \\ \lambda & \kappa \end{pmatrix}\begin{pmatrix} S \\ T \end{pmatrix} + \begin{pmatrix} \theta \\ \phi \end{pmatrix} \quad (1.36)$$

where Λ, γ, λ, and κ depend on t and θ, ϕ depend on (x, t). In place of the condition $\Lambda \neq 0$ one imposes the stipulation $\Delta = \kappa\Lambda - \lambda\gamma \neq 0$ so that Δ is invertible. One models this transformation group $\mathfrak{G}$ on $GL(2, \mathbf{R})$. The variables S and T are treated on an equal footing (see e.g. [**1276**]). One needs here the complexification of (1.31) to (1.33) and can now write this in the form

$$\dot{S} = a_1\nabla^2 S + a_2\nabla^2 T + a_3(\nabla S)^2 + a_4\nabla S\cdot\nabla T + a_5(\nabla T)^2 + a_6 S + a_7 T + \quad (1.37)$$

$$+u_0 + \mathbf{u}_1\cdot\nabla S + \mathbf{u}_2\cdot\nabla T; \ \dot{T} = b_1\nabla^2 S + b_2\nabla^2 T +$$

$$+b_3(\nabla S)^2 + b_4\nabla S\cdot\nabla T + b_5(\nabla T)^2 + b_6 S + b_7 T + v_0 + \mathbf{v}_1\cdot\nabla S + \mathbf{v}_2\cdot\nabla T$$

The relation between (1.31) and (1.37) is given via

- $a_1 = -\mu_1$; $a_2 = -2\mu_2$; $a_3 = -\mu_3$; $a_4 = -2\mu_1 - 2\mu_4$; $a_5 = -4\mu_2 - 4\mu_5$; $a_6 = -\alpha_2$; $a_7 = -2\alpha_1$; $u_0 = -U - \nabla\cdot\mathfrak{A}_1$; $\mathbf{u}_1 = -\mathfrak{A}_2$; $\mathbf{u}_2 = -2\mathfrak{A}_1$
- $b_1 = \nu_1$; $b_2 = 2\nu_2$; $b_3 = \nu_3$; $b_4 = 2\nu_1 + 2\nu_4$; $b_5 = 4\nu_2 + 4\nu_5$; $b_6 = \delta_2$; $b_7 = 2\delta_1$; $v_0 = \mathfrak{T} + \nabla\cdot\mathfrak{A}$; $\mathbf{v}_1 = \mathfrak{D}$; $\mathbf{v}_2 = 2\mathfrak{A}$

The usual SE (as in [**254**]) involves

$$\dot{S} + \frac{1}{2m}S_x^2 + V - \frac{\hbar^2}{2m}\frac{R''}{R} = 0; \ \partial_t R^2 + \frac{1}{m}(R^2 S_x)_x = 0 \equiv \quad (1.38)$$

$$\equiv \dot{S} + \frac{1}{2m}(\nabla S)^2 + V - \frac{\hbar^2}{2m}[\nabla^2 T + (\nabla T)^2] = 0; \ \dot{T} + \frac{1}{2m}(2\nabla S\cdot\nabla T + \nabla^2 S) = 0$$

Hence

$$a_1 = 0; \ a_2 = \frac{\hbar^2}{2m}; \ a_3 = -\frac{1}{2m}; \ a_4 = 0; \ a_5 = \frac{\hbar^2}{2m}; \quad (1.39)$$

$$b_1 = -\frac{1}{2m},\ b_2 = b_3 = b_5 = 0;\ b_4 = -\frac{1}{m}$$

(note the S here corresponds to $\hbar S$ above). In general the coefficients a_j, b_j satisfy ($\Delta = \kappa\Lambda - \lambda\gamma$)

$$(1.40)\qquad \begin{pmatrix} a_1' \\ a_2' \\ b_1' \\ b_2' \end{pmatrix} = \Delta^{-1} \begin{pmatrix} \kappa\Lambda & -\lambda\Lambda & \kappa\gamma & -\lambda\gamma \\ -\gamma\Lambda & \Lambda^2 & -\gamma^2 & \gamma\Lambda \\ \kappa\lambda & \lambda^2 & \kappa^2 & -\kappa\lambda \\ -\lambda\gamma & \lambda\Lambda & \kappa\gamma & \kappa\Lambda \end{pmatrix} \begin{pmatrix} a_1 \\ a_2 \\ b_1 \\ b_2 \end{pmatrix};$$

$$\begin{pmatrix} a_3' \\ a_4' \\ a_5' \\ b_3' \\ b_4' \\ b_5' \end{pmatrix} = \Delta^{-2}\mathfrak{M} \begin{pmatrix} a_3 \\ a_4 \\ a_5 \\ b_3 \\ b_4 \\ b_5 \end{pmatrix}$$

$$\mathfrak{M} = \begin{pmatrix} \kappa^2\Lambda & -\kappa\lambda\Lambda & \lambda^2\Lambda & \kappa^2\gamma & -\kappa\lambda\gamma & \lambda^2\gamma \\ -2\kappa\gamma\Lambda & \Lambda(\kappa\Lambda+\lambda\gamma) & -2\lambda\Lambda^2 & -2\kappa\gamma^2 & \gamma(\kappa\Lambda+\lambda\gamma) & -2\lambda\gamma\Lambda \\ \gamma^2\Lambda & -\gamma\Lambda^2 & \Lambda^3 & \gamma^3 & -\gamma^2\Lambda & \gamma\Lambda^2 \\ \kappa^2\lambda & -\kappa\lambda^2 & \lambda^3 & \kappa^3 & -\kappa^2\lambda & \kappa\lambda^2 \\ -2\kappa\lambda\gamma & \lambda(\kappa\Lambda+\lambda\gamma) & -2\lambda^2\Lambda & -2\kappa^2\gamma & \kappa(\kappa\Lambda+\lambda\gamma) & -2\kappa\lambda\Lambda \\ \lambda\gamma^2 & -\lambda\gamma\Lambda & -\lambda\Lambda^2 & \kappa\gamma^2 & -\kappa\gamma\Lambda & \kappa\Lambda^2 \end{pmatrix}$$

(1.41)

$$\begin{pmatrix} a_6' \\ a_7' \\ b_6' \\ b_7' \end{pmatrix} = \Delta^{-1} \begin{pmatrix} \kappa\Lambda & -\lambda\Lambda & \kappa\gamma & -\lambda\gamma \\ -\gamma\Lambda & \Lambda^2 & -\gamma^2 & \gamma\Lambda \\ \kappa\lambda & \lambda^2 & \kappa^2 & -\kappa\lambda \\ -\lambda\gamma & \lambda\Lambda & -\kappa\gamma & \kappa\Lambda \end{pmatrix} \begin{pmatrix} a_6 \\ a_7 \\ b_6 \\ b_7 \end{pmatrix} + \Delta^{-1} \begin{pmatrix} \kappa\dot{\Lambda} - \lambda\dot{\gamma} \\ \Lambda\dot{\gamma} - \gamma\dot{\Lambda} \\ \kappa\dot{\lambda} - \lambda\dot{\kappa} \\ \Lambda\dot{\kappa} - \gamma\dot{\lambda} \end{pmatrix}$$

The behavior of the external fields under gauge transformation is more complicated and not written out in [**539**].

Now a main point here is that it is possible to write combinations formed from S and T that are invariant under (1.36). Consider e.g. the matrix part of (2.7) (i.e. $\theta = \phi = 0$) and denote the gauge transformation matrix by $\mathcal{A}$. Suppose d_1, d_2 are some coefficients depending on the a_j, b_j and assume $d_1S + d_2T$ is an invariant combination; then it follows that this can happen if and only if $[d_1\,d_2]\mathcal{A}^{-1} = [d_1'\,d_2']$ (e.g. $d_1 = 2a_3 + b_4$ and $d_2 = a_4 + 2b_5$ obey this condition). Next let $L_1 = a_1S + a_2T$ and $L_2 = b_1S + b_2T$; then (L_1, L_2) transforms under $\mathcal{A}$ exactly as does the pair (S, T) so that $d_1L_1 + d_2L_2$ is also an invariant. In fact any combination $d_1(\sigma L_1 + \tau S) + d_2(\sigma L_2 + \tau T)$, where σ and τ are fully invariant combinations of the coefficients, is also invariant. One checks that $a_1 + b_2 = 2\tau_1$ and $a_1b_2 - a_2b_1 = 2\tau_2$ are then gauge invariants under (1.36). One can interpret now $\tau_2 > 0$ as characterizing the class in (1.37) pertaining to quantum mechanics (QM) and $\tau_2 \to 0$ defines the classical limit; this seems interesting to examine in terms of the quantum potential Q.

REMARK 8.1.1. From Section 8.1 with $Z = \partial log(R)$ we can write $Z = \partial T$ of course and since in the QM situation $Z' + Z^2 = -\alpha Q$ with $\alpha = 2m/\hbar^2$ one has

$\partial^2 T + (\partial T)^2 = -\alpha Q$ which in 3-D corresponds to $\nabla^2 T + (\nabla T)^2 = -\alpha Q$. Now $2\tau_2 = a_1 b_2 - a_2 b_1$ which in the QM situation is from (1.39)

$$2\tau_2 = -a_2 b_1 = \frac{\hbar^2}{4m} \Rightarrow \tau_2 \to 0 \iff \hbar \to 0 \tag{1.42}$$

Thus in this context there are gauge theoretic reasons to think of $\hbar \to 0$ as a classical limit. Recall that in the linear theory $\hbar \to 0 \not\equiv Q \to 0$ (cf. [**163, 275**]). In fact in [**390**] one shows more about the gauge invariants τ_i. In particular for the linear SE

$$\tau_1 = 0;\ \tau_2 = \frac{\hbar^2}{8m^2};\ \tau_3 = 1;\ \tau_4 = 0;\ \tau_5 = -\frac{\hbar^2}{16m^2} \tag{1.43}$$

and one can write

$$\left[\frac{\hbar}{m}\right]_{obs} = (8\tau_2)^{1/2};\ \left[\frac{\hbar}{m}\right]_{obs} = (-16\tau_5)^{1/2} \tag{1.44}$$

where $obs \sim$ observed. Since τ_2 and τ_5 are independent variables $[\hbar/m]_{obs}$ could have different values which are the same in the special case of a linear SE. Also for $V = 0$ the SE can contain $\hbar/m$ as a parameter so $\hbar$ and m can not be individually obtained from the equations. For the Ehrenfest family one has

$$\tau_1 = 0;\ \tau_2 = \frac{\hbar^2}{2m^2} - \kappa;\ \tau_3 = -1;\ \tau_4 = 0; \tag{1.45}$$

$$\tau_5 = -\frac{1}{2}\left(\frac{\hbar^2}{8m^2} - \kappa\right);\ \kappa = \frac{\hbar}{2m}D'c_2 + \frac{1}{2}D^2$$

(cf. (1.7)). As long as $\kappa < (\hbar^2/8m^2)$ so that $\tau_2 > 0$ the NLSE is linearizable and the equations in the Ehrenfest family are a 1-parameter family of gauge inequivalent theories differing from one another only in the physical observable $[\hbar/m]_{obs}$. ∎

EXAMPLE 1.1. For a plane wave $\psi = Rexp[i(kx - (\hbar k^2 t/2m))]$ with R constant one has $Q = 0$ and $S_x = k$ ($k \sim$ constant). Here R constant corresponds to $Z = 0$ or $T = log(R) = constant$ and $\partial^2 S = 0$ as well so (1.37) is valid for $\dot{T} = 0$. Also $\dot{S} = -\hbar k^2/2m = -(\hbar/2m)S_x^2$ agress with (1.37). Thus $\hbar \to 0$ corresponds here to $T = S = constant$ (with $U = 0$) so there is no dynamics. Consider now the same SE with two solutions as in [**237**], namely $\psi_{\pm} = Rexp[i \pm kx - (\hbar k^2 t/2m))]$ and set $\psi = (1/2)\psi_+ + \psi_-) = Cos(kx)exp(-i\hbar k^2 t/2m)$. Then $R = Cos(kx)$ with $Q = \hbar k^2/2m$ and $S = -\hbar k^2 t/2m$ which means $\dot{S} = -\hbar k^2/2m$ as before but $S_x = 0$; however again $\dot{T} = 0$ (agreeing with (2.8) and $T = log(R) = log[Cos(kx)]$. Further $T' = -kTan(kx)$ and $T'' = -k^2 Sec^2(kx)$ so (2.8) becomes for $\dot{S}$

$$\dot{S} = -\frac{\hbar k^2}{2m} = -\frac{\hbar k^2}{2m}Sec^2(kx) + \frac{\hbar k^2}{2m}Tan^2(kx) = -\frac{\hbar k^2}{2m} \tag{1.46}$$

as required. Here one sees that $SE \leftrightarrow Q$ is not 1-1 and Q plays the role of kinetic energy with $Q \to 0$ along with $\hbar \to 0$. ∎

EXAMPLE 1.2. An example from Bolivar [**163**] with potential $V = m\omega^2 x^2/2$ and a stationary SE gives solutions

$$\psi_n(x) = c_n H_n(\xi x)e^{-\xi^2 x^2/2};\ \xi = (m\omega\hbar)^{1/2};\ c_n = \frac{\xi}{\sqrt{\pi 2^n n!}} \tag{1.47}$$

Then $Q = \hbar\omega[n + (1/2)] - (1/2)m\omega^2 x^2$ so $\hbar \to 0 \not\Rightarrow Q \to 0$ and $Q = 0$ corresponds to $x = \pm\sqrt{(2\hbar/m\omega)[n + (1/2)]}$. Here one has $T(x) = \psi_n(x)$ with $S = 0$. ■

1.2. COMPLEX NONLINEARITIES. We go now to Kaniadakis-Scarfone [**723**] (cf. also [**254, 722, 726, 727, 728**]); this provides some embellishments to the Doebner-Goldin (DG) theory. One introduces the Lagrangian

$$(1.48)\quad \mathfrak{L} = \frac{i\hbar}{2}(\psi^*\partial_t\psi - \psi\partial_t\psi^*) - \frac{\hbar^2}{2m}|\partial_x\psi|^2 - U[\psi,\psi^*];\ U[a] = U(a_0, a_1, \cdots)$$

where $a_n = \partial^n a/\partial x^n$. One uses

$$(1.49)\qquad \rho = \psi\psi^*;\ S = \frac{i\hbar}{2}log\left(\frac{\psi}{\psi^*}\right);\ \psi = \rho^{1/2}exp[(i/\hbar)S)]$$

The evolution equations for the fields ψ, ψ^*, ρ, S can be obtained via (★) $\delta\mathfrak{A}/\delta a = 0$ with $\mathfrak{A} = \int \mathfrak{L}dxdt$ where a standard formula of the type $(\delta/\delta a)\int U[a]dxdt = \sum_0^N(-1(^n\partial_x^n(\partial U/\partial a_n)[a])$ is used. This leads to the SE

$$(1.50)\qquad i\hbar\partial_t\psi = -\frac{\hbar^2}{2m}\partial_x^2\psi + W[\rho,S]\psi + i\mathfrak{W}[\rho,S]\psi;$$

$$W[\rho,S] = \frac{\delta}{\delta\rho}\int u[\rho,S]dxdt;\ \mathfrak{W}[\rho,S] = \frac{\hbar}{2\rho}\frac{\delta}{\delta S}\int U[\rho,S]dxdt$$

From (★) one has directly (for $a = S$ with S canonically conjugate to $-\rho$)

$$(1.51)\quad \partial_t\rho + \partial_x j_\psi = \frac{\partial}{\partial S}U[\rho,S];\ j_\psi = \frac{S_1}{m}\rho + \sum_0^N(-1)^n\partial_x^n\left(\frac{\partial}{\partial S_{n+1}}U[\rho,S]\right)$$

When the conservation of the number of particles $N = \int \rho dx$ is required the hypothesis that $U[\rho, S]$ does not depend on S but only on its derivative must be introduced so that the continuity equation takes the form (•) $\partial_t\rho + \partial_x j_\psi = 0$. Note that the imaginary part $\mathfrak{W}$ is responsible for the nonlinearity of the expression for the current j_ψ. Now consider the transformation (••) $\psi \to \phi = \mathfrak{U}[\psi,\psi^*]\psi$; this allows one to eliminate the imaginary part of the evolution equation for ψ, which corresponds to linearizing the current j_ψ. The operator generating this transformation is unitary and defined via

$$(1.52)\qquad \mathfrak{U}[\psi,\psi^*] = exp\left[\frac{im}{\hbar}\sum_0^N(-1)^n\int\frac{1}{\rho}\partial_x^n\left(\frac{\partial}{\partial S_{n+1}}U[\rho,S]\right)dx\right]$$

One can rewrite this as

$$(1.53)\qquad \phi = \rho^{1/2}e^{(i/\hbar)\sigma(x,t)};\ \sigma = S + \frac{m}{\hbar}\sum_0^N(-1)^n\int\frac{1}{\rho}\partial_x^n\left(\frac{\partial}{\partial S_{n+1}}U\right)dx$$

The two wave functions ψ and ϕ should then represent the same physical system and one obtains (using (••))

$$(1.54)\quad i\hbar\partial_t\phi = -\frac{\hbar^2}{2m}\partial_x^2\phi + W[\rho,S]\phi - \frac{m}{2}\left[\sum_0^N(-1)^n\frac{1}{\rho}\partial_x^n\left(\frac{\partial U}{\partial S_{n+1}}\right)dx\right]\phi -$$

$$-\sum_0^N(-1)^n\left\{m\partial_t\left[\int\frac{1}{\rho}\partial_x^n\left(\frac{\partial U}{\partial S_{n+1}}\right)dx\right]\phi+\frac{S_1}{\rho}\partial_x^n\left(\frac{\partial U}{\partial S_{n+1}}\right)\phi\right\}$$

Note that the nonlinearity in (3.6) is real and the continuity equation takes the form

$$\partial_t\rho+\partial_x j_\phi=0;\ j_\phi=\frac{\sigma_1}{m}\rho \tag{1.55}$$

Thus the gauge transformation (•• and (3.4) make the complex nonlinearity real and produces a noncanonical new dynamical system (cf. here also [**360, 390**]).

EXAMPLE 1.3. Consider the canonical DG equation in [**360, 390**] where $c_1 = -c_4 = 1$ and $c_2 = c_3 = c_5 = 0$ arising from (3.1) when (♦♦) $U = (D/2)(\rho_1 S_2 - \rho S_2)$ one has a complex nonlinearity in the evolution equation for ψ with

$$W[\rho,S]=-mD\partial_x\left(\frac{j_\psi}{\rho}\right);\ \mathfrak{W}=\frac{\hbar D}{2\rho}\partial_x^2\rho \tag{1.56}$$

(note here $2(R''/R)=(\partial^2\rho/\rho)-(1/2)(\rho'/\rho)^2$ so Q is not present here directly). Further the quantum current $\jmath_\psi$ has a Fokker-Planck form (•••) $j_\psi=(S_1/m)\rho+D\rho_1$ (a drift current plus a Fick current). The generator of the transformation $\mathfrak{U}$ takes the form $\mathfrak{U}[\psi,\psi^*]=exp[(i/\hbar)mDlog(\rho)]$ (as in the DG theory) and one can then write

$$i\hbar\partial_t\phi=-\frac{\hbar^2}{2m}\partial_x^2\phi+2mD^2\frac{1}{\sqrt{\rho}}\frac{\partial^2\sqrt{\rho}}{\sqrt{\rho}}\phi \tag{1.57}$$

and here the quantum potential does make an appearance (note that, after rescaling $\sigma\to\sqrt{1-(2mD/\hbar)^2}\sigma$, it reduces to a linear SE). ■

In the second paper of [**723**] one considers a class of NLSE minimally coupled with an Abelian gauge field A_μ; thus $\mathfrak{L}=\mathfrak{L}_M+\mathfrak{L}_G$ ($M\sim$ matter and $G\sim$ gauge) where

$$\mathfrak{L}_M=\frac{ic\hbar}{2}[\psi^*D_0\psi-\psi(D_0\psi^*)]-\frac{\hbar^2}{2m}|\mathbf{D}\psi|^2-U([\psi],[\psi^*],\mathbf{A}) \tag{1.58}$$

with $D_\mu=\partial_\mu+(ie/\hbar c)A_\mu\equiv(D_0,\mathbf{D})$ a covariant derivative involving $\partial_\mu\equiv(c^{-1}\partial_t,\nabla)$ for $\mu=0,1,\cdots,n$. The nonlinear potential U is assumed to be real and one writes as before $\rho=\psi\psi^*$ and $S=(i\hbar c/2e)log(\psi^*/\psi)$ with $\psi=\rho^{1/2}exp[(ie/\hbar c)S(t,x)]$. For simplicity one assumes the standard form of the EM field (★★) $\mathfrak{L}_G=-(1/4)F^{\mu\nu}F_{\mu\nu}$ where $F_{\mu\nu}=\partial_\mu A_\nu-\partial_\nu A_\mu$. Also here $\eta^{\mu\nu}=diag(1,-1,\cdots,-1)$ and $A^\mu=\eta^{\mu\nu}A_\nu$ (Latin indices take the values $1-n$ and Greek indices the values $0-n$). One uses an action $\mathfrak{A}=\int\mathfrak{L}d^nx$ with fields $a=\psi,\ \psi^*,\ A_\mu,\ \rho,\ S$ and the evolution equations are found from $\delta\mathfrak{A}/\delta a=0$ where

$$\frac{\delta}{\delta a}\int F[a]d^nx=\sum_0^N(-1)^k\sum_I\mathfrak{D}_I\frac{\partial F[a]}{\partial(\mathfrak{D}_I a)} \tag{1.59}$$

where $I = (i_1, \cdots, i_n)$, $0 \leq i_p \leq k$, $\sum i_p = k$, and $\mathfrak{D}_I = \partial^k/\partial x_1^{i_1} \cdots \partial x_n^{i_n}$. When $a = \psi^*$ one obtains (cf. (3.3))

$$(1.60) \qquad ic\hbar D_0\psi = -\frac{\hbar^2}{2m}\mathbf{D}^2\psi + W\psi + i\mathfrak{W}\psi;\; W = \frac{\delta}{\delta\rho}\int U dt d^n x;$$

$$\mathfrak{W}([\rho], [S], \mathbf{A}) = \frac{\hbar c}{2e\rho}\frac{\delta}{\delta S}\int U([\rho], [S], \mathbf{A}) dt d^n x$$

For $a = S$ on the other hand one obtains

$$(1.61) \qquad \partial_t\rho + \nabla\cdot\left[\frac{e}{mc}(\nabla S - \mathbf{A})\rho\right] - \frac{c}{e}\frac{\delta}{\delta S}\int U dt d^n x = 0 \Rightarrow$$

$$\Rightarrow \partial_t\rho + \nabla\cdot\mathbf{J} = \frac{c}{e}\int\frac{\partial}{\partial S}U([\rho], [S], \mathbf{A}) dt d^n x;$$

$$J_i = \frac{e}{mc}(\partial_i S + A_i)\rho + \frac{c}{e}\frac{\delta}{\partial(\partial_i S)}\int U dt d^n x$$

If the nonlinear potential U depends on S only through its derivatives the right side of the second line vanishes and it becomes a continuity equation $\partial_t\rho + \nabla\cdot\mathbf{J} = 0$ which implies a conservation of charge $q = e\int \rho d^n x$. There is also then a symmetry of the action $\mathfrak{A}$ under $\psi \to \psi' = \psi exp(i\epsilon)$. Similarly the evolution equation $\delta\mathfrak{A}/\delta A_\nu = 0$ becomes $\partial^\mu F_{\mu\nu} = (e/c)J_\nu$ where the covariant current is $J_\nu = (c\rho, -\mathbf{J})$.

Next one introduces a nonlinear and nonlocal transformation to eliminate the imaginary part of the nonlinearity in the NLSE; this can be done in two ways, either by a unitary transformation on the field ψ or by a gauge transformation on the field A_μ and we refer to [**723**] for details. Here we only look at the situation for a subclass of DG equations

$$(1.62) \quad i\hbar\partial_t\psi = -\frac{\hbar^2}{2m}\Delta\psi + m\nu\nabla\left(\frac{\mathbf{j}_0}{\rho}\right)\psi - \frac{2\alpha\hbar^2}{m}\left[\frac{\Delta\rho}{\rho} - \frac{1}{2}\left(\frac{\nabla\rho}{\rho}\right)^2\right]\psi + \frac{i\hbar\nu}{2}\frac{\Delta\rho}{\rho}\psi$$

where $\mathbf{j}_0 = (-i\hbar/2m)(\psi^*\nabla\psi - \psi\nabla\psi^*)$ is the standard QM current, ν is a diffusion coefficient, and α a dimensionless coupling constant. This equation is obtainable starting from the nonlinear potential

$$(1.63) \qquad U_{DG}[\rho], [S]) = \frac{\nu}{2}(\rho\Delta S - \nabla\rho\cdot\nabla S) + \alpha\frac{\hbar^2}{m}\frac{(\nabla\rho)^2}{\rho}$$

Here one has also

$$(1.64) \qquad \mathbf{j} = \frac{\nabla S}{m}\rho - \nu\nabla\rho;\; \partial_t\rho + \nabla\cdot\mathbf{j} = 0$$

(Fokker-Planck equation). In the case of interacting charged particles (1.62) must be replaced by

$$(1.65) \qquad ic\hbar D_0\psi = -\frac{\hbar^2}{2m}\mathbf{D}^2\psi + m\nu\nabla\left(\frac{\mathbf{J}_0}{\rho}\right)\psi -$$

$$-\frac{2\alpha\hbar^2}{m}\left[\frac{\Delta\rho}{\rho} - \frac{1}{2}\left(\frac{\nabla\rho}{\rho}\right)^2\right]\psi + \frac{i\hbar\nu}{2}\frac{\Delta\rho}{\rho}\psi$$

whose associated potential is

$$U = \frac{\nu e}{2c}[\rho\nabla(\nabla S - \mathbf{A}) - \nabla\rho\cdot(\nabla S - \mathbf{A})] + \frac{\alpha\hbar^2}{m}\frac{(\nabla\rho)^2}{\rho} \tag{1.66}$$

Note here also

$$\mathbf{J}_0 = -\frac{i\hbar}{2m}\left[\psi^*\left(\nabla - \frac{ie}{\hbar c}\mathbf{A}\right)\psi - \psi\left(\nabla + \frac{ie}{\hbar c}\mathbf{A}\right)\psi^*\right] \tag{1.67}$$

and for $\mathbf{J} = \mathbf{J}_0 - \nu\nabla\rho$ one has $\mathbf{J} = (e/mc)\rho(\nabla S - \mathbf{A}) - \nu\nabla\rho$. Writing now (★★★) $\psi \to \phi = \mathfrak{U}\psi$ where $\mathfrak{U} = exp[-(im/\hbar)\nu log(\rho)]$ (1.65) becomes

$$ic\hbar D_0\phi = -\frac{\hbar^2}{2m}\mathbf{D}^2\phi + \left(m\nu^2 - \frac{2\alpha\hbar^2}{m}\right)\left[\frac{\Delta\rho}{\rho} - \frac{1}{2}\left(\frac{\nabla\rho}{\rho}\right)^2\right]\phi \tag{1.68}$$

We note that the transformation (★★★) is the same as that of the DG theory; further here we see the emergence of Q in (1.68) and (1.65) (recall that $2(\Delta R/R) - (\Delta\rho/\rho) - (1/2)(\nabla\rho/\rho)^2$). One remarks further that (1.68) can also be obtained via the gauge transformation

$$\mathbf{X} = \mathbf{A} + \frac{mc\nu}{e}\frac{\nabla\rho}{\rho};\ X_0 = A_0 + \frac{\nu}{c\rho}\nabla[(\nabla S - \mathbf{X})\rho] \tag{1.69}$$

REMARK 8.1.2. In order to deal with $\tau_2 \to 0$ as a classical limit we have been looking for Q in the DG equations and it does not seem to arise in a natural manner in (1.11)-(1.13) or (1.34) except in the form $Q = -(\hbar^2/2m)[\nabla^2 T + (\nabla T)^2]$ arising from the linear SE. In (1.62), (1.63), and (1.66) however we have a somewhat different picture involving a Q term in the form (cf. also (4.10))

$$-\frac{2\alpha\hbar^2}{m}\left[\frac{\Delta\rho}{\rho} - \frac{1}{2}\left(\frac{\nabla\rho}{\rho}\right)^2\right] = -\frac{2\alpha\hbar^2}{m}\left(2\frac{\Delta R}{R}\right) = 8\alpha Q \tag{1.70}$$

This is claimed to arise from a DG potential (1.63) $U = (\alpha\hbar/m)[(\nabla\rho)^2/\rho]$ but this is not homogeneous and should not be referred to as a DG potential. Rather consider R_2 and R_5 from (2.4) - (2.6); namely

$$R_2 = \frac{\Delta\rho}{\rho};\ R_5 = 4\frac{(\nabla\rho)^2}{\rho^2} = 4\left(\frac{\nabla\rho}{\rho}\right)^2 \tag{1.71}$$

Then $R_2 - (1/8)R_5 = (\Delta\rho/\rho) - (1/2)(\nabla\rho/\rho)^2$ so (1.70) yields

$$8\alpha Q = -\frac{2\alpha\hbar^2}{m}\left(R_2 - \frac{1}{8}R_5\right) \Rightarrow Q = -\frac{\hbar^2}{4m}(R_2 - (1/8)R_5) \tag{1.72}$$

However adding such a potential to the SE gives $i\hbar\psi_t = -(\hbar^2/2m)\psi_{xx} + 8\alpha Q\psi$ in 1-D, leading for $\psi = Rexp(iS/\hbar)$ to

$$R_t + \frac{1}{m}\partial(R^2 S_x) = 0;\ S_t + \frac{1}{2m}(S_x)^2 + Q + 8\alpha Q = 0 \tag{1.73}$$

Then via $Q = -(\hbar^2/2m)[\nabla^2 T + (\nabla T)^2]$ one obtains

$$\dot{S} + \frac{1}{2m}(\nabla S)^2 - (8\alpha + 1)\frac{\hbar^2}{2m}[\nabla^2 T + (\nabla T)^2] = 0 \tag{1.74}$$

leading to $2\tau_2 = (8\alpha+1)\hbar^2/4m^2$ with $\tau_5 = (8\alpha+1)\hbar^2/4m^2$, and there is no progress over Remark 8.1.1. ■

2. NLS AND QP

We discuss now some topics arising in [**78, 152, 808, 807, 998, 999, 1061, 1113, 1119**] involving nonlinear SE (NLS) and QP. There are many physical applications to vortices and hydrodynamics, Chern-Simons theory, solitons, black holes, etc. and a number of interesting points of view. For NLS we refer to [**254, 276, 275, 359, 360, 390, 392, 391, 539, 541, 542, 1134**] and Section 8.1 for a sampling of material. In Auberson-Sabatier [**78**] now one refers to two types of NLS ($s \in \mathbf{R}$)

$$i\psi_t = (-\Delta + V)\psi + s\psi\Delta log(|\psi|);\ i\psi_t = (-\Delta+V)\psi + s\psi\frac{\Delta(|\psi|)}{|\psi|} \tag{2.1}$$

The first equation has linear phase solutions exponentially confined by the nonlinear term when $s > 1$. However it cannot be derived directly from a local Lagrangian and aside from the norm no conserved quantities can be derived from local expressions in ψ. In contrast the second equation in (2.1) does not have exponentially confined solutions but it can be derived from a Lagrangian density

$$\mathfrak{L} = \frac{i}{2}(\psi^*\psi_t - \psi_t^*\psi) - \nabla\psi^*\cdot\nabla\psi - V|\psi|^2 + \tag{2.2}$$
$$+\frac{s}{4}\left[\frac{\psi}{\psi^*}\nabla\psi^*\cdot\nabla\psi^* + \frac{\psi^*}{\psi}\nabla\psi\cdot\nabla\psi + 2\nabla\psi^*\cdot\nabla\psi\right]$$

Moreover setting $\psi(x,t) = \xi(x,t)exp(-i\theta(x,t))$ where $\xi = |\psi|$ one obtains

$$\theta_t = (s-1)\frac{\Delta\xi}{\xi} + \nabla\theta\cdot\nabla\theta + V;\ \xi^{-1}\xi_t = \Delta\theta + 2\nabla\theta\cdot\frac{\nabla\xi}{\xi} \tag{2.3}$$

Now for linearization there are three cases:

(1) ($s < 1$): One writes (**1A**) $t = (1-s)^{-1/2}t'$ and $gt = (1-s)^{1/2}\theta'$ to get

$$\theta'_{t'} = -\frac{\Delta\xi}{\xi} + \nabla\theta'\cdot\nabla\theta' + (1-s)^{-1}V;\ \xi^{-1}\xi_{t'} = \Delta\theta' + 2\nabla\theta'\cdot\frac{\nabla\xi}{\xi} \tag{2.4}$$

which are equivalent for $\psi' = \xi(x,(1-s)^{-1/2}t')exp[-\theta'(x,t')]$ to the linear SE (**1B**) $i\psi'_{t'} = -\Delta\psi' + [V/(1-s)]\psi'$.

(2) ($s > 1$): Set first (**1C**) $t = (s-1)^{-1/2}\tau$ and $\theta(x,t) = (s-1)^{1/2}\psi(x,\tau)$ to arrive at

$$\psi_\tau = \frac{\Delta\xi}{\xi} + \nabla\psi\cdot\nabla\pi + (s-1)^{-1}V;\ \xi^{-1}\xi_\tau = \Delta\psi + 2\nabla\pi\cdot\frac{\nabla\xi}{\xi} \tag{2.5}$$

so that $\psi'(x,\tau) = \xi(x,(s-1)^{-1/2}\tau)exp[-i\psi(x,\tau)]$ is a solution of

$$i\psi'_\tau = -\Delta\psi' + \frac{V}{s-1}\psi' + 2\psi'\frac{\Delta|\psi'|}{|\psi|} \tag{2.6}$$

Note that (2.5) is invariant under the transformation $\tau \to -\tau$ and $\psi \to -\psi$ while (**1D**) $\phi(x,\tau) = \xi(x,(s-1)^{-1/2}\tau)exp[\psi(x,\tau)]$ is a solution of the linear (heat) equation (**1E**) $\phi_\tau = \Delta\phi + [V/(s-1)]\phi$. Also

(1F) $\phi'(x,\tau) = \xi(x,(s-1)^{-1/2}\tau)exp[-\psi(x,\tau)]$ is a solution of the backward (heat) equation **(IG)** $-\phi'_\tau = \Delta\phi' + [V/(s-1)]\phi'$. Conversely given a positive solution ϕ of **(1E)** and a solution ϕ' of **(1G)** there results **(1H)** $\partial_\tau(\phi\phi') = div(\phi'\nabla\phi - \phi\nabla\phi')$. Setting then **(1I)** $\xi = \sqrt{\phi\phi'}$ and $\psi = log(\phi/\xi) = -log(\phi'/\xi)$ one sees that **(J)** is nothing but the second equation of (2.3) and the first one follows from **(1E)** or **(1G)**. Hence $\psi' = \xi exp(-i\psi)$ is a solution of (2.5) and inverting the similarity **(1C)** yields a solution (2.1). Note however that if the second equation in (2.1) is treated by this reduction it will be ill posed because of the occurance of the backward heat equation.

(3) $(s=1)$: Introduce **(1J)** $\rho = \xi^2$ and $v = -2\nabla\theta$ to obtain **(1K)** $v_t = -(v\nabla)v - 2\nabla V$ and $\rho_t = -div(\rho v)$. Then viewing ρ as a density field and v as a velocity field on has in **(1K)** the Euler and continuity equation for the irrotational flow of a gas of pointlike particles of mass 1/2 moving without mutual interactions (and in zero pressure) in the force field $-\nabla V$. Thus when $s = 1$ the effect of the nonlinear part of the second equation in (2.1) is just to remove the quantum correction ($\propto \hbar^2\nabla(\Delta\sqrt{\rho}/\sqrt{\rho})$) appearing in the usual hydrodynamical interpretation of the linear SE.

One shows also in [**78**] that no equation of the form

$$i\psi_t = (-\Delta + V)\psi + F\left(\frac{\nabla|\psi|}{|\psi|}, \frac{\Delta|\psi|}{|\psi|}\right) \tag{2.7}$$

can be linearized by a change of functions and a rescaling of variables (except for the second equation in (2.1)). Some other equations studied are of the form

$$i\psi_t = (-\Delta + V)\psi + [s\Delta log|\psi| + r\nabla log(|\psi|)\cdot\nabla log(|\psi|)]\psi \tag{2.8}$$

and we refer to [**78**] for details. In general equations which can be derived from a local Lagrangian are linearizable but they do not admit regular confined solutions. The equations not generated by a Lagrangian do admit such confined solutions for suitable parameter values but they are perturbatively unstable.

We extract next from Santini [**1119**] (cf. also [**1061, 1113**]). The Lagrangian for a nonrelativistic scalar particle can be taken as

$$\mathfrak{L} = \frac{i\hbar}{2}(\psi^*\partial_t\psi - (\partial_t\psi^*)\psi) - \frac{\hbar^2}{2m}\nabla\psi^*\cdot\nabla\psi - U\psi^*\psi \tag{2.9}$$

By taking the variations of $\mathfrak{L}$ with respect to ψ and ψ^* one gets the SE and its conjugate. By contrast since a generic wave function $\psi = \sqrt{\rho}exp[i\phi\hbar]$ one can take the variations of

$$\mathfrak{L} = -\left[\partial_t\phi + \frac{1}{2m}(\nabla\phi)^2 + \frac{\hbar^2}{8m}\left(\frac{\nabla\rho}{\rho}\right)^2 + U\right]\rho \tag{2.10}$$

with respect to ρ and ϕ. One then obtains the two equations for a Madelung fluid which taken together are equivalent to the SE; thus

$$\partial_t\phi + \frac{1}{2m}(\nabla\phi)^2 + \frac{\hbar^2}{4m}\left[\frac{1}{2}\left(\frac{\nabla\rho}{\rho}\right)^2 - \frac{\Delta\rho}{\rho}\right] + U = 0 \tag{2.11}$$

and (**1L**) $\partial_t\rho + \nabla\cdot(\rho\nabla\phi/m) = 0$ which are the HJ and continuity equation for the quantum fluid, where

$$(2.12)\qquad \frac{\hbar^2}{4m}\left[\frac{1}{2}\left(\frac{\nabla\rho}{\rho}\right)^2 - \frac{\Delta\rho}{\rho}\right] = -\frac{\hbar^2}{2m}\frac{\Delta|\psi|}{|\psi|}$$

is the QP. Such a potential derives from a nonclassical term $(\hbar^2/8m)(\nabla\rho/\rho)^2$ (which involves Fisher information - cf. [**254, 255, 595, 995, 996**] and earlier chapters).

Now one does not need to go to stochastic mechanics even if one recognizes the existence of a Zitterbewegung (ZBW) or diffusive or internal motion (i.e. of a motion observed in the center of mass frame (CMF) where $\mathbf{p} = 0$ by definition) along with the external, drift, or translational or convective motion of the CM. The existence of such an internal motion is announced (in addition to the presence of spin) by the fact that in the Dirac theory $\mathbf{v} \neq \mathbf{p}/m$. Moreover while $[\hat{\mathbf{p}}, \hat{H}] = 0$ so that $\mathbf{p}$ is a preserved quantity the quantity $\mathbf{v}$ is not a constant of the motion (i.e. $[\hat{\mathbf{v}}, \hat{H}] \neq 0$). In order to deal with ZBW one splits the motion variables as (**1K**) $x = \xi + X$ with $\dot{x} = v = w + V$ (bold face is omitted here). Here ξ and $w = \dot{\xi}$ describe the motion of the CM in some chosen reference frame while X and $V = \dot{X}$ describe the internal motion referred to the CMF. From a dynamical point of view the conserved electric current is associated with helical trajectories of the electric charge (i.e. with x and $v = \dot{x}$) while the center of mass Coulomb field is associated with the geometrical center of such trajectories (i.e. with ξ and $w = \dot{\xi} = p/m$). Going back now to (2.10) one can give an interpretation of the nonclassical term (★) $(\hbar^2/8m)(\nabla\rho/\rho)^2$ where the first term in the right side represents the total energy (up to sign) $\partial_t\phi = -E$ while the second term is the kinetic energy $p^2/2m$ of the CM if one assumes (**1M**) $p = -\nabla\phi$. The third term will be shown to represent the kinetic energy in the CMF, i.e. the internal energy due to the ZBW motion.

One starts form the Pauli current (i.e. the Gordon decomposition of the Dirac current in the NR (nonrelativistic) limit, namely

$$(2.13)\qquad \mathbf{j} = \frac{i\hbar}{2m}[(\nabla\pi^\dagger)\psi - \psi^\dagger\nabla\psi] - \frac{e\mathbf{A}}{m}\psi^\dagger\psi + \frac{1}{m}\nabla\wedge(\psi^\dagger\hat{\mathbf{s}}\psi)$$

A spinning NR particle can be factored into (**P**) $\pi = \sqrt{\rho}\phi$ where ϕ is a Pauli 2-component spinor which obeys the normalization constraint $\phi^\dagger\phi = 1$ (in order to have $\|\psi|^2 = \rho$. By definition $\rho\mathbf{s} = \psi^\dagger\hat{\mathbf{s}}\psi = \rho\phi^\dagger\hat{\mathbf{s}}\phi$ and hence using $\psi = \sqrt{\rho}\phi$ in (2.19) one obtains

$$(2.14)\qquad \mathbf{j} = \rho\mathbf{v} = \rho\frac{\mathbf{p} - e\mathbf{A}}{m} + \frac{\nabla\wedge(\rho\mathbf{s})}{m}$$

which is the decomposition of Hestenes [**599, 586**] using Clifford algebra language

$$(2.15)\qquad \mathbf{v} = \frac{\mathbf{p} - e\mathbf{A}}{m} + \frac{\nabla\wedge(\rho\mathbf{s})}{m\rho}$$

where $c = 1$, e is the electric charge, $\mathbf{A}$ is the external electromagnetic vector potential, $\mathbf{s}$ is the spin vector $\mathbf{s} = \rho^{-1}\psi^{\dagger}\hat{\mathbf{s}}\psi$, and $\hat{\mathbf{s}}$ is the spin operator usually represented via Pauli matrices as (**1O**) $\hat{\mathbf{s}} = (\hbar/2)(\sigma_x, \sigma_y, \sigma_z)$. One thinks now of every quantity as a local or field quantity (e.g. $\mathbf{v} = \mathbf{v}(\mathbf{x}, t)$, etc.) and the interal ZBW velocity is then (**1P**) $\mathbf{V} = [\nabla \wedge (\rho\mathbf{s})]/m\rho$. The Schrödinger case in which the vector spin field $\mathbf{s}$ is constant in time and uniform in space corresponds to spin eigenstates so that one needs now a wave function factorizable into the product of a non-spin part $\sqrt{\rho}exp(i\phi)$ (scalar) and a spin part χ where (**1Q**) $\psi = \sqrt{\rho}exp(i\phi/\hbar)\chi$ (with χ constant in time and space). Therefore when $\mathbf{s}$ has no precession (and no external field is present so $\mathbf{A} = 0$) one has $\mathbf{s} = \chi^{\dagger}\hat{\mathbf{s}}\chi = constant$ and (**1R**) $\mathbf{V} = (\nabla\rho \times \mathbf{s})/m\rho \neq 0$ (Schrödinger case). Thus even in the Schrödinger framework the ZBW does not vanish except for plane waves (i.e. for the nonphysical case of $\mathbf{p}$ eigenfunctions when not only $\mathbf{s}$ but also ρ is constant and uniform so $\nabla\rho = 0$). Now one can write

$$\mathbf{V}^2 = \left(\frac{\nabla\rho \wedge \mathbf{s}}{m\rho}\right)^2 = \frac{(\nabla\rho)^2\mathbf{s}^2 \quad (\nabla\rho\cdot\mathbf{s})^2}{(m\rho)^2} \tag{2.16}$$

(note here $(\mathbf{a} \wedge \mathbf{b})^2 = \mathbf{a}^2\mathbf{b}^2 - (\mathbf{a}\cdot\mathbf{b})^2$). Now observe that from the smallness of the negative energy component of the Dirac bispinor follows the smallness also of $\nabla\rho\cdot\mathbf{s} \simeq 0$. This was already known from the Clifford algebra approach to Dirac theory that yielded $\nabla\cdot(\rho\mathbf{s}) = -m\rho Sin(\beta)$ (where β is the Takabayashi angle) which in the NR limit corresponds to $\beta = 0$ (pure electron) or $\beta = \pi$ (pure positron). Hence one gets $\nabla\cdot(\rho\mathbf{s}) = 0$ and in the Schrödinger case where $\mathbf{s}$=constant with $\nabla\cdot\mathbf{s} = 0$ one has (**1S**) $\nabla\rho\cdot\mathbf{s} = 0$ and putting this in (2.16) yields the result (**1T**) $\mathbf{V}^2 = \mathbf{s}^2(\nabla\rho/m\rho)^2$. This allows one to attribute to the nonclassical term (★) $(\hbar^2/8m)(\nabla\rho/\rho)^2$ of the Lagrangian (2.10) a simple meaning of kinetic energy of the internal (ZBW) motion (i.e. the kinetic energy associated with the internal ZBW velocity $\mathbf{V}$) provided (**1U**) $\hbar = 2|\mathbf{s}|$ (i.e. $(1/2(m\mathbf{V}^2 = (|\mathbf{s}|/2m)(\nabla\rho/\rho)^2 = (\hbar^2/8m)(\nabla\rho/\rho)^2)$. Vice versa if one assumes (within a ZBW philosophy) that $\mathbf{V}$ in (**1T**) is the velocity attached to the kinetic energy term (★) then one can deduce (**1U**) in the form $|\mathbf{s}| = \hbar/2$. In the stochastic approaches the diffusion velocity is $\mathbf{V} = \mathbf{v}_{diff} = \nu(\nabla\rho/\rho)$ where ν is the diffusion coefficient, postulated to be $\nu = \hbar/2m$. In the present approach if one adopted for a moment the stochastic language it would follow by comparison of (★), (**1T**), and (**1U**) that $\nu = \hbar/2m$ and hence immediately that $\nu = |\mathbf{s}|/m$. In this spirit one could say that the fundamental relation should be $\hbar = 2|\mathbf{s}|$ in which case for $\mathbf{s} = 0$ the quantum potential would vanish so that quantum effects could be attributed to spin and quantum scalar particles are always composed of spinning objects with ZBW (this is obviously a rough kind of conjecture since $\hbar \to 0$ is not equivalent to $Q \to 0$, etc.).

2.1. REMARKS. We go now to [**201, 406, 407, 571, 630, 631, 683, 696, 808, 807, 859, 998, 999, 1195, 1267**] (cf. also [**233, 321, 352, 369, 434, 435, 485, 576, 577, 633, 715, 716, 798, 824, 1101, 1149, 1226, 1311, 1321, 1339**]). Before plunging into details we want to make a few general comments. In dealing with nonlinear equations such as KdV (**2A**) $u_t + 6uu_x + u_{xxx} = 0$ or the cubic NLS (**2B**) $i\hbar\psi_t = -(\hbar^2/2m)\psi_{xx} + 2g|\psi|^2\psi + V\psi$ ideas of dispersion play an important role (cf. [**238, 252, 434, 1306**]). For KdV (or KP) type equations there is an elaborate systematic dispersionless theory (cf. [**238, 252, 247, 246**]) which we omit here. For NLS type equations the dispersionless aspect is somewhat more subtle and it can arise e.g. as follows. Write $\psi = \sqrt{\rho}exp(iS/\hbar)$ in (**2B**) and compute

$$(2.17)\qquad \psi_t e^{-iS/\hbar} = \frac{1}{2}\rho^{-1/2}\rho_t + \frac{\sqrt{\rho}iS_t}{\hbar};\ \psi_x e^{-iS/\hbar} = \frac{1}{2}\rho^{-1/2}\rho_x + \frac{\sqrt{\rho}iS_x}{\hbar};$$

$$\psi_{xx}e^{-iS/\hbar} = -\frac{\rho_x^2}{4\rho^{3/2}} + \frac{\rho_{xx}}{2\sqrt{\rho}} + \frac{i\rho_x S_x}{\rho^{1/2}\hbar} + \frac{i\rho^{1/2}S_{xx}}{\hbar} - \frac{\rho^{1/2}S_x^2}{\hbar^2}$$

leading to

$$(2.18)\qquad i\hbar\left[\frac{\rho_t}{2\sqrt{\rho}} + \frac{i\sqrt{\rho}S_t}{\hbar}\right] = 2g\rho^{3/2} + V\sqrt{\rho} -$$

$$-\frac{\hbar^2}{2m}\left[-\frac{\rho_x^2}{4\rho^{3/2}} + \frac{\rho_{xx}}{2\sqrt{\rho}} + \frac{i\rho_x S_x}{\sqrt{\rho}\hbar} + \frac{i\sqrt{\rho}S_{xx}}{\hbar} - \frac{\sqrt{\rho}S_x^2}{\hbar^2}\right]$$

Equating real and imaginary parts gives then

$$(2.19)\ S_t + \frac{S_x^2}{2m} + 2g\rho + V + Q = 0;\ Q = -\frac{\hbar^2}{4m}\left[\frac{\rho_{xx}}{\rho} - \frac{\rho_x^2}{2\rho^2}\right];\ \rho_t = -\frac{1}{m}\partial_x(\rho S_x)$$

Thus the quantum potential Q arises as an additional term in the classical HJ equation and this term would be related to dispersion effects as the only term in the HJ equation involving a small parameter $\hbar$. This has led various authors to think of setting $V = U - Q$ to produce a classical type HJ equation (no small parameter) and a "dispersionless" NLS

$$(2.20)\qquad i\hbar\psi_t = -\frac{\hbar^2}{2m}\psi_{xx} + 2g|\psi|^2\psi + U\psi + \frac{\hbar^2}{2m}\frac{\partial|\psi|^2}{|\psi|}\psi$$

(see e.g. [**808, 807, 859, 998, 999**]). In particular for $U = 0$ one connects a linear SE to a dispersionless NLS with corresponding HJ equation (**1C**) $S_t + (1/2m)S_x^2 + U = 0$ of classical type.

REMARK 8.2.1. We recall from [**199, 201, 551, 552, 553, 554, 627**] that Bohmian (quantum) mechanics can be thought of as classical symplectic mechanics applied to the Hamiltonian $H + Q$. ■

REMARK 8.2.2. We note also that the addition of a term of Fisher information type (**1D**) $\nabla|\psi|\nabla|\psi| \sim (\partial\sqrt{\rho}\partial\sqrt{\rho}) = (1/4)(\rho_x^2/\rho)$ to the Lagrangian will give rise to a QP with standard variational methods (cf. [**254, 255, 490, 592, 595, 599, 627, 995, 996**]). Indeed with an integration by parts (**2C**) $\delta\int(\rho_x^2/\rho)dx \sim \int[(\rho'/\rho)^2 - 2(\rho''/\rho)]dx\delta\rho$ and $Q = -(\hbar^2/4m)[(\rho''/\rho) - (1/2)(\rho'/\rho)^2]$. Further one notes (cf. [**255, 258, 259**]) the important relation of the form (**2D**) $\int\rho Q dx = (\hbar^2/4m)\int(\rho_x^2/\rho)dx$ (assuming e.g. that $\rho \to 0$ outside of a compact set - as befitting a probability density). ■

REMARK 8.2.3. We see that the QP can be regarded as

- An object producing a quantization of a classical HJ system by inducing a standard SE or NLS essentially via dispersion
- An object which can transform a SE or NLS into a dispersionless NLS by removing a term Q from the HJ equation and adding it to the SE as $\pm(\hbar^2/4m)(\Delta|\psi|/|\psi|)\psi$. ■

REMARK 8.2.4. If we have a stationary problem with $i\hbar\psi_t = E\psi$ (i.e. $\rho = \rho(x)$ and $= -Et + s(x)$ so $\psi_t = (iS_t/\hbar)\psi = -(iE/\hbar)\psi)$ then $\rho s_x = c$ or $\rho_x s_x + \rho s_{xx} = 0$ so $(\rho_x/\rho) = -(s_{xx}/s_x)$ and $\rho_{xx} = -\rho_x(s_{xx}/s_x) - \rho(s_{xxx}/x_x) + \rho(s_{xx}/_x)^2$. This implies then

$$Q = -\frac{\hbar^2}{4m}\left(\frac{\rho_{xx}}{\rho} - \frac{\rho_x^2}{2\rho^2}\right) = -\frac{\hbar^2}{4m}\left[\frac{3}{2}\left(\frac{s_{xx}}{s_x}\right)^2 - \frac{s_{xxx}}{s_x}\right] = \tag{2.21}$$

$$= \frac{\hbar^2}{4m}\{s,x\} = \frac{\hbar^2}{4m}\mathfrak{S}(s,x)$$

where $\mathfrak{S}$ is the Schwartzian derivative (cf. [**133, 238, 252, 245, 449**]). ■

2.2. WAVES AND THE QUANTUM POTENTIAL. The quantum potential arises of course in studies of nonlinear wave motion and soliton theory although often in a disguised form. In particular this arises in using WKB analysis in the context of inverse scattering theory (cf. [**22, 205, 206, 428, 434, 485, 696, 715, 716, 717, 718**]). Thus following [**428, 715, 717, 718**] one considers a NLS (**2G**) $i\epsilon\psi_t + \epsilon^2\psi_{xx} \mp |\psi|^2\psi = 0$ and looks for a WKB type solution (**2H**) $\psi = \sqrt{\rho}exp[(i/\epsilon)\int^x v(x)dx]$ leading to

$$\rho_t + \partial_x(\rho v) = 0;\ v_t + \partial_x\left\{\frac{1}{2}v^2 \pm \rho - \frac{\epsilon^2}{4}\left[\frac{\rho''}{\rho} - \frac{1}{2}\left(\frac{\rho'}{\rho}\right)^2\right]\right\} = 0 \tag{2.22}$$

Since $Q = -(\hbar^2/4m)[(\rho''/\rho) - (1/2)(\rho'/\rho)^2]$ we can think of $\epsilon = \hbar$ with $m = 1$ and write (3.6) as

$$\rho_t + \partial_x(\rho v) = 0;\ v_t + \partial_x\left[\frac{1}{2}v^2 \pm \rho + Q\right] = 0 \tag{2.23}$$

One thinks here of an initial value problem with $\psi(x,0) \sim \rho_0(x),\ v_0(x)$. For $\epsilon = 0$ (which implies $Q = 0$ here) one has a classical Euler hydrodynamical problem which can be analyzed via Riemann invariants, etc. (cf. [**428, 1306**]). Then ψ is a rapidly oscillating function as $\epsilon \to 0$ assuming $v_0(x) \neq 0$.

Now we go to [**205, 206, 715, 716, 717, 718, 719, 1249**] for some interesting formulas involving Baker-Akhieser (BA) functions and square eigenfunctions (cf. also [**57, 238, 252, 247, 246, 245**]). Thus for the KdV equation $u_t + 6uu_x + u_{xxx} = 0$ one can consider

$$u_t = \frac{1}{2}\mathfrak{B}_{xxx} + 2\mathfrak{B}_x(u+\lambda) + \mathfrak{B}u_x \tag{2.24}$$

for $\mathfrak{B} = 4\lambda - 2u$ as a compatibility condition of two linear equations

$$\psi_{xx} = -(u+\lambda)\psi = \mathfrak{A}\psi;\ \psi_t = -\frac{1}{2}\mathfrak{B}_x\psi + \mathfrak{B}\psi_x \tag{2.25}$$

Then one can write (**2E**) $\psi^{\pm}=\sqrt{g}exp[\pm\sqrt{P(\lambda)}\int^x(dx/g)]$ to obtain two solutions of (3.9) where $g=\psi_+\psi_-$ is a "square eigenfunction" satisfying the Gelfand-Dickey resolvant equation

$$g_{xxx}+2u_xg+4(u+\lambda)g_x=0;\ g_t=\mathfrak{B}g_x-\mathfrak{B}_xg \tag{2.26}$$

One notes that a first integral in (2.26) is

$$\frac{1}{2}gg_{xx}-\frac{1}{4}g_x^2+(u+\lambda)g^2=P(\lambda) \tag{2.27}$$

where $P(\lambda)$ is an odd degree polynomial in λ for example. In fact when (**2F**) $P(\lambda)=\prod_1^{2n+1}(\lambda-\lambda_i)=\lambda^{2n+1}-s_1\lambda^{2n}+s_2\lambda^{2n-1}+\cdots+s_{2n}\lambda-s_{2n+1}$ for example it follows that

$$g=\prod_1^n(\lambda-\mu_i)=\lambda^n-\sigma_1\lambda^{n-1}+\sigma_2\lambda^{n-2}+\cdots+(-1)^n\sigma_n; \tag{2.28}$$

$$\sigma_1=\sum\mu_i=\frac{1}{2}(u+s_1);\ \sigma_2=\sum_{i<j}(\mu_i\mu_j)=\frac{1}{4}(u_{xx}+3u^2)+\frac{3}{2}s_1u+s_2-\frac{1}{4}s_1^2;\cdots$$

Then putting this information into (**2E**) one obtains the BA function in terms of u and its x derivatives.

This can be related to the NLS equation as follows. Consider an AKNS scheme (see e.g. [**238, 252**])

(2.29)

$$\vec{\psi}_x=U\vec{\psi};\ \vec{\psi}_t=V\vec{\psi};\ \vec{\psi}=\begin{pmatrix}\psi_1\\ \psi_2\end{pmatrix};\ U=\begin{pmatrix}F & G\\ H & -F\end{pmatrix};\ V=\begin{pmatrix}A & B\\ C & -A\end{pmatrix}$$

where compatibility $\vec{\psi}_{xt}=\vec{\psi}_{tx}$ requires

(2.30)

$$F_t-A_x+CG-BH=0;\ G_t-B_x+2(BF-AG)=0;\ H_t-C_x+2(AH-CF)=0$$

The NLS arises for the choice (**2G**) $F=-i\lambda$, $G=iu$, $H=iu^*$, $B=2iu\lambda-u_x$, $A=-2i\lambda^2+i|u|^2$, $C=2i\lambda u^*+u_x^*$ and one has two basis solutions $\vec{\psi}^{\pm}$ from which square eigenfunctions are constructed via

$$f=-\frac{i}{2}(\psi_1^+\psi_2^-+\psi_1^-\psi_2^+);\ g=\psi_1^+\psi_1^-;\ h=-\psi_2^+\psi_2^- \tag{2.31}$$

They satisfy

$$f_x=-iHg+iGh;\ g_x=2iGf+2Fg;\ h_x=-2iHf-2Fh; \tag{2.32}$$

$$f_t=-iCg+iBh;\ g_t=2iBf+2Ag;\ h_t=-2iCf-2Ah$$

Constancy of the Wronskian for (2.29) implies then

$$-\frac{1}{4}(\psi_1^+\psi_2^--\psi_1^-\psi_2^+)^2=f^2-gh=P(\lambda) \tag{2.33}$$

where periodic solutions are characterized by $P(\lambda)$ being a polynomial and some calculation shows that (**2H**) $gh=(f-\sqrt{P(\lambda)})(f+\sqrt{P(\lambda)})$ which leads to

$$\psi_1^{\pm}=\sqrt{g}e^{\pm i\sqrt{P(\lambda)}\int^x(G/g)dx};\ \psi_2^{\pm}=\sqrt{-h}e^{\pm i\sqrt{P(\lambda)}\int^x(H/h)dx} \tag{2.34}$$

It is shown in [**715, 717**] how to make formulas (2.34) for say $\psi_1^{\pm}$ correspond to (**2I**). This would hold for the same $P(\lambda)$ if $g_{NLS} = Gg_{KdV}$ and $\psi_1^{\pm} = \sqrt{G}\psi^{\pm}$ for example. Further the $P(\lambda)$ would be the same for

$$\mathfrak{A} = \left(F - \frac{G_x}{2G}\right)^2 + GH + \left(F - \frac{G_x}{2G}\right)_x ;\ \mathfrak{B} = \frac{B}{G} \tag{2.35}$$

In particular e.g. the NLS (**2I**) $iu_t + u_{xx} \pm 2|u|^2 u = 0$ can be presented as a compatibillity condition for

$$\psi_{xx} = \mathfrak{A}\psi;\ \psi_t = -\frac{1}{2}\mathfrak{B}_x\psi + \mathfrak{B}\psi_x \tag{2.36}$$

provided

$$\mathfrak{A} = -\left(\lambda - \frac{iu_x}{2}\right)^2 \mp |u|^2 - \left(\frac{u_x}{2u}\right)_x ;\ \mathfrak{B} = \pm 2\lambda + \frac{iu_x}{u} \tag{2.37}$$

and (**2E**) would again be applicable. In this situation however $\mathfrak{A}$ is complex and (2.26) will not have the same form.

REMARK 8.2.5. It is interesting to note that (2.27) can be written as

$$u + \lambda = \frac{P(\lambda)}{g^2} - \frac{1}{2}\left[\frac{g_{xx}}{g} - \frac{1}{2}\left(\frac{g_x}{g}\right)^2\right] \tag{2.38}$$

where the last term has the form of a quantum potential (**2J**) $Q_g = -(1/2)[(g''/g) - (1/2)(g'/g)^2]$. Further (**2E**) suggests $g \sim \rho$ as a density and since $P(\lambda) = -P(-\lambda)$ we consider a tentative wave function based on (**2E**) in the form

$$\psi = \sqrt{g}e^{\sqrt{P(\lambda)}\int^x \frac{dx}{g}} = \sqrt{g}e^{i\sqrt{P(-\lambda)}\int^x \frac{dx}{g}}$$

corresponding to a phase $S \sim \sqrt{P(-\lambda)}\int^x (dx/g)$ (imagine here $\hbar = 1$ and e.g. λ is negative with $P(-\lambda)$ positive). Then $S_x = \sqrt{P(-\lambda)}/g$ and $S_x g_x + S_{xx} g = 0$ which implies

$$Q_q = \frac{1}{2}\left[\frac{3}{2}\left(\frac{S_{xx}}{S_x}\right)^2 - \frac{S_{xxx}}{S_x}\right] = \frac{1}{2}\mathfrak{S}(S, x) \tag{2.39}$$

(cf. Remark 8.2.4) where $\mathfrak{S}$ is a Schwartzian derivative (see here also [**238, 252, 245, 449**]) where the square eigenfunction term corresponds to a prepotential). The resulting HJ equation (for $m = 1/2$ and $\hbar = 1$) would have the form $S_t + S_x^2 + V + Q_g = 0$ which for $S_t = 0$ and $\lambda < 0$ takes the form $P(-\lambda)/g^2 + Q_g + V = 0$. For $\lambda \to -\lambda$ this corressponds to (3.22) with $V = -u - \lambda$ or one could simply use $P(-\lambda)/g^2 + Q_g = u - \lambda$. Thus (**2E**) suggests a WKB (or Madelung) type solution to a quantum mechanical problem and in some sense promotes BA square eigenfunctions to QM wave functions. A question remains of how to connect kind of QM problem to NLS, which is already a QM type problem for u as in (**2I**) (note that the connection of $\psi^{\pm}$ to say $\psi_1^{\pm}$ is already in place and in addition (2.36)-(2.37) hold). In terms of dispersionless KP or KdV the methods of [**238, 252, 245**], which connect to the (x, ψ) duality theme of [**447, 448**], provide a somewhat refined WKB type theory for the SE which is connected to say KdV.

Hence this could be of some use here. ■

REMARK 8.2.6. From [**719**] one sees that g for KdV also satisfies a linearized KdV equation (**2K**) $g_t + 6ug_x + g_{xxx} = 0$. Further one recalls the Bloch eigenfunctions (of $\psi_{xx} + (u+\lambda)\psi = 0$ and the translation operator) in the form (**2L**) $\psi_\pm = \sqrt{g}exp(\pm i\sqrt{\lambda}\int^x (dx/g)$. ■

We recall also the original Zakharov-Shabat form for NLS, leading to (**2I**), namely $\psi_x = U\psi$ and $\psi_t = V\psi$ with

$$U = \begin{pmatrix} -i\lambda & iu \\ \pm iu^* & i\lambda \end{pmatrix};\; V = \begin{pmatrix} -2i\lambda^2 \pm i|u|^2 & 2iu\lambda - u_x \\ \pm 2iu^*\lambda \pm u_x^* & 2i\lambda^2 \mp i|u|^2 \end{pmatrix} \tag{2.40}$$

Now (**2I**) with a $\pm$ sign is respectively focusing (defocusing). Going to [**715**] we consider now (**2I**) with an ϵ inserted and a $\pm$ sign as indicated. Then (**2M**) $i\epsilon u_t + \epsilon^2 u_{xx} \pm 2|u|^2 u = 0$ and as in (2.36)-(2.37) one can use formulas

$$\epsilon^2\psi_{xx} = \mathfrak{A}\psi;\; \psi_t = -\frac{1}{2}\mathfrak{B}_x\psi + \mathfrak{B}\psi_x \tag{2.41}$$

where now for the defocusing case one has

$$\mathfrak{A} = -\left(\lambda + \frac{i\epsilon}{2}\frac{u_x}{u}\right)^2 + |u|^2 - \frac{\epsilon^2}{2}\left(\frac{u_{xx}}{u} - \frac{u_x^2}{u^2}\right);\; \mathfrak{B} = -2\lambda + \frac{i\epsilon u_x}{u} \tag{2.42}$$

(note here $(u_x/u)_x = (u_{xx}/u) - (u_x^2/u^2)$). Again for $u = \sqrt{\rho}exp[(i/\epsilon)\int^x vdx]$ one obtains (2.6) and is led to consider BA functions as in (**2E**), but now with an ϵ term (i.e. (**2N**) $\psi^\pm = \sqrt{g}exp[\pm(i/\epsilon)\int^x(\sqrt{P(\lambda)}/g)dx]$) where (**2O**) $(\epsilon^2/2)[(g_{xx}/g) - (1/2)(g_x^2/g^2)] - \mathfrak{A} = P(\lambda)/g^2$ (cf. (2.38)). Periodic solutions are distinguished by $P(\lambda)$ being a polynomial and the one phase periodic solution corresponds to (**2P**) $P(\lambda) = \prod_1^4(\lambda - \lambda_i) = \lambda^4 - s_1\lambda^3 + s_2\lambda^2 - s_3\lambda + s_4$ with (**2Q**) $g = \mu(x,t) - \lambda$ leading to (**2R**) $\epsilon\mu_x = 2\sqrt{-P(\mu)}$ (via $\mu = \lambda$). Further $g_t = \mathfrak{B}g_x - \mathfrak{B}_x g$ implies $\mu_t = -s_1\mu_x$ (via $\mu = (s_1/2) + i\epsilon\mu_x/2u$ from the coefficient of λ^3). Consequently $\mu(x,t)$ depends only on the phase $\xi = x - s_1 t$ and (**2S**) $\epsilon(d\mu/d\xi) = 2\sqrt{-P(\mu)}$. There is much more on this in [**715**]. ■

3. VORTICES

We go here to [**55, 138, 136, 140, 139, 141, 419, 1056, 1099, 1149, 1256, 1283, 1311, 1314**] and will start with [**138**]. Thus consider $\psi = Rexp(iS/\hbar)$ in 3-D with

$$\rho(\mathbf{r},t) = |\psi|^2 = R^2;\; \mathbf{v} = \frac{1}{m}\frac{\Re[\psi^*(-i\hbar\nabla - e\mathbf{A})\psi]}{|\psi|^2} = \frac{1}{m}[\nabla S - e\mathbf{A}] \tag{3.1}$$

The gradient term ∇S does not contribute to vorticity and thus the bulk vorticity $\nabla\times\mathbf{v}$ in the probability fluid is completely determined by the magnetic field except at points where the phase S is singular; this may occur only at points where the wave function vanishes. The vanishing of a complex wave function involves the vanishing of two real functions, defining two surfaces, whose intersection is then a (vorticity) line. Thus in addition to a given distribution of vorticity in the fluid (uniquely determined by the magnetic field) one may also have isolated vortex lines embedded in the fluid. Along these lines the vorticity has a 2-D delta funnction singularity δ^2 in the plane perpendicular to each vortex line and it is the dynamics of these lines which is studied here. Thus to insure that the wave function is single valued the strength of every vortex as measured via circulation along any closed contour encircling the vortex must satisfy (**3A**) $\Gamma = \oint_C d\mathbf{l}\cdot\mathbf{v}(\mathbf{r},t) = (2\pi\hbar/m)n$ where $n = 0, \pm 1, \cdots$. As one approaches the vortex line $\mathbf{v}$ must tend to infinity in order to satisfy the quantization. Vortex lines may be created in the form of a closed vortex line that springs from a point or in pairs of opposite circulation. In QM the line vortices occur in exact solutions of the SE without any limiting procedure; the fact that the velocity becomes infinite near the vortex is acceptable because it is only the probability and not real matter that flows with that velocity.

One can consider a broader class of vortices which do not exhibit cylindrical symmetry even when they form a straight line. Thus consider an arbitrary vortex line $\vec{\xi}(s)$ where s is length along the line. The local properties of such a vortex are determined by the wave function near this point (**3B**) $\psi(x,y,z) \approx [\mathbf{r} - \vec{\xi}(s)]\cdot\nabla\psi(\vec{\xi}(s))$ (keeping only the lowest terms in a Taylor expansion). Define a complex vector (**3C**) $\mathbf{w}(s) = \nabla\psi(\mathbf{r})|_{\mathbf{r}=\vec{\xi}(s)}$. One shows that this vector describes the basic properties of the vortex. Note that since the derivative in s of the wave function along the vortex line is zero the real and imaginary parts of $\mathbf{w}(s)$ lie in the plane perpendicular to the tangent vector $\mathbf{t}(s) = d\vec{\xi}(s)/ds$, i.e. (**3D**) $(d/ds)\psi(\vec{\xi}(s)) = \mathbf{t}(s)\cdot\mathbf{w}(s) = 0$. The general complex vector $\mathbf{w}$ will have four real components but only two parameters contain the information about the vortex since multiplication of the wave function by a complex number does not change the velocity field, i.e.

$$\mathbf{v}(x,y,z) = \frac{\hbar}{2mi}\left(\frac{\mathbf{w}}{\mathbf{w}\cdot\mathbf{r}} - \frac{\mathbf{w}^*}{\mathbf{w}^*\cdot\mathbf{r}}\right) = \frac{\hbar}{2mi}\frac{\mathbf{r}\times(\mathbf{w}\times\mathbf{w}^*}{|\mathbf{w}\cdot\mathbf{r}|^2} \tag{3.2}$$

is a homogeneous fuction of $\mathbf{w}$. Therefore $\mathbf{v}$ does not change when $\mathbf{w}$ is multiplied by a complex number (here the origin has been chosen at $\vec{\xi}(s)$). Thus vortex lines may appear without any special conditions. The vector $\mathbf{w}$ also plays a crucial role in determining the motion of the vortex line. Indeed the velocity $\mathbf{u} = (d/dt)\vec{\xi}(s,t)$ of a point $\vec{\xi}(s,t)$ lying on the vortex line can be obtained from (**3E**) $(d/dt)\psi[\vec{\xi}(s,t)] = \mathbf{u}\cdot\mathbf{w} + \partial_t\psi = 0$. Here $\mathbf{u}$ is determined only up to a vector parallel to the vortex line and solving (**3E**) gives

$$\mathbf{u} = \frac{\mathbf{w}\times\mathbf{w}^*}{|\mathbf{w}\times\mathbf{w}^*|^2}\times(\partial_t\psi^*\mathbf{w} - \partial_t\psi\mathbf{w}^*) \tag{3.3}$$

Using the SE

$$i\hbar\partial_t\psi = \left[-\frac{\hbar^2}{2m}\Delta + V\right]\psi \tag{3.4}$$

one obtains

$$\mathbf{u} = \frac{\hbar}{2mi}\frac{\mathbf{w}\times\mathbf{w}^*}{|\mathbf{w}\times\mathbf{w}^*|}\times(\mathbf{w}\Delta\psi^* + \mathbf{w}^*\Delta\psi) \tag{3.5}$$

Thus the motion of the vortex line at a given point is completely determined by the local properties of the wave function, namely its gradient and its Laplacian at this point. There is no direct relation between $\mathbf{u}$ and $\mathbf{v}$. One notes also that using an initial condition for the SE $\phi_k(\mathbf{r}) = exp(i\mathbf{k}\cdot\mathbf{r})\phi_0(\mathbf{r})$ one can differentiate in k any number of times to obtain new wave functions. Then setting $\mathbf{k} = 0$ one can arrive at an expression $[W_R(\mathbf{r}) + iW_I(\mathbf{r})]\phi_0(\mathbf{r}) = 0$ where W_R, W_I are real polynomials in (x, y, z). Then $W_R = 0 = W_I$ determines two 2-D surfaces intersecting in a vortex line if the circulation is different from 0. This procedure leads to wave functions that have initial vortex lines of almost arbitrary shape and topology embedded in them.

We go now to Bialynicki-Birula et al [**140**] and recall (3.1) and (**3A**) in the QM situation where a recovery formula for the phase can also be given, namely $S(\mathbf{r}) = m\int_{r_0}^{r} \mathbf{d}l\cdot\mathbf{v}$. One wants to generalize matters now to EM fields and defines a replacement for the wave function via the Riemann-Silberstein (RS) vector (**3F**) $\mathbf{F} = (\mathbf{E} + i\mathbf{B})/\sqrt{2}$. The Maxwell equations in free space are then (for $c = 1$) (**3G**) $i\partial_t\mathbf{F} = \nabla\times\mathbf{F}$ and $\nabla\cdot\mathbf{F} = 0$. This presents a strong analogy to a SE and one can think of $\mathbf{F}$ as a photon wave function with a view to analyzing the vortex lines and their motion. There are however 3 independent phases ϕ_k corresponding to the 3 components of $\mathbf{F}$ and one thinks first of a natural generalization of (5.1) in the form

$$\mathbf{v} = \frac{1}{2i}\frac{\sum_k [F_k^*\nabla F_k - (\nabla F_k^*)F_k]}{\sum F_k^* F_k} \tag{3.6}$$

However this does not work and one loses some of the simplicity of the scalar situation. Thus note that the reconstruction of the EM field from geometry may be described via the Einstein equations (**3H**) $R_{\mu\nu} - (1/2)g_{\mu\nu}R = \kappa T_{\mu\nu}$. The knowledge of the energy momentum tensor of the EM field is however alone not sufficient to determine the EM field. This can be seen via

$$T_{00} = \mathbf{F}^*\cdot\mathbf{F};\ T_{0i} = \epsilon_{ijk}(1/i)F_j^*F_k;\ T_{ij} = -F_i^*F_j - F_j^*F_i + \delta_{ij}\mathbf{F}^*\cdot\mathbf{F} \tag{3.7}$$

All components of the EM tensor are invariant under (**3I**) $\mathbf{E}' = \mathbf{E}Cos(\phi) - \mathbf{B}Sin(\phi)$ and $\mathbf{B}' = \mathbf{E}Sin(\phi) + \mathbf{B}Cos(\phi)$ so the overall phase cannot be determined from the EM tensor. However if one assumes the Maxwell equations then the phase can be determined from $T_{\mu\nu}$. This was discovered by Rainich [**1057**] and an alternative (equivalent) proposal appears here. Thus one defines the phase of the EM field ϕ via (**3J**) $\mathbf{F}^2(x) = exp[2i\phi(x)]|\mathbf{F}|^2$. In full analogy with (3.1) for nonrelativistic wave mechanics one defines a velocity 4-vector u^μ via

$$u_\mu = \frac{(\mathbf{F}^2)^*\partial_\mu\mathbf{F}^2 - \mathbf{F}^2\partial_\mu(\mathbf{F}^2)^*}{4i|\mathbf{F}|^2} \tag{3.8}$$

Since $\mathbf{F}^2$ can be written as (**3K**) $\mathbf{F}^2 = \mathfrak{S} + i\mathfrak{P} = (1/2)(\mathbf{E}^2 - \mathbf{B}^2) + i\mathbf{E}\cdot\mathbf{B}$ one sees that u_μ is a relativistic 4-vector

$$u_\mu = \frac{\mathfrak{S}\partial_\mu\mathfrak{P} - \mathfrak{P}\partial_\mu\mathfrak{S}}{2(\mathfrak{S}^2 + \mathfrak{P}^2)} \tag{3.9}$$

In this formulation the square of the RS vector $\mathbf{F}$ plays the role of the wave function ψ and vortex lines are to be found at the intersection of the surfaces $\mathfrak{S} = 0$ and $\mathfrak{P} = 0$. As for the Schrödinger wave function ψ at all points where $\mathbf{F}^2 \neq 0$ the vector u_μ is by construction a pure gradient (**3L**) $u_\mu = \partial_\mu\phi(x)$ and therefore one can recover the phase of **F** via (**3M**) $\phi(x) = \int_{x_0}^{x} d\xi^\mu u_\mu(\xi)$. Since the RS vector is univalued the phases obtained by choosing different paths between x_0 and x can differ only by a multiple of π and consequently (**3N**) $\oint d\xi^\mu u_\mu(\xi) = 2\pi n$. The phase defined by (**5M**) is determined up to a global phase ϕ_0 (the value of $\phi(x)$ at the lower limit of integration) and this value ϕ_0 cannot be obtained from the EM tensor. Under duality rotations (**5I**) when ϕ varies from 0 to 2π the vector $\mathbf{E}'$ at each spacetime point draws an ellipse in the $\mathbf{E} - \mathbf{B}$ plane and the same ellipse is drawn by $\mathbf{B}'$. These ellipses become circles on the vortex line since then the vectors **E** and **B** are orthogonal and of equal length. The denominator in (3.9) can also be written as

$$\mathfrak{S}^2 + \mathfrak{P}^2 = \left(\frac{\mathbf{E}^2 + \mathbf{B}^2}{2}\right)^2 - (\mathbf{E} \times \mathbf{B})^2 \tag{3.10}$$

Hence the vanishing of $\mathfrak{S}^2 + \mathfrak{P}^2$ at a point also means that the EM field at this point is pure radiation; the energy density and the Poynting vector form a null 4-vector so that one can say that on vortex lines the energy of the EM field moves locally with the speed of light. Note that the geometric properties of the Poynting vector and the space part of u_μ are different; since $\mathfrak{S}$ is a scalar and $\mathfrak{P}$ is a pseudoscalar, the vector u_μ is really a pseudovector. In the simplest case of a constant EM field the Poynting vector is $\mathbf{E} \times \mathbf{B}$ while $u_\mu = 0$ identically. There does not seem to be a physical quantity whose flow can be identified with u_μ (in contrast to nonrelativistic wave mechanics where the gradient of the phase determines the velocity of the probability flow. Note also that the EM field does not have to vanish identically along the lines where the phase is singular; it is only that the field is null, i.e. the two invariants $\mathfrak{S}$ and $\mathfrak{P}$ vanish. Nevertheless lines where the field is null will be called vortex lines.

In Bialynicki-Birula and Radozycki [**142**] another method of introducing vortex lines of the EM field is outlined. Here it is observed that when a null solution of the Maxwell equations is taken as the background field an extra scalar multiplier may imprint on this solution a rich vortex structure. Then all components of the EM field will vanish on vortex lines (but the vortex lines obtained in this manner do not have a generic character since they arise only from null fields). First one notes that a complex scalar function $\phi(\mathbf{r}, t)$ multiplying the RS vector will control the zeros of the EM field. Assume now that the background field **F**, as well as the product $\phi\mathbf{F}$ satisfy the Maxwell equations

$$i\partial_t\mathbf{F} = \nabla \times \mathbf{F};\ \nabla \cdot \mathbf{F} = 0;\ i\partial_t(\phi\mathbf{F}) = \nabla \times (\phi\mathbf{F});\ \nabla \cdot (\phi\mathbf{F}) = 0 \tag{3.11}$$

This leads to conditions

$$\mathbf{F}i\partial_t\phi = -\mathbf{F} \times \nabla\phi;\ \mathbf{F} \cdot \nabla\phi = 0 \tag{3.12}$$

These equations have nontrivial solutions only when the background field is null, i.e. $\mathbf{F}^2 = (1/2)(\mathbf{E}^2 - \mathbf{B}^2) + i\mathbf{E} \cdot \mathbf{B} = 0$ (see [**182**] for proof). Further one obtains two equations for ϕ, namely (**5O**) $\partial_t\phi + \mathbf{n} \cdot \nabla\phi = 0$ and $\mathbf{F} \cdot \nabla\phi = 0$ where **n** is the normalized Poynting vector

$$\mathbf{n} = \frac{-i\mathbf{F}^* \times \mathbf{F}}{\mathbf{F}^* \cdot \mathbf{F}} = \frac{\mathbf{E} \times \mathbf{B}}{(1/2)(\mathbf{E}^2 + \mathbf{B}^2)} \tag{3.13}$$

The solutions of (**5O**) will generally have zeros and this will lead to vortex lines of the EM field riding atop the background solution **F**. Near the vortex line the EM field vectors followed around a closed curve rotate by $2\pi m$ where m is the topological charge of the vortex. There is no interaction between vortex lines (in contrast to nonrelativistic wave mechanics). Examples and more discussion are provided in [**142**].

REMARK 8.3.1. The RS vector **F** in the form $\mathbf{F} = (\mathbf{D}/\sqrt{2\epsilon_0}) + i(\mathbf{B}/\sqrt{2\mu_0})$ is used in the first paper of [**142**] to describe photon states. ■

REMARK 8.3.2. The paper Radozycki [**1056**] extends the analysis of [**136**] to show how quantum (field theoretic) effects influence the nodal lines of the EM wave function. Here pair creation leads to photon-photon interaction and the Maxwell equations are supplemented by additional terms. An Euler-Heisenberg Lagrangian

(3.14) $$L(\mathbf{r},t) = \mathfrak{S} + \frac{2\alpha^2}{45m^4}[4\mathfrak{S}^2 + 7\mathfrak{P}^2]$$

is used where $\alpha \sim$ the fine structure constant and m is the electron mass. The α^2 term corresponds to one loop accuracy and e.g.

(3.15) $$\mathbf{E} = [1 + \alpha^2 K(\mathbf{r},t)]\mathbf{D} + \alpha^2 M(\mathbf{r},t)\mathbf{B}$$

where

(3.16) $$K = -\frac{16}{45m^4}[\mathbf{D}^2 - \mathbf{B}^2];\ M = -\frac{28}{45m^4}\mathbf{D}\cdot\mathbf{B}$$

We refer to [**1056**] for more details and see also [**142, 708**]. ■

We go now to van Holten [**1256**] where there is some unifying discussion concerning particles fluids, fields, vortices, etc. We enumerate via equations and formulas.

(1) First for particles

$$L = (1/2)g_{ij}(x)\dot{x}^i\dot{x}^j;\ D^2x^i/Dt^2 = \ddot{x}^i + \Gamma^{\ i}_{jk}\dot{x}^j\dot{x}^k = 0$$

(2)

$$p_i = \frac{\partial L}{\partial \dot{x}^i} = g_{ij}\dot{x}^j;\ H = \frac{1}{2}g^{ij}p_ip_j$$

(3)

$$\partial_t F = \{F,H\} = \frac{\partial F}{\partial x^i}\frac{\partial H}{\partial p_i} - \frac{\partial F}{\partial p_i}\frac{\partial H}{\partial x^i};\ \dot{x}^i = \frac{\partial H}{\partial p_i};\ \dot{p}_i = -\frac{\partial H}{\partial x^i}$$

(4)

$$\frac{\partial S}{\partial t} = -H(x, p = \nabla S);\ \partial_t S = -\frac{1}{2}g^{ij}\nabla_i S\nabla_j S;\ S = \int_0^t d\tau L(x,\dot{x})|_{x^i(\tau)}$$

EXAMPLE 3.1. Consider a particle on the surface of the unit sphere S^2 with $m = 1$

$$L(\theta,\phi) = \frac{1}{2}(\dot{\theta}^2 + Sin^2(\theta)\dot{\phi}^2;\ H = \frac{1}{2}\left(p_\theta^2 + \frac{p_\phi^2}{Sin^2(\theta)}\right) = \frac{\mathbf{J}^2}{2}$$

with $p_\theta = \dot{\theta}$ and $p_\phi = Sin^2(\theta)\dot{\phi}$. Here $\mathbf{J}$ is a Casimir invariant given via

$$J_x = -Sin(\phi)p_\theta - Cos(\phi)Ctn(\theta)p_\phi;\ J_y = Cos(\phi)p_\theta - Sin(\phi)Ctn(\theta)g_\phi$$

with $J_z = p_\phi$ and the geodesics on the sphere are great circles parametrized via

$$Cos[\theta(\tau)] = Sin(\alpha)Sin[\omega(\tau - \tau_*)];\ Tan[\phi(\tau) - \phi_*] = Cos(\alpha)Tan[\omega(\tau - \tau_*)]$$

where α is a constant and τ_* ϕ_* refer to the time and longitude at which the orbit crosses the equator $\theta_* = \pi/2$. On these orbit the angular frequency is given via $\omega^2 = 2H = \mathbf{J}^2$. ■

Next for fluids

(1)

$$A(\rho, S) = \int dt \int d^n x \sqrt{g}\rho(\partial_t S + (1/2)g^{ij}\nabla_i S\nabla_j)$$

(2)

$$\frac{1}{\sqrt{g}}\frac{\delta A}{\delta\rho} = \partial_t S + \frac{1}{2}g^{ij}\nabla_i S\nabla_j S = 0;\ -\frac{1}{\sqrt{g}}\frac{\delta A}{\delta S} = \partial_t\rho + \nabla_i(g^{ij}\rho\nabla_j S) = 0$$

(3) First $v_i = \nabla_i S \Rightarrow \partial_t\rho + \nabla_i(\rho v^i) = 0$ leading to

$$\partial_t v_i + v^j\nabla_j v_i = 0;\ \nabla_j v_i = \frac{\partial v_i}{\partial x^j} - \Gamma^{\ k}_{ji} v_k$$

(4) This flow is of potential type and in the absence of torsion $\Gamma^{\ k}_{ij}$ is symmetric which makes the vorticity zero (i.e. $\nabla_i v_j - \nabla_j v_i = 0$ while for the fluid to be incompressible one must have $\nabla \cdot \mathbf{v} = \Delta S = 0$ where $\Delta = g^{ij}\nabla_i\nabla_j$. Hence the number of incompressible modes of flow on the manifold equals the number of zero-modes of the scalar Laplacian; on S^2 or any compact Riemann surface there is only one incompressible flow ($v^i = 0$).

(5) Generally to solve the equations in item 3 the equation of continuity takes the form $\partial_t\rho + \nabla_i(\rho\nabla^i S) = 0$ and a stationary flow (ρ independent of time) is possible if $\nabla \cdot (\rho\nabla S) = 0$. In addition to $\rho = c$ with $v = \nabla S/m$ one can find nontrivial solutions here for spatially varying ρ. As an example in 2-D space one can introduce a stream function T, dual to the fluid momentum, and write

$$\rho\nabla^i S = \frac{1}{\sqrt{g}}\epsilon^{ij}\nabla_j T;\ \rho = \frac{\epsilon^{ij}\nabla_i S\nabla_j T}{\sqrt{g}(\nabla S)^2} = \frac{\epsilon^{ij}\nabla_i S\nabla_j T}{2H\sqrt{g}}$$

With H constant one can write $\tilde{T} = T/2H$ and hence

$$\rho = \frac{1}{\sqrt{g}}\epsilon^{ij}\nabla_i S\nabla_j\tilde{T} = \frac{1}{\sqrt{g}}\epsilon^{ij}v_i\nabla_j\tilde{T}$$

One thinks here of T as a pseudoscalar field dual to $\rho\nabla S$ and one notes that $\nabla S \cdot \nabla T = v \cdot \nabla T = 0$.

For vortices one has now

(1) The duality between S and T suggests studying the dynamics of a fluid for which T is the velocity potential, namely $v_i = (1/\rho_*)\nabla_i T$ (where ρ_* is included for dimensional reasons). This velocity field is stationary ($\partial_t v_i = 0$) but it is not geodesic. Indeed

$$v \cdot \nabla v_i = \frac{1}{2}\nabla v^2 = \frac{1}{2\rho_*^2}\nabla_i(\nabla T)^2 = \frac{1}{2\rho_*^2}\nabla_i(\rho\nabla S)^2$$

Here ρ and S denote the previous functions with $(\nabla S)^2 = 2H = \omega^2 = c$ and hence

$$v \cdot \nabla v_i = \frac{\omega^2}{2\rho_*^2}\nabla_i\rho^2 = \nabla_i h;\ \frac{1}{2}v^2 = -(h - h_0) = \frac{\omega^2\rho^2}{2\rho_*^2}$$

where h denotes the external potential. Since $v_i = (1/\rho_*)\nabla_i T$ the local vorticity vanishes (i.e. $\nabla_i v_j - \nabla_j v_i = 0$) but this is not necessarily true globally. Indeed at singular points of the original geodesic flow (with sources/sinks) the dual flow generally has vortices/anti-vortices. We refer to [**1255**] for further discussion of the dual particle model.

REMARK 8.2.9. We refer to [**55, 1267**] for vortices in superfluids and non-Riemannian spacetimes and we will make some preliminary remarks about Chern-Simons vortices; the latter topic will be picked up later in more detail. ■

4. CHERN-SIMONS

We refer here especially to [**406, 407, 408, 409, 410, 419, 420, 682, 683, 808, 807, 859, 998, 999, 1149, 1195**] and will extract from Dunne [**410**] for a brief introduction (many important points will be ignored). First recall the Yang-Mills action (**4A**) $S_{YM} = \int d^4x Tr(F_{\mu\nu}F_{\mu\nu})$ where $F_{\mu\nu} = \partial_\mu A_\nu - \partial_\nu A - \mu + [A_\mu, A_\nu]$ is the gauge field curvature. The Euler-Lagrange (EL) equations have the form (**4B**) $D_\mu F_{\mu\nu} = 0$ where $D_\mu = \partial_\mu + [A_\mu, \cdot]$ is the covariant derivative. In 4-D space the YM action is minimized by solutions of the self dual (or anti self dual) YM equations $F_{\mu\nu} = \pm\tilde{F}_{\mu\nu}$ where $\tilde{F}_{\mu\nu} = (1/2)\epsilon_{\mu\nu\rho\sigma}F_{\rho\sigma}$ is the dual field strength. Solutions of the self duality equations are automatically solutions to the EL equations since $D_\mu\tilde{F}_{\mu\nu} = 0$ (Bianchi identity). One recalls also the (self dual) Bogolmolny equations (**4C**) $D_i\Phi = -\epsilon_{ijk}F_{jk}$ where Φ is a scalar field; these arise in the theory of magnetic monopoles in 3+1. The abelian Higgs model in 2+1 involves a complex scalar field ϕ interacting with a $U(1)$ gauge field with conventional Maxwell dynamics. This is a relativistic analogue of the Ginsburg-Landau model and has the static energy functional

$$E = \int d^2x \left[\frac{1}{2}B^2 + |\vec{D}\phi|^2 + V(|\phi|)\right] = \int d^2x \left[|(D_j - i\epsilon_{jk}D_k)\phi|^2 + \right. \tag{4.1}$$

$$\left. + \frac{1}{2}(B + |\phi|^2 - v^2)^2 + v^2 B - \frac{1}{2}(|\phi|^2 - v^2)^2 + V(|\phi|)\right]$$

Thus for a special self dual quartic potential $V_{SD} = (1/2)(|\phi|^2 - v^2)^2$ the static energy functional is bounded below by v^2 times the magnetic flux B and this Bogomolny bound is saturated by solutions of the self duality equations

$$D_j\phi = i\epsilon_{ijk}D_k\phi;\ F_{12} = v^2 - |\phi|^2 \tag{4.2}$$

The term self duality arises from the appearance of the duality operation in the first equation of (4.2). The self dual CS theories in [**355**] describe charged scalar fields in 2+1 minimally coupled to a gauge field whose dynamics is given by a CS Lagrangian of the form (**4D**) $L_{CS} = \epsilon^{\mu\nu\rho}Tr(\partial_\mu A_\nu A_\rho + (2/3)A_\mu A_\nu A_\rho)$. The gauge field A_μ takes values in a finite dimensional representation of the gauge Lie algebra $\mathfrak{g}$ and the totally antisymmetric symbol $\epsilon^{\mu\nu\rho}$ is normalized with $\epsilon^{012} = 1$. In an abelian theory the gauge fields A_μ commute and the trilinear term in (**4D**) vanishes. The EL equations here are $F_{\mu\nu} = 0$ which follows from (**4E**) $\delta L_{CS}/\delta A_\mu = \epsilon^{\mu\nu\rho}F_{\nu\rho}$. There are pure gauge solutions $A_\mu = g^{-1}\partial_\mu g$ of $F_{\mu\nu} = 0$ but putting in topological effects and external sources makes the situation more interesting and important; one notes that these equations are first order in the space time derivatives. Note also the gauge invariance of $F_{\mu\nu} = 0$ under $A_\mu \to A^g_\mu = g^{-1}A_\mu g + g^{-1}\partial_\mu g$ but the Lagrangian is not gauge invariant since

$$(4.3)\quad L_{CS}(A) \to L_{CS}(A) - \epsilon^{\mu\nu\rho}\partial_\mu Tr(\partial_\nu g g^{-1} A_\rho - \frac{1}{3}\epsilon^{\mu\nu\rho}(g^{-1}\partial_\mu g g^{-1}\partial_\nu g g^{-1}\partial_\rho g)$$

For an abelian CS theory the last term in (4.3) vanishes so the change is a total spacetime derivative and the action is gauge invariant; hence a sensible quantum theory should be possible. For a nonabelian CS theory the last term is proportional to the winding number of the group element g and to insure that a quantum amplitude $exp(iS)$ remain invariant the CS density L_{CS} must be multiplied by a dimensionless coupling parameter $\kappa' = n/4\pi$ for n an integer.

One notes also that the CS term describes a topological gauge field theory in the sense that there is explicit dependence on the spacetime metric since the Lagrange density can be written as a 3-form $L_{CS} = Tr(AdA + A^3)$. Hence the action is independent of the metric and the CS density does not contribute to the energy momentum tensor. One couples the CS Lagrangian go an external matter field via (**4F**) $L_{CS} = \epsilon^{ij}Tr(A_i\dot{A}_j) + Tr(A_0F_{12})$ where A_1, A_2 are canonically conjugate. The A_0 part of the Lagrangian produces the Gauss law constraint and there is no contribution to the Hamiltonian; this implies that the CS gauge field does not have any real dynamics of its own and is a nonpropagating field whose dynamics comes from the fields to which it is minimally coupled. One can couple to an external matter current J^μ via (**4G**) $L = (\kappa/2)L_{CS} - Tr(A_\mu J^\mu)$ leading to equations of motion $F_{\mu\nu} = -(1/\kappa)\epsilon_{\mu\nu\rho}J^\rho$ (where $D_\mu J^\mu = 0$) with $J^0 = \kappa F_{12}$. The spatial component J^i are perpendicular to the electric field $J^i = \kappa\epsilon^{ij}F_{j0}$ $(E_i = F_{0i})$.

The abelian nonrelativistic selfdual CS system describes a complex scalar field ψ minimally coupled to an abelian gauge field A_μ where $A_\mu = (A_0, \mathbf{A})$ and $g_{\mu\nu} = diag(-1,1,1)$ $(c = 1)$. Thus the gauge field does not have any dynamics of its own and the Lagrangian is

$$(4.4)\qquad L = \frac{\kappa}{2}\epsilon^{\mu\nu\rho}A_\mu\partial_\nu A_\rho + i\psi^* D_0\psi - \frac{1}{2m}|\mathbf{D}\psi|^2 + \frac{g}{2}|\psi|^4$$

where m is the mass of the scalar field ψ, κ is a coupling constant determining the strength of the CS term $L_{CS} = \epsilon^{\mu\nu\rho}A_\mu\partial_\nu A_\rho$ and g is a coupling constant determining the strength of the π nonlinearity. Here $\epsilon^{\mu\nu\rho}$ is normalized via $\epsilon^{012} = 1$

and $D_\mu = \partial_\mu + iA_\mu$. The EL equations are then

(4.5) $$iD_0\psi = -\frac{1}{2m}\mathbf{D}^2\psi - g|\psi|^2\psi;\ F_{\mu\nu} = -\frac{1}{\kappa}\epsilon_{\mu\nu\rho}J^\rho$$

where $J^\mu = (\rho, \mathbf{J})$ with $\rho = |\psi|^2$ and $J^i = -\frac{i}{2m}(\psi^* D^j\psi - (D^j\psi)^*\psi)$. The equations (4.5) are called the planar gauged nonlinear SE (cf. [**686**]). The equations of motion (6.5) are invariant under gauge transformations $\psi \to exp(-i\lambda)\psi$ and $A_\mu \to A_\mu + \partial_\mu\lambda$ but the Lagrange density is not invariant since $L_{CS} \to L_{CS} + \partial_\mu(\lambda\epsilon^{\mu\nu\rho}\partial_\nu A_\rho)$. Nevertheless the classical theory is sensible since the equations of motion are gauge invariant. One can also write ($\epsilon^{12} = 1$)

(4.6) $$G = \frac{1}{\kappa}\rho;\ E^i = -\frac{1}{\kappa}\epsilon^{ij}J_j;\ B = \partial_1 A_2 - \partial_2 A_1 = F_{12};\ E^i = \partial_i A_0 - \partial_0 A_i = F_{i0}$$

(thus $\mathbf{J} \perp \mathbf{E}$). From (4.5) one has

(4.7) $$\dot{\rho} + \vec{\nabla}\cdot\mathbf{J} = 0 \equiv \partial_\mu J^\mu = 0;\ \Phi = \int d^2xB = \frac{1}{\kappa}Q$$

and Q is conserved. There is a Hamiltonian formulation via

(4.8) $$L = i\psi^*\dot{\psi} + \kappa\dot{A}_1 A_2 - \frac{1}{2m}|\mathbf{D}\psi|^2 + \frac{g}{2}|\psi|^4 + A_0(\kappa B - \rho)$$

The last term enforces the CS Gauss law $\kappa B = \rho$ and this constraint may be solved by writing

(4.9) $$\mathbf{A} = -\frac{1}{\kappa}\int d^2x'\mathbf{G}(\mathbf{x},\mathbf{x}')\rho(\mathbf{x}');\ \mathbf{G} = \mathbf{G}(\mathbf{x}-\mathbf{x}');\ \vec{\nabla}\times\mathbf{G} = \epsilon^{ij}\partial_i G_j = -\delta^2(\mathbf{x})$$

with $G^i(\mathbf{x}) = -(1/2\pi)\epsilon^{ij}\partial_j log(\mathbf{r})$. This leads to a contrained Hamiltonian system with

(4.10) $$H = \int d^2x\left(\frac{1}{2m}|\mathbf{D}\psi|^2 - \frac{g}{2}|\psi|^4\right);\ i\partial_0\psi = \frac{\delta H}{\delta\psi^*}$$

where the $\mathbf{A}$ appearing in the covariant derivative is given by (4.9). Here in computing $\delta H/\delta\psi^*$ one must also vary $\mathbf{A}$ with respect to ψ^* leading to the A_0 for (4.6) in the form (**4H**) $A_0 = -(1/\kappa)\int d^2\mathbf{G}(\mathbf{x},\mathbf{x}')\cdot\mathbf{J}(\mathbf{x}')$. To find solutions of (4.5) one can make a selfdual Ansatz for the matter fields. Thus suppose the matter field ψ satisfies (**4I**) $D_j\psi \pm i\epsilon_{jk}D_k\psi$ or equivalently $D_\mp\psi = 0$. Writing $x^\pm = (1/2)(x^1 \mp x^2)$ with $\partial_\pm = \partial_1 \pm i\partial_2$ and $D_\pm = D_1 \pm iD_2$ one finds

(4.11) $$J^j = \pm\frac{1}{2m}\epsilon^{jk}\partial_k(|\psi|^2);\ \partial_i A_0 - \partial_0 A_i = \pm\frac{1}{2m\kappa}\partial_i(|\pi|^2)$$

Note also the identity (**4J**) $\mathbf{D}^2\psi = D_\pm D_\mp\psi \mp F_{12}\psi$ and using $B = (1/\kappa)\rho$ the gauge NLS becomes

(4.12) $$i\partial_0\psi = -\frac{1}{2m}D_\pm D_\mp\psi - \left(g \mp \frac{1}{2m\kappa}\right)|\psi|^2 + A_0\psi$$

With the self dual Ansatz the first term on the right vanishes and one has

(4.13) $$i\partial_0\psi = -\left(\left(g \mp \frac{1}{2m\kappa}\right)|\psi|^2 - A_0\right)\psi$$

One sees that (4.11) and (4.13) are solved by static solutions $\partial_0\psi = 0$ and $\partial_0 A_i = 0$ with A_0 chosen as $A_0 = \pm(1/2m\kappa)|\psi|^2$ provided $g = \pm(1/m\kappa)$ (called self dual coupling). The corresponding self dual Lagrangian is then

(4.14) $$L = \frac{\kappa}{2}\epsilon^{\mu\nu\rho}A_\mu\partial_\nu A_\rho + i\psi^* D_0\psi - \frac{1}{2m}|\mathbf{D}\psi|^2 \pm \frac{1}{2m\kappa}|\psi|^4$$

This leads to the nonrelativistic selfdual CS equations

(4.15) $$D_\mp\psi = 0;\ F_{12} = \frac{1}{\kappa}|\psi|^2$$

One remarks that (**4J**) can be rewritten as (**4K**) $|\mathbf{D}\psi|^2 = |D_\mp\psi|^2 \pm F_{12}|\psi|^2 \mp m\epsilon^{ij}\partial_i J_j$ leading to

(4.16) $$H = \int d^2x\left(\frac{1}{2m}|D_\mp\psi|^2 - \frac{1}{2}\left(g \mp \frac{1}{m\kappa}\right)|\psi|^4\right)$$

When $g = \pm(1/m\kappa)$ this gives (**4L**) $H = (1/2m)\int d^2x|D_\mp\psi|^2$. Note that self dual solutions are necessarily static and in fact all static solution are solutions of the self duality equations.

The particular self dual form of the potential in (4.14) may also be understood as a Pauli like magnetic interaction. Thus consider a two component spinor χ and let (**4M**) $S = \vec{\sigma}\cdot(\vec{\nabla} + i\mathbf{A})\chi$ where the Pauli matrices satisfy $\sigma^i\sigma^j = \delta^{ij} + i\epsilon^{ij}\sigma^3$. Then the Pauli energy is

(4.17) $$E = \frac{1}{2m}\int d^2x S^\dagger S = \frac{1}{2m}\int d^2x|\mathbf{D}\chi|^2 + \frac{1}{2m}\int d^2x B\chi^\dagger\sigma^3\chi$$

Taking χ to be an eigenstate of σ^3 with eigenvalue ±1, i.e. $\chi_+ = \begin{pmatrix}\psi\\0\end{pmatrix}$ or $\chi_- = \begin{pmatrix}0\\\psi\end{pmatrix}$ the Pauli energy becomes

(4.18) $$E = \frac{1}{2m}\left(\int d^2x|\mathbf{D}\psi|^2 \pm \int d^2x B|\psi|^2\right) = \frac{1}{2m}\int d^2x\left[|\mathbf{D}\psi|^2 \pm \frac{1}{\kappa}|\psi|^4\right]$$

(using $\rho = B\kappa$). Thus the self dual nonlinear interaction term in (4.14) may be alternatively viewed as a magnetic moment interaction with a magnetic field that is given self consistently in terms of the charge density ρ by the CS equation $\rho = B\kappa$.

The nonrelativistic field theory with Lagrangian (4.4) possesses the kinematical symmetry of Galilean invariance, namely

- Time translation $t \to t' = t + a$ and $\mathbf{x} \to \mathbf{x}'$.
- Space translation $t \to t' = t$ and $\mathbf{x} \to \mathbf{x}' = \mathbf{x} + \mathbf{a}$
- Space rotation $t \to t' = t$ and $x_i \to x_i' = \mathcal{R}_{ij}(\omega)x_j$
- Galilean boost $t \to t' = t$ and $\mathbf{x} \to \mathbf{x}' = \mathbf{x} + \mathbf{v}t$

Under the first three transformations $\psi'(\mathbf{x}', t') = \psi(\mathbf{x}, t)$ but under a Galilean boost $\psi(\mathbf{x}', t') = exp[im\mathbf{v}\cdot(\mathbf{x} + \mathbf{v}t)/2]\psi(\mathbf{x}, t)$ (i.e. a cocycle term appears). The nonrelativistic CS system with Lagrangian (4.4) has additional invariance beyond Galilean symmetrization corresponding to conformal reparametrization of the time coordinate. Consider e.g. (**4N**) $t \to t' = T(t)$ and $\mathbf{x} \to \mathbf{x}' = \sqrt{\dot{T}}\mathbf{x}$

and define the transformed field as $\psi'(\mathbf{x}',t') = (1/\sqrt{\dot{T}})exp[imr^2\ddot{T}/(4\dot{T})]\psi(\mathbf{x},t)$. An interesting thing happens here in that the transformed field is $\psi'(\mathbf{x}',t') = (1/\sqrt{\dot{T}})exp[imr^2\ddot{T}/(4\dot{T})]\psi(\mathbf{x},t)$ and if ψ' satisfies the original gauge NLS (4.5) in terms of primed coordinates it follows that the field $\psi(\mathbf{x},t)$ satisfies

$$iD_0\psi = -\frac{1}{2m}\mathbf{D}2\psi - g|\psi|^2\psi + \frac{1}{2}m\omega^2(t)\mathbf{x}^2\psi; \tag{4.19}$$

$$\omega^2(t) = \frac{\dddot{T}}{2\dot{T}} - \frac{3}{4}\frac{(\ddot{T})^2}{(\dot{T})^2} = -\frac{1}{2}\mathfrak{S}(T,t)$$

where $\mathfrak{S}$ is the Schwartzian (other transformations are also discussed in [**410**]).

5. REMARKS ON TIME

Time is perhaps the least understood of all physical concepts and we make here a few preliminary remarks (to be extended later).

5.1. MULTIFINGERED TIME. The discussion of the multifingered time (MFT) of Tomonaga in Gerlach [**521**] (cf. Remark 8.5.1) can be improved as in Nikolić [**930, 932, 934**] (cf. also [**254, 1141, 1233**]). Let $x = \{x^\mu\} = (x^0, \mathbf{x})$ be spacetime coordinates. A timelike Cauchy hypersurface Σ can be defined via a function $T(x)$ via the equation ((**5A**) $x^0 = T(\mathbf{x})$. If $T(\mathbf{x})$ is given then $\mathbf{x} \in \Sigma$ is correct and if $\sigma \subset \Sigma$ then e.g. T_σ denotes the set of values for $\mathbf{x} \in \sigma$. For a scalar field ϕ one describes its dynamics via

$$\hat{H}(\mathbf{x})\psi[\phi,T] = i\frac{\delta\psi[\phi,T]}{\delta T(\mathbf{x})} \tag{5.1}$$

A wave functional $\psi[\phi,T]$ can be viewed as a functional of ϕ_Σ and (5.1) shows how ψ changes for an infinitesimal change $\delta T(x)$ (we will ocassionally omit boldface on x now). Thus (5.1) is a generalized SE but it does not involve any preferred foliation of spacetime. Since Σ is determined by T one can say that $\rho[\phi,T] = |\psi[\phi,T]|^2$ is the probability density for the field to have the value ϕ at time T but remember that T is a collection of real parameters with one real parameter for each point $\mathbf{x}$. Consider now a free scalar field with Hamiltonian density (**5B**) $\hat{H}(x) = -(1/2)(\delta^2/\delta\phi^2(x)) + (1/2)[(\nabla\phi(x))^2 + m^2\phi^2(x)]$. Then writing $\psi = Rexp(iS)$ one obtains

$$\frac{1}{2}\left(\frac{\delta S}{\delta\phi(x)}\right)^2 + \frac{1}{2}[(\nabla\phi(x))^2 + m^2\phi^2(x)] + \mathfrak{Q}(x,\phi,T) + \frac{\delta S}{\delta T(x)} = 0 \tag{5.2}$$

$$\frac{\delta\rho}{\delta T(x)} + \frac{\delta}{\delta\phi(x)}\left(\rho\frac{\delta S}{\delta\phi(x)}\right) = 0;\ \mathfrak{Q} = -\frac{1}{2R}\frac{\delta^2 R}{\delta\phi^2(x)} \tag{5.3}$$

The Bohmian interpretation involves a deterministic time dependent hidden variable such that the time evolution of this variable is consistent with the probabilistic interpretation of ρ. This is naturally achieved by introducing a MFT field $\Phi(x,T)$ satisfying the MFT Bohmian equation of motion

$$\frac{\delta\Phi(x,T)}{\delta T(x')} = \delta^3(x-x')\left.\frac{\delta S}{\delta\phi(x)}\right|_{\phi=\Phi};\ \int_{\sigma_x} d^3x'\frac{\delta\Phi(x,T)}{\delta T(x')} = \left.\frac{\delta S}{\delta\phi(x)}\right|_{\Phi=\phi} \tag{5.4}$$

where σ_x is an arbitrarily small region around x. The second equation in (5.4) is the MFT version of the usual single-time Bohmian equation of motion $\partial_t\Phi(x,t) = (\delta S/\delta\phi(x))|_{\phi=\Phi}$ whereas the first equation is more fundamental since no σ_x is involved. For comparison purposes however integration within σ_x is useful; e.g. using (5.2) and (5.4) one has

$$\left[\left(\int_{\sigma_x} d^3x' \frac{\delta}{\delta T(x')}\right)^2 - \nabla_x^2 + m^2\right]\Phi(x,T) = -\int_{\sigma_x} d^3x' \frac{\delta\mathfrak{Q}(x',\phi,T)}{\delta\phi(x)}\bigg|_{\phi=\Phi} \tag{5.5}$$

This can be viewed as an MFT Klein-Gordon equation with a quantum term added. Note that officially one should write $\Phi(\mathbf{x},T(\mathbf{x})) = \phi(\mathbf{x},x^0) = \Phi(x)$ and we assume this is understood throughout.

Now to provide a manifestly covariant QFT one introduces $\mathbf{s} = (s^1,s^2,s^3)$ which serve as coordinates on a 3-D manifold; then write $x^\mu = X^\mu(\mathbf{s})$ leading to one equation $f(x^0,x^1,x^2,x^3) = 0$ determining a 3-D hypersurface in spacetime. Assume a background metric $g_{\mu\nu}(x)$ is given with induced metric

$$h_{ij}(s) = g_{\mu\nu}(X(s))\frac{\partial X^\mu(s)}{\partial s^i}\frac{\partial X^\nu(s)}{\partial s^j} \tag{5.6}$$

on the hypersurface. A normal and unit normal to this surface is then

$$\tilde{n}_\mu(s) = \epsilon_{\mu\alpha\beta\gamma}\frac{\partial X^\alpha}{\partial s^1}\frac{\partial X^\beta}{\partial s^2}\frac{\partial X^\gamma}{\partial s^3};\ n^\mu(s) = \frac{g^{\mu\nu}\tilde{n}_\nu}{\sqrt{|g^{\alpha\beta}\tilde{n}_\alpha\tilde{n}_\beta}} \tag{5.7}$$

Now the equations above can be written in a covariant form via

$$\mathbf{x}\to\mathbf{s};\ \frac{\delta}{\delta T(\mathbf{x})}\to\frac{\delta}{\delta\tau(\mathbf{s})}\equiv n^\mu(\mathbf{s})\frac{\delta}{\delta X^\mu(\mathbf{s})} \tag{5.8}$$

The Tomonaga-Schwinger equation (5.1) becomes then

$$\hat{H}(\mathbf{s})\psi[\phi,X] = in^\mu(\mathbf{s})\frac{\delta\psi[\phi,X]}{\delta X^\mu(\mathbf{s})} \tag{5.9}$$

and for free fields the Hamiltonian density operator in curved spacetime is

$$\hat{H} = \frac{-1}{2|h|^{1/2}}\frac{\delta^2}{\delta\phi^2(\mathbf{s})} + \frac{|h|^{1/2}}{2}[-h^{ij}(\partial_i\phi)(\partial_j\phi) + m^2\phi^2] \tag{5.10}$$

The Bohmian equations of motion (5.4) become

$$\frac{\delta\Phi(s,X)}{\delta\tau(\mathbf{s}'} = \frac{\delta^3(\mathbf{s}-\mathbf{s}')}{|h(\mathbf{s})|^{1/2}}\frac{\delta S}{\delta\phi(\mathbf{s})}\bigg|_{\phi=\Phi} \tag{5.11}$$

and (5.5) becomes

(5.12)
$$\left[\left(\int_{\sigma_x} d^3s' \frac{\delta}{\delta\tau(s')}\right)^2 + \nabla^i\nabla_i + m^2\right]\Phi(s,X) = -\int_{\sigma_x}\frac{d^3s'}{\sqrt{|h|}}\frac{\delta\mathfrak{Q}(s',\phi,X)}{\delta\phi(s)}\bigg|_{\phi=\Phi}$$

where ∇_i is the covariant derivative in s^i and

$$\mathfrak{Q}(\mathbf{s},\phi,X) = -\frac{1}{\sqrt{|h(\mathbf{s})|}}\frac{1}{2R}\frac{\delta^2 R}{\delta\phi^2(\mathbf{s})} \tag{5.13}$$

There is a sort of gauge freedom associated related to the covariance due to the freedom in choosing the $X^\mu(\mathbf{s})$. For a timelike hypersurface the simplest choice of gauge is $X^i(\mathbf{s}) = s^i$. This choice implies $\delta X^i(\mathbf{s}) = 0$ which leads to some of the previous equations prior to covariance. For example (5.11) becomes

$$(g^{00}(\mathbf{x}))^{1/2}\frac{\delta\Phi(x,X^0)}{\delta X^0(x')} = \frac{\delta^3(\mathbf{x}-\mathbf{x}')}{|h(\mathbf{x})|^{1/2}}\frac{\delta S}{\delta\phi(\mathbf{x})}\Big|_{\phi=\Phi} \tag{5.14}$$

which is the curved spacetime version of (5.4). The covariant formulation of QFT leads to a covariant MFT Bohmian interpretation of quantum fields which also does not involve a preferred foliation of spacetime. The covariant Bohmian dynamics does not depend on the choice of coordinates but when a choice is made then the solution of the MFT Bohmian equations of motion can be written so that the MFT nature of the field is not manifest. However the Bohmian equation of motion retains its covariant form.

REMARK 8.5.1. We remark that the Einstein equations can be derived from quantum geometrodynamics following Gerlach [**521**]. He works with the Einstein HJ (EHJ) equation in the Perez form (cf. [**1012**])

$$^3R + h^{-1}\left(\frac{1}{2}h_{ij}h_{k\ell} - h_{ik}h_{j\ell}\right)\frac{\delta S}{\delta h_{ij}}\frac{\delta S}{\delta h_{k\ell}} \tag{5.15}$$

where h_{ij} is the metric of the spatial hypersurface Σ. One defines (**5C**) $\delta S = \int[\delta S/\delta h_{ij}(x)]\delta h_{ij}(x)d^3x$ with integration over Σ and assumes that S is a function of the 3-geometry only, namely (**5D**) $S = S[^3\mathfrak{G}]$ (i.e. S is coordinate independent). A principle of constructive interference is assumed (cf. [**258, 521**]) and Σ is taken to be either finite with no boundary or Σ asymptotically flat. Under these conditions one proves that there are 4 functions N, N_i ($i = 1,2,3$) which together with h_{ij} give a spacetime metric

$$ds^2 = h_{ij}(N^i dx^0 + dx^i)(N^j dx^0 + dx^j) - N^2(dx^0)^2 = \tag{5.16}$$
$$= h_{ij}dx^i dx^j + 2N_i dx^i dx^0 + (N_jN^j - N^2)(dx^0)^2$$

which satisfies the Einstein field equations. Further the manifestly covariant equations of geometrodynamics

$$\frac{\delta h_{ij}}{\delta\sigma} = \frac{\delta H}{\delta\pi^{ij}(x)};\ \frac{\delta\pi^{ij}(x)}{\delta\sigma} = -\frac{\delta H}{\delta h_{ij}(x)} \tag{5.17}$$

hold where $H = H[h_{ij}]$ and $\pi^{ij} = (\delta S/\delta h_{ij})$. Here σ is the Tomonaga-Schwinger many fingered time parameter (cf. [**254, 930, 932, 934, 1141, 1233**]). One notes that the dynamical phase $S = S[h_{ij}]$ is required to be a functional of the 3-geometry alone, regardless of coordinates so one writes (**5E**) $S = S[^3\mathfrak{G}]$ which means that (**5F**) $\nabla_j[\delta S/\delta h_{ij}] = 0$. To see this consider $h_{ij}(x)$ with $x^i \to x'^i = x^i + \epsilon\xi^i(x)$ while preserving the geometry where

$$h'_{ij}(x) = h_{ij}(x) + \delta h_{ij}(x);\ \delta h_{ij}(x) = -\epsilon(\nabla_j\xi_i + \nabla_i\xi_j) \tag{5.18}$$

The ostensible change in S would be

$$\delta S = \int\frac{\delta S}{\delta h_{ij}(x)}\delta h_{ij}(x)d^3x = \tag{5.19}$$

$$= -2\epsilon \int \frac{\delta S}{\delta h_{ij}(x)} \nabla_j \xi_i d^3x = 2\epsilon \int \nabla_j \left(\frac{\delta S}{\delta h_{ij}(x)} \right) \xi_i d^3x$$

However the $^3\mathfrak{G}$ itself is not changed so δS must vanish which means that (**5F**) holds. ∎

5.2. CONNECTIONS TO GRAVITY. We have seen how MFT arises in QFT and we want to examine this further in connection with gravity. We begin with remarks based on [**47, 48, 89, 95, 207, 396, 431, 518, 519, 521, 740, 741, 751, 753, 930, 931, 934, 935, 988, 1176, 1297, 1319, 1327, 1328, 1329**]. We note first that time can arise naturally for WDW when using the dDW theory but a MFT approach seems to require the semiclassical approach and some interaction with matter (see however [**521**] as discussed in [**254**] and Remark 8.6.1 above). Weakening the Hamiltonian constraint as in [**932**] also provides a time and the semiclassical approach is illustrated in [**95, 521, 751**] for example. and we will sketch some of this here following Giulini-Kiefer [**534**].

We forgo the sandwich ordering here for convenience - it remains our principal ordering candidate however. Thus consider ($c = 1$)

$$H\psi[h_{ab}, \phi] = \left(-16\pi G\hbar^2 G_{abcd} \frac{\delta^2}{\delta h_{ab} \delta h_{cd}} - \frac{\sqrt{h}}{16\pi G}(R - 2\Lambda) + H_m \right) \psi = 0 \tag{5.20}$$

The integrated form of (5.20) is

$$\int d^3x N H\psi \equiv H^N \psi = (H_G^N + H_m^N)\psi = 0 \tag{5.21}$$

One uses now an Ansatz ($M = 32\pi G)^{-1}$)

$$\psi = exp\left[i(MS_0 + S_1 + M^{-1}S_2 + \cdots)/\hbar \right] \tag{5.22}$$

leading to a set of equations of consecutive orders in M. The highest order M^2 shows that S_0 depends only on the 3-metric h (cf. [**751**]) and the next order M gives the HJ equation for the gravitational field

$$H_x = \frac{1}{2} G_{abcd} \frac{\delta S_0}{\delta h_{ab}} \frac{\delta S_0}{\delta h_{cd}} - 2\sqrt{h}(R - 2\Lambda) = 0 \tag{5.23}$$

Note that these depend on the lapse function $N(x)$. At the next order M^0 it is convenient to introduce a functional (**5G**) $\psi = D(h_{ab}) exp(iS_1/\hbar)$ and require that D satisfies

$$G_{abcd} \frac{\delta S_0}{\delta h_{ab}} \frac{\delta D}{\delta h_{cd}} - \frac{1}{2} G_{abcd} \frac{\delta^2 S_0}{\delta h_{ab} \delta h_{cd}} D = 0 \tag{5.24}$$

(note D corresponds to the vanVleck determinant). The important observation here is that ψ obeys the equation

$$i\hbar G_{abcd} \frac{\delta S_0}{\delta h_{ab}} \frac{\delta \psi}{\delta h_{cd}} = H_m \psi \tag{5.25}$$

which can be rewritten in terms of vector fields

$$\chi(x) = G_{abcd} \frac{\delta S_0}{\delta h_{ab}(x)} \frac{\delta}{\delta h_{cd}} = -2K_{cd} \frac{\delta}{\delta h_{cd}(x)};\ K_{cd} = -\frac{1}{2} G_{abcd} \frac{\delta S_0}{\delta h_{ab}} \tag{5.26}$$

where K_{cd} has the meaning of an extrinsic curvature. If one now writes (**5H**) $\chi(x) = (\delta/\delta\tau(x))$ then (7.6) would be a Tomonaga-Schwinger equation with respect to the MFT $\tau(x)$ (note τ is really a function on $Riem(\Sigma)$). However this leads to a contradiction since $[(\delta/\delta\tau(x), \delta/\delta\tau(y)] = 0$ of necessity but $[H_m(x), H_m(y)] \neq 0$. One writes then (**5I**) $i\hbar\chi^N = H_m^N\psi$ and (with some argument) shows that in fact

$$(5.27)\qquad [\chi^N, \chi^M] = -2\int_x (N\partial_a M - M\partial_a N)\nabla_b\left(\frac{\delta}{\delta h_{ab}}\right) = \int \mathfrak{L}_K h_{ab}\frac{\delta}{\delta h_{ab}}$$

where (**5J**)) $K^a = h^{ab}(N\partial_b M - M\partial_b N)$. Hence $[\chi^N, \chi^M] \neq 0$ and time functions as above can never be introduced (because the Ricci scalar R is not ultralocal in h_{ab}). The vector fields χ^N are generators of a hypersurface deformation normal to itself and the commutator generates stretchings of the hypersurface. A proper understanding of this and its compatibility with (5.25) is obtained however if one expands the diffeomorphism constraints in powers of G (or M) which gives (**5K**) $2h_{bc}D_a(\delta S_0/\delta h_{ab}) = 0$ (cf. [**773**] for notation - $D_a \sim$ covariant derivative). The highest order M yields (since S_0 does not depend on the scalar field ϕ) (**5L**) $2h_{bc}D_a(\delta S_0/\delta h_{ab}) = 0$ (diffeomorphism invariance of S_0). The next order M^0 leads to a condition on ψ, namely

$$(5.28)\qquad 2h_{bc}D_a\left(\frac{\delta\psi}{\delta h_{ab}} - \frac{\psi}{D}\frac{\delta D}{\delta h_{ab}}\right) = \phi_{,c}\frac{\delta\psi}{\delta\phi}$$

Since D depends only on the 3-metric it is appropriate to demand that it be diffeomorphism invariant by itself, i.e. (**5M**) $h_{bc}D_a(\delta D/\delta h_{ab}) = 0$. One finds then (**5N**) $2h_{bc}D_a(\delta\psi/\delta h_{ab}) = \phi_{,x}(\delta\psi/\delta\phi)$ which is of the same form as the general solution (**5K**). Thus it expresses the invariance of the wave functional $\psi[h_{ab}, \phi]$ with respect to simultaneous diffeomorphisms of the metric and matter field. The consistency condition condition for (**5I**) is (**5O**) $[\chi^N, \chi^M]\psi = [H_m^M, H_m^N]\psi$. This however is nothing but the momentum constraint in this order of approximation, namely (**5N**), since $[\chi^N, \chi^M]$ generates a diffeomorphism of the metric and $[H_m^M, H_m^N]$ closes on the momentum density of matter which generates a diffeomorphism of the matter field. Thus in the full theory the momentum constraints provide the integrability conditions for the Tomonaga-Schwinger equations (**6I**).

In the explicit case of a scalar field one has e.g.

$$(5.29)\qquad [H_m^M, H_m^N] = -\int_x (N\partial_a M - M\partial_a N)h^{ab}\phi_{,b}\frac{\delta}{\delta\phi}$$

Although a family of time functions $\tau(x)$ on $Riem(\Sigma)$ does not exist one can integrate (**5I**) along the vector field χ^N for one particular choice of N and this defines a global time parameter t with respect to which one global SE can be written down. It is in this sense that QFT with respect to a chosen foliation emerges from full quantum gravity. If there are no such general time functions on $Riem(\Sigma)$ what about $S(\Sigma) = Riem(\Sigma)/Diff(\Sigma)$? To answer this one projects the vector fields χ^N to S which is possible since χ^N is invariant under diffeomorphisms - referring to [**534**] for details one arrives at

$$(5.30)\qquad \pi_*[\chi^N, \chi^M] = [\pi_*\chi^N, \pi_*\chi^M] = 0$$

and there exist functions $\bar{\tau}^N$ on S such that (**5P**) $\bar{\chi}^N = \delta/\delta\bar{\tau}^N$ where $\bar{\chi} = \pi_*\chi$, etc. However the WDW operator is only defined on $Riem(\Sigma)$ and some of the intervening calculations do not make sense on S. There is further discussion of anomalies, etc. that is worth reading. This paper corrects some confusion about the existence of Tomonaga-Schwinger times on $Riem(\Sigma)$ in other papers (e.g. [**95, 751, 521**]) and one should also exercise caution in this respect relative to the calculations from [**521**] (cf. also [**888**]).

5.3. EXTRINSIC CURVATURE AND TIME. We go now to some papers [**47, 48, 518, 519, 740, 1020, 1327, 1328, 1329**] where from Anderson-York [**47**] one recalls that it is not N but the slicing density $\alpha(x,t) = Nh^{-1/2}$ is the freely specifiable quantity for the lapse. One writes then

$$ds^2 = -N^2dt^2 + h_{ij}(dx^i + \beta^i dt)(dx^j + \beta^j dt) \tag{5.31}$$

(in what follows $R \sim {}^3R$). The momentum conjugate to to the metric is a density of weight one $\pi^{ij} = h^{1/2}(Kh^{ij} - K^{ij})$ where K_{ij} is the extrinsic curvature with trace K. The natural time derivative for evolution $\hat{\partial}_0$ acts in the normal future direction to the spacelike slice Σ and is denoted by an over-dot; one has $\hat{\partial}_0 = \partial_t - \mathfrak{L}_\beta$ where $\mathfrak{L}_\beta$ is the Lie derivative along the shift β. Every foliation is described by a wave equation for N for some value of α thus making N a dynamical variable. The Hamiltonian constraint does not fix the time but does fix the proper time rate $d\tau/dt = \alpha h^{1/2} = N$ along the normal $\hat{\partial}_0$. Using α has the effect of altering the Hamiltonian density from H to

$$\tilde{H} = h^{1/2}H = \pi^{ij}\pi_{ij} - \frac{1}{2}\pi^2 - hR \tag{5.32}$$

which is of scalar weight 2 and a rational function of the metric. $\tilde{H}$ will be referred to as the Hamiltonian density and may not vanish. This leads to a modification of the ADM action as in [**9, 1217**], namely ($16\pi G = c = 1$)

$$S(h,\pi,\alpha,\beta) = \int d^4x(\pi^{ij}\dot{h}_{ij} - \alpha\tilde{H}) \tag{5.33}$$

(one assumes $N \sim 1 + O(r^{-1})$). Explicitly the Lie derivative term in $\dot{\pi}^{ij}$ is, up to a divergence, (**5Q**) $2\beta^i\nabla_j\pi_i^j = -\beta^i H_i$.

Consider now a general variation of the modified Hamiltonian density

$$\delta\tilde{H} = (2\pi_{ij} - h_{ij}\pi)\delta\pi^{ij} + (2\pi^{ik}\pi_k^j - \pi\pi^{ij} + \tag{5.34}$$

$$+hR^{ij} - hh^{ij}R)\delta h_{ij} - h(\nabla^i\nabla^j\delta h_{ij} - h^{ij}\nabla^k\nabla_k\delta h_{ij})$$

Note that this does not involve either the Hamiltonian or momentum densities; in contrast the variation of the ADM Hamiltonian density $\delta H = \delta(h^{-1/2}\tilde{H})$ does contain a term proportional to the Hamiltonian density. Requiring that S above be stationary under a variation with respect to π^{ij} gives the definition of the extrinsic curvature

$$\dot{h}_{ij} = \alpha\frac{\delta\tilde{H}}{\delta\pi^{ij}} = \alpha(2\pi_{ij} - h_{ij}\pi) \equiv -2NK_{ij} \tag{5.35}$$

Requiring stationarity under a variation in h_{ij} gives the equation of motion

(5.36) $$\dot{\pi}^{ij} = -\alpha\frac{\delta\tilde{H}}{\delta h_{ij}} = -\alpha h(R^{ij} - h^{ij}R) - \alpha(2\pi^{ik}\pi^j_k - \\ -\pi\pi^{ij}) + h(\nabla^j\nabla^i\alpha - h^{ij}\nabla^k\nabla_k\alpha)$$

The slicing density α and the shift β^j are not to be varied; instead the constraints are imposed on initial data and are preserved dynamically as shown below. Thus consider the familiar 3+1 identities

(5.37) $$\dot{h}_{ij} \equiv -2NK_{ij};\ \dot{K}_{ij} \equiv N(R_{ij} - {}^4R_{ij} + KK_{ij} - K_{ik}K^k_j - N^{-1}\nabla_i\nabla_j N)$$

One recalls also that $h^{-1}\dot{h} = h^{ij}\dot{h}_{ij} = -2NK$. Now pass to canonical variables and use (7.18) to arrive at

(5.38) $$\dot{\pi}^{ij} \equiv Nh^{1/2}(Rh^{ij} - R^{ij}) - Nh^{-1/2}(2\pi^{ik}\pi^j_k - \pi\pi^{ij}) + \\ +h^{1/2}(\nabla^i\nabla^j N - h^{ij}\nabla^k\nabla_k N) + Nh^{1/2}\mathfrak{R}^{ij};\ \mathfrak{R}_{ij} = {}^4R_{ij} - h_{ij}{}^4R^k_k$$

One sees that the equations of motion (5.35)-(5.36) derived from the action principle are (5.37)-(5.38) when ${}^4R^{ij} - h^{ij4}R^k_k = 0$. Thus to say that (5.36) holds is to assert that ${}^4R^{ij} = 0$. In fact the equations of motion hold strongly independent of whether the constraints are satisfied or not and this is not true in the ADM formulation because of the presence of the Hamiltonian density in the equations of motion for π^{ij}. This difference can be explained more fully as follows. Given $G_{\mu\nu} = {}^4R_{\mu\nu} - (1/2)g_{\mu\nu}{}^4R^\sigma_\sigma$ and the observation that $2G^0_0 = {}^4R^0_0 - {}^4R^k_k$ one has (**5R**) $G_{ij} + h_{ij}G^0_0 \equiv {}^4R_{ij} - h_{ij}{}^4R^k_k$. The vanishing of the right side does not depend on either the Hamiltonian or momentum densities and is equivalent to ${}^4R_{ij} = 0$ or $G_{ij} = -h_{ij}G^0_0$. Thus while ${}^4R_{\mu\nu} = 0$ and $G_{\mu\nu} = 0$ are equivalent $R_{ij} = 0$ and $G_{ij} = 0$ are not equivalent as equations of motion - unless the Hamiltonian density $H = 2h^{1/2}G^0_0$ vanishes exactly (i.e. unless the Hamiltonian constraint holds). The ADM action principle is equivalent to $G_{ij} = 0$ and one recalls that the use of R_{ij} instead of G_{ij} has always been preferred by the French school. This raises the important principle that a constrained Hamiltonian theory should be well behaved even when the constraints are violated. There is much further calculation in this direction which we omit here (cf. [**1020, 1329**]).

REMARK 8.5.2. There is a great deal of material now available on general relativity in terms of Ashtekar variables (see e.g. [**76, 74, 86, 533, 689, 749, 870, 1085, 1087, 1186, 1192, 1222**] for a very incomplete list of references on loop quantum gravity, etc.). In [**1192**] for example one recasts the WDW equation in the new variables in terms of the 3-geometry elements C and $\mathfrak{K}$ where C is the Chern-Simons functional and $\mathfrak{K}$ is the integral of the trace of the extrinsic curvature (cf. also [**1186**]). ■

6. EXACT UNCERTAINTY REVISITED

In [**592, 599, 594, 1065**] an exact uncertainty principle was enunciated (cf. also [**185, 254, 955, 1115, 1116**]) which has the form $\Delta x\Delta p \geq \delta x\Delta p \geq \delta x\Delta p_{nc} = (\hbar/2)$ (here $p \sim \nabla S + p_{nc}$ and Δ refers to variance). Then one arrives at a formula for the $p_{nc} \sim \delta p \sim c(\nabla\rho/\rho)$ where $\rho = |\psi|^2$. This in turn leads to a quantum potential (QP) $Q = -(\hbar^2/2)(\nabla^2\sqrt{\rho}/\sqrt{\rho})$ in a standard form. We reverse the argument in [**255**] (cf. also [**258, 259**]) and observe that any quantization via a QP as above can be identified with a quantization obtained via $\delta p \sim c\nabla log(\rho)$ (see Theorem below). A similar (but more complicated) procedure is then developed for the famous Wheeler-deWitt (WDW) equation, which has been a corner stone of research in geometrodynamics for many years and which plays an important role in various ideas of quantum gravity (see e.g. [**44, 254, 374, 534, 749, 903, 1085, 1297**]). Applications of exact uncertainty ideas to WDW arises in [**587**] and we show here again (using necessarily different arguments) how the procedure can be reversed to exhibit any quantization (6.15) using Q as arising from momentum fluctuations as in (6.8) (indicated in another Theorem)). Then heuristically one arrives at an entropy connection as in Theorem 3.1. There are other types of entropy connections related to [**955**] which we sketched in Section 7.8. The suggested formulas involving Ashtekar variables etc. are mainly heuristic in nature.

In [**258, 259, 255**] we sketched some heuristic results concerning WDW and exact uncertainty following [**254, 598, 592, 599, 594, 1065**]. Basically following e.g. Hall, Kumar, Reginatto [**592, 598, 599**] one defines Fisher information via an equation (**6A**) $F_x = \int dx P(x)[\partial_x log(P(x))]^2$ and a Fisher length by $\delta x = F_x^{-1/2}$ where $P(x)$ is a probability density for a 1-D observable x. The Cramer-Rao inequality says $Var(x) \geq F_x^{-1}$ or simply $\Delta x \geq \delta x$. For a quantum situation with $P(x) = |\psi(x)|^2$ and ψ satisfying a SE one finds immediatly

$$F_X = \int dx|\psi|^2 \left[\frac{\psi'}{\psi} + \frac{\bar{\psi}'}{\bar{\psi}}\right]^2 dx = \frac{4}{\hbar^2}\left[< p^2 >_\psi - < p_{cl}^2 >_\psi\right] \tag{6.1}$$

where $p_{cl} = (\hbar/2i)[(\psi'/\psi) - (\bar{\psi}'/\bar{\psi})]$ is the classical momentum observable conjugate to x ($\sim S_x$ for $\psi = Rexp(iS/\hbar)$). Setting now $p = p_{cl} + p_{nc}$ one obtains after some calculation (**6B**) $F_x = (4/\hbar^2)(\Delta p_{nc})^2 = 1/(\delta x)^2 \Rightarrow \delta x \Delta p_{nc} = \hbar/2$ as a relation between nonclassicality and Fisher information. Note $< p >_\psi = < p_{cl} >_\psi$, $\partial_t|\psi|^2 + \partial_x[|\psi|^2 m^{-1} p_{cl}] = 0$ from the SE, and $(\Delta x)(\Delta p) \geq (\delta x)(\Delta p) \geq (\delta x)(\Delta p_{nc})$.

We recall also that from (6.1) F_x is proportional to the difference of a quantum and a classical kinetic energy. Thus $(\hbar^2/4)F_x(1/2m) = (1/2m) < p^2 >_\psi -(1/2m) < p_{cl}^2 >_\psi$ and $E_F = (\hbar^2/8m)F_x$ is added to E_{cl} to get E_{quant}. By deBroglie-Bohm (dBB) theory there is a quantum potential

$$Q = \frac{\hbar^2}{8m}\left[\left(\frac{P'}{P}\right)^2 - 2\frac{P''}{P}\right];\ P = |\psi|^2 \tag{6.2}$$

and evidently (**6C**) $< Q >_\psi = \int PQdx = (\hbar^2/8m)F_x$ (upon neglecting the boundary integral term at $\pm\infty$ - i.e. $P' \to 0$ at $\pm\infty$).

Now the exact uncertainty principle (cf. [**592, 594, 599, 1065**]) looks at momentum fluctuations (**6D**) $p = \nabla S + f$ with $< f > = \bar{f} = 0$ and replaces a classical ensemble energy $< E >_{cl}$ by $(P \sim |\psi|^2)$

$$(6.3)\qquad < E >= \int dx P\left[(2m)^{-1}\overline{|\nabla S + f|^2} + V\right] =< E >_{cl} + \int dx P\frac{\overline{f\cdot f}}{2m}$$

The background action principle here involves (★) $A = \int dt[-H + \int dx PS_t t]$ and $\delta A = 0 \Rightarrow P_t = \delta H/\delta S$ with $S_t = -\delta H/\delta P$. Upon making an assumption of the form (**6E**) $\overline{f\cdot f} = \alpha(x, P, S, \nabla P, \nabla S, \cdots)$ one looks at a modified Hamiltonian (**6F**) $\tilde{H}_q[P,S] = \tilde{H}_{cl} + \int dx P(\alpha/2m)$. Then, assuming

(1) Causality - i.e. α depends only on S, P and their first derivatives
(2) Independence for fluctuations of noninteracting uncorrelated ensembles
(3) $f \to L^T f$ for invertible linear coordinate transformations $x \to L^{-1}x$
(4) Exact uncertainty - i.e. $\alpha = \overline{f\cdot f}$ is determined solely by uncertainty in position

one arrives at

$$(6.4)\qquad \tilde{H}_q = \tilde{H}_{cl} + c\int dx \frac{\nabla P\cdot\nabla P}{2mP}$$

and putting $\hbar = 2\sqrt{c}$ with $\psi = \sqrt{P}exp(\imath S/h)$ a SE is obtained. Note that a similar conclusion based on a more purely information theoretical argument appears in [**996**].

As pointed out in [**258, 259, 255**] in the SE situation with Q as in (6.2), in 3-D one has

$$(6.5)\qquad \int PQd^3x \sim -\frac{\hbar^2}{8m}\int\left[2\Delta P - \frac{1}{P}(\nabla P)^2\right]d^3x = $$
$$= \frac{\hbar^2}{8m}\int\frac{1}{P}(\nabla P)^2 d^3x$$

since $\int_\Omega \Delta P d^3x = \int_{\partial\Omega}\nabla P\cdot\mathbf{n}d\Sigma$ can be assumed zero for $\nabla P = 0$ on $\partial\Omega$.

THEOREM 6.1. Given that any quantum potential for the SE has the form (6.2) (with $\nabla P = 0$ on $\partial\Omega$) it follows that the quantization via Q can be identified with momentum fluctuations of the type studied in Hall-Reginatto [**596**] and thus has information content as described by the Fisher information. Thus any SE described via a probability distribution $P\ (= |\psi|^2)$ gives rise to a quantum potential Q and one can then identify this SE as a quantum model arising from a classical quadratic Hamiltonian $\tilde{H}_{cl}$ perturbed by a Fisher information term as in (6.4). Thus the SE, for any wave function ψ, involves an information content with entropy significance (cf. below and Chapter 6 for entropy connections).

6.1. WDW. The same sort of arguments can be applied for the WDW equation following [**258, 259, 255, 1065**] (cf. also [**374, 749, 786, 1023, 1026, 1085, 1160, 1163, 1152, 1168**] for WDW). Thus take an ADM situation

$$(6.6)\qquad ds^2 = -(N^2 - h^{ij}N_iN_j) + 2N_i dx^i dt + h_{ij}dx^i dx^j$$

Assume dynamics generated by an action (**6G**) $A = \int dt[\tilde{H} + \int \mathfrak{D}hP\partial_t S]$. One will have equations of motion (**6H**) $\partial_t P = \delta\tilde{H}/\delta S$ and $\partial_t S = -\delta\tilde{H}/\delta P$ (cf. [**254,**

599]). A suitable "classical" Hamiltonian is

$$\tilde{H}_c[P,S] = \int \mathfrak{D}hPH_0\left[h_{ij}, \frac{\delta S}{\delta h_{ij}}\right]; \tag{6.7}$$

$$H_0 = \int dx\left[N\left(\frac{1}{2}G_{ijk\ell}\pi^{ij}\pi^{k\ell} + V(h_{ij})\right) - 2N_i\nabla_j\pi^{ij}\right]$$

where $G_{ijk\ell}$ is the deWitt (super)metric (**6I**) $G_{ijk\ell} = (1/\sqrt{h})(h_{ik}h_{j\ell} + h_{i\ell}h_{jk} - h_{ij}h_{k\ell})$ and $V \sim \hat{c}\sqrt{h}(2\Lambda - {}^3R)$. Then thinking of $\pi^{ij} = \delta S/\delta h_{ij} + f^{ij}$ and e.g. $\tilde{H}_q = \tilde{H}_c + (1/2)\int \mathfrak{D}hP\int dxNG_{ijk\ell}\overline{f^{ij}f^{k\ell}}$ one arrives via exact uncertainty at a Fisher information contribution (cf. [**254, 258, 259, 255, 592, 599**] and see (•) with (6.20) below)

$$\tilde{H}_q[P,S] = \tilde{H}_{cl} + \frac{c}{2}\int \mathfrak{D}h\int dxNG_{ijk\ell}\frac{1}{P}\frac{\delta P}{\delta h_{ij}}\frac{\delta P}{\delta h_{k\ell}} \sim \tag{6.8}$$

$$\sim \tilde{H}_{cl} + \frac{c}{2}\int\int \mathfrak{D}h\, dxNPQ$$

where $\hbar = 2\sqrt{c}$ and $\psi = \sqrt{P}exp(iS/\hbar)$ resulting in (for $N = 1$ and $N_i = 0$)

$$\left[-\frac{\hbar^2}{2}\frac{\delta}{\delta h_{ij}}G_{ijk\ell}\frac{\delta}{\delta h_{k\ell}} + V\right]\psi = 0 \tag{6.9}$$

with a sandwich ordering ($G_{ijk\ell}$ in the middle - cf. also below). In general there are also constraints

$$\frac{\delta\psi}{\delta N} = \frac{\delta\psi}{\delta N_i} = \partial_t\psi = 0;\ \nabla_j\left(\frac{\delta\psi}{\delta h_{ij}}\right) = 0 \tag{6.10}$$

Now take any WDW equation of the form (6.9) and compute

$$\frac{\delta}{\delta h_{ij}}\left(G_{ijk\ell}\frac{\delta}{\delta h_{k\ell}}\sqrt{P}e^{iS/\hbar}\right) = \left[\frac{\delta G_{ijk\ell}}{\delta h_{ij}}\left(\frac{1}{2}P^{-1/2}\frac{\delta P}{\delta h_{k\ell}} + \frac{iP^{1/2}}{\hbar}\frac{\delta S}{\delta h_{k\ell}}\right)\right. \tag{6.11}$$

$$+G_{ijk\ell}\left\{-\frac{1}{4}P^{-3/2}\frac{\delta P}{\delta h_{k\ell}}\frac{\delta P}{\delta h_{ij}} + \frac{1}{2}P^{-1/2}\frac{\delta^2 P}{\delta h_{k\ell}\delta h_{ij}} - \frac{P^{1/2}}{\hbar^2}\frac{\delta S}{\delta h_{k\ell}}\frac{\delta S}{\delta h_{ij}} +\right.$$

$$\left.\left.+\frac{i}{2\hbar}P^{-1/2}\left(\frac{\delta P}{\delta h_{k\ell}}\frac{\delta S}{\delta h_{ij}} + \frac{\delta S}{\delta h_{k\ell}}\frac{\delta P}{\delta h_{ij}}\right) + \frac{iP^{1/2}}{\hbar}\frac{\delta^2 S}{\delta h_{k\ell}\delta h_{ij}}\right\}\right]e^{iS/\hbar}$$

Therefore writing out the WDW equation gives (cf. [**349**])

$$-\frac{\hbar^2}{4P}\frac{\delta}{\delta h_{ij}}\left[G_{ijk\ell}\frac{\delta P}{\delta h_{k\ell}}\right] + \tag{6.12}$$

$$+\frac{\hbar^2}{8P^2}G_{ijk\ell}\frac{\delta P}{\delta h_{k\ell}}\frac{\delta P}{\delta h_{ij}} + G_{ijk\ell}\left[\frac{\hbar^2}{8P}\frac{\delta^2 P}{\delta h_{ij}\delta h_{ij}} + \frac{1}{2}\frac{\delta S}{\delta h_{k\ell}}\frac{\delta S}{\delta h_{ij}}\right] + V = 0;$$

$$2P\frac{\delta G}{\delta h_{ij}}\frac{\delta S}{\delta h_{k\ell}} + G\left(\frac{\delta P}{\delta h_{k\ell}}\frac{\delta S}{\delta h_{ij}} + \frac{\delta S}{\delta h_{k\ell}}\frac{\delta P}{\delta h_{ij}}\right) + 2PG\frac{\delta^2 S}{\delta h_{k\ell}\delta h_{ij}} = 0$$

It is useful here to compare with $-(\hbar^2/2m)\psi'' + V\psi = 0$ which for $\psi = Rexp(iS/\hbar)$ yields

$$\frac{1}{2m}S_x^2 + V + Q = 0;\ Q = -\frac{\hbar^2}{4m}\frac{R''}{R} = \frac{\hbar^2}{8m}\left[\frac{2P''}{P} - \left(\frac{P'}{P}\right)^2\right] \tag{6.13}$$

along with $\partial(R^2S') = \partial(PS') = 0$ (leading to (6.5)). The analogues here are then in particular

(6.14) $$\frac{1}{2m}S_x^2 \sim \frac{1}{2}G_{ijk\ell}\frac{\delta S}{\delta h_{k\ell}}\frac{\delta S}{\delta h_{ij}};\ Q = \frac{\hbar^2}{8m}\left[\frac{2P''}{P} - \left(\frac{P'}{P}\right)^2\right] \sim$$

$$\sim -\frac{\hbar^2}{4P}\frac{\delta}{\delta h_{ij}}\left[G_{ijk\ell}\frac{\delta P}{\delta h_{k\ell}}\right] + G_{ijk\ell}\left\{\frac{\hbar^2}{8P^2}\frac{\delta P}{\delta h_{k\ell}}\frac{\delta P}{\delta h_{ij}} + \frac{\hbar^2}{4P}\frac{\delta^2 P}{\delta h_{ij}\delta h_{k\ell}}\right\}$$

We note that the Q term arises directly from

(6.15) $$Q = -\frac{\hbar^2}{2}P^{-1/2}\frac{\delta}{\delta h_{ij}}\left(G_{ijk\ell}\frac{\delta P^{1/2}}{\delta h_{k\ell}}\right)$$

and

(6.16) $$\int \mathfrak{D}h\, PQ = -\frac{\hbar^2}{2}\int \mathfrak{D}h P^{1/2}\frac{\delta}{\delta h_{ij}}\left(G_{ijk\ell}\frac{\delta P^{1/2}}{\delta h_{k\ell}}\right)$$

But from (•) $\int \mathfrak{D}h\delta[\ ] = 0$ (cf. [**587**] and see below) one has

(6.17) $$\int \mathfrak{D}hP^{1/2}\frac{\delta}{\delta h_{ij}}\left(G_{ijk\ell}\frac{\delta P^{1/2}}{\delta h_{k\ell}}\right) = -\int \mathfrak{D}h\frac{\delta P^{1/2}}{\delta h_{ij}}G_{ijk\ell}\frac{\delta P^{1/2}}{\delta h_{k\ell}}$$

Thus let us assume there exists a suitable $\mathfrak{D}f$ indicated below which is a translation invariant measure in the (super)space of fields h. Then there is an integration by parts formula (•) (cf. (6.20) below) leading to (6.17). Consequently

THEOREM 6.2. Given a WDW equation of the form (6.9) with associated quantum potential given via (6.15) and $\psi = \sqrt{P}exp(iS/\hbar)$ it follows that the quantum potential can be modeled as arising from momentum fluctuations as in (6.8) (for $N = 1$) since it gives rise to a corresponding Fisher information type perturbation of a classical $\tilde{H}_c$ as in (6.8) (cf. here [**254, 258, 259, 255, 599**]).

Here P represents a probability density of fields h_{ij} which determine $G_{ijk\ell}$ and one should perhaps stipulate that the perturbations of h_{ij} generated by the $f^{ij} \sim (1/P)(\delta P/\delta h_{ij})$ are symmetric (note $\overline{f^{ij}} \sim \int \mathfrak{D}hP(1/P)(\delta P/\delta h_{ij}) = 0$). In general one would expect not necessarily symmetric perturbations of a Riemannian metric but in order to use the functional integral developed, involving $\mathfrak{D}h$ etc., it seems required to remain within the Riemannian framework where $G_{ijk\ell}$ is defined. Thus the very existence of a quantum WDW in sandwich form seems to require entropy type input via Fisher information fluctuation of fields.

REMARK 8.6.1. We go here to [**254, 599, 616**] and will sketch briefly the derivation of (•) following [**592, 599**] (cf. also [**254**]). The relevant functional calculus goes as follows. One defines a functional F of fields f and sets

(6.18) $$\delta F = F[f + \delta f] - F[f] = \int dx\frac{\delta F}{\delta f_x}\delta f_x$$

Here e.g. $dx \sim d^4x$ and in the space of fields there is assumed to be a measure $\mathfrak{D}f$ such that $\int \mathfrak{D}f \equiv \int \mathfrak{D}f'$ for $f' = f + h$ (cf. Hatfield [**616**] for product type measures). Then evidently (**6J**) $\int \mathfrak{D}f(\delta F/\delta f) = 0$ when $\int \mathfrak{D}f\, F[f] < \infty$. Indeed

$$0 = \int \mathfrak{D}f(F[f + \delta f] - F[f]) = \int dx \delta f_x \left(\int \mathfrak{D}f \frac{\delta F}{\delta f_x} \right) \tag{6.19}$$

and this provides an integration by parts formula

$$\int \mathfrak{D}f\, P\left(\frac{\delta F}{\delta f}\right) = -\int \mathfrak{D}f \left(\frac{\delta P}{\delta f}\right) F \tag{6.20}$$

for $P[f]$ a probability density functional. ■

7. ENTROPY AND TIME

In Nikolić [**932, 934**] the problem of time in quantum gravity is addressed by weakening the Hamiltonian constraint $\hat{H} = 0$ to $<\psi|\hat{H}|\psi>= 0$ which is consistent with the classical Hamiltonian constraint. This can be written as (we shift $g \to h$ here in thinking of applications below to the deWitt metric and $^3h \sim h$)

$$\int \mathfrak{D}h \psi^* \hat{H} \psi = 0 \tag{7.1}$$

and for $\psi = Rexp(iS/\hbar)$, $\hat{H}\psi = i\hbar\partial_t\psi$ and a condition $(d/dt)\int \mathfrak{D}h\psi^*\psi = 0$ one finds that (7.1) holds if $\partial_t S = 0$. Hence the weak Hamiltonian constraint (7.1) is consistent with $\psi = R(h,t)exp(iS(h)/\hbar)$. The idea now is to allow $i\hbar\partial_t\psi = \hat{H}\psi$ but insist that this not contradict $\hat{H} = 0$ in the classical limit. Consider then $H = \tilde{G}_{AB}(h)\pi^A\pi^B + V(h)$ ($h = \{h_A\}$ and $\tilde{G}_{AB} = \tilde{G}_{BA}$) or explicitly

$$\tilde{G}_{AB}\pi^A\pi^B \equiv \kappa \int d^3x \tilde{G}_{ijk\ell}\pi^{ij}\pi^{k\ell};\ V = -\kappa^{-1}\int d^3x \sqrt{|h|}^3 R \tag{7.2}$$

where $\kappa = 8\pi G$ and

$$\tilde{G}_{ijk\ell} = \frac{\sqrt{|h|}}{2}(h_{ik}h_{j\ell} + h_{jk}h_{i\ell} - h_{ij}h_{k\ell}) \tag{7.3}$$

($\tilde{G}_{ijk\ell}$ differs from $G_{ijk\ell}$ by a factor of $\sqrt{h}$ and this can be absorbed in $\mathfrak{D}h$ as needed yielding $\tilde{\mathfrak{D}}h$). In the quantum case π^A becomes $\hat{\pi}^A = -i\hbar(\delta/\delta h_A) \equiv -i\hbar\partial^A$ and different orderings of the $\hat{\pi}^A$ in $\hat{H}$ become important. Some argument shows that a form (**7A**) $\hat{H} = \hat{\pi}^A\tilde{G}_{AB}\hat{\pi}^B + V$ implies $<\psi|\psi>$ as well as all $<\psi_1|\psi_2>$ are time independent since

$$\frac{d}{dt}\int \tilde{\mathfrak{D}}h\psi_1^*\psi_2 = \hbar \int \tilde{\mathfrak{D}}h\partial^A[\tilde{G}_{AB}(\psi_1^* i \overleftrightarrow{\partial}^B\psi_2] \tag{7.4}$$

which vanishes because the integral over a total derivative vanishes (thus unitary time evolution implies the sandwich ordering). Moreover for $\hbar \to 0$ one obtains densities

$$\tilde{G}_{AB}\partial^A S\partial^B S + V = 0;\ \partial_t R^2 + \partial^A[2R^2\tilde{G}_{AB}\partial^B S] = 0 \tag{7.5}$$

which is the classical HJ equation (via $\pi^A = \partial^A S$) and

$$\dot{h}_A = \partial_t h_A = \frac{\partial H}{\partial \pi^A} = 2\tilde{G}_{AB}\pi^B;\ \partial_t\rho + \partial^A(\rho\dot{h}_A) = 0 \tag{7.6}$$

($\sim (d\rho/dt) = 0$) for $\rho = R^2$. Hence in fact the conventional strong form of the Hamiltonian constraint (leading to $\partial_t \rho = 0$) does not have the correct classical limit, but the weaker form does.

One recalls now (cf. [**254, 258, 259, 255, 509**]) that with the SE (under certain circumstances) one has a differential entropy $\mathfrak{S} = -\int dx\rho log(\rho)$ (1-D for simplicity here) with $\partial_t \rho = -\partial(v\rho)$ and $v = -u = -D\partial log(\rho)$ (diffusion current) leading to

$$\partial_t \mathfrak{S} = -\int dx\rho_t(log(\rho)+1) = \int dx\,(log(\rho)+1)\,\partial(v\rho) = \tag{7.7}$$

$$= -\int \partial\rho D\partial log(\rho) = D\int \frac{(\partial\rho)^2}{\rho}$$

Thus the Fisher information is the time derivative of the differential entropy and there should be some analogue of this for WDW. There is not a priori a natural time evolution for WDW but we have indicated a way around this above. In any case one might look for a formula of the form

$$\delta\int \mathfrak{D}hF(S,P,h_{ij}) = \int \mathfrak{D}h\left[\frac{\delta F}{\delta S}\delta S + \cdots\right] = \tag{7.8}$$

$$= \int \mathfrak{D}h\frac{\delta P^{1/2}}{\delta h_{ij}}G_{ijk\ell}\frac{\delta P^{1/2}}{\delta h_{k\ell}}$$

where F represents some kind of entropy term. Hence one thinks of $R^2 = \rho$ ($= P$) and looks at (7.5). The second equation is in fact fixed by the sandwich ordering as

$$\partial_t\rho + \partial^A[2\rho\tilde{G}_{AB}\partial^B S] = 0 \tag{7.9}$$

where $\partial^B S = -i\hbar(\delta/\delta h_B)S$. Now recall from [**254, 509**] that in a Brownian motion situation the use of a drift velocity $u = D\nabla log(\rho) = -v = -(1/m)\nabla S$ is natural ($D = \hbar/2m$). Another context involving the SE with statistical geometry and a Weyl space produces a Weyl vector $\phi_i = -\partial_i log(\rho)$ related to an osmotic velocity field (cf. [**254, 237, 1115, 1116**]). Thus a relation $\mathbf{u} = -c\phi = c\nabla log(\rho)$ can be envisioned with $\rho = P \sim R^2$ so that, instead of dealing with $\delta S/\delta h_{ij} = \pi^{ij} - (1/P)(\delta P/\delta h_{ij})$ one is motivated to consider

$$\frac{\delta S}{\delta h_B} \sim -\frac{\hat{c}}{P}\frac{\delta P}{\delta h_B} \tag{7.10}$$

provided one is only interested in metric fluctuations (there is no particle mass here to impede this). In this case on could work with (7.6) as ($-(i\hbar)^2 = -\hbar^2$)

$$\partial_t P - \frac{\delta}{\delta h_A}\left[2P\tilde{G}_{AB}\frac{\hbar^2\hat{c}}{P}\frac{\delta P}{\delta h_B}\right] = 0 \tag{7.11}$$

Then for a differential entropy defined via (**7B**) $\mathcal{S} = -(1/\hat{c})\int\tilde{\mathfrak{D}}hPlog(P)$ one would have

$$\mathcal{S}_t = -\frac{\hbar^2}{\hat{c}}\int dx\int\tilde{\mathfrak{D}}hP_t[1+log(P)] = \tag{7.12}$$

$$= -\hbar^2 \int dx \int \tilde{\mathfrak{D}}h[1 + log(P)]\left[\frac{\delta}{\delta h_A}\left(2\tilde{G}_{AB}\frac{\delta P}{\delta h_B}\right)\right] =$$
$$= \hbar^2 \int dx \int \tilde{\mathfrak{D}}h\frac{2}{P}\tilde{G}_{AB}\frac{\delta P}{\delta h_B}\frac{\delta P}{\delta h_A} \sim 16\int \tilde{\mathfrak{D}}h\, PQ$$

(cf. (6.16), (6.17), and (6.32)). One arrives then at a very heuristic

THEOREM 7.1. Assuming only metric fluctuations satisfying (7.10) and the weak form of the Hamiltonian constraint one can define a differential entropy (**7B**) and express the Fisher information (expressed via the quantum potential Q) as a time derivative $\partial_t\mathcal{S}$.

REMARK 8.7.1. We note from [**44, 473, 533, 903, 1297**] that, for Σ a 3-manifold of fixed topology, the "superspace" $S(\Sigma) = Riem(\Sigma)/Diff)\Sigma)$ with the deWitt metric $G_{ijk\ell} = G_{k\ell ij}$, has a complicated structure. ■

REMARK 8.7.2. We refer to [**431, 685, 976, 978**] for some direct connections of thermodynamics and entropy to the Einstein equations (cf. also Chapter 6). ■

REMARK 8.7.3. The multifingered time (MFT) of Schwinger, Tomonaga, Wheeler, et al is used in a field theoretic context to create a time variable (cf. [**95, 254, 521, 534, 932, 1141, 1233**]). ■

REMARK 8.7.4. Covariant quantum field theory based on deDonder-Weyl ideas is very powerful and can be used in gravitational situations (see e.g. [**254, 258, 259, 255, 721, 930, 931, 932**]; in particular and one obtains a theorem analogous the Theorem above. Let us spell this out here and discuss again the covariant QFT of Nikolić [**930, 931**], expanding somewhat the presentation in [**254**] (for more on deDonder-Weyl theory see also [**254, 721**]). One shows here following Nikolić [**930, 931, 932, 934, 935**] that the deterministic evolution of quantum fields is a covariant version of the Bohmian hidden variable interpretation of quantum field theory (QFT). The deDonder-Weyl (dDW) covariant canonical formalism is exploited in a novel manner and a covariant Bohmian formulation is not postulated but derived; this suggests that the Bohmian interpretation could be the missing link between QM and GR. The dDW formalism treats space and time variables on an equal footing. Thus given a Lagrangian $L(y^a, \partial_\mu y^a, x^\nu)$ with field variables y^a and $\mu, \nu = 1, \cdots, n)$ one defines polynomials $p^\mu_a = \partial L/\partial(\partial_\mu y^a)$ and a dDW Hamiltonian $H = \partial_\mu y^a p^\mu_a - L$ such that the Euler-Lagrange (EL) field equations take the form

$$\partial_\mu y^a = \frac{\partial H}{\partial p^\mu_a};\ \partial_\mu p^\mu_a = -\frac{\partial H}{\partial y^a} \tag{7.13}$$

The fields are treated as a multitime dDW system evolving in space and time (not just in time) and everything is manifestly covariant. Consequently this is an ideal framework for quantum gravity. Following now Nikolić [**930, 931**] (cf. also [**254**]) one writes (using only one field ϕ for illustration)

$$\mathfrak{A} = \int d^4x\mathfrak{L};\ \mathfrak{L} = \frac{1}{2}(\partial^\mu\phi)(\partial_\mu\phi) - V(\phi);\ \pi^\mu = \frac{\partial\mathfrak{L}}{\partial(\partial_\mu\phi)} = \partial^\mu\phi \tag{7.14}$$

The covariant canonical equations of motion and dDW Hamiltonian (not related to the energy density) are

$$(7.15)\qquad \partial_\mu\phi = \frac{\partial\mathfrak{H}}{\partial\pi^\mu};\ \partial_\mu\pi^\mu = -\frac{\partial\mathfrak{H}}{\partial\phi};\ \mathfrak{H}(\pi^a,\phi) = \pi^\mu\partial_\mu\phi - \mathfrak{L} = \frac{1}{2}\pi^\mu\pi_\mu + V$$

By introducing the local vector $S^\mu(\phi(x),x)$ the dynamics can also be described by the covariant dDW Hamilton-Jacobi equation and equation of motion

$$(7.16)\qquad \mathfrak{H}\left(\frac{\partial S^a}{\partial\phi},\phi\right) + \partial_\mu S^\mu = 0;\ \partial^\mu\phi = \pi^\mu = \frac{\partial S^\mu}{\partial\phi}$$

Note here that ∂_μ acts only on the second argument of $S^\mu(\phi(x),x)$ and the corresponding total derivative is $d_\mu = \partial_\mu + (\partial_\mu\phi)(\partial/\partial\phi)$. To describe the relation between the covariant HJ equation and the conventional HJ equation one writes from (7.15) - (7.16)

$$(7.17)\qquad \frac{1}{2}\frac{\partial S_\mu}{\partial\phi}\frac{\partial S^\mu}{\partial\phi} + V + \partial_\mu S^\mu = 0;\ \frac{1}{2}\frac{\partial S_\mu}{\partial\phi}\frac{\partial S^\mu}{\partial\phi} = \frac{1}{2}\frac{\partial S^0}{\partial\phi}\frac{\partial S^0}{\partial\phi} + \frac{1}{2}(\partial_i\phi)(\partial^i\phi)$$

where $i = 1,2,3$ are the space indices and one notes also that (**7C**) $\partial_\mu S^\mu = \partial_0 S^0 + d_i S^i - (\partial_i\phi)(\partial^i\phi)$. Now introduce the quantity $\mathfrak{S} = \int d^3x S^0$ leading to

$$(7.18)\qquad \frac{\partial S^0(\phi(x),x)}{\partial\phi(x)} = \frac{\delta\mathfrak{S}([\phi(x,t)],t)}{\delta\phi(x,t)};\ \frac{\delta}{\delta\phi(x,t)} = \left.\frac{\delta}{\delta\phi(x)}\right|_{\phi(x)=\phi(x,t)}$$

Putting the second equation of (7.17) and (7.18) into the first equation of (7.17) yields upon integration then (cf. (7.8))

$$(7.19)\qquad \int d^3x\left[\frac{1}{2}\left(\frac{\delta\mathfrak{S}}{\delta\phi(x,t)}\right)^2 + \frac{1}{2}(\nabla\phi)^2 + V(\phi)\right] + \partial_t\mathfrak{S} = 0$$

which is the standard non-covariant HJ equation (recall here $\partial^i\phi = \partial S^i/\partial\phi$). The time evolution of the field $\phi(x,t)$ is now given via (**7D**) $\partial_t\phi(x,t) = \delta\mathfrak{S}/\delta\phi(x,t)$ (from the time component in (7.16)) and one notes that in deriving (7.19) it was necessary to use the space part of the equations of motion in (7.16); this will be important in the quantum extension below.

We recall that QFT can be formulated in the Schrödinger picture via

$$(7.20)\qquad \hat{H}\psi = i\hbar\partial_t\psi;\ \hat{H} = \int d^3x\left[-\frac{\hbar^2}{2}\left(\frac{\delta}{\delta\phi(x)}\right)^2 + \frac{1}{2}(\nabla\phi)^2 + V(\phi)\right]$$

Write now (**7E**) $\psi([\phi(x)],t) = \mathfrak{R}([\phi(x)],t)exp(i\mathfrak{S}([\phi(x)],t)/\hbar)$ and (7.20) will be equivalent to a set of two real equations

$$(7.21)\qquad \int d^3x\left[\frac{1}{2}\left(\frac{\delta\mathfrak{S}}{\delta\phi(x)}\right)^2 + \frac{1}{2}(\nabla\phi)^2 + V(\phi) + \mathfrak{Q}\right] + \partial_t\mathfrak{S} = 0;$$

$$\int d^3x\left[\frac{\delta\mathfrak{R}}{\delta\phi(x)}\frac{\delta\mathfrak{S}}{\delta\phi(x)} + \mathfrak{J}\right] + \partial_t\mathfrak{R} = 0;\ \mathfrak{Q} = -\frac{\hbar^2}{2\mathfrak{R}}\frac{\delta^2\mathfrak{R}}{\delta\phi^2(x)};\ \mathfrak{J} = \frac{\mathfrak{R}}{2}\frac{\delta^2\mathfrak{S}}{\delta\phi^2(x)}$$

The second equation is equivalent to

$$\partial_t \mathfrak{R}^2 + \int d^3x \frac{\delta}{\delta\phi(x)}\left(\mathfrak{R}^2 \frac{\delta\mathfrak{S}}{\delta\phi(x)}\right) = 0 \tag{7.22}$$

and this represents the unitarity of the theory since it provides a norm of the form (**7F**) $\int[d\phi(x)]\psi^*\psi = \int[d\phi(x)]\mathfrak{R}^2$ that does not depend on time (some argument is needed here). One must also stipulate that the quantity $exp(i\mathfrak{S}/\hbar)$ be single valued. This formulation also suggests an interesting Bohmian interpretation stating that the quantum fields have a deterministic time evolution given by the classical equation (**7D**) and the statistical predictions will be equivalent to those of the conventional interpretation (cf. [**254**] for discussion). Comparing now (7.21) with (7.19) we see that the quantum field satisfies an equation similar to the classical one except for the additional nonlocal quantum potential $\mathfrak{Q}$. There are no contradictions here with the Bell theory (which specifies local hidden variables) and the quantum equation of motion will be

$$\partial^\mu\partial_\mu\phi + \frac{\partial V(\phi)}{\partial\phi} + \frac{\delta Q}{\delta\phi(x,t)} = 0 \tag{7.23}$$

where $Q = \int d^3x\mathfrak{Q}$.

We now indicate a covariant version of the Bohm theory following [**930, 931**]. One wants first a quantum version of the classical covariant dDW HJ equation in (7.17) and one formulates the classical version first in a somewhat different way. Thus let $A([\phi],x)$ be a functional of ϕ and a function of x; define then (★) $dA/d\phi(x) = \int d^4x'(\delta A([\phi],x')/\delta\phi(x))$ where $\delta/\delta\phi(x)$ is a spacetime functional derivative. If $A([\phi],x) = A(\phi(x),x)$ (local functional) then $dA([\phi],x)/d\phi(x) = \int d^4x'(\delta A(\phi(x'),x')/\delta\phi(x)) = \partial A(\phi(x),x)/\partial\phi(x)$. An example of particular interest here is a functional nonlocal in space but local in time so that

$$\frac{\delta A([\phi],x')}{\delta\phi(x)} = \frac{\delta A([\phi],x')}{\delta\phi(x,x^0)}\delta(x'^0 - x^0);\quad \frac{dA([\phi],x)}{d\phi(x)} = \tag{7.24}$$

$$= \frac{\delta}{\delta\phi(x,x^0)}\int d^3x' A([\phi],x',x^0)$$

One can write the HJ equation in (7.17) as

$$\frac{1}{2}\frac{dS_\mu}{d\phi}\frac{dS^\mu}{d\phi} + V + \partial_\mu S^\mu = 0 \tag{7.25}$$

which is appropriate for the quantum modification. Similarly the classical equations of motion in (7.16) can be written as (**7G**) $\partial^\mu\phi = dS^\mu/d\phi$. This leads now to the quantum analogue of the classical covariant equation, namely

$$\frac{1}{2}\frac{dS_\mu}{d\phi}\frac{dS^\mu}{d\phi} + V + \mathfrak{Q} + \partial_\mu S^\mu = 0 \tag{7.26}$$

Here (7.26) is manifestly covariant provided that Q in (7.21) can be written in a covariant form (see below for this). One can then show that (7.26) implies (7.21) provided S^0 is local in time (so that (7.24) can be used - cf. (7.18)) and S^i must be completely local so that $dS^i/d\phi = \partial S^i/\partial\phi$ and hence $d_iS^i = \partial_iS^i + (\partial_i\phi)(dS^i/d\phi)$

(cf. (7.16)). Thus in the covariant quantum theory based on the dDW formalism one must require the validity of (**7G**) and this is nothing but a covariant version of the Bohmian equations of motion written for an arbitrarily nonlocal S^{μ}. To produce covariant versions of the remaining terms in (7.21) introduce a vector $R^{\mu}([\phi],x)$ which generates a preferred foliation of spacetime with R^{μ} normal to the leaves of the foliation. Then introduce (**7H**) $\mathfrak{R}([\phi],\Sigma)=\int_{\Sigma}d\Sigma_{\mu}R^{\mu}$ where Σ is a 3-D leaf generated by R^{μ}. Similarly a covariant version of $\mathfrak{S}$ is (**7I**) $\mathfrak{S}([\phi],\Sigma)=\int_{\Sigma}d\Sigma_{\mu}S^{\mu}$ with Σ again generated by R^{μ}. The covariant version of (**7E**) is then (**7J**) $\psi([\phi],\Sigma)=\mathfrak{R}([\phi],\Sigma)exp(i\mathfrak{S}([\phi],\Sigma)/\hbar)$ and for R^{μ} one postulates the equation

$$\frac{dR^{\mu}}{d\phi}\frac{dS_{\mu}}{d\phi}+\mathfrak{J}+\partial_{\mu}R_{\mu}=0 \tag{7.27}$$

In this manner a preferred foliation emerges dynamically as a foliation generated by the solution R^{μ} of (7.27) and (7.26). Note that R^{μ} plays no classical role and the existence of a preferred foliation is a purely quantum effect. Now the relation betweeen (7.27) and (7.21B) is obtained by assuming that nature has chosen a solution of the form $R^{\mu}=(R^0,0,0,0)$ where R^0 is local in time and by integration of (7.27) over d^3x with S^0 local one sees that (7.27) is truely a covariant substitute for (7.21B). Finally one has covariant versions of Q and J in the form

$$\mathfrak{Q}=-\frac{\hbar^2}{\mathfrak{R}}\frac{\delta^2\mathfrak{R}}{\delta_{\Sigma}\phi^2(x)};\ \mathfrak{J}=\frac{\mathfrak{R}}{2}\frac{\delta^2\mathfrak{S}}{\delta_{\Sigma}\phi^2(x)} \tag{7.28}$$

where $\delta/\delta_{\Sigma}\phi(x)$ is a version of (7.18) in which Σ is generated by R^{μ}. Here Σ depends on x ($x\in\Sigma$) and Σ is kept fixed in the variation $\delta_{\Sigma}\phi(x)$. Thus (7.26)-(7.27) with (7.28) represent a covariant substitute for the functional SE (7.20) equivalent to (7.21). The covariant Bohmian equations (**7G**) imply a covariant version of (7.23), namely

$$\partial^{\mu}\partial_{\mu}\phi+\frac{\partial V}{\partial\phi}+\frac{d\mathfrak{Q}}{d\phi}=0 \tag{7.29}$$

Since the last term can also be written as $\delta(\int d^4x\mathfrak{Q}/\delta\phi(x)$ the equation of motion (7.29) can be obtained by varying the quantum action (**7K**) $\mathfrak{A}_Q=\int d^4x\mathfrak{L}_Q=\int d^4x(\mathfrak{L}-\mathfrak{Q})$. Generalizations are included in [**918**] dealing with a larger number of fields and curved spacetimes.

Recall now that the convenentional SE (and WDW equation) corresponds to a special class of solutions of the covariant canonical quantization of fields given by (7.26), (7.27), and (7.28) for which $R^i=0$, S^i is local, and R^0, S^0 are local in time. Let us sketch then how to derive a version of Theorem 7.1 using this restricted situation. First generalize to Hamiltonians involving $G_{AB}\pi^A\pi^B$ as before and use (7.21B) $\equiv$ (7.22) for calculations based on the covariantly equivalent (7.27) with $R^i=0$ etc. It is convenient to choose $\mathfrak{Q}$ and $\mathfrak{J}$ in a form leading to a SE with Hermitian Hamiltonian, namely

$$\mathfrak{Q}=-\frac{\hbar^2}{2\mathfrak{R}}\frac{\delta}{\delta_{\Sigma}h_A}G_{AB}\frac{\delta}{\delta_{\Sigma}h_B}\mathfrak{R};\ \mathfrak{J}=\frac{\mathfrak{R}}{2}\frac{\delta}{\delta_{\Sigma}h_A}\left(G_{AB}\frac{\delta\mathfrak{S}}{\delta_{\Sigma}h_B}\right) \tag{7.30}$$

insted of (7.28). Then one can essentially repeat the proof of Theorem 7.1 by writing (7.22) as

$$\partial_t \mathfrak{R}^2 + \int d^3x \frac{\delta}{\delta h_A}\left(\mathfrak{R}^2 G_{AB}\frac{\delta \mathfrak{S}}{\delta h_B}\right) = 0 \tag{7.31}$$

and setting $\mathcal{S} \sim -(1/\hat{c})\int \tilde{\mathfrak{D}}hPlog(P)$ with $P \sim \mathfrak{R}^2$. From (7.7) and (**7L**) $(\delta\mathfrak{S}/\delta h_B) \sim -(\hat{c}/P)(\delta P/\delta h_B)$ (cf. (7.10)) we get then

$$\partial_t \mathcal{S} = \frac{1}{\hat{c}}\int \mathfrak{D}h(1+log(P))\int d^3x \frac{\delta}{\delta h_A}\left(PG_{AB}\frac{\delta\mathfrak{S}}{\delta h_B}\right) = \tag{7.32}$$

$$= -\frac{1}{\hat{c}}\int \mathfrak{D}h\int d^3x G_{AB}\frac{\delta\mathfrak{S}}{\delta h_B}\frac{\delta P}{\delta h_A} = \int \mathfrak{D}h \int d^3x G_{AB}\frac{1}{P}\frac{\delta P}{\delta h_B}\frac{\delta P}{\delta h_A}$$

(cf. (7.12)) whereas (recall $P = \mathfrak{R}^2$)

$$\int \mathfrak{D}hPQ = \int \mathfrak{D}h\int d^3x P\mathfrak{Q} = -\frac{\hbar^2}{2}\int d^3x \int \mathfrak{D}h\mathfrak{R}\frac{\delta}{\delta h_A}\left(G_{AB}\frac{\delta\mathfrak{R}}{\delta h_B}\right) = \tag{7.33}$$

$$= \frac{\hbar^2}{2}\int d^3x\int \mathfrak{D}hG_{AB}\frac{\delta\mathfrak{R}}{\delta h_A}\frac{\delta\mathfrak{R}}{\delta h_B} = \frac{\hbar^2}{2}\partial_t\mathcal{S}$$

Consequently

THEOREM 7.2. Using the covariant formulation of QFT related to WDW (where $R^i = 0$, S^i is local, and R^0, S^0 are local in time) one can use (7.22) with a differential entropy $\mathcal{S} = (1/\hat{c})\int \mathfrak{D}hPlog(P)$ ($P \sim \mathfrak{R}^2$) to obtain $\partial_t\mathcal{S} \sim (2/\hbar^2)\int d^3x\int \mathfrak{D}hP\mathfrak{Q} = (2/\hbar^2)\int \mathfrak{D}hPQ$ where $\mathfrak{Q}$ is given via (7.31). This provides a connection of metric fluctuations as in (**3K**) to Fisher information via the quantum potential and leads to a relation between a differential entropy and Fisher information as in Theorem 7.1. ■

REMARK 8.7.5. In [**207**] one develops a theory of a time direction hidden in quantum mechanics based on $Q(t) > 0$ where Q is the quantum potential. The idea is that $\int^t Q(\tau)d\tau$ is a monotone increasing function of time which can be useful to characterize the direction of time. We will not go into the idea of a knowledge functional $\mathcal{K}$ here except to remark that $\mathcal{K} \sim Q$ (up to a factor of $\hbar^2/2$). In any event this also seems to be compatible with entropy change as envisioned in (7.7), (7.12) and Theorem 7.1. ■

REMARK 8.7.6. In all these approaches Bohmian equations of motion emerge along with the quantum potential and provides more support for treating the quantum potential as the unifying link betweem classical and quantum physics (see e.g. [**254, 258, 259, 255, 930, 931, 932, ?, 934, 935, 936, 937, 938, 1023, 1024, 1025, 1160, 1163, 1152, 1168**]). ■

REMARK 8.7.6. Following [**254, 255, 599**] one can use the classical geometrodynamical framework of deWitt [**374**] to describe quantum metric fluctuations (in momentum) in terms of Fisher information ideas. Here one is starting with a Riemannian space V_4 and symmetric metric fluctuations should perhaps be stipulated with the Levi-Civita connection imposed. In more general geometries somewhat more flexibility is allowed (cf. [**150**]). ■

REMARK 8.7.7. Heuristically one can also formulate a version of Theorem 6.1 based on Ashtekar variables (cf. [**76, 86, 102, 749, 1085, 1278**]). One begins with densitized triads $E^a_i(x) = \sqrt{h}e^a_i(x)$ and extrinsic curvature $K^i_a(x) = K_{ab}(x)e^{bi}(x)$ (where in the ADM context $K_{ab} = (1/2N)(\dot{h}_{ab} - D_aN_b - D_bN_a)$). An arbitrary vector has the form $v^a = v^ie^a_i$ and a covariant derivative is defined via $D_av^i = \omega_a{}^i{}_jv^j$ where $\omega_a{}^i{}_j = \Gamma^i_{k\,j}e^k_a$ (spin connection) with $\Gamma^i_{k\,j} = e^d_ke^f_je^i_c\Gamma^c_{d\,f} - e^d_ke^f_j\partial_de^i_f$. Parallel transport is defined via $dv^i = -\omega_a{}^i{}_jv^jdx^a$ and one defines $\Gamma^i_a = -(1/2)\omega_{ajk}\epsilon^{ijk}$ where ϵ^{ijk} is the Levi-Civita symbol. Then a connection is defined via $GA^i_a(x) = \Gamma^i_a(x) + \beta K^i_a(x)$ where β is the Barbero-Immirzi parameter and G is the gravitational constant. In a quantum context one will have

$$\hat{E}^b_j\Psi(A) = 8\pi\beta(\hbar/i)(\delta/\delta A^j_b)\Psi(A) \tag{7.34}$$

The Hamiltonian constraint is then $\epsilon^{ijk}F_{abk}E^a_iE^b_j \approx 0$ where $\beta = i$ for the Lorentzian case and F_{abk} refers to the curvature of the A^i_b. In this spirit one can envision a WDW type equation with a sandwich ordering in the form (cf. 6.9 where $V \sim \hat{c}\sqrt{h}(2\Lambda - {}^3R)$)

$$\hbar^2\frac{\delta}{\delta A^i_a}\left(\epsilon^{ij}_kF^k_{ab}\frac{\delta\Psi}{\delta A^j_b}\right) - V(\Psi) = 0 \tag{7.35}$$

For $\psi = Rexp(iS/\hbar)$ one obtains then for $\mathfrak{G}^{ij}_{ab} = \epsilon^{ij}_kF^k_{ab}$

$$Q + \mathfrak{G}^{ij}_{ab}\frac{\delta S}{\delta A^i_a}\frac{\delta S}{\delta A^j_b} + V = 0;\ Q = -\frac{\hbar^2}{R}\frac{\delta}{\delta A^i_a}\left(\mathfrak{G}^{ij}_{ab}\frac{\delta R}{\delta A^j_b}\right) \tag{7.36}$$

One can then envision "momentum" fluctuations $\delta log(R^2)/\delta A^i_a$ for $R = P^{1/2}$ and essentially repeat earlier arguments, which upon inserting a field integral $\int\mathfrak{D}A$ (with suitable properties) leads to a Fisher information type expression

$$\int\mathfrak{D}A\int d^3xPQ = -\hbar^2\int\mathfrak{D}A\int d^3xP^{1/2}\frac{\delta}{\delta A^i_a}\left(\mathfrak{G}^{ij}_{ab}\frac{\delta P^{1/2}}{\delta A^j_b}\right) \sim \tag{7.37}$$

$$\sim \hbar^2\int\mathfrak{D}A\int d^3x\mathfrak{G}^{ij}_{ab}\frac{\delta P^{1/2}}{\delta A^i_a}\frac{\delta P^{1/2}}{\delta A^j_b}$$

which perturbs a classical Hamiltonian. ∎

REMARK 8.7.8. In [**1278**] Wang develops an interesting treatment for classical conformal geometrodynamics and the resulting system of first class constraints has the form

$$\tilde{C}_\perp = \phi^2\tilde{\mu}^{-1}\left[\epsilon_{ijk}\tilde{F}^k_{ab} - \frac{4\phi^8+1}{2\phi^8}\tilde{K}^i_{[a}\tilde{K}^j_{b]}\right]\tilde{E}^a_i\tilde{E}^b_j + 8\tilde{\mu}\phi\tilde{\Delta}\phi; \tag{7.38}$$

$$\tilde{C}_a = \tilde{F}^k_{ab}\tilde{E}^b_k - \tilde{A}^k_a\tilde{C}_k + \pi\phi_{,a};\ \tilde{C}_k = \tilde{D}_a\tilde{E}^a_k;\ \tilde{C} = -\tilde{K}^i_a\tilde{E}^a_i - \frac{1}{4}\phi\pi$$

where (**7M**) $\tilde{E}^a_i = \phi^{-4}E^a_i$ (with E^a_i already densitized) and $\tilde{A}^i_a = \tilde{\Gamma}^i_a + \phi^4K^i_a$ are conjugate variables. Hence in quantization one would expect a formula of the form (**7N**) $\tilde{E}^a_i \sim \kappa(\delta/\delta\tilde{A}^i_a)$ and we recall that $\tilde{\Delta} = \tilde{g}^{ab}\tilde{\nabla}_a\tilde{\nabla}_b$ where $\tilde{g}_{ab}$ refers to the conformal metric and $\tilde{\mu}^2 = det(\tilde{g})_{ab}$. In view of the somewhat complicated structure of $\tilde{C}_\perp$ one does not exclude the posibility that κ depends on ϕ. Hence in analogy to (7.14) we might think here very heuristically of a WDW type equation in sandwich ordering of the form (a potential term $\tilde{V}$ is gratuitously added)

$$\kappa\frac{\delta}{\delta\tilde{A}^i_a}\left(\kappa\mathcal{H}^{ij}_{ab}\frac{\delta\Psi}{\delta\tilde{A}^j_b}\right)+W\Psi-\tilde{V}\Psi=0;\; W=8\tilde{\mu}\phi\tilde{\Delta}\phi \tag{7.39}$$

$$\mathcal{H}^{ij}_{ab}=\phi^2\tilde{\mu}\left[\epsilon_{ijk}\tilde{F}^k_{ab}-\frac{4\phi^8+1}{2\phi^8}\tilde{K}^i_{[a}\tilde{K}^j_{b]}\right] \tag{7.40}$$

(here e.g. $\tilde{V}\sim\hat{c}\sqrt{\tilde{h}}(2\Lambda-{}^3\tilde{R})$); the ϕ evolution, and hence W, is determined in principle by $\tilde{C}_a$ and $\tilde{C}$ (but simultaneous solution of all constraints may be indicated). Now writing (7.18) in the form (**7O**) $\kappa(\delta/\delta\tilde{A}^i_a)(\tilde{\mathcal{H}}^{ij}_{ab}(\delta\Psi/\delta\tilde{A}^j_b)+W\Psi-\tilde{V}\Psi=0$ and putting $\Psi=Rexp(iS/\hbar)$ leads to (cf. (7.3))

$$\kappa\delta_1\left[\tilde{\mathcal{H}}\delta_2(Re^{iS/\hbar})\right]+W(\Psi)-\tilde{V}(\Psi)=0=W(\Psi)+ \tag{7.41}$$
$$+\left\{\kappa\left[\delta_1(\tilde{\mathcal{H}}R_2)-\frac{1}{\hbar^2}\tilde{\mathcal{H}}RS_1S_2\right]+\right.$$
$$\left.+\frac{i}{\hbar}\left[\tilde{\mathcal{H}}R_2S_1+\delta_1(\tilde{\mathcal{H}}RS_2)\right]\right\}e^{iS/\hbar}-\tilde{V}\Psi$$

Consequently

$$\kappa\left[\frac{1}{R}\frac{\delta}{\delta\tilde{A}^i_a}\left(\tilde{\mathcal{H}}\frac{\delta R}{\delta\tilde{A}^j_b}\right)-\frac{\tilde{\mathcal{H}}}{\hbar^2}\frac{\delta S}{\delta\tilde{A}^i_a}\frac{\delta S}{\delta\tilde{A}^j_b}\right]++W-\tilde{V}=0 \tag{7.42}$$

This implies

$$Q=-\frac{\hbar^2}{R}\left[\kappa\frac{\delta}{\delta\tilde{A}^i_a}\left(\tilde{\mathcal{H}}^{ij}_{ab}\frac{\delta R}{\delta\tilde{A}^j_b}\right)\right];\; Q+\kappa\mathcal{H}^{ij}_{ab}\frac{\delta S}{\delta\tilde{A}^i_a}\frac{\delta S}{\delta\tilde{A}^j_b}-W+\tilde{V}=0 \tag{7.43}$$

One could now manipulate Q as in (7.16) to produce a Fisher information type perturbation to a classical Hamiltonian. We emphasize again the speculative nature of (7.18)-(7.22) and note the difficulties involved in solving $\tilde{C}$ and $\tilde{C}_a$ to determine ϕ. ∎

CHAPTER 9

PHASE SPACE ASPECTS

We gather here material on various topics.

1. MATRIX MODELS AND EIGENVALUES

We begin in a different vein with Smolin [**1182**] connecting string theory (and hence gravity) via matrix models to quantum mechanics (QM) in terms of hidden variables (cf. Banks et al [**94**] for matrix models and Adler [**13**] for the statistical mechanics of matrix models). Given the dynamics postulated by matrix models one shows that the finite temperature classical statistical mechanics of the matrix elements reproduces the quantum theory of the matrix eigenvalues - when the temperature is scaled as $T \approx 1/N$ with $N \to \infty$. Thus matrix model theories contain not only a non-perturbative definition of string theory but also a reformulation of quantum theory in terms of the ordinary statistical mechanics of matrix elements as a set of non-local hidden variables. The term hidden variables here refers to the existence of degrees of freedom beyond the observables of the quantum theory whose statistical fluctuations are the source of the quantum fluctuations and uncertainties of quantum theory. Here the hidden degrees of freedom are elements of matrices while the observables are the eigenvalues (see e.g. Holland [**637**] for hidden variables in the Bohmian theory). Here instead of trying to quantize the classical matrix model in some standard manner one simply assumes that the off diagonal elements of the model are in a classical thermal state and hence the formulation is automatically background independent. The interaction of diagonal elements with a large number of off-diagonal elements introduces a Brownian motion which is transferred to a Brownian motion of the eigenvalues at low temperatures. With T scaled as $1/N$ the off diagonal elements are of order $1/\sqrt{N}$ and their collective effects on the eigenvalues remain as $N \to \infty$ while the diffusion coefficient for the Brownian motion of the eigenvalues remains finite. This is very similar to Nelson's formulation of QM as reviewed in [**254, 907, 908, 926**].

Thus one goes to a bosonic matrix model with degrees of freedom given by d $N \times N$ real symmetric matrices X^j_{ai} where $a = 1, \cdots, d$ and $i, j = 1, \cdots, N$; the action is

$$S = \mu \int dtTr[\dot{X}_a^2 + \omega^2[X_a, X_b][X^a, X^b]] \tag{1.1}$$

The matrices X^a are dimensionless, ω is a frequency, and μ has dimensions of $mass \cdot length^2$; $\hbar$ is not yet meaningful and will be introduced later while the parameters define an energy $\epsilon = \mu\omega^2$. The off diagonal matrix elements of X^a

wil be the non local hidden variables and the physical observables will be the eigenvalues λ_i^a of the matrices. The system will be put at a small but finite temperature the result of which will be that the matrix elements undergo Brownian motion as they oscillate in the potential. When $T = 0$ the matrices must commute and can be simultaneously diagonalized; when T is finite but small compared to ϵ the off diagonal elements will on average be small and one writes (•) $X_{ai}^j = D_{ai}^j + Q_{ai}^j$ where $D^a = diag(d_1^q, \cdots, d_N^a)$ is diagonal and Q_{ai}^j has no diagonal elements; the Q_{ai}^j are expected to scale like a power of $T/\mu\omega^2$ and one writes

$$S = \int dt[\mathfrak{L}^d + \mathfrak{L}^Q + \mathfrak{L}^{int}];\ \mathfrak{L}^d = \mu \sum_{a,i} (\dot{d}_i^a)^2; \tag{1.2}$$

$$\mathfrak{L}^Q = \mu \left[\sum_{a,i,j} (\dot{Q}_{ai}^j)^2 + \omega^2 [Q_a, Q_b][Q^a, Q^b] \right]$$

The interaction terms are

$$\mathfrak{L}^{int} = 2\mu\omega^2 \sum_{a,b,i,j} \left[-(d_i^a - d_j^a)^2 (Q_{bi}^j)^2 - (d_i^a - d_j^a)(d_i^b - d_j^b) Q_{ai}^j Q_{bj}^i + \right. \tag{1.3}$$

$$\left. +2(d_i^a - d_j^a) Q_i^{bj} [Q_a, Q_b]_j^i \right]$$

One notes that the model has a translation symmetry (♦) $d_i^a a \to d_i^a + v^a$ so that the center of mass momentum of the system is conserved.

Now recall that a dynamical system in n-dimensional configuration space x^α can be described via a Hamilton-Jacobi (HJ) function $S(x,t)$ with

$$\dot{S} + \frac{1}{2m}(\partial_a S)^2 + U = 0;\ p_a(x) = \partial_a S \tag{1.4}$$

There are many solutions and a statistical description of the system can be given via a probability density $\rho(x,t)$ and a probability current $v^a(s,t)$ such that (★) $\dot{\rho} + \partial_a(\rho v^a) = 0$ (note $\int d^n x \rho(x,t) = 1$). Now restrict attention to an ensemble of classical trajectories whose evolution is determined by a particular solution S of the HJ equation (an S ensemble). These will have a probability density $\rho_S(x,t)$ and since the momentum is determined by S so is the probability current so

$$m v_a = \partial_a S;\ \dot{\rho} + \frac{1}{m}\partial_a(\rho \partial^a S) = 0 \tag{1.5}$$

The total probability density is then $\rho_{tot} = \sum_S \rho_S$ over all solutions to the HJ equation and one notes that such ensembles are given by the solution of a variational problem

$$I[\rho, S] = \int dt \int d^n x \rho(x,t) \left[\dot{S} + \frac{1}{2m}(\partial_a S)^2 + U(x) \right] \tag{1.6}$$

The equations that arise from varying ρ and S are respectively the HJ equation (1.4) and the probability conservation equation (1.5). One notes that the action and equations of motion are invariant under time reversal with $t \to -t,\ S \to -S$, and $\rho \to \rho$.

One now has to adjust for nondifferentiable paths via the language of stochastic differential equations. Thus write

$$(1.7)\qquad Dx^a = b^a(x(t),t)dt + \Delta w^a \ (dt > 0);$$
$$D^*x^a = -b^{*a}(x(t),t)dt + \Delta w^{*a} \ (dt < 0)$$

The forward and backward drift velocities b^a and b^{*a} can be defined via

$$(1.8)\qquad b^a(x,t) = lim_{\Delta t \to 0}\left\langle \frac{x^a(t+\Delta t) - x^a(t)}{\Delta t}\right\rangle_{x(t)=x};$$
$$b^{*a}(x,t) = lim_{\Delta t \to 0}\left\langle \frac{x^a(t) - x^a(t+\Delta t)}{\Delta t}\right\rangle_{x(t)=x}$$

The different elements of the ensemble are distinguished by their Brownian motion, given via a Markov process

$$(1.9)\qquad < \Delta w^a, \Delta w^b >= - < \Delta w^{*a}\Delta w^{*b} >= \nu dt q^{ab};\ < \Delta w^a \Delta w^{*b} >= 0$$

Here q^{ab} is a metric on the configuration space and ν is the diffusion constant; the averages $< \ >$ are relative to the ensemble. Thus $< F >= \int d^n x \rho(x,t)F(x)$ and one obtains the forward and backward Fokker-Planck (FP) equations

$$(1.10)\qquad \dot{\rho} = -\partial_a(\rho b^a) + \nu\nabla^2\rho;\ \dot{\rho} = -\partial_a(\rho b^{*a}) - \nu\nabla^2\rho$$

It follows that $v^a = (1/2)(b^a + b^{*a})$ and the difference between forward and backward drift velocities is called the osmotic velocity (cf. [**234**]) which satisfies

$$(1.11)\qquad u^a = (1/2)(b^a - b^{*a}) = \nu\partial^a log(\rho)$$

Note in (1.8) the limit $\Delta t \to 0$ is taken after averaging over the ensemble. Now one modifies the statistical variational principle to include the effects of Brownian motion and we refer to [**1182**] for details; the result will be

$$(1.12)\qquad I^\nu[\rho,S] = \int dt \int d^n x \rho(x,t)\left[\dot{S} + \frac{1}{2m}(\partial_a S)^2 + \frac{m\nu^2}{2}(\partial_a log(\rho))^2 + U(x)\right]$$

The new HJ equation following from (1.12) is

$$(1.13)\qquad \dot{S} + \frac{1}{2m}(\partial_a S)^2 - 2m\nu^2\frac{1}{\sqrt{\rho}}\nabla^2\sqrt{\rho} + U = 0$$

and voila the quantum potential appears in the third term. As usual now (1.13) and (1.5) with (♠) $\psi = \sqrt{\rho}exp(iS/\hbar)$ and $\hbar = 2\nu m$ lead immediately to the Schrödinger equation (SE)

$$(1.14)\qquad i\hbar\partial_t\psi(x,t) = \left[-\frac{\hbar^2}{2m}\nabla^2 + U(x)\right]\psi(x,t)$$

(cf. [**234**]). From the present point of view a quantum state is nothing more nor less that an ordinary statistical ensemble of Brownian motion trajectories which share a single HJ solution S of (1.13).

This can be formulated in Hamiltonian form by writing

$$(1.15)\qquad I^\nu[\rho,S] = \int dt \int d^n x[S\dot{\rho} - \mathfrak{H}[\rho,S,x]];$$

$$\mathfrak{H} = \rho\left[\frac{1}{2m}(\partial_a S)^2 + \frac{m\nu^2}{2}(\partial_a log(\rho))^2 + U(x)\right]$$

To see this note that ρ can be considered as a conjugate variable to S and one has

(1.16) $$\{\rho(x), S(x')\} = \delta^n(x', x)$$

so that the Hamiltonian $H = \int d^n x \mathfrak{H}$ is conserved in time. To see through all this write out the conserved Hamiltonian

(1.17) $$H = \int d^n x \left[\frac{1}{2m}(\partial_a S)^2 + \frac{m\nu^2}{2}(\partial_a log(\rho))^2 + U(x)\right]$$

This seems to be nonlinear but it is nothing but the expectation value of a linear operator; indeed one can write

(1.18) $$H = \int d^n x \sqrt{\rho} e^{-iS/\hbar}\left[\frac{1}{2m}(\partial_a S)^2 - 2m\nu^2\frac{\nabla^2\sqrt{\rho}}{\sqrt{\rho}} + U(x)\right]\sqrt{\rho}e^{iS/\hbar}$$

and using (♠) this is seen to be (♣) $H = \int d^n x \bar{\psi}\hat{\mathcal{H}}\psi$ with $\hat{\mathcal{H}} = -(\hbar^2/2m)\nabla^2 + U(x)$. Thus $\dot{H} = 0$ is equivalent to the SE.

Now go to the matrix model and one will show that the ordinary statistical mechanics of this model has a critical behavior when the off diagonal sector is heated to finite temperature and the large N limit is taken with the temperature scaled so that $T \approx 1/N$. One feature of the critical behavior is QM itself, i.e. to leading order in $1/N$ the evolution of the probability density and current for the eigenvalues of the matrices is equivalent to that given by the free SE. One first studies the behavior of the matrix model at low temperature and large N; then picking $T \approx 1/N$ the off diagonal elements scale as $1/\sqrt{N}$ and they move harmonically in an average field given by the average values of all the other off diagonal elements. The diagonal elements remain of order unity and move in a random potential given by the oscillations of all the off diagonal elements; they pick up a random Brownian motion on top of their free motion and this is transmitted to the eigenvalues. Under this critical scaling the diffusion coefficients for the diagonal elements and eigenvalues go to constant limits as $N \to \infty$ and $T \to 0$. Then one studies the Brownian motion of the eigenvalues using an effective statistical action and the Brownian motion in (1.12) emerges naturally as a term in the effective statistical action for the eigenvalues as a result of the induced Brownian motion indicated. Finally in the large N limit Nelson's stochastic formulation of QM emerges as a description of the statistical behavior of the eigenvalues. We omit the details here (cf. [**1182**]) and go directly to the final stages. Thus one defines an S-ensemble for the matrix elements via a variational principle (with some abuse of notation)

(1.19) $$I[\rho, S] = \int dt \int d^n x (dd)(dQ)\rho(d, Q, t)\left[\dot{S}(d, Q) + \frac{1}{2\mu}\left(\frac{\delta S(d, Q, t)}{\delta d_i^a}\right)^2 + \right.$$
$$\left. + \frac{1}{2\mu}\left(\frac{\delta S(d, Q, t)}{\delta Q_{ij}^a}\right)^2 + U(d, Q)\right]$$

For physical assumptions connecting ρ and S one has

- The Q system is in a distribution that is to leading order in $1/N$ statistically independent of the distribution of the eigenvalues. This means that to leading order $\rho(d,Q) = \rho_d(d)\rho_Q(Q) + O(1/N)$
- The Q subsystem is in thermal equilibrium at a temperature T so $\rho_Q(Q) = (1/Z)exp(-H(Q)/T)$ where

$$H(Q) = \mu\left[\sum_{aij}(\dot{Q}^j_{ai})^2 - \omega^2[Q_a,Q_b][Q^a,Q^b]\right];\ Z = \int dQ e^{-H(Q)/T} \tag{1.20}$$

Consequently the variational principle involves

$$I[\rho_d,S,T] = \int dt \int d^n x (dd)(dQ)\rho_d(d)\rho_Q(Q)\left[\dot{S}(d,Q)+ \right. \tag{1.21}$$
$$\left. +\frac{1}{2\mu}\left(\frac{\delta S}{\delta d^a_i}\right)^2 + \frac{1}{2\mu}\left(\frac{\delta S}{\delta Q^a_{ij}}\right)^2 + U(d,Q)\right]$$

- Next for an effective variational principle one does some averaging (1.21) over the values of the matrix elements and extracting the leading order behavior. Thus one uses

$$1 = \int \prod_{a,i} d\lambda^a_i \delta\left(\lambda^a_i - d^a_i - \sum_j \frac{Q^a_{ij}Q^a_{ji}}{d^a_i - d^a_j} + \cdots\right) \tag{1.22}$$

and inserts this in I to get

$$I[\rho_d,S,T] = \frac{1}{Z}\int dt \int dddQ \int d\lambda\delta\left(\lambda^a_i - d^a_i - \sum_j \frac{Q^a_{ij}Q^a_{ji}}{\lambda^a_i - \lambda^a_j} + \cdots\right)\times \tag{1.23}$$
$$\times\rho_d e^{-H(Q)/T}\left[\dot{S} + \frac{1}{2\mu}\left(\frac{\delta S}{\delta d^a_i}\right)^2 + \frac{1}{2\mu}\left(\frac{\delta S}{\delta Q^a_{ij}}\right)^2 + U(d,Q)\right]$$

Further calculation involves

$$S_\lambda(\lambda) = \int (dQ)\rho_Q(Q)S(\lambda,Q);\ \mu v^a_i = \frac{\delta S_\lambda(\lambda)}{\delta \lambda^a_i} \tag{1.24}$$
$$\psi(\lambda,t) = \sqrt{\rho}e^{S_\lambda/\hbar};\ \hbar = \mu\nu_\lambda$$

with a renormalized ψ and SE

$$\psi_r(\lambda) = e^{iE'_Q t/\hbar}\psi(\lambda);\ i\hbar\partial_t\psi_r = \left[-\frac{\hbar^2}{2\mu}\frac{\delta^2}{\delta(\lambda^a_i)^2}\right]\psi_r \tag{1.25}$$

There is also a quantum potential of the form

$$U^{qp} = -\mu\nu^2_\lambda \frac{1}{\sqrt{\rho_\lambda(\lambda)}}\nabla^2\sqrt{\rho_\lambda(\lambda)} \tag{1.26}$$

The term E'_Q is best forgotten since it involves infinite constants but in any case one is able to obtain a SE from the matrix model in a reasonable manner.

2. BOHMIAN MECHANICS AND PHASE SPACE

2.1. PHASE SPACE AND OPERATORS. We go first to [**201, 199, 551, 552, 553, 554, 627**] (for phase space considerations see also [**65, 252, 239, 343, 380, 1317, 1331, 1332**]). In Brown-Hiley [**201**] one argues that if the quantum potential (QP) reflects the quantum aspects of a sysem it should be possible to to identify such aspects within the QP and in particular one shows how the balance between localisation and dispersion energies suggests a link between the QP and the Heisenberg uncertainty principle. Recall first that from the SE (**2A**) $i\hbar\partial_t\psi = [-(\hbar^2/2m)\nabla^2 + V]\psi$ there follows

$$\partial_t S + \frac{(\nabla S)^2}{2m} - \frac{\hbar^2}{2m}\frac{\nabla^2 R}{r} + V = 0;\ \partial_t\rho + \nabla\cdot\left(\rho\frac{p}{m}\right) = 0 \tag{2.1}$$

where $\psi = Rexp(iS/\hbar)$, $\rho = |\psi|^2$, and $p = \nabla S$. The QP is manifestly of the form $Q = -(\hbar^2/2m)(\nabla^2 R/R)$ and one writes $F = -\nabla(Q+V)$ and $v = j/\rho = p/m$. Now consider a more general derivation of the QP by writing the SE in the form (**2B**) $i\hbar\partial_t\psi = (T(\hat{p}) + V(\hat{x}))\psi$. Setting again $\psi = Rexp(iS/\hbar)$ one obtains

$$\partial_t S + \Re\left(\frac{T\psi}{\psi}\right) + V(x) = 0;\ \partial_t\rho - \frac{2\rho}{\hbar}\Im\left(\frac{T\psi}{\psi}\right) = 0 \tag{2.2}$$

Correspondingly in the momentum space with $\hat{x} = i\hbar\nabla_p$ and $\hat{p} = p$ the real and imaginary parts of the SE are

$$\partial_t S + T(p) + \Re\left(\frac{V\psi}{\psi}\right) = 0;\ \partial_t\rho - \frac{2\rho}{\hbar}\Im\left(\frac{V\psi}{\psi}\right) = 0 \tag{2.3}$$

Then expanding exponentials one writes

$$\Re\left(\frac{\psi^* T\psi}{\rho}\right) = \Re\left(\frac{R[1-(iS/\hbar)-\cdots)T(\hat{p})R(1+(iS/\hbar)-\cdots]}{\rho}\right) \tag{2.4}$$

(note the formal equivalence $T\psi/\psi = \psi^* T\psi/\rho$). If now $T(\hat{p})$ is a general but analytic function of $\hat{p}$ one can expand in a power series in $\hat{p} = -i\hbar\nabla$ and the kinetic term may be separated into the sum of two parts

$$\Re\left(\frac{T\psi}{\psi}\right) = T_h(x) + T_0(x);\ T_0(x) = T(\nabla S) \tag{2.5}$$

where $T_h(x)$ is an expansion in even positive powers of $\hbar$ and $T_0(x)$ is independent of $\hbar$ and identifies $p = \nabla S$. The same line of argument allows the potential term of the HJ equation in (2.3) to be separated as

$$\Re\left(\frac{V\psi}{\psi}\right) = V_h(p) + V_0(p);\ V_0(p) = V(-\nabla_p S) \tag{2.6}$$

where V_h is an expansion in even positive powers of $\hbar$ and $V_0(p)$ is independent of $\hbar$ and identifies $x = -\nabla_p S$.

REMARK 9.2.1. To see how this goes one thinks of $T(\hat{p}) = \sum a_{\ell m}\hat{p}_\ell^m$ and assuming only one momentum component the index ℓ is dropped. Then writing

$\Re(T\psi/\psi) = (1/R)\Re[exp(-iS/\hbar)T(\hat{p})Rexp(iS/\hbar)]$ with $\hat{p} = -i\hbar\partial_x$ and expanding one obtains

$$\Re\left(\frac{T\psi}{\psi}\right) = \frac{1}{R}\sum_{j,k=0}^{\infty}\sum_{m}^{M} a_m(-1)^{j+m}\Re(i^{j+k+m})\hbar^{m-k-j}\frac{S^j}{j!k!}\frac{\partial^m(RS^k)}{\partial x^m} \tag{2.7}$$

Thus $j+k+m \in Z_e$ (even integers) and the expression must not change when one adds an arbitrary constant to S; this means all powers of S must sum to zero so only the $j=0$ terms will contribute to the summation. Similarly those parts of the partial derivative in (2.7) which have non-zero powers of S as factors cannot contribute. Thus $k \leq m$ and finally eliminating all factors involving S alone one obtains

$$\Re\left(\frac{t\psi}{\psi}\right) = \frac{1}{R}\sum_{m}^{M}\sum_{k=0,k+m\in Z_e}^{m} a_m(-1)^m(-1)^{\frac{k+m}{2}}\frac{\hbar^{m-k}}{k!}\frac{\partial^m(RS^k)}{\partial x^m}\bigg|_{S:0} \tag{2.8}$$

where $S:0$ denotes the process of eliminating terms with S as a factor. This leads to

$$T_h(x) = \frac{1}{R}\sum_{m}^{M}\sum_{k=0,k+m\in Z_e}^{<m} a_m(-1)^{\frac{3m+k}{2}}\frac{\hbar^{m-k}}{k!}\frac{\partial^m(RS^k)}{\partial x^m}\bigg|_{S:0} \tag{2.9}$$

while $T_0(x) = \sum_m^M a_m(\partial_x S)^m$ (which corresponds to a one dimensional classical form of energy for $p = \partial_x S$). The same procedure is also used for the momentum representation using $\dot{x} = i\hbar\partial_p$ and $x = -\nabla_p S$. ■

REMARK 9.2.2. In the x-representation it is simply $T(\hat{p}) = p^2/2m$ that leads to the simple 1-term expression for the QP, namely $T_h(x) = -(\hbar^2/2m)(\nabla^2 R/R)$ while $T_0(x) = (\nabla S)^2/2m$ is the same as the classical kinetic energy. In the momentum representation the QP does not in general have the simple one term form but for the harmonic oscillator $V(x) = (1/2)m\omega^2$ the QP has the simple form $V_h(p) = -(\hbar^2/2)m\omega^2(\nabla_p^2 R/R)$ while $V_0(p) = [(\nabla_p S)^2/2]m\omega^2$. ■

Now one indicates the symplectic structure for the quantum situation. From (2.2) and (2.5) there is an effective Hamiltonian **(2C)** $H = T(p) + T_h(x) + V(x)$ where $p = \nabla S$ is understood. One finds

$$\dot{x} = \left(\frac{\partial H}{\partial p}\right)_{p=\nabla S} = \left(\frac{\partial T(p)}{\partial p}\right)_{p=\nabla S}; \tag{2.10}$$

$$\dot{p} = -\partial_x H = -\frac{\partial(T_h(x) + V(x))}{\partial x}$$

Similarly in the momentum representation $H_p = T(p) + V_h(x) + V(x)$ with $x = -\nabla_p S$ and one obtains

$$\dot{p} = -\left(\frac{\partial H_p}{\partial x}\right)_{x=-\nabla_p S} = -\left(\frac{\partial V(x)}{\partial x}\right)_{x=-\nabla_p S}; \tag{2.11}$$

$$\dot{x} = \frac{\partial H_p}{\partial p} = \frac{\partial(T(p) + V_h(p))}{\partial p}$$

This illustrates the breaking of precise classical symplectic symmetry and one notes the difference between e.g. the x trajectories in configuration space and momentum space.

Given that the QP is a uniquely quantum energy one inquires about the source of this energy. Unlike classical systems a quantum system has intrinsic internal energies associated with spatial localization and momentum dispersion. To look further into this write

$$T_h(x) = \frac{-\hbar^2}{2mR}\nabla^2 R = M_d(x) + \Sigma_\ell(x); \tag{2.12}$$

$$M_d = \frac{-\hbar^2}{8m}(\nabla^2 log(\rho));\ \Sigma_\ell = \frac{-\hbar^2}{8m}\left(\frac{\nabla^2\rho}{\rho}\right)$$

In [**491**] M_d is identified as the Wigner potential and one recalls here the moments of the Wigner density are

$$\rho(x) = \int f(x,p)dp;\ P^1(x) = \int pf(x,p)dp;\ P^2(x) = \int p^2 f(x,p)dp \tag{2.13}$$

Defining the mean moments as $\overline{p(x)} = P^1(x)/\rho$ and $P^2(x)/\rho = \overline{p(x)^2}$ the lowest order condition on the moments is the dispersion relation $\overline{p(x)^2} - \overline{p(x)}^2 = -(\hbar^2/4)\nabla^2 log(\rho)$. Thus M_d in (2.12) can be identified as a momentum dispersion energy. On the other hand Σ_ℓ is a measure of the local curvature of the probability density ρ and is called the localization energy (cf. [**201**] for discussion of this). In the momentum representation only the harmonic oscillator has a QP of the form

$$V_h(p) = \frac{-\hbar^2}{2R}m\omega^2\nabla_p^2 R = \Sigma_d(p) + M_\ell(p); \tag{2.14}$$

$$\Sigma_d(p) = \frac{-\hbar^2}{8}m\omega^2(\nabla_p^2 log(\rho));\ M_\ell(x) = \frac{-\hbar^2}{8}m\omega^2\left(\frac{\nabla_p^2\rho}{\rho}\right)$$

One finds then

$$\overline{x(p)^2} - \overline{x(p)}^2 = \frac{-\hbar^2}{4}\nabla_p^2 log(\rho) \tag{2.15}$$

where $\Sigma_d(p)$ is identified as the spatial dispersion energy and $M_\ell(p)$ is the momentum localisation energy (proportional to the local curvature of the probability density in momentum space). These facts are evidently related to the Heisenberg uncertainty principle.

We go now to [**201, 199, 551, 552, 553, 554, 628**] where the ideas revolve around a medly of themes related to symplectic geometry and algebraic constructions of the SE (cf. also [**252, 239, 343, 1317, 1331, 1332**] for the Weyl calculus and phase space, etc.). The clearest presentation of the algebraic theory seems to be in [**201, 199, 627**] where one main idea is to express the Schrödinger picture of QM through two representation independent algebraic forms. First recall the standard SE in the form

$$i\partial_t\psi(a_i,t) = H\psi(a_i,t)\ (\hbar = 1\ at\ times) \tag{2.16}$$

where the a_i are the eigenvalues of an operator A in the algebra $\mathfrak{A}$ of observables. In the Heisenberg icture one writes the Hamiltonian flow of the operator A in the form (**2D**) $dA/dt = (1/i)[A, H]_-$ or more generally $dA/dt = \partial_t A + (1/i)[A, H]_-$ where $[\ ,\ ]_-$ is the standard commutator. In the algebraic approach the state function is introduced via the density operator $\hat{\rho}$ which is in the operator algebra, namely one write $\hat{\rho} = |\psi><\psi|$ or better $\hat{\rho} =><= \epsilon$ with $\epsilon^2 = \epsilon$. Given ρ or $\hat{\rho}$ as an operator one will then first the Liouville equation $d\rho/dt = 0$ which implies via (**2D**) the Heisenberg equation

(2.17) $$\partial_t \rho = \frac{1}{i}[H, \rho]_-$$

(here both H and ρ should have hats but we expect no confusion will arise). The algebraic equivalent of the vector space of ket vectors is a left ideal I_L in the algebra which one can take in the form $B\epsilon$; similarly $\epsilon C \sim I_R$ is a right ideal representing bra vectors. In particular one can write $\rho = B\epsilon\epsilon C = B\epsilon C$ so a pure state density operator corresponds to a two sided ideal with $\rho^2 = \rho$ and $Tr(\rho) = 1$. Now to find an algebraic SE one can write

(2.18) $$i\hbar(\partial_t B)\epsilon C + i\hbar B\epsilon(\partial_t C) = HB\epsilon C - B\epsilon CH$$

One assumes there are operator elements $B^\dagger,\ C^\dagger$ such that $B^\dagger B = 1$ and $C^\dagger C = 1$ which leads then to

(2.19) $$B^\dagger(i\hbar\partial_t B - HB)\epsilon = -\epsilon(i\hbar\partial_t C + CH)C^\dagger$$

Since $B,\ C$ are arbitrary nonzero elements of the algebra $\mathfrak{A}$ one can write

(2.20) $$i(\partial_t B)\epsilon = HB\epsilon;\ -i\epsilon(\partial_t C) = \epsilon CH$$

These have the same form as the SE and its conjugate (H is assumed to be Hermitian).

We essentially rewrite this now in the framework of [**627**] which is somewhat cleaner (although [**201, 199**] is much more complete and easier reading). Thus one takes $\hat{\psi}_L \sim B\epsilon$ and $\hat{\psi}_R \sim \epsilon C$ so $\rho = \hat{\psi}_L\hat{\psi}_R$ (ρ should also have a hat but we see no problems arising). Then we have ($\hbar = 1$ now)

(2.21) $$i\partial_t\hat{\psi}_L = \hat{H}\hat{\psi}_L;\ -i\partial_t\hat{\psi}_R = \hat{\psi}_R\hat{H}$$

Taking the difference yields then

(2.22) $$i\partial_t\hat{\rho} + [\hat{\rho}, \hat{H}]_- = 0;\ i\left[(\partial_t\hat{\psi}_L)\hat{\psi}_R - \hat{\psi}_L(\partial_t\hat{\psi}_R)\right] = [\hat{\rho}, \hat{H}]_+$$

This can be simplified via polar coordinates (**2E**) $\hat{\psi}_L = \hat{R}\hat{U}$ and $\hat{\psi}_R = \hat{U}^\dagger\hat{R}$ where $\hat{R}$ is positive definite and $\hat{U}$ is unitary. There results then

(2.23) $$i[(\partial_t\hat{\psi}_L)\hat{\psi}_R - \hat{\psi}_L(\partial_t\hat{\psi}_R)] = i\hat{R}[(\partial_t\hat{U})\hat{U}^\dagger - \hat{U}(\partial_t\hat{U}^\dagger)]\hat{R} = [\hat{\rho}, \hat{H}]_+$$

Now write $\hat{U} = exp(i\hat{S})$ where $\hat{S} = \hat{S}^\dagger$ and assume $[\hat{R}, \partial_t\hat{S}] = 0$; then

(2.24) $$\hat{R}(\partial_t\hat{S})\hat{R} + \frac{1}{2}[\hat{\rho}, \hat{H}]_+ = 0 \Rightarrow \hat{\rho}(\partial_t\hat{S}) + \frac{1}{2}[\hat{\rho}, \hat{H}]_+ = 0$$

Thus the two Schrödinger type equations in (2.21) are replaced by

$$i\partial_t\hat{\rho} + [\hat{\rho}, \hat{H}]_- = 0;\ \hat{\rho}(\partial_t\hat{S}) + \frac{1}{2}[\hat{\rho}, \hat{H}]_+ = 0 \tag{2.25}$$

(cf. also [**201**]).

Now to relate this to the Bohm approach one notes first the similarity between the pair of defining equations (2.25) and the pair of equations forming the classical Bohm approach, namely ($v = p/m,\ \rho \sim R^2 \sim P,\ \nabla S = p,\ \hbar = 1$, etc.)

$$\partial_t\rho + \nabla(v\rho) = 0;\ \partial_t S + \frac{(\nabla S)^2}{2m} - \frac{1}{2m}\frac{\nabla^2 R}{R} + V = 0 \tag{2.26}$$

Note here the current is usually

$$J = \frac{\hbar}{m}\Im(\psi^*\psi') = \frac{\hbar}{m}(p\rho) \tag{2.27}$$

and in Hiley [**627**] one writes $[Q, P] = i$ so $P \sim -i\nabla$ (the equation (6.31) in [**627**] is however curious). The first difference now is that (2.25) involves operator equations and secondly they are representation free (whereas (2.26) is tied to the x-representation). Actually (2.23) will produce the classical equations if one goes to the x-representation even though there is nothing like a quantum potential visible in (2.25). Thus consider a general representation defined by $\hat{A}|a>= a|a>$ so that (2.23) becomes

$$i\partial_t P(a) - \left\langle [\hat{H}, \hat{\rho}]_- \right\rangle_a = 0 \tag{2.28}$$

where $P(a) \sim R^2 \sim \rho(a)$ is the probability of finding the particle in the state $|a>$. (2.28) is just the Liouville equation expressing the conservation of probability. If one replaces a by x the x-representation is chosen and (2.28) becomes (2.24). In a general representation we have

$$P(a)\partial_t S(a) + \frac{1}{2}\left\langle [\hat{H}, \hat{\rho}]_+ \right\rangle_a = 0 \tag{2.29}$$

which is clearly an equation for the time development of the phase function S. To compare with (2.11b) however requires in particular a quantum potential and this is in fact implicit in the commutator. Indeed pick e.g. a Hamiltonian for the harmonic oscillator, namely $H = (p^2/2m) + (Kx^2/2)$ and put this in (2.29) to get

$$\partial_t S_x + \frac{1}{2m}(\partial_x S_x)^2 + \frac{Kx^2}{2} - \frac{1}{2mR_x}(\partial^2 R_x^2/\partial x^2) = 0 \tag{2.30}$$

and the quantum potential arises, giving one an expression for the conservation of energy. Doing the same thing in the p-representation yields

$$\partial_p S_p + \frac{p^2}{2m} + \frac{K}{2}(\partial_p S_p)^2 - \frac{K}{2R_p}\left(\frac{\partial^2 R_p}{\partial p^2}\right) = 0 \tag{2.31}$$

and a quantum potential appears again; thus there is also a Bohm interpretation in the p-representation and (2.31) is again an expression for conservation of energy. Consequently one can construct a Bohm interpretation for any representation and one has not lost the symmetry of the Heisenberg approach; moreover the full symplectic symmetry is still there (cf. remarks below). One chooses the x-representation in general simply because it is more "natural" and easier to visualize.

In order to interpret (2.11b) as a conservation equation one has to identify $Q_x \sim \pm(1/2mR_x)(\partial^2 R_x^2/\partial x^2)$ as an energy and also $\partial S_x/\partial x$ needs to be interpreted as the momentum p. The latter is sometimes known as the "guidance" condition of Bohm. We omit mention here of connections with a classical limit $Q \to 0$ (or $\hbar^2 Q \to 0$) since there is not a clear cut meaning to this (see e.g. [**163, 257**]). However we note that in the x-representation (resp. p-representation) for $P \sim R^2$ one has

$$\frac{\Re[\psi^*(x)P\psi(x)]}{|\psi|^2(x)} = \frac{\partial S_x}{\partial x} = p; \quad \frac{\Re[\psi^*(p)X\psi(p)]}{|\psi|^2(p)} = -\frac{\partial S_p}{\partial p} = x \tag{2.32}$$

This all makes it clear that the quantum potential is not an ad hoc object but a necessary feature of QM. It is needed in order to ensure that both energy and momentum are conserved. Indeed the kinetic energy used in (2.30) is calculated from the real part of $(1/2m)[\psi^*(x,t)\hat{P}\psi(x,t)]^2$ which is clearly not the quantum kinetic energy calculated from $(1/2m)\psi^*(x,t)\hat{P}^2\psi(x,t)$; the difference is the quantum potential giving the conservation of energy as in (2.30). In the p-representation the potential energy which is calculated from the real part of $[\psi^*(p,t)\hat{X}\psi(p,t)]^2$ cannot be the total potential energy which must be calculated from $\psi^*(p,t)\hat{X}^2\psi(x,t)$ leading to (2.31).

Let us redo some of this following [**109**]. Thus go back to (2.20) and to see how these have the same general form as the SE and its conjugate one writes $B\epsilon = B(X,t) \in I_L$ and projects $B(X,t)$ to a function $B(x,t) = B(X,t)(x) = \psi(x,t)$. Then the first equation in (2.20) becomes the SE

$$i\partial_t\psi(x,t) = H(x)\psi(x,t) \tag{2.33}$$

The conjugate equation follows similarly via $C(X,t)(x) = C(x,t)$ so the second equation in (2.20) becomes the conjugate SE. Now write $B = exp[iS_Q(t)]$ and $C = exp[-S_Q(t)]$ where $S_Q = S - ilog(R)$. Then (2.20) becomes

$$-\partial_t S_Q B\epsilon = HB\epsilon; \ -\epsilon C\partial_t S_Q^\dagger = \epsilon CH \tag{2.34}$$

which are algebraic equivalents of the HJ equation $\partial_t S_{cl} + H = 0$; here S_{cl} is the classical action and one will call S_Q the quantum action. Now to pull these equations out of their respective ideals one post and pre mulltiplies by ϵC and $B\epsilon$ to get

$$-(\partial_t S_Q)\rho + \rho(\partial_t S_Q^\dagger) = [H,\rho]_-; \ -(\partial_t S_Q)\rho - \rho(\partial_t S_Q^\dagger) = [H,\rho]_+ \tag{2.35}$$

Putting the first equation in (2.35) in Hermitian form one recovers (2.17). The second equation in (2.35) cannot be simplified algebraically but may be recognized as a symmetrized operator form of the HJ equation. If one writes now $[\ ](a) =< a|[\ ]|a>$ it takes the form ($\rho_R = R^2$)

$$\partial_t\rho_R(a) + \frac{1}{i}[\rho,H]_- = -; \ \rho_R(a)\partial_t S(a) + \frac{1}{2}[\rho,H]_+(a) = 0 \tag{2.36}$$

(cf. (2.28)-(2.29)). Choosing A to be the position operator X the first equation in (2.36) becomes $\partial_t \mathcal{P} + \nabla \cdot \mathbf{j} = 0$ where $\mathcal{P} = \mathcal{P}(x) = \rho_R(X) =< x|\rho|x >$ and $\mathbf{j}$ is a probability current ($\mathbf{j} \sim P\mathbf{v}\rho$ - cf. (2.26)).

3. SYMPLECTIC MECHANICS

To be more complete here we go to [**65, 252, 551, 552, 553, 554, 1049, 1238**] and first in connection with (2.26) we note that one can define a Bohmian (**3A**) $H^\psi = H + Q^\psi$ where $Q^\psi = -(1/2m)(\nabla^2 R/R)$ (here $\hbar = 1$). Then putting this in the HJ equation, written as

$$\partial_t S + H^\psi(\mathbf{r}, \nabla_r S) = 0 \tag{3.1}$$

one can show that there is a symplectomorphism f^ψ_{t,t_0} given via $(\mathbf{r}^\psi(t), \mathbf{p}^\psi(t)) = f^\psi_{t,t_0}(\mathbf{r}_0, \mathbf{p}_0)$ which can be written as

$$\frac{d\mathbf{r}^\psi}{dt} = \nabla_p H^\psi, \ \frac{d\mathbf{p}^\psi}{dt} = -\nabla_r H^\psi = -\nabla_r(V + Q^\psi) \tag{3.2}$$

The first equation in (3.2) is simply the guidance condition $\mathbf{p} = \nabla S$ written in an unusual form. The second equation is the generalization of Newton's equation of motion where the classical potential is supplemented with the quantum potential and this ensures that momentum is always conserved. Now write

$$S(\mathbf{r}^\psi(t), t) = S_0(x_0) + \int_0^t (\mathbf{p} \cdot d\mathbf{r} - H^\psi dt') \tag{3.3}$$

Then one can easily show that

$$\psi(\mathbf{r}^\psi(t), t)|d^n \mathbf{r}^\psi|^{1/2} = exp\left[\frac{i}{\hbar} S((\mathbf{r}^\psi(t), t)\right] \psi_0(x_0)|d^n x_0|^{1/2} \tag{3.4}$$

where $\psi(\mathbf{r}^\psi(t), t)$ is a solution of the SE (cf. de Gosson [**551, 552, 553, 554**] and below for proofs). This demonstrates a key role played by the quantum potential in the relation between the symplectic space and its double cover! In this direction think of the SE as describing the time evolution of the flow in the covering space; then for each initial point of the covering space one can project a distribution of initial points on the phase space lying directly below. As the SE develops a series of trajectories unfold in the underlying manifold determining classical type symplectic flows which are defined by the Bohmian. It is a property of covering spaces that ensures the non-crossing of the underlying trajectories (well known for Bohmian trajectories). Thus the Bohm flows seem to be a necessary structure of the quantum formalism when looked at in terms of the underlying geometry. We note also a clarification of the uncertainty principle when one realizes that in the x-representation the momentum used in the Bohm approach is not the eigenvalue of the momentum operator; and it is the eigenvalues which are measured quantities!

Note that the notation is sometimes noncanonical (in the spirit of [**65**]) but seems to be consistent and correct. Thus (from [**551, 552, 553, 554**]) we sketch some background for the symplectic theory. One recalls classical Hamiltonian theory based on $\dot{\mathbf{r}} = \nabla_p H(\mathbf{r}, \mathbf{p}, t)$ $\dot{\mathbf{p}} = -\nabla_r H(\mathbf{r}, \mathbf{p}, t)$, and e.g. $H = (1/2m)(\mathbf{p} - \mathbf{A}(\mathbf{r}, t))^2 + U(\mathbf{r}, t)$. Further there are gauge transformations $\mathbf{A}' = \mathbf{A} + \nabla_r \chi$ and

$U' = U - \partial_t\chi$ with $H'(\mathbf{r},\mathbf{p},t) = H(\mathbf{r},\mathbf{p}-\nabla_r\chi,t) - \chi_t$. One writes now $z \sim (\mathbf{r},\mathbf{p})$ with $\dot{z} = X_H(z,t)$ where $X_H = (\nabla_p H, -\nabla_r H)$ is a Hamiltonian vector field. Set $f_{t,t'} : z' = (\mathbf{r}',\mathbf{p}') \to z_t = (\mathbf{r}_t,\mathbf{p}_t)$ and there are Chapman-Kolmogorov equations

$$f_{t,t'} \circ f_{t',t''} = f_{t,t''};\ (f_{t,t'})^{-1} = f_{t',t};\ f_{t,t} = I \tag{3.5}$$

Generally on a real vector space E a map $\Omega:\ E \times E \to \mathbf{R}$ is symplectic if it is bilinear, antisymmetric, and nondegenerate. The basic example here is for $E = \mathbf{R}^3_r \times \mathbf{R}^3_p$ (phase space) with $\Omega(z,z') = \mathbf{p}\cdot\mathbf{r}' - \mathbf{p}'\cdot\mathbf{r}$. Note Ω here can be identified with $d\mathbf{p} \wedge d\mathbf{r} = dp_x \wedge dx + dp_y \wedge dy + dp_z \wedge dz$ where e.g. $dp_x \wedge dx)(\mathbf{r},\mathbf{p},\mathbf{r}',\mathbf{p}') = p_x x' - p'_x x$. Then one has

$$\Omega(z,z') = {}^T\!z' J z;\ J = \begin{pmatrix} 0 & I \\ -I & 0 \end{pmatrix};\ X_H = J\nabla_z H;\ \dot{z} = J\nabla_z H(z,t) \tag{3.6}$$

Further one has

$$\Omega(X_H(z,t),z') = z'\cdot\nabla_z H(z,t) \equiv i_{X_H}\Omega + dH = 0; \tag{3.7}$$

$$i_{X_H}\Omega(z)(z') = \Omega(X_H(z),z')$$

Note also that symplectic matrices s are defined via $\Omega(sz,sz') = \Omega(z,z')$ or $sJs^T = s^TJs = J$. Next one looks at the HJ equation $\partial_t\Phi + H(\mathbf{r},\nabla_r\Phi,t) = 0$ via the action $\mathfrak{A}(\Gamma) = \int_\Gamma \mathbf{p}\cdot d\mathbf{r} - Hdt$ where Γ is an arc of curve $s \to f_{s,t'}(z')$ for $t' \leq s \leq t$ joining z' to $z = f_{t,t'}(z')$ (cf. [**65**]). For small $t - t'$, Γ will project diffeomorphically onto a curve $\gamma \subset \mathbf{R}^3_r \times \mathbf{R}_t$ with $A(\gamma) = \int_\gamma \mathbf{p}\cdot d\mathbf{r} - Hdt = W(\mathbf{r},\mathbf{r}',t,t')$ and $dW(\mathbf{r},t) = \mathbf{p}d\dot{\mathbf{r}} - H(\mathbf{r},\mathbf{p},t)dt$ (see [**65, 380, 551, 857**] for proofs). One can then look at Cauchy problems $\Phi_t + H(\mathbf{r},\nabla_r\Phi,t) = 0$ with solutions $\Phi(\mathbf{r},t) = \Phi_0(\mathbf{r}_0) + W(\mathbf{r},\mathbf{r}_0,t,0)$.

Now the symplectic group $Sp(3)$ has a double cover $Sp_2(3)$ and the metaplectic group $Mp(3)$ is generated by generalized Fourier transforms

$$S_W\psi(\mathbf{r}) = \left(\frac{1}{2\pi i\hbar}\right)^{3/2}\sqrt{Hess(-W)}\int e^{(i/\hbar)W(\mathbf{r},\mathbf{r}')}\psi(\mathbf{r};)d^3\mathbf{r}' \tag{3.8}$$

(cf. (3.23) below) associated to quadratic forms W for which $Hess(-W) \neq 0$ (see below). $Mp(3)$ is a unitary representation for $Sp_2(3)$ and, by the path lifting property for coverings, the one parameter subgroup (f_t) of $Sp(3)$ can be lifted to a unique 1-parameter subgroup of its coverings and hence in particular to $Mp(3)$. This lift is in fact the quantum evolution group U_t. This means that the classical Hamiltonian flow (f_t) and the quantum evolution group U_t are identical (but of course with different physical meanings). However this only works for quadratic Hamiltonians and will be discussed below.

For the Bohmian theory one can go to Dürr et al [**415**] (cf. also [**254, 234, 235**]) where for a given wave function ψ satisfying the SE there is a trajectory velocity

$$\mathbf{v}^\psi = \dot{\mathbf{r}}^\psi = \frac{\hbar}{m}\Im\frac{\nabla_r\psi(\mathbf{r}^\psi,t)}{\psi(\mathbf{r}^\psi,t)} \tag{3.9}$$

The Bohmian form, where $\psi = Rexp(iS/\hbar)$ leads to the HJ equation ($S \sim \Phi$)

$$\partial_t\Phi + \frac{1}{2m}(\nabla_r\Phi)^2 + U + Q^\psi = 0;\ Q^\psi = -\frac{\hbar^2}{2m}\frac{\nabla^2 R}{R} \tag{3.10}$$

Now for $H^\psi = H + Q^\psi$ one can represent Φ as the unique solution of

$$\partial_t\Phi + H^\psi(\mathbf{r}, \nabla_r\Phi, t) = 0;\ \Phi(\mathbf{r}, 0) = \Phi_0(\mathbf{r}) \tag{3.11}$$

and one has a new Hamiltonian theory

$$\dot{\mathbf{r}}^\psi = \nabla_p H^\psi(\mathbf{r}^\psi, \mathbf{p}^\psi, t);\ \dot{\mathbf{p}}^\psi = -\nabla_r H^\psi(\mathbf{r}^\psi, \mathbf{p}^\psi, t) \tag{3.12}$$

and the initial condition involves $\mathbf{r}^\psi(0) = \mathbf{r}_0$, $\mathbf{p}^\psi(0) = \nabla_r\Phi_0(\mathbf{r})$. Via $\mathbf{p}^\psi = \nabla_r\Phi$ this can be expressed via

$$\dot{\mathbf{r}}^\psi = \nabla_p H^\psi(\mathbf{r}^\psi\Phi(\mathbf{r}^\psi(t), t) \tag{3.13}$$

and evidently $\mathbf{p}^\psi = \hbar\Im(\nabla_r\psi/\psi)$ is identical with (3.8). In general the Bohmian trajectories may differ from the classical trajectories.

In $\mathbf{R}^{2n} = \mathbf{R}^n_x \times \mathbf{R}^n_p$ one writes coordinates $z = (x,p)$ with $\Omega(x,p;x',p') = p\cdot x' - p'\cdot x$ as before (but now without vector notation). Thus $\Omega(z,z')$ is the value on (z,z') of $dp\wedge dx = dp_1\wedge dx_1 + \cdots + dp_n \wedge dx_n$ and

$$\Omega(x,p;x',p') = -\sum_1^n \begin{vmatrix} x_j & x'_j \\ p_j & p'_j \end{vmatrix} \tag{3.14}$$

Consider now $s_{t,t'}(z') = f'_{t,t'}(z')$ where f' denotes the Jacobian matrix so

$$s_{t,t'}(z') = \begin{pmatrix} \partial x/\partial x' & \partial x/\partial p' \\ \partial p/\partial x' & \partial p/\partial p' \end{pmatrix} \tag{3.15}$$

One shows that $s_{t,t'}$ is symplectic (i.e. belongs to $Sp(n)$) and satisfies

$$\dot{s}(t) = JH''_{x,p}(s(t),t)s(t);\ H''_{x,p} = \left(\frac{\partial^2 H}{\partial x_i \partial p_j}\right) \tag{3.16}$$

Now for a few facts from deGosson [**551, 552, 553, 554**]. The complex unitary group $U(n,\mathbf{C})$ and its real subgroup $O(n,\mathbf{R})$ are subgroups of $Sp(n)$. Here

$$r = \begin{pmatrix} A & -B \\ B & C \end{pmatrix}\ (symplectic) \sim R = A + iB\ (unitary) \tag{3.17}$$

In fact $Sp(n)\cap O(2n,\mathbf{R}) = U(n)$ and the Lie algebra $\mathfrak{sp}(n)$ consists of matrices

$$X = \begin{pmatrix} \alpha & \beta \\ \gamma & -\alpha^T \end{pmatrix};\ \beta = \beta^T;\ \gamma = \gamma^T \tag{3.18}$$

There is a 1-1 correspondence between quadratic polynomials in (x,p) and elements of $\mathfrak{sp}(n)$. Indeed write

$$H = \frac{1}{2}\alpha p^2 + \beta x\cdot p + \frac{1}{2}\gamma x^2;\ \dot{x} = \beta x + \alpha p;\ \dot{p} = -\gamma x - \beta^T p;\ z(t) = e^{tX}z(0) \tag{3.19}$$

Note that one parameter subgroups of $Sp(n)$ do not cover $Sp(n)$; one can show that $Sp(n) \simeq U(n)\times\mathbf{R}^{n(n+1)/2}$ and dim $Sp(n)$ = n(2n+1).

REMARK 9.3.1. We will omit discussion of the uncertainty principle in classical mechanics and Gromov's symplectic camel theorem. ■

One defines a symplectomorphism f on a subset of phase space to be free if given (x', x) the equation $(x, p) = f(x', p')$ uniquely determines (p, p'). This will be the case if and only if

$$det\frac{\partial(x,x')}{\partial(p',x')} = det\begin{pmatrix} \partial x/\partial p' & \partial x/\partial x' \\ 0 & I \end{pmatrix} = det\left(\frac{\partial x}{\partial p'}\right) \neq 0 \tag{3.20}$$

REMARK 9.3.2. We also omit background discussion of Lagrangian manifolds, Maslov quantization, and Maslov-Leray indices. ■

In Chapter 5 of de Gosson [**551**] (first book) one theme is that the time evolution of the QM wave function is essentially the Bohmian motion of half densities. We will omit technical details involving Lagrangian manifolds, caustics, etc. Now a density is defined as a function ρ satisfying (cf. (2.1)) $\partial_t\rho + \nabla_x(\rho v) = 0$ and a half density is the square root of a density. With suitable smoothness one will have $\rho(x(t),t)dx(t) = \rho(x,0)dx$ (showing that ρdx is constant in time - or that total mass is conserved under a flow $x(t)$). For the SE with $\psi = Rexp(i\Phi/\hbar)$ one has (2.1) so (**3B**) $R^2(x(t,t)dx(t) = R^2(x,0)dx$. Writing $H^\psi = H + Q^\psi$ again one has

$$\Phi(x(t),t) = \Phi(x,0) + \int_{(x,0)}^{(x(t),t)} pdx - H^\psi dt;\ (x(t),p(t)) = f^\psi_{t,0}(x,p) \tag{3.21}$$

(here f^ψ is the flow determined by H^ψ). Multiply now (**3B**) by the function $exp[i\Phi(x(t),t)/\hbar]$ to get (motion of wave form)

$$\psi(x(t),t)\sqrt{dx(t)} = exp\left(\frac{i}{\hbar}\int_{(x.0)}^{(x(t),t)} pdx - H^\psi dt\right)\psi(x,0)\sqrt{dx} \tag{3.22}$$

(here $\sqrt{dx}$ is defined via the Maslov index - see below). Equation (3.22) shows that the evolution of ψ is unitary ($\int|\psi(x(t),t)|^2dx(t) = \int|\psi(x,0)|^2dx$).

In Chapter 6 of [**551**] (first book) one looks at the double cover of $Sp(n)$ realized as a group of unitary operators in $L^2(\mathbf{R}^n)$ (and identified with the metaplectic group $Mp(n)$). Here $Sp_2(n)$ is the only covering of $Sp(n)$ that can be represented as a group of unitary operators acting on $L^2(\mathbf{R}^n_x)$. This realization $Mp(n)$ is generated by objects (cf. (3.7))

$$F\psi(x) = \left(\frac{1}{2\pi i\epsilon}\right)^{n/2}\int e^{-(i/\epsilon)x\cdot x'}\psi(x')dx' \tag{3.23}$$

where $x\cdot x'$ is replaced by non-degenerate quadratic forms which are generating functions of free linear symplectomorphisms ($\epsilon \sim \hbar$). This representation will be reducible but the subrepresentations on L^2_{odd} and L^2_{even} are irreducible. Symplectic matrices are involved here where

(3.24)

$$s = \begin{pmatrix} A & B \\ C & D \end{pmatrix};\ A^TC,\ D^TB,\ AB^T,\ DC^T,\ AC^T,\ DB^T\ \textit{are symmetric};$$

$$A^TD - C^TB = I;\ DA^T - CB^T = I;\ AD^T - BC^T = I$$

Notations here involve $Px^2 = x^T Px$ for symmetric $n \times n$ matrices P and $Lx \cdot x' = {}^T x' Lx$ for any $n \times n$ matrix L. A symplectic s as above is free when $det(B) \neq 0$. The set of free symplectic matrices $Sp_0(n)$ is a submanifold of $Sp(n)$ with dimension $(n+1)(2n-1)$ (codimension 1 in $Sp(n)$ with measure 0) and a generator for s as in (3.24) is

$$W(x, x') = \frac{1}{2} DB^{-1}s^2 - B^{-1}x \cdot x' + \frac{1}{2} B^{-1}A(x')^2 \tag{3.25}$$

Conversely if (**3C**) $W(x, x') = (1/2)Px^2 - Lx \cdot x' + (1/2)Q(x')^2$ with $P = P^T$, $Q = Q^T$, and $det(L) \neq 0$ then

$$S_W = \begin{pmatrix} L^{-1}Q & L^{-1} \\ PL^{-1}Q - L^T & PL^{-1} \end{pmatrix} \tag{3.26}$$

is a free symplectic matrix with generating function (**6C**) (note for $(x, p) = s(x', p')$ one has $W(x, x') = (1/2)(p \cdot x - p' \cdot x')$).

Now one writes $ISp(n)$ for the inhomogeneous symplectic group where $\tau(z_0) \circ s = s \circ \tau(s^{-1}z_0)$ ($\tau : z \to z_0$) and $ISp(n)$ consists of matrices

$$< s, z_0 >= \begin{pmatrix} s & z_0 \\ 0_{1\times 2n} & 1 \end{pmatrix} \tag{3.27}$$

Then an affine symplectomorphism $< s, z_0 >$ is free if and only if s is free and a free generator of $f = \tau(z_0) \circ s_W$ ($z_0 = (x_0, p_0)$) is

$$W_{z_0}(x, x') = W(x - x_0, x') + p_0 \cdot x \tag{3.28}$$

where W is a free generator for s. Various facts about all this can be found in [**551**]. Now let s_W be a free symplectomorphism with $W = (P, L, Q)$ and recall $Hess_{x,x'}(-W) = det(L) \neq 0$. One defines two quadratic Fourier transforms for $\psi \in \mathcal{S}(\mathbf{R}^n_x)$ (where $\mathcal{S}$ is the standard Schwartz space). Thus

$$S_{W,m}\psi(x) = \left(\frac{1}{2\pi i}\right)^{n/2} \Delta(W) \int_{\mathbf{R}^n_x} e^{iW(x,x')}\psi(x')d^n x' \tag{3.29}$$

where $\Delta(W) = i^m\sqrt{|det(L)|}$, $m = arg(Hess_{x,x'}(-W))$ ($arg(det(L)) = m\pi \; mod(2\pi)$, and m is even (resp. odd) if $det(L) > 0$ $(resp. < 0)$. Then $m = m(S_{W,m})$ is called a Maslov index of W and exactly 2 indices mod 4 are associated to each W, namely m and $m + 2$. One can show that $Mp(n)$ is the set of all products $S = S_{W_1,m_1} \cdots S_{W_k,m_k}$ and it is a connected Lie group with a group isomorphism $\pi : Mp(n) \to SP(n)$ whose kernel is $\pm I$. Actually $Mp(n)$ can be generated by operators $M_{L,m}$ and V_P (with J) where

$$M_{L,m}\psi(x) = i^m\sqrt{|det(L)|}\psi(Lx); \; V_P\psi(x) = e^{(i/2)Px^2}\psi(x) \tag{3.30}$$

where $arg(det(L)) = m\pi mod(2\pi)$ and P is a symmetric $n \times n$ matrix. It is shown that if $W = (P, L, Q)$ and $arg\, det(L) = m\pi \; mod(2\pi)$ then $S_{W,m} = V_P M_{L,m} F V_Q$. Further every $S \in Mp(n)$ is the product of two quadratic transforms $S_{W,m}$ and $S_{W',m'}$ (note also that every symplectic matrix is the product of two free symplectic matrices).

Now $Sp(n)$ arises from symplectic matrices and thre is a larger group $Symp(n)$

of symplectomorphisms (whose Jacobian matrices are symplectic). If f and g are symplectomorphisms then $(f\circ g)'(z) = f'(g(z))g'(z)$ so $(f\circ g)'$ is a symplectic matrix and this gives a space $Symp(n)\subset diff(n)$ of symplectomorphisms on phase space. Further if $t\to f_t$ is a continuous path in $Symp(n)$ with $f_0 = Id$ then there exists a function $H(z,t)$ such that (f_t) is the time dependent flow of the Hamiltonian vector field $X_H = (\nabla_p H, -\nabla_x H)$. Thus $(d/dt)f_t(z) = X_t f_t(z)$ and $i_{X_t}\Omega = -dH_t$ where Ω is the standard symplectic form.

We treat some special situations following deGosson [**551**].

EXAMPLE 3.1. Take a free particle in $\mathbf{R}^3_r$ to which is associated a plane wave $\Theta(t,\mathbf{r}) = \mathbf{k}\cdot\mathbf{r} - \omega(\mathbf{k})t + c$ where $\mathbf{k} = m\mathbf{v}/\hbar$ and $\omega(\mathbf{k}) = mc^2/\hbar$ (following deBroglie). Noting that $mc^2 = m_0c^2 + (1/2)m_0v^2 + O(v^4/c^2)$ one writes

$$\Theta(\mathbf{r},t) = \frac{1}{\hbar}\Phi(\mathbf{r},t) - \frac{m_0c^2}{\hbar}t + O(v^4/c^2);\ \Phi = \mathbf{p}_0\cdot\mathbf{r} - \frac{p_0^2 t}{2m} + c\hbar \tag{3.31}$$

(here $\mathbf{p}_0 = m_0\mathbf{v}$ and $p_0 = |\mathbf{p}_0|^2$). For small v one approximates by $\Theta' = (1/\hbar)\Phi - (m_0c^2/\hbar)t + c$ and dropping $m_0c^2t/\hbar$ (which affects neither the phase nor the group velocity) the phase can be defined as $\Theta = (1/\hbar)(\Phi + c)$. Fix c by requiring $\Theta = 0 \sim \mathbf{p}\cdot\mathbf{r} = \mathbf{p}_0\cdot\mathbf{r}_0$ to get

$$\Phi = \mathbf{p}_0\cdot(\mathbf{r}-\mathbf{r}_0) - \frac{p_0^2}{2m}(t-t_0) \tag{3.32}$$

Then $\partial_t\Phi + (1/2m)(\nabla_r\Phi)^2 = 0$ with $\Phi(\mathbf{r},t_0) = \mathbf{p}_0\cdot(\mathbf{r}-\mathbf{r}_0)$. Replacing $(\mathbf{r}_0,\mathbf{p}_0,t_0)$ by $(\mathbf{r}',\mathbf{p}',t')$ one defines a propagator via

$$G(\mathbf{r},\mathbf{r}';t,t') = \frac{1}{(2\pi\hbar)^3}\int e^{(i/\hbar)\Phi_{p'}(\mathbf{r},\mathbf{r}';t,t')}d^3p' \tag{3.33}$$

This is a Fresnel integral and one has

$$i\hbar\partial_t G = -\frac{\hbar^2}{2m}\nabla_r^2 G;\ lim_{t\to t'}G = \delta(\mathbf{r}-\mathbf{r}') \tag{3.34}$$

and for $\psi'\in\mathcal{S}(\mathbf{R}^3_r)$

$$\psi(\mathbf{r},t) = \int G(\mathbf{r},\mathbf{r}';t,t')\psi'(\mathbf{r}')d^3\mathbf{r}'; \tag{3.35}$$

$$i\hbar\partial_t\psi = -\frac{\hbar^2}{2m}\nabla_r^2\psi;\ lim_{t\to t'}\Psi(\mathbf{r},t) = \psi'(\mathbf{r})$$

$$G = \left(\frac{m}{2\pi i\hbar(t-t')}\right)^{3/2} exp\left(\frac{i}{\hbar}W_f(\mathbf{r},\mathbf{r}';t,t')\right);\ W_f = m\frac{(\mathbf{r}-\mathbf{r}')^2}{2(t-t')} \tag{3.36}$$

(W_f is the free particle generating function). This can be related to the metaplectic representation as follows. Consider $H = p^2/2m,\ p = \mathbf{p}^2$ on $\mathbf{R}^3_r\times\mathbf{R}^3_p$. The flow here involves

$$s_{t,t'} = \begin{pmatrix} I & \frac{t-t'}{m}I \\ 0 & I\end{pmatrix} \tag{3.37}$$

with W_f as above. Let $\Pi^\hbar : M^\hbar p(3) \to Sp(3)$ be the covering map which to every quadratic Fourier transform

$$(3.38)\qquad S^\hbar_{W,m}\psi(\mathbf{r}) = \left(\frac{1}{2\pi i\hbar}\right)^{3/2}\Delta(W)\int e^{(i/\hbar)W(\mathbf{r},\mathbf{r}';t,t')}\psi(\mathbf{r}')d^3\mathbf{r}'$$

associates the free symplectic matrix s_W. Let $\pm S^\hbar_{t,t'}$ be the two quadratic Fourier transforms with projections $\Pi^\hbar(\pm S^\hbar_{t,t'}) = s_{t,t'}$ given via

$$(3.39)\qquad S^\hbar_{t,t'}\psi'(\mathbf{r},t) = \pm\left(\frac{m}{2\pi i\hbar(t-t')}\right)^{3/2}\int e^{(i/\hbar)W(\mathbf{r},\mathbf{r}';t,t')}\psi'(\mathbf{r}')d^3\mathbf{r}'$$

where the $\pm$ factor is determined by $sgn(t-t')$ (see [**551, 552**] for details). Then (for the + sign in (**??**)), writing $\psi(\mathbf{r},t) = S^\hbar_{t,t'}\psi'(\mathbf{r})$ it follows that $i\hbar\partial_t\psi = -(\hbar^2/2m)\nabla^2_r\psi$ with $\psi(\mathbf{r},t)\to\psi'(\mathbf{r})$ as $t\to t'$. ■

EXAMPLE 3.2. Suppose (dropping some vector notation)

$$(3.40)\qquad H = \frac{1}{2m}(p-Ax)^2 + \frac{1}{2}Kx^2 + a\cdot x$$

where A, K are $n\times n$ matrices with K symmetric. More generally one could assume H is time independent (not essential) and of Maxwell type, $H = \sum_1^n (1/2m_j)(p_j - A_j(x,t))^2 + U(x,t)$. Then the flow determined by H is a one parameter subgroup $s_t \in Sp(n)$ and for t sufficiently small s_t will be a free symplectic matrix. Set $W(t) = W(x,x';t,0)$ with $s_t = s_{W(t)}$ and associate this to two elements $\pm S^\epsilon_{W(t),m(t)}$ of $M^\epsilon p(n)$ (this essentially means $S_{W,m} \sim S^\epsilon_{W,m}$ as in (3.38) with $\epsilon\sim\hbar$ - see [**551**], first book, Section 6.4.2). Let S_t be a choice of $S_{W(t),m(t)}$ and $\psi(x,t) = S_t\psi_0(x)$ is then a solution of the SE (for small t); the formulas are (cf. (**??**)-(**??**))

$$(3.41)\qquad \psi = \left(\frac{1}{2\pi i\hbar}\right)^{n/2}\Delta(W)\int e^{(i/\hbar)W(x,x';t)}\psi_0(x')d^nx'$$

Actually $\psi = S_t\psi_0$ leads to a solution for all t by using $\psi(x,t) = (S_{t/N})^N\psi_0$ where N is chosen large enough so that $s_{t/N}$ is a free symplectic matrix. ■

The situation of Example 6.2 is developed further in [**520**] using the vanVleck determinant. Thus take a time dependent Maxwell Hamiltonian as in Example 4.2 with associated Hamiltonian equations

$$(3.42)\qquad \dot{x}_j = \frac{1}{m_j}(p_j - A_j);\ \dot{p}_j = \frac{1}{m_j}(p_j - A_j)\frac{\partial A_j}{\partial x_j} - \frac{\partial U}{\partial x_j}$$

leading to

$$(3.43)\qquad \ddot{x}_j + \frac{1}{m_j}\left(\frac{\partial U}{\partial t}(x,t) + \frac{\partial A_j}{\partial t}(x,t)\right) = 0$$

For $|t-t'|$ small enough there is a unique trajectory in configuration space joining x', x in a time $t-t'$ with $\mathbf{v}$ also determined (the initial data are $x(t') = x'$ and

$\mathbf{v}(t') = \mathbf{v}'$). The question is then posed about how much Δx is produced from initial $\Delta p'$. If one takes $\Delta x = (t-t')v'_x$ and $\Delta y = (t-t')v'_y$ then

$$\det\frac{\partial(v'_x, v'_y)}{\partial(x,y)} = (t-t')^{-2} \tag{3.44}$$

measures the variation of the number of trajectories (or "density" of trajectories). This leads one to the vanVleck determinant

$$det\frac{\Delta p'}{\Delta x} = det\left(\frac{\Delta p'_j}{\Delta x_j}\right)_{1\leq i,j\leq n};\ lim_{\Delta x\to 0}det\frac{\Delta p'}{\Delta x} = det\frac{\partial p'}{\partial x} = \rho(x,x';t,t') \tag{3.45}$$

Note ρ can be negative but the sign can be adjusted via a Maslov index and for $t-t'$ small (or whenever $f_{t,t'}$ is free) one can write

$$\rho(x,x';t,t') = Hess_{x,x'}(-W) \tag{3.46}$$

To see this imagine $det(\partial x/\partial p') \neq 0$ and since $p = \nabla_x W$ with $p' = -\nabla_{x'}W$ we have

$$p'_i = -\partial_{x'_i}W(x,x';t,t') \Rightarrow \frac{\partial p'}{\partial x} = \left(-\frac{\partial^2 W}{\partial x'_i \partial x_j}\right)_{1\leq i,j\leq n} \tag{3.47}$$

If H is a Maxwell Hamiltonian quadratic in the position and momentum variables (so $W = (1/2)Px^2 - Lx\cdot x' + Q(x')^2 + \alpha\cdot x + \alpha'\cdot x'$) then $\rho(t,t') = det(L(t,t'))$ and (in general)

$$\partial_t\rho + div(\rho v) = 0 \tag{3.48}$$

Referring back to (3.36) for the free particle one can show that there is an analogue for the general Maxwell Hamiltonian, namely

$$G^{sh} = \left(\frac{1}{2\pi i\hbar}\right)^{n/2}\sqrt{\rho}e^{(i/\hbar)W} \tag{3.49}$$

(short time propagator). Moreover

$$i\hbar\partial_t G^{sh} = (\hat{H}+Q)G^{sh};\ Q = -\frac{\hbar^2}{2m}\frac{\nabla_x^2\sqrt{|\rho|}|}{\sqrt{|\rho|}} \tag{3.50}$$

and this involves a quantum potential type object in yet another fascinating manner! Equation (3.49) leads then to a solution of the SE

$$\psi(x,t) = \int G^{sh}(x,x';t,t')\psi'(x')d^n x';\ i\hbar\partial_t\psi = (\hat{H}+\hat{Q})\psi \tag{3.51}$$

$$\hat{Q}\psi(x,t) = \int Q(x,x';t,t')G^{sh}(x,x';t,t')\psi'(x')d^n x' \tag{3.52}$$

For quadratic Hamiltonians $H = (1/2m)(p-Ax)^2 + (1/2)Kx^2 + \alpha\cdot x$ one has $Qexp[(i/\hbar)W] = 0$. We note that this is exactly what one expects using a Bohmian H^ψ in a "classical" symplectic theory.

4. ELABORATION

We sketch here a number of topics following [**14, 93, 211, 239, 303, 476, 552, 553, 554, 582, 862, 890, 1049, 1077, 1238, 1258, 1264, 1331, 1332**]. There are several questions which need elucidation, in particular one can ask

(1) What connection is there between the van Vleck determinant, the quantum potential, and space time (or quantum) geometry? In particular there is a natural ρ in phase space connected to hydrodynamics for example and this is related to the quantum potential so what would be the meaning of a quantum potential term in phase space?

The first question is partially confused and partially covered already in de Gosson [**552**] (first paper - cf. also [**637**]). Thus Bohmian mechanics can be completely characterized as the theory of half densities on Lagrangian manifolds. The underlying idea is that it is the square roots of densities which intervene in QM (going back to van Vleck [**1258**]). It turns out that the inclusion of half densities in Bohmian mechanics (BM) leads to a QM in phase space totally different from the traditional Wigner-Weyl-Moyal formalism (cf. [**239, 1331, 1332**]). In fact if one transports an initial half density $\psi_0(q)|d^nq|^{1/2}$ using the Hamiltonian flow associated to the Bohmian $H + U^\psi$ (we use U^ψ for the quantum potential here instead of Q since Q will be needed for a coordinate) then, up to a phase factor κ, one obtains the half density $\psi(Q,t)|d^nQ|^{1/2}$, where $(q,0) \to (Q,t)$ is the flow and $\psi(t,\cdot)$ is the solution to the SE associated with H. The phase factor is $\kappa = exp[(i/\hbar)\int pdq - (H + U^\psi)dt]$ where the action integral is calculated along the Bohmian trajectory in phase space starting from $(q, \nabla_q S_0(q))$ at $t = 0$ and arriving at $(Q, \nabla_Q S_0(Q,t))$ at time t. The Bohmian trajectory is determined by the usual Hamiltonian equations associated however to $H + U^\psi$ (rather than H). If one neglects U^ψ the classical trajectories are given by the usual formulas of the semiclassical approximation.

Recall, for a polynomial Hamiltonian at most quadratic in q,p, that a flow (f_t) consists of linear symplectic transforms and can be lifted to a group F_t of unitary operators acting on $L^2(\mathbf{R}^n)$ (a subgroup of the metaplectic group $Mp(n)$ which is a two fold covering of the symplectic group $Sp(n)$). Then $i\hbar\partial_t F_t = \hat{H}F_t$ and one gives first here a direct derivation of the SE for Hamiltonians of the form $H(q,p,t) = (1/2)p^2 + (1/2)R\cdot q$ where R is a real symmetric $n \times n$ matrix ($p^2 = \sum p_i^2$, $m = 1$). One knows first

$$\text{(4.1)} \quad \dot{q} = p;\ \dot{p} = -Rq;\ f_t = exp\left[t\begin{pmatrix} 0 & tI \\ -R & 0 \end{pmatrix}\right] = \begin{pmatrix} 0 & tI \\ -tR & 0 \end{pmatrix} + O(t^2)$$

Let S be the two point free generating function (cf. [**65**]) determined via

$$\text{(4.2)} \quad (Q,P) = f_t(q,p) \iff \begin{cases} P = \partial_Q S(Q,q,t) \\ p = -\partial_q S(Q,q,t) \end{cases};\ S(Q,q,t) = \int_0^t pdq - Hdt'$$

$(Q = q(t))$. It follows from this that $\partial_t S + H(Q, \nabla_Q S) = 0$ and

$$\text{(4.3)} \qquad J(Q,q,t) = det\frac{\partial^2 S}{\partial Q \partial q};\ \partial_t J + \nabla_Q \cdot J\nabla_p H = 0$$

(cf. [**862**]). This property holds in fact for the free generating function determined by an arbitrary Hamiltonian. However in the present case (quadratic Hamiltonian) the f_t are linear and the generating function S is itself quadratic in (Q, q) so that J does not contain the variables (q, Q). One has then $\partial_t J + J\nabla_Q^2 S = 0$, noting that $\nabla_p H = \nabla_Q S$ via (3.1) - (3.2). Let now ψ_0 be an initial wave function (in the Schwartz space $\mathcal{S}$ for convenience only). Then a solution to the SE $i\hbar\partial_t\psi = H(q, -i\hbar\nabla_q)\psi$ is given for $0 < |t| < T$ by

$$\psi(q,t) = (2\pi i\hbar)^{-n/2}a(t)\int exp\left(\frac{i}{\hbar}S(q,q',t)\right)\psi_0(q')d^nq' \tag{4.4}$$

where $a(t)$ is a conveniently chosen square root of $J(t)$ (actually the square root can be chosen so that $lim_{t\to 0}\psi(q,t) = \psi_0(q)$). To see this note

$$i\hbar\frac{\partial\psi}{\partial t} = (2\pi i\hbar)^{-n/2}\int e^{(i/\hbar)S}\left[H(q, -i\hbar\nabla_q a) + i\hbar\partial_t a\right]\psi_0(q')d^nq'; \tag{4.5}$$

$$H(q, -i\hbar\nabla_q)\psi = (2\pi i\hbar)^{-n/2}\int e^{iS/\hbar}\left[H(q, -i\hbar\nabla_q a) + \frac{1}{2}a\nabla_q^2 S\right]\psi_0(q')d^nq'$$

Now $a = \sqrt{J}$ satisfies the equation $\partial_t a + (1/2)\nabla_q^2 S \cdot a = 0$ via (3.3) and hence ψ satisfies the SE as in (5.4). There are well known procedures for determining the correct sign of the square root in defining a (e.g. Maslov indices, etc.) and one can write ($S = S(q, q', t)$ and A, B, C, D are functions of t)

$$f_t = \begin{pmatrix} A & B \\ C & D \end{pmatrix};\ S = (1/2)DB^{-1}q\cdot q - B^{-1}q\cdot q' + (1/2)B^{-1}Aq'\cdot q \tag{4.6}$$

$$\psi(q,t) = (2\pi i\hbar)^{-n/2}i^{m(t)}|det(B(t))|^{-1/2}\int e^{(i\hbar)S(q,q',t)}\psi_0(q')dq'$$

where $m(t)$ is the inertia (number of less than zero eigenvalues of $B(t)$); this recovers a formula of Robbin-Salmon [**1075**].

Now one goes to moving half densities by the Bohmian flow. Let $H = (1/2)p^2 + U(q, t)$ be a more general Hamiltonian (U a suitable function). Assume that the solution ψ of the SE $i\hbar\partial_t\psi = H(q, -i\hbar\nabla_q)\psi$ exists and is unique for t in some interval $[-T, T]$ and write $\psi = exp(iS/\hbar)\sqrt{\rho(q,t)}$ so that

$$\partial_t\rho + \nabla_q(\rho S_q) = 0;\ \partial_t S + H(q, \nabla_q S) = \frac{\hbar^2}{2}\frac{\nabla_q^2(\sqrt{\rho})}{\sqrt{\rho}} = -U^\psi \tag{4.7}$$

Thus $\partial_t S + H^\psi(q, \nabla_q S) = 0$ where $H^\psi = H + U^\psi$ is the Bohmian. One notes here that it can very well happen that $\hbar \to 0$ while U^ψ becomes infinite (cf. Hiley [**627**]) so $\hbar \to 0$ is not equivalent to $U^\psi \to 0$ (cf. also [**163, 257**]).

(1) Setting $t = 0$ one has $\psi_0(q) = exp[(i/\hbar)S_0(q)]\sqrt{\rho_0(q)}$ and we write L_0 for the graph of the gradient of S_0, i.e. $L_0 = \{(q,p);\ p = \nabla_0 S_0(q);\ q \in D\}$ where D is some domain $D \subset \mathbf{R}_q^n$ (note $L_0 \subset \mathbf{R}_q^n \times \mathbf{R}_p^n$ - phase space).
(2) The canonical symplectic form $\Omega = dp_1 \wedge dq_1 + \cdots + dp_n \wedge dq_n$ vanishes on any pair of tangent vectors to L_0 and the restriction π_0 to L_0 of the projection $\pi : (q,p) \to p$ is a diffeomorphism with (L_0, π_0) a global chart.

(3) Denote now by $(f^{\psi}_{t,t'})$ the time dependent flow associated with the Bohmian H^{ψ}; $f^{\psi}_{t,t'}$ is thus the symplectic transformation of phase space defined by $(q^{\psi}(t), p^{\psi}(t)) = f^{\psi}_{t,t'}(q,p)$ if and only if $\dot{q}^{\psi} = \nabla_p H^{\psi}$, $\dot{p}^{\psi} = -\nabla_q H^{\psi}$, and $q^{\psi}(t') = q$ with $p^{\psi}(t') = p$.
(4) Using the Bohmian flow thus defined we can carry the manifold L_0 in phase space to another manifold L_t and defining $L_t = f^{\psi}_t(L_0)$ with $f^{\psi}_t = f^{\psi}_{t,0}$ we have $f^{\psi}_{t,t'}(L_{t'}) = L_t$.

This is summarized via

PROPOSITION 4.1. The manifold L_t is a Lagrangian submanifold of phase space which projects diffeomorphically on $D_t = f^{\psi}_t(D_0)$. In fact it is the graph $L_t = \{(Q,P);, P = \nabla_Q S(Q,t)\}$ and one has $S(q^{\psi}(t),t) = S_0(q) + \int_0^t pdq - H^{\psi}dt'$ where the integration is over the phase space trajectory from $(q, \nabla_q S_0)$ to $(Q,P) = (q^{\psi}(t), \nabla_q S(q^{\psi}(t),t)$.

The proof follows standard lines with $\Phi(Q,t) = S_0(q) + \int_0^t pdq - H^{\psi}dt'$ and $d\Phi(Q,t) = PdQ - H^{\psi}dt$ which yields $P = \partial_Q \Phi$ and $\partial_t \Phi = -H^{\psi}(Q,P,t)$ leading to $\Phi = S$.

REMARK 9.4.1. Note that L_t is a graph, unlike classical flows which deform graphs; there are no caustics here to worry about the phase $S(q,t)$ is globally defined a priori. ∎

PROPOSITION 4.2. With notation as above one has

$$(4.8)\qquad (q^{\psi}(t),t) = exp\left[\frac{i}{\hbar}\int_0^t pdq - H^{\psi}dt'\right]\left|det\frac{\partial q^{\psi}(t)}{\partial q}\right|^{1/2}\psi_0(q)$$

where $\partial q^{\psi}/\partial q$ is the Jacobian matrix. Denoting by $d^n q$ the Lebesgue measure on $\mathbf{R}^n_q$ and by $|d^n q|^{1/2}$ the associated half density, (5.8) is equivalent to

$$(4.9)\qquad \psi(Q,t)|d^n Q|^{1/2} = exp\left[\frac{i}{\hbar}\int_0^t pdq - H^{\psi}dt'\right]\psi_0(q)|d^n q|^{1/2}$$

Proof. All that remains to be proved is $\rho_0(q) = \rho(q^{\psi}(t),t)|det(\partial q^{\psi}(t)/\partial q)|$ which is an immediate consequence of (4.7). Indeed from fluid dynamics if a function ρ satisfies $\partial_t\rho + \nabla_q(\rho v)$ where $v = v(q,)$ is a velocity field then $\rho(q,0) = \rho(q,t)|det(\partial q(t)/\partial q)|$; thus set $v(q,t) = \partial S(q,t)/\partial q)$. ∎

4.1. PHASE SPACE AND DEFORMATION QUANTIZATION. We go to [**560**] (paper 2) and refer to [**239, 252, 1238, 1331, 1332**] for Wigner-Weyl-Moyal background. In [**1238**] a family of SE of the form

$$(4.10)\qquad i\hbar\partial_t\psi = H\left(\frac{x}{2} + i\hbar\partial_p, \frac{p}{2} - i\hbar\partial_x\right)\psi$$

was proposed (cf. also [**14, 295, 1049**]). It is shown in de Gosson [**552**] that (4.10) is equivalent to the usual SE provided one restricts the set of solutions to a closed subspace of $L^2(\mathbf{R}^2_{x,p})$ and it corresponds to the choice of an irreducible unitary representation of the Heisenberg group. Further one reveals that the theory of Torres-Vega and Frederick [**1238**] is in fact a Doppelgänger of deformation quantization (cf. [**239, 252**]). The idea is to start with an action form

$$\beta_H = \frac{1}{2}(pdx - xdp) - Hdt \tag{4.11}$$

instead of the traditional $\alpha_H = pdx - Hdt$ (note $d\alpha_H = d\beta_H = dp \wedge dx - H_p dp \wedge dt - H_x dx \wedge dt$). Let the canonical symplectic form on the phase space $\mathbf{R}_z = \mathbf{R}_x \times \mathbf{R}_p$ be

$$\sigma(z, z') = px' - p'x \;\; (z = (x, p),\; z' = (x', p'),\; x = (x_1, \cdots, x_n),\; etc. \tag{4.12}$$

The generalized gradients are denoted by ∂_x, ∂_p and one recalls that the Heisenberg group is defined via

$$(z, t) \cdot (z', t') = (z + z', t + t' + (1/2)\sigma(z, z')) \tag{4.13}$$

Recall also on $\mathbf{R}^{2n}$, $T(tz_0)\Psi_0(z) = \Psi_0(z - tz_0)$ and for $\psi_0 \in L^2(\mathbf{R}^n_x)$ one defines Heisenberg-Weyl operators via $(z - tz_0 \to z)$

$$\hat{T}(tz_0)\psi_0(x) = e^{(i/\hbar)\phi(z,t)}T(tz_0)\psi_0(x) = e^{(i/\hbar)(p_0xt - (t^2/2)p_0x_0)}\psi_0(x - tx_0); \tag{4.14}$$

$$\phi(z, t) = p_0xt - \frac{t^2}{2}p_0x_0 = \int_{-t}^{0} pdx - H_{x_0}dt$$

The Schrödinger representation is then $T_{sch} : \mathbf{H}_n \to \mathfrak{U}(L^2(\mathbf{R}^n_x))$ given via

$$T_{sch}(z_0, t_0)\psi_0(x) = e^{(i/\hbar)t_0}\hat{T}(z_0)\psi_0(x) \tag{4.15}$$

and this is a unitary and irreducible representation (in fact the only such). However one can also construct non-trivial irreducible representations in other Hilbert spaces. First note that the standard quantization involves $x_j \to x_j$ and $p_j \to -i\hbar\partial_{x_j}$ and from (4.14) one sees immediately that $\psi(x, t) = \hat{T}(tz_0)\psi_0(x)$ is a solution of

$$i\hbar\partial_t\psi = H_{z_0}(x, -i\hbar\partial_x)\psi;\;\; \psi(x, 0) = \psi_0(x) \tag{4.16}$$

Now let $\hat{T}(tz_0)$ act on $L^2(\mathbf{R}^{2n}_z)$ via

$$\hat{T}_{ph}(tz_0)\Psi_0(z) = e^{(i/\hbar)\phi'(z,t)}T(tz_0)\Psi_0(z); \tag{4.17}$$

$$\phi'(z, t) = -\frac{1}{2}H_{z_0}(z)t = -\frac{1}{2}\sigma(z, z_0)t$$

(using here $\beta_{H_{z_0}} = (1/2)(pdx - xdp) - H_{z_0}dt$). Thus one has defined $\hat{T}_{ph}(tz_0)\Psi_0(z) = exp[-(i/2\hbar)\sigma(z, z_0)t]\Psi_0(z - tz_0)$ and this will satisfy

$$i\hbar\partial_t\Psi = H\left(\frac{x}{2} + i\hbar\partial_p, \frac{p}{2} - i\hbar\partial_x\right)\Psi \tag{4.18}$$

One then proves that $\hat{T}_{ph}(tz_0)$ correspond to a new irreducible unitary representation of $\mathbf{H}_n$ on a closed subspace of $L^2(\mathbf{R}^{2n}_z)$, unitarily equivalent to the Schrödinger representation) and there is a relation to deformation quantization indicated below. Thus in analogy with (4.15) one has

$$\hat{T}_{ph}(z_0, t_0)\Psi_0(z) = e^{(it_0/\hbar)}\hat{T}_{ph}(tz_0)\Psi_0(z); \tag{4.19}$$

$$\hat{T}_{ph}(z_0, t_0)\hat{T}_{ph}(z_1, t_1) = e^{(i/2\hbar)\sigma(z_0,z_1)}\hat{T}_{ph}(z_0 + z_1, t_0 + t_1 + (1/2)\sigma(z_0, z_1))$$

To show equivalence to the Schrödinger representation let $\phi \in \mathcal{S}(\mathbf{R}^n_x)$ be normalized to 1 and consider the operator $L^2(\mathbf{R}^n_x) \to L^2(\mathbf{R}^{2n}_z)$ defined via

$$V_\phi \psi(z) = \left(\frac{\pi\hbar}{2}\right)^{n/2} W(\psi, \bar{\phi})(z/2); \tag{4.20}$$

$$W(\psi, \bar{\phi})(x,p) = \left(\frac{1}{2\pi\hbar}\right)^n \int e^{-(i/\hbar)(p,y)} \psi(x + (y/2))\phi(x - (y/2)) d^n y$$

In fact V_ϕ is an extension of the coherent state representation, to which it reduces (up to a factor $exp(-ipx/\hbar)$) if ϕ is taken to be a Gaussian $\phi_0(x) = (1/\pi\hbar)^{n/4} exp(-|x|^2/2\hbar)$. Indeed $V_\phi \psi(z) = exp(-ipx/2\hbar) U_\phi(z)$ where

$$U_\phi \psi(z) = \left(\frac{1}{2\pi\hbar}\right)^{n/2} \int e^{(i/\hbar)(p, x - x')} \phi(x - x') \psi(x') d^n x' \tag{4.21}$$

One can show that V_ϕ is an isometry and extends to an isometric operator with $V_\phi^* V_\phi = I$; further $P = V_\phi V_\phi^*$ is the orthogonal projection onto the closed range $H_\phi \subset L^2(\mathbf{R}^{2n}_z)$. Further some calculation shows that $\hat{T}_{ph}(z_0) V_\phi = V_\phi \hat{T}_{sch}(z_0)$ and

$$\left(\frac{x}{2} + i\hbar\partial_p\right) V_\phi \psi = V_\phi(x\psi); \; \left(\frac{p}{2} - i\hbar\partial_x\right) V_\phi \psi = V_\phi(-i\hbar\partial_x \psi) \tag{4.22}$$

Hence the transform V_ϕ takes the usual quantization rules to the phase space quantization rules $x \to (x/2) + i\hbar\partial_p$ and $p \to (p/2) - i\hbar\partial_x$.

Finally one writes

$$\tilde{a}(z) = \mathfrak{F}_\sigma a(z) = \left(\frac{1}{2\pi\hbar}\right)^n \int e^{(i/\hbar)\sigma(z,z')} a(z') d^{2n} z'; \tag{4.23}$$

$$\hat{A}_{ph} \Psi(z) = \left(\frac{1}{2\pi\hbar}\right)^n \int \tilde{a}(z_0) \hat{T}_{ph}(z_0) \Psi(z) d^{2n} z_0 =$$

$$= \left(\frac{1}{2\pi\hbar}\right)^n \int e^{-(i/2\hbar)\sigma(z,z_0)} \mathfrak{F}_\sigma a(z_0) \Psi(z - z_0) d^{2n} z_0$$

and the notation $\hat{A} = a^w$ (Weyl operators with symbols a) is used. Then for $\hat{C} = \hat{A} \circ \hat{B}$ one has

$$\tilde{c} = \left(\frac{1}{2\pi\hbar}\right)^n \int e^{-(i/2\hbar)\sigma(z,z')} a(z - z') b(z') d^{2n} z'; \tag{4.24}$$

$$c(z) = \left(\frac{1}{4\pi\hbar}\right)^{2n} \int e^{(i/2\hbar)\sigma(z',z'')} a(z + (z'/2)) b(z - (z''/2)) d^{2n} z' d^{2n} z''$$

Finally one can write

$$c(z) = a(z) exp\left[\frac{i\hbar}{2}(\overleftarrow{\partial_x} \cdot \overrightarrow{\partial_p} - \overleftarrow{\partial_p} \cdot \overrightarrow{\partial_x}\right] b(z); \; \hat{A}_{ph}\Psi = \mathfrak{F}_\sigma(a \bigstar \Psi) \tag{4.25}$$

where $\bigstar$ is the standard Moyal star.

4.2. METAPLECTIC OPERATORS. We go now to de Gosson [**552, 553, 554**] where in particular one looks at the Weyl representation of metaplectic operators. First some standard facts about $Mp(n)$ are as follows (with some repetition). Every $\hat{S} \in Mp(n)$ is the product of two quadratic Fourier transforms (defined on $\mathcal{S}(\mathbf{R}^n)$ via

$$\hat{S}_{W,m}f(X) = \left(\frac{1}{2\pi i}\right)^{n/2} i^m \sqrt{|det(L)|} \int e^{iW(x,x')}f(x')d^n x' \tag{4.26}$$

$$W(x,x') = \frac{1}{2} < Px, x > - < Lx, x' > + \frac{1}{2} < Qx', x' > \tag{4.27}$$

with $P = P^T$, $Q = Q^T$, $det(L) \neq 0$. The integer m (Maslov index) is given via $m\pi \equiv arg(det(L))\ mod(2\pi)$ and there are two choices of m $(mod(2\pi))$, namely m and $m+2$. The projection $\pi : Mp(n) \to Sp(n)$ is entirely specified by the $\pi(\hat{S}_{W,m}) = S_W$ where

$$(x,p) = S_W(x',p') \iff p = \partial_x W(x,x')\ and\ p' = -\partial_{x'}W(x,x') \tag{4.28}$$

Rewriting in terms of P, L, Q one gets $p = Px - L^T x'$ and $p' = Lx - Qx'$ leading to

$$x = L^{-1}(p' + Qx');\ p = (PL^{-1}Q - L^T)x' + PL^{-1}p' \tag{4.29}$$

$$S_W = \begin{pmatrix} L^{-1}Q & L^{-1} \\ PL^{-1}Q - L^T & PL^{-1} \end{pmatrix} \tag{4.30}$$

Note here that if S is a free symplectic matrix

$$S = \begin{pmatrix} A & B \\ C & D \end{pmatrix} \in Sp(n);\ det(B) \neq 0 \tag{4.31}$$

then $S = S_W$ with $P = DB^{-1}$, $L = B^{-1}$, and $Q = B^{-1}A$ (the free symplectic matrices form a dense subset of $Sp(n)$). The inverse $\hat{S}_{W,m}^{-1} = (\hat{S}_{W,m})^* = \hat{S}_{W^*,m^*}$ where $W^*(x,x') = -W(x',x)$ and $m^* = n - m\ (mod(4))$. Recall that the operators $\hat{T}(z_0)$ satisfy the metaplectic covariance relation (**4A**) $\hat{S}\hat{T}(z) = \hat{T}(Sz)\hat{S}$ where $S = \pi(\hat{S})$. Further for every $S \in Sp(n)$ there exists a unitary transformation $\hat{U}$ acting in $L^2(\mathbf{R}^n)$ satisfying (**4A**) (up to a constant factor of modulus one). The Heisenberg-Weyl operators satisfy moreover

$$\hat{T}(z_0)\hat{T}(z_1) = e^{i\sigma(z_0,z_1)}\hat{T}(z_1)\hat{T}(z_0);\ \hat{T}(z_1 + z_1) = e^{-(i/2)\sigma(z_0,z_1)}\hat{T}(z_0)\hat{T}(z_1) \tag{4.32}$$

We recall also the Weyl symbols and operators with

$$a^w f(x) = \left(\frac{1}{2\pi}\right)^n \int\int e^{i(p,x-y)}a((1/2)(x+y),p)f(y)d^n y d^n p \equiv \tag{4.33}$$

$$\equiv a^w = \left(\frac{1}{2\pi}\right)^n \int a_\sigma(z_0)\hat{T}(z_0)d^{2n}z_0$$

where the twisted symbol a_σ is the symplectic Fourier transform $\mathfrak{F}_\sigma a$ given via $(1/2\pi)^n \int exp[-i\sigma(z,z')]a(z')d^{2n}z'$. For the composition $c^w = a^w \circ b^w$ one has (cf.

[**1317**])

$$a *_\sigma b(z) = \int e^{(i/2)\sigma(z,u)} a(z-u)b(u)d^{2n}u \tag{4.34}$$

We recall also a generalized Fresnel formula

$$\left(\frac{1}{2\pi}\right)^{m/2} \int e^{-i<v,u>} e^{(i/2)<Mu,u>} d^m u = \tag{4.35}$$

$$= |det(M)|^{-1/2} e^{(i\pi/4)sgn(M)} e^{-(i/2)<M^{-1}v,v>}$$

(here $sgn(M)$ is the number of positive eigenvalues of M minus the number of negative ones).

We go now to the Mehlig-Wilkinson (M-W) formula (cf. [**869**]). First note that the matrix $M_S = (1/2)J(S+I)(S-I)^{-1}$ is symmetric since $S \in Sp(n) \iff S^TJS = J \iff SJS^T - J$. Further for every M with $det(M-(1/2)J) \neq 0$ the equation $M = (1/2)J(S+I)(S-I)^{-1}$ can be solved for S to get $S = (M-(1/2)J)^{-1}(M+(1/2)J)$. The relation $S \in Sp(n)$ is then equivalent to M being real and symmetric and one proves then that the MW operator

$$\hat{R}_\nu(S) = \left(\frac{1}{2\pi}\right)^n \frac{i^\nu}{\sqrt{|det(S-I)|}} \int e^{(i/2)<M_S z,z>} \hat{T}(z)d^{2n}z \tag{4.36}$$

can be written as

$$\hat{R}_\nu(S) = \left(\frac{1}{2\pi}\right) i^\nu \sqrt{|det(S-I)|} \int e^{-(i/2)\sigma(Sz,z)} \hat{T}((S-I)z)d^{2n}z = \tag{4.37}$$

$$= \left(\frac{1}{2\pi}\right) i^\nu \sqrt{|det(S-I)|} \int \hat{T}(Sz)\hat{T}(-z)d^{2n}z$$

(ν can be identified with the Conley-Zehnder index). Further $\hat{R}_\nu(S) = c_S \hat{S}_{W,m}$ with $|c_S| = 1$ due to the relation $\hat{R}_\nu(S)\hat{T}(z) = \hat{T}(Sz)\hat{R}_\nu(S)$ (cf. (**4A**) and note that $\hat{R}(S)$ is unitary).

One shows next that the MW operators coincide with the metaplectic operators $\hat{S}_{W,m}$ when $S = S_W$. First for S_W a free symplectic matrix as in (4.31) one can write $S_W - I$ as

$$\begin{pmatrix} A-I & B \\ C & D-I \end{pmatrix} = \begin{pmatrix} 0 & B \\ I & D-I \end{pmatrix} \begin{pmatrix} C-(D-I)B^{-1}(A-I) & 0 \\ B^{-1}(A-I) & I \end{pmatrix} \tag{4.38}$$

leading to $det(S_W - I) = (-1)^n det(B)det(B^{-1}A + DB^{-1} - B^{-1} - (B^T)^{-1})$ and for S_W as in (4.30) this means $det(S_W - I) = (-1)^n det(L^{-1})det(P+Q-L-L^T)$. Now the Hessian matrix of the transformation $x \to W(x,x)$ will be (**4B**) $W_{xx} = P+Q-L-L^T$ and $\hat{R}_\nu(S) = \hat{S}_{W,m}$ provided that $\nu = m - Inert(W_{xx})\ mod(4)$ in which case $(1/\pi)arg(det(S-I)) \equiv -\nu + n\ mod(2)$ (see [**579**] for proofs). One remarks here also that the MW formula is related to a famous formula of Gutzwiller (cf. [**579**]). In general now one has via (4.38)

$$det(S_W - I) = (-1)^n det(L^{-1})det(P+Q-L-L^T) \tag{4.39}$$

Some further calculation then shows that every $\hat{S} \in Mp(n)$ can be written as a product

$$\hat{S} = \hat{R}_\nu(S_W)\hat{R}_{\nu'}(S_{W'}) \tag{4.40}$$

Finally one shows that if $\hat{S} \in Mp(n)$ with $det(S-I) \neq 0$ and $\hat{S} = \hat{R}_\nu(S_W)\hat{R}_{\nu'}(S_{W'})$ then $\hat{S} = \hat{R}_{\nu(S)}(S)$ with $\nu(S) = \nu + \nu' + n - Inert(M + M')$ (where $M, M' \sim S_W, S_{W'}$ via e.g. $M_S = (1/2)J(S+I)(S-I)^{-1}$) and the MW operators thus generate $Mp(n)$.

4.3. THE VAN VLECK DETERMINANT. This was discussed briefly above and we expand this here following [**295, 343, 372, 373, 785, 853, 990, 1188, 1234, 1264**]. We begin with Toms [**1234**] which deals with QM on a curved background and the resulting SE via the path integral. This involves a van Vleck determinant $\Delta^p(x,x')$ in the measure with various powers p and we refer to Remark 8.2.5 for further general comments. In particular it is shown that the Schwinger action principle is equivalent to the Feynman path integral when $p = 1$ (cf. also [**372**]). Now for the Schwinger action principle let q^i be local coordinates on a manifold M and denote by $< q_2, t_2|q_1, t_1 >$ the transition amplitude ($|q_i, t_i >$ represents a quantum state at time t_i and here $t_2 \geq t_1$). The position operator is $\hat{q}^i|q_\alpha, t_\alpha >= q^i(t_\alpha)|q_\alpha, t_\alpha >$ $(\alpha = 1, 2)$ and the Schwinger principle says that

$$\delta < q_2, t_2|q_1, t_1 >= \frac{i}{\hbar} < q_2, t_2|\delta S|q_1, t_1 > \tag{4.41}$$

where S represents the action obtained by the replacement of q^i in the classical action by $\hat{q}^i$, along with an operator ordering making S self adjoint. δ in (6.1) refers to any possible variation (times, dynamical variables, structure of the Lagrangian, etc.). A source J is often added to the theory to generate $n-point$ functions and for a flat space this is (**4C**) $S_J[q] = S[q] + \int_{t_1}^{t_2} dt J_i(t)q^i(t)$. The covariant generalization of this for curved space is obtained via

$$S_J[q, q_*] = S[q] + \int_{t_1}^{t_2} dt J_i(t)(q^i(t) - q_*^i(t)) \tag{4.42}$$

This gives the same classical theory and indicates that the natural replacement for $(q^i - q_*^i)$ is the tangent vector at q_*^i to the geodesic connecting q^i and q_*^i. The latter can be introduced via the geodetic interval $\sigma(q_*; q)$ where (**4D**) $\sigma(q_*; q) = (1/2)\ell^2(q_*; q)$ with ℓ is the length of the geodesic in question. The tangent vector to the geodesic at q_* is (**4E**) $\sigma^i(q_*; q) = g^{ij}(q_*)(\partial/\partial q_*^j)\sigma(q_*; q)$ where M has metric tensor g_{ij}. If M is flat and the q^j are Cartesian coordinates (so $g_{ij} = \delta_{ij}$) then $\sigma^i(q_*; q) = -(q^i - q_*^i)$ leading to (**4F**) $S_J[q, q_*] = S[q] - \int_{t_1}^{t_2} dt J_i(t)\sigma^i(q_*; q)$. If M is flat but the q^i are not Cartesian coordinates one can also derive (**4F**) from (6.2). It is convenient now to write (**4G**) $S_J[q, q_*] = S[q] - J_i\sigma^i(q_*; q)$ where $i \sim$ time label and a repeated index means integration over time. Further one can regard $S_J[q, q_*]$ as a functional $\tilde{S}_J[q_*, \sigma^i(q_*, q)]$ defined using a covariant Taylor expansion

$$S[q] = \sum_0^\infty \frac{(-1)^n}{n!} S_{;(i_1, \cdots, i_n)}[q_*]\sigma_1^i(q_*; q) \cdots \sigma_n^i(q_*; q) \tag{4.43}$$

in (**4G**) (the semicolon means covariant derivative using the Christoffel connection based on g_{ij}). Let then $<q_2,t_2|q_1,t_1>[J]$ be the transition amplitude with action $S_J[q;q_*]$ in (**4G**) and the Schwinger principle gives

(4.44) $$\delta <q_2,t_2|q_1,t_1>[J] = \frac{i}{\hbar}<q_2,t_2|\delta S_J|q_1,t_1>[J]$$

If the variation involves q^i with values fixed at times t_j then the amplitude does not change and one has $0 = <q_2,t_2|\delta S|q_1,t_1>$ leading to an equation of motion $(\delta \tilde{S}_J/\delta \sigma^i) - J_i = 0$. One can also write

(4.45) $$<q_2,t_2|q_1,t_1>[J] = \sum_0^\infty \frac{1}{n!}J_{i_1}\cdots J_{i_n}\frac{\delta^n <q_2,t_2|q_1,t_1>[J]}{\delta J_{i_1}\cdots \delta J_{i_n}}\bigg|_{J=0}$$

Suppose the variation in (6.4) to be relative to J_i; then

(4.46) $$\frac{\delta <q_2,t_2|q_1,t_1>[J]}{\delta J_i} = -\frac{i}{\hbar}<q_2,t_2|\sigma^i(q_*;q)|q_1,t_1>[J]$$

A further variation is then taken with respect to the source and terms $1 = \int dv'|q',t'><q',t'|$ are inserted ($t_1 < t' < t_2$ and $dv' = d^n q' g^{1/2}(q')$ - the invariant volume element on M). One considers time ordering with i,t',t_1,t_2 and concludes that

(4.47) $$\delta <q_2,t_2|\sigma^i|q_1,t_1>[J] = -\frac{i}{\hbar}\delta J_j <q_2,t_2|T(\sigma^i\sigma^j)|q_1,t_1>[J]$$

where T is the time ordering symbol. It follows that

(4.48) $$\frac{\delta^2 <q_2,t_2|q_1,t_1>[J]}{\delta J_j \delta J_i} = \left(-\frac{i}{\hbar}\right)^2 <q_2,t_2|T(\sigma^i\sigma^j)|q_1,t_1>[J]$$

By induction one then sees that

(4.49) $$\frac{\delta^n <q_2 t_2|q_1,t_1>[J]}{\delta J_{i_1}\cdots \delta J_{i_n}} = \left(-\frac{i}{\hbar}\right)^n <q_2,t_2|T(\sigma^{i_1}\cdots \sigma^{i_n})|q_1,t_1>[J]$$

This leads to (cf. [**1239**])

(4.50) $$<q_2,t_2|q_1,t_1>[J] = <q_2,t_2|T\left[exp\left(-\frac{i}{\hbar}J_i\sigma^i\right)\right]|q_1,t_1>[J=0]$$

Define then (**4H**) $E_i[q_*;\sigma^i(q_*;q)] = \delta \tilde{S}/\delta \sigma^i$ so that the operator equation of motion above becomes $E_i[q_*;\sigma^i(q_*;q)] = J_i$. Consider $E_i[q_*;-(\hbar/i)(\delta/\delta J_i)]$ where σ^i is replaced by $-(\hbar/i)(\delta/\delta J_i)$. Now view E_i as defined in terms of the Taylor series obtained by differentiating (6.3) and use (6.10) to get

(4.51) $$E_i\left[q_*;-\frac{\hbar}{i}\frac{\delta}{\delta J_i}\right]<q_2,t_2|q_1,t_1>[J] = J_i <q_2,t_2|q_1,t_1>[J] =$$

$$<q_2,t_2|T\left[E_i[q_*;\sigma^i]exp\left(-\frac{i}{\hbar}J_i\sigma^i\right)\right]|q_1,t_1>|[J=0]$$

This provides a functional differential equation for $<q_2,t_2|q_1,t_1>|[J]$ and integration of (4.51) gives the link between the Schwinger action principle and the Feynman path integral.

Now in order to deal with (6.11) write

$$(4.52)\quad <q_2,t_2|q_1,t_1>[J]=\int\left(\prod_i d\sigma^i(q_*;q)\right)F[q_*;\sigma^i(q_*;q)]exp\left(-\frac{i}{\hbar}J_i\sigma^i\right)$$

for some function F and some calculation leads to

$$(4.53)\qquad F[q_*;\sigma^i(q_*;q)]=f(q_*)exp\left(\frac{i}{\hbar}\tilde{S}(q_*;\sigma^i)\right)$$

for arbitrary $f(q_*)$. The condition for the surface term in integration to vanish is that $S[q=q_1]=S[q=q_2]$ (assuming that J_i is only nonzero for $t_1<t<t_2$)). Then one has

$$(4.54)\qquad <q_2,t_2|q_1,t_1>[J]=f(q_*)\int\left(\prod_i d\sigma^i\right)exp\left[\frac{i}{\hbar}(\tilde{S}-J_i\sigma^i)\right]$$

One can change the integration variables via

$$(4.55)\qquad \left(\prod_i d\sigma^i(q_*;q)\right)=\left|det\frac{\delta}{\delta q^j}\sigma^i(q_*;q)\right|\left(\prod_i dq^i\right)$$

Note from (**7E**) that $\sigma^i(q_*;q)=g^{ik}(q_*)\delta\sigma(q_*;q)/\delta q_*^k$ and the vanVleck determinant is

$$(4.56)\qquad \Delta(q_*;q)=|g(q_*)|^{-1/2}|g(q)|^{-1/2}det\left(-\frac{\delta^2\sigma(q_*;q)}{\delta q^i\delta q_*^j}\right)$$

Then (4.55) becomes

$$(4.57)\qquad \left(\prod_i d\sigma^i(q_*;q)\right)=\left(\prod_i dq^i\right)|g(q)|^{1/2}|\Delta(q_*;q)||g(q_*)|^{-1/2}$$

The factors of $g(q)$ and $g(q_*)$ have been chosen to make $\Delta(q_*;q)$ a scalar in each argument. The transition amplitude now becomes

$$(4.58)\qquad <q_2,t_2|q_1,t_1>[J]=$$

$$=|g(q_*)|^{-1/2}f(q_*)\int\left(\prod_i dq^i\right)|g(q)|^{1/2}|\Delta(q_*;q)|exp\left[\frac{i}{\hbar}(\tilde{S}-J_i\sigma^i)\right]$$

The amplitude must be invariant under a change of coordinates $q_*\to q'_*$ and this constrains $|g(q_*)|^{-1/2}f(q_*)$ to transform as a scalar, and in fact for the Feynman expression it must be a constant, so one simply takes

$$(4.59)\quad <q_2,t_1|q_1,t_1>[J]=\int\left(\prod_i dq^i\right)|g(q)|^{1/2}|\Delta(q_*;q)|exp\left[\frac{i}{\hbar}(\tilde{S}-J_i\sigma^i)\right]$$

as the path integral representation for the amplitude.

REMARK 9.4.2. We go here to Parvate-Gangal [**994**] and consider the SE (cf. also [**234**])

$$(4.60)\qquad i\hbar\partial_s<x,s|\psi>=\left[-\frac{\hbar^2}{2\mu}g^{\alpha\beta}(x)\nabla_\alpha\nabla_\beta+\frac{\hbar^2}{2\mu}\xi R\right]<x,s|\psi>$$

(this can also be related to equations $(-\nabla^\alpha\nabla_\alpha + \xi R + m^2)\phi = 0$). One shows here that the probability amplitude $< x, s|x', 0 >$ satisfying (6.20) and the boundary condition (**4I**) $lim_{s\to 0} < x, s|x', 0 >= [g(x)]^{-1/2}\delta(x - x')$ can be written in the path integral form

(4.61) $$< x, s|x', 0 >=$$

$$= \int d[x(s')][\Delta^p]exp\left(\frac{i}{\hbar}\int_0^s ds'\left[\frac{\mu}{2}g_{\alpha\beta}\frac{dx^\alpha}{ds'}\frac{dx^\beta}{ds'} - \frac{\hbar^2}{2\mu}[\xi + \frac{p-1}{3}]R(x)\right]\right)$$

Note for $p = 0$ and $\xi = 1/3$ the scalar curvature term vanishes (cf. [**288, 383**]). The path integral in (4.61) can be defined by breaking the time interval $[0, s]$ into $N + 1$ equal increments of length ϵ and writing

(4.62) $$< x, s|x', 0 >= lim_{N\to\infty}\left(\frac{\mu}{2\pi i\epsilon}\right)^{n(N+1)/2}\int\prod_{j=1}^N [d^n x_j[g(x_j)]^{1/2}]\times$$

$$\times exp\left\{\sum_{j=0}^N\left[i\int_{j\epsilon}^{(j+1)\epsilon}\left(\frac{\mu}{2}g_{\alpha\beta}\frac{dx^\alpha}{ds'}\frac{dx^\beta}{ds'} - \frac{\lambda R}{2\mu}\right)ds' + plog(\Delta(x_{j+1}, x_j)\right]\right\}$$

where $d^n x_j = dx_j^1 dx_j^2 \cdots dx_j^n$, $g = det(g_{\alpha\beta}), s = (N+1)\epsilon$, $x_0 = x'$, $x_{N+1} = x$, and $\lambda = \xi + (1/3)(p-1)$ $(c = \hbar = 1$ and the metric signature is $(+, +, \cdots))$. If the signature were $(-, +, +, \cdots)$ a further fator of $(1/i)^{N+1}$ would be present and $g^{1/2}$ would be replaced by $(-g)^{1/2}$. Here

(4.63) $$\Delta(x_{j+1}, x_j) = [g(x_{j+1})]^{-1/2}det\left[-\frac{\partial^2\sigma(x_{j+1}, x_j)}{\partial x_{j+1}\partial x_j}\right][g(x_j)]^{-1/2};$$

$$\sigma(x_{j+1}, x_j) = \frac{1}{2}\int_{j\epsilon}^{(j+1)\epsilon} ds'\left[g_{\alpha\beta}\frac{dx^\alpha}{ds'}\frac{dx^\beta}{ds'}\right]^{1/2}$$

Thus $\sigma(x_{j+1}, x_j)$ is (1/2) of the proper arc length along the geodesic from x_j to x_{j+1}. The symbol $[\Delta^p]$ in (4.61) indicates that terms $exp[i\int_{j\epsilon}^{(j+1)\epsilon} ds'(\,)]$ in (4.62) are multiplied by a factor of $[\Delta(x_{j+1}, x_j)]^p$. In flat space $\Delta(x_{j+1}, x_j) = 1$ and $R = 0$. Further calculation in the general case shows that (4.60) holds and in fact the path integral of (4.61) (for any p) satisfies (4.60) and (**4I**). By choosing $p = 0$ one can eliminate the $[\Delta^p]$ term or by choosing $p = 1 - 3\xi$ the scalar curvature term. ∎

REMARK 9.4.3. We have seen earlier that a term Q arises in (3.50) whiich has the form of a quantum potential with ρ the vanVleck density and this is consistent with the Bohmian approach. On the other hand the SE (3.51) seems to arise in a geometric form in(4.60) for example with a scalar curvature term R somehow related to Q. In fact this is consistent with results indicated in [**234, 235, 250, 254**] (based on [**1115, 1116**] for example) where such a situation arises naturally (cf. also [**239, 339**]). Note explicitly in (3.51)

(4.64) $$i\hbar\psi_t = (\hat{H} + \hat{Q})\psi;\ \hat{Q} = \int QG^{sh}\psi' d^n x';\ Q = -\frac{\hbar^2}{2m}\frac{\nabla_x^2\sqrt{|\rho|}}{\sqrt{|\rho|}}$$

while in (4.60) one has (for $\mu \sim m$)

$$i\hbar\psi_t = -\frac{\hbar^2}{2m}g^{\alpha\beta}\nabla_\alpha\nabla_\beta\psi + \frac{\hbar^2}{2m}\xi R\psi \tag{4.65}$$

Recall also in [**990**] there is a term $(\hbar^2/2m)[\xi+(1/2)(p-1)]R$ and (4.65)corresponds to $p = 1$. In [**1115, 1116**] for a flat Riemannian space one arrives at a SE $i\hbar\psi = H\psi$ with

$$\psi = \sqrt{\rho}e^{iS/\hbar};\ Q = -\frac{\hbar^2}{2m}\frac{\nabla^2\sqrt{\rho}}{\sqrt{\rho}} = -\frac{\hbar^2}{12m}R \tag{4.66}$$

where R is the Weyl-Ricci scalar curvature and ρ is an invariant probability density for particle motion. Actually in a general 3-D Weyl space one gets the standard SE with a Hamilton-Jacobi equation

$$\partial_t S + H(x, \nabla S, t) - \gamma\frac{\hbar^2}{2m}R = 0;\ R = \dot{R} + \left(\frac{1}{2\gamma\sqrt{\rho}}\right)\frac{1}{\sqrt{g}}\partial_i(\sqrt{g}g^{ik}\partial_k\sqrt{\rho}) \tag{4.67}$$

where $\gamma = 1/12$ and $\dot{R}$ is the Riemann scalar curvature. We note that if $\dot{R} \neq 0$ then the resulting SE in [**1115, 1116**] will have the form

$$\textbf{(7J)}\ \ i\hbar\psi_t = -\frac{\hbar^2}{2m}\frac{1}{\sqrt{g}}\partial_i\sqrt{g}g^{ik}\partial_k\psi + V\psi - \gamma\frac{\hbar^2}{m}\dot{R}\psi$$

The Weyl geometry term disappears but a Riemannian curvature term persists (as in (4.65)) and this suggests that in (4.65) $\xi = -1$ with $\dot{R} = R$. ■

5. PHASE SPACE QUANTUM MECHANICS

We follow here Nasiri et al [**785, 915, 917, 1188**] for a somewhat more "hands on" approach to phase space quantization and will concentrate here on [**915, 916, 917, 1188**] which summarizes and extends the earlier work of Sobouti-Nasiri [**1188**]. The theme here argues that in classical dynamics and classical statistical mechanics the generalized coordinates q and momenta p play symmetric and independent roles whereas in the operator based quantum theory (QT) one or the other loses its identity at the expense of the other. One could avoid this by carrying the q and p formalisms concomitantly and arrive at state functions in $q - p$ space to get $q - p$ representations of the mixed states of quantum statistical mechanics. The operator theory emerges as a special case with a definite ordering rule for non-commutative operators. We concentrate on equations and begin with $H(q,p)$ and find $L^q(q,\dot{q})$ as a solution of

$$H\left(q, \frac{\partial L^q}{\partial\dot{q}}\right) - \dot{q}\frac{\partial L^q}{\partial\dot{q}} + L^q = 0 \tag{5.1}$$

Similarly one could replace q by p and arrive at $L^p(p,\dot{p})$ as a solution of

$$H\left(\frac{\partial L^p}{\partial\dot{p}}, p\right) + \dot{p}\frac{\partial L^p}{\partial\dot{p}} - L^p = 0 \tag{5.2}$$

Following [**1188**] one defines an extended Lagrangian (**5A**) $\mathfrak{L}(q,\dot{q};p,\dot{p}) = -\dot{q}p - q\dot{p} + L^q(q,\dot{q}) + L^p(p,\dot{p})$ and obtain Euler-Lagrange (EL) equations

$$(5.3)\qquad \frac{d}{dt}\frac{\partial\mathfrak{L}}{\partial\dot{q}} - \frac{\partial\mathfrak{L}}{\partial q} = \frac{d}{dt}\frac{\partial L^q}{\partial\dot{q}} - \frac{\partial L^q}{\partial q} = 0;$$
$$\frac{d}{dt}\frac{\partial\mathfrak{L}}{\partial\dot{p}} - \frac{\partial\mathfrak{L}}{\partial p} = \frac{d}{dt}\frac{\partial L^p}{\partial\dot{p}} - \frac{\partial L^p}{\partial p} = 0$$

With preassigned values at say $t = t_0$ each equation can be solved separately; the condition for q and p orbits to represent the same state of motion are $p(t_0) = (\partial L^q/\partial\dot{q})|_{t=t_0}$ and $q(t_0) = (\partial L^p/\partial\dot{p})|_{t=t_0}$. Such a state will be called a pure state - otherwise a mixed state. One can now define "extended momenta"

$$(5.4)\qquad \pi_q = \frac{\partial\mathfrak{L}}{\partial\dot{q}} = \frac{\partial L^q}{\partial\dot{q}} - p;\ \pi_p = \frac{\partial\mathfrak{L}}{\partial\dot{p}} = \frac{\partial L^p}{\partial\dot{p}} - q$$

These in turn produce an "extended Hamiltonian" (**5B**) $\mathfrak{H}(q,\pi_q;p,\pi_p) = \dot{q}\pi_q + \dot{p}\pi_q - \mathfrak{L}(q,\dot{q};p,\dot{p})$. Eliminating $\dot{q}$ and *dotp* from $\mathfrak{H}$ yields

$$(5.5)\quad \mathfrak{H}(q,\pi_q;p,\pi_p) = H(q,p+\pi_q) - H(q+\pi_p,p) = \sum_{n=0}\frac{1}{n!}\left[\frac{\partial^n H}{\partial p^n}\pi_q^n - \frac{\partial^b H}{\partial q^n}\pi_p^n\right]$$

The condition for pure state motions is $\pi_q(t_0) = \pi_p(t_0) = 0$ and these terms then remain zero. Poisson brackets can be extended via

$$(5.6)\qquad \{F,G\} = \frac{\partial F}{\partial q}\frac{\partial G}{\partial\pi_q} - \frac{\partial F}{\partial\pi_q}\frac{\partial G}{\partial q} + \frac{\partial F}{\partial p}\frac{\partial G}{\partial\pi_p} - \frac{\partial F}{\partial\pi_p}\frac{\partial G}{\partial p}$$

Now for dynamics in $p - q$ space:

(1) Let $\mathfrak{X}$ be the space of integrable complex functions $\chi(p,q)$ with q,p,π_q,π_p operators satisfying ((**5C**) $[q,\pi_q] = [p\pi_p] = i\hbar$; $[q,p] = [q,\pi_p] = [p,\pi_q] = [\pi_q,\pi_p] = 0$.

(2) By virtue of (**5C**) $\mathfrak{H}$ is now an operator on **X** and one looks at state functions satisfying (**5D**) $i\hbar(\partial\chi/\partial t) = \mathfrak{H}\chi = [H(q,p-i\hbar(\partial_q)) - H(q - i\hbar(\partial_p),p)]\chi = 0$.

(3) For a (real) observable $O(q,p)$ the expectation on χ is (**5E**) $<O(q,p)>= \int O(q,p)\Re(\chi)dqdp = (1/2)\int O(q,p)(\chi+\chi^*)dqdp \in \mathbf{R}$

One can think of χ in the form $\chi(q,p) = F(q,p)exp(-ipq/\hbar)$ (the exponential factor comes from $-d(qp)/dt$ in (**5A**). One checks

$$(5.7)\qquad (p - i\hbar\partial_q)\chi = i\hbar\partial_q F e^{-ipq/\hbar};\ (q - i\hbar\partial_p)\chi = i\hbar\partial_p F e^{-ipq/\hbar}$$

Solutions of (**5F**) and (8.7) in (**8D**) give

$$(5.8)\qquad i\hbar\partial_t F = [H(q,-i\hbar\partial_q) - H((-i\hbar\partial_p,p)]F = 0$$

This leads to superposition of the separable solutions in the form (**5G**) $\chi(q,p,t) = \sum_{\alpha\beta}A_{\alpha\beta}\psi_\alpha(q,t)\phi_\beta^*(p,t)exp(-ipq/\hbar)$ where

$$(5.9)\qquad i\hbar\partial_t\psi_\alpha = H(q,-i\hbar\partial_q)\psi_\alpha;\ i\hbar\partial_t\phi_\beta = H(i\hbar\partial_p,p)\phi_\beta$$

Further to each $\psi_\alpha(q)$ there corresponds a Fourier transform $\phi_\alpha(p)$ with (**5H**) $\psi_\alpha(q) = (1/(2\pi\hbar)^{N/2})\int\phi_\alpha(p)exp(ipq/\hbar)dp$ (where N is the number of degrees of freedom of the system).

Let now $U(q)$ be a real polynomial or series observable in q with matrix representation

$$(5.10) \qquad U_{\beta\alpha} = \int \psi_\alpha(q)U(q)\phi^*_\beta(p)e^{-ipq/\hbar}dpdq =$$

$$= \int \psi^*_\beta(q)U(q)\psi_\alpha(q)dq = \int \phi^*_\beta(p)U(i\hbar\partial_p)\phi_\alpha(p)dp = U^*_{\alpha\beta}$$

(the coefficient $(2\pi\hbar)^{N/2}$ is suppressed for brevity). Now from (**5E**) one has (**5I**) $<U>= (1/2)\int U(\chi+\chi^*)dpdq = (1/2)Tr[\hat{U}(\hat{A}+\hat{A}^\dagger)]$ where $\hat{A}$ is the matrix $A_{\alpha\beta}$ of (**5G**). This gives the freedom of choosing $\hat{A} = \hat{A}^\dagger$ and simplifying (**5E**) to read $<U>= \int U\chi dpdq = Tr(\hat{U}\hat{A})$. Choosing $U = 1$ requires $Tr(\hat{A}) = 1$ and for averages of all positive definite functions of q to be positive one needs also that $\hat{A}$ be a positive definite matrix. Thus χ of (**5G**) is a physically acceptable solution if ($\bigstar$) $\hat{A} = \hat{A}^\dagger$, $Tr\hat{A} = 1$, and $\hat{A}$ is positive definite. Then the averaging rule (**5E**) for $U(q)+V(p)$ reduces to (**5J**) $<U(q)+P(p)>= \int(U+V)\chi dpdq$. For a product via (**5E**) and ($\bigstar$) one has (**5K**) $<UV>= \Re Tr(\hat{U}\hat{V}\hat{A}) = (1/2)Tr(\hat{U}\hat{V}\hat{A}+\hat{A}\hat{V}\hat{U}) = Tr[(1/2)(\hat{U}\hat{V}+\hat{V}\hat{U})\hat{A}]$ and translation of this to q space language gives

$$(5.11) \qquad <UV>= \frac{1}{2}A_{\alpha\beta}\int \pi^*_\beta[U(q)P(-i\hbar\partial_q)+P(-i\hbar\partial_q)U(q)]\psi_\alpha dq$$

This means that the ordering rule associated with a product $U(q)V(p)$ is the symmetric ordering.

One shows next that the classical limit of this theory is the Liouville equation governing the dynamics of classical ensembles - hence the dynamics here is essentially ensemble dynamics. Thus in (**5D**), expanding about (q,p) and retaining only the first terms gives

$$(5.12) \qquad \partial_t\chi + \partial_p H\partial_q\chi - \partial_q H\partial_p\chi = \frac{d\chi}{dt} = 0$$

The most general solutions have the form $\chi(q(t)p(t))$ where $q(t)$ and $p(t)$ are the classical trajectories in q and p space; they will represent the same state of motion if (8.3) holds. For the Schrödinger situation one allows for only one term in (**5G**). Thus, for ψ and ϕ Fourier transforms of each other with $\int\chi dpdq = \int\psi^*\psi dq = \int\phi^*\phi dp = 1$ and

$$(5.13) \qquad \chi = \psi(q)\phi^*(p)e^{-ipq/\hbar};\ i\hbar\psi_t = H(q,-i\hbar\partial_q)\psi;$$

$$<U(q)V(p)>= \frac{1}{2}\int\psi^*[V(-i\hbar\partial_q)U(q)+U(q)V(-i\hbar\partial_q)]\psi dq$$

This leads immediately to the Heisenberg uncertainty principle with the added feature of specifying the ordering rule in the last equation.

REMARK 9.5.1. The state function (**5G**) represents an ensemble in a mixed state. If $\hat{A}$ is diagonalized via $A_{\alpha\beta} = A_\alpha\delta_{\alpha\beta}$ then one can write $\chi = \sum A_\alpha\psi)\alpha\phi^*_\alpha exp(-ipq/\hbar)$. Integrating over q or p shows A_α to be the probability of the system to be in the state $\psi_\alpha(q,t)$ or $\phi_\alpha(p,t)$. However let $\{\psi_n(q)\}$ be a complete orthornormal time independent basis set and $\{\phi_\alpha(p)\}$ be the Fourier replica. Let

Latin subscripts denote the members of the basis set and Greek subcripts denote the solutions of (5.9). Then $\chi(q,p,t) = A_{mn}(t)\psi_n(q)\phi^*_m(p)exp(-ipq/\hbar)$. Putting this in (**5D**) and multiplying by $\psi^*_n(q)\phi_m(p)exp(ipq/\hbar)$ gives upon integration a vonNeumann equation (**5L**) $i\hbar(d\hat{A}/dt) = [\hat{A},\hat{H}]$ for $\hat{A} = \hat{A}^\dagger$ positive definite and $Tr(\hat{A}) = 1$ where $\hat{A}$ is the matrix of the expansion coefficients and $\hat{H}$ refers to $H(q,p)$ in either a χ, ψ, or ϕ basis. This gives the evolution of the density matrix and $Tr(\hat{A}^2) = Tr(\hat{A}) = 1$ involves an ensemble in a pure state while if $Tr(\hat{A}^2) < 1$ the ensemble is in a mixed state. ∎

Concerning canonical transformations consider (**5M**) $q = Q - \delta\alpha\Pi_P$, $\pi_q = \Pi_Q$, $p = P - \delta\alpha\Pi_Q$, and $\pi_p = \Pi_P$. The generator is $G = \pi_p\pi_q$ and to this (for a finite α) there corresponds the unitary operator (**5N**) $U_\alpha = exp(-i\alpha G/\hbar) = exp[i\hbar\alpha(\partial^2/\partial_q\partial_p)]$ with $U^\dagger_\alpha U_\alpha = 1$. Operating by U_α on a pure state function $\chi = \psi(q)\phi^*(p)exp(-ipq/\hbar)$ generates another state function (α-representation)

$$(5.14)\qquad \chi_\alpha(q,p,t) = U_\alpha\chi = \left(\frac{1}{2\pi\hbar}\right)^N \int \psi(q-\alpha\tau)\psi^*(q+(1-\alpha)\tau)e^{ip\tau/\hbar}d\tau$$

(see [**917**] for details). For $\alpha = 1/2$ (8.11) gives Wigner's standard function $\chi_{1/2} = W(q,p,t)$ and the cases $\alpha = 0$ and $\alpha = 1$ simply give back χ and χ^*. Similarly operating on (**5D**) with U_α gives an evolution equation for χ_α (see [**913**] for details).

Next one looks at the ordering of different factors of noncommuting q and π_q in a given α-representation and how averages change from one α-representation to another. Let $\hat{F}_\alpha(q,\pi_q)$ be the q space operator corresponding to q^np^m in phase space when averaged by χ_α. The defining equation is (**5O**) $<q^np^m>_\alpha = \int q^np^m\chi_\alpha dpdq = \int \psi^*(q)\hat{F}_\alpha(q,\pi_q)\psi(q)dq$. For the combination of $\alpha = 0$ and $\alpha = 1$ corresponding to $(\chi+\chi^*)$ of (**5E**) this is already worked out in (8.13) and is the symmetric ordering (**5P**) $q^np^m \to (1/2)q^n\pi_q^m + \pi_q^mq^n)$. For a general α one gets (cf. [**913**] for details)

$$(5.15)\qquad q^np^m \to \sum_0^m \binom{m}{r}((1-\alpha)\pi_q)^rq^n(\alpha\pi_q)^{m-r}$$

For $\alpha = 1/2$ this reduces to Weyl's ordering which is known to go with Wigner's function. One shows also that q^np^m averaged by χ is the same as $U_\alpha(q^np^m)$ averaged by χ_α (cf. [**917**]). Setting further (**5Q**) $F(q,p) = \sum F_{mn}\chi_{mn} =< q|\hat{F}p > exp(-ipq/\hbar)$ where $F_{mn} =< n|\hat{F}|m >$ i the matrix element of $\hat{F}$ in the basis of eigenstates of $\hat{H}(q,\pi_q)$; some product formulas are also worked out.

5.1. QUANTUM POTENTIAL AND SYMMETRIES. We go now to Nasiri [**915**] and show how the QP behaves as a representation dependent quantity in the phase space. Look back at (5.1)-(5.4) and (**5A**) and corresponding to (5.5) one has

$$(5.16)\qquad \mathfrak{H}(\pi_p,\pi_q,p,q) = \dot{p}\pi_p + \dot{q}\pi_q - \mathfrak{L} = H(p+\pi_q,q) - H(p,q+\pi_p) =$$

$$= \sum \frac{1}{n!}\left[\left(\frac{\partial^n H}{\partial p^n}\right)_{\pi_q=0}\pi_q^n - \left(\frac{\partial^n H}{\partial q^n}\right)_{\pi_p=0}\pi_p^n\right]$$

The following postulates are outlined:

(1) Let p, q, π_p, and π_q be operators in a Hilbert space $\mathbf{X}$ of complex functions satisfying $\pi_q = -i\hbar\partial_q$ and $\pi_p = -i\hbar\partial_p$ with

(5.17) $$[\pi_q, q] = -i\hbar;\ [\pi_p, p] = -i\hbar;\ [p, q] = [\pi_p, \pi_q] = 0$$

This implies that $\mathfrak{H}$ is also an operator in $\mathbf{X}$.

(2) A state function $\chi(p, q, t) \in \mathbf{X}$ is assumed to satisfy (cf. (**5D**))

(5.18) $$i\hbar\partial_t\chi = \mathfrak{H}\chi = [H(p - i\hbar\partial_q, q) - H(p, q - i\hbar\partial_p)]\,\chi =$$

$$= \sum \frac{(-i\hbar)^n}{n!}\left[\frac{\partial^n H}{\partial p^n}\frac{\partial^n}{\partial q^n} - \frac{pp^n H}{\partial q^n}\frac{pp^n}{\partial p^n}\right]\chi$$

The evolution operator on the right side of (5.18) is Hermitian as far as H is Hermitian and the extension operation is done by the canonical transformations corresponding to unitary transformations at quantum level. Note that probability conservation may seem to be violated but χ is not a probability distribution function; it may become negative or complex in some regions of phase space.

(3) The averaging rule for $O(p, q)$ is given as in (**5E**)).

To find solutions for(5.18) assume (**5R**) $\chi(p, q, t) = F(p, q, t)exp(-ipq/\hbar)$ and one checks that

(5.19) $$(p - i\hbar\partial_q)\chi = i\hbar\partial_q F e^{-ipq/\hbar};\ (q - i\hbar\partial_p)\chi = i\hbar\partial_p F e^{-ipq/\hbar}$$

Substituting (5.19) into (5.18) and eliminating the exponential factor gives the equation (**5S**) $H(-i\hbar\partial_q, q) - H(p, -i\hbar\partial_p)F = i\hbar\partial_t F$ which has separable solutions of the form (**5T**) $F(p, q, t) = \psi(q, t)\phi^*(p, t)$ where ψ and ϕ are solutions of the SE in q and p representations respectively.

Now the canonical transformations that keep the extended Hamiltonian equations of motion form invariant are obtained as follows. Consider an infinitesimal extended canonical transformation on p, q, π_p, π_q via (**5U**) $p \to p + \alpha\pi_q$, $q \to q + \beta\pi_p$, $\pi_p \to \pi_p + \gamma q$, and $\pi_q \to \pi_q + \eta p$. This transformation is chosen so that an ordering problem does not arise (cf. (5.17)). To be canonical one requires (**5V**) $-\beta + \alpha = 0$, $-\beta\eta + 1 = 1$, and $-1 + \alpha\gamma = -1$ leading to $\beta = \alpha$ and $\eta = \gamma = 0$ so the generator is (**5W**) $G = \pi_p\pi_q = \hbar^2(\partial^2/\partial p\partial q)$. Then the unitary transformation for finite α becomes (**5X**) $U = exp(\alpha G/i\hbar) = exp{-}i\alpha\hbar\partial^2/\partial p\partial q) = exp(-i\alpha\hbar\partial^2/\partial P\partial Q)$ with $UU^\dagger = 1$. Using (5.16) and (**5X**) for $\alpha = -1/2$ the new extended Hamiltonian becomes

(5.20) $$\mathfrak{H}' = U\mathfrak{H}U^{-1} = H\left(p + \frac{1}{2}\pi_q, q - \frac{1}{2}\pi_p\right) - H\left(p - \frac{1}{2}\pi_q, q + \frac{1}{2}\pi_p\right)$$

For $H = (1/2m)p^2 + V(q)$ (5.20) becomes

(5.21) $$\mathfrak{H}' = \frac{1}{2m}[p + (1/2)\pi_q]^2 + V(q - (1/2)\pi_p) - \frac{1}{2m}[p - (1/2)\pi_q]^2 - V(q + (1/2)\pi_p)$$

For the linear potential $H = (1/2m)p^2 + bq$ one gets (•) $\mathfrak{H}' = (p/m)\pi_q - b\pi_p$ while for an harmonic potential $H = (1/2m)p^2 + (1/2)kq^2$ (with $k = m\omega^2$) one has (••) $\mathfrak{H}' = (p/m)\pi_q - kq\pi_p$. Via (5.21) one has (5.18) in the form

$$(5.22)\quad i\hbar\partial_t W = \mathfrak{H}'W = -i\hbar\frac{p}{m}\partial_q W + \sum \frac{i\hbar}{(2n+1)!}\left(\frac{\hbar}{2i}\right)^{2n}\frac{\partial^{2n+1}V}{\partial q^{2n+1}}\frac{\partial^{2n+1}W}{\partial p^{2n+1}}$$

$$W(p,q,t) = U\chi = \int \psi[q + (1/2)\hbar\tau, t]\psi^*[q - (1/2)\hbar\tau, t]e^{-ip\tau}d\tau$$

Here the first equation is the Wigner function and one calls the new form of the extended phase space (EPS) obtained via (**8U**) for $\beta = \alpha = -1/2$ the Wigner representation.

Finally, getting to the quantum potential (QP), one writes (**5Y**) $\psi(q,t) = R^q(q,t)exp(iS^q(q,t)/\hbar)$ with (♦) $S^q = \int^t \mathfrak{L}^q(q,\dot{q},t')dt'$. Using (**5Y**)k in the SE gives

$$(5.23)\quad i\hbar\left(\partial_t R^q + \frac{i}{\hbar}R^q\frac{\partial S^q}{\partial t}\right) = V(q)R^q -$$
$$-\frac{\hbar^2}{2m}\left[\frac{\partial^2 R^q}{\partial q^2} + \frac{2i}{\hbar}\partial_q R^q\partial_q S^q - \frac{R^q}{\hbar^2}(\partial_q S^q)^2\right]$$

The real part of (5.23) gives

$$(5.24)\quad \partial_t S^q - \frac{\hbar^2}{2m}\frac{1}{R^q}\frac{\partial^2 R^q}{\partial q^2} + \frac{1}{2m}(\partial_q S^q)^2 + V(q) = 0$$

This exhibits the QP in the usual form in q-space. In p-space (referring to [**199**] for some examples) one takes some special cases:

(1) A linear potential $V(q) = bq$ gives

$$(5.25)\quad i\hbar\partial_t\phi(p,t) = \frac{p^2}{2m}\phi + i\hbar b\partial_p\phi$$

One assumes now

$$(5.26)\quad \phi(p,t) = R^p(p,t)e^{-iS^p(p,t)/\hbar};\ S^p = \int^t \mathfrak{L}^p(p,\dot{p},t')dt'$$

(cf. (8.2). This yields then

$$(5.27)\quad i\hbar\left(\partial_t R^p + \frac{i}{\hbar}R^p\partial_t S^p\right) = \frac{p^2}{2m}R^p + i\hbar b\left(\partial_p R^p + \frac{iR^p}{\hbar}\partial_p S^p\right)$$

The real part gives then

$$(5.28)\quad \partial_t S^p + \frac{p^2}{2m} - b\partial_p S^p = 0 \Rightarrow \partial_t S^p + H = 0$$

Thus there is no QP for the linear potential in p-space.

(2) For the harmonic potential $V = (1/2)kq^2$ one obtains

$$(5.29)\quad i\hbar\left(\partial_t R^p + \frac{i}{\hbar}R^p\partial_t S^p\right) = \frac{p^2}{2m}R^p -$$

$$-\frac{k\hbar^2}{2}\left[\frac{\partial^2 R^p}{\partial p^2}+\frac{2i}{\hbar}\partial_p R^p\partial_p S^p+\frac{i}{\hbar}R^p\frac{\partial^2 S^p}{\partial p^2}-\frac{T^p}{\hbar^2}(\partial_p S^p)^2\right]$$

The real part is then

(5.30) $$\partial_t S^p+\frac{p^2}{2m}-\frac{k\hbar^2}{2R^p}\frac{\partial^2 R^p}{\partial p^2}+\frac{k}{2}(\partial_p S^p)^2=0$$

The p-space version of the QP is (**5Z**) $\mathfrak{Q}=-(\hbar^2/2)(k/R^p)(\partial^2 R^p/\partial p^2)$.

(3) For a linear potential (5.16) becomes

(5.31) $$\mathfrak{H}=\frac{\pi_p^2}{2m}+\frac{p}{m}\pi_q-b\pi_p$$

and the state function χ from (8.18) is

(5.32) $$\chi(p,q,t)=\mathcal{R}(p,q,t)e^{i\mathfrak{S}(p,q,t)/\hbar};\ \mathfrak{S}(p,q,t)=\int^t\mathfrak{L}(p,q,\dot p,\dot q,t')dt'$$

where $\mathfrak{L}$ is given via (**5A**). Using (**5A**) one gets then (♦♦) $\mathfrak{S}=S^p+S^q-pq$ where S^p and S^q are given by (♦) and (5.26) respectively. An alternative approach to (♦♦) is to use (**8Y**) and (5.26) in (**5R**) and (**5T**) and compare with (5.32). One also sees that $\mathcal{R}(p,q,t)=R^q(q,t)R^{*p}(p,t)$. Now (8.18) gives

(5.33) $$i\hbar\left(\partial_t\mathcal{R}+\frac{i\mathcal{R}}{\hbar}\partial_t\mathfrak{S}\right)=-\frac{\hbar^2}{2m}\left[\frac{\partial^2\mathcal{R}}{\partial q^2}+\frac{2i}{\hbar}\partial_q\mathcal{R}\partial_q\mathfrak{S}+\frac{i\mathcal{R}}{\hbar}\frac{\partial^2\mathfrak{S}}{\partial q^2}-\right.$$

$$\left.-\frac{\mathcal{R}}{\hbar^2}(\partial_q\mathfrak{S})^2\right]-\frac{i\hbar p}{m}\left(\partial_q\mathcal{R}+\frac{i\mathcal{R}}{\hbar}\partial_q S\right)+i\hbar b\left(\partial_p\mathcal{R}+\frac{i\mathcal{R}}{\hbar}\partial_p\mathfrak{S}\right)$$

Consequently

(5.34) $$\pi_p=\partial_p\mathfrak{S}(p,q,t)\ and\ \pi_q=\partial_q\mathfrak{S}(p,q,t)\Rightarrow\partial_t\mathfrak{S}-\frac{\hbar^2}{2m}\frac{1}{\mathcal{R}}\frac{\partial^2\mathcal{R}}{\partial q^2}+\mathfrak{H}=0$$

This is the modified HJ equation for the linear potential in EPS.

(4) For the harmonic potential one has $|mfH=(\pi_q^2/2m)+(p/m)\pi_q-(1/2)k\pi_p^2-kq\pi_q$ and there results

(5.35)
$$i\hbar\left(\partial_t\mathcal{R}+\frac{i\mathcal{R}}{\hbar}\partial_t\mathfrak{S}\right)=-\frac{\hbar^2}{2m}\left[\frac{\partial^2\mathcal{R}}{\partial q^2}+\frac{2i}{\hbar}\partial_q\mathcal{R}\partial_q\mathfrak{S}+\frac{i\mathcal{R}}{\hbar}\frac{\partial^2\mathfrak{S}}{\partial q^2}-\frac{\mathcal{R}}{\hbar^2}(\partial_q S)^2\right]-$$

$$-\frac{i\hbar p}{m}\left(\partial_q\mathcal{R}+\frac{i\mathcal{R}}{\hbar}\partial_q\mathfrak{S}\right)+\frac{k\hbar^2}{2}\left[\frac{\partial^2\mathcal{R}}{\partial p^2}+\frac{2i}{\hbar}\partial_p\mathcal{R}\partial_p\mathfrak{S}+\frac{i\mathcal{R}}{\hbar}\frac{\partial^2\mathfrak{S}}{\partial p^2}-\frac{\mathcal{R}}{\hbar^2}(\partial_p\mathfrak{S})^2\right]$$

The real part of (8.35) gives then

(5.36) $$\partial_t\mathfrak{S}-\frac{\hbar^2}{2m}\frac{1}{\mathcal{R}}\frac{\partial^2\mathcal{R}}{\partial q^2}+\frac{\hbar^2 k}{2\mathcal{R}}\frac{\partial^2\mathcal{R}}{\partial p^2}+\mathfrak{H}=0$$

This is the modified HJ equation for the harmonic potential in EPS.

One looks now for a possible extended canonical transformation that could remove the QP from (5.34) and (5.36). Thus consider

(5.37) $$p\to p+\alpha\pi_q;\ q\to q+\alpha\pi_p;\ \pi_p\to\pi_p;\ \pi_q\to\pi_q$$

As indicated above this is canonical for arbitrary α and one obtains

$$\mathfrak{H}' = \frac{\pi_q^2}{2m} + \frac{(p+\alpha\pi_q)}{m}\pi_q - \frac{1}{2}k\pi_p^2 - k(q+\alpha\pi_p)\pi_p \tag{5.38}$$

Then (8.34) and (8.36) transform into

$$\partial_t\mathfrak{S}' - \frac{\hbar^2}{m}\left(\frac{1}{2}+\alpha\right)\frac{1}{\mathcal{R}'}\frac{\partial^2\mathcal{R}'}{\partial q^2} + \frac{p}{m}\pi_q - b\pi_p = 0; \tag{5.39}$$

$$\partial_t\mathfrak{S}' - \frac{\hbar^2}{m}\left(\frac{1}{2}+\alpha\right)\frac{1}{\mathcal{R}'}\frac{\partial^2\mathcal{R}'}{\partial q^2} + \hbar^2 k\left(\frac{1}{2}+\alpha\right)\frac{1}{\mathcal{R}'}\frac{\partial^2\mathcal{R}'}{\partial p^2} + \frac{p}{m}\pi_q - kq\pi_p = 0$$

With the assumption $\alpha = -1/2$ and using (•) and (••) these both become (★★) $\partial_t\mathfrak{S}' + \mathfrak{H}' = 0$ which looks like a classical HJ equation (no QP). Recall here that (**5U**) transforms (5.18) into the Wigner equation for $\alpha = -1/2$ and $\eta = \gamma = 0$. Hence one concludes that (★★) is the Wigner representation.

6. STRINGS

We go here to some work of Nikolić in [**930, 932, 933**] and consider the derivation of a Bohmian interpretation of relativistic quantum particles from world sheet covariance in string theory. Thus for bosonic strings one goes back to [**930, 931**] and considers strings with $\alpha,\,\beta = 0,1,\cdots,D-1$ for target indices and $\mu,\,\nu = 0,1$ for world sheet indices. Let $\sigma(\sigma_0,\sigma_1)$ and write the action for a bosonic string as

$$\mathfrak{A} = \int d^2\sigma\mathfrak{L};\ \mathfrak{L} = -\frac{1}{2}|h|^{1/2}h^{\mu\nu}\eta_{\alpha\beta}(\partial_\mu X^\alpha(\partial_\nu X^\beta) \tag{6.1}$$

Here $\eta_{\alpha\beta}$ is a flat Minkowski metric in D dimensions, $h^{\mu\nu}(\sigma)$ is an arbitrary metric on the world sheet, and $h = det(h_{\mu\nu})$. By requiring that the variation of the first equation with respect to $h^{\mu\nu}$ should vanish one concludes that $h_{\mu\nu}$ must be proportional to the induced metric on the world sheet (cf. [**1340**]), i.e. (★) $h_{\mu\nu})\sigma) = f(\sigma)(\partial_\mu X^\alpha)(\partial_\nu X_\alpha)$ where $f(\sigma)$ is an arbitrary positive function. The canonical momentum world sheet vector density is defined via

$$P^\mu_\alpha = \frac{\partial\mathfrak{L}}{\partial(\partial_\mu X^\alpha)} = -|h|^{1/2}\partial^\mu X_\alpha \tag{6.2}$$

and the covariant dDW Hamiltonian density is given by the Legendre transform

$$\mathcal{H} = P^\mu_\alpha\partial_\mu X^\alpha - \mathfrak{L} = -\frac{1}{2}\frac{h_{\mu\nu}}{|h|^{1/2}}\eta^{\alpha\beta}P^\mu_\alpha P^\nu_\beta \tag{6.3}$$

When (6.2) is satisfied then $\mathcal{H} = \mathfrak{L}$ and the covariant Hamiltonian equations of motion are

$$\partial_\mu X^\alpha = \frac{\partial\mathcal{H}}{\partial P^\mu_\alpha};\ \partial_\mu P^\mu_\alpha = -\frac{\partial\mathcal{H}}{\partial X^\alpha} \tag{6.4}$$

Using (6.2) one sees that the first equation in (6.4) is equivalent to (6.2). Since $\mathcal{H}$ does not depend on X^α the second equation in (6.4) leads to the covariant string-wave equation (♣) $\partial_\mu(|h|^{1/2}\partial^\mu X_\alpha) = 0$. Thus the classical dDW covariant canonical formalism is equivalent to the classical Lagrangian formalism which also leads to the covariant equation of motion (6.4). Similarly it is also equivalent to the ordinary non-covariant Hamiltonian formalism in which the Hamilitonian is defined such that only $\mu = 0$ contributes in the first line of (6.2).

Next one introduces the covariant dDW HJ formalism via a vector density function $\mathcal{S}^\mu(X(\sigma),\sigma)$ satisfying the dDW HJ equation (♠) $\mathcal{H} + \partial_\mu \mathcal{S}^\mu = 0$. Here $\mathcal{H}$ is given by (6.3) with the replacement $P^\mu_\alpha \to (\partial\mathcal{S}^\mu/\partial X^\alpha)$ where ∂_μ acts only on the second argument of $\mathcal{S}^\mu(X(\sigma),\sigma)$ and the corresponding total derivative is (••) $d_\mu = \partial_\mu + (\partial_\mu X^\alpha(\partial/\partial X^\alpha)$. For a given solution $\mathcal{S}^\mu$ of the dDW HJ equation the σ-dependence of $X^\alpha(X,\sigma)$ is determined by the equation of motion (♦♦) $-|h|^{1/2}\partial^\mu X_\alpha = \partial\mathcal{S}^\mu/\partial X^\alpha$. The classical dDW HJ formalism has a manifest world sheet covariance and one looks for an analogous quantum formalism. For the ordinary non- covariant HJ formalism one knows how to quantize via the Schrödinger equation (SE) and hence one seeks to obtain the ordinary HJ formalism from the covariant one. Choosing $h_{\mu\nu} = \eta_{\mu\nu}$ (♠) can be written as

$$-\frac{1}{2}\frac{\partial\mathcal{S}^0}{\partial X^\alpha}\frac{\partial\mathcal{S}^0}{\partial X_\alpha} + \frac{1}{2}\frac{\partial\mathcal{S}^1}{\partial X^\alpha}\frac{\partial\mathcal{S}^1}{\partial X_\alpha} + \partial_0\mathcal{S}^0 + \partial_1\mathcal{S}^1 = 0 \tag{6.5}$$

Then via (••) and (♦♦) the last term can be written as (■■) $\partial_1\mathcal{S}^1 = d_1\mathcal{S}^1 - (\partial_1 X^\alpha)(\partial_1 X_\alpha)$. Also the second term in (6.5) can be written $(1/2)(\partial_1 X^\alpha)(\partial_1 X_\alpha)$. One introduces now (★★) $S = \int d\sigma^1 \mathcal{S}^0$ so that

$$\frac{\partial\mathcal{S}^0(X(\sigma),\sigma)}{\partial X^\alpha(\sigma)} = \frac{\delta S([X(\sigma^0,\sigma^1)],\sigma^0)}{\delta X^\alpha(\sigma^1;\sigma^0)}; \tag{6.6}$$

$$\frac{\delta}{\delta X^\alpha(\sigma^1;\sigma^0)} \equiv \left.\frac{\delta}{\delta X^\alpha(\sigma^1)}\right|_{X(\sigma^1)=X(\sigma)}$$

By integrating (6.5) over $d\sigma^1$ one obtains the ordinary non-covariant HJ equation (■) $H + \partial_0 S = 0$ where

$$H = -\int d\sigma^1\left[\frac{1}{2}\frac{\delta S}{\delta X^\alpha(\sigma^1;\sigma^0)}\frac{\delta S}{\delta X_\alpha(\sigma^1;\sigma^0)} + \frac{1}{2}(\partial_1 X^\alpha)(\partial_1 X_\alpha)\right] \tag{6.7}$$

is written for the σ^0 dependent string coordinate $X^\sigma(\sigma^0,\sigma^1)$. The integral of a total derivative $\int d\sigma^1 d_1\mathcal{S}^1$ is ignored since it is a constant without physical significance. The σ^0 evolution of $X^\alpha(\sigma^0,\sigma^1)$ is given by

$$-\partial^0 X_\alpha(\sigma^0,\sigma^1) = \frac{\delta S}{\delta X^\alpha(\sigma^1;\sigma^0)} \tag{6.8}$$

which is a consequence of the $\mu = 0$ component of (♦♦). The covariant constraint (★) implies the non-covariant Hamiltonian constraint $H = 0$ (cf. [**1340**]). To anticipate the implications for the quantum case it is crucial to observe the following. First, to derive (■) from (♠) it was necessary to use the $\mu = 1$ component of (♦♦). Second if the world sheet covariance is required then the validity of the $\mu = 1$ component of (♦♦) also impolies the validity of the $\mu = 0$ component. Third the validity of the $\mu = 0$ component implies the classical determinism incoded in (10.19). Thus the determinism in classical string theory can be derived from the world sheet covariance and the requirement that the covariant HJ equation (♠) and the noncovariant one (■) should both be valid.

Now one coniders how to quantize strings so that the world sheet covariance is covariant. The standard method goes via path integral quantization (with $D=26$ to avoid anomalies - cf. Hatfield [**616**]). One would try to use the σ^0 dependent quantum states $\psi([X(\sigma^1)],\sigma^0)$ satisfying (♣♣) $\hat{H}\psi = i\hbar\partial_0\psi$

$$\hat{H} = -\int d\sigma^1\left[-\frac{\hbar^2}{2}\frac{\delta}{\delta X^\alpha(\sigma^1)}\frac{\delta}{\delta X_\alpha(\sigma^1)} + \frac{1}{2}(\partial_1 X^\alpha)(\partial_1 X_\alpha)\right] \tag{6.9}$$

but this runs into the problem that physical states satisfy (♠♠) $(\hat{H}+a)\psi = 0$ where a arises from operator ordering (cf. [**1340**]). Thus (6.9) needs a substitute and one imagines that a covariant substitute for (6.9) might be an equation similar to the covariant dDW HJ equation (♠) (as developed in [**918**]). One applies now these general methods of [**918**] to bosonic strings. Let $\psi = Rexp(iS/\hbar)$ where R, S are real functionals and then (♣) is equivalent to

$$-\int d\sigma^1\left[\frac{1}{2}\frac{\delta S}{\delta X^\alpha(\sigma^1)}\frac{\delta}{\delta X_\alpha(\sigma^1)} + \frac{1}{2}(\partial_1 X^\alpha)(\partial_1 X_\alpha) - \mathfrak{Q}\right] + \partial_0 S = 0; \tag{6.10}$$
$$-\int d\sigma^1\left[\frac{1}{2}\frac{\delta R}{\delta X^\alpha(\sigma^1)}\frac{\delta S}{\delta X_\alpha(\sigma^1)} - \mathfrak{J}\right] + \partial_o R = 0$$

$$\mathfrak{Q} = \frac{\hbar^2}{2R}\frac{\delta^2 R}{\delta X^\alpha(\sigma^1)\delta X_\alpha(\sigma^1)};\ \mathfrak{J} = -\frac{R}{2}\frac{\delta^2 S}{\delta X^\alpha(\sigma^1)\delta X_\alpha(\sigma^1)} \tag{6.11}$$

One sees that (6.10) is very similar to (■) with (6.7) (differing only in the $\mathfrak{Q}$ term). Now following Nikolić one replaces the classical dDW HJ equation (♠) with the quantum one

$$-\frac{1}{2}\frac{h_{\mu\nu}}{|h|^{1/2}}\eta^{\alpha\beta}\frac{d\mathcal{S}^\mu}{dX^\alpha}\frac{d\mathcal{S}^\nu}{dx^\beta} + \mathfrak{Q} + \partial_\mu\mathcal{S}^\mu = 0 \tag{6.12}$$

Here $\mathcal{S}^\mu([X],\sigma)$ is a functional of $X(\sigma)$ and a function of σ which incorporates quantum nonlocalities in a covariant manner. The quantum potential is defined as in (6.11) but with the replacement $\delta/\delta X^\alpha \to \delta/\delta_C X^\alpha$ where $\delta/\delta_C X^\alpha$ is a covariant version of (6.6) with C denoting a curve on the world sheet that generalizes the curve $\sigma^0 = constant$ in (6.6). The foliation of the world sheet into curves C is induced by the dynamical vector density $\mathfrak{R}^\mu([X],\sigma)$ with $\mathfrak{R}^\mu$ orthogonal to the curves at each point and satisfying

$$-\frac{1}{2}\frac{h_{\mu\nu}}{|h|^{1/2}}\eta^{\alpha\beta}\frac{d\mathfrak{R}^\mu}{dX^\alpha}\frac{d\mathcal{S}^\nu}{dX^\beta} + \mathfrak{J} + \partial_\mu\mathfrak{R}^\mu = 0 \tag{6.13}$$

where $\mathfrak{J}$ is defined as above with now $\delta/\delta X^\alpha \to \delta/\delta_C X^\alpha$. One defines then, covariantly,

$$R = \int_C d\Sigma_\mu R^\mu;\ S = \int_C d\Sigma_\mu S^\mu;\ R^\mu = |h|^{-1/2}\mathfrak{R}^\mu;\ S^\mu = |h|^{-1/2}\mathcal{S}^\mu \tag{6.14}$$

The non-covariant SE is then derived via $\mathfrak{R}^\mu = (\mathfrak{R}^0, 0)$ for $\mathcal{S}^1$ a local functional in σ^0, along with $\mathfrak{R}^0$ and $\mathcal{S}^0$. Choosing $h_{\mu\nu} = \eta_{\mu\nu}$ and integrating (6.12) and (6.13) over $d\sigma^1$ one recovers (6.10) which is equivalent to the SE (♣♣). We refer to [**919**] for more details and discussion. One conclusion is that in order that the covariant method of quantization should lead to the standard non-covariant quantization without violating covariance, it turns out that the quantization method should be supplemented with an equation that corresponds to the Bohmian deterministic hidden variable formulation of QM, namely (6.8).

CHAPTER 10

ENTROPY AND GEOMETRY

1. INTRODUCTION

In [**254**] and Chapter 6 we outlined a number of thermodynamic aspects of quantum mechanics (QM) and information theory (see also [**509, 510, 1121**]). Further in a number of papers some strong connections of thermodynamics and Einstein gravity have emerged (cf. [**21, 213, 258, 259, 431, 517, 685, 690, 768, 841, 842, 974, 975, 976, 977, 978, 979, 980, 982, 983, 1124, 1194, 1232**]). We will sketch some of this material in connection with general ideas about Bohmian mechanics (BM), the quantum potential (QP), and Schrödinger equations linear (SE) and nonlinear (NLS), following [**103, 124, 163, 207, 254, 261, 276, 333, 389, 570, 543, 624, 706, 711, 714, 729, 730, 877, 937**]. Subsequently we add some remarks connecting GL theory, the EH action, and Ricci-Yamabe flows following Kholodenko [**742, 743, 744, 745**]; this brings the Perelman entropy into the picture.

2. REMARKS ON INFORMATION DYNAMICS

We follow here Garbaczewski [**510**] (second paper) and mainly indicate the equations. For $\psi \in L^2(\mathbf{R})$ one writes $\rho(x) = |\psi^2(x)|$ and $\tilde{\rho} = |(\mathfrak{F}\psi)(p)|^2$ (Fourier transform) with

$$S(\rho) = S_q = - < log(\rho) >= - \int \rho(x) log(\rho(x)) dx; \tag{2.1}$$

$$S(\tilde{\rho}) = S_p = - < log(\tilde{\rho}) >= - \int \tilde{\rho}(p) log(\tilde{\rho}(p)) dp$$

where S is the Shannon (or differential) entropy. It is known (cf. [**139**]) that (**2A**) $S_q + S_p \geq (1 + log(\pi))$ and, writing $(\Delta A)^2 = (\psi, [A- < A >]^2\psi)$ with $< A >= (\psi, A\psi)$, for position and momentum operators X and P, one has for $\hbar = 1$ (cf. [**952**])

$$\Delta X \cdot \Delta P \geq \frac{1}{2\pi e} exp[S(\rho) + A(\tilde{\rho}) \geq \frac{1}{2} \tag{2.2}$$

Writing $\sigma^2 =< (X- < X >)^2 >= \Delta x^2$ gives (**2B**) $S(\rho) \leq (1/2) log(2\pi e \sigma^2)$ with maximum if and only if ρ is a Gaussian with variance σ^2. The Fisher information is defined via (cf. [**254, 509**]) (**2C**) $\mathfrak{F}(\rho) =< [\nabla log(\rho)]^2 >= \int [(\nabla\rho)^2/\rho] dx$ and one has

$$\mathfrak{F}(\rho) \geq (2\pi e) exp[-2S(\rho)] \geq \frac{1}{\sigma^2} \tag{2.3}$$

Evidently $\mathfrak{F}(\rho) \geq (1/\sigma^2)$ with equality allowed only if ρ is a Gaussian with variance σ^2. Standard Fourier analysis with (**2A**) yields then

$$(2.4) \qquad 4\tilde{\sigma}^2 \geq 2(e\pi)^{-1}exp[-2 < log(\tilde{\rho} >] \geq (2\pi e)exp[2 < log(\rho) >] \geq \sigma^{-2}$$

Now let P be conjugate to X, $\hbar = 1$, and $P \sim -i(d/dx)$ with all averages finite; then $[< P^2 > - < P >^2] = (\Delta P)^2 = \tilde{\sigma}^2$ leading to the standard $\sigma \cdot \tilde{\sigma} \geq (1/2)$. All of the above formulas hold if $\psi = \psi(x,t)$ with the parameter t properly inserted.

Now look at

$$(2.5) \qquad i\partial_t\psi = -D\Delta\psi + \frac{\mathcal{V}(x,t)}{2mD}\psi;\ D \sim \frac{\hbar}{2m}$$

Using (**2D**) $\psi = \sqrt{\rho}exp(is/2D)$ with $v = \nabla S$ one has as usual (**2E**) $\partial_t\rho = -\nabla(v\rho)$ and the QHJ equation (**2F**) $\partial_t s + (1/2)(\nabla s)^2 + (\Omega - \tilde{Q}) = 0$ where $\Omega = \mathcal{V}/m$. Introducing the velocity field $u = D\nabla log(\rho(x,t))$ this leads to

$$(2.6) \qquad Q = 2D^2\frac{\Delta\rho^{1/2}}{\rho^{1/2}} = \frac{1}{2}u^2 + D\nabla \cdot u$$

Note that $\tilde{Q}$ here is the negative of the usual QP (usually denoted by Q - cf. [**254, 261**]) and we will use $\mathcal{Q}$ to denote "heat" as in standard thermodynamics. If now (**2G**) $< \psi|\hat{H}|\psi >= E < \infty$ then unitary QM gives

$$(2.7) \qquad \mathfrak{H} = \frac{1}{2}[< v^2 > + < u^2 >] + < \Omega >= - < \partial_t s >= \mathcal{E} = \frac{E}{m} = const.$$

Further since $< u^2 >= -D < \nabla u >$ one has

$$(2.8) \qquad \frac{D^2}{2}\mathfrak{F} = \frac{D^2}{2}\int \frac{1}{\rho}(\partial_x\rho)^2 dx = \int \frac{\rho u^2}{2} = - < \tilde{Q} >$$

(one observes that (**2H**) $(\Delta u)^2 = \sigma_u^2 =< [u- < u >]^2 >=< u^2 >= D^2\mathfrak{F}$ here with (**2I**) $(\Delta v)^2 = \sigma_v^2 =< v^2 > - < v >^2$). Then with the definition $P = -i(2mD)(d/dx)$ and $< P >= m < v >$ one writes (**2J**) $\tilde{\sigma}^2 = (\Delta P)^2 =< P^2 > - < P >^2$ leading to (cf. Hall [**595**])

$$(2.9) \qquad \mathfrak{F}(\rho) = \frac{1}{D^2}\sigma_u^2 = \int dx|\psi|^2\left[\frac{\psi'}{\psi} + \frac{\psi^{*'}}{\psi^*}\right]^2 = 4\int dx\psi^{*'}\psi' +$$

$$+ \int dx|\psi|^2\left[\frac{\psi'}{\psi} - \frac{\psi^{*'}}{\psi^*}\right]^2 = \frac{1}{m^2D^2}[< P^2 > -m^2 < v^2 >] =$$

$$\frac{1}{m^2D^2}[(\Delta P)^2 - m^2\sigma_v^2] \Rightarrow (\mathbf{2K})\ m^2(\sigma_u^2 + \sigma_v^2) = \tilde{\sigma}^2$$

Note here that $< (P - mv) >= 0$ while (**2L**) $< (P - mv)^2 >=< P^2 > -m^2 < v^2 >= m^2D^2\mathfrak{F}$. This permits a sharpening of (2.3) by passing to dimensionless quantities as in (2.9) (e.g. $2mD = 1$) and writing $p_{cl} = (arg(\psi(x,t))'$ one obtains

$$(2.10) \qquad \mathfrak{F} = 4[< P^2 > - < p_{cl}^2 >] = 4[(\Delta P)^2 - (\Delta p_{cl})^2] = 4[\tilde{\sigma}^2 - \sigma_{cl}^2]$$

leading to

$$(2.11) \qquad 4\tilde{\sigma}^2 \geq 4]\tilde{\sigma}^2 - \tilde{\sigma}_{cl}^2] = \mathfrak{F} \geq (2\pi e)exp[-2S(\rho)] \geq \frac{1}{\sigma^2}$$

3. THERMODYNAMICS, DIFFUSION, AND QUANTUM MOTION

We go next to the first paper in Garbaczewski [**510**] and again mainly indicate equations (cf. also [**509**]). It turns out that direct analogs of the first and second laws of thermodynamics are faithfullly reproduced in the quantum situation and here asymptotic behavior of the Helmholz free energy is important in providing a tool to discriminate between classical and quantum regimes. One shows that the Schrödinger picture evolution attributes to a quantum system in its pure state a never vanishing work rate term whose interpretation is either (i) The time rate of work externally performed upon the system - if positive or (ii) The time rate of work performed by the system - if negative. Thermodynamic systems will be distinguished as either isolated with no energy or matter exchange with the environment, or closed with energy but no matter exchange, or open with energy-matter exchange unrestricted. Generally the nonequilibrium thermodynamics (NET) of closed systems involves the first law (**3A**) $\dot{U} = \dot{\mathcal{Q}} + \dot{W}$ and the second law (**3B**) $\dot{S} = \dot{S}_{int} + \dot{S}_{ext}$ where $\dot{S}_{int} \geq 0$ and $\dot{S}_{ext} = \dot{\mathcal{Q}}/T$. The time derivative notation does not imply that one may describe $\mathcal{Q}$ and W as legitimate functions - i.e. the issue of imperfect differentials arises as in classical thermodynamics (cf. Reif [**1067**]) - and one recalls that extremum principles are usually invoked in connection with the large time behavior of irreversible processes. In particular if T and the available volume V are kept constant then the minimum of the Helmholz free energy (**3C**) $F = U - TS$ is often preferred in describing system evolution and one has (**3D**) $\dot{F} = -T\dot{S}_{int} \leq 0$.

First (as an example) one considers a phase-space diffusion process governed by the Langevin equation $m\ddot{x} + m\gamma\dot{x} = -\nabla V(x,t) + \xi(t)$ with standard assumptions about the while noise, namely $<\xi(t)>=0$ and $<\xi(t)\xi(t')>=\sqrt{2m\gamma k_B T}\delta(t-t')$. The pertinent phase space density $f(x,u,t)$ is a solution of the Fokker-Planck-Kramers equation (with suitable initial data)

$$\partial_t f = \left[-\frac{\partial}{\partial x}u + \frac{\partial}{\partial u}\left(\gamma u + \frac{1}{m}\nabla V\right) + \frac{\gamma k_B T}{m}\frac{\partial^2}{\partial x^2}\right] f \tag{3.1}$$

One defines the Shannon or differential entropy $S(t)$ as usual via (**3E**) $S(t) = -\int dxduflog(f) = -<log(f)>$ (a dimensional factor is omitted) and the internal energy U of the stochastic process is $U = <E>$ with $E = (1/2)mu^2 + V(x,t)$. The first law takes the form (**3F**) $\dot{U} = \dot{\mathcal{Q}} + \dot{W}$ as before with $\dot{W} = <\partial_t V>$ interpreted as the work externally performed upon the system (so it is positive). Further as an analogue of the free energy one takes

$$F = <E + k_B T log(f)> = U - TS;\ \dot{F} - \dot{W} = \dot{\mathcal{Q}} - T\dot{S} = -T\dot{S}_{int} \leq 0 \tag{3.2}$$

via (3.1). Assuming suitable behavior of f at integration boundaries and sufficiently rapid decay at ∞ one has then

$$\dot{\mathcal{Q}} = \gamma(k_B T - <mu^2>);\ \dot{S} = \gamma\left[\frac{k_B T}{m}\left\langle\left(\frac{\partial log(f)}{\partial u}\right)^2\right\rangle - 1\right] \tag{3.3}$$

Writing $\mathfrak{S} = k_B S$ one arrives at $\dot{\mathcal{Q}} \leq T\dot{\mathfrak{S}}$ and as a byproduct of all this $\dot{F} \leq \dot{W}$. For time independent V one has the standard Helmholz extremum principle which amounts to minimizing the free energy F in the course of random motion so (**3G**) $\dot{F} = \dot{\mathcal{Q}} - T\dot{\mathfrak{S}} = -T\dot{\mathfrak{S}} \leq 0$. This discussion encompasses both forced and unforced Brownian motion. The Smoluchowski process is also considered in [**510**] but we omit this here.

Now for the thermodynamics of quantum motion one goes again to (2.5), namely $(i\partial_t\psi = -D\Delta\psi + (\mathcal{V}/2mD)\psi$, where $\mathcal{V}(x,t)$ is a continuous and say bounded below function with dimensions of energy and $D = \hbar/2m$. One introduces (**3H**) $k_B T_0 = \hbar\omega_0 = mc^2$ and hence $D = \hbar/2m = k_B T_0/m\beta_0$ for $\beta_0 = 2\omega_0 = 2mc^2/\hbar$. Using now $\psi = \sqrt{\rho}exp(is/2D)$ with $v = \nabla s$ one obtains in the standard manner $\partial_t\rho = -\nabla(v\rho)$ and (**2F**) again, namely $\partial_t s + (1/2)(\nabla s)^2 + (\Omega - \tilde{Q}) = 0$, and (2.6) holds, i.e. $\tilde{Q} = 2D^2(\Delta\sqrt{\rho}/\sqrt{\rho}) = (1/2)u^2 + D\nabla\cdot u$. The probability density $\rho = |\psi|^2$ is propagated by a Fokker-Planck equation of the form (3.2) with the drift $b = v - u = \nabla(s - Dlog(\rho))$ where $u = D\nabla log(\rho)$. The Shannon entropy $S = -< log(\rho) >$ satisfies

$$D\dot{S} =< v^2 > - < b\cdot v >= D(\dot{S}_{int} + \dot{S}_{ext}) \tag{3.4}$$

This is an analogue of the second law in the QM sense, namely

$$\dot{S}_{ing} = \dot{S} - \dot{S}_{ext} = \frac{1}{D} < v^2 > \geq 0 \Rightarrow \dot{S} \geq \dot{S}_{ext} \tag{3.5}$$

For the first law one wants expressions for U and $F = U - TS$ so write $D = k_B T_0/m\beta_0$ with $\beta_0 = 2\omega_0$ to arrive at $k_B T_0\dot{S}_{ext} = \dot{\mathcal{Q}}$. Then (cf. (**3G**))

(3.6)

$$v = \nabla s = \nabla(s + Dlog(\rho) - D\nabla log(\rho) = -\frac{1}{m\beta}\nabla(V + k_B T_0 log(\rho)) = -\frac{1}{m\beta_0}\nabla\Psi$$

where the time dependent potential (**3I**) $V = V(x,t) = -m\beta_0(s + Dlog(\rho))$ is defined in order to remain in conformity with the Smoluchowski process definition $b = -\nabla V/m\beta_0$. This yields then (**3J**) $-m\beta < s >=< \Psi >=< V > -TS \Rightarrow F = U - TS$ where $U =< V >$ and $F =< \Psi >$. Recalling now the explicit time dependence of $b(x,t) = -(1/m\beta_0)\nabla V(x,t)$ one arrives at the first law in the quantum context as (**3K**) $\dot{U} =< \partial_t V > -m\beta_0 < bv >= \dot{W} + \dot{\mathcal{Q}}$. The externally performed work entry is $\dot{W} =< \partial_t V >$ but

$$V = -m\beta s - k_B T log(\rho) \Rightarrow < \partial_t V >= m\beta_0 < \partial_t s >= \dot{W} \tag{3.7}$$

and consequently

$$-\frac{d}{dt} < s >= - < v^2 > - < \partial_t s > \Rightarrow \dot{F} = -T_0\dot{S}_{int} + \dot{W} \tag{3.8}$$

where $\dot{S}_{int} \geq 0$. If $< \psi|H|\psi >= E < \infty$ now then (2.7) holds (cf. (**2F**)) and consequently in the thermodynamical description of the quantum motion there is a never vanishing work term (**3L**) $\dot{W} = m\beta_0\mathcal{E} = E/m = const.$. The associated Helmholz extremum principle reads then (**3M**) $\dot{F} - m\beta_0\mathcal{E} = -T_0\dot{S}_{int} \leq 0$. One notes that (**3N**) $T\dot{S}_{int} = T\dot{S} - \dot{\mathcal{Q}} \geq 0 \Leftrightarrow \dot{\mathcal{Q}} \leq T\dot{S}$ goes in parallel with (**3O**) $\dot{F} \leq \dot{W} = \beta_0 < \hat{H} >$. It is emphasized that the nonvanishing external work term is generic to the quantum motion; if a stationary state is considered $<\hat{H}>$ is equal to a corresponding energy eigenvalue. For negative eigenvalues the work term receives an interpretation of "work performed" by the system (on its hypothetical environment); then $\dot{F}$ is negative and F may possibly have a chance to attain a minimum. Since bounded from below Hamiltonians can be replaced by positive operators one can in principle look at $m\beta_0\mathcal{E} = \beta_0 <\hat{H}>$ as a positive (constant and nonvanishing) time rate of work externally performed upon the system. In the present situation one does not expect monotonic behavior of F in time and even if so F may be an increasing function; consequently the standard Helmholz minimum principle is not a generic property of the (non-dissipative) quantum motion.

Thus some basic features of nonequilibrium thermodynamics of closed systems have been seen to occur in the quantum Schrödinger picture evolution. There are direct analogues of the first and second laws of thermodynamics as well as the involved notions of $\dot{S}_{int} \geq 0$ and $\dot{S}_{ext} = (1/T)\dot{Q}$. The real distinction between dissipative and nondissipative motions can be attributed to violation of the Helmholz extremum principle by the quantum dynamics. Examples are given in [**510**] and we refer to this paper, as well as [**509**], for an excellent aperçu of these matters.

4. ENTROPY AND GRAVITY

The derivation of the Einstein equations via an entropy functional and its holographic aspects in Padmanabhan [**976, 977**] (and sketched in [**258**] and Chapter 6) has been developed much further and should be understood as embedded in a whole array of papers (see e.g. [**431, 679, 685, 690, 768, 904, 976, 977, 978, 979, 980, 981, 982, 983, 984, 986, 1124, 1194, 1232, 1273, 1327**]). We omit reference here to the classic Beckenstein and Hawking papers (and many other references) on black hole entropy and radiation.

REMARK 10.4.1. There is an important concept of horizon which we only sketch here (see e.g. Padmanabhan [**977, 978, 979**] and cf. also e.g. [**77, 356, 1072**]). Perhaps an example from [**978**] will illustrate the matter best. Given a timelike curve $X^a(t)$ with past light cone $C(t)$ for the event $P[X^a(t)]$ one looks at the union U of all the past light cones for $(-\infty < t < \infty)$. If U has a nontrivial boundary then there will be regions in the spacetime from which this observer cannot receive signals. The boundary of the union of the causal pasts of all the observers in a congruence of such timelike curves (essentially the boundary of the union of backward light cones) will define a causal horizon for this congruence. A general class of metrics with such a static horizon can be described via ($\bigstar$) $ds^2 = -N^2(x^\alpha)dt^2 + \gamma_{\alpha\beta}(x^\alpha)dx^\alpha dx^\beta$ with conditions (i) $g_{00}(\mathbf{x}) = -N^2(\mathbf{x})$ vanishes on some 2-surface $\mathcal{H}$, (ii) $\partial_\alpha N$ is finite and nonzero on $\mathcal{H}$, and (iii) All other metric components and curvature remain finite and regular on $\mathcal{H}$. The natural congruence of observers with $\mathbf{x} = const.$ will perceive $\mathcal{H}$ as a horizon. The 4-velocity $u_a = -N\delta_a^0$ of such observers has a 4-acceleration $a^i = u^j\nabla_j u^i = (0, \mathbf{a})$ with $a_\alpha = (\partial_\alpha N/N)$. These static spacetimes have a more natural coordinate system defined locally in terms of the level surfaces of N. Thus transform from the

x^μ to the set (N, y^A), $A = 2,3$ where the y^A are transverse coordinates on the $N = constant$ surface. The metric can be transformed to

$$ds^2 = -N^2 dt^2 + \frac{dN^2}{(Na)^2} + dL_\perp^2 \tag{4.1}$$

where $dL_\perp^2$ is the metric on the transverse plane. Near the $N = 0$ surface $Na \to \kappa$ (surface gravity) and the metric reduces to the Rindler form

$$ds^2 \simeq -N^2 dt^2 + \frac{dN^2}{\kappa^2} + dL_\perp^2 = -\kappa^2 x^2 dt^2 + dx^2 + dL_\perp^2 \tag{4.2}$$

with $x = N/\kappa$. In classical theory the horizon at $N = 0$ acts as a 1-way membrane and shields the observers at $N > 0$ from anything taking place on the other side of the horizon ($N < 0$) (ignoring entanglement and tunneling in QM). ■

In Padmanabhan [**977, 978**] for example one develops the idea that the true degrees of freedom for gravity in a volume V live on the boundary ∂V. In this case one should be able to obtain the dynamics of gravity using only the surface term of an Einstein-Hilbert (EH) type action. Starting with a spacetime manifold V part of whose boundary is made up of the horizon $\mathcal{H}$ one posits a surface term arising from integrating a 4-divergence term in a Lagrangian with generic form (**4A**) $A_{sur} = \int_V d^4x\sqrt{g}\nabla_a U^a$ where U^a is to be built out of (i) The normal u^i to the boundary ∂V (ii) The metric g_{ab} and (iii) The covariant derivative operator ∇_j acting at most once. The normal u^i is defined only on ∂V but can be extended in any manner to V as a vector field. Allowing the action to depend on u^i as well as g_{ab} introduces a foliation (observer) dependence, although A_{sur} is still generally covariant; the resulting dynamical equations turn out to be independent of u^i as desired. There are then only four possible choices for U^i, namely $u^j\nabla^i u_j$, $u^j\nabla_j u^i$, $u^i\nabla_j u^j$, and u^i. The first vanishes since u^j has constant norm and the second (which is the acceleration $a^i = u^j\nabla^i u_j$) gives zero upon integration, since the dot product with the normal on ∂V vanishes, i.e. $u_i U^i = a^i u_i = 0$. Hence the most general U^i that need be considered is a linear combination of u^i and Ku^i where $K = -\nabla_i u^i$ is the trace of the extrinsic curvature of ∂V. Of these $U^i = u^i$ will lead to the volume of the bounding surface and can be ignored (it only adds a constant to K and does not alter any conclusions to follow). Thus the surface term (arising from Ku^i) must have the form

$$A_{sur} \propto \int_V d^4x\sqrt{g}\nabla_i(Ku^i) = \frac{1}{8\pi G}\int_{\partial V} d^3x\sqrt{h}K \tag{4.3}$$

where G is a constant to be determined (with dimensions of area in natural units with $c = \hbar = 1$) and 8π is introduced with hindsight. In fact A_{sur} is familiar but it is introduced here without recourse to the EH action. More importantly $-A_{sur}$ has the physical interpretation of the entropy attributed to the horizon by observers (cf. [**976, 977, 979**]). This follows by an argument using Rindler coordinates which yields (**4B**) $A_{sur} \sim -(1/4)(A_\perp/G)$ which is minus (1/4) of the transverse area $A_\perp$ of the horizon; here the surface contribution is due to removing the inaccessible region it makes to identify $-A_{sur}$ with an entropy (see [**976, 977, 979**] for details and discussion - matters are only sketched here).

To obtain the dynamics for the spacetime continuum one takes the total action $A_{tot} = A_{sur} + A_M[\phi_i, g]$ where $A_M[\phi_i, g]$ is the standard matter action in a spacetime with metric g_{ab} (the ϕ_i denote some generic matter degrees of freedom and varying the ϕ_i leads to standard equations of motion for matter in a background metric with energy-momentum (EM) tensor T^a_b satisfying $\nabla_a T^a_b = 0$. One shows now in [**976, 977**] that the Einstein equations arise from the demand that A_{tot} should be invariant under virtual displaements of the horizon normal to itself. Let $\mathcal{H} \subset \partial V$ be the horizon and consider an infinitesimal coordinate transformation $x^a \to \bar{x}^a = x^a + \xi^a(x)$ where $\xi^a(x)$ is nonzero only on $\mathcal{H}$ and is in the direction of the normal to the horizon (making it a null vector). This induces a (virtual) displacement of the horizon normal to itself, leaving the other parts of the boundary intact. The metric changes via $\delta g^{ab} = \nabla^a \xi^b + \nabla^b \xi^a$ the

$$\delta A_M = -\frac{1}{2}\int_V d^4x\sqrt{-g}T_{ab}\delta g^{ab} = -\int_V d^4x\sqrt{-g}\nabla_a(T^a_b\xi^b \tag{4.4}$$

where one uses $\nabla_a T^a_b = 0$ (arising from the equations of motion for the matter). Next one needs $\delta A_{sur}/\delta g^{ab}$ under infinitesimal coordinate transformation which is given via

$$\delta A_{sur} = \frac{1}{8\pi G}\int_V d^4x\sqrt{-g}\nabla_a(R^a_b\xi^b) \tag{4.5}$$

A derivation is provided since this result is not found in standard text books. Thus varying (4.3) directly is possible but more clever is to recall that A_{sur} in (4.3) is the usual extrinsic curvature term which is added to the EH action in order to cancel the variation in the term involving the second derivatives of the metric (cf. Padmanabhan [**979**]). Hence the variation of $(-A_{sur})$ is the same as that of the second derivative term in the EH action (mentioned and usually ignored in standard textbooks when deriving the Einstein equations). Therefore one has

$$\delta(-A_{sur}) = \frac{1}{16\pi G}\int_V d^4x\sqrt{-g}g^{ab}\delta R_{ab} \tag{4.6}$$

Now under $x^a \to \bar{x}^a$ the integrand in (4.6) is $\sqrt{-g}g^{ab}\delta R_{ab} = -2\sqrt{-g}\nabla^a(R_{ab}\xi^b)$ which establishes (4.5). Indeed to see this note

$$\delta[\sqrt{-g}R] = -\sqrt{-g}\nabla_a(R\xi^a) = \sqrt{-g}[G_{ab}\delta g^{ab} + g^{ab}\delta R_{ab}] \tag{4.7}$$

The first equality follows from the fact that the local functional variation $\delta[\sqrt{-g}Q(x)]$ of any scalar density made from a generally covariant scalar $Q(x)$ is $\delta[\sqrt{-g}Q(x)] = -\sqrt{-g}\nabla_a(Q\xi^a)$. The second equality in (4.7) is a standard text book result for $\delta[\sqrt{-g}R]$. Using $\delta g^{ab} = \nabla^a\xi^b + \delta^b\xi^a$ and $\delta^a G_{ab} = 0$ one obtains then

$$\begin{aligned} -\sqrt{-g}\nabla_a(R\xi^a) &= 2\sqrt{-g}\nabla^a[(R_{ab} - (1/2)g_{ab}R)\xi^b] + \sqrt{-g}g^{ab}\delta R_{ab} = \\ &= 2\sqrt{-g}\nabla^a(R_{ab}\xi^b) - \sqrt{-g}\nabla_a(R\xi^a) + \sqrt{-g}g^{ab}\delta R_{ab} \end{aligned} \tag{4.8}$$

and this gives immediately $\sqrt{-g}g^{ab}\delta R_{ab} = -2\sqrt{-g}\nabla^a(R_{ab}\xi^b)$ needed to prove (4.5).

The rest is straightforward. The integration of the divergences in (4.4) and (4.5) leads to surface terms which contribute only on $\mathcal{H}$ since ξ^a is nonzero only on $\mathcal{H}$. Further since ξ^a is in the direction of the normal the demand $0 = \delta A_{tot}$

leads to the result $(R^a_b - 8\pi G T^a_b)\xi^b\xi_a = 0$. But ξ^a is arbitrary (except for being a null vector) and hence $R^a_b - 8\pi G T^a_b = F(g)\delta^a_b$ where F is arbitrary. Finally since $\nabla_a T^a_b = 0$ identically, $R^a_b - F(g)\delta^a_b$ must have identically zero divergence, so F must have the form $F = (1/2)R + \Lambda$ where R is the scalar curvature and λ is an (unfortunately) undetermined cosmological constant. The resulting equation is then (**4C**) $R^a_b - (1/2)R\delta^a_b + \Lambda\delta^a_b = 8\pi G T^a_b$ which is Einstein's equation without any bulk term involved in its derivation.

In this approach the action functional for the continuum spacetime is

$$A_{tot} = A_{sur} + A_M = \frac{1}{8\pi G}\int_{\partial V} d^3x\sqrt{h}K + \int_V d^x\sqrt{-g}L_M \tag{4.9}$$

where matter lives in the bulk V while the gravity contributes on the boundary ∂V. Since S_{sur} is related to entropy, its variation when the horizon is moved infinitesimally is equivalent to the change in entropy dS. The variation of the matter term contributes the PdV and dE terms so that the entire variational principle is equivalent to the thermodynamic identity $TdS = dE + PdV$ (cf. also [**981**] for examples). This also shows that the Einstein theory has an intrinsic holography and it is noted that

$$\sqrt{-g}L_{sur} = -\partial_a\left(g_{ik}\frac{\partial\sqrt{-g}L_{bulk}}{\partial(\partial_a g_{ik})}\right) \tag{4.10}$$

This gives some structure and meaning but the subject is worthy of much more elaboration and discussion.

We give now an ingress into some expansion of the above arguments following Padmanabhan [**978, 981**] (cf. also [**904, 979, 981**]). The papers in [**978, 981**] deal with fundamental matters such as dark energy, the cosmological constant, and vacuum energy in connection with holographic gravity. Also [**904, 982**] for example deal with Lanczos-Lovelock type gravities. We would like to write out all of the details (in order to properly understand all this) but there are too many so we continue to only sketch some of the material and begin with [**978**]. Thus the issue of cosmological constant seems to be related to questions such as: (i) How is the microscopic structure of the vacuum modified by gravity and (ii) What kind of macroscopic gravitational field is produced by the vacuum. In dealing with horizons and boundaries one inserts a level of observer dependence in the action functional via the existence of horizons, namely the physics of the region blocked by the horizzon will be encoded in a boundary term in the action. In fact, as we have seen above, it is possible to obtain the dynamics of gravity using only the surface term of the EH action (without using the bulk term). In this approach

$$A_{tot} = A_{sur} + A_M = \frac{1}{16\pi G}\int_{\partial V} d^3x\sqrt{-g}n_c Q_a{}^{bcd}\Gamma^a{}_{bd} + \int_V d^4x\sqrt{-g}L_m(g,\phi) \tag{4.11}$$

where $Q_a{}^{bcd} = (1/2)(-\delta^c_a g^{bd} + \delta^d_a g^{bc})$ (cf. [**904**]). Matter degrees of freedom live in the bulk V with gravity contributing via ∂V. When the boundary has a part which acts as a horizon for a class of observers one demands that the action should be invariant under virtual displacements of this horizon and this leads to the Einstein theory with a cosmological constant arising as an integration constant. This is called then the holographic dual description of Einstein gravity. We have seen also that this has a thermodynamic interpretation as in (**4D**) ($TdS = dE + PdV$). In this approach the continuum spacetime is like an elastic solid (the Sakharov paradigm) with the Einstein equations providing the macroscopic description. The horizons in the spacetime are similar to defects in the solid so that their displacement costs entropy.

The general setting goes as follows. Consider a generalized theory of gravity in D dimensions based on a covariant Lagrangian $L[g^{ab}, R^a_{\ bcd}]$ and use $(g^{ab}, \Gamma^i_{\ k\ell}, R^a_{\ bcd})$ as the independent variables. The curvature tensor $R^a_{\ bcd}$ can be expressed entirely in terms of $\Gamma^i_{\ k\ell}$ and $\partial_j \Gamma^i_{\ k\ell}$; it is independent of g^{ab}. One defines also the tensor $P_a^{\ bcd} = (\partial L/\partial R^a_{\ bcd})$ which has the same symmetries as $R^a_{\ bcd}$. Variation of the action functional gives

$$\delta A = \delta \int_V d^D x \sqrt{-g} L = \int_V d^D x \sqrt{-g} E_{ab} \delta g^{ab} + \int_V d^D x \sqrt{-g} \nabla_j \delta v^j \quad (4.12)$$

$$E_{ab} = \left(\frac{\partial \sqrt{-g} L}{\partial g^{ab}} - 2\sqrt{-g} \nabla^m \nabla^n P_{amnb} \right) \quad (4.13)$$

where $\delta v^j = [2P^{ibjd}(\nabla_b \delta g_{di}) - 2\delta g_{di}(\nabla_c P^{ijcd})]$. One usually assumes that $n_a \delta v^a = 0$ on ∂V where n_a is the normal to the boundary. This would require however conditions on the dynamical variables and their derivatives on the boundary and to simplify this one concentrates on a subset of Lagrangians for which P^{abcd} is divergence free, i.e. (**4E**) $\nabla_c P^{ijcd} = 0$, and because of the symmetries this means P^{abcd} is divergence free in all indices. Then from (4.13) the source free equations of motion $E_{ab} = 0$ reduce to (**4F**) $(\partial \sqrt{-g} L / \partial g^{ab}) = 0$. It turns out in fact that this condition encompasses all the gravitational theories in D dimensions in which the field equations are of no higher than second degree. The idea here is that possible fourth rank tensors P^{abcd} which have the symmetries of the curvature tensor, are divergence free, and are made up from g^{ab} alone; hence we can write (**4G**) $P_a^{\ bcd} = (1/2)(\delta^c_a g^{bd} - \delta^d_z g^{bc})$ leading to (**4H**) $L \equiv P_a^{\ bcd} R^a_{\ bcd} + c \equiv R - 2\Lambda$. Then the standard field equations arise from the ordinary derivative as in (**4F**) with $\sqrt{-g} L = \sqrt{-g}[g^{ab} R_{ab} - 2\Lambda]$. If P^{abcd} is allowed depend linearly on curvature there are additional Gauss-Bonnet type terms possible in P^{abcd} which lead to a pure divergence for the action in 4-D. In fact the holographic dual description of gravity, in which the same equations arise from a surface term, exists for many related theories (see [**978**] for details).

Going now to [**981**] one notes that most interactions of nature are invariant under shifting L_M by a constant which shifts the energy by $T_{ab} \to T_{ab} + \Lambda g_{ab}$; however this does not occur for gravity because of the $\sqrt{-g}$ term. Hence it is desirable to change the field equations to become invariant under $T_{ab} \to T_{ab} + \Lambda g_{ab}$ which is the same as working with the trace free part of the equations or alternatively with (**4I**) $(G_{ab} - 8\pi T_{ab}) \xi^a \xi^b = 0$ for all null vectors ξ^a. Either formulation, when combined with the Bianchi identity leads to $G_{ab} = 8\pi T_{ab} + C g_{ab}$ with an integration constant C which changes the nature of the cosmological constant term. This cannot be obtained from a bulk action which is an integral of a local Lagrangian

L with a measure $\sqrt{-g}d^Dx$ but one considers here

$$L_{EH} \sim R \sim (\partial g)^2 + \partial^2 g \equiv \sqrt{-g}L_{bulk} - L_{sur}; \tag{4.14}$$

$$L_{bulk} = 2Q_{ab}^{dc} g^{bi} \Gamma^a{}_{dk} \Gamma^k{}_{ic};\ L_{sur} = 2Q_{ak}^{cd} \partial_c [\sqrt{-g} g^{bk} \Gamma^a{}_{bd}$$

where $2Q_{ab}^{cd} = (\delta_a^d \delta_b^c - \delta_a^c \delta_b^d)$ is the alternating (determinant) tensor. The minus sign provides that L_{sur} is the Lagrangian that should be added to L_{EH} to get an action which is quadratic in the first derivatives of the metric. The surface term obtained by integrating $L_{sur} \propto \partial^2 g$ should be ignored or cancelled by an extrinsic curvature term (cf. York [**1327**]) to obtain a well defined variational derivative that will lead to the Einstein equations. Thus the (covariant) field equations essentially arise from the variation of the noncovariant (or foliation dependent) bulk term $L_{bulk} \propto (\partial g)^2$ - called the Γ^2 Lagrangian. On the other hand via [**977, 978, 979**] one knows that (as above)

$$[(D/2) - 1]\sqrt{-g}L_{sur} - \partial_a \left(g_{ik} \frac{\partial \sqrt{-g} L_{bulk}}{\partial(\partial_a g_{ik})} \right) \tag{4.15}$$

(for $D = 4$ the coefficient on the left is unity). In this spirit the transition from L_{bulk} to $L_{EH} = L_{bulk} - L_{sur}$ can be thought of as a transition to momentum representation. Thus given any $L_q(\dot{q}, q)$ one can always construct a $L_p(\ddot{q}, \dot{q}, q)$ which depends on the second derivatives $\ddot{q}$ but gives the same equation of motion by using (**4J**) $L_p = L_q - (d/dt)\left(q\frac{\partial L_q}{\partial \dot{q}}\right)$ (cf. [**979**]). Here keeping $\delta p = 0$ at the end points and varying L_p leads to the same equations of motion as keeping $\delta q = 0$ at the end points and varying L_q. Relations of this kind clearly indicate that both L_{sur} and L_{bulk} contain the same information content. Using this perspective one provides again in [**981**] a derivation of the Einstein equations via an action principle using L_{sur}. It is more instructive to read the original exposition here so we refrain from an attempt at sketching this.

5. GINZBURG-LANDAU, EINSTEIN-HILBERT, AND RICCI-YAMABE

We sketch here some material from Kholodenko and Ballard [**742, 743**] involving Ginzburg-Landau equations, the Einstein-Hilbert action, Ricci and Yamabe flows, and Perelman entropy. These topics go very deep and are connected to number theoretic ideas (cf. [**742**]); in particular one looks at statements of the form "black holes make the universe arithmetic" (cf. paper one in [**742**]) although we make no attempt to discuss black holes or algebraic number theory here. There is also an excellent survey of background ideas in [**742, 743**]. The Ginzburg-Landau equations have been mentioned in Chapter 8 and we refer to [**1029, 1267**] for much more on that in connection with Bose-Einstein condesation and superfluidity (cf. also [**38, 677**]). One begins paper two of Kholodenko [**742**] with the ϕ^4 model and an action functional in d dimensions $S[\phi] = \int d^dx \mathfrak{L}(x)$ (cf. [**677**]). The requirement that $S[\phi]$ be independent of a scale factor λ, i.e. $\int d^dx \mathfrak{L}(x) = \int d^dx \lambda^d \mathfrak{L}(\lambda x)$ leads to the constraint

$$\int d^dx \left(x \cdot \frac{\partial}{\partial x} + d \right) \mathfrak{L}(x) = 0 \tag{5.1}$$

(obtained by differentiation of S in λ and setting $\lambda = 1$). For (**5A**) $\mathfrak{L}(x) = (1/2)(\nabla\phi)^2 + (m^2/2)\phi^2 + (\hat{G}/4!)\phi^4$ one obtains (for $\epsilon \sim log(\lambda)$)

(5.2) $$\frac{\delta\mathfrak{L}}{\delta\epsilon} = \left(x \cdot \frac{\partial}{\partial x} + d\right)\mathfrak{L}(x) + (d-4)\frac{\hat{G}}{4!}\phi^4 - m^2\phi^2$$

(cf. [**677**]). Consequently the action $S[\phi]$ is scale invariant if $d = 4$ and $m^2 = 0$. One looks then for conformal invariance and noting from [**1070**] that for a metric $\tilde{g} = exp(f)g$ one has, with some sign adjustment (cf. [**677**]),

(5.3) $$\Delta_{\tilde{g}} = e^{-f}\Delta_g - \frac{1}{2}\left(\frac{d}{2} - 1\right)[fe^{-f}\Delta_g + e^{-f}(\Delta_g f) - e^{-f}\Delta_g \circ f]$$

(here $\Delta_g\psi = -(det(g)^{-1/2}\partial_i[g^{ij}det(g)^{1/2}\partial_j\psi]$ for a scalar function ψ). This means that for $d > 2$ the conformal invariance of the ϕ^4 model is absent. However using [**38**] one considers first scaling of the noninteracting (free) G-L theory with action functional (**5B**) $S[\phi] = \int d^dx[(\nabla\phi)^2 + m^2\phi^2]$ via (**5C**) $\tilde{\phi}(Lx) = L^{\omega}\phi(x)$. Then if one requires

(5.4) $$\int d^dx[(\nabla\phi)^2 + m^2\phi^2] = \int d^dxL^d[(\tilde{\nabla}\tilde{\phi})^2 + \tilde{m}^2\tilde{\phi}^2]$$

and uses (**5C**) there results

(5.5) $$S[\phi] = \int d^dxL^d[(\nabla\phi)^2L^{2\omega-2} + \tilde{m}^2\phi^2L^{2\omega}]$$

Thus for $S[\phi]$ to be scale invariant the mass m^2 should scale via $\tilde{m}^2 = m^2L^{-2}$ with $\omega = 1 - (d/2)$ (cf. also [**38, 742**] for variations).

One notes next that to have the mass scaling as scalar curvature R for some Riemannian manifold, i.e. $\tilde{m}^2 = m^2L^{-2}$, is the same as scaling R via (**5D**) $\tilde{R} = L^{-2}R$ (cf. Appendix D in [**1274**]). In general the scalar curvature $R(g)$ changes under conformal transformation $\hat{g} = exp(2f)g$ via

(5.6) $$\hat{R}(\hat{g}) = e^{-2f}[R(g) - 2(d-1)\Delta_g f - (d-1(d-2)|\nabla_g f|^2]$$

Thus for constant f the scaling is in accord with (**5D**). In fact one can do more following [**805**] by writing $exp(2f) = \phi^{p-2}$ where $p = [2d/(d-2)]$ leading to $\hat{g} = \phi^{p-2}g$; it follows that (5.6) becomes (**5E**) $\hat{R}(\hat{g}) = \phi^{1-p}[\alpha\Delta_g\phi + R(g)\phi]$ with $\alpha = 4[(d-1)/)d-2)]$. Consequently one can deal with actions which are both scale and conformally invariant for $d \geq 3$.

Now let $\tilde{R}(\tilde{g})$ in (5.6) ($\tilde{g} \sim \hat{g}$ etc.) be some constant (shown below); then (**5E**) takes the form (**5F**) $\alpha\Delta_g\phi + R(g)\phi = \hat{R}(\hat{g})\phi^{p-1}$. Since $p - 1 = [(d+2)/(d-2)]$ one obtains $p - 1 = 3$ for $d = 4$ and $p - 1 = 5$ for $d = 3$. These are the familiar Ginzburg-Landau values for critical and tricritical GL theories (cf. [**744**]). The action functional can then be constructed (in place of (5.5)) in a manifestly covariant form, namely

(5.7) $$S[\phi] = \frac{1}{(\int_M d^dx\sqrt{g}\phi^p)^{2/p}}\int_M d^dx\sqrt{g}[\alpha(\nabla_g\phi)^2 + R(g)\phi^2] = \frac{E[\phi]}{\|\phi\|_p^2}$$

Minimization of this gives the Euler-Lagrange equations (**5G**) $\alpha\Delta_g\tilde{\phi} + R(g)\tilde{\phi} - \lambda\tilde{\phi}^{p-1} = 0$ with λ denoting the extremum value of the ratio $\lambda = (E[\tilde{\phi}]/\|\tilde{\phi}\|_p^p) = inf\{S[\phi];\ \hat{g}\ conformal\ to\ g\})$. In accord with the Landau theory of phase transitions (cf. [**792**]) one expects that the conformal factor ϕ is a smooth nonnegative function on M with extreme value $\tilde{\phi}$. Comparison of (**5F**) and (**5G**) yields $\lambda = \tilde{R}(\tilde{g})$ as required. These results are due to Yamabe and λ is known as the Yamabe invariant (cf. [**183, 805**]) which is an invariant of the conformal class (M, g). One recalls here

DEFINITION 5.1. The Yamabe problem is to find a compact Riemannian manifold (M, g) of dimension $n \geq 3$ whose metric is conformal to a metric $\hat{g}$ of constant scalar curvature. ■

Further developments (see e.g. [**436**]) extended this problem to manifolds with boundaries and to noncompact manifolds. One can prove that the (Yamabe-Ginzburg-Landau-like) functional is manifestly conformally invariant and to see this one can write (**5E**) in the form (**5H**) $\phi^p\hat{R}(\hat{g}) = (\alpha\phi\Delta_g\phi + R(g)\phi^2)$. Then rewrite $E[\phi]$ as $E[\phi] = \int d^dx\sqrt{\hat{g}}\hat{R}(\hat{g})$ and note that $\int d^dx\sqrt{\hat{g}} = \int d^dx\sqrt{g}\phi^p$ leading to an equation of the form (**5I**) $S[\phi] = [\int d^dx\sqrt{\hat{g}}\hat{R}(\hat{g})/(\int d^dx\sqrt{\hat{g}})^{2/p}]$ (Einstein-Hilbert form) where both the numerator and denominator are invariant with respect to conformal changes in the metric (e.g. scale change).

REMARK 10.5.1. In order to use such results in statistical mechanics one has to show that the extremum of the Yamabe functional $S[\phi]$ is realized for manifolds M whose scalar curvature $R(g)$ in (**5G**) is also constant and this is the essence of the Yamabe problem solved by several authors (cf. [**79, 805, 1130**]). Next in view of the relation $\int d^dx\sqrt{\hat{g}} = \int d^dx\sqrt{g}\phi^p$ it is clear that for the fixed background metric g the result (**5G**) can be obtained also using the functional

$$\tilde{S}[\phi] = \int d^dx\sqrt{g}[\alpha(\nabla_g\phi)^2 + R(g)\phi^2] - \tilde{\lambda}\int d^dx\sqrt{g}\phi^p \tag{5.8}$$

where $\tilde{\lambda} = (2/p)\lambda = \lambda[(d-2)/d]$ is responsible for the volume constraint. Such a functional is used for compatibility with other formulations in [**742**]. Apart from the normalizing denominator (**5I**) represents the Einstein-Hilbert (EH) action for pure gravity defined for a Riemannian d dimensional space and the denominator (volume to power $2/p$) serves to make $S[\phi]$ manifestly conformally invariant (cf. [**1130**]). ■

One considers next (following Dirac [**378**]) the extended EH action functional $S^c[g]$ for pure gravity with a cosmological constant C defined for a pseudo Riemannian manifold M of total space-time dimension d without boundary, namely

$$S^c[g] = \int_M R\sqrt{g}d^dx + C\int_M d^dx\sqrt{g} \tag{5.9}$$

Here one takes R_{ij} for the Ricci curvature tensor with Einstein equation (**5J**) $R_{ij} = \lambda g_{ij}$ (λ constant). It follows that in the Riemannian case (•) $R = d\lambda$ while in the pseudo-Riemannian case (♦) $R = (d-2)\lambda$. From [**378**] variation of the action

$S^c[g]$ yields then

$$R_{ij} - \frac{1}{2}g_{ij}R + \frac{1}{2}Cg_{ij} = 0 \tag{5.10}$$

Combining (•), (♦), and (5.10) gives also (★) $C = \lambda(d-2)$ (Riemannian) and $C = \lambda d$ (pseudo-Riemannian). One can then write (5.10) as

$$R_{ij} - \frac{1}{2}g_{ij}R + \frac{1}{2d}(d-2)Rg_{ij} = 0 \ \ or \ \ R_{ij} - \frac{1}{2}g_{ij}R + \frac{d}{2(d-2)}Rg_{ij} \tag{5.11}$$

The latter equation is for the pseudo-Riemannian case but Kholodenko reports that it is sufficient here to deal with the Riemannian situation (a result to appear in a paper in preparation). Consequently one can argue that (5.10) can be obtained as well by varying the Yamabe functional (**5I**) and, following [**79, 1130**], let t be a small parameter labeling a family of metrics $g_{ij}(t) = g_{ij} + th_{ij}$. It was shown there that

$$\left(\frac{dR_t}{dt}\right)_{t=0} = \nabla^i\nabla^j h_{ij} - \nabla^j\nabla_j h^i_i - R^{ij}h_{ij};\ \left(\frac{d}{dt}\sqrt{|g_t|}\right)_{t=0} = \frac{1}{2}\sqrt{|g|}g^{ij}h_{ij} \tag{5.12}$$

Now consider the Yamabe functional (**5I**) written for the family of metrics which belong to the same conformal class, namely

$$\mathcal{R}(g(t)) = (V(t))^{-2/p}\int_M R(g(t))DV(t) \tag{5.13}$$

where $V(t) = \int_M d^dx\sqrt{g(t)}$ and $DV(t) = d^dx\sqrt{g(t)}$. Using (5.12) and (5.13) and noting that $\nabla^i\nabla^j h_{ij} - \nabla^j\nabla_j h^i_i$ is a total divergence one obtains the formula

$$\left(\frac{d}{dt}\mathcal{R}(g(t))\right)_{t=0} = V(0)^{-\frac{2-d}{d}}\left[\int_M\left(\frac{1}{2}Rg^{ij} - R^{ij}\right)h_{ij}DV(0)\int DV(0) - \right. \tag{5.14}$$
$$\left. -\left(\frac{1}{2}-\frac{1}{d}\right)\int_M DV(0)\int_M h_{ij}g^{ij}DV(0)\right]$$

If the metric g is the critical point for $\mathcal{R}(g(t))$ then

$$\left[R_{ij} - \frac{R}{2}g_{ij}\right]\int_M DV(0) + \left(\frac{1}{2}-\frac{1}{d}\right)[\int_M RDV(0)]g_{ij} = 0 \tag{5.15}$$

Multiplying by g^{ij} and summing yields (**5K**) $R - (d/2)R + [(1/2)-(1/d)] < R > d = 0$ where $< R >= (1/V(0))\int RDV(0)$ is the average scalar curvature. Note that (**5K**) can be written as $R =< R >$ and this condition is formally equivalent to the Einstein condition (**5J**) in view of (•); hence under such circumstances (5.15) and (**5K**) are equivalent.

REMARK 5.10.2. One note that in general the condition $R =< R >$ is more restrictive than the condition (**5J**); the constant Ricci tensor R_{ij} always leads to the constant scalar curvature R while the Riemannian spaces of constant scalar curvature R are not necessarily spaces for which the Einstein condition (**5J**) holds and they may or may not be of Einstein type (cf. [**1133**]). In [**742**] one discusses this problem dynamically following work of Hamilton and Perelman and it is shown that the equation $R =< R >$ is compatible with the Einstein condition and hence describes Einstein spaces of Riemannian type (see remarks below). ■

In paper two of [**742**] there follows a discussion of GL theory in 2-D and connections to string theory and conformal field theory (CFT). This is followed by a section on designing higher dimensional CFT where use of the conformal Yamabe Laplacian (**5L**) $\Box_g = \Delta_g + \tilde{\alpha}R(g)$ ($\tilde{\alpha} = \alpha^{-1} = (1/4)[(d-2)/(d-1)]$) is productive. We refer to [**742**] for this material, from which a few formulas are extracted below, and go to Section 7 of [**742**] (paper two) where one goes to critical dynamics and Yamabe and Ricci flows. The time dependent GL theory of Landau and Khalatnikov (cf. [**794**]) is based on the assumption that an order parameter satisfies a relaxation equation of the type (**5M**) $\partial_t\phi = -\gamma(\delta\mathfrak{F}(\phi)/\delta\phi) \equiv -\gamma grad\mathfrak{F}(\phi)$ where $\mathfrak{F}$ is a free energy of the form (cf. also [**38**])

$$\mathfrak{F}(\phi) = \mathfrak{F}_0 + \frac{1}{2}\int d\mathbf{x}\left[c(\nabla\phi(\mathbf{x}))^2 + a\phi^2(\mathbf{x}) + \frac{b}{2}\phi^4(\mathbf{x})\right] \tag{5.16}$$

(γ is a friction coefficient which can be eliminated by rescaling). Note that (**5M**) is an example of gradient flow popularized recently by Perelman's solution of the Poincaré conjecture (cf. [**221, 609, 891, 1009, 1235**]). Following [**609**] one writes

DEFINITION 5.2. The normalized Ricci flow is described by a dynamical equation (**5N**) $\partial_t g_{ij} = (2/d)g_{ij} < R(g) > -2R_{ij}(g)$ where $< R(g) >= (\int_M R\sqrt{g}d^dx)/\int_M \sqrt{g}d^d \equiv (\int_M Rd\mu)/\int_M d\mu$ with $g_{ij}(t=0) = \hat{g}_{ij}$. ■

REMARK 10.5.3. In two dimensions $R_{ij}(g) = (1/2)Rg_{ij}$ so that all two dimensional spaces are Einsteinian and (**5N**) becomes (**5O**) $\partial_t g_{ij} = (< R(g) > -R(g))g_{ij}$. This last equation defines the Yamabe flow and it can be considered for any $d \geq 2$ and the fixed points (or equilibrium situation) should correspond to (★) $< R(g) > -R(g) = 0$. It is only in 2 dimensions that the Yamabe and the normalized Ricci flow are essentially equivalent (cf. [**609**]) and this is due to the fact that only in two dimensions one can write $R_{ij}(g) = (1/2)Rg_{ij}$. ■

REMARK 10.5.4. Even though Yamabe and Ricci flows are different in higher dimensions one can ask whether some Ricci flows can be used in order to obtain results for Yamabe type flows and this is demonstrated to be possible. To this end using (**5O**) one has (**5P**) $(1/2)g^{ij}\partial_t g_{ij} =< R(g) > -R(g)$. In addition, since in any dimension $\partial_t d\mu \equiv \partial_t\sqrt{det(g_{ij})}d^dx = (1/2)g^{ij}\partial_t g_{ij}d\mu$ one obtains as well (**5Q**) $\partial_t \int_M d\mu = \int_M \partial_t d\mu = \int_M(< R > -R(g))d\mu = 0$ which is compatible with (**5N**), valid for $d \geq 2$. This follows via $\partial_t d\mu = (1/2)g^{ij}\partial_t g_{ij}d\mu$ and (**5N**). Hence after some manipulation, (**5N**), describing Ricci flow and valid for $d \geq 2$, can be brought into a form identical with (**5P**) which is a Yamabe flow (obtained originally for $d = 2$). This does not mean that the flows are equivalent but the Yamabe flow can be looked upon as a special case of Ricci flow. The fixed points for such flows are given via (★). Further since the fixed point (★) is the trace form of the fixed point for the Ricci flow (**5N**), i.e. of (**5S**) $(1/d)g_{ij} < R(g) > -R_{ij}(g) = 0$, which is equivalent to (**5I**), this means that the fixed point solutions of the normalized Ricci flow always produce Einstein spaces, provided these fixed points are stable (see Remark 10.5.2 and below). ■

The Yamabe flow (**5O**) should be supplemented by the initial condition $g_{ij}(t =$

$0) = g_{ij}(0) \equiv \hat{g}$ and in order to study the evolution of this metric it is convenient to write (**5T**) $g_{ij}(t) = |\phi(t)|^{4/(d-2)} g_{ij}(0)$. Putting this in (**5O**) and using (**5E**) gives an equivalent form

$$\partial_t \phi = -\phi^{2-p}(\alpha\Delta_{\hat{g}}\phi + R(\hat{g})\phi) + < R(g) > \phi \tag{5.17}$$

This can be compared with the phenomenological result (**5M**) and although different in appearance their fixed points (if stable) are the same. It is suggested that the equations of critical dynamics as in [**290**] should be replaced by (5.17) in accordance with the requirement of conformal invariance at criticality (cf. [**745**]). Using then (**5P**) one introduces (**5U**) $\eta = [4/(d-2)](< R(g) > -R(g))$ which in view of (**5Q**) upon integration leads to $\int_M \eta d\mu = 0$. Combining (**5Q**) and (5.17) yields, after calculations as in [**1139**], the equation (**5W**) $\partial_t < R(g)(t) >= [(d-2)/2]\int_M R(g)\eta d\mu$. Then combining (**5U**) and (**5W**) one gets
(5.18)

$$\partial_t < R(g(t)) >= 2\int_M R(g)(< R(g) > -R(g)) = -2\int_M [< R(g) > -R(g)]^2$$

where the second equality comes from (**5V**). The last expression has an entropic meaning following Boltzman; his H-function (**5X**) $H = -\int f log(f) d\mathbf{v}$ where $f(\mathbf{v}, t)$ satisfies the Boltzman equation

$$\partial_t f = \int [f(\mathbf{v}', t)f(\mathbf{v}_1', t) - f(\mathbf{v}, t)f(\mathbf{v}_1, t)]|\mathbf{v}_1 - \mathbf{v}|I(\theta, |\mathbf{v}_1 - \mathbf{v}|)d\mathbf{v}_1 d\Omega \tag{5.19}$$

with $I(\theta, |\mathbf{v}_1 - \mathbf{v}|)$ some known function and $d\Omega = Sin(\theta)d\theta d\phi$ while $\mathbf{v}' = \mathbf{v} + \mathbf{n}(\mathbf{n}\cdot\mathbf{g})$ and $\mathbf{v}_1' = \mathbf{v}_1 - \mathbf{n}(\mathbf{n}\cdot\mathbf{g})$ are respective velocities of the colliding particles after (primed) and before (unprimed) scattering on each other ($\mathbf{g} = \mathbf{v}_1 - \mathbf{v}$ here and $\mathbf{n}$ is a unit normal - see [**667**] for details). It can be shown that $d\mathbf{v}d\mathbf{v}_1 = d\mathbf{v}'d\mathbf{v}_1'$ and combining (**5X**) and (5.21) one can show that $\partial_t H \geq 0$ leading to the Maxwell distribution of particle velocities in the equilibrium situation. Given this background one is motivated to choose (**5Y**) $- < R(g(t)) >$ as an entropy (given a normalized volume of unity). Thus one identifies (**5O**) (or (5.17)) with a Boltzman type equation and the result (**5K**) provides physically meaningful equilibrium solutions. One arrives at the conclusion that the fixed point solutions for the Yamabe flow are described by the Einstein equation (**5J**) leading to Einstein manifolds of constant scalar curvature in accord with Remark 10.5.2.

Next one demonstrates that some Ricci flows produce results which are in agreement with those obtained for the Yamabe flows and one finds an entropy for the Ricci flows following Perelman [**1009**]. Thus the Ricci flow is not gradient (cf. [**905**]) and one cannot therefore immediately find an entropy to study its stability; but there is a way around this sketched in [**742**]. First for the Yamabe flow the choice of $- < R(g(t)) >$ as an entropy is somewhat artificial since there was no entropic procedure behind this. One begins now with the observation that the heat operator $\Box = \partial_t - \Delta_g$ has an associated operator $\Box^* = -\partial_t - \Delta_g + R(g)$ as an adjoint with respect to Ricci flow (cf. [**905**]). The proof is based on the fact that upon some rescaling (see [**609**]) it is possible to suppress the normalization of the Ricci flow (**5N**), i.e. one can eliminate the factor $(2/d)g_{ij} < R(g) >$. Consequently

(**5P**) acquires the form

$$\frac{1}{2}g^{ij}\partial_t g_{ij} = -R(g) \tag{5.20}$$

Let now g_{ij} be a solution of the non-normalized Ricci flow equation with $u = exp(-f)$ the solution of the equation $\Box^* u = 0$ for some function f such that $\int_M exp(-f)dV = 1$. By analogy with Boltzman entropy (**5X**), following [**905**], one defines the Nash entropy (**5Z**) $N(u) = \int_M ulog(u)dV$. One can then calculate

$$\partial_t N(u) = \int_M [(\partial_t u)log(u)dV + (\partial_t u)dV + ulog(u)\partial_t dV] = \tag{5.21}$$

$$= \int_M [(\partial_t - R(g))ulog(u) + \partial_t u]dV = \int_M [(-\Delta_g u)log(u) + (R(g) - \Delta_g)u]dV$$

On a closed manifold the integral of $\Delta_g u$ vanishes and the last equation acquires the form

(5.22)

$$\partial_t N(u) = \int_M \left[\frac{|\nabla_g u|^2}{u} + R(g)u\right] dV = \int_M (|\nabla_g f| + R(g))\, e^{-f} dV \equiv \mathfrak{F}(g_{ij}, f)$$

The functional $\mathfrak{F}(g_{ij}, f)$ is Perelman's entropy (cf. [**1009**]) and is also an action for the dilaton gravity [**499**]. In case f is a constant, taking into account the sign differences between the Boltzman entropy (**5X**) and the Nash entropy (**5Z**), one obtains the earlier result for the entropy $- < R(g(t)) >$ as required.

REMARK 10.5.5. The entropic nature of the EH and dilaton gravity actions could be useful fr application to black hole dynamics (cf. [**619**]). ■

6. ENHANCEMENT

Given now an entropy for the Ricci flow one wants to determine the analogue of (5.20) and following the previous logic Kholodenko begins with $\phi = exp(-f/2)$ in (5.21). This causes $\mathfrak{F}(g_{ij}, f)$ to acquire the form

$$\mathfrak{F}(g_{ij}, \phi) = \int_M [4|\nabla\phi|^2 + R(g)\phi^2]dV \tag{6.1}$$

which is to be compared to $E[\phi]$ in (5.7). By analogy with (**5H**) one can define the constant λ_g in terms of the Raleigh quotient

$$\lambda_g = \inf_\phi \frac{\mathfrak{F}(g_{ij}, \phi)}{\int_M \phi^2 dV} \tag{6.2}$$

In accord with (**5G**) λ_g serves as an eigenvalue in the equation (**6A**) $4\Delta_g\tilde{\phi} + R(g)\tilde{\phi} - \lambda_g\tilde{\phi} = 0$ where $\tilde{\phi}$ is the minimizer for (6.2). Equivalently

$$\lambda_g = inf\left\{\int_M [4|\nabla\phi|^2 + R(g)\phi^2]dV;\ \int_M \phi^2 dV = 1\right\} \tag{6.3}$$

By analogy with (5.12)-(5.14) one introduces the family of metrics $g_{ij}(s) = g_{ij} + sh_{ij}$ and instead of (5.12) one obtains

$$\frac{d}{ds}\lambda(g_{ij}, s)) = \int_M [(-h_{ij})(R_{ij} + \nabla_i\nabla_j f)e^{-f}dV \tag{6.4}$$

This is analogous to (**5W**) and one follows the same course of action. Thus following [**1009**] and also [**761**] one sees that the variation of $\mathfrak{F}(g_{ij}, f)$ leads to coupled equations for the Ricci flow

$$\partial_t g_{ij} = -2(R_{ij} + \nabla_i \nabla_j f); \ \partial_t f = -R + \Delta f \tag{6.5}$$

DEFINITION 6.1. The flow defined by (6.5) is called the generalized Ricci flow.

After making the identification $\partial_t g_{ij} = h_{ij}$ and using (6.5A) in (6.4) one obtains Perelman's monotonicity result

$$\frac{d}{ds}\lambda(g_{ij}(s)) = 2\int_M |R_{ij} + \nabla_i \nabla_j f|^2 e^{-f} dV \tag{6.6}$$

to be compared with (5.20). One concludes that $(d/ds)\lambda(g_{ij}(s)) = 0$ only when (**6B**) $R_{ij} + \nabla_i \nabla_j f = 0$.

DEFINITION 6.2. Equation (**6B**) is known as the equation for the gradient steady soliton (cf. [**221, 222, 891, 1009**]). A metric $g_{ij}(t)$ evolving via the Ricci flow is called a breather if for some $t_1 < t_2$ and $\alpha > 0$ the metric $\alpha g_{ij}(t_1)$ and $g_{ij}(t_2)$ differ only by a diffeomorphism. The case when $\alpha = 1$, $\alpha < 1$, and $\alpha > 1$ correspond to steady, shrinking, and expanding breathers respectively. Solitons are trivial breathers for which the above relation holds for any pair (t_1, t_2).

REMARK 10.6.1. If one considers the Ricci flow as a dynamical system on the space of Riemannian metrics modulo diffeomorphisms then the breathers and solitons are respectively the periodic orbits and the fixed points for such a system. One concludes that there are no breathers for the Ricci flows in agreement with Perelman (cf. [**742**] for more detail). ■

Going back to (**6B**) and multiplying by g^{ij} yields (**6C**) $R = \Delta_g f$ where one has used $-\Delta_g f = \nabla^i \nabla_i f$. Taking into account the arguments leading from (5.21) to (5.22) one sees that for compact manifolds $f = const$ and hence $R = 0$ (cf. [**609, 678**]). One looks now for physical meaning for some of this and recalls that the fundamental solution of the heat equation in d-dimensional Euclidean space is

$$u(x, y, \tau) = (4\pi\tau)^{-d/2} exp(-|x - y|^2/4\tau) \tag{6.7}$$

where $\tau = t - T$ or $T - t = -\tau$ depending on whether one is dealing with the forward or backward equation. Since $\int_M udV = 1$ one looks for a solution of the heat equation on M via an Ansatz $u = (4\pi\tau)^{-d/2} exp(-f)$ which amounts to redefining the earlier f via $f = \tilde{f} - (d/2)log(4\pi\tau)$ so that $exp(-\tilde{f}) = (4\pi\tau)^{-d/2} exp(-f)$. Then however $\nabla f = \nabla \tilde{f}$ and $\Delta_g f = \Delta_g \tilde{f}$ and one begins in the flat case. Using (6.7) one has $N_{flat} = (d/2) - (d/2)log(4\pi\tau)$ and the normalized Nash entropy is defined via $N(u) - N_{flat}$ yielding

$$\tilde{N}(u) = N(u) - N_{flat} = \int_M \left(-f + \frac{d}{2}\right) udV \tag{6.8}$$

Kholodenko then clarifies the difference between this and a partition function which caused some confusion in the interpretation of [**1009**]. Thus one defines a free energy $\mathfrak{F}$ via $log(\tilde{N}(u)) = -\beta\mathfrak{F}$ with β eventually identified with an inverse temperature (when $k_B = 1$). Here the role of temperature is played by τ and one can define an "energy" $U =< E >$ via $U = -(\partial/\partial\beta)log(\tilde{N}(u)) = \tau(\partial/\partial\tau)log(\tilde{N}(u))$. Explicitly one obtains

$$(6.9)\quad U = \tau^2\frac{\partial}{\partial\tau}[\int_M ulog(u)dV - N_{flat}] = \int_M \tau^2\partial_\tau[ulog(u)dV] + \frac{d}{2}\tau\int_M udV$$

Proceeding, via (5.21) and (5.22) (with $\int_M udV = 1$) one arrives at

$$(6.10)\quad U = \tau^2\frac{\partial}{\partial\tau}\left[\int_M ulog(u)dV - N_{flat}\right] = \int_M\left[\tau^2(R+(\nabla f)^2) + \frac{d}{2}\tau\right]udV$$

This result differs slightly from [**1009**] but this is useful since thermodynamically (**6D**) $\beta U - \beta\mathfrak{F} = S$ determines the entropy S from which the results above give (**6E**) $S_+ = \int_M[\tau(R+(\nabla f)^2) - f + d]udV$; this coincides with the entropy result for Ricci expanders obtained in [**905**]. In order to obtain the entropy for Ricci shrinkers one changes signs when taking time derivatives to get

$$(6.11)\quad S_- = \int_M[\sigma(R+(\nabla f)^2 + f - d]udV$$

again in accord with [**905**]. At this point one could proceed either by computing the heat capacity $C_v = (\partial_T U)_V$ for constant volume or repeat the argument for steady solitons adapted to the present case (this was discussed in [**222**]). More physically attractive however is to follow the logic of Perelman and note that

$$(6.12)\quad C_v = \left(\frac{\partial U}{\partial\beta}\right)\frac{\partial\beta}{\partial\tau} = -\left(\frac{\partial^2}{\partial\beta^2}log(\tilde{N}(u)\right)\frac{\partial\beta}{\partial\tau}\ \ or\ \ \tau^2C_v = \frac{\partial^2}{\partial\beta^2}log(\tilde{N}(u)$$

Recall from statistical mechanics (**6F**) $C_b = \beta^2(< E^2 > - < E >^2)$ where again $k_B = 1$ so Perelman's fluctuations are really heat capacities. Next straightforward calculations analogous to that used in (6.10) finally leads to

$$(6.13)\quad \tau^2C_v = \tau^4\int_M\left|R_{ij} + \nabla_i\nabla_j f - \frac{1}{2\tau}g_{ij}\right|^2 dV$$

This agrees with (6.6) for steady solitons (corresponding to $|\tau| \to \infty$). The gradient shrinking (or expanding) solitons are respectively solutions to

$$(6.14)\quad R_{ij} + \nabla - i\nabla_j f - \frac{1}{2\sigma}g_{ij} = 0\ \ or\ \ R_{ij} + \nabla_i\nabla_j f + \frac{1}{2\tau}g_{ij} = 0$$

By analogy with (**6B**) one can multiply these equations by g^{ij} and sum to obtain

$$(6.15)\quad R - \Delta_g f - \frac{d}{2\sigma} = 0;\ R - \Delta_g f + \frac{1}{2\tau} = 0$$

For compact manifolds $\Delta_g f = 0$ as before so in both cases one obtains again (•) and combining with (**6C**) gives Ivey's result that there are no 3-dimensional solitons or breathers on a compact connected 3-manifold other than those of constant curvature metrics.

REMARK 10.6.2. This result is due to Ivey [**678**] (cf. also [**609**]) and the derivation here is inspired by [**1009**]. It thus appears that, at least for compact connected 3-manifolds, one can extract needed physical information for the Yamabe flow from that of the Ricci flow. This provides support for claims in Remark 10.5.3. The generality of arguments used in proving (5.20) can be applied in principle to manifolds of any dimension $d \geq 2$. In particular the Euclidean dilaton gravity described by (5.22) for any $d \geq 2$ is reduced to a more familiar Euclidean gravity. ■

6.1. PERELMAN'S APPROACH. In order to expand on this and perhaps clarify some points we go now to the first paper in [**1009**] and sketch some of the material. The notation will differ at times from that of Kholodenko [**742**] exhibited already above in Sections 5 and 6. Thus the Ricci flow equation is written as $(d/dt)g_{ij}(t) = -2R_{ij}$ and Hamilton [**610**] proved that this equation has a unique short time solution for an arbitrary (smooth) metric on a closed manifold. The evolution equation for the metric tensor implies the evolution equation for the curvature tensor leading to $R_t = \Delta R + 2|Ric|^2$ so by the maximum principle its minimum is nondecreasing along the flow. Hamilton [**610**] also proved that the Ricci flow preserves the positivity of the Ricci tensor in 3-D and of the curvature tensor in all dimensions. This let to the result (for example) that evolving metrics (on a closed manifold) of positive Ricci curvature in dimension 3 converge modulo scaling to metrics of constant positive curvature. We will not discuss pinching, surgery, or singularities here. One notes that the Ricci flow arises also in QFT as an approximation to the renormalization group (cf. [**516**]).

Consider first the functional $\mathfrak{F} = \int_M (R+|\nabla f|^2)exp(-f)dV$ for a closed manifold M. Its first variation is ($\delta f = h$ and $\delta g_{ij} = v_{ij}$ where $v = g^{ij}v_{ij}$)

$$\delta\mathfrak{F}(g_{ij},h) = \int_M e^{-f}[-\Delta v + \nabla_i\nabla_j v_{ij} - R_{ij} - \tag{6.16}$$

$$-v_{ij}\nabla_i f\nabla_j f + 2 < \nabla f, \nabla h > +(R+|\nabla f|^2)((1/2)v - h)] =$$

$$= \int_M e^{-f}[-v_{ij}(R_{ij} + \nabla - i\nabla_j f) + ((1/2)v - h)(2\Delta f - |\nabla f|^2 + R]$$

Note that $(1/2)v{-}h$ vanishes identically if and only if the measure $dm = exp(-f)dV$ is kept fixed. Therefore the symmetric tensor $-(R_{ij} + \nabla_i\nabla_j f)$ is the L^2 gradient of the functional $\mathfrak{F}^m = \int_M (R+|\nabla f|^2)dm$ where now $f \sim log(dV/m)$. Hence given a measure m one may consider the gradient flow $(g_{ij})_t = -2(R_{ij} + \nabla_i\nabla_j f)$ for $\mathfrak{F}^m$. For general m this flow may not exist, even for a short time; however when it exists it is just the Ricci flow, modified by a diffeomorphism. In particular different choices of m lead to the same flow up to a diffeomorphism, i.e. the choice of m is analogous to a choice of gauge. One proves then (cf. [**1009**])

PROPOSITION 6.1. Suppose that the gradient flow for $\mathfrak{F}^m$ exists for $t \in [0,T]$; then at $t = 0$ one has $\mathfrak{F}^m \leq (n/2T)\int_M dm$.

Proof: One can assume $\int_M dm = 1$ and the evolution equations for the gradient flow of $\mathfrak{F}^m$ are then (**6G**) $(g_{ij})_t = -2(R_{ij} + \nabla_i\nabla_j f)$ with $f_t = -R - \Delta f$ and $\mathfrak{F}^m$ satisfies (**6H**) $\mathfrak{F}^m_t = 2\int |R_{ij} + \nabla_i\nabla_j f|^2 dm$. Modifying by an appropriate diffeomorphism one obtains evolution equations

$$(g_{ij})_t = -2R_{ij};\ f_t = -\Delta f + |\nabla f|^2 - R \tag{6.17}$$

and **(6H)** is retained in the form)**6I)** $\mathfrak{F}_t = 2\int |R_{ij} + \nabla_i\nabla_j f|^2 exp(-f)dV$. One computes then

$$\mathfrak{F}_t \geq \frac{2}{n}\int (R+\Delta f)^2 e^{-f}dV \geq \frac{2}{n}\int (R+\Delta f)e^{-f}dV)^2 = \frac{2}{n}\mathfrak{F}^2 \tag{6.18}$$

proving the proposition. ■

REMARK 10.6.3. The functional $\mathfrak{F}^m$ is related to some formulas of Bochner-Lichnerowicz. Thus one knows $\nabla^*\nabla u_i = (d^*d + dd^*)u_i - R_{ij}u_j$ and $\nabla^*\nabla\psi = \delta\psi^2 - (1/4)R\psi$ for 1-forms and spinors. Here ∇^* and d^* are defined via the Riemannian volume form and the Dirac operator satisfies $\delta^* = \delta$. If one substitutes $dm = exp(-f)dV$ for dV one obtains modified formulas $\nabla^{*m}\nabla u_i = (d^{*m}d + dd^{*m})u_i - R^m_{ij}u_j$ and $\nabla^{*m}\nabla\psi = (\delta^m)^2\psi - (1/4)R^m\psi$ where $\delta^m\psi = \delta\psi - (1/2)(\nabla f)\cdot\psi$, $R^m_{ij} = R_{ij} + \nabla_i\nabla_j f$, and $R^m = 2\Delta f - |\nabla f|^2 + R$. Note that $g^{ij}R^m_{ij} = R + \Delta f \neq R^m$ but one does have the Bianchi identity $\nabla^{*m}_i R^m_{ij} = \nabla_i R^m_{ij} - R_{ij}\nabla_i f = (1/2)\nabla_j R^m$ leading to $\mathfrak{F}^m = \int_M R^m dm = \int_M g^{ij}R^m_{ij}dm$. ■

A metric $g_{ij}(t)$ evolving via the Ricci flow is called a breather if, as above, for some $t_1 < t_2$ and $\alpha > 0$ the metrics $\alpha g_{ij}(t_1)$ and $g_{ij}(t_2)$ differ only by a diffeomorphism. The cases $\alpha = 1$, $\alpha < 1$, and $\alpha > 1$ correspond respectively to steady, shrinking, and expanding breathers. Trivial breathers, for which $g_{ij}(t_1)$ and $g_{ij}(t_2)$ differ only by a diffeomorphism and scaling for each pair of (t_i) are called Ricci solitons. Again if one considers Ricci flow as a dynamical system on the space of Riemannian metrics modulo diffeomorphism and scaling, then the breathers and solitons correspond to periodic orbits and fixed points respectively. At each time the Ricci soliton satisfies an equation $R_{ij} + cg_{ij} + \nabla_i b_j + \nabla_j b_i = 0$ where c is a number and b_i is a 1-form; in particular when $b_i = (1/2)\nabla_i a$ for some function a we get a gradient Ricci soliton. An important example of a gradient shrinking soliton is the Gaussion soliton where g_{ij} is just the Euclidean metric on $\mathbf{R}^n$ with $c = 1$ and $a = -|x|^2/2$.

One defines the **(6J)** $\lambda(g_{ij}) = inf\,\mathfrak{F}(g_{ij}, f)$ over all smooth functions f satisfying $\int_M exp(-f)dV = 1$. Then $\lambda(g_{ij})$ is the lowest eigenvalue of the operator $-4\Delta + R$ and **(6J)** implies that $\lambda(g_{ij}(t))$ is nondecreasing in t; also if $\lambda(t_1) = \lambda(t_2)$ then for $t \in [t_1, t_2]$ one has $R_{ij} + \nabla_i\nabla_j f = 0$ for the minimizing f. Thus a steady breather is necessarily a steady soliton. To deal with the expanding situation consider a scale invariant version $\bar{\lambda}(g_{ij}) = \lambda(g_{ij})V^{2/n}(g_{ij}$ and one proves that $\bar{\lambda}$ is nondecreasing along the Ricci flow whenever it is nonpositive; moreover the monotonicity is strict unless one is on a gradient soliton. To see this note first (intuitively) that on an expanding breather one would have $dV/dt > 0$ for some $t \in [t_1, t_2]$ while, for every t, $-(d/dt)log(V) = (1/V)\int RdV \geq \lambda(t)$ so $\bar{\lambda}$ cannot be negative everywhere. More formally consider

$$\frac{d\bar{\lambda}}{dt} \geq 2V^{2/n}\int |R_{ij} + \nabla_i\nabla_j f|^2 e^{-f}dV + \frac{2}{n}V^{(2-n)/n}\lambda\int(-R)dV \geq \tag{6.19}$$

$$\geq 2V^{2/n}\left[\int |R_{ij} + \nabla_i\nabla_j f - (1/n)(R+\Delta f)g_{ij}]^2 e^{-f}dV + \right.$$

$$+\frac{1}{n}\left(\int(R+\Delta f)^2e^{-f}dV-(\int(R+\Delta f)e^{-f}dV)^2\right)\Bigg]\geq 0$$

where f is the minimizer for $\mathfrak{F}$. These arguments also show that there are no non-trivial (i.e. with non-constant Ricci curvature) steady or expanding Ricci solitons (on closed manifolds M). Indeed the equality case in the chain of inequalities above requires that $R+\Delta f$ be constant on M; however the Euler-Lagrange equation for the minimizer f is $2\Delta f = |\nabla f|^2 + R = const.$. Therefore $\Delta f - |\nabla f|^2 = const. = 0$ because $\int(\Delta f - |\nabla f|^2)exp(-f)dV = 0$ and hence f is constant by the maximum principle (cf. also [**609**]).

In order to deal with the shrinking case when $\lambda > 0$ one replaces $\mathfrak{F}$ by its generalization which contains insertion of the scale parameter τ. Thus one considers

$$\mathfrak{W}(g_{ij}, f, \tau) = \int_M [\tau(|\nabla f|^2 + R) + f - n](4\pi\tau)^{-n/2}e^{-f}dV \tag{6.20}$$

restricted to f satisfying (**6K**) $\int_M (4\pi\tau)^{-n/2}exp(-f)dV = 1$ $(\tau > 0)$. Evidently $\mathfrak{W}$ is invariant under simultaneous scaling of τ and g_{ij} and the evolution equations generalizing (6.17) are

$$(g_{ij})_t = -2R_{ij};\ f_t = \Delta f + |\nabla f|^2 - R + \frac{n}{2\tau};\ \tau_t = -1 \tag{6.21}$$

The equation for f can also be written as $\Box^* u = 0$ where $u = (4\pi\tau)^{-n/2}exp(-f)$ and $\Box^* = -\partial_t - \Delta + R$ is the conjugate heat operator. Then one has

$$\frac{d\mathfrak{W}}{dt} = \int_M 2\tau\left[R_{ij} + \nabla_i\nabla_j f - \frac{1}{2\tau}g_{ij}\right]^2 (4\pi\tau)^{-n/2}e^{-f}dV \tag{6.22}$$

Hence if one lets $\mu(g_{ij}, \tau) = inf\,\mathfrak{W}(g_{ij}, f, \tau)$ over smooth f satisfying (**6K**), and $\nu(g_{ij}) = inf\,\mu(g_{ij}, \tau)$ over positive τ, then ν is nondecreasing along the Ricci flow. One can see that there always exists a smooth minimizer f (on a closed M) and that $lim_{\tau\to\infty}\mu = +\infty$ whenever the first eigenvalue of $-4\Delta + R$ is positive. Hence the statement that there is no shrinking breather other than gradient solitons is implied by

CLAIM: For an arbitrary metric g_{ij} on a closed manifold M the function $\mu(g_{ij}, \tau)$ is negative for small $\tau > 0$ and tends to zero as $\tau \to 0$. The proof is sketched in [**1009**].

We sketch next a few remarks on statistical mechanics from [**1009**] and refer to the previous extracts from [**742**] for comparison. One recalls that the partition function for the canonical ensemble at temperature β^{-1} is given by $Z = \int exp(-\beta E)d\omega(E)$ where $\omega(E)$ is a density of states measure not depending on β. One computes the average energy $< E >= -(\partial/\partial\beta)log(Z)$, the entropy $S = \beta < E > +log(Z)$, and the fluctuation $\sigma =< (E- < E >)^2 >= (\partial^2/\partial\beta^2)log(Z)$. Then fix a closed manifold M with a probability measure m and suppose that a system is described by a metric $g_{ij}(\tau)$ which depends on temperature τ via $(g_{ij})_\tau = 2(R_{ij} + \nabla_i\nabla_j f)$ and set $dm = udV$ with $u = (4\pi\tau)^{-n/2}exp(-f)$ with partition function now given by $log(Z) = \int(-f+(n/2))dm$ (assuming this possible

with suitable assumptions on g_{ij}). Then consider
(6.23)

$$< E >= -\tau^2 \int_M \left(R + |\nabla f|^2 - \frac{n}{2\tau}\right) dm;\ S = -\int_M [\tau(R + |\nabla f|^2) + f - n] dm;$$

$$\sigma = 2\tau^4 \int_M \left| R_{ij} + \nabla_i \nabla_j f - \frac{1}{2\tau} g_{ij} \right|^2 dm$$

Alternatively one could prescribe the evolution equations by replacing the t- derivatives by minus τ-derivatives in (6.21) and get the same formulas for Z, $> | < E >$, S, σ via Proposition 2.1. Further if (A) u tends to a delta function as $\tau \to 0$ or (B) u is a limit of a sequence u_i such that each u_i tends to a delta function as $\tau \to \tau_i > 0$ and $\tau_i \to 0$ then S is also nonnegative. In case (A) all the quantities $< E >$, S, σ tend to zero as $\tau \to 0$ while in case (B) S may tend to a positive limit (of interest concerning singularities). One observes that heuristically this statistical analogy is related to renormalization group flow and we refer to [?] for a more complete discussion.

7. CONNECTIONS TO THE QP

We recall now from e.g. [**254, 261, 1160**] that a relativistic Klein-Gordon (KG) equation in Bohmian form leads to a quantum HJ equation ($\psi = Rexp(iS/\hbar)$)

$$g^{ab}\nabla_a S \nabla_b S = \mathfrak{M}^2 c^2;\ \mathfrak{M}^2 = m^2 exp(Q);\ Q = \frac{\hbar^2}{m^2c^2}\frac{\Box_g |\psi|}{|\psi|} \tag{7.1}$$

along with the standard conservation law (**7A**) $\nabla_a(\rho \nabla^b S) = 0$ where $\rho \sim R^2$ with $R \sim |\psi|$. Equation (7.1) becomes (**7B**) $(m^2/\mathfrak{M}^2) g^{ab} \nabla_a S \nabla_b S = m^2 c^2$ with a conformal transform of the metric introduced via (**7C**) $g_{ab} \to \tilde{g}_{ab} = (\mathfrak{M}^2/m^2) g_{ab}$. Then writing $\tilde{\nabla}_a$ for the covariant derivatives with respect to $\tilde{g}$ one has

$$\tilde{g}^{ab} \tilde{\nabla}_a S \tilde{\nabla}_b S = m^2 c^2;\ \tilde{g}^{ab} \tilde{\nabla}_a (\rho \tilde{\nabla}_b S) = 0 \tag{7.2}$$

Thus the presence of a quantum potential (QP) Q is equivalent to a conformal metric change as in (**7C**). We recall also the standard geodesic equation (in g_{ab} terms)

$$\mathfrak{M} \frac{d^2 x^a}{ds^2} + \mathfrak{M} \Gamma^a_{bc} u^b u^c = (c^2 g^{ab} - u^a u^b) \nabla_b \mathfrak{M} \tag{7.3}$$

It was suggested by L. Crowell (private communication) that this transformation to a quantum mass $\mathfrak{M} \sim m_Q$, with $\mathfrak{M}^2 = m^2 exp(Q) \sim m^2(1 + Q)$ for small Q, might be related to Ricci flow and renormalization theory (cf. also [**56, 90, 339, 432, 474, 560, 578, 810, 829**]). We append to this the remark that the idea of quantum mass was used also by Faraggi-Matone and by Floyd in implementing their work (cf. [**133, 241, 254, 446, 478, 479**]) and this is also implicit in any approach to Bohmian mechanics. Hence one might perhaps think of Bohmian mechanics itself as a renormalization of classical mechanics via mass or geometry where here we concentrate on the geometry. Since a small parameter

$\hbar \sim \tau$ appears already in Q it would be perhaps the natural parameter τ to imploy in renormalization so let us write **(7D)** $Q = \tau(1/c^2)(\Box_g R/R) = \tau\Xi$ with

$$(7.4)\qquad \tilde{g}_{ab} = \left(\frac{\mathfrak{M}^2}{m^2}\right) g_{ab} = (e^Q) g_{ab};\ Q = Q(\tau, x, t)$$

REMARK 10.7.1. In 4-D however we are not using a Riemannian metric so it might be more appropriate to rephrase matter in 3-D with a time parameter (cf. Remark 10.7.2). There is a clear exposition of some of the Ricci features of Riemannian geometry relative to the Hamilton-Perelman et al theory in [**223, 229, 474, 560, 691, 905, 1235**] for example and we extract from that a few items. First in connection with the quantum potential we note that in Topping [**1235**] Perelman's functional **(7E)** $\mathfrak{F}(g, f) = \int_M (\mathcal{R} + |\nabla f|^2) exp(-f) dV$ is described as a form of Fisher information (cf. (5.22)) and this equation is generally related to an entropy formula $\partial_t \mathfrak{E} \propto \mathfrak{F}$ where $\mathfrak{E}$ is a differential entropy (whose role is played here by the so-called Nash entropy **(7F)** $N(u) = \int_M u log(u) dV$ with $u = exp(-f)$). The result $\partial_t \mathfrak{E} \propto \mathfrak{F}$ goes back at least to Garbaczewski [**508**] (cf. here also [**254, 255, 509, 510**]) for more details and other references). In a "classical" construction one can write for $P(x) \sim R^2(x) \sim |\psi|^2(x)$ a probability density (in 1-D for convenience with $P' \sim P_x$ etc.)

$$(7.5)\qquad F = \int P[\partial_x log(P)]^2 dx = \int \frac{P_x^2}{P} dx;\ Q = \frac{\hbar^2}{8m}\left[\left(\frac{P'}{P}\right)^2 - 2\left(\frac{P''}{P}\right)\right];$$

$$< Q >_\psi = \int PQ dx = \frac{\hbar^2}{8m} F$$

and this is consistent with momentum fluctuations proportional to $(\nabla P/P) \sim (P'/P)$. The action term $< Q >_\psi \sim F$ can be added to a classical Hamiltonian to quantize it and, for example when $|\psi|$ depends only on x, any such quantization can be modeled on momentum fluctuations of the type indicated (see here [**254, 261, 592, 598, 599, 1062**] for details and discussion). ■

REMARK 10.7.2. Following [**254, 261, 270, 275, 276, 1115, 1116**] we consider the SE in a Weyl space based on a Riemannian metric g_{ab} over a 3-D manifold M and a probability density $P(x,t) \sim |\psi|^2(x,t)$ with density $\rho = mP$ and $\hat{\rho} = \rho/\sqrt{g}$. The Weyl vector is $\phi_i = -\partial_i log(\hat{\rho})$ and the SE is

$$(7.6)\qquad i\hbar\psi_t = -\frac{\hbar^2}{2m}\frac{1}{\sqrt{g}}\left[\partial_i\left(\sqrt{g}g^{ik}\partial_k\right)\right]\psi + \left[V - \gamma\left(\frac{\hbar^2}{m}\right)\dot{\mathcal{R}}\right]\psi = 0$$

with corresponding HJ equation via $\psi = R exp(iS/\hbar)$

$$(7.7)\qquad \partial_t S + \frac{1}{2m} g^{ik}\partial_i S \partial_k S + V - \gamma\frac{\hbar^2}{m}\mathcal{R} = 0$$

where $\mathcal{R}$ is the Weyl-Ricci curvature

$$(7.8)\qquad \mathcal{R} = \dot{\mathcal{R}} + \frac{1}{2\gamma\sqrt{\hat{\rho}}}\left[\frac{1}{\sqrt{g}}\partial_i\left(\sqrt{g}g^{ik}\partial_k\sqrt{\hat{\rho}}\right)\right] = \dot{\mathcal{R}} + \mathcal{R}_w$$

(here $\dot{\mathcal{R}}$ the Riemann curvature for g). Further there is a standard equation **(7G)** $\partial_t\hat{\rho} + (1/\sqrt{g})\partial_i(\sqrt{g}v^i\hat{\rho}) = 0$ where $mv^i \sim p^i \sim \partial_i S$. Moreover there is a lovely relation for the quantum potential via (**7H**) $Q = -\gamma(\hbar^2/m)\mathcal{R} = -(\hbar^2/16m)\mathcal{R}$ ($\gamma = 1/16$ for $n = 3$). Let us recall here some formulas for covariant derivatives, namely (cf. [**263**] for a good reference to spacetime geometry)

$$\nabla_m\phi = \partial_m\phi;\ \nabla^m\phi = g^{mn}\partial_n\phi;\ \nabla_m v^m = \frac{1}{\sqrt{g}}\partial_m(\sqrt{g}v^m); \tag{7.9}$$

$$\nabla^m v_m = \frac{1}{\sqrt{g}}\partial_m(\sqrt{g}g^{mn}v_m);\ \Delta\phi = \nabla_m\nabla^m\phi = \frac{1}{\sqrt{g}}\partial_m(\sqrt{g}g^{mn}\partial_n\phi)$$

Hence we can rewrite (7.6)-(7.8) in the conventional form

$$i\hbar\psi_t + \Delta\psi + \left[V - \gamma\left(\frac{\hbar^2}{m}\right)\dot{\mathcal{R}}\right]; \tag{7.10}$$

$$\partial_t S = \frac{1}{2m}(\nabla S)^2 + V - \gamma\frac{\hbar^2}{m}\mathcal{R} = 0;\ \mathcal{R} = \dot{\mathcal{R}} + \frac{1}{2\gamma\sqrt{\hat{\rho}}}\Delta\sqrt{\hat{\rho}}$$

and since $\gamma = 1/16$ for 3-D one obtains (**7H**) $Q = -(\hbar^2/16m)\mathcal{R}$. Let us remark also in passing (cf. [?]) that if (**7I**) $\tilde{g}_{ab} = \omega^2 g_{ab}$ then

$$\tilde{\Gamma}^a_{bc} = \Gamma^a_{bc} + C^a_{bc};\ C^a_{bc} = \omega^{-1}(\delta^a_b\nabla_c\omega + \delta^a_c\nabla_b\omega - g_{bc}g^{ad}\nabla_d\omega) \tag{7.11}$$

$$\tilde{R}_{ab} = R_{ab} - [(n-2)\delta^c_a\delta^d_b + g_{ab}g^{cd}]\omega^{-1}(\nabla_c\nabla_d\omega) + \tag{7.12}$$

$$+[2(n-2)\delta^c_a\delta^d_b - (n-3)g_{ab}g^{cd}]\omega^{-2}(\nabla_c\omega)(\nabla_d\omega)$$

$$\tilde{R} = \omega^{-2}R - 2(n-1)g^{ab}\omega^{-3}(\nabla_a\nabla_b\omega) - (n-1)(n-4)g^{ab}\omega^{-4}(\nabla_a\omega)(\nabla_b\omega) \tag{7.13}$$

$$\tilde{\nabla}_a\phi = \nabla_a\phi = \partial_a\phi;\ \tilde{\Box}\phi = \omega^{-2}\Box\phi + (n-2)g^{ab}\omega^{-3}(\nabla_a\omega)(\nabla_b\omega) \tag{7.14}$$

(see [**302, 905, 1235**] for more formulas). ■

We see clearly from (7.4) for example that the effect of quantum mass is codified completely in a conformal change of geometry so that somehow renormalizing the mass corresponds to renormalizing the geometry. On the other hand from [**254, 261, 1160**] and Section 3, in a Weyl geometry with Dirac-Weyl action the Dirac field $\beta \sim \mathfrak{M} =$ the quantum (Bohmian) mass and the Weyl vector is a function of β. Following [**1159, 1160**] both ϕ_μ and $\mathfrak{M}$ determine the Weyl geometry of spacetime and we refer here to [**81, 82, 254, 261, 270, 275, 276, 668, 661, 669, 963, 1115, 1116, 1131, 1132, 1153, 1159, 1160, 1299**] for more on Weyl geometry.

The 3-D situation of Remark 10.7.2 has a different flavor. It is developed in a 3-D weyl space with time parameter t and arises in a specifically quantum mechanical context. Note $Q = 0 \sim \mathcal{R} = 0$ or $\sqrt{\hat{\rho}}\dot{\mathcal{R}} + 8\Delta\sqrt{\hat{\rho}} = 0$ (so $\dot{\mathcal{R}} = 0$ and $\hat{\rho} = const.$ would do). The Weyl geometry is defined via $\phi_i = -\partial_i log(\hat{\rho})$ with $\hat{\rho} = \rho/\sqrt{g}$ and $Q = -(\hbar^2/16m)\mathcal{R}$ as in (**7H**) where $\mathcal{R}_w$ is given in (7.8).

Now as indicated in (7.5) there is a strong relation between Fisher information, entropy, and the quantum potential. This is developed in [**258, 261, 255**] (based on [**254, 508, 509, 592, 598, 599, 1062**]) for the WDW equation and discussed in Sections 7.7, 7.8, and 8.6 of [?]. Thus we sketch here from [**255**]. Thus let $\psi = \sqrt{P}exp(iS/\hbar)$ and take an ADM situation

$$ds^2 = -(N^2 - h^{ij}N_iN_j) + 2N_idx^idt + h_{ij}dx^idx^j \tag{7.15}$$

Assume dynamics generated by an action (**7J**) $A = \int dt[\tilde{H} + \int \mathfrak{D}hP\partial_tS]$. One will have equations of motion (**7K**) $\partial_tP = \delta\tilde{H}/\delta S$ and $\partial_tS = -\delta\tilde{H}/\delta P$ (cf. [**254, 592**]). A suitable "classical" Hamiltonian is

$$\tilde{H}_{cl}[P,S] = \int \mathfrak{D}hPH_0\left[h_{ij}, \frac{\delta S}{\delta h_{ij}}\right]; \tag{7.16}$$

$$H_0 = \int dx\left[N\left(\frac{1}{2}G_{ijk\ell}\pi^{ij}\pi^{k\ell} + V(h_{ij})\right) - 2N_i\nabla_j\pi^{ij}\right]$$

where $G_{ijk\ell}$ is the deWitt (super)metric (**7L**) $G_{ijk\ell} = (1/\sqrt{h})(h_{ik}h_{j\ell} + h_{i\ell}h_{jk} - h_{ij}h_{k\ell})$ and $V \sim \hat{c}\sqrt{h}(2\Lambda - {}^3R)$. Then thinking of $\pi^{ij} = \delta S/\delta h_{ij} + f^{ij}$ and e.g. $\tilde{H}_q = \tilde{H}_{cl} + (1/2)\int \mathfrak{D}hP\int dxNG_{ijk\ell}\overline{f^{ij}f^{k\ell}}$ one arrives via exact uncertainty at a Fisher information contribution

$$\tilde{H}_q[P,S] = \tilde{H}_{cl} + \frac{c}{2}\int \mathfrak{D}h\int dxNG_{ijk\ell}\frac{1}{P}\frac{\delta P}{\delta h_{ij}}\frac{\delta P}{\delta h_{k\ell}} \sim \tag{7.17}$$

$$\sim \tilde{H}_{cl} + \frac{c}{2}\int\int \mathfrak{D}h\,dxNPQ$$

where $\hbar = 2\sqrt{c}$ and $\psi = \sqrt{P}exp(iS/\hbar)$ resulting in (for $N = 1$ and $N_i = 0$)

$$\left[-\frac{\hbar^2}{2}\frac{\delta}{\delta h_{ij}}G_{ijk\ell}\frac{\delta}{\delta h_{k\ell}} + V\right]\psi = 0 \tag{7.18}$$

with a sandwich ordering ($G_{ijk\ell}$ in the middle). Now compute in (7.18) to obtain

$$-\frac{\hbar^2}{4P}\frac{\delta}{\delta h_{ij}}\left[G_{ijk\ell}\frac{\delta P}{\delta h_{k\ell}}\right] + \tag{7.19}$$

$$+\frac{\hbar^2}{8P^2}G_{ijk\ell}\frac{\delta P}{\delta h_{k\ell}}\frac{\delta P}{\delta h_{ij}} + G_{ijk\ell}\left[\frac{\hbar^2}{8P}\frac{\delta^2 P}{\delta h_{ij}\delta h_{ij}} + \frac{1}{2}\frac{\delta S}{\delta h_{k\ell}}\frac{\delta S}{\delta h_{ij}}\right] + V = 0;$$

$$2P\frac{\delta G}{\delta h_{ij}}\frac{\delta S}{\delta h_{k\ell}} + G\left(\frac{\delta P}{\delta h_{k\ell}}\frac{\delta S}{\delta h_{ij}} + \frac{\delta S}{\delta h_{k\ell}}\frac{\delta P}{\delta h_{ij}}\right) + 2PG\frac{\delta^2 S}{\delta h_{k\ell}\delta h_{ij}} = 0$$

It is useful here to compare with $-(\hbar^2/2m)\psi'' + V\psi = 0$ which for $\psi = Rexp(iS/\hbar)$ yields $\partial(R^2S') = \partial(PS') = 0$ along with

$$\frac{1}{2m}S_x^2 + V + Q = 0;\ Q = -\frac{\hbar^2}{4m}\frac{R''}{R} = \frac{\hbar^2}{8m}\left[\frac{2P''}{P} - \left(\frac{P'}{P}\right)^2\right] \tag{7.20}$$

The analogues here are then in particular

$$\frac{1}{2m}S_x^2 \sim \frac{1}{2}G_{ijk\ell}\frac{\delta S}{\delta h_{k\ell}}\frac{\delta S}{\delta h_{ij}};\ Q = \frac{\hbar^2}{8m}\left[\frac{2P''}{P} - \left(\frac{P'}{P}\right)^2\right] \sim \tag{7.21}$$

$$\sim -\frac{\hbar^2}{4P}\frac{\delta}{\delta h_{ij}}\left[G_{ijk\ell}\frac{\delta P}{\delta h_{k\ell}}\right] + G_{ijk\ell}\left\{\frac{\hbar^2}{8P^2}\frac{\delta P}{\delta h_{k\ell}}\frac{\delta P}{\delta h_{ij}} + \frac{\hbar^2}{4P}\frac{\delta^2 P}{\delta h_{ij}\delta h_{k\ell}}\right\}$$

We note that the Q term arises directly from

$$Q = -\frac{\hbar^2}{2}P^{-1/2}\frac{\delta}{\delta h_{ij}}\left(G_{ijk\ell}\frac{\delta P^{1/2}}{\delta h_{k\ell}}\right) \tag{7.22}$$

and

$$\int \mathfrak{D}h\, PQ = -\frac{\hbar^2}{2}\int \mathfrak{D}hP^{1/2}\frac{\delta}{\delta h_{ij}}\left(G_{ijk\ell}\frac{\delta P^{1/2}}{\delta h_{k\ell}}\right) \tag{7.23}$$

But from (•) $\int \mathfrak{D}h\delta[\ ] = 0$ (cf. below) one has

$$\int \mathfrak{D}hP^{1/2}\frac{\delta}{\delta h_{ij}}\left(G_{ijk\ell}\frac{\delta P^{1/2}}{\delta h_{k\ell}}\right) = -\int \mathfrak{D}h\frac{\delta P^{1/2}}{\delta h_{ij}}G_{ijk\ell}\frac{\delta P^{1/2}}{\delta h_{k\ell}} \tag{7.24}$$

Thus one assumes there exists a suitable $\mathfrak{D}f$ which is a translation invariant measure in the (super)space of fields h leading to (•) (see below) and hence to (7.24). Consequently

THEOREM 7.1. Given a WDW equation of the form (7.18) with associated quantum potential given via (7.22) and $\psi = \sqrt{P}exp(iS/\hbar)$ it follows that the quantum potential can be modeled as arising from momentum fluctuations as in (7.17) (for $N = 1$) since it gives rise to a corresponding Fisher information type perturbation of a classical $\tilde{H}_c$ as in (7.17).

Here P represents a probability density of fields h_{ij} which determine $G_{ijk\ell}$ and one should perhaps stipulate that the perturbations of h_{ij} generated by $f^{ij} \sim (1/P)(\delta P/\delta h_{ij})$ are symmetric (note $\overline{f^{ij}} \sim \int \mathfrak{D}hP(1/P)(\delta P/\delta h_{ij}) = 0$). In general one would expect not necessarily symmetric perturbations of a Riemannian metric but in order to use the functional integral developed, involving $\mathfrak{D}h$ etc., it seems required to remain within the Riemannian framework where $G_{ijk\ell}$ is defined. Thus the very existence of a quantum WDW in sandwich form seems to require entropy type input via Fisher information fluctuation of fields (see below for entropy).

REMARK 10.7.3. We go here to [**254, 592, 616**] and will sketch briefly the derivation of (•) following [**592, 599**] (cf. also [**254**]). The relevant functional calculus goes as follows. One defines a functional F of fields f and sets

$$\delta F = F[f+\delta f] - F[f] = \int dx \frac{\delta F}{\delta f_x}\delta f_x \tag{7.25}$$

Here e.g. $dx \sim d^4x$ and in the space of fields there is assumed to be a measure $\mathfrak{D}f$ such that $\int \mathfrak{D}f \equiv \int \mathfrak{D}f'$ for $f' = f + h$ (cf. [**599**] and e.g. [**616**] for product type measures). Then evidently **(7M)** $\int \mathfrak{D}f(\delta F/\delta f) = 0$ when $\int \mathfrak{D}f\, F[f] < \infty$. Indeed

$$0 = \int \mathfrak{D}f(F[f+\delta f] - F[f]) = \int dx\delta f_x \left(\int \mathfrak{D}f\frac{\delta F}{\delta f_x}\right) \tag{7.26}$$

and this provides an integration by parts formula

$$\int \mathfrak{D}f\, P\left(\frac{\delta F}{\delta f}\right) = -\int \mathfrak{D}f\left(\frac{\delta P}{\delta f}\right)F \tag{7.27}$$

for $P[f]$ a probability density functional (we refer to [**254, 598**] for further details of the functional analysis). ■

REMARK 10.7.4. In [**932**] the problem of time in quantum gravity is addressed by weakening the Hamiltonian constraint $\hat{H} = 0$ to $<\psi|\hat{H}|\psi>= 0$ which is consistent with the classical Hamiltonian constraint. This can be written as (we shift $g \to h$ here in thinking of applications below to the deWitt metric and ${}^3h \sim h$) (**7N**) $\int \mathfrak{D}h\psi^*\hat{H}\psi = 0$ and for $\psi = Rexp(iS/\hbar)$, $\hat{H}\psi = i\hbar\partial_t\psi$ and a condition $(d/dt)\int \mathfrak{D}h\psi^*\psi = 0$ one finds that (**7N**) holds if $\partial_t S = 0$. Hence the weak Hamiltonian constraint (**7N**) is consistent with $\psi = R(h,t)exp(iS(h)/\hbar)$. The idea now is to allow $i\hbar\partial_t\psi = \hat{H}\psi$ but insist that this not contradict $\hat{H} = 0$ in the classical limit and we refer to [**254, 932**] for details. One recalls now (cf. [**254, 257, 508**]) that with the SE (under certain circumstances) one has a differential entropy $\mathfrak{S} = -\int dx\rho log(\rho)$ (1-D for simplicity here) with $\partial_t\rho = -\partial(v\rho)$ and $v = -u = -D\partial log(\rho)$ (diffusion current) leading to

$$\partial_t\mathfrak{S} = -\int dx\rho_t(log(\rho)+1) = \int dx\,(log(\rho)+1)\,\partial(v\rho) = \tag{7.28}$$
$$= -\int \partial\rho D\partial log(\rho) = D\int\frac{(\partial\rho)^2}{\rho}$$

Consequently the Fisher information is the time derivative of the differential entropy. ■

There should be some analogue of this related to the Wheeler-deWitt equation (WDW) where the action is minimized via perturbations of the metric. There is not a priori a natural time evolution for WDW but but this is handled as in [**255, 932**]. One considers

$$\partial_t\rho + \partial^A[2\rho\tilde{G}_{AB}\partial^B S] = 0 \tag{7.29}$$

where $\partial^B S = -i\hbar(\delta/\delta h_B)S$. Then one is motivated to posit

$$\frac{\delta S}{\delta h_B} \sim -\frac{\hat{c}}{P}\frac{\delta P}{\delta h_B} \tag{7.30}$$

provided one is only interested in metric fluctuations (there is no particle mass here to impede this). Thus consider (cf. [**255, 932**] for more details)

$$\partial_t P - \frac{\delta}{\delta h_A}\left[2P\tilde{G}_{AB}\frac{\hbar^2\hat{c}}{P}\frac{\delta P}{\delta h_B}\right] = 0 \tag{7.31}$$

and for a differential entropy defined via (**7O**) $\mathcal{S} = -(1/\hat{c})\int\tilde{\mathfrak{D}}hPlog(P)$ we would have

$$\mathcal{S}_t = -\frac{\hbar^2}{\hat{c}}\int dx\int\tilde{\mathfrak{D}}hP_t[1+log(P)] = \tag{7.32}$$
$$= -\hbar^2\int dx\int\tilde{\mathfrak{D}}h[1+log(P)]\left[\frac{\delta}{\delta h_A}\left(2\tilde{G}_{AB}\frac{\delta P}{\delta h_B}\right)\right] =$$
$$= \hbar^2\int dx\int\tilde{\mathfrak{D}}h\frac{2}{P}\tilde{G}_{AB}\frac{\delta P}{\delta h_B}\frac{\delta P}{\delta h_A} \sim 16\int\tilde{\mathfrak{D}}h\,PQ$$

(cf. (2.16), (2.17), and (2.32)) leading to a heuristic

THEOREM 7.2. Assuming only metric fluctuations satisfying (7.30) one can define a differential entropy (**7O**) and express the Fisher information (expressed via the quantum potential Q) as a time derivative $\partial_t \mathcal{S}$.

It should now be possible to provide a version of this procedure with WDW in the format of gradient Ricci flows. the analogies clearly involve using (**7P**) $\mathfrak{F} = \int_M (\mathcal{R} + |\nabla f|^2) exp(-f) dV$ (cf. (★) before (6.16)) corresponding in some way to $(c/2) \int \mathcal{D}hdxPQ$ in (7.17) and the Nash entropy $N(u)$ of (**7F**) with $u = exp(-f)$ as a form of differential entropy. What then is Q and does this relate to quantization? The idea is basically that in the WDW analysis it is perturbations in h_{ij} ($\sim g_{ij}$) on a 3-D M) which produces variation in an action integral of Fisher information type. The same thing happens in varying $\mathfrak{F}$ of (**7P**), i.e. perturbations of g_{ij} produce variations in $\mathfrak{F}$ leading to Ricci flow equations of the form (6.17) (to make $(d/ds)\mathfrak{F} \geq 0$ - i.e. this is not an extremization problem). We sketch this here again following [**905, 1235**] (for a "cleaner" derivation). Thus if $g(t) = a(t)\phi_t^*(g(0))$ is a pullback of $g(0)$ by a family of diffeomorphisms $\phi_t : M \to M$ with $\phi_0 = Id$ and $a(t) \in \mathbf{R}$ one calls $g(t)$ a Ricci soliton where $\partial_t g = -2Ric$. If further (••) $\partial_t \phi_t(p) = X \circ \phi_t(p)$ then $g(t)$ is called a gradient steady soliton (if $X = \nabla f$ one writes $X^i = g^{ij} X_j = g^{ij} \nabla_j f$). Now one wants to evaluate how $\mathfrak{F}$ is affected by variations in g_{ij} and f while retaining $\int_M exp(-f) dV = 1$. One thinks of a variation of $\mathfrak{F}$ via perturbations h_{ij} and considers e.g. (**7Q**) $\delta_h \mathfrak{F} = \partial_s \mathfrak{F}(g_{ij} + sh_{ij}, f)|_{s=0}$. In this direction one proves easily that (**7R**) $\delta(dV) = \partial_s(dV) = (1/2)Tr_g h)dV$ where $dV = \sqrt{det(g)}dx$ (with $dx \sim \prod dx^i$). Such a calculation applied to (**7S**) $\mathfrak{E}(g) = \int_M RdV$ for example yields the Einstein tensor E_{ik} via (cf. [**905**])

$$(7.33)\quad \partial_s \mathfrak{E} = \int_M (\partial_s RdV + R\partial_s(dV)) = -\int_M h^{ik} E_{ik} dV;\ E_{ik} = R_{ik} - (1/2)Rg_{ik}$$

and $\partial_s \mathfrak{E} = 0 \sim E_{ik} = 0$ (the energy momentum tensor is absent here); note also that the Einstein equations involve an indefinite or Lorentzian metric - cf. [**474**] for more general situations). One shows in [**905, 1235**] that $(d/ds)\mathfrak{F}(g, f) \geq 0$ provided

$$(7.34)\qquad \partial_s g = -2Ric;\ \partial_s f = -\Delta f + |\nabla f|^2 - R$$

(for variations of f, g preserving $exp(-f)dV$) and for such f, g it follows that $\mathfrak{F}$ is in fact monotone increasing in s (such solutions exist - cf. [**1235**], p. 159). However the Ricci flow solutions needed in the Perelman theory are not involved in extremizing an "action" $\mathfrak{F}$ so we don't have immediately an analogue to the quantization situation.

Hence let us write out $\partial_t \mathfrak{F}$ (using (••)) and set it equal to zero which can be written in many forms (collected below following [**705, 905, 1235**]). Thus first from [**905**] suppose $g(t)$ is a gradient steady soliton as in (••) leading to

$$(7.35)\qquad Ric + Hess(f) = 0;\ R + \Delta f = 0;\ |\nabla f|^2 + R = const.;\ \partial_t f = |\nabla f|^2$$

(see [**905**], p. 24). Here for $Hess(f)$ we recall (see [**705, 1235**]) (**7T**) $Hess = \nabla df$ for $f : M \to M$. Note that combining (ii) and (iv) in (7.35) gives $\partial_t f + \Delta f =$

$|\nabla f|^2 - R$ which is (ii) in (7.34). To obtain (7.35) we follow [**905**] and write from $g(t) = \phi_t^*(g(0))$ and $\partial_t \phi_t(p) = X \circ \phi_t(p)$

$$(7.36) \qquad -2R_{ij} = \partial_t g_{ij} = (L_X g)_{ij} = -2div^* X = \nabla_i X_j + \nabla_j X_i$$

Note here from [**905**], p. 15 that (♦) $div(\eta) = \nabla \eta_i,\ div(T_{ik}) = \nabla_i T_{ik}$, and $div^*\eta = -(1/2)(\nabla_i \eta_k + \nabla_k \eta_i)$. Then from (7.36) for $X = \nabla f$ we have $X_i = \nabla_i f$ and $-R_{ij} = \nabla_i \nabla_j f = Hess(f)$ ((i) in (7.35)). For more detail we have from [**705**] for $g = g_{ij} dx^i \otimes dx^j$

$$(7.37) \qquad L_X g = \partial_k g_{ij} dx^i \otimes dx^j + g_{ij} \frac{\partial X^i}{\partial x^k} dx^k \otimes dx^j + g_{ij} \frac{\partial X^j}{\partial x_k} dx^i \otimes dx^k =$$

$$= \left[\partial_k g_{ij} + g_{ki} \frac{\partial X^k}{\partial x^i} + g_{ik} \frac{\partial X^k}{\partial x^j} \right] dx^i \otimes dx^j$$

Now (ii) in (7.35) is the trace of the first (recall the trace is $g^{ij} R_{ij} = g^{ij} \nabla_i \nabla_j f = \nabla^j \nabla_j f = \Delta f$). For (iii) one can write from (i) (cf. [**1235**], p. 24)

$$(7.38) \qquad \nabla_k R_{ij} + \nabla_k \nabla_i \nabla_j f = 0;\ \nabla_i R_{kj} + \nabla_i \nabla_k \nabla_j f = 0$$

It is then concluded that (★★) $\nabla_k R_{ij} - \nabla_i R_{kj} + R_{kijp} \nabla_p f = 0$ and this needs some argument. We recall that $\nabla_i = \partial_i$ acting on functions but on indexed objects connection coefficients should arise. We recall the Bianchi identities

$$(7.39) \qquad R_{ijk\ell} + R_{i\ell jk} + R_{ik\ell j} = 0;\ \nabla_m R_{ijk\ell} + \nabla_k R_{ij\ell m} + \nabla_\ell R_{ijmk} = 0$$

and the last formula gives

$$(7.40) \qquad g^{im} \nabla_m R_{ijk\ell} = \nabla^i R_{ijk\ell} = \nabla_k R_{j\ell} - \nabla_\ell R_{jk}$$

which traced again with $g^{j\ell}$ yields (after an index change)

$$(7.41) \qquad \nabla^i R_{ik} = \nabla_k R - \frac{1}{2} \nabla^j R_{jk} \Rightarrow \nabla^i R_{ik} = \frac{1}{2} \nabla_k R$$

REMARK 7.10.5. This is all fine but we need some old fashioned formulas and extract from [**12, 80, 254, 474, 705, 905, 950, 1235, 1274**]. Thus (cf. [**1235**], p. 340)

$$(7.42) \qquad R^a_{bmn} = -\partial_n \Gamma^a_{bm} + \partial_m \Gamma^a_{bn} + \Gamma^s_{bn} \Gamma^a_{sm} - \Gamma^s_{bm} \Gamma^a_{sn};$$

$$\nabla_i T_{ab} = \partial_i T_{ab} - \Gamma^s_{ai} T_{sb} - \Gamma^s_{bi} T_{as}$$

Further recall (cf. [**12, 80, 705**]) (**7U**) $\Gamma^k_{ij} = \Gamma^k_{ji}$ and $R_{ik} = R_{ki}$ while from [**705**], p. 108 one has

$$(7.43) \qquad R^k_{\ell ij} = \partial_i \Gamma^k_{j\ell} - \partial_j \Gamma^k_{i\ell} + \Gamma^k_{im} \Gamma^m_{j\ell} - \Gamma^k_{jm} \Gamma^m_{i\ell}$$

(cf. also [**80**], p. 102). There are also standard formulas of the form (**7V**) $\Gamma^k_{ijj} = (1/2) g^{k\ell} (\partial_j g_{i\ell} + \partial_i g_{j\ell} - \partial_\ell g_{ij})$ (cf. [**705**], p. 128). ■

We consider then

$$(7.44) \quad \nabla_k R_{ab} - \nabla_a R_{kb} = \partial_k R_{ab} - \Gamma^s_{ak} R_{sb} - \Gamma^s_{bk} R_{as} - (\partial_a R_{kb} - \Gamma^s_{ka} R_{sb} - \Gamma^s_{ba} R_{kx}) =$$

$$= \partial_k R_{ab} - \partial_a R_{kb} + \Gamma^s_{ba} R_{ks} - \Gamma^s_{bk} R_{as}$$

since $\Gamma^s_{ak} = \Gamma^s_{ka}$. This is not too instructive but going to [**80**], pp. 106, 116 we see that

$$(7.45)\qquad (\nabla_i\nabla_j - \nabla_j\nabla_i)Z^k = R^k_{\ell ij}Z^\ell - T^\ell_{ij} = R^k_{\ell ij}Z^\ell$$

since the torsion $T^\ell_{ij} = 0$ for the Levi-Civita connection. Consequently we can write from (7.38)

$$(7.46)\qquad \nabla_k R_{j\ell} - \nabla_\ell R_{jk} = \nabla_k R_{\ell j} - \nabla_\ell R_{kj} = -\nabla_k\nabla_\ell\nabla_j f + \nabla_\ell\nabla_k\nabla_j f =$$
$$= (\nabla_\ell\nabla_k - \nabla_k\nabla_\ell)\nabla_j f = R^j_{m\ell k}\nabla_j f \sim R_{jm\ell k}\nabla^j f$$

in agreement with (★★) and [**905**], p. 24. Note that $\nabla_\ell f = \nabla^\ell f$ so (7.46) is OK for indexing purposes.

Now consider Perelman's $\mathfrak{F}$ (cf. [**1009**]) as in (★) of Section 2.1, i.e. (★) $\mathfrak{F} = \int_M (R + |\nabla f|^2)exp(-f)dV$ with first variation $\delta g_{ij} = h_{ij}$. Insstead of (6.16) we use now formulas from [**905**]. Note for $u = exp(-f)$ and (♦♦) $\Box^* u = -\partial_t u - \Delta u + Ru = 0$ (or equivalently $f_t + \Delta f = |\nabla f|^2 - R$) it follows that $\int_M exp(-f)dV = \int_M udV$ is preserved. In addition we know that

$$(7.47)\qquad \partial_t N = \int_M (|\nabla f|^2 + R)exp(-f)dV$$

where $N(u) = \int_M ulog(u)dV$. Let now $\partial_s g_{ij} = h_{ij}$ (i.e. (• • •) $\delta g_{ij} = h_{ij}$) and we use s and t interchangeably at times. Recall also $\partial_s det(g) = (Tr_g h)det(g)$. A crucial lemma of Perelman ([**1009**]) is (cf. [**905**], pp. 50-51 and (**7Q**) and (**7R**))
(7.48)

$$\partial_t\mathfrak{F} = \int_M \left[-h^{ij}(R_{ij} + \nabla_i\nabla_j f) + \left(\frac{1}{2}Tr_g h - \ell\right)(2\Delta f - |\nabla f|^2 + R)\right]e^{-f}dV$$

and if $dm = exp(-f)dV$ is pointwise fixed one has (**7W**) $\partial_s(exp(-f)dV) = [(1/2)Tr_g h - \ell]exp(-f)dV = 0$ (this defines ℓ) leading to (**7X**) $\delta_h\mathfrak{F} = -\int_M h^{ij}(R_{ij} + \nabla_i\nabla_j f)exp(-f)dV$. Some equations needed here are (cf. [**905**], pp. 50-51)

$$(7.49)\qquad \partial_t R = -h^{ik}R_{ik} - \Delta(Tr_g h) + div^2 h;\ \partial_t dV = \frac{1}{2}Tr_g h dV$$

$$(7.50)\qquad \partial_t|\nabla f|^2 = \delta < \nabla f, \nabla f >_g = 2 < \nabla f, \nabla \ell > -h^{ik}\nabla_i f\nabla_k f;$$
$$\partial_t exp(-f) = -\ell exp(-f)$$

Some further calculation gives then the result (cf. [**905, 1009**]) that if $\partial_t g = -2Ric$ and $u = exp(-f)$ satisfies the adjoint heat equation (♦♦) $\Box^* u = 0$ then

$$(7.51)\qquad \partial_t\mathfrak{F} = \partial_t^2 N = \int_M 2u|Hess(f) + Ric|^2 dV \geq 0$$

Evidently then $\delta\mathfrak{F} \sim \partial_t\mathfrak{F} = 0$ when (**7Y**) $Ric + Hess(f) = 0$ which becomes an equation for extreme action, namely (**7Z**) $R_{ij} + \nabla_i\nabla_j f = 0$. In some sense this should correspond to a WDW equation as in (7.18) where the h_{ij} correspond to metric terms in a 3-D geometry (one is dealing with a space $Riem(\Sigma)$ of Riemannian metrics in the WDW theory). We recall that time can be developed in the WDW framework in various ways (cf. [**254, 258, 255, 521, 592, 599, 931, 932**]) and here we can perhaps imagine time as entering e.g. via $f(t)$, $g(t)$ (cf. (♦♦) and

($\bullet\bullet\bullet$)) where one could think of general $g_{ij}(x,t)$ with $\partial_t g_{ij} = h_{ij}(x,t)$. In any case the question now is to find an analogue of Q such that $\int_M PQ \sim \mathfrak{F} = \int_M [|\nabla f|^2 + R]exp(-f)dV$ where P is a probability distribution indicated below for the SE in Weyl space. Recall that the WDW framework (or in fact the general Fisher information framework) involves (in 1-D for simplicity) $Q = cP^{-1/2}\partial(G\partial P^{1/2})$ with $\int QP = c\int P^{1/2}\partial(G\partial P^{1/2}) \to -c\int \partial P^{1/2} G\partial P^{1/2}$.

8. RICCI FLOW AND THE QP

Now this is very interesting since it means we are interested in analogues (modulo constant multipliers and in 1-D for convenience)
(8.1)

$$P \sim e^{-f};\ P' \sim P_x \sim -f'e^{-f};\ Q \sim e^{f/2}\partial(G\partial e^{-f/2});\ PQ \sim e^{-f/2}\partial(G\partial e^{-f/2};$$

$$\int PQ \to -\int \partial e^{-f/2} G\partial e^{-f/2} \sim -\int \partial P^{1/2} G\partial P^{1/2}$$

and this reproduces exactly the format sketch in (7.22)-(7.24) for the WDW equation. Moreover we could imagine ourselves here in the context of a Schrödinger equation (SE) in a Weyl space as in the work of [**1115, 1116**] (cf. [**254, 261, 237, 270, 276**]) in which case one would have $P \sim \hat{\rho} = \rho/\sqrt{g} \sim exp(-f)$ and there would be a Weyl vector $\vec{\phi} = -\nabla log(\hat{\rho}) \sim \nabla f$ in 3-D (also a differential entropy exists as in [**254, 261**] playing the role of the Nash entropy). Thus in 3-D

$$P^{1/2} \sim \sqrt{\hat{\rho}} \sim e^{-f/2}; Q \sim \frac{1}{\sqrt{\hat{\rho}}}\nabla(G\nabla\sqrt{\hat{\rho}});\ G \sim (R + |\vec{\phi}|^2) \tag{8.2}$$

Note also $\int_M exp(-f)dV = 1$ corresponds to $\int_M PdV = 1$ as befits a quantum mechanical situation. We recall however that for a SE in a Weyl space à la [**1115, 1116**] one has (cf. [**254, 261**])

$$Q \sim -\frac{\hbar^2}{16m}\left[\dot{\mathcal{R}} + \frac{8}{\sqrt{\hat{\rho}}}\frac{1}{\sqrt{g}}\partial_i\left(\sqrt{g}g^{ik}\partial_k\sqrt{\hat{\rho}}\right)\right] = -\frac{\hbar^2}{16m}\left[\dot{\mathcal{R}} + \frac{8}{\sqrt{\hat{\rho}}}\Delta\sqrt{\hat{\rho}}\right] \tag{8.3}$$

(recall $div\, grad\, v = \Delta v = (1/\sqrt{g})\partial_m(\sqrt{g}g^{mn}\partial_n v)$). Thus in a flat space where $\dot{\mathcal{R}} = 0$ we obtain the standard form of Q (and if $\dot{\mathcal{R}} \neq 0$ one notes that it will appear in the corresponding SE). We recall also that the Weyl-Ricci curvature is $\dot{\mathcal{R}} + \mathcal{R}_w$ where

$$\mathcal{R}_w = 2|\vec{\phi}|^2 - 4\nabla\cdot\vec{\phi} = 4\frac{\Delta\sqrt{\hat{\rho}}}{\sqrt{\hat{\rho}}} \tag{8.4}$$

Thus the quantum potential is purely a geometrical object of curvature form, having to do with probability and information, and the quantum mechanics only arises via an equation of the type

$$\partial_t\hat{\rho} + \frac{1}{m}div(\hat{\rho}\nabla S) = 0 \tag{8.5}$$

arising from the SE. This was sketched in [**262**] (somewhat hastily) and we reorganize this here. Thus one looks at the Perelman entropy functional $\mathfrak{F}$ of ($\bigstar$) with $\int_M exp(-f)dV = 1$ resulting form ($\blacklozenge\blacklozenge$) $\partial_t f + \Delta f - |\nabla f|^2 + \dot{R} = 0$ ($\dot{R}$ is the Riemannian curvature). Note first that we think of $exp(-f) = P$ ($\sim u$) as a probability distribution so that $\int_M PdV = 1$ is a desideratum. One thinks then of $P \sim \hat{\rho} = \rho/\sqrt{g}$ as in the Santamato theorey with $\vec{\phi} = -\nabla log(\hat{\rho}) \sim \nabla f$. We note that (♦♦) becomes (recall $u \sim P$)

(8.6) $$\Box^* P = -\partial_t P - \Delta P + RP = 0 \simeq \partial_t \hat{\rho} + \Delta\hat{\rho} - R\hat{\rho} = 0$$

REMARK 1.8.1. The quantum mechanics relation associated with wave function $\psi = \sqrt{\hat{\rho}}exp(iS\hbar)$ involves

(8.7) $$\partial_t\hat{\rho} + \frac{1}{m}div(\hat{\rho}\nabla S) = 0;\ p = mv = \nabla S$$

and this has no obvious relation to (8.6). However we can imagine it as being an equation for S and there seems to be no contradictions involved. One is faced with an equation

(8.8) $$R\hat{\rho} - \Delta\hat{\rho} = -\frac{1}{m}div(\hat{\rho}\nabla S)$$

which is not immediately threatening. ■

Now one knows from [**254, 261, 270, 276, 1115, 1116**] for 3-D

(8.9) $$\mathcal{R} = \dot{\mathcal{R}} + \mathcal{R}_w;\ \vec{\phi} = -\nabla log(\hat{\rho});$$

$$\mathcal{R}_w = 2|\vec{\phi}|^2 - 4\nabla\cdot\vec{\phi} = \frac{8}{\sqrt{\hat{\rho}}}\partial_i(\sqrt{g}g^{ik}\partial_k\sqrt{\hat{\rho}})$$

and (**8A**) $Q = -(\hbar^2/16m)\mathcal{R}$ which for $\dot{\mathcal{R}} = 0$ becomes the standard (**8B**) $Q = -(\hbar^2/2m)(\Delta\sqrt{\hat{\rho}}/\sqrt{\hat{\rho}})$ (cf. [**254, 261, 270, 276, 1115, 1116**]) (note for $\dot{\mathcal{R}} \neq 0$ one can put this in the SE directly and treat $Q \sim Q_w$). Set then $Q_w = -(\hbar^2/16m)\mathcal{R}_w = -(\hbar^2/2m)(\Delta\sqrt{\hat{\rho}}/\sqrt{\hat{\rho}})$ and note that the main object here is to show that the functional $\mathfrak{F}$ is related to $\int_M PQdV$ where Q corresponds to a quantum potential for some (undetermined) quantum process. Thus consider the quantum potential Q_w of (**8B**). In 1-D the calculations are easy and one has (use ρ for $\hat{\rho}$ for convenience)

(8.10) $$\frac{\partial^2\rho^{1/2}}{\rho^{1/2}} = \frac{1}{2}\frac{\rho''}{\rho} - \frac{1}{4}\frac{(\rho')^2}{\rho^2}$$

Consequently **8C**) $\hat{\rho}Q_w \sim (1/2)\hat{\rho}'' - (1/4)[(\hat{\rho}')^2/\hat{\rho}]$. The first term integrates out to zero and integrating the second leads to a Fisher information term corresponding to $\int|\vec{\phi}|^2\hat{\rho}dV \sim \int[(\hat{\rho}')^2/\hat{\rho}]dV$. More directly one can write (**8D**) $\dot{\mathcal{R}} + |\vec{\phi}|^2 = \dot{\mathcal{R}} + \mathcal{R}_w + (|\vec{\phi}|^2 - \mathcal{R}_w) = \alpha Q + (4\nabla\cdot\vec{\phi} - |\vec{\phi}|^2)$ which leads to (**8E**) $\alpha\int_M QPdV + \beta\int_M|\vec{\phi}|^2PdV$ putting Q directly into the picture and suggesting some sort of quantum connection.

APPENDIX A

DeDONDER WEYL THEORY

We refer to [**254**] for the full Appendix A where material is extracted from [**265, 707, 788, 812**]. It is better here to read the original sources.

APPENDIX B

RELATIVITY AND ELECTROMAGNETISM

We extract first from Adler et al [**12**], which still appears to be the best book ever written on classical general relativity, and will sketch some of the essential features. Four criteria for field equations are stated as:

(1) Physical laws do not distinguish between accelerated systems and inertial systems. This will hold if all laws are written in tensor form.
(2) Both gravitational forces and fictitious forces appear as Christoffel symbols (connection coefficients) in a mathematically similar form. This is desirable since they should be indistinguishable in the small.
(3) The gravitational equations should be phrased in covariant tensor form and should be of second order in the components of the metric tensor.
(4) For unique solutions one wishes the field equations to be quasi-linear (i.e. the second derivatives enter linearly).

Now the signature for a Lorentz metric is taken to be $(1,-1,-1,-1)$ and one writes $f_{|\alpha} = \partial f/\partial x^\alpha$ while $f_{||\alpha}$ is the covariant derivative (defined below). We assume known here the standard techniques of differential geometry as used in [**12**]. Then e.g. for a contravariant (resp. covariant) vector ξ^i (resp. η_m) one writes

(B.1) $$\xi^i_{||k} = \xi^i_{|k} - \Gamma^i_{k\ell}\xi^\ell;\ \Gamma^i_{k\ell} = -\left\{\begin{matrix} i \\ k\ \ell \end{matrix}\right\};\ \eta_{m||\ell} = \eta_{m|\ell} + \left\{\begin{matrix} r \\ m\ \ell \end{matrix}\right\}$$

(the bracket notation is used for Christofel symbols which are connection coefficients). One defines the Riemann curvature tensor via

(B.2) $$R^\alpha_{\ \eta\beta\gamma} = \left\{\begin{matrix} \alpha \\ \beta\ \eta \end{matrix}\right\}_{|\gamma} - \left\{\begin{matrix} \alpha \\ \eta\ \gamma \end{matrix}\right\}_{|\beta} + \left\{\begin{matrix} \alpha \\ \tau\ \gamma \end{matrix}\right\}\left\{\begin{matrix} \tau \\ \beta\ \eta \end{matrix}\right\} - \left\{\begin{matrix} \alpha \\ \tau\ \beta \end{matrix}\right\}\left\{\begin{matrix} \tau \\ \gamma\ \eta \end{matrix}\right\}$$

and a necessary (and sufficient) condition for a Riemann space to have a Lorentz metric is $R^\alpha_{\ \eta\beta\gamma} = 0$ (i.e. the space is flat); this equation is in fact a field equation for a flat and gravity free space (with Lorentz metric as a solution). Note here $\xi^\alpha_{\ ||\beta||\gamma} - \xi^\alpha_{\ ||\gamma||\beta} = R^\alpha_{\ \eta\beta\gamma}\xi^\eta$ which implies $\xi_{\alpha||\beta||\gamma} - \xi_{\alpha||\gamma||\beta} = R_{\alpha\rho\beta\gamma}\xi^\rho$ in a metric space (since the metric is used to lower indices, i.e. $T^\alpha_\gamma = g_{\gamma\beta}T^{\alpha\beta}$ etc.). Also one has generally

(B.3) $$T^{\alpha\delta}_{\ \ ||\beta||\gamma} - T^{\alpha\delta}_{\ \ ||\gamma||\beta} = R^\alpha_{\ \tau\beta\gamma}T^{\tau\delta} + R^\delta_{\ \tau\beta\gamma}T^{\alpha\tau}$$

The notation $\{R_{\alpha\eta\beta\gamma}\xi^\eta\}_{(\alpha,\beta,\gamma)} = 0 = \{R_{\alpha\eta\beta\gamma}\}_{(\alpha,\beta,\gamma)}\xi^\eta$ involves an antisymmetrization in α, β, γ and since ξ^η is arbitrary this means $\{R_{\alpha\eta\beta\gamma}\}_{(\alpha,\beta,\gamma)} = 0$. Written out (with some relabeling and combination) this means that there are symmetries

(B.4) $$R_{\alpha\eta\beta\gamma} = -R_{\alpha\eta\gamma\beta},\ R_{\alpha\eta\beta\gamma} = -R_{\eta\alpha\beta\gamma},\ R_{\alpha\eta\beta\gamma} = R_{\beta\gamma\alpha\eta}$$

and $R_{1023} + R_{2031} + R_{3012} = 0$. The Bianchi identities are $\{R_{\alpha\eta\beta\gamma||\delta}\}_{(\beta,\gamma,\delta)} = 0$. Next via parallel transport one has $d\xi^\alpha = -\left\{ \begin{matrix} \alpha \\ \beta\ \gamma \end{matrix} \right\} \xi^\beta dx^\gamma$ and displacements along paths dx, $d\hat{x}$ and $d\hat{x}$, dx respectively leads to a vector transport difference $\Delta\xi^\alpha = R^\alpha{}_{\beta\eta\gamma}\xi^\beta dx^\eta d\hat{x}^\gamma$. Now the only meaningful contraction of $R_{\alpha\beta\gamma\delta}$ is given by $R_{\eta\gamma} = R^\alpha{}_{\eta\alpha\gamma} = g^{\alpha\beta}R_{\beta\eta\alpha\gamma} = g^{\alpha\beta}R_{\alpha\gamma\beta\eta} = R_{\gamma\eta}$ with 10 independent components. The equation $R_{\beta\delta} = 0$ satisfies conditions 1-4 above and has the Lorentz metric for one solution. Note here

(B.5)
$$R_{\beta\delta} = \left\{ \begin{matrix} \alpha \\ \beta\ \alpha \end{matrix} \right\}_{|\delta} - \left\{ \begin{matrix} \alpha \\ \beta\ \delta \end{matrix} \right\}_{|\alpha} + \left\{ \begin{matrix} \alpha \\ \tau\ \delta \end{matrix} \right\}\left\{ \begin{matrix} \tau \\ \beta\ \alpha \end{matrix} \right\} - \left\{ \begin{matrix} \alpha \\ \tau\ \alpha \end{matrix} \right\}\left\{ \begin{matrix} \tau \\ \beta\ \delta \end{matrix} \right\} = 0$$

can be written out in terms of the metric tensor $g_{\alpha\gamma}$ and this is the free space Einstein field equation. Some calculation shows that this can be written also in terms of the zero divergence Ricci tensor $G^{\beta\delta} = R^{\beta\delta} - (1/2)g^{\beta\delta}R = 0$ where $R = R^\eta{}_\eta$ is the Riemann scalar). Finally for a one parameter family $\Gamma(v)$ of geodesics given by $x^\mu = x^\mu(u,v)$ one has geodesic equations

(B.6) $$\frac{\partial^2 x^\mu}{\partial u^2} = -\left\{ \begin{matrix} \alpha \\ \beta\ \gamma \end{matrix} \right\} \frac{\partial x^\beta}{\partial u}\frac{\partial x^\gamma}{\partial u}$$

Now one can write out (B.5) in the form

(B.7) $$R_{\mu\nu} = \frac{1}{2}g^{\rho\sigma}\left[-g_{\mu\sigma|\nu|\rho} - g_{\nu\rho|\mu|\sigma} + g_{\mu\nu|\rho|\sigma} + g_{\rho\sigma|\mu|\nu}\right] + K_{\mu\nu}$$

where $K_{\mu\nu}$ contains only metric potentials and their first derivatives. One thinks of solving an initial value problem with data on a 3-dimensional hypersurface S described locally via $x^0 = 0$ (so $g_{00} > 0$). On S one gives $g_{\alpha\beta}$ and all first derivatives (only $g_{\mu\nu}$ and $g_{\mu\nu|0}$ need to be prescribed). Then one can calculate that

(B.8) $$R_{ij} = (1/2)g^{00}g_{ij|0|0} + M_{ij} = 0;\ R_{i0} = -(1/2)g^{0j}g_{ij|0|0} + M_{i0} = 0;$$

$$R_{00} = (1/2)g^{ij}g_{ij|0|0} + M_{00} = 0$$

where the $M_{\mu\nu}$ can be computed from data on S. A change of coordinates will make all $g_{\lambda 0|0|0} = 0$ on S (these are not contained in (B.8)) and thus (B.8) consists of 10 equations for six unknowns $g_{ij|0|0}$ on S which is overdetermined and leads to compatability conditions for the data $M_{\mu\nu}$ on S. This can be reduced to the form

(B.9) $$R_{ij} = 0;\ G^0_\lambda = 0$$

The first set of 6 equations determines the six unknowns $g_{ij|0|0}$ from initial data. The additional 4 equations in terms of the Ricci tensor G^0_λ represent necessary conditions on the initial data in order to insure a solution. There is much more material in [**12**] about the Cauchy problem which we omit here. For the Einstein equations in nonempty space one needs an energy momentum tensor for which a typical form is

(B.10) $$T^{\mu\nu} = \rho_0 u^\mu u^\nu + (p/c^2)(u^\mu u^\nu - g^{\mu\nu})$$

where ρ_0 is a density, p a pressure term, and u^μ a 4-velocity field. One assumes $T^{\mu\nu}$ has zero divergence or $T^{\mu\nu}_{\ \ ||\nu} = 0$ (which is a covariant formulation of fluid flow under the effect of its own internal pressure force).

REMARK B.1 One can also include EM fields in $T^{\mu\nu}$ by dealing with a Lorentz force $f^i = \sigma(\mathbf{E} + (\mathbf{v}/c) \times \mathbf{H})^i \sim -\sigma_0 F^{i\nu} u_\nu$ where $F^{i\nu}$ is the EM field tensor. However we will work in electromagnetism later in more elegant fashion via the Dirac-Weyl theory and omit this here (cf. [**12**] for more details). ■

In any event the Einstein field equations for nonempty space involve a zero divergence $T^{\mu\nu}$ so one uses the Ricci tensor $G_{\mu\nu} = R_{\mu\nu} - (1/2)g_{\mu\nu}R$ and notes that the most general second order tensor $B^{\alpha\gamma}$ of zero divergence can be written as $B^{\alpha\gamma} = G^{\alpha\gamma} + \Lambda g^{\alpha\gamma}$ (a result of E. Cartan). Hence one takes the Einstein field equations to be $G^{\alpha\gamma} + \Lambda g^{\alpha\gamma} = cT^{\alpha\gamma}$.

The exposition in Section 5.2 suggests the desirability of having a differential form discription of EM fields and we supply this via [**950**]. Thus one thinks of tensors $T = T^\sigma_{\ \mu\nu}\partial_\sigma \otimes dx^\mu \otimes dx^\nu$ with contractions of the form $T(dx^\sigma, \partial_\sigma) \sim T_\nu dx^\nu$. For $\eta = \eta_{\mu\nu} dx^\mu \otimes dx^\nu$ one has $\eta^{-1} = \eta^{\mu\nu}\partial_\mu \otimes \partial_\nu$ and $\eta\eta^{-1} = 1 \sim diag(\delta^\mu_\mu)$. Note also e.g.

(B.11) $$\eta_{\mu\nu} dx^\mu \otimes dx^\nu(\mathbf{u}, \mathbf{w}) = \eta_{\mu\nu} dx^\mu(\mathbf{u}) dx^\nu(\mathbf{w}) =$$
$$= \eta_{\mu\nu} dx^\mu(u^\alpha \partial_\alpha) dx^\nu(w^\tau \partial_\tau) = \eta_{\mu\nu} u^\mu w^\nu$$

(B.12) $$\eta(\mathbf{u}) = \eta_{\mu\nu} dx^\mu \otimes dx^\nu(\mathbf{u}) = \eta_{\mu\nu} dx^\mu(\mathbf{u}) dx^\nu =$$
$$= \eta_{\mu\nu} dx^\mu(u^\alpha \partial_\alpha) dx^\nu = \eta_{\mu\nu} u^\mu dx^\nu = u_\nu dx^\nu$$

for a metric η. Recall $\alpha \wedge \beta = \alpha \otimes \beta - \beta \otimes \alpha$ and

(B.13) $$\alpha \wedge \beta = \alpha_\mu dx^\mu \wedge \beta_\nu dx^\nu = (1/2)(\alpha_\mu \beta_\nu - \alpha_\nu \beta_\mu) dx^\mu \wedge dx^\nu$$

The EM field tensor is $F = (1/2) F_{\mu\nu} dx^\mu \wedge dx^\nu$ where

(B.14) $$F_{\mu\nu} = \begin{pmatrix} 0 & E_x & E_y & E_z \\ -E_x & 0 & -B_z & B_y \\ -E_y & B_z & 0 & -B_x \\ -E_z & -B_y & B_x & 0 \end{pmatrix};$$

$$F = E_x dx^0 \wedge dx^1 + E_y dx^0 \wedge dx^2 + E_z dx^0 \wedge dx^3 -$$
$$-B_z dx^1 \wedge dx^2 + B_y dx^1 \wedge dx^3 - B_x dx^2 \wedge dx^3$$

The equations of motion of an electric charge is then $d\mathbf{p}/d\tau = (e/m)\mathbf{F}(\mathbf{p})$ where $\mathbf{p} = p^\mu \partial_\mu$. There is only one 4-form, namely $\epsilon = dx^0 \wedge dx^1 \wedge dx^2 \wedge dx^3 = (1/4!)\epsilon_{\mu\nu\sigma\tau} dx^\mu \wedge dx^\nu \wedge dx^\sigma \wedge dx^\tau$ where $\epsilon_{\mu\nu\sigma\tau}$ is totally antisymmetric. Recall also for $\alpha = \alpha_{\mu\nu\ldots} dx^\mu \wedge dx^\nu \cdots$ one has $d\alpha = d\alpha_{\mu\nu\ldots} \wedge dx^\mu \wedge dx^\nu \cdots = \partial_\alpha \alpha_{\mu\nu\ldots} dx^\sigma \wedge dx^\mu \wedge dx^\nu \cdots$ and $dd\alpha = 0$. Define also the Hodge star operator on F and j via

${}^*F = (1/4)\epsilon_{\mu\nu\sigma\tau}F^{\sigma\tau}dx^\mu \wedge dx^\nu$ and ${}^*j = (1/3!)\epsilon_{\mu\nu\sigma\tau}j^\tau dx^\mu \wedge dx^\nu \wedge dx^\sigma$; these are called dual tensors. Now the Maxwell equations are

(B.15) $$\partial_\mu F^{\mu\nu} = \frac{4\pi}{c}j^\nu;\ \partial^\alpha F^{\mu\nu} + \partial^\mu F^{\nu\alpha} + \partial^\nu F^{\alpha\mu} = 0$$

and this can now be written in the form

(B.16) $$dF = 0;\ d^*F = \frac{4\pi}{c} * j$$

and $0 = d^*j = 0$ is automatic. In terms of $A = A_\mu dx^\mu$ where $F = dA$ the relation $dF = 0$ is an identity $ddA = 0$.

A few remarks about the tensor nature of j^μ and $F^{\mu\nu}$ are in order and we write $n = n(x)$ and $\mathbf{v} = \mathbf{v}(x)$ for number density and velocity with charge density $\rho(x) = qn(x)$ and current density $\mathbf{j} = qn(x)\mathbf{v}(x)$. The conservation of particle number leads to $\nabla\cdot\mathbf{j} + \rho_t = 0$ and one writes $j^\nu = (c\rho, j_x, j_y, j_z) = (c\rho n, qnv_x, qnv_y, qnv_z)$ or equivalently $j^\nu = n_0 q u^\nu = j^\nu = \rho_0 u^\nu$ where $n_0 = n\sqrt{1-(v^2/c^2)}$ and $\rho_0 = qn_0$ (ρ_0 here is charge density). Since j^ν consists of u^ν multiplied by a scalar it must have the transformation law of a 4-vector $j'^\beta = a^\beta_{\ \nu} j^\nu$ under Lorentz transformations. Then the conservation law can be written as $\partial_\nu j^\nu = 0$ with obvious Lorentz invariance. After some argument one shows also that $F^{\mu\nu} = a_\beta^{\ \nu} a_\alpha^{\ \mu} F'^{\alpha\beta}$ under Lorentz transformations so $F^{\mu\nu}$ is indeed a tensor. The equation of motion for a charged particle can be written now as $(d\mathbf{p}/dt) = q\mathbf{E} + (q/c)\mathbf{v}\times\mathbf{B}$ where $\mathbf{p} = m\mathbf{v}/\sqrt{1-(v^2/c^2)}$ is the relativistic momentum. This is equivalent to $dp^\mu/dt = (q/m)p_\nu F^{\mu\nu}$ with obvious Lorentz invariance. The energy momentum tensor of the EM field is

(B.17) $$T^{\mu\nu} = -(1/4\pi)[F^{\mu\alpha}F^\nu_{\ \alpha} - (1/4)\eta^{\mu\nu}F^{\alpha\beta}F_{\alpha\beta}]$$

(cf. Ohanian-Ruffini [**950**] for details) and in particular $T^{00} = (1/8\pi)(\mathbf{E}^2 + \mathbf{B}^2)$ while the Poynting vector is $T^{0k} = (1/4\pi)(\mathbf{E}\times\mathbf{B})^k$.

One can equally well work in a curved space where e.g. covariant derivatives are defined via

(B.18) $$\nabla_n T = lim_{d\lambda\to 0}[(T(\lambda + d\lambda) - T(\lambda) - \delta T]/d\lambda$$

where δT is the change in T produced by parallel transport. One has then the usual rules $\nabla_u(T\otimes R) = \nabla_u T \otimes R + T \otimes \nabla_u R$ and for $\mathbf{v} = v^\nu\partial_\nu$ one finds $\nabla_\mu\mathbf{v} = \partial_\mu v^\nu \partial_\nu + v^\nu\nabla_\mu\partial_\nu$. Now if $\mathbf{v}$ was constructed by parallel transport its covariant derivative is zero so, acting with the dual vector dx^α gives

(B.19) $$\frac{\partial x^\nu}{\partial x^\mu}dx^\alpha(\partial_\nu) + v^\nu dx^\alpha(\nabla_\mu\partial_\nu) = 0 \equiv \partial_\mu v^\alpha + v^\nu dx^\alpha(\nabla_\mu\partial_\nu) = 0$$

Comparing this with the standard $\partial_\mu v^\alpha + \Gamma^\alpha_{\ \mu\nu}v^\nu = 0$ gives $dx^\alpha(\nabla_\mu\partial_\nu) = \Gamma^\alpha_{\ \mu\nu}$. One can show also for vectors u, v, w (boldface omitted) and a 1-form α

(B.20)
$$(\nabla_u\nabla_v - \nabla_v\nabla_u - uv + vu)\alpha(w) = R(\alpha, u, v, w);\ R = F^\sigma_{\ \beta\mu\nu}\partial_\sigma \otimes dx^\beta \otimes dx^\mu \otimes dx^\nu$$

so R represents the Riemann tensor.

For the nonrelativistic theory first we go to [**879**] we define a transverse and longitudinal component of a field F via

(B.21) $$F^{||}(r) = -\frac{1}{4\pi}\int d^3r'\frac{\nabla'\cdot F(r')}{|r-r'|};\ F^{\perp}(r) = \frac{1}{4\pi}\nabla\times\nabla\times\int d^3r'\frac{F(r')}{|r-r'|}$$

For a point particle of mass m and charge e in a field with potentials A and ϕ one has nonrelativistic equations $m\ddot{x} = eE + (e/c)v\times B$ (boldface is suppressed here) where one recalls $B = \nabla\times A$, $v = \dot{x}$, and $E = -\nabla\phi - (1/c)A_t$ with $H = (1/2m)(p-(e/c)A)^2 + e\phi$ leading to

(B.22) $$\dot{x} = \frac{1}{2m}\left(p - \frac{e}{c}A\right);\ \dot{p} = \frac{e}{c}[v\times B + (v\cdot\nabla)A] - e\nabla\phi$$

Recall here also

(B.23) $$B = \nabla\times A,\ \nabla\cdot E = 0,\ \nabla\cdot B = 0,\ \nabla\times E = -(1/c)B_t,$$
$$\nabla\times B = (1/c)E_t,\ E = -(1/c)A_t - \nabla\phi$$

(the Coulomb gauge $\nabla\cdot A = 0$ is used here). One has now $E = E^{\perp} + E^{||} \sim E^T + E^L$ with $\nabla\cdot E^{\perp} = 0$ and $\nabla\times E^{||} = 0$ and in Coulomb gauge $E^{\perp} = -(1/c)A_t$ and $E^{||} = -\nabla\phi$. Further

(B.24) $$H \sim \frac{1}{2m}\left(p - \frac{e}{c}A\right)^2 + e\phi + \frac{1}{8\pi}\int d^3r((E^{\perp})^2 + B^2)$$

(covering time evolution of both particle and fields).

For the relativistic theory one goes to the Dirac equation $i(\partial_t + \alpha\cdot\nabla)\psi = \beta m\psi$ which, to satisfy $E^2 = \mathbf{p}^2 + m^2$ with $E \sim i\partial_t$ and $\mathbf{p} \sim -i\nabla$, implies $-\partial_t^2\psi = (-i\alpha\cdot\nabla + \beta m)^2\psi$ and ψ will satisfy the Klein-Gordon (KG) equation if $\beta^2 = 1$, $\alpha_i\beta + \beta\alpha_i \equiv \{\alpha_i, \beta\} = 0$, and $\{\alpha_i, \alpha_j\} = 2\delta_{ij}$ (note $c = \hbar = 1$ here with $\alpha\cdot\nabla \sim \sum\alpha_\mu\partial_\mu$ and cf. [**879, 855**] for notations and background). This leads to matrices

(B.25) $$\sigma_1 = \begin{pmatrix} 0 & 1 \\ 1 & 0 \end{pmatrix};\ \sigma_2 = \begin{pmatrix} 0 & -i \\ i & 0 \end{pmatrix};\ \sigma_3 = \begin{pmatrix} 1 & 0 \\ 0 & -1 \end{pmatrix};$$
$$\alpha_i = \begin{pmatrix} 0 & \sigma_i \\ \sigma_i & 0 \end{pmatrix};\ \beta = \begin{pmatrix} 1 & 0 \\ 0 & -1 \end{pmatrix}$$

where α_i and β are 4×4 matrices. Then for convenience take $\gamma^0 = \beta$ and $\gamma^i = \beta\alpha_i$ which satisfy $\{\gamma^\mu, \gamma^\nu\} = 2g^{\mu\nu}$ (Lorentz metric) with $(\gamma^i)^\dagger = -\gamma^i$, $(\gamma^i)^2 = -1$, $(\gamma^0)^\dagger = \gamma^0$, and $(\gamma^0)^2 = 1$. The Dirac equation for a free particle can now be written

(B.26) $$\left(i\gamma^\mu\frac{\partial}{\partial x^\mu} - m\right)\psi = 0 \equiv (i\partial\!\!\!/ - m)\psi = 0$$

where $A\!\!\!/ = g_{\mu\nu}\gamma^\mu A^\nu = \gamma^\mu A_\mu$ and $\partial\!\!\!/ = \gamma^\mu\partial_\mu$. Taking Hermitian conjugates in, noting that α and β are Hermitian, one gets $\bar{\psi}(i\overleftarrow{\partial\!\!\!/} + m) = 0$ where $\bar{\psi} = \psi^\dagger\beta$. To define a conserved current one has an equation $\bar{\psi}\gamma^\mu\partial_\mu\psi + \gamma^\mu\bar{\psi}_\mu\psi = \partial_\mu(\bar{\psi}\gamma^\mu\psi) = 0$ leading to the conserved current $j^\mu = \bar{\psi}\gamma^\mu\psi = (\psi^\dagger\psi, \psi^\dagger\alpha\psi)$ (this means $\rho = \psi^\dagger\psi$ and $\mathbf{j} = \psi^\dagger\alpha\psi$ with $\partial_t\rho + \nabla\cdot\mathbf{j} = 0$). The Dirac equation has the Hamiltonian form

(B.27) $$i\partial_t\psi = -i\alpha\cdot\nabla\psi + \beta m\psi = (\alpha\cdot\mathbf{p} + \beta m)\psi \equiv H\psi$$

($\alpha \cdot \mathbf{p} \sim \sum \alpha_\mu p_\mu$). To obtain a Dirac equation for an electron coupled to a prescribed external EM field with vector and scalar potentials A and ϕ one substitutes $p^\mu \to p^\mu - eA^\mu$, i.e. $\mathbf{p} \to \mathbf{p} - e\mathbf{A}$ and $p^0 = i\partial_t \to i\partial_t - e\Phi$, to obtain

$$i\partial_t \psi = [\alpha \cdot (\mathbf{p} - e\mathbf{A}) + e\Phi + \beta m]\psi \tag{B.28}$$

This identifies the Hamiltonian as $H = \alpha \cdot (\mathbf{p} - e\mathbf{A}) + e\Phi + \beta m = \alpha \cdot \mathbf{p} + \beta m + H_{int}$ where $H_{int} = -e\alpha \cdot \mathbf{A} + e\Phi$, suggesting α as the operator corresponding to the velocity v/c; this is strengthened by the Heisenberg equations of motion

$$\dot{\mathbf{r}} = \left(\frac{1}{i\hbar}\right)[\mathbf{r}, H] = \alpha;\ \dot{\pi} = \left(\frac{1}{i\hbar}\right)[\pi, H] = e(\mathbf{E} + \alpha \times \mathbf{B}) \tag{B.29}$$

Another bit of notation now from [**874**] is useful. Thus (again with $c = \hbar = 1$) one can define e.g. $\sigma_z = -i\alpha_x\alpha_y$, $\sigma_x = -i\alpha_y\alpha_z$, $\sigma_y = -i\alpha_z\alpha_x$, $\rho_3 = \beta$, $\rho_1 = \sigma_z\alpha_z = -i\alpha_x\alpha_y\alpha_z$, and $\rho_2 = i\rho_1\rho_3 = \beta\alpha_x\alpha_y\alpha_z$ so that $\beta = \rho_3$ and $\alpha^k = \rho_1\sigma^k$. Recall also that the angular momentum $\vec{\ell}$ of a particle is $\vec{\ell} = \mathbf{r} \times \mathbf{p}$ ($\sim (-i)\mathbf{r} \times \nabla$) with components ℓ_k satisfying $[\ell_x, \ell_y] = i\ell_z$, $[\ell_y, \ell_z] = i\ell_x$, and $[\ell_z, \ell_x] = i\ell_y$. Any vector operator L satisfying such relations is called an angular momentum. Next one defines $\sigma_{\mu\nu} = (1/2)i[\gamma_\mu, \gamma_\nu] = i\gamma_\mu\gamma_\nu$ ($\mu \neq \nu$) and $S_{\alpha\beta} = (1/2)\sigma_{\alpha\beta}$. Then the 6 components $S_{\alpha\beta}$ satisfy

$$\begin{gathered} S_{10} = (i/2)\alpha_x;\ S_{20} = (i/2)\alpha_y;\ S_{30} = (i/1)\alpha_z; \\ S_{23} = (1/2)\sigma_x;\ ,\ S_{31} = (1/2)\sigma_y;\ S_{12} = (1/2)\sigma_z \end{gathered} \tag{B.30}$$

The $S_{\alpha\beta}$ arise in representing infinitesimal rotations for the orthochronous Lorentz group via matrices $I + i\epsilon S_{\alpha\beta}$. Further one can represent total angular momentum J in the form $J = L + S$ where $L = \mathbf{r} \times \mathbf{p}$ and $S = (1/2)\sigma$ (L is orbital angular momentum and S represents spin). We recall that the gamma matrices are given via $\gamma = \beta\alpha$. Now from

$$[(i\partial_t - e\phi) - \alpha \cdot (-i\nabla - e\mathbf{A}) - \beta m]\psi = 0 \tag{B.31}$$

one gets

$$[i\gamma^\mu D_\mu - m)\psi = [\gamma^\mu(i\partial_\mu - eA_\mu) - m]\psi = 0 \tag{B.32}$$

where $D_\mu = \partial_\mu + ieA_\mu \equiv (\partial_0 + ie\phi, \nabla - ie\mathbf{A})$. Working on the left with $(-i\gamma^\lambda D_\lambda - m)$ gives then $[\gamma^\lambda\gamma^\mu D_\lambda D_\mu + m^2]\psi = 0$ where $\gamma^\lambda\gamma^\mu = g^{\lambda\mu} + (1/2)[\gamma^\lambda, \gamma^\mu]$. By renaming the dummy indices one obtains

$$[\gamma^\lambda, \gamma^\mu]D_\lambda D_\mu = -[\gamma^\lambda, \gamma^\mu]D_\mu D_\lambda = (1/2)[\gamma^\lambda, \gamma^\mu][D_\lambda, D_\mu] \tag{B.33}$$

leading to

$$[D_\lambda, D_\mu] = ie[\partial_\lambda, A_\mu] + ie[A_\lambda, \partial_\mu] = ie(\partial_\lambda A_\mu - \partial_\mu A_\lambda) = ieF_{\lambda\mu} \tag{B.34}$$

This yields then

$$\gamma^\lambda\gamma^\mu D_\lambda D_\mu = D_\mu D_\lambda + eS^{\lambda\mu}F_{\lambda\mu} \tag{B.35}$$

where $S^{\lambda\mu}$ represents the spin of the particle. Therefore $[D_\mu D^\mu + eS^{\lambda\mu}F_{\lambda\mu} + m^2]\psi = 0$. Comparing with the standard form of the KG equation we see that this differs by the term $eS^{\lambda\mu}F_{\lambda\mu}$ which is the spin coupling of the particle to the EM field and has no classical analogue.

APPENDIX C

REMARKS ON QUANTUM GRAVITY

We refer here to [**69, 70, 71, 72, 73, 74, 75, 86, 87, 231, 379, 505, 663, 749, 853, 898, 1085, 1086, 1183, 1184, 1185, 1220, 1221, 1222, 1223**] and will mainly follow [**86, 749**] for basic material. First (cf. Baez-Muniain [**86**] recall that upon assuming the spacetime manifold M to be diffeomorphic to $\mathbf{R}\times S$ where S is a 3-dimensional manifold one can choose spacelike slices $\Sigma\subset M$ with a space $Met(\Sigma)$ of Riemannian metrics. Writing $\phi : M\to\mathbf{R}\times S$ one defines a time coordinate via $\tau=\phi^*t$ and $\Sigma\subset M$ is determined via $\tau = constant$. The extrinsic curvature K of Σ provides Cauchy data $({}^3g,K)$ for the metric and regarding Einstein's 10 equations, 4 are constraint equations for the Cauchy data and 6 are evolution equations saying how the 3-metric changes in time. This is the Arnowitt-Deser-Misner (ADM) formulation. Now in more detail, one takes $g(v,v)>0$ for $v\in T\Sigma$ and $g(n,n)=-1$ (where n is the unit normal vector to Σ). One can write $v=-g(v,n)n+(v+g(v,n)n)$ in terms of orthogonal vectors and, for ∇ corresponding to the covariant derivative for the Levi-Civita connection, one can write

$$\text{(C.1)}\qquad \nabla_u v=-g(\nabla_u v,n)+(\nabla_u v+g(\nabla_u v,n)n);\ K(u,v)n=-g(\nabla_u v,n)n$$

where K defines the extrinsic curvature. Since ∇ is torsion free one has $K(u,v)=K(v,u)$ and

$$\text{(C.2)}\qquad K_{ij}u^iv^j=K(\partial_i,\partial_j)u^iv^j;\ K(u,v)=g(\nabla_u n,v)$$

Write now $\partial_\tau=Nn+\vec{N}$ whee $\vec{N}$ is the shift field and N the lapse function. Then one has

$$\text{(C.3)}\qquad N=-g(\partial_\tau,n);\ \vec{N}=\partial_\tau+g(\partial_\tau,n)n$$

Write the Christoffel symbols of the connection ∇ on Σ as ${}^3\Gamma^i{}_{jk}$ and the Riemann tensor of ${}^3g$ as ${}^3R^m_{ijk}$. Then some calculation (cf. [**86**]) gives the Gauss-Codazzi equations

$$\text{(C.4)}\qquad R(\partial_i,\partial_j)\partial_k=({}^3\nabla_iK_{jk}-{}^3\nabla_jK_{ik})n+({}^3R^m_{ijk}+K_{jk}K^m_i-K_{ik}K^m_j)\partial_m$$

Now the Einstein tensor has the form $G_{\mu\nu}=R_{\mu\nu}-(1/2)g_{\mu\nu}R$ where $R_{\mu\nu}=R^{\alpha}{}_{\mu\alpha\nu}$ with $R=R^{\alpha\beta}_{\alpha\beta}$ and for vectors ∂_j tangent to Σ at a point in question this leads to (cf. [**86**]

$$\text{(C.5)}\qquad G_{\mu\nu}n^\mu n^\nu=-\frac{1}{2}({}^3R+(TrK)^2-Tr(K^2));\ G_{\nu i}n^\mu={}^3\nabla_jK^j_i-{}^3\nabla_iK^j_j$$

The remaining 6 Einstein equations $G_{ij} = 0$ are dynamical in nature and describe the time evolution of ${}^3g$. Now looking only at the vacuum Einstein equations one writes $q_{ij} \sim {}^3g_{ij}$ with q for $det(q_{ij})$. It can be shown then (cf. [**86**]) that

$$K_{ij} = \frac{1}{2}N^{-1}(\dot{q}_{ij} - {}^3\nabla_i N_j - {}^3\nabla_j N_i) \tag{C.6}$$

The Lagrangian density for the Einstein-Hilbert action is $R\sqrt{-detg}d^4x$ and one writes here $\mathfrak{L} = R\sqrt{-g}$ which in terms of the 3-metric and lapse function becomes $\mathfrak{L} = q^{1/2}NR$. Discarding terms that give total divergences (and would integrate to zero for compact Σ at least) one has then

$$\mathfrak{L} = q^{1/2}N({}^3R + Tr(K^2) - (trK)^2) \tag{C.7}$$

Now the conjugate momenta are determined via

$$p^{ij} = \frac{\partial\mathfrak{L}}{\partial\dot{q}_{ij}} = q^{1/2}(K^{ij} - Tr(K)q^{ij}) \tag{C.8}$$

The Hamiltonian structure involves now

$$\mathfrak{H}(p^{ij}, q_{ij}) = p_{ij}\dot{q}^{ij} - \mathfrak{L};\ H = \int \mathfrak{H}d^3x;\ \mathfrak{H} = q^{1/2}(NC + N^iC_i) \tag{C.9}$$

where

$$C = -{}^3R + q^{-1}\left(Tr(p^2) - \frac{1}{2}Tr(p)^2\right);\ C_i = -2\,{}^3\nabla^j(q^{-1/2}p_{ij}) \tag{C.10}$$

Note one must specify the lapse and shift to know the meaning of time evolution. However one can compute that

$$C = -2G_{\mu\nu}n^\mu n^\nu;\ C_i = -2G_{\mu i}n^\mu \tag{C.11}$$

Now the equations $C = C_i = 0$ are precisely the constraint Einstein equations and hence $\mathfrak{H} = 0$ is a constraint on the phase space in $T^*Met(\Sigma)$ yet the dynamics is not trivial. To formulate Hamilton's equations one can define Poisson brackets on phase space via

$$\{f, g\} = \int_\Sigma \left\{\frac{\partial f}{\partial p^{ij}(x)}\frac{\partial g}{\partial q_{ij}(x)} - \frac{\partial f}{\partial q_{ij}(x)}\frac{\partial g}{\partial p^{ij}(x)}\right\} q^{1/2}d^3x \tag{C.12}$$

This is often written in a functional derivative notation but it simply amounts to defining say $\partial f/\partial q_{ij}(x)$ for example via

$$\int_\Sigma h_{ij}(x)\frac{\partial f}{\partial q_{ij}(x)}q^{1/2}d^3x = \frac{d}{ds}f(q + sh)|_{s=0} \tag{C.13}$$

for every symmetric $(0, 2)$ tensor field h (for $h \sim \delta g$ one writes the right side of (C.13) as δf. In this spirit one arrives at

$$\{p^{ij}(x), q_{k\ell}(y)\} = (\delta^i_k\delta^j_\ell + \delta^i_\ell\delta^j_k)\delta^3(x - y); \tag{C.14}$$
$$\{p^{ij}(x), p^{k\ell}(y)\} = 0;\ \{q_{ij}(x), q_{k\ell}(y)\} = 0$$

The equations $G_{ij} = 0$ in this disguise are now simply $\dot{q}^{ij} = \{H, q^{ij}\}$ and $\dot{p}_{ij} = \{H, p_{ij}\}$ which take the form

$$\dot{q}_{ij} = 2q^{-1/2}N\left(p_{ij} - \frac{1}{2}p^k_k q_{ij}\right) + 2\,{}^3\nabla_{[i}N_{j]}; \tag{C.15}$$

$$\dot{p}^{ij} = -Nq^{1/2}\left({}^3R^{ij} - \frac{1}{2}{}^3Rq^{ij}\right) + \frac{1}{2}Nq^{-1/2}q^{ij}\left(p_{ab}p^{ab} - -\frac{1}{2}(p^a_a)^2\right) -$$
$$-2Nq^{-1}\left(p^{ia}p^j_a - \frac{1}{2}p^a_a q^{ij}\right) + q^{1/2}(\nabla^i\nabla^j N - q^{ij}\,{}^3\nabla^a\,{}^3\nabla_a N) +$$
$$+q^{1/2}\nabla_a(q^{-1/2}N^a p^{ij}) - 2p^{a[i3}\nabla_a N^{j]}$$

This is a horror story but it does show that the time evolution given by Hamilton's equations is nontrivial. One notes in passing that the lapse and shift measure how much the time evolution push the slice Σ in the normal or tangent direction respectively. For example for shift (resp. lapse) zero one has

$$C(N) = \int_\Sigma NCq^{1/2}d^3x \ (resp.\ C(\vec{N}) = \int_\Sigma N^i C_i q^{1/2}d^3x) \tag{C.16}$$

Here $C(\vec{N})$ or C_i (resp. $C(N)$ or C) is called the diffeomorphism (resp. Hamiltonian) constraint and one can calculate

$$\{C(\vec{N}), C(\vec{N}')\} = C([\vec{N}, \vec{N}']);\ \{C(\vec{N}), C(N')\} = C(\vec{N}N'); \tag{C.17}$$
$$\{C(N), C(N')\} = C((N\partial^i N' - N'\partial^i N)\partial_i)$$

where $\vec{N}N'$ is the derivative of N' in the direction $\vec{N}$ and $(N\partial^i N' - N'\partial^i N)\partial_i$ is the result of converting the 1-form $NdN' - N'dN$ into a vector field by raising indices. These formulas are known as the Dirac algebra and one notes that the constraints are closed under Poisson brackets.

Now for quantization one proceeds formally for various reasons (cf. [**86**]) and takes the operator corresponding to the 3-metric (and momentum) to be

$$(\hat{q}_{ij}(x)\psi)(q) = g_{ij}(x)\psi(q);\ (\hat{p}^{ij}(x)\psi)(q) = -i\frac{\partial}{\partial q_{ij}(x)}\psi(q) \tag{C.18}$$

where $g \in Met(\Sigma)$. These operators satisfy then

$$[\hat{p}^{ij}(x), \hat{q}_{k\ell}(y)] = -i(\delta^i_k\delta^j_\ell + \delta^i_\ell\delta^j_k)\delta^3(x,y); \tag{C.19}$$
$$[\hat{p}^{ij}(x), \hat{p}^{k\ell}(y)] = 0;\ [\hat{q}_{ij}(x), \hat{q}_{k\ell}(y)] = 0$$

In order to obtain quantum versions $\hat{C}$ and $\hat{C}_i$ one encounters operator ordering problems; ideally one would like

$$[\hat{C}(\vec{N}), \hat{C}(\vec{N}')] = -i\hat{C}(\vec{N}, \vec{N}']);\ [\hat{C}(\vec{N}, \hat{C}(N')] = -i\hat{C}(\vec{N}N'); \tag{C.20}$$
$$[\hat{C}(N), \hat{C}(N')] = -i\hat{C}((N\partial^i N' - N'\partial^i N)\partial_i)$$

but this seems virtually impossible to achieve. Suppose nevertheless that one obtained somehow satisfactory operators $\hat{C}$ and $\hat{C}_i$; then one could write

$$\hat{H} = \int_\Sigma (N\hat{C} + N^i\hat{C}_i)q^{1/2}d^3x \tag{C.21}$$

It could then be said that a vector $\psi \in L^2(Met(\Sigma))$ (whatever that may mean) is a physical state if it satisfies $\hat{C}(N)\psi = \hat{C}(\vec{N})\psi = 0$ or alternatively $\hat{H}\psi = 0$ for all choices of lapse and shift and this is the WDW equation. There are many problems here and even if one could find solutions there arises the problem of time, namely the Hamiltonian vanishes on the space of physical states so any operator

A on $\mathfrak{H}_{phys}$ must satisfy $(d/dt)A_t = i[\hat{H}, A_t] = 0$ and the dynamics disappears. Recent developments using the Ashtekar variables have made some progress in this area and will be discussed briefly below.

We describe briefly now the Ashtekar variables following [**86**] which are based on a modification of the Palatini formalism. First for background the Palatini action is $S(g) = \int_M R \cdot vol$ rewritten so that it is not a function of the metric but rather a function of a connection and a frame field. Thus a trivialization of TM is a vector bundle isomorphism $e : M \times \mathbf{R}^n \to TM$ sending each fiber $\{p\} \times \mathbf{R}^n$ of the trivial bundle $M \times \mathbf{R}^n$ to the corresponding tangent space T_pM. A trivialization of TM is also called a frame field sending the standard basis of $\mathbf{R}^n$ to a basis of tangent vectors at p or a frame. If M is 3 (resp. 4) dimensional a frame field is called a triad (resp. tetrad). One goes back and forth now using the frame field e and its inverse $e^{-1} : TM \to M \times \mathbf{R}^n$. Given a basis of sections of $M \times \mathbf{R}^n$ of the form $\xi_i = (0, \cdots, 0, 1, 0, 0, \cdots)$ one writes any section as $s = s^I \xi_I$ where I denotes an internal index (whereas ∂_μ refers to coordinate vector fields on a chart). Thus $e(\xi_I) = e_I^\alpha \partial_\alpha$ and $e(\xi_I) \sim e_I$. Now given sections s, s' one can define $\eta(s, s') = \eta_{IJ} s^I {s'}^J$ where η_{IJ} is copied after a Minkowski metric $(-1, 1, \cdots, 1)$ (internal metric). One can raise and lower indices via η_{IJ} and set $g(v, v') = g_{\alpha\beta} v^\alpha {v'}^\beta$. The frame field is said to be orthonormal if $g(e_I, e_J) = \eta_{IJ}$; in this case one has $g(e(s), e(s')) = \eta(s, s')$ since

(C.22) $$g(e(s), e(x')) = g(e(S^I \xi_I), e(s^J \xi_J)) = s^I s^J g(e_I, e_J) = \eta_{IJ} s^I s^J =$$
$$= \eta(s^I \xi_I, s^J \xi_J) = \eta(s, s')$$

Note that one can write $\eta_{IJ} = g(e_I, e_J) = g_{\alpha\beta} e_I^\alpha e_J^\beta$ and hence $\delta_J^I = e_\alpha^I e_J^\alpha$. Further $e^{-1}v = e_\alpha^I v^\alpha \xi_I$ since if $v = e(s)$ one has $e^{-1}v = e_\alpha^I v^\alpha \xi_I = e_\alpha^I e_J^\alpha s^J \xi_I = \delta_J^I s^J \xi_I = s^I \xi_I = s$ This leads to a formula for the metric g in terms of the coframe field e_α^I, namely

(C.23) $$g_{\alpha\beta} = g(\partial_\alpha, \partial_\beta) = \eta(e^{-1}\partial_\alpha, e^{-1}\partial_\beta) = \eta(e_\alpha^I \xi_I, e_\beta^J \xi_J) = \eta_{IJ} e_\alpha^I e_\beta^J$$

The other ingredient in the Palatini formalism is a connection on the trivial bundle $M \times \mathbf{R}^n$. One says a connection D here is a Lorentz connection if $v\eta(x, s') = \eta(D_v s, s') + \eta(s, D_v s')$ (standard $S(O(n,1)$ connection). Note torsion free is meaningless and there is no Levi-Civita connection on $M \times \mathbf{R}^n$; however the standard flat connection $D^0 s = v(s^I)\xi_I$ is nice and any connection can be written as $D = D^0 + A$ for some potential A, which is an $End(\mathbf{R}^n)$-valued 1-form on M. Thus

(C.24) $$D_v s = (v(s^J) + A_{\mu I}^J v^\mu s^I)\xi_J; \; F_{\alpha\beta}^{IJ} = \partial_\alpha A_\beta^{IJ} - \partial_\beta A_\alpha^{IJ} + [A_\alpha, A_\beta]^{IJ}$$

where $F_{\alpha\beta}^{IJ}$ is the curvature of D. If A defines a Lorentz connection then $F_{\alpha\beta}^{IJ} = -F_{\beta\alpha}^{IJ} = -F_{\alpha\beta}^{JI}$. Next given a frame field e and a Lorentz connection D one can transfer the Lorentz connection from $M \times \mathbf{R}^N$ to TM to obtain a connection $\bar{\nabla}$ given by

(C.25) $$\bar{\nabla}_\alpha \partial_\beta = \bar{\Gamma}^\gamma{}_{\alpha\beta} \partial_\gamma; \; \bar{\Gamma}^\gamma{}_{\alpha\beta} = A_{\alpha I}^J e_\beta^I e_J^\gamma$$

Here $\bar{\nabla}$ is called the imitation Levi-Civita connection and $\bar{\Gamma}^{\gamma}{}_{\alpha\beta}$ are the imitation Christoffel symbols, leading to an imitation Riemann tensor

$$\text{(C.26)} \qquad \bar{R}_{\alpha\beta}{}^{\delta} = F^{IJ}_{\alpha\beta} e^{\gamma}_I e^{\delta}_J;\ \bar{R}_{\alpha\beta} = \bar{R}^{\gamma}{}_{\alpha\gamma\beta};\ \bar{R} = \bar{R}^{\alpha}_{\alpha}$$

Now the Palatini action is basically the Einstein-Hilbert action in disguise, being a function of the frame field given by $g_{\alpha\beta} = \eta_{IJ} e^I_\alpha e^J_\beta$ in the form

$$\text{(C.27)}\ S(A,e) = \int_M e^{\alpha}_I e^{\beta}_J F^{IJ}_{\alpha\beta} \cdot vol;\ \delta S = 2\int_M (\bar{R}_{\alpha\beta} - (1/2)\bar{R} g_{\alpha\beta}\eta^{IJ} e^{\beta}_J (\delta e^{\alpha}_I) \cdot vol$$

Thus $\delta S = 0$ for an arbitrary variation of the frame field when $\bar{R}_{\alpha\beta} - (1/2)\bar{R} g_{\alpha\beta} = 0$ which is of course Einstein's equation when $\bar{\nabla} = \nabla$. We refer to [**86**] for further computations, formulas, and discussion.

Now for the Ashtekar variables themselves define the complexified tangent bundle $\mathbf{C}TM$ to have fibers $\mathbf{C} \times T_pM$ and an imitation complexified tangent bundle $M \times \mathbf{C}^4$. A complex frame field is an isomorphism $e: M \times \mathbf{C}^4 \to \mathbf{C}TM$. A connection A on $M \times \mathbf{C}^4$ is an $End(\mathbf{C}^4)$ valued 1-form on M with components $A^J_{\alpha I}$ or A^{IJ}_α; it is Lorentz if $A^{IJ}_\alpha = -A^{JI}_\alpha$. Recall the Hodge $*$ operator maps 2-forms to 2-forms in 4 dimensions and define it here via

$$\text{(C.28)} \qquad *T^{IJ} = (1/2)\epsilon^{IJ}{}_{KL} T^{KL};\ \epsilon^{i_1,\cdots,i_n} = \begin{cases} sgn(i_1,\cdots,i_n) & i_j\ distinct \\ 0 & otherwise \end{cases}$$

for any T with two antisymmetric raised internal indices. In particular

$$\text{(C.29)} \qquad (*A)^{IJ}_\alpha = (1/2)\epsilon^{IJ}{}_{KL} A^{KL}_\alpha$$

Now write any Lorentz connection A as a sum of self-dual and anti-self-dual parts

$$\text{(C.30)} \qquad A = {}^+A + {}^-A;\ *{}^{\pm}A = \pm i {}^{\pm}A;\ {}^{\pm}A = (1/2)(A \mp i * A)$$

In the self dual formulation of GR one of the two basic fields is a self-dual Lorentz connection, i.e. a Lorentz connection on $M \times \mathbf{C}^4$ with $*{}^+A = i{}^+A$. The other basic field is a complex frame field $e: M \times \mathbf{C}^4 \to \mathbf{C}TM$ and the action is built using the curvature ${}^+F$ of ${}^+A$ via

$$\text{(C.31)} \qquad {}^+F^{IJ}_{\alpha\beta} = \partial_\alpha {}^+A^{IJ}_\beta - \partial_\beta {}^+A^{IJ}_\alpha + [{}^+A_\alpha, {}^+A_\beta]^{IJ}$$

As in the Palatini formalism one writes

$$\text{(C.32)} \qquad g_{\alpha\beta} = \eta_{IJ} e^I_\alpha e^J_\beta;\ e^{-1}\partial_\alpha = e^I_\alpha \xi_I$$

(note the metric is now complex). The self dual action is

$$\text{(C.33)} \qquad S_{SD}({}^+A, e) = \int_M e^{\alpha}_I e^{\beta}_J {}^+F^{IJ}_{\alpha\beta} \cdot vol;\ vol = \sqrt{-g}d^4x$$

Now define the internal Hodge dual of the curvature of a connection F on $M \times \mathbf{C}^4$ via

$$\text{(C.34)} \qquad (*F)^{IJ}_{\alpha\beta} = (1/2)\epsilon^{IJ}{}_{KL} F^{KL}_{\alpha\beta}$$

and call the curvature self dual if $*F = iF$. It turns out that the curvature of a self dual Lorentz connection is self dual (computation needed) and this has Lie algebraic meaning (cf. [**86**] for details). Next compute $\delta S_{SD} = 0$ and following [**86**] one obtains two equations. First by varying the self dual connection there is an equation saying that ^+A is the self dual part of a Lorentz connection A for which the self dual part of the Riemann tensor of g is related to ^+F, namely

$$(C.35)\qquad {}^+R^{\alpha}{}_{\beta\gamma}{}^{\delta} = (1/2)(R^{\alpha}{}_{\beta\gamma}{}^{\delta} - (i/2)\epsilon^{\alpha\delta}{}_{\mu\nu}R^{\mu}{}_{\beta\gamma}{}^{\nu});\ {}^+R^{\alpha}{}_{\beta\gamma}{}^{\delta} = {}^+F^{IJ}_{\beta\gamma}e^{\alpha}_{I}e^{\delta}_{J}$$

Second by varying the frame field there arises a self dual analogue of Einstein's equation

$$(C.36)\qquad {}^+R_{\alpha\beta} - (1/2)g_{\alpha\beta}{}^+R = 0;\ {}^+R_{\alpha\beta} = {}^+R^{\gamma}{}_{\alpha\gamma\beta};\ {}^+R = {}^+R^{\alpha}_{\alpha}$$

Using symmetries of the Riemann tensor this is equivalent to the vacuum Einstein equation. Note however that we have complex metrics here; some reality conditions are needed and this is not exactly a trivial matter. Thus let Σ be a spacelike slice and work in coordinates such that ∂_0 is normal to Σ and ∂_i is tangent for spacelike indices. Given a self dual Lorentz connection $^+A^{IJ}_{\alpha}$ on $M \times \mathbf{C}^4$ one can restrict it to a connection A^{IJ}_i on $\Gamma \times \mathbf{C}^4$ with $A^{IJ}_i = -A^{JI}_i$ and $*A = iA$ (the + sign is gratuitously omitted here). Since $sl(2,\mathbf{C})$ has a basis in terms of Pauli matrices one can also write this as $-(i/2)A^a_i\sigma_a$ $(a = 1,2,3)$. The field playing the role analogous to position is the self dual Lorentz conection A^a_i with conjugate momentum $\tilde{E}^i_a = q^{1/2}e^i_a$ and one has Poisson brackets

$$(C.37)\quad \{\tilde{E}^i_a(x), A^b_j(y)\} = -i\delta^b_a\delta^i_j\delta^3(x,y);\ \{\tilde{E}^i_a(x), \tilde{E}^j_b(y)\} = 0 = \{A^a_i(x), A^b_j(y)\}$$

The Hamiltonian and diffeomorphism constraints are given via

$$(C.38)\qquad \tilde{C} = \epsilon^{abc}\tilde{E}^i_a\tilde{E}^j_bF_{ijc};\ C_j = \tilde{E}^k_aF^a_{jk};\ G_a = D_i\tilde{E}^i_a$$

(the latter being a Gauss law constraint). The tilde appears because of densitation, i.e. $\tilde{C}$ is $q^{1/2}$ times the earlier C. To quantize now one writes

$$(C.39)\qquad (\hat{A}^a_i(x)\psi)(A) = A^a_i(x)\psi(A);\ (\hat{E}^i_a(x)\psi)(A) = \frac{\partial}{\partial A^a_i(x)}\psi(A)$$

$$(C.40)\qquad [\hat{E}^i_a(x), \hat{A}^b_j(y)] = \delta^b_a\delta^i_j\delta^3(x,y);\ [\hat{E}^i_a(x), \hat{E}^j_b(y)] = 0 = [\hat{A}^a_i(x), \hat{A}^b_j(y)]$$

A convenient choice of operator orderings is now
(C.41)

$$\hat{C} = \epsilon^{abc}\hat{E}^i_a\hat{E}^j_b\hat{F}_{ijc};\ \hat{C}_j = \hat{E}^k_a\hat{F}^a_{jk};\ \hat{G}_a = \hat{D}_i\hat{E}^i_a;\ (\hat{F}^a_{jk}(x)\psi)(A) = F^a_{jk}(x)\psi(A)$$

It seems that these operators satisfy commutation relations analogous to the Poisson brackets of the classical constraints. The physical state space $\mathfrak{H}_{phys}$ then consists of functions $\psi(A)$ satisfying the constraints in quantum form, i.e.

$$(C.42)\qquad \mathfrak{H}_{phys} = \{\psi : \hat{C}\psi = \hat{C}_j\psi = \hat{G}_a\psi = 0\}$$

Here $\hat{C}_j\psi = 0$ means $\psi(A) = \psi(A')$ whenever A' is obtained from A by applying a diffeomorphism connected to the identity by a flow. Similarly $\hat{G}_a\psi = 0$ says that $\psi(A) = \psi(A')$ whenever A' is obtained from A by a small gauge transformation. The Hamiltonian constraint $\hat{C}\psi = 0$ contains the dynamics of the theory and finding solutions is difficult.

There is an interesting relation between Chern-Simons (CS) theory and quantum gravity however that provides some solutions. First if the cosmological constant is nonzero the Hamiltonian constraint becomes

$$\hat{C} = \epsilon^{abc}\hat{E}^i_a\hat{E}^j_b\hat{F}_{ijc} - \frac{\Lambda}{6}\epsilon_{ijk}\epsilon^{abc}\hat{E}^i_a\hat{E}^j_b\hat{E}^k_c \tag{C.43}$$

The CS state ψ_{CS} is defined as $\psi_{CS}(A) = exp[-(6/\hbar)S_{CS}(A)]$ where $S_{CS}(A) = \int_\Sigma Tr(A \wedge dA + (2/3)A \wedge A \wedge A)$. It is then shown in [**86**] that $\hat{C}_j\psi_{CS} = \hat{G}_a\psi_{CS} = \hat{C}\psi_{CS} = 0$ with some discussion. The book [**86**] was written in 1994 and there has since been enormous activity in loop quantum gravity for which we refer to [**67, 71, 73, 231, 379, 505, 749, 1085, 1086, 1183, 1220, 1221, 1222, 1223**].

APPENDIX D

DIRAC ON WEYL GEOMETRY

First we give some background on Weyl geometry and Brans-Dicke theory following Adler et al [**12**]; for differential geometry we use the tensor notation of [**12**] and refer to e.g. [**179, 457, 618, 950, 962, 1274, 1309**] for other notation (see also [**1300**] for interesting variations). For general background see [**189, 225, 293, 450, 451, 460, 757, 815, 1041, 1059, 1229, 1230, 1231, 1318, 1335**] and note that for our purposes the most important background features appear already in Sections 3.2, 3.2.2, and 4.1. One thinks of a differential manifold $M = \{U_i, \phi_i\}$ with $\phi : U_i \to \mathbf{R}^4$ and metric $g \sim g_{ij}dx^i dx^j$ satisfying $g(\partial_k, \partial_\ell) = g_{k\ell} =< \partial_k, \partial_\ell >= g_{\ell k}$. This is for the bare essentials; one can also imagine tangent vectors $X_i \sim \partial_i$ and dual cotangent vectors $\theta^i \sim dx^i$, etc. Given a coordinate change $\tilde{x}^i = \tilde{x}^i(x^j)$ a vector ξ^i transforming via $\tilde{\xi}^i = \sum \partial_i \tilde{x}^j \xi^j$ is called contravariant (e.g. $d\tilde{x}^i = \sum \partial_j \tilde{x}^i dx^j$). On the other hand $\partial\phi/\partial\tilde{x}^i = \sum(\partial\phi/\partial x^j)(\partial x^j/\partial\tilde{x}^i$ leads to the idea of covariant vectors $A_j \sim \partial\phi/\partial x^j$ transforming via $\tilde{A}_i = \sum(\partial x^j/\partial\tilde{x}^i)A_j$ (i.e. $\partial/\partial\tilde{x}^i \sim (\partial x^j/\partial\tilde{x}^j)\partial/\partial x^j$). Now define connection coefficients or Christoffel symbols via (strictly one writes $T^\gamma{}_\alpha = g_{\alpha\beta}T^{\gamma\beta}$ and $T_\alpha{}^\gamma = g_{\alpha\beta}T^{\beta\gamma}$ which are generally different; we use that notation here but it is not used in subsequent sections since it is unnecessary)

$$\text{(D.1)} \qquad \Gamma^r{}_{ki} = -\left\{ \begin{array}{c} r \\ k\ i \end{array} \right\} = -\frac{1}{2}\sum(\partial_i g_{k\ell} + \partial_k g_{\ell i} - \partial_\ell g_{ik})g^{\ell r} = \Gamma^r{}_{ik}$$

(note this differs by a minus sign from some other authors). Note also that (D.1) follows from equations

$$\text{(D.2)} \qquad \partial_\ell g_{ik} + g_{rk}\Gamma^r{}_{i\ell} + g_{ir}\Gamma^r{}_{\ell k} = 0$$

and cyclic permutation; the basic definition of $\Gamma^i{}_{mj}$ is found in the transplantation law

$$\text{(D.3)} \qquad d\xi^i = \Gamma^i{}_{mj}dx^m\xi^j$$

Next for tensors $T^\alpha_{\beta\gamma}$ define derivatives

$$\text{(D.4)} \qquad T^\alpha_{\beta\gamma|k} = \partial_k T^\alpha_{\beta\gamma};\ T^\alpha_{\beta\gamma||\ell} = \partial_\ell T^\alpha_{\beta\gamma} - \Gamma^\alpha{}_{\ell s}T^s_{\beta\gamma} + \Gamma^s{}_{\ell\beta}T^\alpha_{s\gamma} + \Gamma^s{}_{\ell\gamma}T^\alpha_{\beta s}$$

In particular covariant derivatives for contravariant and covariant vectors respectively are defined via

$$\text{(D.5)} \qquad \xi^i_{||k} = \partial_k\xi^i - \Gamma^i{}_{k\ell}\xi^\ell = \nabla_k\xi^i;\ \eta_{m||\ell} = \partial_\ell\eta_m + \Gamma^r{}_{m\ell}\eta_r = \nabla_\ell\eta_m$$

Now to describe Weyl geometry one notes first that for Riemannian geometry (D.3) holds along with

(D.6) $$\ell^2 = \|\xi\|^2 = g_{\alpha\beta}\xi^\alpha\xi^\beta;\ \xi^\alpha\eta_\alpha = g_{\alpha\beta}\xi^\alpha\eta^\beta$$

However one does not demand conservation of lengths and scalar products under affine transplantation (D.3). Thus assume

(D.7) $$d\ell = (\phi_\beta dx^\beta)\ell$$

where the covariant vector ϕ_β plays a role analogous to $\Gamma^\alpha_{\beta\gamma}$. Combining one obtains

(D.8) $$d\ell^2 = 2\ell^2(\phi_\beta dx^\beta) = d(g_{\alpha\beta}\xi^\alpha\xi^\beta) =$$

$$= g_{\alpha\beta|\gamma}\xi^\alpha\xi^\beta dx^\gamma + g_{\alpha\beta}\Gamma^\alpha_{\rho\gamma}\xi^\rho\xi^\beta dx^\gamma + g_{\alpha\beta}\Gamma^\beta_{\rho\gamma}\xi^\alpha\xi^\rho dx^\gamma$$

Rearranging etc. and using (D.6) again gives

(D.9) $$(g_{\alpha\beta|\gamma} - 2g_{\alpha\beta}\phi_\gamma) + g_{\sigma\beta}\Gamma^\sigma{}_{\alpha\gamma} + g_{\sigma\alpha}\Gamma^\sigma{}_{\beta\gamma} = 0$$

leading to

(D.10) $$\Gamma^\alpha_{\beta\gamma} = -\left\{ \begin{array}{c} \alpha \\ \beta\ \gamma \end{array} \right\} + g^{\sigma\alpha}[g_{\sigma\beta}\phi_\gamma + g_{\sigma\gamma}\phi_\beta - g_{\beta\gamma}\phi_\sigma]$$

Thus we can prescribe the metric $g_{\alpha\beta}$ and the covariant vector field ϕ_γ and determine by (D.10) the field of connection coefficients $\Gamma^\alpha_{\beta\gamma}$ which admits the affine transplantation law (D.3). If one takes $\phi_\gamma = 0$ the Weyl geometry reduces to Riemannian geometry. This leads one to consider new metric tensors via the gauge transformation $\hat{g}_{\alpha\beta} = f(x^\lambda)g_{\alpha\beta}$ and it turns out that $(1/2)\partial log(f)/\partial x^\lambda$ plays the role of ϕ_λ in (D.7). The ordinary connections $\left\{ \begin{array}{c} \alpha \\ \beta\ \gamma \end{array} \right\}$ constructed from $g_{\alpha\beta}$ are equal to the more general connections $\hat{\Gamma}^\alpha_{\beta\gamma}$ constructed according to (D.10) from $\hat{g}_{\alpha\beta}$ and $\hat{\phi}_\lambda = (1/2)\partial log(f)/\partial x^\lambda$. The generalized differential geometry is conformal in that the ratio

(D.11) $$\frac{\xi^\alpha\eta_\alpha}{\|\xi\|\|\eta\|} = \frac{g_{\alpha\beta}\xi^\alpha\eta^\beta}{[(g_{\alpha\beta}\xi^\alpha\xi^\beta)(g_{\alpha\beta}\eta^\alpha\eta^\beta)]^{1/2}}$$

does not change under the gauge transformation above. Again if one has a Weyl geometry characterized by $g_{\alpha\beta}$ and ϕ_α with connections determined by (D.10) one may replace the geometric quantities by use of a scalar field f with

(D.12) $$\hat{g}_{\alpha\beta} = f(x^\lambda)g_{\alpha\beta},\ \hat{\phi}_\alpha = \phi_\alpha + (1/2)(log(f)_{|\alpha};\ \hat{\Gamma}^\alpha_{\beta\gamma} = \Gamma^\alpha_{\beta\gamma}$$

without changing the intrinsic geometric properties of vector fields; the only change is that of local lengths of a vector via $\hat{\ell}^2 = f(x^\lambda)\ell^2$. Note that one can reduce $\hat{\phi}_\alpha$ to the zero vector field if and only if ϕ_α is a gradient field, namely $F_{\alpha\beta} = \phi_{\alpha|\beta} - \phi_{\beta|\alpha} = 0$ (i.e. $\phi_\alpha = (1/2)\partial_\alpha log(f) \equiv \partial_\beta\phi_\alpha = \partial_\alpha\phi_\beta$). In this case one has length preservation after transplantation around an arbitrary closed curve and the vanishing of $F_{\alpha\beta}$ guarantees a choice of metric in which the Weyl geometry becomes Riemannian; thus $F_{\alpha\beta}$ is an intrinsic geometric quantity for Weyl geometry - note $F_{\alpha\beta} = -F_{\beta\alpha}$ and

(D.13) $$\{F_{\alpha\beta|\gamma}\} = 0;\ \{F_{\mu\nu|\lambda}\} = F_{\mu\nu|\lambda} + F_{\lambda\mu|\nu} + F_{\nu\lambda|\mu}$$

Similarly the concept of covariant differentiation depends only on the idea of vector transplantation. Indeed one can define

(D.14) $$\xi^{\alpha}_{||\beta} = \xi^{\alpha}_{|\beta} - \Gamma^{\alpha}_{\beta\gamma}\xi^{\gamma}$$

In Riemann geometry the curvature tensor is

(D.15) $$\xi^{\alpha}_{||\beta|\gamma} - \xi^{\alpha}_{||\gamma|\beta} = R^{\alpha}_{\eta\beta\gamma}\xi^{\eta}$$

Hence here we can write

(D.16) $$R^{\alpha}_{\ \beta\gamma\delta} = -\Gamma^{\alpha}_{\ \beta\gamma|\delta} + \Gamma^{\alpha}_{\ \beta\delta|\gamma} + \Gamma^{\alpha}_{\ \tau\delta}\Gamma^{\tau}_{\ \beta\gamma} - \Gamma^{\alpha}_{\ \tau\gamma}\Gamma^{\tau}_{\ \beta\delta}$$

Using (D.11) one then can express this in terms of $g_{\alpha\beta}$ and ϕ_{α} but this is complicated. Equations for $R_{\beta\delta} = R^{\alpha}_{\ \beta\alpha\delta}$ and $R = g^{\beta\delta}R_{\beta\delta}$ are however given in [**12**]. One notes that in Weyl geometry if a vector ξ^{α} is given, independent of the metric, then $\xi_{\alpha} = g_{\alpha\beta}\xi^{\beta}$ will depend on the metric and under a gauge transformation one has $\hat{\xi}_{\alpha} = f(x^{\lambda})\xi_{\alpha}$. Hence the covariant form of a gauge invariant contravariant vector becomes gauge dependent and one says that a tensor is of weight n if, under a gauge transformation $\hat{T}^{\alpha\cdots}_{\beta\ldots} = f(x^{\lambda})^n T^{\alpha\cdots}_{\beta\ldots}$. Note ϕ_{α} plays a singular role in (D.12) and has no weight. Similarly $(\sqrt{-\hat{g}} = f^2\sqrt{-g}$ (weight 2) and $F^{\alpha\beta} = g^{\alpha\mu}g^{\beta\nu}F_{\mu\nu}$ has weight -2 while , $\mathfrak{F}^{\alpha\beta} = F^{\alpha\beta}\sqrt{-g}$ has weight 0 and is gauge invariant; further $F_{\alpha\beta}F^{\alpha\beta}\sqrt{-g}$ is gauge invariant. Now for Weyl's theory of electromagnetism one wants to interpret ϕ_{α} as an EM potential and one has automatically the Maxwell equations $\{F_{\alpha\beta|\gamma}\} = 0$ along with a gauge invariant complementary set $\mathfrak{F}^{\alpha\beta}_{|\beta} = \mathfrak{s}^{\alpha}$ (source equations). These equations are gauge invariant as a natural consequence of the geometric interpretation of the EM field. For the interaction between the EM and gravitational fields one sets up some field equations as indicated in [**?**] and the interaction between the metric quantities and the EM fields is exhibited there.

REMARK D.1. As indicated earlier in [**12**] $R^{i}_{\ jk}$ is defined with a minus sign compared with Ohanian-Ruffini and Willmore e.g. [**950, 1309**]. There is also a difference in definition of the Ricci tensor which is taken to be $G^{\beta\delta} = R^{\beta\delta} - (1/2)g^{\beta\delta}R$ in [**12**] with $R = R^{\delta}_{\delta}$ so that $G_{\mu\gamma} = g_{\mu\beta}g_{\gamma\delta}G^{\beta\delta} = R_{\mu\gamma} - (1/2)g_{\mu\gamma}R$ with $G^{\gamma}_{\eta} = R^{\eta}_{\eta} - 2R \Rightarrow G^{\eta}_{\eta} = -R$ (recall $n = 4$). In [**950**] the Ricci tensor is simply $R_{\beta\mu} = R^{\alpha}_{\beta\mu\alpha}$ where $R^{\alpha}_{\beta\mu\nu}$ is the Riemann curvature tensor and $R = R^{\eta}_{\eta}$ again. This is similar to [**1309**] where the Ricci tensor is defined as $\rho_{j\ell} = R^{i}_{\ ji\ell}$. To clarify all this we note that $R_{\eta\gamma} = R^{\alpha}_{\ \eta\alpha\gamma} = g^{\alpha\beta}R_{\beta\eta\alpha\gamma} = -g^{\alpha\beta}R_{\beta\eta\gamma\alpha} = -R^{\alpha}_{\ \eta\gamma\alpha}$ which confirms the minus sign difference. ■

For completeness it is worthwhile to reflect on the comments of a master craftsman and hence we refer her to Dirac [**377**] where first there are two papers on a new classical theory of the electron and in the third paper of [**377**] the original Dirac-Weyl action is developed (cf. also Sections 3.2.1 and 4.1) which we sketch here in some detail. The main point is to think of EM fields as a property of spacetime rather than something occuring in a gravity formed spacetime. This seems to be in the spirit of considering a microstructure of the vacuum (or an ether) and we find it attractive. The solution proposed by Weyl involved a length change

$\delta\ell = \ell\kappa_\mu \delta x^\mu$ under parallel transport $x^\mu \to x^\mu + \delta x^\mu$. The κ_μ are field quantities occuring along with the $g_{\mu\nu}$ in a fundamental role. Suppose ℓ gets changed to $\ell' = \ell\lambda(x)$ and $\ell + \delta\ell$ becomes

$$\ell' + \delta\ell' = (\ell + \delta\ell)\lambda(x + \delta x) = (\ell + \delta\ell)\lambda(x) + \ell\lambda_{,\mu}\delta x^\mu \tag{D.17}$$

with neglect of second order terms (here $\lambda_{,\mu} \equiv \partial\lambda/\partial x^\mu$). Then

$$\delta\ell' = \lambda\delta\ell + \ell\lambda_{,\mu}\delta x^\mu = \lambda(\kappa_\mu + \phi_{,\mu})\delta x^\mu;\ \phi = log(\lambda) \tag{D.18}$$

Hence

$$\delta\ell' = \ell'\kappa'_\mu \delta x^\mu;\ \kappa'_\mu = \kappa_\mu + \phi_{,\mu} \tag{D.19}$$

If the vector is transported by parallel displacement around a small closed loop the total change in length is

$$\delta\ell = \ell F_{\mu\nu}\delta S^{\mu\nu};\ F_{\mu\nu} = \kappa_{\mu,\nu} - \kappa_{\nu,\mu} \tag{D.20}$$

and $\delta S^{\mu\nu}$ is the element of area enclosed by the small loop. this change is unaffected by (D.19). It will be seen that the field quantities κ_μ can be taken to be EM potentials, subject to the transformations (D.19) which correspond to no change in the geometry but a change only in the choice of artificial standards of length. The derived quantities $F_{\mu\nu}$ have a geometrical meaning independent of the length standard and correspond to the EM fields. Thus the Weyl geometry provides exactly what is needed for describing both gravitational and EM fields in geometric terms. There was at first some apparent conflict with atomic standards and the theory was rejected, leaving only the idea of gauge transformation for length standard changes.

Dirac's approach however helped to resurrect the Weyl theory; since we feel that this theory is not perhaps sufficiently appreciated a sketch is given here (cf. also Blagojević [**150**]). Dirac first goes into a discussion of large numbers, e.g. e^2/GMm (proton and electron masses), e^2/mc^2 (age of universe), etc. and the Einsteinian theory requires that G be constant which seems in contradiction to $G \sim t^{-1}$ where t represents the epoch time, assumed to be increasing. Dirac reconciles this by assuming the large numbers hypothesis (all dimensionless large numbers are connected) and stipulating that the Einstein equations refer to an interval ds_E which is different from the interval ds_A measured by atomic clocks. Then the objections to Weyl's theory vanish and it is assumed to refer to ds_E. In this spirit then one deals with transformations of the metric gauge under which any length such as ds is multiplied by a factor $\lambda(x)$ depending on its position x, i.e. $ds' = \lambda ds$ and a localized quantity Y may get transformed according to $Y' = \lambda^n Y$, in which case Y is said to be of power n and is called a co-tensor. If $n = 0$ then Y is called an **in-tensor** and it is invariant under gauge transformations. The equation $ds^2 = g_{\mu\nu}dx^\mu dx^\nu$ shows that $g_{\mu\nu}$ is a co-tensor of power 2, since the dx^μ are not affected by a gauge transformation. Hence $g^{\mu\nu}$ is a co-tensor of power -2 and one writes $\mathfrak{g}$ for $\sqrt{-g}$ with $T_{:\mu}$ denoting the covariant derivative ($\nabla_\mu T$ would be better). We see that the covariant derivative of a co-tensor is not generally a co-tensor. However there is a modifed covariant derivative $T_{*\mu}$ which is a co-tensor. Consider first a scalar S of power n; then $S_{:\mu} = S_{,\mu} \equiv S_\mu$; under a change of gauge it transforms to

(D.21) $$S'_\mu = (\lambda^n S)_{,\mu} = \lambda^n S_\mu + n\lambda^{n-1}\lambda_\mu S = \lambda^n[S_\mu + n(\kappa'_\mu - \kappa_\mu)S]$$

(via (D.19)). Thus

(D.22) $$(S_\mu - n\kappa_\mu S)' = \lambda^n(S_\mu - n\kappa_\mu S)$$

so $S_\mu - n\kappa_\mu S$ is a covector of power n and is defined to be the co-covariant derivative of S, i.e.

(D.23) $$\S_{*\mu} = S_\mu - n\kappa_\mu S$$

To obtain the co-covariant derivative of co-vectors and co-tensors we need a modified Christoffel symbol

(D.24) $${}^*\Gamma^\alpha_{\mu\nu} = \Gamma^\alpha_{\mu\nu} - g^\alpha_\mu\kappa_\nu - g^\alpha_\nu\kappa_\mu + g_{\mu\nu}\kappa^\alpha$$

(the notation $\Gamma^\alpha_{\mu\nu}$ for the more correct form $\Gamma^{\alpha}_{\ \mu\nu}$ is used in [**377**]). This is known to be invariant under gauge transformations. Let now A_μ be a co-vector of power n and form $A_{\mu,\nu} - {}^*\Gamma^\alpha_{\mu\nu}A_\alpha$ which is evidently a tensor since it differs from the covariant derivative $A_{\mu:\nu}$ by a tensor and under gauge transformations one has (cf. (D.19) where $\phi_{,\mu} = \kappa'_\mu - \kappa_\mu$)

(D.25) $$(A_{\mu,\nu} - {}^*\Gamma^\alpha_{\mu\nu}A_\alpha)' = \lambda^n A_{\mu,\nu} + n\lambda^{n-1}\lambda_\nu A_\mu - {}^*\Gamma^\alpha_{\mu\nu}\lambda^n A_\alpha =$$
$$= \lambda^n[A_{\mu,\nu} + n(\kappa'_\nu - \kappa_\nu)A_\mu - {}^*\Gamma^\alpha_{\mu\nu}A_\alpha]$$

Thus

(D.26) $$(A_{\mu,\nu} - n\kappa_\nu A_\mu - {}^*\Gamma^\alpha_{\mu\nu}A_\alpha)' = \lambda^n[A_{\mu,\nu} - n\kappa_\nu A_\mu - {}^*\Gamma^\alpha_{\mu\nu}A_\alpha]$$

so take

(D.27) $$A_{\mu*\nu} = A_{\mu,\nu} - n\kappa_\nu A_\mu - {}^*\Gamma^\alpha_{\mu\nu}$$

as the co-covariant derivative of A_α. this can be written via (D.24) as

(D.28) $$A_{\mu*\nu} = A_{\mu:\nu} - (n-1)\kappa_\nu A_\mu + \kappa_\mu A_\nu - g_{\mu\nu}\kappa^\alpha A_\alpha$$

Similarly for a vector B^μ of power n one has

(D.29) $$B^\mu_{*\nu} = B^\mu_{:\nu} - (n+1)\kappa_\nu B^\mu + \kappa^\mu B_\nu - g^\mu_\nu\kappa_\alpha B^\alpha$$

For a co-tensor with various suffixes up and down one can form the co-covariant derivative via the same rules; one notes that the co-covariant derivative always has the same power as the original. Next observe

(D.30) $$(TU)_{*\sigma} = T_{*\sigma}U + TU_{*\sigma}$$

while

(D.31) $$g_{\mu\nu*\sigma} = 0;\ G^{\mu\nu}_{*\sigma} = 0$$

so one can raise and lower suffixes freely in a co-tensor before carrying out co-covariant differentiation. Thus one can raise the μ in (D.28) giving (D.29) with A^μ replacing B^μ and $n-2$ in place of n. The potentials κ_μ do not form a co-vector because of the wrong transformation laws (D.19) but the $F_{\mu\nu}$ defined by (D.19) are unaffected by gauge transformations so they form an in-tensor. One obtains the co-covariant divergence of a co-vector B^μ by putting $\nu = \mu$ in (D.29) to get

$$B^\mu_{*\mu} = B^\mu_{:\mu} - (n+4)\kappa_\mu B^\mu \tag{D.32}$$

(for $n = -4$ this is the ordinary covariant divergence).

We list some formulas for second co-covariant derivatives now with a sketch of derivation. Thus for a scalar of power n

$$S_{*\mu*\nu} = S_{*\mu:\nu} - (n-1)\kappa_\nu S_{*\mu} + \kappa_\mu S_{*\nu} - g_{\mu\nu}\kappa^\sigma S_{*\sigma} \tag{D.33}$$

Putting $S_{*\mu} = S_\mu - n\kappa_\mu S$ on gets

$$S_{*\mu*\nu} = S_{\mu:\nu} - n\kappa_{\mu:\nu} - n\kappa_\mu S_\nu - n\kappa_\nu(S_\mu - n\kappa_\mu S) + \kappa_\nu S_{*\mu} + \kappa_\mu S_{*\nu} - g_{\mu\nu}\kappa^\sigma S_\sigma \tag{D.34}$$

Now $S_{\mu:\nu} = S_{\nu:\mu}$ so

$$S_{*\mu*\nu} - S_{*\nu*\mu} = -n(\kappa_{\mu:\nu} - \kappa_{\nu:\mu})S = -nF_{\mu\nu}S \tag{D.35}$$

This is tedious but instructive and we continue. Let A_μ be a co-vector of power n so

$$A_{\mu*\nu*\sigma} = A_{\mu*\nu:\sigma} - n\kappa_\sigma A_{\mu*\nu} + (g^\rho_\mu\kappa_\sigma + g^\rho_\sigma\kappa_\mu - g_{\mu\sigma}\kappa^\rho)A_{\rho*\nu} + \\ +(g^\rho_\nu\kappa_\sigma + g^\rho_\sigma\kappa_\nu - g_{\sigma\nu}\kappa^\rho)A_{\mu*\rho} \tag{D.36}$$

A lengthy calculation then yields

$$A_{\mu*\nu*\sigma} - A_{\mu*\sigma*\nu} = {}^*\mathfrak{B}_{\mu\nu\sigma\rho}A^\rho - (n-1)F_{\nu\sigma}A_\mu \tag{D.37}$$

where

$${}^*\mathfrak{B}_{\mu\nu\sigma\rho} = B_{\mu\nu\sigma\rho} + g_{\rho\nu}(\kappa_{\mu:\sigma} + \kappa_\mu\kappa_\sigma) + g_{\mu\sigma}(k_{\rho:\nu} + \kappa_\rho\kappa_\nu) - g_{\rho\sigma}(\kappa_{\mu:\nu} + \kappa_\mu\kappa_\nu) - \\ -g_{\mu\nu}(\kappa_{\rho:\sigma} + (\kappa_\rho\kappa_\sigma) + (g_{\rho\sigma}g_{\mu\nu} - g_{\rho\nu}g_{\mu\sigma})\kappa^\alpha\kappa_\alpha \tag{D.38}$$

One can consider ${}^*\mathfrak{B}$ as a generalized Riemann-Christoffel tensor but it does not have the usual symmetry properties for such a tensor; however one can write

$${}^*\mathfrak{B}_{\mu\nu\sigma\rho} = {}^*B_{\mu\nu\sigma\rho} + (1/2)(g_{\rho\nu}F_{\mu\sigma} + g_{\mu\sigma}F_{\rho\nu} - g_{\rho\sigma}F_{\mu\nu} - g_{\mu\nu}F_{\rho\sigma}) \tag{D.39}$$

and then ${}^*B_{\mu\nu\sigma\rho}$ has all the usual symmetries, namely

$${}^*B_{\mu\nu\sigma\rho} = -{}^*B_{\mu\sigma\nu\rho} = -{}^*B_{\rho\nu\sigma\mu} = {}^*B_{\nu\mu\rho\sigma};\ {}^*B_{\mu\nu\sigma\rho} + {}^*B_{\mu\sigma\rho\nu} + {}^*B_{\mu\rho\nu\sigma} = 0 \tag{D.40}$$

Thus is is appropriate to call ${}^*B_{\mu\nu\sigma\rho}$ the Riemann-Christoffel (RC) tensor for Weyl space; it is a co-tensor of power 2. The contracted RC tensor is

$${}^*R_{\mu\nu} = {}^*B^\sigma_{\mu\nu\sigma} = R_{\mu\nu} - \kappa_{\mu:\nu} - \kappa_{\nu:\mu} - g_{\mu\nu}\kappa^\sigma_{:\sigma} - 2\kappa_\mu\kappa_\nu + 2g_{\mu\nu}\kappa^\sigma\kappa_\sigma \tag{D.41}$$

and is an in-tensor. A further contraction gives the total curvature

$${}^*R = {}^*R^\sigma_\sigma = R - 6\kappa^\sigma_{:\sigma} + 6\kappa^\sigma_\sigma \tag{D.42}$$

which is a co-scalar of power -2.

One gets field equations from an action principle with an in-invariant action, hence one of the form $I = \int \Omega \mathfrak{g} d^4x$ where Ω must be a co-scalar of power -4 to compensate $\mathfrak{g}$ having power 4. Ths usual contribution to Ω from the EM field is $(1/4)F_{\mu\nu}F^{\mu\nu}$ (of power -4 since it can be written as $F_{\mu\nu}F_{\rho\sigma}g^{\mu\rho}g^{\nu\sigma}$ with F factors of power zero and g factors of power -2). One also needs a gravitational term and the standard $-R$ could be ${}^*R$ but this has power -2 and will not do. Weyl proposed $({}^*R)^2$ which has the correct power but seems too complicated to be satisfactory. Here one takes ${}^*R = 0$ as a constraint and puts the constraint into the Lagrangian via γ^*R with γ a co-scalar field of power -2 in the form of a Lagrange multiplier. This leads to a scalar-tensor theory of gravitation and one can insert other terms involving γ. For convenience one takes $\gamma = -\beta^2$ with β as the basic field variable (co-scalar of power -1) and adds terms $k\beta^{*\sigma}\beta_{*\sigma}$ (co-scalar of power -4); terms $c\beta^4$ can also be added to get

(D.43) $$I = \int [(1/4)F_{\mu\nu}F^{\mu\nu} - \beta^{2*}R + k\beta^{*\mu}\beta_{*\mu} + c\beta^4]\mathfrak{g}d^4x$$

as a vacuum action. Now $\beta^{*\mu}\beta_{*\mu} = (\beta^\mu + \beta\kappa^\mu)(\beta_\mu + \beta\kappa_\mu)$ and using (D.42) one obtains

(D.44) $$-\beta^{2*}R + k\beta^{*\mu}\beta_{*\mu} = -\beta^2 R + k\beta^\mu\beta_\mu + (k-6)\beta^2\kappa^\mu\kappa_\mu + \\ +6(\beta^2\kappa^\mu)_{:\mu} + (2k-12)\beta\kappa^\mu\beta_\mu$$

The term involving $(\beta^2\kappa^\mu)_{:\mu}$ can be discarded since its contribution to the action density is a perfect differential, namely $(\beta^2\kappa^\mu)_{:\mu}\mathfrak{g} = (\beta^2\kappa^\mu\mathfrak{g})_{,\mu}$ and for the simplest vacuum equations one chooses $k = 6$ so that (D.43) becomes

(D.45) $$I = \int [(1/4)F_{\mu\nu}F^{\mu\nu} - \beta^2 R + 6\beta^\mu\beta_\mu + c\beta^4]\mathfrak{g}d^4x$$

Thus I no longer involves the κ_μ directly but only via $F_{\mu\nu}$ and I is invariant under transformations $\kappa_\mu \to \kappa_\mu + \phi_{,\mu}$ so the equations of motion that follow from the action principle will be unaffected by such transformations (i.e. they have no physical significance). Now consider three kinds of transformation:

(1) Any transformation of coordinates.
(2) Any transformation of the metric gauge combined with the appropriate transformation of potentials $\kappa_\mu \to \kappa_\mu + \phi_{,\mu}$.
(3) In the vacuum one may make a transformation of potentials as above without changing the metric gauge or alternatively one may transform the metric gauge without changing the potentials. This works only where there is no matter.

For the field equations one makes small variations in all the field quantities $g_{\mu\nu}$, κ_μ, and β, calculates the change in I and sets it equal to zero. Thus write

(D.46) $$\delta I = \int [(1/2)P^{\mu\nu}\delta g_{\mu\nu} + Q^\mu\delta\kappa_\mu + S\delta\beta)\mathfrak{g}d^4x$$

and drop the $c\beta^4\mathfrak{g}$ term since it is probably only of interest for cosmological purposes. One has

(D.47) $$\delta[(1/4)F_{\mu\nu}F^{\mu\nu}\mathfrak{g}] = (1/2)E^{\mu\nu}\mathfrak{g}\delta g_{\mu\nu} - J^\mu\mathfrak{g}\delta\kappa_\mu$$

with neglect of a perfect differential. Here $E^{\mu\nu}$ is the EM stress tensor

(D.48) $$E^{\mu\nu} = (1/4)g^{\mu\nu}F^{\alpha\beta}F_{\alpha\beta} - F^{\mu\alpha}F^\nu_\alpha$$

and J^μ is the charge current vector

(D.49) $$F^\mu = F^{\mu\nu}_{:\nu} = \mathfrak{g}^{-1}(F^{\mu\nu}\mathfrak{g})_{,\nu}$$

Considerable calculation and neglect of perfect differentials leads finally to

(D.50) $P^{\mu\nu} = E^{\mu\nu} + \beta^2[2R^{\mu\nu} - g^{\mu\nu}R] - 4g^{\mu\nu}\beta\beta^{\rho}_{:\rho} + 4\beta\beta^{\mu:\nu} + 2g^{\mu\nu}\beta^{\sigma}\beta_{\sigma} - 8\beta^{\mu}\beta^{\nu};$

$$Q^{\mu} = -J^{\mu};\ S = -2\beta R - 12\beta^{\mu}_{:\mu}$$

and the field equations for the vacuum are

(D.51) $$P^{\mu\nu} = 0,\ Q^{\mu} = 0;\ S = 0$$

These are not all independent since

(D.52) $$P^{\sigma}_{\sigma} = -2\beta^2 R - 12\beta\beta^{\sigma}_{:\sigma} = \beta S$$

so the S equation is a consequence of the P equations. If one omits the EM term from the action it becomes the same as the Brans-Dicke action except that the latter allows an arbitrary value for k; with $k \neq 6$ the vacuum equations are independent so the BD theory has one more vacuum field equation, namely $\Box(\beta^2) = 0$.

Now the action integral is invariant under transformations of the coordinate sysem and transformations of gauge; each of these leads to a conservation law connecting the quantities $P^{\mu\nu}$, Q^{μ}, S defined via (D.46). For coordinate transformations $x^{\mu} \to x^{\mu} + b^{\mu}$ one gets

(D.53) $-\delta g_{\mu\nu} = g_{\mu\sigma}b^{\sigma}_{,\nu} + g_{\nu\sigma}b^{\sigma}_{,\mu} + g_{\mu\nu,\sigma}b^{\sigma};\ -\delta\beta = \beta_{\sigma}b^{\sigma};\ -\delta\kappa_{\mu} = \kappa_{\sigma}b^{\sigma}_{,\mu} + \kappa_{\mu,\sigma}b^{\sigma}$

Putting these variations in (D.46) yields

(D.54) $$\delta I = -\int[(1/2)P^{\mu\nu}(g_{\mu\sigma}b^{\sigma}_{,\nu} + g_{\nu\sigma}b^{\sigma}_{,\mu} + g_{\mu\nu,\sigma}b^{\sigma}) +$$

$$+Q^{\mu}(\kappa_{\sigma}b^{\sigma}_{,\mu} + \kappa_{\mu,\sigma}b^{\sigma}) + S\beta_{\sigma}b^{\sigma}]\mathfrak{g}d^4x =$$

$$= \int[(P^{\mu}_{\sigma}\mathfrak{g})_{,\mu} - (1/2)P^{\mu\nu}g_{\mu\nu,\sigma}\mathfrak{g} + (Q^{\mu}\kappa_{\sigma}\mathfrak{g})_{,\mu} - Q^{\mu}\kappa_{\mu,\sigma}\mathfrak{g} - S\beta_{\sigma}\mathfrak{g}]b^{\sigma}d^4x$$

This δI vanishes for arbitrary b^{σ} so one puts the coefficient of b^{σ} equal to zero; using

(D.55) $(P^{\mu}_{\sigma}\mathfrak{g})_{,\mu} - (1/2)P^{\mu\nu}g_{\mu\nu,\sigma}\mathfrak{g} = P^{\mu}_{\sigma:\mu}\mathfrak{g};\ (Q^{\mu}\kappa_{\sigma}\mathfrak{g})_{,\mu} = \kappa_{\sigma}Q^{\mu}_{:\mu}\mathfrak{g} + \kappa_{\sigma,\mu}Q^{\mu}\mathfrak{g}$

this reduces to

(D.56) $$P^{\mu}_{\sigma:\mu} + \kappa_{\sigma}Q^{\mu}_{:\mu} + F_{\sigma\mu}Q^{\mu} - S\beta_{\sigma} = 0$$

Next consider a small transformation in gauge

(D.57) $$\delta g_{\mu\nu} = 2\lambda g_{\mu\nu},\ \delta\beta = -\lambda\beta;\ \delta\kappa_{\mu} = [log(1+\lambda)]_{,\mu} = \lambda_{\mu}$$

Putting this in (D.46) yields

(D.58) $$\delta I = \int[P^{\mu\nu}\lambda g_{\mu\nu} + Q^{\mu}\lambda_{\mu} - S\lambda\beta]\mathfrak{g}d^4x = \int[P^{\mu}_{\mu}\mathfrak{g} - (Q^{\mu}\mathfrak{g})_{,\mu} - S\beta\mathfrak{g}]\lambda d^4x$$

Putting the coefficient of λ equal to zero gives $P^{\mu}_{\mu} - Q^{\mu}_{:\mu} - S\beta = 0$ which with (D.56) comprise the conservation laws. For the vacuum one sees that (D.58) is the same as (D.52) since $Q^{\mu}_{:\mu} = 0$ from (D.50); also (D.56) reduces to

(D.59) $$P^{\mu}_{\sigma:\mu} + F_{\sigma\mu}Q^{\mu} - \beta^{-1}\beta_{\sigma}P^{\mu}_{\mu} = 0$$

which may be considered as a generalization of the Bianchi identities. The conservation laws (D.56) and (D.58) hold more generally than for the vacuum, namely whenever the action integral can be constructed from the field variables $g_{\mu\nu}$, κ_μ, β alone.

Now let the coordinates of a particle be z^μ, functions of the proper time s measured along its world line. Put $dz^\mu/ds = v^\mu$ for velocity so $v_\mu v^\mu = 1$ and v^μ is a co-vector of power -1. One adds to the action the further terms

(D.60) $$I_1 = -m\int \beta ds;\ I_2 = e\int \beta^{-1}\beta_{*\mu}v^\mu ds$$

(m and e being constants). Then these terms are in-invariants with

(D.61) $$I_2 = e\int(\beta^{-1}\beta_\mu + \kappa_\mu)v^\mu ds = e\int[(d/ds)(log(\beta)) + \kappa_\mu v^\mu]ds$$

and the first term contributes nothing to the action principle. Thus $I_2 = e\int \kappa_\mu v^\mu ds$ which is unchanged when $\kappa_\mu \to \kappa_\mu + \phi_{,\mu}$ since the extra term is $e\int(d\phi/ds)ds$. Thus for a particle with action $I_1 + I_2$ the transformations (3) above are still possible. Now some calculation yields

(D.62) $$m[g_{\mu\sigma}d(\beta v^\mu)/ds + \beta\Gamma_{\sigma\mu\nu}v^\mu v^\nu - \beta_\sigma] =$$
$$= -ev^\mu F_{\mu\sigma} \equiv m[d(\beta v^\mu)/ds + \Gamma^\mu_{\rho\sigma}v^\rho v^\sigma - \beta^\mu] = eF^{\mu\nu}v_\nu$$

This is the equation of motion for a particle of mass m and charge e; if $e = 0$ it could be called an in-geodesic. If one works with the Einstein gauge then the case $e = 0$ gives the usual geodesic equation. Next one considers the influence the of particle on the field and this is done by generating a dust of particles and a continuous fluid leading to an equation

(D.63) $$\rho[(\beta v^\mu)_{,\nu} + \Gamma^\mu_{\alpha\sigma}v^\alpha v^\sigma - \beta^\mu] = \sigma^{\mu\nu}$$

where ρ and σ refer to mass and charge density respectively.

APPENDIX E

BICONFORMAL GEOMETRY

We sketch here some beautiful work of J. Wheeler [**1302, 1304**] which has led to a number of important developments in mathematical physics (cf. also Anderson, Wehner, and Wheeler [**42, 43, 1284, 1295, 1300, 1301, 1303**]). The paper [**1302**] (not published) gives a very nice discussion of normal biconformal spaces and lays some of the foundation for some later work of Wheeler et al for which [**1304**] is apparently the best starting point. We will therefore extract here from [**1304**] and remark that this work alone transcends earlier approaches to unifying GR and EM. One works over an 8-D base space where in a flat situation the 4-D biconformal cospace to 4-D Minkowski space corresponds to a standard tangent space (or momentum space) with variables p_μ transforming via L^{-1} under Lorentz transformations L of x^μ (or dx^μ). The symplectic form given by the exterior derivative of the Weyl 1-form provides a typical Hamiltonian dynamical structure and one gives general necessary and sufficient conditions for curved 8-D geometry to be in 1-1 correspondence with 4-D Einstein-Maxwell spacetime; further a consistent unified geometrical theory of gravity and electromagnetism is obtained. This is very powerful stuff and we can only sketch a few items here. Some connections of biconformal geometry to QM are given in Section 3.5.1 and we refer to [**1295, 1304**] for history and philosophy.

The conformal group is the most general set of transformations preserving ratios of infinitesimal lengths. On a 4-D spacetime this group is 15 dimensional, with Lorentz transformations (6), translations (4), 4 inverse translations (special conformal transformations), and dilations (1). One concentrates on flat situations and develops biconformal structure as a conformal fiber bundle. This is constructed as the quotient $\mathcal{C}/\mathcal{C}_0$ of the conformal group by its isotropy subgroup (7-D homogeneous Weyl group) producing a conformal Cartan connection on an 8-D manifold; this is then generalized to a curved 8-D manifold with the 7-D homogeneous Weyl group as fiber by the addition of horizontal curvature 2-forms to the group structure equations. The resulting 8-D base manifold is called a biconformal space and the full 15-D fiber bundle the biconformal bundle. We will try to illustrate this via the equations. One uses the $O(4,2)$ representation of the conformal group for notation where $(A, B, \cdots) = (0, 1, \cdots, 5)$. Then the $O(4.2)$ metric is $\eta_{ab} = diag(1, 1, 1, -1)$ $(a, b = 1, \cdots, 4)$ with $\eta_{05} = \eta_{50} = 1$ and all other components are zero. Introducing a connection 1-form ω^A_B one has covariant constancy via

$$D\eta_{AB} = d\eta_{AB} - \eta_{CB}\omega_A^C - \eta_{AC}\omega_B^C = 0 \tag{E.1}$$

The conformal connection may be broken into 4 independent Weyl invariant parts, the spin connection ω_b^a, the solder form ω_0^a, the co-solder form ω_a^0, and the Weyl vector ω_0^0 where the spin connection satisfies $\omega_b^a = -\eta_{bc}\eta^{ad}\omega_d^c$ and the remaining components of ω_{AB} are related via

$$\omega_0^5 = \omega_5^0 = 0;\ \omega_5^5 = -\omega_0^0;\ \omega_5^a = -\eta^{ab}\omega_b^0;\ \omega_a^5 = -\eta_{ab}\omega_0^b \tag{E.2}$$

These constraints reduce the number of independent connection forms ω_B^A to the required 15 and one can run $A, B\cdots$ from 0 to 4 (with 5 implicit). The structure constants of the conformal Lie algebra now lead immediately to the Maurer-Cartan (MC) equations of the conformal group as $d\omega_B^A = \omega_B^C\omega_C^A$ (wedge product is assumed) or written out

$$d\omega_b^a = \omega_b^c\omega_c^a + \omega_b^0\omega_0^a - \eta_{bc}\eta^{ad}\omega_d^0\omega_0^c; \tag{E.3}$$

$$d\omega_0^a = \omega_0^0\omega_0^a + \omega_0^b\omega_b^a;\ d\omega_a^0 = \omega_a^0\omega_0^0 + \omega_a^b\omega_b^0;\ d\omega_0^0 = \omega_0^g o_a^0$$

Note that d in (E.2) includes partial derivatives in all eight of the base space directions and when using coordinates one will write (x^μ, y_ν) corresponding to index positions on (ω_0^b, ω_b^0). Also $\partial_\mu\phi = \partial\phi/\partial x^\mu$ while $\partial^\mu\phi = \partial\phi/\partial y_\mu$. The generalization of (E.3) to a curved base space is obtained via

$$d\omega_b^a = \omega_b^c\omega_c^a + \omega_b^0\omega_0^a - \eta_{bc}\eta^{ad}\omega_d^0\omega_0^c + \Omega_b^a = \omega_b^c\omega_c^a + \Delta_{bc}^{da}\omega_d^0\omega_0^c + \Omega_b^a; \tag{E.4}$$

$$d\omega_0^a = \omega_0^0\omega_0^a + \omega_0^b\omega_b^a + \Omega_0^a;\ d\omega_a^0 = \omega_a^0\omega_0^0 + \omega_a^b\omega_b^0 + \Omega_a^0;\ d\omega_0^0 = \omega_0^a\omega_a^0 + \Omega_0^0$$

One calls the four types of curvature Ω_b^a, Ω_0^a, Ω_a^0, Ω_0^0 the Riemann curvature, torsion, co-torsion, and dilational curvature respectively. Horizontality requires each of the curvatures to take the form

$$\Omega_B^A = (1/2)\Omega_{Bcd}^A\omega_0^c\omega_0^d + \Omega_{Bd}^{Ac}\omega_0^d\omega_c^0 + (1/2)\Omega_B^{Acd}\omega_c^0\omega_d^0 \tag{E.5}$$

The connection of a flat biconformal space is in the standard flat form when written as

$$\omega_0^0 = \alpha_a(x)dx^a - y_a dx^a \equiv W_a dx^a;\ \omega_0^a = dx^a; \tag{E.6}$$

$$\omega_a^0 = dy_a - (\alpha_{a,b} + W_a W_b - (1/2)W^2\eta_{ab})dx^b;\ \omega_b^a = (\eta^{ac}\eta_{bd} - \delta_d^a\delta_b^c)W_c dx^d$$

Note that the Weyl vector $W_a = \alpha_a(x) - y_a$ depends on an arbitrary 4-vector α_a and also on the 4 coordinates y_a; the presence of α_a gives the generality required for the EM vector potential while the y_a keeps the dilational curvature zero. In general the dilational gauge vector of biconformal space is of the form

$$\omega_0^0 = \omega_{0\mu}^0(x,y)dx^\mu + \omega_0^{0\mu}(x,y)dy_\mu \tag{E.7}$$

(i.e. an 8-D vector field depending on 8 independent variables) but constraining the biconformal geometry to have vanishing curvatures forces $\omega_0^0 = (\alpha_\mu(x) - y_\mu)dx^\mu$ which is precisely the form required to give the Lorentz force law. Then one proves in [**1304**] that when the curvatures of biconformal space $\Omega_B^A = 0$ there exist global coordinates x^a, y_a) such that the connection takes the standard flat form.

Note in (E.6) if one holds the y coordinates fixed then equations 1,2,4 are the connection forms for a 4-D Weyl spacetime with conformally flat metric η_{ab}; the remaining equation 3 is then simply a 1-form constructed from the Weyl-Ricci tensor. However the dilational curvature of a 4-D Weyl geometry is given by the curl of the Weyl vector, equivalent here to the curl of the arbitrary α_a, and viewed from the Weylian 4-D perspective the solution gives unphysical size change; it is only with the inclusion of the additional momentum variables proportional to y_a that the dilational curvature can be seen to vanish. It is thus seen that the actual motion of a particle in biconformal space is 8 dimensional and one can in fact interpret biconformal space, and therefore conformal gauge theory, as a generalization of phase space (with symplectic structure). Indeed for a given Hamiltonian system one can specify a unique flat biconformal space by judicious choice of the solder and co-solder forms (see [**1304**] for details); in particular the extra 4 dimensions are identified with momenta and the integral of the Weyl vector is identified with action. In order to make a full identification of a biconformal space with an Einstein-Maxwell spacetime one can proceed as follows. The idea is that the solder form should satisy the Einstein equations with arbitrary matter as source and the vector potential obtained via $\alpha_a = q(\phi, -A_i) = -qA_a$ should satisfy the Maxwell field equations with arbitrary EM currents. One assumes for example that the torsion is zero (but not the cotorsion) leading to constraints $\Omega^0_0 = 0$ and $\Omega^a_0 = 0$ and, setting $\omega^a_0 = e^a$, one wants a completion f_a to the e^a basis in which $\Omega^a_{bac} = 0$. Further one posits two field equations $*d * d\omega^0_0 = J = J_a(x)e^a$ and $\omega^0_a = T_a + \cdots$ where $T_a = -(1/2)(T_{ab} - (1/3)\eta_{ab}T)e^b$. This can all be achieved and leads to the general identification stated above (cf. [**1304**]).

Now to set the stage for [**42**] and connections to QM we gather the material from the appendices to [**42**] (where the formulation is different). The conformal group generators include Lorentz transformations $M^a_b = -M_{ba}\eta_{ac}M^c_b$, translations P_a, special conformal transformations K^a, and dilatations D satisfying the commutation relations

$$[M^a_{\ b}, M^c_{\ d}] = -(\delta^c_b M^a_{\ d} + \eta_{df}\eta^{ac}M^f_{\ b} + \eta_{bd}\eta^{ae}M^c_{\ e} - \delta^a_d M^c_{\ b}); \tag{E.8}$$

$$[M^a_{\ b}, P_c] = -(\eta_{cb}\eta^{ad}P_d - \delta^a_c P_b);\ [M^a_{\ b}, K^d] = -(\delta^d_b\delta^a_c - \eta^{ad}\eta_{bc})D^c;$$

$$[P_a, K^b] = 2M^b_{\ a} - 2\delta^b_a D;\ [D, K^b] = K^b;\ [D, P_a] = -P_a$$

(note dilatation corresponds to dilation and both terms seem to be popular). The conformal Lie algebra has two independent involutive automorphisms; the first is

$$\sigma_1 : (M^a_b, P_a, K^a, D) \to (M^a_b, -P_a, -K^a, D) \tag{E.9}$$

and this identifies the invariant subgroup used as the isotropy subgroup in the bicomformal gauging. The second is

$$\sigma_2 : (M^a_b, P_a, K^a, D) \to (M^a_b, -\eta_{ab}K^b, -\eta^{ab}P_b, -D) \tag{E.10}$$

and may be chosen to be complex conjugation in order to define what are called σ_C representations of the algebra. Thus if we assume the generators to be complex, σ_C representations have P_a and K_a as complex conjugates, while M^a_b is real and D pure imaginary. As an illustration note that while both $so(3)$ and $su(2)$ have involutive automorphisms the existence of a σ_C representation singles out $su(2)$. Thus while

(E.11) $$[J_i, J_j] = \epsilon_{ijk} J_k; \; [\tau_i, \tau_j] = \epsilon_{ijk} \tau_k$$

are both invariant under

(E.12) $$\rho: \; (J_1, J_2, J_3) \to (-J_1, J_2, -J_3); \; (\tau_1, \tau_2, \tau_3) \to (-\tau_1, \tau_2, -\tau_3)$$

where $[J_j]_{ik} = \epsilon_{ijk}$ and $\tau_j = -(i/2)\sigma_j$ (σ_j being the Pauli matrices - cf. (B.25)) it is only with the complex representation that $\rho = \rho_C : \; \overline{(\tau_1, \tau_2, \tau_3)} = (-\tau_1, \tau_2, -\tau_3) = \rho(\tau_1, \tau_2, \tau_3)$. As examples of conformal representations with this property first consider the covering group $SU(2,2)$, whose Lie algebra is isomorphic to that of $O(4,2)$. Due to the local isomorphism between $Spin(4,2)$ and $SU(2,2)$ this algebra can be represented via spinors. Thus using 4×4 Dirac matrices γ^a one can write $su(2,2)$ via

(E.13) $$\{\gamma^a, \gamma^b\} = 2\eta^{ab} = 2diag(-1, 1, 1, 1) \; (a, b = 0, , 1, 2, 3)$$

One also defines

(E.14) $$\sigma^{ab} = -(1/8)[\gamma^a, \gamma^b]; \; \gamma_5 = i\gamma^0\gamma^1\gamma^2\gamma^3$$

where the full Clifford algebra has the basis

(E.15) $$\Gamma = \{1, i1, \gamma^a, \sigma^{ab}, i\sigma^{ab}, \gamma_5\gamma^a, i\gamma_5\gamma^a, \gamma_5, i\gamma_5\}$$

The conformal Lie algebra may be obtained from this set be demanding invariance of a spinor metric Q given by $Q = i\gamma^0$. Then if one requires $Q\Gamma + \Gamma^\dagger Q = 0$ the generators of the conformal Lie algebra are (cf. [**42**])

(E.16)
$$M^a_b = \eta_{bc}\sigma^{ac}; \; P_a = (1/2)\eta_{ab}(1 + \gamma_5)\gamma^b; \; K^a = (1/2)(1 - \gamma_5)\gamma^a; \; D = -(1/2)\gamma_5$$

Choosing any real representation for the Dirac matrices γ_5 is necessarily imaginary and it follows that

(E.17) $$\bar{M}^a_b = M^a_b; \; \bar{P}_a = \eta_{ab} K^b; \; \bar{D} = -D$$

so the action of σ_C is realized. Alternatively one may consider a complex function space representation of the conformal algebra via

(E.18) $$M^\mu_\nu = -\frac{1}{2}\left(z^\mu \frac{\partial}{\partial z^\nu} + \bar{z}^\mu \frac{\partial}{\partial \bar{z}^\nu} - z_\nu \frac{\partial}{\partial z_\mu} - \bar{z}^\nu \frac{\partial}{\partial \bar{z}^\mu}\right);$$

$$D = z^\mu \frac{\partial}{\partial z^\mu} - \bar{z}^\nu \frac{\partial}{\partial \bar{z}^\nu}; \; P_\mu = \frac{\partial}{\partial z^\mu} + \left(\bar{z}_\mu \bar{z}^\nu - \frac{1}{2}\bar{z}^2 \delta^\nu_\mu\right) \frac{\partial}{\partial \bar{z}^\nu};$$

$$K_\mu = \frac{\partial}{\partial \bar{z}^\mu} + \left(z_\mu z^\nu - \frac{1}{2} z^2 \delta^\nu_\mu\right) \frac{\partial}{\partial z^\nu}$$

Note the generators are complex but the group manifold is real. In either of these representations the Maurer-Cartan (MC) equations inherit the same symmetry under σ_C and in particular the gauge vector of dilatations (the Weyl vector) is imaginary. To clarify this one shows that the dilatations generated by an imaginary

D nevertheless give a real factor as expected. Thus first consider $su(2,2)$ with a basis of Dirac matrices in which

(E.19) $$D = -\frac{1}{2}\gamma_5 = -\frac{1}{2}\begin{pmatrix} -\sigma_y & \\ & \sigma_y \end{pmatrix};\ \sigma_y = \begin{pmatrix} & -i \\ i & \end{pmatrix}$$

Define the definite conformal weight spinors χ^A, ψ^B via

(E.20) $$D\chi = (1/2)\chi;\ D\psi = -(1/2)\psi;\ e^{\lambda D}\chi = e^{\lambda/2}\chi;\ e^{\lambda D}\psi = e^{-\lambda/2}\psi$$

For the complex function space representation of the conformal group (E.18) one has (for one variable $z = rexp(i\phi)$)

(E.21) $$D \sim -i\frac{\partial}{\partial\phi}$$

so D measures the phase of a complex number. Homogeneous functions of z and $\bar{z}$ are then eigenfuncitons and D measures the degree of homogeneity. Thus if $f(z,\bar{z}) = z^a\bar{z}^b$ there results $exp(\lambda D)f(z,\bar{z}) = exp[(a-b)\lambda]f(z,\bar{z})$ so there are dilatations with the weight of the function encoded into the total phase. Similarly in multiple complex dimensions eigenfunctions can be built up from powers of the norms

(E.22) $$f_{\alpha-\beta} = (\sqrt{z^2})^\alpha(\sqrt{\bar{z}^2})^\beta;\ Df_{\alpha-\beta} = D(z^2)^{\alpha/2}(\bar{z}^2)^{\beta/2} = (\alpha-\beta)f_{\alpha-\beta}$$

Note that $z^a\bar{z}_a$ is of weight zero with $D(z^a\bar{z}_a) = 0$.

Gauge transformations will remain real even though there is a complex valued connection. A local gauge transformation is given via

(E.23) $$\Lambda = M^a_b\Lambda^b{}_a + D\Lambda^0$$

Note Λ is complex, since Λ^b_a, Λ^0 are real parameters used to exponentiate the generators M (real) and D (imaginary), and it follows that a gauge transformation of the Weyl vector is $\delta\omega = -d\Lambda^0$ where Λ^0; one can then define a scale covariant derivative of a definite weight scalar field via

(E.24) $$Df = df + k\omega f$$

where k is the conformal weight of f. To see that this is a gauge invariant expression one takes a dilatational gauge transformation

(E.25) $$f' = fexp(k\Lambda^0);\ \omega' = \omega + \delta\omega = \omega - d\Lambda^0$$

which implies

(E.26) $$D'f' = d(fexp(k\Lambda^0)) + k(\omega - d\Omega^0)f = exp(k\Lambda^0)Df$$

Thus the equation is covariant and the MC structure equations are invariant under real scalings. Of course in generic gauges the Weyl vector is complex but the invariance fo the structure equations under gauge transformations guarantees consistency. Note also that whether the Weyl vector is complex or pure imaginary $exp(\oint\omega)$ remains a pure phase since the Weyl vector is pure imaginary in at least one gauge and the above expression is gauge invariant. Finally one writes the Cartan structure equations for flat σ_C biconformal space via

(E.27) $$d\omega^a_b = \omega^c_b\omega^a_c + 2\omega_b\omega^a;\ d\omega^a = \omega^c\omega^a_c + \omega\omega^a;\ d\omega = 2\omega^a\omega_a$$

(and their conjugates); here ω^a corresponds to translation generators, ω is the Weyl vector, and ω^a_b is the spin connection. A first order perturbative solution is

$$\omega^a_b(\delta^a_e\eta_{cb} - \delta^a_c\eta_{eb})x^c dx^e + (\delta^a_e\eta_{cb} - \delta^a_c\eta_{eb})y^c dy^e;$$
$$\omega = i(y_a dx^a - x_a dy^a);\ \omega^a = \{dx^a + idy^a +$$
$$+\left(-\frac{1}{2}x^a x_e + \frac{i}{2}(\delta^a_e x_c y^c - x^a y_e) + \frac{1}{2}y^a y_e\right)(dx^e - idy^e)\} \tag{E.28}$$

APPENDIX F

A FEW BASIC FORMULAS

We gather here some formulas involving tensor analysis and differential geometry from a "classical" point of view (see e.g. [**12, 150, 351, 487, 1199, 1274**]). We follow mainly Wald [**1274**] but have always found Adler et al [**12**] an enjoyable source of information. Thus let $\psi : M \to \mathbf{R}^n$ be a diffeomorphism and define the tangent space $T_p(M)$ to M at p via (**1A**) $X_\mu(f) = (\partial/\partial x^\mu)(f \circ \psi^{-1})|_{\psi(p)}$ for say $f \in C^\infty(p)$. Then vectors in $T_p(M)$ have the form $v = \sum v^\mu X_\mu$ with $V^\mu = v(x^\mu \circ \psi)$ (i.e. $v(f) = \sum v^\mu X_\mu(f) = \sum v^\mu (\partial/\partial x^\mu)(f \circ \psi^{-1}$ and for $f = x^\nu \circ \psi$ one has (**1B**) $(\partial/\partial x^\mu)(x^\nu) = v^\nu$). Given a different chart ψ' one obtains (**1C**) $X_\mu = \sum (\partial x'^\nu/\partial x^\mu)_{\psi(p)} X'_\nu$ with (**1D**) $v'^\nu = \sum v^\mu (\partial x'^\nu/\partial x^\mu)$. Given then $dx^\mu(\partial/\partial x^\nu) = \delta^\mu_\nu$ one has dual vectors (or covectors) (**1E**) $\omega = \sum \omega_\mu dx^\mu$ with $\omega'_\nu = \sum \omega_\mu (\partial x^\mu/\partial x'^\nu)$ generating the cotangent space $T^*_p(M)$. The transformation laws (**1D**) and (**1E**) indicate that v is contravariant and ω is covariant. Tensors of type $(k\ell)$ are written via (**1F**) $T = \sum T^{\mu_1 \cdots \mu_k}{}_{\nu_1 \cdots \nu_\ell} v_{\mu_1} \otimes \cdots \otimes v_{\mu_k} \otimes \omega_1 \otimes \cdots \otimes \omega_\ell$ where the v_μ (resp. ω_ν) transform contravariantly (resp. covariantly). Thus

$$T^{\mu'_1 \cdots \mu'_k}{}_{\nu'_1 \cdots \nu'_\ell} = \sum T^{\mu_1 \cdots \mu_k}{}_{\nu_1 \cdots \nu_\ell} \left(\frac{\partial x_1'^{\mu}}{\partial x^{\mu_1}}\right) \cdots \left(\frac{\partial x^\nu_\ell}{\partial x_\ell'^{\nu}}\right) \tag{F.1}$$

where e.g. $x^{\nu'}_\ell \sim x'^{\nu}_\ell$. A metric g has the form $g = \sum g_{\mu\nu} dx^\mu \otimes dx^\nu$ and one writes

$$T_{ab} = \frac{1}{2}(T_{ab} + T_{ba});\ T_{[ab]} = \frac{1}{2}(T_{ab} - T_{ba}) \tag{F.2}$$

More generally

$$(\bigstar)\ T_{(a_1 \cdots a_n)} = (1/n!) \sum T_{a_{\pi(1)} \cdots a_{\pi(n)}};\ T_{[a_1 \cdots a_n]} = (1/n!) \sum \delta(\pi) T_{a_{\pi(1)} \cdots a_{\pi(n)}}$$

where $\delta(\pi) = 1$ (resp. -1) for even (resp. odd) permutations of $(1, \cdots, n)$. A covariant derivative ∇ satisifies

(1) Linearity: $\nabla_c(\alpha A^{\mathbf{a}}{}_{\mathbf{b}} + \beta B^{\mathbf{a}}{}_{\mathbf{b}}) = \alpha \nabla_c A^{\mathbf{a}}{}_{\mathbf{b}} + \beta \nabla_c B^{\mathbf{a}}{}_{\mathbf{b}}$
(2) Leibnitz rule: $\nabla_e [A^{\mathbf{a}}{}_{\mathbf{b}} B^{\mathbf{c}}{}_{\mathbf{d}}] = [\nabla_e A^{\mathbf{a}}{}_{\mathbf{b}}] B^{\mathbf{c}}{}_{\mathbf{d}} + A^{\mathbf{a}}{}_{\mathbf{b}} [\nabla_e B^{\mathbf{c}}{}_{\mathbf{d}}]$
(3) Commutativity with contraction: $\nabla_d (A^{\mathbf{acb}}{}_{\mathbf{ecg}}) = \nabla_d A^{\mathbf{acb}}{}_{\mathbf{ecg}}$
(4) Consistency with the idea of tangent vectors as directional derivatives on scalar fields: For $T^a \in T_p(M)$ one has $T(f) = T^a \nabla_a f$
(5) Torsion free: $\nabla_a \nabla_b f = \nabla_b \nabla_a f$. This condition is dropped in some theories of gravitation.

One notes that #4 and #5 imply (**1G**) $[v,w](f) = v(w(f)) - w(v(f)) = v^a\nabla_a(w^b\nabla_b f) - w^a\nabla_a(v^b\nabla_b f) = [v^a\nabla_a w^b - w^a\nabla_a v^b]\nabla_b f$ which implies $[v,w]^b = v^a\nabla w^b - w^a\nabla_a v^b$ when e.g. $v \sim v^a\nabla_a$. In particular one has (**1H**) $\nabla_a v^b = \partial_a v^b + \Gamma^b_{ac}v^c$ which is of course coordinate independent. Parallel transport of v^a corresponds to $T^a\nabla_a v^b = 0$ along a curve with tangent vector T^a and given a metric g_{ab} there is a unique ∇_a satisfying $\nabla_a g_{bc} = 0$ defined by

$$\Gamma^c{}_{ab} = \frac{1}{2}g^{cd}[\partial_a g_{bd} + \partial_b g_{ad} - \partial_d g_{ab}] \tag{F.3}$$

(Levi-Civita connection). The Riemann curvature tensor is defined via

$$\nabla_a\nabla_d\omega_c - \nabla_b\nabla_a\omega_c = R_{abc}{}^d\omega_d \tag{F.4}$$

for all covector fields ω_c. Some properties of the Riemann curvature tensor are (see [**1274**] for details)

(1) $R_{abc}{}^d = -R_{bac}{}^d$
(2) $R_{[abc]}{}^d = 0$
(3) For ∇ satisfying $\nabla_a g_{bc} = 0$ one has $R_{abcd} = -R_{abdc}$
(4) The Bianchi identity $\nabla_{[a}R_{bc]d}{}^e = 0$ holds and to prove this one uses e.g.

$$(\nabla_a\nabla_b - \nabla_b\nabla_a)\nabla_c\omega_d = R_{abc}{}^e\nabla_e\omega_d + R_{abd}{}^f\nabla_c\omega_f \tag{F.5}$$

Next the Ricci tensor is (**1I**) $R_{ac} = R_{abc}{}^b$ with $R_{ac} = R_{ca}$ and $R = R_a{}^a$ is the scalar curvature. Further, contraction of the Bianchi identity in #4 yields

$$\nabla R_{bcd}{}^a + \nabla_b R_{cd} - \nabla_c R_{bd} = 0 \tag{F.6}$$

leading to the Einstein tensor (**1J**) $G_{ab} = R_{ab} - (1/2)Rg_{ab}$ with $\nabla^a G_{ab} = 0$. The Einstein equations are then $G_{ab} = 8\pi T_{ab}$ where T_{ab} denotes the stress-energy tensor (with $\nabla^a T_{ab} = 0$) and this is discussed at various points in the book. One notes that frequently formulas for T_{ab} will contain the metric explicitly so one needs then to solve simultaneously for the metric and the matter distribution.

Geodesics are defined as curves whose tangent vector is parallel transported along itself, i.e. (**1K**) $T^a\nabla_a T^b = 0$. In a coordinate system ψ the geodesic is mapped into a curve $C \sim x^\mu(t)$ and we know first that $T^\mu = dx^\mu/dt$ via (**1L**) $\mathbf{T} = (d/dt)(f\circ C) = \sum\partial_\mu(f\circ\psi^{-1})(dx^\mu/dt) = \sum(dx^\mu/dt)X_\mu(f)$. From (**1H**) now (**1K**) becomes $T^a(\partial_a T^b + \Gamma^b{}_{bc}T^c) = 0$. But $(d/dt)T^b = \sum\partial_a T^b(dx^a/dt) = T^a\partial_a T^b$ so one obtains

$$\frac{d^2x^\mu}{dt^2} + \sum\Gamma^\mu{}_{\sigma\nu}\frac{dx^\sigma}{dt}\frac{dx^\nu}{dt} = 0 \tag{F.7}$$

It is noted that the most efficient way of computing the $\Gamma^\mu{}_{\sigma\nu}$ is to start with the Lagrangian (**1M**) $L = \sum g_{\mu\nu}(dx^\mu/dt)(dx^\nu/dt)$ and write down the corresponding Euler-Lagrange equations to compare with (F.7).

REMARK F.1. From (F.3) one has $\Gamma^c{}_{ab} = (1/2)[\partial_a g_{bd} + \partial_b g_{ad} - \partial_d g_{ab}]$ which leads immediately to

$$\sum\Gamma^c{}_{cb} = \frac{1}{2}\sum g^{cd}\partial_b g_{cd} = \frac{1}{2g}\frac{\partial g}{\partial x^b} = \partial_b log(\sqrt{|g|}) \tag{F.8}$$

where $g = det(g_{cd})$ (recall $g_{cd} = g_{cd}$). Further the natural volume element on M induced by g_{cd} is $\sqrt{|g|}\prod dx^{\mu}$. Consequently from (**1H**)

(F.9) $$\nabla_a v^a = \partial_a v^a + \Gamma^a_{\ ab} v^b = \sum \frac{1}{\sqrt{|g|}}\partial_a(\sqrt{|g|}v^a)$$

provides a divergence formula. ■

REMARK F.2. It is appropriate to make a few comments about Killing vector fields. If $\phi_t : M \to M$ is a 1-parameter group of isometries $\phi^* g_{ab} = g_{ab}$ then the vector field ξ^a generating ϕ_t is called a Killing field. By definitions a necessary and sufficient condition for $\phi_t^* g_{ab} = g_{ab}$ is $\mathfrak{L}_\xi g_{ab} = 0$ where $\mathfrak{L}$ is the Lie derivative (**1N**) $\mathfrak{L}_\xi T = lim_{t\to 0}(1/t)(\phi^*_{-t}T - T)$ for tensors T evaluated at the same point p. Note here $(\phi_t^* v)(f) = v(f \circ \phi_t)$ and $(\phi_t\mu)_a v^a = \mu_a(\phi^* v)^a$ which extends to T of all types. Then one shows easily that (**1**)) $\mathfrak{L}_\xi v^a = [\xi, v]^a$ and $\mathfrak{L}_\xi \mu_a = \xi^b\nabla_b\mu_a + \mu_b\nabla_a\xi^b$ so extending this as in [**1274**] yields

(F.10) $$\mathfrak{L}_\xi g_{ab} = \xi^c\nabla_c g_{ab} + g_{cb}\nabla_a\xi^c + g_{ac}\nabla_b\xi^c = \nabla_a\xi_b + \nabla_b\xi_a$$

(with standard index manipulation). In particular one can say that ξ is a Killing field if and only if (**1P**) $\nabla_a\xi_b + \nabla_v\xi_a = 0$ where ∇_a is the covariant derivative associated with g_{ab}. ■

REMARK F.3. We extract here a few formulas from [**351, 487, 1274**]. Thus one can write following (F.9)

(F.11) $$\nabla^m v_m = \frac{1}{\sqrt{|g|}}\partial_m(\sqrt{|g|}g^{mn}v_n);$$

$$\Delta\phi = \nabla_m\nabla^m\phi = \nabla_m(g^{mn}\partial_n\phi) = \frac{1}{\sqrt{|g|}}\partial_m(\sqrt{|g|}g^{mn}\partial_n\phi)$$

Next for $\sqrt{|g|}d\Omega$ an invariant volume element and S a surface bounding Ω one has

(F.12) $$\oint_S v^m dS_m = \int_\Omega \nabla_m v^m\sqrt{|g|}d\Omega$$

Here dS_k can be defined as $(1/2)\epsilon_{kmn}dx^m dx^n$ where ϵ_{kmn} is the totally antisymmetric 3-D Ricci tensor. ■

Bibliography

[1] L. Abbot and M. Wise, Amer. Jour. Phys., 49 (1981), 37-39

[2] M. Abdalla, A. Gadelha, and I. Vancea, hep-th 0002217

[3] S. Abe, Phys. Rev. A, 48 (1993), 4102-4106; Phys. Lett. A, 224 (1997), 326-330

[4] S. Abe and N. Suzuki, Phys. Rev. A, 41 (1990), 4608-4613

[5] S. Abe, Phys. Lett. A, 271 (2000), 74

[6] S. Abe, Phys. Rev. E, 66 (2002), 046134

[7] S. Abe and A. Rajagopal, cond-mat 0304066

[8] M. Abolhasani and M. Golshani, gr-qc 9709005; quant-ph 9808015

[9] E. Abreu and M. Hott, Phys. Rev. D, 62 (2000), 027702

[10] C. Adami, quant-ph 0405005

[11] F. Ben Adda and J. Cresson, Quantum derivatives and the Schrödinger equation, Chaos, Solitons, and Fractals, 19 (2004), 1323-1334

[12] R. Adler, M. Bazin, and M. Schiffer, Introduction to general relativity, McGraw-Hill, 1965

[13] S. Adler, Quantum theory as an emergent phenomenon, Cambridge Univ. Press, 2004

[14] G. Agarwal and E. Wolf, Phys. Rev. D, 2 (1970), 2161-2186, 2187-2205, and 2206-2225

[15] A. Agnese and R. Festa, Phys. Lett. A, 227 (1997), 165-171

[16] M. Agop, P. Ioannou, and P. Nica, Chaos, Solitons, and Fractals, 19 (2004), 1057-1070

[17] M. Agop, P. Ioannou, C. Buzea, and P. Nica, Chaos, Solitons, and Fractals, 16 (2003), 321-338

[18] M. Agop, P. Ioannou, and C. Buzea, Chaos, Solitons, and Fractals, 13 (2002), 1137

[19] Y. Aharonov and D. Bohm, Phys. Rev. 115 (1959), 485

[20] K. Akama, Y. Chikashige, T. Matsuki, and H. Terazawa, Prog. Theor. Phys., 60 (1978), 868-877

[21] M. Akbar and R. Cai, gr-qc 0609128 and 612089

[22] M. Alber, G. Luther, and J. Marsden, Nonlinearity, 10 (1999), 223-241

[23] R. Aldrovandi, J. Almeida, and J. Pereira, gr-qc 0403099 and 0405104

[24] E. Alfinito, R. Manda, and G. Vitiello, gr-qc 9904027

[25] R. Alicki, quant-ph 0201012

[26] G. Allemandi, M. Capone, S. Capozziello, and M. Francaviglia, hep-th 0409198

[27] V. Allori, D. Dürr, S. Goldstein, and N. Zanghi, quant-ph 0112005

[28] V. Allori and N. Zanghi, quant-ph 0112008 and 0112009

[29] M. Altaie, gr-qc 0104100 and 0212123

[30] E. Alvarez and C. Gomez, hep-th 9807226

[31] R. Alvargonzalez and L. Soto, physics 0312096

[32] R. Alvargonzalez, Physics 0311139 and 0311027

[33] G. Amelino-Camelia, gr-qc 0309054 and 0412136

[34] J. Almeida, physics 0211056, 0303034, and 0403058
[35] M. Almeida, Physica A, 300 (2001), 424
[36] M. Almeida, F. Potiguar, and U. Costa, cond-mat 0206243
[37] A. Ozorio de Almeida, Hamiltonian systems: Chaos and quantization, Cambridge Univ. Press, 1988
[38] D. Amit, Field theory; The renormalization group and critical phenomena, World Scientific, 2005
[39] W. Amrein, Helv. Phy. Acta, 42 (1969), 149
[40] J. Anandan, quant-ph 0012011; gr-qc 9505011, 9506011, 9712015; Found. Phys., 21 (1991), 1265-1284; Phys. Lett. A, 147 (1990), 3-8; 19 (1994), 284-292; Ann. Inst. H. Poincaré, 49 (1988), 271-286
[41] J. Anandan and Y. Aharonov, Phys. Rev. Lett., 65 (1990), 1697-1700; Phys. Rev. D, 38 (1988), 1863-1870
[42] L. Anderson and J. Wheeler, hep-th 0406159
[43] L. Anderson and J. Wheeler, hep-th 0305017 and 0412229
[44] E. Anderson, gr-qc 0205118, 0302035, 0409123, and 0511070
[45] E. Anderson, J. Barbour, B. Foster, and N. Ó Murchadha, gr-qc 0211022
[46] A. Anderson and J. Halliwell, gr-qc 9304025
[47] A. Anderson and J. York, gr-qc 9807041
[48] A. Anderson, A. Abrahams, and C. Lea, gr-qc 9710041
[49] L. deAndrade, gr-qc 9810010, 0301013, 0302042,0405062, and 0410036
[50] L. deAndrade, hep-th 0104070; cond-mat 0501657 and 0501741; astro-ph 0005519
[51] M. deAndrade and I. Vancea, gr-qc 9907059
[52] M. deAndrade, M. Santos, and I. Vancea, hep-th 0308169 and 0104154
[53] V. deAndrade, A. Barbosa, and J. Pereira, gr-qc 0501037
[54] V. deAndrade, H. Arcos, and J. Pereira, gr-qc 0412034
[55] L. Garcia deAndrade, gr-qc 0502106; cond-mat 9812263
[56] I. Antoniadis and E. Mottola, Phys. Rev. D, 45 (1992), 2013-2025
[57] H. Aratyn, E. Nissimov, and S. Pacheva, solv-int 9701017
[58] D. Arbatsky, math-ph 0402003
[59] O. Arias, T. Gonzalez, Y. Leyva, and I. Quiros, gr-qc 0307016
[60] O. Arias and I. Quiros, gr-qc 0212006
[61] O. Arias, T. Gonzalez, and I. Quiros, gr-qc 0210097
[62] T. Arimitsu, math-ph 0206015; quant-ph 0206062
[63] T. Arimitsu and N. Arimitsu, cond-mat 0210027 and 0306042
[64] M. Arminjon, gr-qc 0401021 and 0409092
[65] V. Arnold, Mathematical methods of classical mechanics, Springer, 1989
[66] R. Arnowitt, S. Deser, and C. Misner, gr-qc 0405109
[67] A. Ashtekar, gr-qc 0112038 and 9901023
[68] A. Ashtekar and T. Schilling, gr-qc 9706069
[69] A. Ashtekar and J. Lewandowski, hep-th 9603083
[70] A. Ashtekar, A. Corichi, and J. Zapata, gr-qc 9806041
[71] A. Ashtekar and J. Lewandowski, gr-qc 0404018
[72] A. Ashtekar and C. Isham, Class. Quantum Grav., 9 (1992), 1433
[73] A. Ashtekar, New perspectives in quantum gravity, Bibliopolis, 1988
[74] A. Ashtekar and R. Tate, Lectures on nonperturbative canonical gravity, World Scientific, 1991
[75] A. Ashtekar, gr-qc 0410054
[76] A. Ashtekar, Phys. Rev. D, 36 (1987), 1587-1602; Phys. Rev. Lett., 57 (1986), 2244-2247

[77] A. Ashtekar, Phys. Rev. D, 68 (2003), 104030
[78] G. Auberson and P. Sabatier, Jour. Math. Phys., 35 (1994), 4028-4040
[79] T. Aubin, Some nonlinear problems in Riemannian geometry, Springer, 1998
[80] T. Aubin, A course in differential geometry, Amer. Math. Soc., 2001
[81] J. Audretsch, Phys. Rev. D, 27 (1983), 2872-2884
[82] J. Audretsch, F. Gähler, and N. Straumann, Comm. Math. Phys., 95 (1984),41-51
[83] A. ben-Avraham and S. Havlin, Diffusion and reactions in fractal and disordered systems, Cambridge Univ. Press, 2000
[84] A. Babakhani and V. Dattardar-Gejji, Jour. Math. Anal. Appl., 270 (20020, 66-79
[85] G. Bacciagaluppi, quant-ph 9711048, 9811040 and 0302099
[86] J. Baez and J. Muniain, Gauge fields, knots, and gravity, World Scientific, 1994
[87] J. Baez (editor), Knots and quantum gravity, Oxford Univ. Press, 1994
[88] D. Bahns, S. Doplicher, K. Fredenhagen, and G. Piacitelli, hep-th 0301100
[89] R. Baierlein, D. Sharp,. and J.A. Wheeler, Phys. Rev. 126 (1962), 1864-1865
[90] I. Bakas, hep-th 0511057 and 0702034
[91] R. Balian, cond-mat 0501322
[92] F. Baldovin and A. Robledo, cond-mat 0304410
[93] M. Ban, Jour. Math. Phys., 39 (1998), 1744-1765; 40 (1999), 3718-3722
[94] T. Banks, W. Fishler, S. Shenker, and L. Susskind, Phys. Rev. D, 55 (1997), 5112 (hep-th 9610043)
[95] T. Banks, Nucl. Phys. B, 249 (1985), 332-360
[96] O. Barabash and Y. Shtanov, astro-ph 9904144
[97] D. Barabash and Y. Shtanov, hep-th 9807291
[98] J. Barbero and E. Villasenor, Phys. Rev. D, 68 (2003), 087501
[99] G. Barbosa and N. Pinto-Neto, hep-th 0304105 and 0407111; Phys. Rev. D, 69 (2004), 065014
[100] G. Barbosa, hep-th 0408071; Jour. High Energy Phys., 05 (2003), 024
[101] G. Barbosa, Phys. Rev. D, 69 (2004), 065014
[102] C. Barcelo, S. Liberati, S. Sonego, and M. Visser, gr-qc 0408022
[103] G. Beretta, quant-ph 0402180
[104] M. Barnsley, Fractals everywhere, Academic Press, 1988
[105] J. Acacio de Barros and N. Pinto-Neto, gr-qc 9611029; Inter. Jour. Mod. Phys. D, 7 (1998), 201
[106] J. Acacio de Barros, J. de Mendonca, and N. Pinto-Neto, quant-ph 0307193
[107] J. Acacio de Barros, G. Oliveira-Neto, and T. Vale, gr-qc 0404073
[108] J. Acacio de Barros, N. Pinto-Neto, and A. Sagioro-Leal, General Relativity and Gravitation, 32 (2000), 15; Phys. Lett. A, 241 (1998), 229
[109] A. Barut and N. Zanghi, Phys. Rev. Lett., 52 (1984), 2009-2012
[110] A. Barut and M. Pavsic, Class. Quant. Grav., 4 (1987), L41-L45
[111] A. Barvinsky and C. Kiefer, gr-qc 9711037
[112] H. Bateman, Proc. Royal Soc. London A, 125 (1929), 598-618
[113] C. Beck and F. Schlögl, Thermodynamics of chaotic systems, Cambridge Univ. Press, 1997
[114] J. Beckenstein, astro-ph 0403694 and 0412652
[115] J. Beckenstein and R. Sanders, astro-ph 0509519
[116] J. Bekenstein and L. Parker, Phys. Rev. D, 23 (1981), 2850
[117] V. Belavkin, quant-ph 0208087
[118] J. Bell, Speakable and unspeakable in quantum mechanics, Cambridge Univ. Press, 1987

[119] J. Bell, Phys. Repts., 137 (1986), 49-54; Physics (1964), 195-200; Rev. Mod. Phys., 38 (1966), 447-452

[120] G. Benenti, G. Casati, I. Guarneri, and M. Terraneo, cond-mat 0104450

[121] G. Benenti and G. Casati, quant-ph 0112060

[122] F. Berezin and M. Shubin, The Schrödinger equation, Kluwer, 1991

[123] I. Bengtsson and K. Zyczkowski, Geometry of quantum states, Cambridge Univ. Press, 2006

[124] J. Berger, quant-ph 0309143

[125] S. Bergliaffa, K. Hibberd, M. Stone, and M. Visser, cond-mat 0106255

[126] K. Berndl, quant-ph 9509009

[127] K. Berndl, M. Daumer, and D. Dürr, quant-ph 9504010

[128] K. Berndl, D. Dürr, S. Goldstein, G. Peruzzi, and N. Zanghi, Comm. Math. Phys., 173 (1995), 647-673 (quant-ph 9503013)

[129] K. Berndl, D. Dürr, S. Goldstein, and N. Zanghi, Phys. Rev. A, 53 (1996), 2062 (quant-ph 9510027)

[130] K. Berndl, M. Daumer, D. Dürr, S. Goldstein, and N. Zanghi, Nuovo Cimento, 110B (1995), 737-750

[131] M. Berry, Proc. Royal Soc. London A, 392 (1984), 45; Jour. Phys. A, 29 (1996), 6617

[132] M. Bertola, B. Eynard, and J. Harnad, nlin.SI 0108049, nlin.SI 0204054, nlin.SI 0208002

[133] G. Bertoldi, A. Faraggi, and M. Matone, Class. Quant. Grav., 17 (2000), 3965 (hep-th 9909201)

[134] I. Bialynicki-Birula and J. Mycielski, Annals Phys., 100 (1976), 62-93; Comm. Math. Phys., 44 (1975), 129-132

[135] I. and Z. Bialynicki-Birula, physics 0305012; quant-ph 0110145

[136] I. and Z. Bialynicki-Birula, Phys. Rev. D, 10 (1971), 2410-2412; Phys. Rev. A, 67 (2003), 062114; 65 (2002), 063606

[137] I. Bialynicki-Birula, Acta Phys. Pol. A, 86 (1994), 97

[138] I. and Z. Bialynicki-Birula and C. Sliwa, quant-ph 9911007

[139] I. Bialynicki-Birula and J. Mycielski, Annals Phys., 100 (1976), 62-93; Comm. Math. Phys., 44 (1976), 129

[140] I. Bialynicki-Birula, P. Gornicki, and J. Rafelski, Phys. Rev. D, 44 (1991), 1825

[141] I. Bialynicki-Birula, Phys. Rev. Lett., 80 (1998), 5247-5250; Jour. Optics A, 6 (2004), 5181-5183; quant-ph 0608116

[142] I. Bialynicki-Birula and T. Radozycki, quant-ph 0602049

[143] I. Bialynicki-Birula, T. Mloduchowski, T. Radozycki, and C. Sliwa, Acta Phys. Polon. A, 100 (2001), 1-10

[144] N. Bilić, gr-qc 9908022

[145] N. Birrell and P. Davies, Quantum fields in curved space time, Cambridge Univ. Press, 1982

[146] Y. Bisabr and H. Salehi, gr-qc 0109087

[147] Y. Bisabr, gr-qc 0302102; hep-th 0306092

[148] S. Biswas, A. Shaw, and D. Biswas, gr-qc 9906009

[149] J. Bjorken and S. Drell, Relativistic quantum mechanics, McGraw-Hill, 1964; Relativistic quantum fields, McGraw-Hill, 1965

[150] M. Blagojević, Gravitation and gauge symmetry, IOP Press, 2002

[151] M. Blassone, P. Jizba, and G. Vitiello, hep-th 0007138

[152] M. Blasone, E. Graziano, O. Pashaev, and G. Vitiello, hep-th 9603092

[153] A. Blaut and J. Kowalski-Glikman, gr-qc 9506081, 9509040, and 9509039

[154] A. Blaut and J. Kowalski-Glikman, gr-qc 9710136, 9706076, 9710039, and 9607004
[155] K. Bleuler, Helv. Phys. Acta, 23 (1950), 567
[156] L. Bogdanov, B. Konopelchenko, and L. Martinez-Alonso, nlin.SI 0111062
[157] Y. Bogdanov, quant-ph 0303013, 0303014, 0310011, 0312042, 0509213, 05070312, 0512062, 0605208, and 0612025
[158] D. Bohm, B. Hiley, and P. Kaloyerou, Phys. Rept. 144 (1987), 323-375
[159] D. Bohm and B. Hiley, The undivided universe, Routledge, Chapman and Hall, 1993
[160] D. Bohm and B. Hiley, Phys. Repts., 144 (1987), 323-348
[161] D. Bohm, Phys. Rev. 85 (1952), 166-179, 180-193; 84 (1951), 166
[162] D. Bohm and J. Vigier, Phys. Rev., 96 (1954), 208
[163] A. Bolivar, Quantum classical correspondence, Springer, 2004
[164] R. Bonal, I. Quiros, and R. Cardenas, gr-qc 0010010
[165] W. Boothby, An introduction to differentiable manifolds and Riemannian geometry, Acad. Press, 1975
[166] E. Borges and I. Roditi, Phys. Lett. A, 246 (1998), 399-402
[167] J. Borgman and L. Ford, gr-qc 0406066
[168] A. Bouda and T. Djama, quant-ph 0103071, 0108022, and 0206149
[169] A. Bouda, quant-ph 0004044 and 0210193
[170] A. Bouda and F. Hammad, quant-ph 0111114
[171] A. Bouda and A. Meziane, quant-ph 0701159
[172] D. Boyanovsky, e. Newman, and C. Rovelli, Phys. Rev. D, 45 (1992), 1210
[173] A. Boyarsky and P. Gora, chaos, Solitons, and Fractals, 9 (2001), 1611-1618; 7 (1996), 611-630, 939-954
[174] A. Boyarsky and O. Ruchayskiy, hep-th 0211010
[175] T. Boyer, Phys. Rev. D, 11 (1975), 790-808; 29 (1984), 1096-1098 and 21 (1980), 2137-2148; Phys. Rev., 182 (1969), 1374-1383 and 186 (1969), 1304-1318
[176] T. Boyer, Phys. Rev. 180 (1969), 19; Phys. Rev. A, 7 (1973), 1832; physics 0206033 and 0210129
[177] M. Bożejko, B. Kümmerer, and R. Speicher, funct-an 9604010
[178] H. Brandt, quant-ph 0303054
[179] C. Brans, gr-qc 9705069
[180] C. Brans and R. Dicke, Phys. Rev., 124 (1961), 925
[181] O. Bratteli and D. Robinson, Operator algebras and quantum statistical mechanics, Springer, 1979
[182] S. Braunstein and C. Caves, Phys. Rev. Lett., 72 (1994), 3439-3443
[183] H. Bray and A. Neves, Ann. Math., 159 (2004), 407-424
[184] R. Breban, quant-ph 0205177
[185] L. Brenig, quant-ph 0608025 and 0610142
[186] D. Brody and L. Hughston, quant-ph 9701051, 9706030, 9706037,9906085, and 9906086; gr-qc 9708032
[187] L. deBroglie, Electrons et photons, Solvay Conf., Paris, pp. 105-141
[188] L. deBroglie, Problemes de propagatons guideés des ondes electromagnetiques, Paris, 1941
[189] K. Bronnikov, gr-qc 0110125 and 0204001
[190] C. Brouder and W. Schmitt, hep-th 0210097
[191] C. Brouder, math-ph 0201033
[192] C. Brouder, hep-th 9904014, 9906111, 0202025, 0208131, and 0307212
[193] C. Brouder and R. Oeckl, hep-th 0206054 and 0208118
[194] C. Brouder and A. Frabetti, hep-th 0003202 and 0011161

[195] C. Brouder, B. Fauser, A. Frabetti, and R. Oeckl, hep-th 0312158
[196] C. Brouder, A. Frabetti, and C. Krattenthaler, math.QA 0406117
[197] H. Brown and P. Holland, quant-ph 0302062
[198] H. Brown, E. Sjöqvist, and G. Bacciagaluppi, quant-ph 9811054
[199] M. Brown, quant-ph 9703007 and 0102102
[200] M. Brown and D. Wallace, quant-ph 0403094
[201] M. Brown and B. Hiley, quant-ph 0005026
[202] L. Brown, Quantum field theory, Cambridge Univ. Press, 1992
[203] C. Brukner and A. Zeilinger, quant-ph 0212084
[204] T. Brun and J. Hartle, Phys. Rev. D, 60 (1999), 123503
[205] Y. Brezhnev, nlin.SI 0106024 and 0505003
[206] Y. Brezhnev and S. Leble, math-ph 0502052
[207] A. Budiyono, quant-ph 0512235 and 0601212; physics 0609216
[208] A. Bunde and S. Havlin, Fractals and disordered systems, Springer, 1991
[209] V. Burdyuzha, J. de Freitas-Pacheco, and G. Vereshkov, gr-qc 0312072
[210] J. Butterfield, quant-ph 0210140
[211] V. Buzek, C. Keitel, and P. Knight, Phys. Rev. A, 51 (1995), 2575-2593
[212] M. Cadoni, hep-th 9803257
[213] R. Cai and L. Cao, gr-qc 0611071
[214] C. Callan, S. Giddings, and A. Strominger, Phys. Rev. D, 45 (1992), 1005
[215] C. Callan and F. Wilczek, hep-th 9401072
[216] X. and J. Calmet, cond-mat 0410452
[217] J. and X. Calmet, math-ph 0403043
[218] X. Calmet, M. Graesser, and S. Hsu, hep-th 0405033
[219] F. Calogero, Phys. Lett. A, 228 (1997), 335-346
[220] R. Campos, physics 0401044
[221] H. Cao and X. Zhu, Asian Jour. Math., 10 (2006), 165-492
[222] H. Cao, Chinese Ann. Math., 27B (2006), 121-142
[223] H. Cao and R. Hamilton, math.DG 0807009
[224] E. Capelas de Olveira and R. de Rocha, math-ph 0502011
[225] S. Capozziello, A. Feoli, G. Lambiase, and G. Papini, gr-qc 0007029
[226] A. Capri and S. Roy, Mod. Phys. Lett. A, 7 (1992), 2317; Inter. Jour. Mod. Phys. A, 9 (1994), 1239
[227] R. Cardenas, T. Gonzalez, O. Martin, and I. Quiros, astro-ph 0210108
[228] R. Cardenas, T. Gonzalez, Y. Leiva, O. Martin, and I. Quiros, astro-ph 0206315
[229] M. Carfora, math.DG 0507309
[230] S. Carlip, Phys. Rev. A, 47 (1993), 3452-3453
[231] S. Carlip, Quantum gravity in 2+1 dimensions, Cambridge Univ. Press, 1998
[232] L. Caron, H. Jirari, H. Kröger, G. Melkonyan, X. Lu, and K. Moriarty, quant-ph 0106159
[233] L. Carr and C. Clark, cond-mat 0408460
[234] R. Carroll, quant-ph 0401082, Inter. Jour. Evol. Eqs., 1 (2005), 23-56
[235] R. Carroll, quant-ph 0403156, Applicable Anal., 84 (2005), 1117-1149
[236] R. Carroll, gr-qc 0406004
[237] R. Carroll, quant-ph 0406203, Foundations of Physics, 35 (2005), 131-154
[238] R. Carroll, Quantum theory, deformation, and integrability, North-Holland, 2000
[239] R. Carroll, Calculus revisited, Kluwer, 2002
[240] R. Carroll, Proc. Conf. Symmetry, Kiev, 2003, Part I, pp. 356-367
[241] R. Carroll, Canadian Jour. Phys., 77 (1999), 319-325

[242] R. Carroll, Direct and inverse problems of mathematical physics, Kluwer, 2000, pp. 39-52
[243] R. Carroll, Generalized analytic functions, Kluwer, 1998, pp. 299-311
[244] R. Carroll, quant-ph 0309023 and 0309159
[245] R. Carroll, Nucl. Phys. B, 502 (1997), 561-593; Springer Lect. Notes Physics, 502, 1998, pp. 33-56
[246] R. Carroll, Acta Applic. Math., 60 (2000), 225-316
[247] R. Carroll and Y. Kodama, Jour. Phys. A, 28 (1995), 6373-6387
[248] R. Carroll and B. Konopelchenko, Inter. Jour. Mod. Phys. A, 11 (1996), 1183-1216
[249] R. Carroll, Mathematical physics, North-Holland, 1988
[250] R. Carroll, Fluctuations, gravity, and the quantum potential, gr-qc 0501045, In book Quantum gravity, Nova Science Publ., to appear
[251] R. Carroll, Abstract methods in partial differential equations, Harper-Row, 1969
[252] R. Carroll, Topics in soliton theory, North-Holland, 1991
[253] R. Carroll, hep-th 9607219, 9610216, and 9702138
[254] R. Carroll, Fluctuations, information, gravity and the quantum potential, Springer, 2006
[255] R. Carroll, Proc. Conf. Nonlin. Physics, Gallipoli, 2006, Teor. i Matem. Fizika, to appear
[256] R. Carroll, physics 0507027
[257] R. Carroll, quant-ph 0506075
[258] R. Carroll, physics 0511076 and 0602036
[259] R. Carroll, gr-qc 0512146
[260] R. Carroll, Proc. Internat. Symposium Quantum Mechanics and Symmetry, Varna, Bulgaria, 2005, Bulg. Jour. Phys., 33 (s1b), 2006, pp. 698-708
[261] R. Carroll, math-ph 0701007
[262] R. Carroll, math-ph 0703065
[263] S. Carroll, Spacetime and geometry, Addison-Wesley, 2004
[264] S. Carroll and E. Lim, hep-th 0407179
[265] E. Cartan, Les systemes differentielles exterieur et leurs applications, Hermann, Paris, 1945
[266] M. Carvahlo and A. Oliveira, hep-th 0212319
[267] M. Carvahlo, A. Oliveira, and C. Rabaca, astro-ph 0212234
[268] M. Casas, F. Pennini, and A. Plastino, Phys. Lett. A, 235 (1997), 457-463
[269] G. Casati and B. Chirikov, Quantum chaos, Cambridge Univ. Press, 1995
[270] C. Castro, Found. Phys., 22 (1992), 569-615; Found. Phys. Lett., 4 (1991), 81
[271] C. Castro, Jour. Math. Phs., 31 (1990), 2626-2633
[272] C. Castro, Chaos, Solitons, and Fractals, 10 (1999), 295-309
[273] C. Castro and A. Granik, Chaos, Solitons, and Fractals, 11 (2000), 2167-2178
[274] C. Castro, physics 0011040; hep-th 9912113, 0001023, 0001134, 0205065, and 0206181
[275] C. Castro, J. Mahecha, and B. Rodriguez, quant-ph 0202026
[276] C. Castro and J. Mahecha, Prog. in Phys., 1 (2006), 38-45
[277] C. Castro and J. Mahecha, hep-th 0009014
[278] C. Castro, Chaos, Solitons, and Fractals, 11 (2000), 1663-1670; 12 (2001), 101-104, 1585-1606
[279] C. Castro, hep-th 9512044, 0203086
[280] C. Castro, hep-th 9512044; physics 0010072
[281] C. Castro, physics 0011040; hep-th 0206181, 0210061, and 0211053
[282] C. Castro and M. Pavsić, hep-th 0203194

[283] C. Castro, On dark energy, and different derivations of the vacuum energy density, Found. Phys., to appear
[284] A. Caticha, gr-qc 0109008, 0301061, and 0508108; math-ph 0008018
[285] A. Caticha, Phys. Rev. A, 57 (1998), 1572-1582; Found. Phys., 30 (2000), 227-251
[286] M. Cavaglia, gr-qc 9497029; Mod. Phys. Lett. A, 9 (1994), 1897
[287] M. Célérier and L. Nottale, hep-th 0112213 and 0210027
[288] W. Chagas-Filho, hep-th 0401091
[289] M. Chaichian, M. Sheikh-Jabbari, and A. Tureanu, Phys. Rev. Lett., 86 (2001), 2716
[290] P. Chaikin and T. Lubensky, Principles of condensed matter physics, Cambridge Univ. Press, 2000
[291] C. Chakrabarti and I. Chakrabarty, quant-ph 0511171
[292] I. Chakrabarty, quant-ph 0511169
[293] S. Chakraborty, N. Chakroborty, and N. Debnath, gr-qc 0306040
[294] O. Chavoya-Aceves, quant-ph 0304133, 0304195, 0305137, and 0409012
[295] K. Cheng, Jour. Math. Phys., 13 (1972), 1723-1726
[296] I. Cherednikov, hep-th 0206245
[297] A. Chervov, math.RT 9905005
[298] L. Chimento, F. Pennini, and A. Plastino, cond-mat 0005006
[299] L. Chimento, F. Pennini, and A. Plastino, Phys. Lett. A, 257 (1999), 275-282; Physica A, 256 (1998), 197-210
[300] M. Choi, B. Kim, B. Yoon, and H. Park, cond-mat 0412156
[301] Y. Choquet-Bruhat and J. York, gr-qc 0511032
[302] B. Chow and D. Knopf, The Ricci flow: An introduction, Amer. Math. Soc., 2004
[303] D. Chruscinski and K. Mlodawski, quant-ph 9810164 and 050163
[304] K. Chung and Z. Zhao, From Brownian motion to the Schrödinger equation, Springer, 2001
[305] K. Chung and J. Zambrini, Introduction to random time and quantum randomness, World Scientific, 2003
[306] K. Chung and R. Williams, Introduction to stochastic integration, Birkhäuser, 1990
[307] R. Cirelli, M. Gatti, and A. Mania, quant-ph 0202076
[308] R. Cirelli, A. Mania, and L. Pizzocchero, Jour. Math. Phys., 31 (1990), 2891-2897 and 2898-2903
[309] R. Cirelli, P. Lanzavecchia, and A. Mania, Jour. Phys. A, 16 (1983), 3829-3835
[310] R. Cirelli and L. Pizzocchero, Nonlinearity, 3 (1990), 1057-1080
[311] R. Cirelli and P. Lanzavecchia, Nuovo Cimento B, 79 (1984), 271-283
[312] I. Ciufolini and J. Wheeler, Gravity and inertia, Princeton Univ. Press, 1995
[313] C. Ciuhu and I. Vancea, gr-qc 9807011
[314] C. Cohen-Tanoudji, J. Dupont-Roc, and G. Grynberg, Photons and atoms, Wiley, 1989
[315] D. Cole and H. Puthoff, Phys. Rev. E, 48 (1993), 1562-1565
[316] D. Cole, Phys. Rev. A, 42 (1990), 1847-1862; 45 (1992), 8953-8956
[317] D. Cole, A. Rueda, and K. Danley, Phys. Rev. A, 63 (1997), O54101
[318] D. Cole and Y. Zou, quant-ph 0307154
[319] R. Colistete Jr., J. Fabris, and N. Pinto-Neto, Phys. Rev. D, 57 (1998), 4707; 62 (2000), 83507
[320] S. Colin, quant-ph 0301119 and 0310056; Phys. Lett. A, 317 (2003), 349-358
[321] J. Colliander and R. Jerrard, math.MP 9712278
[322] A. Compte, Phys. Rev. E, 53 (1996), 4191-4193
[323] A. Connes, Noncommutative geometry, Academic Press, 1994

[324] A. Connes and D. Kreimer, Lett. Math. Phys., 48 (1999), 85-96; Comm. Math. Phys., 210 (2000), 249-273 and 216 (2001), 215-241; hep-th 9912092 and 0003188
[325] H. Conradi, gr-qc 9412049
[326] M. Consoli, hep-th 0109215, 0306070 and 0002098; gr-qc 0306105
[327] M. Consoli and E. Costanzo, hep-th 0311317; gr-qc 0406065
[328] M. Consoli and F. Siringo, hep-th 9910372
[329] M. Consoli P. Stevenson, hep-th 9905427
[330] R. Cook, Phys. Rev. A, 25 (1982), 2164-2167; 26 (1982), 2754-2760
[331] A. Corichi, quant-ph 0407242
[332] N. Costa Dias and J. Prata, quant-ph 0208156
[333] C. Coste, Euro. Phys. Jour., 1 (1998), 245-253
[334] S. Cotsakis, gr-qc 0502008
[335] R. Cremona and J. Lacroix, Spectral theory of random Schrödinger operators, Birkhäuser, 1990
[336] J. Cresson, math.GM 0211071
[337] J. Cresson, Scale calculus and the Schrödinger equation; Scale geometry, I, preprints 2003
[338] J. Cresson, Nondifferentiable variational principles, preprint 2003
[339] L. Crowell, Quantum fluctuations of spacetime, World Scientific, 2005
[340] H. Culetu, hep-th 0410133
[341] W. Curtis and P. Miller, Differentiabe manifolds and theoretical physics, Acad. Press, 1985
[342] T. Curtright and C. Zachos, hep-th 9810164
[343] T. Curtright, A. Polychronakos, and C. Zachos, hep-th 0111173
[344] , J. Cushing, A. Fine, and S. Goldstein (editors), Bohmian mechanics and quantum theory, Kluwer, 1996
[345] J. Cushing, Quantum mechanics, Univ. Chicago Press, 1994
[346] H. Cycon, R. Froese, W. Kirsch, and B. Simon, Schrödinger operators with applications to quantum mechanics and global geometry, Springer, 1987
[347] M. Czachor, quant-ph 9711053
[348] M. Czachor and H. Doebner, quant-ph 0106051 and 0110008
[349] R. Czopnik and P. Garbaczewski, quant-ph 0203018; cond-mat 0202463
[350] Y. Dabaghian, quant-ph 0407239
[351] M. and N. Dalarsson, Tensors, relativity, and cosmology, Elsevier, 2005
[352] B. Damski, Phys. Rev. A, 69 (2004), 043610
[353] R. Dandoloff, quant-ph 0212115
[354] F. Darabi and P. Wesson, gr-qc 0003045
[355] R. Darling, math.PR 0210109
[356] G. Date, Class. Quant. Gravity, 18 (2001), 5219-5225
[357] M. Daumer, S. Goldstein, and N. Zanghi, Jour. Stat. Phys., 88 (1997), 967-977 (quant-ph 9601013)
[358] G. Dautcourt, Acta Phys., Polon. B, 29 (1998), 1047
[359] M. Davidson, Physica A, 96 (1979), 465-487
[360] M. Davidson, Jour. Math. Phys., 20 (1979), 1865-1869; Lett. Math. Phys., 5 (1981), 523-529; quant-ph 0106124
[361] A. Davidson, gr-qc 0409059
[362] M. Davidson, quant-ph 0110050 and 0112157
[363] M. Davidson, quant-ph 0106124
[364] P. Davies, T. Dray, and C. Manogue, Phys. Rev. D, 53 (1996), 4382-4387
[365] P. Davies, Jour. Phys. A, 8 (1975), 609

[366] D. Delphenich, gr-qc 0211065, 0209091, and 0702115
[367] E. Deotto and C. Ghirardi, Found. Phys., 28 (1998), 1-30 and quant-ph 9704021
[368] S. Deser, R. Jackiw, and S. Templeton, Ann. Phys., 140 (1982), 372
[369] A. Desyatnikov, L. Torner, and Y. Kivshar, nlin.PS 0501026
[370] S. Deutsch, physics 9803039
[371] C. Dewdney and G. Horton, quant-ph 0202104
[372] C. DeWitt-Morette, K. Elworthy, B. Nelson, and G. Sammelman, Ann. Inst. H. Poincaré, A, 32 (1980), 327
[373] B. DeWitt, Rev. Mod. Phys., 29 (1957), 377-397
[374] B. DeWitt, Phys. Rev. D, 160 (1967), 1113-1148
[375] D. Diakonov and v. Petrov, hep-th 0108097
[376] R. Dicke, Phys. Rev. 125 (1962), 2163
[377] P. Dirac, Proc. Royal Soc. London A, 209 (1951), 291-296, 212 (1952), 330-339, 333 (1973), 403-418
[378] P. Dirac, General theory of relativity, Princeton Univ. Press., 1996
[379] B. Dittrich and T. Thiemann, gr-qc 0411138, 0411139, 0411140, 0411141, and 0411142
[380] W. Dittrich and M. Reuter, Classical and quantum dynamics, Springer, 1994
[381] T. Dittrich, C. Viviescas, and L. Sandoval, quant-ph 0508057
[382] T. Djama, quant-ph 0111121 and 0201003
[383] T. Djama, quant-ph 0311057 and 0311059
[384] T. Djama, quant-ph 0111142
[385] T. Djama, hep-th 0406255; quant-ph 0404175 and 0407044
[386] T. Djama, quant-ph 0404098
[387] Y. Dobyns, A. Rueda, and B. Haisch, g-qc 0002069
[388] V. Dodonov and S. Mizrahi, Physica A, 214 (1995), 619-628
[389] H. Doebner, G. Goldin, and P. Natterman, quant-ph 9502014 and 9709036; Jour. Math. Phys., 40 (1999), 49-63
[390] H. Doebner and G. Goldin, Phys. Rev. A, 54 (1996), 3764-3771
[391] H. Doebner and G. Goldin, Phys. Lett. A, 162 (1992), 397-401
[392] H. Doebner and G. Goldin, Jour. Phys. A, 27 (1994), 1771-1780
[393] H. Doebner and R. Zhdanov, quant-ph 0304167
[394] D. Dohrn and F. Guerra, Phys. Rev. E, 31 (1985), 2521-2594
[395] J. Doob, Stochastic processes, Wiley, 1953
[396] S. Doplicher, hep-th 0105251; Phys. Rev. D, 70 (2004), 064037
[397] S. Doplicher, K. Fredenhagen, and J. Roberts, hep-th 0303037
[398] M. Douglas and N. Nekrasov, hep-th 0106048
[399] J. Dowker, Jour. Phys. A, 3 (1970), 451
[400] J. Dowker and R. Critchley, Phys. Rev. D, 15 (1977), 1484
[401] W. Drechsler, gr-qc 9901030
[402] W. Drechsler and H. Tann, gr-qc 9802044
[403] I. Drozdov, hep-th 0311199
[404] B. Dubrovin, S. Novikov, and A. Fomenko, Geometrie, contemporaine, Moscow, 1979
[405] A. Dukkipati, M. Murty, and S. Bhatnagar, math-ph 0501025
[406] G. Dunne, hep-th 9204056, 9310182, and 9410065
[407] G. Dunne, A. Kovar, and B. Tekin, hep-th 0008139
[408] G. Dunne, R. Jackiw, and C. Trugenberger, Ann. Phys., 194 (1989)m 197; Phys. Rev. D, 41 (1990), 661-666

[409] G. Dunne, R. Jackiw, S. Pi, and C. Trugenberger, Phys. Rev. D, 43 (1999), 1332-1345

[410] G. Dunne, Self dual Chern-Simons theories, Springer, 1995

[411] G. Dunne and C. Trugenberger, Phys. Rev. D, 43 (1991), 1323-1331

[412] D. Dürr, S. Goldstein, and N. Zanghi, quant-ph 9511016 and 0308039

[413] D. Dürr, S. Goldstein, and N. Zanghi, quant-ph 0308038

[414] D. Dürr, S. Goldstein, R. Tumulka, and N. Zanghi, quant-ph 0208072; 0303156; 0303056; 0311127

[415] D. Dürr, S. Goldstein, and S. Zanghi, Jour. Stat. Phys., 67 (1992), 843-907; Phys. Lett. A, 172 (1992), 6-12

[416] D. Dürr, S. Goldstein, K. Münch-Berndl, and N. Zanghi, Phys. Rev. A, 60 (1999), 2729 (quant-ph 9801070)

[417] D. Dürr, S. Goldstein, and N. Zanghi, quant-ph 9512031; Bohmian mechanics and quantum theory, J. Cushing, et al (editors), Kluwer, 1996, pp. 21-44

[418] D. Dürr, S. Goldstein, R. Tumulka, and N. Zanghi, quant-ph 0407116

[419] C. Duval and P. Horvathy, hep-th 0307025

[420] C. Duval, P. Horvathy, and L. Palla, Phys. Lett. B, 325 (1994), 39-44; Phys. Rev. D, 50 (1994), 6658-6661; 52 (1995), 4700-4703

[421] V. Dvoeglazov, Jour. Phys. A, 33 (2000), 5011-5016; math-ph 0102001

[422] A. Dvurecenskij and S. Pulmannova, New trends in quantum structures, Kluwer, 2000

[423] V. Dzhunushaliev and H. Schmidt, gr-qc 9908049

[424] J. Eberly, quant-ph 0508019

[425] A. Edery and M. Paranjape, astro-ph 9808345

[426] E. Effros and M. Popa, math.FA 0303045

[427] A. Einstein, Annalen d. Physik, 17 (1905), 104

[428] G. El and A. Krylov, Phys. Lett. A, 203 (1993), 402-408

[429] C. Eling and T. Jacobson, gr-qc 0310044

[430] C. Eling, T. Jacobson, and D. Mattingly, gr-qc 0410001

[431] C. Eling, R. Guedens, and t. Jacobson, gr-qc 0602001

[432] E. Elizalde and S. Odintsov, hep-th 9403132

[433] N. Ercolani and R. Montgomery, Phys. Lett. A, 180 (1993), 402-408

[434] N. Ercolani, I. Gabitov, C. Levermore, and D. Serre, Singular limits of dispersive waves, Plenum Press, 1994

[435] L. Erdos, B. Schlein, and h. Yau, math-ph 0606017

[436] J. Escobar, Ann. Math., 136 (1992), 1-50

[437] L. Evans, Partial differential equations, Amer. Math. Soc., 1998

[438] J. Fabris, N. Pinto-Neto, and A. Velasco, gr-qc 9903111

[439] L. Faddeev and R. Jackiw, Phys. Rev. Lett., 60 (1988), 1692

[440] M. Fannes and B. Haegeman, math-ph 0206037

[441] A. Faria, H. Franca, C. Malta, and R. Sponchiado, quant-ph 0409117

[442] A. Faria, H. Franca, and R Sponchiado, quant-ph 0409119

[443] A. Fariborz and D. McKeon, hep-th 9607010

[444] K. Falconer, Fractal geometry, Wiley, 1990; The geometry of fractal sets, 1988; Techniques in fractal geometry, Wiley, 1997

[445] A. Faraggi and M. Matone, Phys. Rev. Lett., 78 (1997), 163-166 (hep-th 9606063)

[446] A. Faraggi and M. Matone, Inter. Jour. Mod. Phys. A, 15 (2000), 1869-2017 (hep-th 9809127)

[447] A. Faraggi and M. Matone, hep-th 9801033

[448] A. Faraggi and M. Matone, hep-th 9705108 and 9809125

[449] A. Faraggi and M. Matone, Phys. Lett. B, 437 (1998), 369; 445 (1999), 77 and 357; 150 (1999), 34; Phys. Lett. A, 249 (1998), 180; Phys. Rev. Lett., 78 (1997), 163-166
[450] V. Faraoni, Phys. Lett. A, 245 (1998), 26 (gr-qc 9805057)
[451] V. Faraoni, E. Gunzig, and P. Nardone, gr-qc 9811047
[452] V. Faraoni, gr-qc 9902083
[453] V. Faraoni and E. Gunzig, astro-ph 9910176
[454] B. Fauser, math.QA 0202059 and 0007137; math-ph 0208018; hep-th 9611069 and 0011026
[455] M. Feldman, T. Ilmanen, and L. Ni, math.DG 0405036
[456] P. Fedichev and U. Fischer, cond-mat 0307200
[457] F. deFelice and C. Clarke, Relativity on curved manifolds, Cambridge Univ. Press, 1995
[458] D. Feng and B. Hu, Quantum classical correspondence, Internat. Press, 1997
[459] J. Fenyes, Zeit. d. Phys., 132 (1952), 81-106
[460] A. Feoli, W. Wood, and G. Papini, gr-qc 9805035
[461] L. Ferandez Jambrina and F. Chinea, gr-qc 0403118
[462] D. Ferry and J. Zhou, Phys. Rev. B, 48 (1993), 7944-7950
[463] R. Feynman and A. Hibbs, Quantum mechanics and path integrals, McGraw-Hill, 1965
[464] R. Feynman, Rev. Mod. Phys., 20 (1948), 367-387; Phys Rev., 80 (1950), 440-457
[465] J. Field, physics 0403076
[466] J. Field, physics 0012011
[467] J. Field, physics 0307133, 0409103, and 0410262
[468] J. Field, physics 0501130 and 0501043
[469] J. Field, quant-ph 0503026
[470] A. Figotin and J. Schenker, Jour. Stat. Phys., 118 (2005), 199-263; math-ph 0608003
[471] U. Fischer, cond-mat 0406086, 9907457, and 0409201
[472] U. Fischer and M. Visser, cond-mat 0110211, 0205139 and 0211029
[473] A. Fischer, Jour. Math. Phys., 27 (1986), 718-738
[474] A. Fisher, math.DG 0312519
[475] R. Fisher, Proc. Cambridge Philos. Soc., 22 (1925), 700
[476] D. Fivel, quant-ph 0311145
[477] E. Flanagan, gr-qc 0403063
[478] E. Floyd, Inter. Jour. Mod. Phys. A, 14 (1999), 1111-1124; 15 (2000), 1363-1378; Found. Phys. Lett., 13 (2000), 235-251; quant-ph 0009070, 0302128 and 0307090
[479] E. Floyd, Phys. Rev. D, 29 (1984) 1842-1844; 26 (1982), 1339-1347; 34 (1986), 3246-3249; 25 (1982), 1547-1551; Jour. Math. Phys., 20 (1979), 83-85; 17 (1976), 880-884; Phys. Lett. A, 214 (1996), 259-265; Inter. Jour. Theor. Phys., 27 (1998), 273-281
[480] E. Floyd, quant-ph 0206114
[481] E. Floyd, quant-ph 9907092
[482] L. Ford, gr-qc 9707062, 0210096, and 0501081
[483] L. Ford and N. Svaiter, quant-ph 9804056, 0003129, and 0204126
[484] L. Ford and C. Wu, gr-qc 0102063
[485] M. Forest and K. McLaughlin, Jour. Nonlin. Sci., 7 (1998), 43-62
[486] B. Foster, gr-qc 0502066
[487] J. Foster and J. Nightingale, A short course in general relativity, Springer, 1995
[488] E. Fradkin, Field theories of condensed matter systems, Addison-Wesley, 1991
[489] H. Franca, A. Maia, and C. Malta, quant-ph 9512007

[490] B. Frieden, Physics from Fisher information, Cambridge Univ. Press, 1998
[491] B. Frieden, A. Plastino, A.R. Plastino, and B. Soffer, cond-mat 0206107; Phys. Rev. E, 60 (1999), 48-53
[492] B. Frieden, Jour. Mod. Opt., 35 (1988), 1297; Amer. Jour. Phys., 57 (1989), 1004
[493] B. Frieden and B. Soffer, Phys. Rev. E, 52 (1995), 2274-2286
[494] B. Frieden and H. Rosu, gr-qc 9703051
[495] B. Frieden and A. Plastino, quant-ph 0006012
[496] J. Friedman and A. Higuchi, Phys. Rev. D, 41 (1990), 2479-2486
[497] H. Frisk, Phys. Lett. A, 227 (1997), 139-142
[498] J. Frölich, M. Merkli, S. Schwarz, and D. Ueltschi, math-ph 0410013
[499] Y. Fujii and K. Maeda, The scalar tensor theory of gravitation, Cambridge Univ. Press, 2003
[500] A. Fujiwara and N. Nagaoka, Phys. Lett. A, 201 (1995), 119-124
[501] S. Fulling, Aspects of quantum field theory in curved space time, Cambridge Univ. Press, 1989
[502] T. Fülöp and S. Katz, quant-ph 9806067
[503] G. Gaeta, Inter. Jour. Theor. Phys., 39 (2000), 1339-1350
[504] C. Galvao, M. Henneux, and C. Titelboim, Jour. Math. Phys., 21 (1980), 1863
[505] R. Gambini and J. Pullin, Loops, knowts, gauge theories and quantum gravity, Cambridge Univ. Press, 1996
[506] R. Garattini, gr-qc 9508060 and 9604004
[507] L. Garay, J. Anglin, J. Cirac, and P. Zoller, gr-qc 0002015
[508] P. Garbaczewski, cond-mat 0211362 and 0301044
[509] P. Garbaczewski, quant-ph 0408192; Jour. Stat. Phys., 123 (2006) 315-355
[510] P. Garbaczewski, cond-mat 0604538; quant-ph 0612151
[511] R. Garcia Compean, O. Obregon, and C. Ramirez, hep-th 0107250
[512] E. Garcia-Rio and D. Kupeli, Semi-Riemannian maps and their applications, Kluwer, 1999
[513] C. Gardiner, Quantum noise, Springer, 1991
[514] C. Gardiner and P. Drummond, Phys. Rev. A, 38 (1988), 4897-4898
[515] M. Gasperini, Class. Quantum Gravity, 41 (1987), 485; Generel Relativ. and Gravitation, 30 (1998), 1703
[516] K. Gawedzki, Lectures on conformal field theory. Quantum fields and strings: A course for mathematicians, Princeton (1996-1997), pp. 727-805
[517] M. Gell-Mann and J. Hartle, quant-ph 0609190
[518] A. Gentle, N. George, A. Kheyfets, and W. Miller, gr-qc 0302044
[519] N. George, A. Gentle, a. Kheyfets, and W. Miller, gr-qc 0302051
[520] H. Georgii and R. Tumulka, math.PR 0312294
[521] U. Gerlach, Phys. Rev., 117 (1969), 1929-1941
[522] A. Gersten, Found. Phys. Lett., 12 (1999), 291-298; 13 (2000), 185-192; Ann. Fond. L. deBroglie, 21 (1996), 67
[523] G. Ghirardi, A. Remini, and T. Weber, Phys. Rev. D, 34 (1986), 470
[524] P. Ghose, quant-ph 0103126
[525] G. Gibbons, Jour. Geom. and Phys., 8 (1992), 147-162
[526] G. Gibbons and H. Pohle, Nucl. Phys. B, 140 (1993), 117-142
[527] G. Gibbons and S. Hawking (editors), Euclidean quantum gravity, World Scientific, 1993
[528] P. Gibilisco and T. Isola, math-ph 0509046 and 0701062
[529] P. Gibilisco, D. Imparato, and T. Isola, math-ph 0702058
[530] P. Gilkey, Jour. Diff. Geom, 10 (1975), 601; Composito Math., 38 (1975), 201

[531] M. Giona and H. Roman, Jour. Phys. A, 25 (1992), 2093-2105
[532] D. Giulini, E. Joos, C. Keifer, J. Kupsch, I. Stamatescu, and H. Zeh, Decoherence and the appearance of a classical world in quantum mechanics, Springer, 1996 and 2002
[533] D. Giulini, gr-qc 9311017 and 9312032
[534] D. Giulini and C. Keifer, gr-qc 9409014 and 9505040
[535] D. Giulini, gr-qc 9312032
[536] R. Glauber, Phys. Rev., 131 (1963), 2766-2788
[537] Y. Goldfarb, I. Degani, and D. Tannor, quant-ph 0604150
[538] G. Goldin and V. Shtelen, quant-ph 0006067
[539] G. Goldin, quant-ph 0002013
[540] G. Goldin and G. Svetlichny, Jour. Math. Phys., 35 (1994), 3322-3332
[541] G. Goldin, Nonlin. Math. Phys., 4 (1997) 6-11
[542] G. Goldin, Inter. Jour Mod. Phys. B, 6 (1992), 1905-1916
[543] Y. Goldfarb, I. Degani, and D. Tannor, quant-ph 0604150
[544] S. Goldstein, J. Taylor, R. Tumulka, and N. Zanghi, quant-ph 0405039 and 0407134
[545] S. Goldstein, (2001) Bohmian mechanics, Stanford Encycl. Philos., http://plato.stanford.edu/archives/win2002/entries/qm-bohm/
[546] S. Goldstein and R. Tumulka, quant-ph 0105040
[547] S. Goldstein and R. Tumulka, quant-ph 9902018
[548] S. Goldstein, quant-ph 9512027 and 9901005
[549] S. Goldstein and J. Lebowitz, quant-ph 9512028
[550] S. Goldstein and S. Teufel, quant-ph 9902018
[551] M. deGosson, math.SG 0504013 and 0411053; The principles of Newtonian and quantum mechanics, Imperial Coll. Press, 2001; Maslov classes, metaplectic representation, and Lagrangian quantization, Wiley-VCH, Berlin, 1997
[552] M. de Gosson, math.SG 0504013; quant-ph 0404072; math-ph 0505073, 0503078, and 0504013
[553] M. deGosson, Jour. Phys. A, 37 (2004), 7297-7314 and 31 (1998), 4239-4247
[554] M. deGosson, math.SG 0411453; Jour. Phys. A, 38 (2005), L324-L329
[555] I. Gottlieb, G. Ciobanu, and C. Buzea, Chaos, Solitons, and Fractals, 17 (2003), 789-796
[556] I. Gottlieb, M. Agop, and M. Jarcau, Chaos, Solitons, and Fractals, 19 (2004), 705-730
[557] J. Gouyet, Physique et structures fractales, Masson, 1992
[558] J. Gracia-Bondi and S. Lazzarini, hep-th 0006106
[559] J. Gracia-Bondi, hep-th 0202023
[560] W. Graf, gr-qc 0602054;; Phys. Rev. D, 67 (2003), 023002
[561] A. Granik, quant-ph 0409018; physics 0309059
[562] A. Granik and G. Chapline, quant-ph 0302013
[563] M. Grasselli and R. Streater, math-ph 0006030
[564] R. Griffiths, Phys. Rev. Lett., 70 (1993), 2201-2204; Phys. Rev. A, 54 (1996), 2759-2774 and 57 (1998), 1604-1618; quant-ph 9902059
[565] P. Grigolini, M. Pala, and L. Palatella, cond-mat 0007323
[566] D. Gron and S. Hervik, gr-qc 0011059
[567] G. Grössing, quant-ph 0311109
[568] B. Grossman, Phys. Rev. Lett., 65 (1990), 3230-3232
[569] Z. Gu and X. Wen, gr-qc 0606100
[570] P. Guerat and J. Vigier, Found. Phys., 12 (1982), 1057-1083; Lett. Nuovo Cimento, 38 (1983), 125-128

[571] F. Guerra, Phys. Reports, 77 (1981), 263-312
[572] F. Guerra and R. Marra, Phys. Rev. D, 28 (1983), 1916-1921
[573] F. Guerra and L. Morato, Phys. Rev. D, 27 (1983), 1784-1786
[574] F. Guerra and M. Pusteria, Lett. Nuovo Cimento, 34 (1982), 351-356
[575] V. Guilleman and S. Sternberg, Symplectic techniques in physics, Cambridge Univ. Press, 1984
[576] H. Gümral, solv-int 9801007
[577] S. Gustafson, K. Nakanishi, and T. Tsai, math.AP 0605655
[578] M. Gutierrez and B. Olea, math.DG 0701067
[579] M. Gutzwiller, Chaos in classical and quantum mechanics, Springer, 1990
[580] E. Guay and L. Marchildon, quant-ph 0407077
[581] I. Guk, Tech. Phys., 43 (1998), 353-357
[582] V. Guillemin and S. Sternberg, Symplectic techniques in physics, Cambridge Univ. Press, 1984
[583] S. Gupta, Proc. Royal Soc. London A, 63 (1950), 681
[584] Z. Haba and H. Kleinert, quant-ph 0106096 and 0101006
[585] C. Hagen, Ann. Phys., 157 (1984), 342; Phys. Rev. D, 31 (1985), 957
[586] B. Haisch, A. Rueda, and H. Puthoff, Phys. Rev. A, 49 (1994), 678-694; physics 9807023
[587] B. Haisch, A. Rueda, and H. Puthoff, Spec. Science and Tech., 20 (1997), 99-114
[588] B. Haisch and A. Rueda, gr-qc 9906084, 9908057, and 0106075; Phys. Lett. A, 268 (2000), 224-227
[589] B. Haisch, A. Rueda, and Y. Dobyns, gr-qc 0009036
[590] B. Haisch, A. Rueda, L. Nichisch, and J. Mollere, gr-qc 0209016
[591] R. Hakim, Jour. Math. Phys., 8 (1967), 1315-1344
[592] M. Hall, gr-qc 0408098
[593] M. Hall, Jour. Phys. A,, 37 (2004), 9549 (quant-ph 0406054)
[594] M. Hall, quant-ph 9806013, 9903045, 9912055, 0103072, 0302007, 0107149
[595] M. Hall, Gen. Relativ. Gravit., 37 (2005), 1505-1515; Phys. Rev. A, 62 (2000), 012107 (gr-qc 0408098; quant-ph 9912055)
[596] M. Hall and M. Reginatto, quant-ph 0201084; Jour. Phys. A, 35 (2002), 3289-2903 (quant-ph 0102069)
[597] M. Hall and M. Reginatto, Jour. Phys. A, 35 (2002), 3289-3303; Fortschr. Phys., 50 (2002), 646-651
[598] M. Hall, K. Kumar, and M. Reginatto, quant-ph 0103041
[599] M. Hall, K. Kumar, and M. Reginatto, Jour. Phys. A, 36 (2003), 9779-9794 (hep-th 0206235 and 0307259)
[600] M. Hall, Jour. Phys. A, 37 (2004), 7799 (quant-ph 0404123)
[601] M. Hall, M. Reineker, and W. Schleich, Jour. Phys. A, 32 (1999), 8275
[602] M. Hall, quant-ph 0007116
[603] J. Hallin and P. Lidjenbeg, Phys. Rev. D, 52 (1995), 1150
[604] J. Halliwell, Phys. Rev. Lett., 83 (1999), 2481-2485; Phys. Rev. A, 58 (1998), 105015 and 60 (1999), 105031; quant-ph 0305084, 0507136, and 0607132
[605] J. Halliwell, Phys. Rev. D, 38 (1988), 2468-2481
[606] H. Halvorson and R. Clifton, quant-ph 0103041
[607] H. Hamber and R. Williams, Phys. Rev. D, 59 (1999), 064014
[608] I. Hamilton, R. Mosna, and L. Delle Site, physics 0609180
[609] R. Hamilton, Jour. Diff. Geom, 17 (1985), 318-419; Contemp. Math., 71 (1988), 237-262; Comm. Anal. Geom., 7 (1999), 695-729
[610] R. Hamilton, Jour. Diff. Geom., 17 (1986), 255-306; 24 (1986), 153-179

[611] R. Hamilton, Jour. Diff. Geom., 37 (1993), 225-243; Comm. Anal. Geom., 1 (1993), 113-126 and 127-137
[612] J. Harriman, Jour. Chem. Phys., 100 (1994), 3651-3661
[613] J. Hartle, quant-ph 0209104; gr-qc 9404017
[614] J. Hartle, R. Laflamme, and D. Marolf, Phys. Rev. D, 51 (1995), 7007-7016
[615] A. Hasegawa and Y. Kodama, Solitons in optical communication, Oxford Univ. Press, 1978
[616] B. Hatfield, Quantum field theory of point particles and strings, Addison-Wesley, 1992
[617] A. Havare, M. Korunar, O. Aydogdu, M. Salti, and T. Yetkin, gr-qc 0506062
[618] S. Hawking and G. Ellis, The large scale structure of spacetime, Cambridge Univ. Press, 1973
[619] M. Headrick and T. Wiseman, hep-th 0606086
[620] C. Heinicke, P. Baekler, F. Hehl, gr-qc 0504005
[621] S. Helgason, Differential geometry, Lie groups, and symmetric spaces, Academic Press, 1978
[622] M. Henneux and C. Titelboim, Ann. Phys., 143 (1982), 127
[623] M. Henneux and C. Titelboim, Quantization of gauge systems, Princeton Univ. Press, 1992
[624] A. Heslot, Phys. Rev. D, 31 (1985), 1341-1348
[625] D. Hestenes Space time algebra, Gordon-Breach, 1966; New foundations for classical mechanics, Kluwer, 1986
[626] C. Hewitt-Horsman, quant-ph 0210204 and 0310014
[627] B. Hiley, Quo vadis quantum mechanics, Springer, 2005, pp. 299-324
[628] B. Hiley, R. Callaghan, and O. Maroney, quant-ph 0010020
[629] R. Hilfer and L. Anton, Phys. Rev. E, 51 (1995), R848-R851
[630] J. Hirschfelder, Jour. Chem. Phys., 67 (1977), 5477-5483
[631] J. Hirschfelder, C. Goebel, and L. Bruch, Jour. Chem. Phys., 61 (1974), 5456-5459
[632] V. Ho and M. Morgan, Jour. Phys. A, 29 (1996), 1497
[633] M. Hoefer, M. Ablowitz, I. Codington, E. Cornell, P. Engels, and V. Schweikhard, cond-mat 0603389
[634] W. Hofer, quant-ph 9610009, 9797024, 9801044, and 9910036
[635] W. Hofer, quant-ph 0001012, 0003061, and 0101091
[636] S. Hojman, K. Kuchar, and C. Teitelboim, Ann. Phys., 96 (1976), 88
[637] P. Holland, The quantum theory of motion, Cambridge Univ. Press, 1997
[638] P. Holland, Foundations of Physics, 38 (1998), 881-911; Nuovo Cimento B, 116 (2001), 1043 and 1143; Phys. Lett. A, 128 (1988), 9; quant-ph 0305175
[639] P. Holland, Phys. Repts., 224 (1993), 95; quant-ph 0302076
[640] P. Holland, quant-ph 0405145 and 0411141
[641] G. 't Hooft, gr-qc 9903084
[642] T. Horiguchi, K. Maeda, and M. Sakamoto, hep-th 9409152
[643] T. Horiguchi, Mod. Phys. Lett. A, 9 (1994), 1429
[644] G. Horton and C. Dewdney, quant-ph 0110007 and 0407089; Jour. Phys. A, 37 (2004), 11935
[645] G. Horton, C. Dewdney, and U. Neeman, quant-ph 0109059
[646] G. Horton, C. Dewdney, and H. Nesteruk, quant-ph 0103114; Jour. Phys. A, 33 (2000), 7337-7352
[647] A. Hosoya, T. Buchert, and M. Morita, gr-qc 0402076
[648] B. Hu, gr-qc 0503067
[649] K. Huang, Statistical mechanics, Wiley, 1989

[650] D. Huard, H. Kröger, G. Melkonyan, K. Moriarty, and L. Nadeau, quant-ph 0407074
[651] D. Huard, H. Kröger, G. Melkonyan, and L. Nadeau, quant-ph 0406131
[652] G. Hunter, M. Kowalski, and C. Alexandrescu, quant-ph 0506231
[653] G. Hunter and R. Wadlinger, Physics Essays, 2 (1989), 158-172
[654] R. Hyman, S. Caldwell, and E. Dalton, quant-ph 0401008
[655] M. Ibison and B. Haisch, Phys. Rev. A, 54 (1996), 2737
[656] M. Ibison, physics 0106046; Found. Phys. Lett., 16 (2003), 83-90
[657] B. Illiev, quant-ph 9902067
[658] Y. Imry, quant-ph 0501156
[659] M. Inaba, Inter. Jour. Mod. Phys. A, 16 (2001), 2965-2973
[660] K. Inoue, A. Kossakowski, and M. Ohya, quant-ph 0406227
[661] A. Iorio, L. O'Raifertaigh, I. Sachs, and C. Wiesendanger, hep-th 9607110
[662] P. Isaev, physics 0111072
[663] C. Isham, R. Penrose, and D. Sciama, Quantum gravity (symposium), Oxford Univ. Press, 1975 and 1981
[664] Y. Ishimori, Prog. Theor. Phy., 72 (1984), 33
[665] J. Isidro, hep-th 0110151
[666] J. Isidro, hep-th 0112032 and 0304143
[667] A. Isihara, Statistical physics, Academic Press, 1971
[668] M. Israelit, The Weyl-Dirac theory and our universe, Nova Science Pub., 1999
[669] M. Israelit, Found. Phys., 29 (1999), 1303-1322; 32 (2002), 295-321 and 945-961
[670] M. Israelit, gr-qc 9608035, 9608047, and 9611060
[671] M. Israelit and N. Rosen, Found. Phys., 22 (1992), 555; 24 (1994), 901
[672] M. Israelit and N. Rosen, Astrophys. Jour., 342 (1989), 627; Astrophys. Space Sci., 204 (1993), 317
[673] M. Israelit and N. Rosen, Found. Phys., 25 (1995), 763; 26 (1996), 585
[674] M. Israelit, Found. Phys., 19 (1989), 33-55
[675] M. Israelit and N. Rosen, Found. Phys., 13 (1983), 1023-1045
[676] K. Ito and H. McKean, Diffusion processes and their sample paths, Springer, 1974
[677] C. Itzykson and J. Zuber, Quantum field theory, Dover, 2005
[678] T. Ivey, Diff. Geom. Appl., 3 (1993), 301-307
[679] V. Iyer and R. Wald, Phys. Rev. D, 52 (1995), 4430-4439
[680] I. Jack and L. Parker, Phys. Rev. D, 31 (1985), 2439-2451
[681] R. Jackiw, physics 0010042
[682] R. Jackiw, Ann. Phys., 201 (1990), 83
[683] R. Jackiw and S. Pi, Phys. Rev. D, 42 (1990), 3500-3515; Phys. REv. Lett., 64 (1990), 2969 and 66 (1990), 2682; Prog. Theor. Phys., Supp. 107 (1992), 1; gr-qc 0308071
[684] J. Jackson, Classical electrodynamics, Wiley, 1999
[685] T. Jacobson, Phys. Rev. Lett., 75 (1995), 1260-1263 (gr-qc 9504004)
[686] T. Jacobson and D. Mattingly, Phys. Rev. D, 64 (2002), 024028; gr-qc 0402005
[687] T. Jacobson, S. Liberati, and D. Mattingly, gr-qc 0404067; astro-ph 0309681; hep-th 0407370
[688] T. Jacobson and G. Volovik, Phys. Rev. D, 58 (1998), 064021
[689] T. Jacobson and L. Smolin, Nucl. Phys. B, 299 (1988), 295
[690] T. Jacobson and R. Parentani, Found. Phys., 33 (2003), 323-348
[691] D. Jane, math.DG 0609624
[692] W. Janke, D. Johnston, and R. Kenna, cond-mat 0401092
[693] J. Jauch and C. Piron, Helv. Phys. Acta, 40 (1967), 550

[694] F. Jegerlehner, hep-th 9803021
[695] V. Jejjala, D. Minic, and C. Tze, gr-qc 0406137
[696] S. Jin, C. Levermore, and D. McLaughlin, Comm. Math. Phys., 52 (1999), 613-654
[697] H. Jirari, H. Kröger, X. Luo, G. Melkonyan, and K. Morairty, hep-th 0103027; quant-ph 0108094
[698] H. Jirari, H. Kröger, X. Luo, and K. Moriarty, quant-ph 0102032
[699] M. John, quant-ph 0102087 and 0109093
[700] R. Johal, cond-mat 0207268
[701] L. Johansen, quant-ph 0402105, 0402050, and 0309025
[702] G. Johnson and M. Lapidus, The Feynman integral and Feynman's operational calculus, Oxford Univ. Press, 1998
[703] D. Johnston, W. Janke, and R. Kenna, cond-mat 0308316
[704] V. Jones, Bull. Amer. Math. Soc., 12 (1985), 103; Ann. Math., 126 (1987), 335
[705] J. Jost, Riemannian geometry and geometric analysis, Springer, 2002
[706] A. Jurisch, quant-ph 0612217
[707] E. Kähler, Einführung in die Theorie der Systeme von Differentialgleichungen, Chelsea, 1949
[708] G. Kaiser, math-ph 0309010
[709] M. Kaku, Quantum field theory, Oxford Univ. Press, 1993
[710] M. Kaku, P. Townsend, and P. van Nieuwenhuizen, Phys. Rev. D, 19 (1979), 3166
[711] G. Kälbermann, quant-ph 0307018
[712] O. Kallenberg, Foundations of modern probability, Springer, 1997
[713] P. Kaloyerou, quant-ph 0311035
[714] P. Kaloyerou and J. Vigier, Phys. Lett. A, 130 (1988), 260-266
[715] A. Kamchatnov, R. Kraenkel, and B. Umarov, Phys. Lett. A, 287 (2001), 223-232; Phys. Rev. E, 66 (2002), 063609
[716] A. Kamchatnov, A. Gammel, and R. Kraenkel, cond-mat 0310457); Phys. Rev. A, 69 (2004), 63605
[717] A. Kamchatnov and R. Kraenkel, Jour. Phys. A, 35 (2002), L13-L18
[718] A. Kamchatnov, Jour. Phys. A, 34 (2001), L441-L445; Phys. Lett. A, 186 (1994), 387-390
[719] A. Kamchatnov, Nonlinear periodic waves and their modulation, World Scientific, 2000
[720] I. Kanatchikov, Phys. Lett. A, 25 (2001), 25
[721] I. Kanatchikov, gr-qc 9810076, 99090332, 9912094, 0004066, 0012038, and 0012074; hep-th 9911175, 0012084, 9811016,and 9410238
[722] G. Kaniadakis and A. Scarfone, Rep. Math. Phys., 51 (2003), 225 (cond-mat 0303334); Jour. Phys. A, 35 (2002), 1943 (quant-ph 0202032)
[723] G. Kaniadakis and A. Scarfone, quant-ph 0209075 and 0209130
[724] G. Kaniadakis, Physica A, 307 (2002), 172 (quant-ph 0112049)
[725] G. Kaniadakis, Phys. Lett. A, 310 (2003), 377 (quant-ph 0303159); Found. Phys. Lett., 16 (2003),99 (quant-ph 0209033)
[726] G. Kaniadakis, cond-mat 0210467 and 0507311; quant-ph 0112049
[727] G. Kaniadakis, P. Quarati, and A. Scarfone, hys. Lett. A, 255 (1998), 474-482; Phys. Rev. E, 58 (1998), 5574-5585
[728] G. Kaniadakis, quant-ph 0209033 and 0303159
[729] G. Kaniadakis, M. Lissia, and A. Scarfone, cond-mat 0409683
[730] G. Kaniadakis, E. Miraldi, and A. Scarfone, quant-ph 0210016
[731] G. Kar, M. Sinha, and S. Roy, Inter. Jour. Theor. Phys., 32 (1993), 593-607
[732] S. Kar, hep-th 9705062

[733] R. Kastner, Amer. Jour. Phys., 61 (1993), 852
[734] H. Kastrup, Phys. Repts., 101 (1983), 1
[735] L. Kaufman, quant-ph 0204007; math.QA 0105255
[736] P. Kazinski and A. Sharapov, hep-th 0212286
[737] B. Kelleher, Class. Quant. Grav., 21 (2004), 483-495
[738] M. Kenmoku, H. Kubortani, E. Takasugi, and Y. Yamazaki, gr-qc 9906056
[739] S. Khaemi and S. Nasiri, quant-ph 0511124
[740] A. Kheyfets and W. Miller, gr-qc 0006001 and 9412037; Phys. Rev. D, 51 (1995), 43-501
[741] A. Kheyfets, D. Holz, and W. Miller, Inter. Jour. Mod. Phys. A, 11 (1996), 2977-3002
[742] A. Kholodenko, gr-qc 0010064; hep-th 0701084
[743] A. Kholodenko and E. Ballard, gr-qc 0410029
[744] A. Kholodenko and K. Freed, Jour. Chem. Phys., 80 (1984), 900-924
[745] A. Kholodenko, Jour. Geom. Phys., 35 (2000), 193-238
[746] A. Khrennikov, G. Adenier, and T. Nieuwenhuizen, quant-ph 0610052
[747] T. Kibble, Comm. Math. Phys., 65 (1979), 189-201
[748] R. Kidd, J. Ardini, and A. Anton, Amer. Jour. Phys., 57 (1989), 27-35
[749] C. Kiefer, Quantum gravity, Oxford Univ. Press, 2004
[750] C. Kiefer, gr-qc 9906100
[751] C. Kiefer, gr-qc 9405039 and 9312015
[752] C. Kiefer and T. Lück, gr-qc 0505158
[753] C. Kiefer and T. Singh, Phys. Rev. D, 44 (1991), 1067-1076
[754] R. Kiehn, physics 9802033, 0101101, and 0102001; math-ph 0101032; gr-qc 0101109
[755] R. Kiehn and P. Baldwin, math-ph 0101033
[756] J. Kigami, Analysis on fractals, Cambridge Univ. Press, 2001
[757] S. Kim, qr-qc 9703065; Phys. Lett. A, 236 (1977), 11; Phys. Rev. D, 55 (1997), 7511
[758] J. Klauder, quant-ph 0112010; gr-qc 0411054 and 0411055
[759] J. Klauder, hep-th 0112010
[760] J. Klauder and B. Skagerstam, Coherent states, World Scientific, 1985
[761] B. Kleiner and J. Lott, math.DG 0605667
[762] L. Kobelev, hep-th 0002005; physics 0003036 and 0011038; math.CA 0002008
[763] A. Kobryn, T. Hayashi, and T. Arimitsu, math-ph 0304023
[764] K. Kolwankar and A. Gangal, cond-mat 9801138; physics 9801010
[765] B. Konopelchenko, solv-int 9905005
[766] B. Konopelchenko, L. Martinez-Alonso, and E. Medina, nlin.SI 0202013
[767] M. Kostin, Jour. Chem. Phys., 57 (1972), 3589-3591
[768] D. Kothawala, S. Sarkar, and T. Padmanabhan, gr-qc 0701002
[769] K. Kowalski, J. Rembielinski, and L. Papalucas, Jour. Phys. A, 29 (1996), 4149
[770] J. Kowalski-Glikman, gr-qc 9511014 and 9805015
[771] J. Kowalski-Glikman and K. Meissner, hep-th 9601062
[772] J. Kowalski-Glikman and J. Vink, Classical Quantum Gravity, 7 (1990), 901-918
[773] M. Kozlowski and J. Marciak-Kozlowska, astro-ph 0307168; cond-mat 0304052 and 0306699
[774] J. Marcial-Kozlowska and M. Kozlowski, astro-ph 0303256; quant-ph 0402069; cond-mat 0402159
[775] M. Kozlowski and J. Marcial-Kozlowska, From quarks to bulk matter, Hadronic Press, 2001
[776] A. Kracklauer, quant-ph 9711013

[777] S. Kreidl, quant-ph 0406045
[778] D. Kreimer, hep-th 9912290, 0005279, 0202110, and 0211188
[779] D. Kreimer, Adv. Theor. Math. Phys., 21 (1998), 303-334; q-alg 9707029; hep-th 0010059
[780] D. Kreimer, Knots and Feynman diagrams, Cambridge Univ. Press, 2000
[781] H. Kröger, Phys. Rev. A, 55 (1997), 951-966
[782] H. Kröger, quant-ph 0106087 and 0212093
[783] S. Kruglov, math-ph 0110008
[784] M. Kruskal, Phys. Rev., 119 (1960), 1743-1745
[785] K. Kuchar, Jour. Math. Phys., 212 (1983), 2122-2141
[786] K. Kuchar, Jour. Math. Phys., 15 (1974), 708-715; 17 (1976), 792-800; 18 (1977), 1589-1597; 19 (1978), 390-400; and 22 (1981), 2640-269l; Quantum gravity II: A second Oxford symposium, 1981, pp. 329-374
[787] M. Kuna and J. Naudts, quant-ph 0201055
[788] A. Kyprianidis, Phys. Repts., 155 (1987), 1-27
[789] F. Laloë, Amer. Jour. Phys., 69 (2001), 655
[790] L. Lamata and J. Leon, quant-ph 0410167
[791] D. Lamb, A. Capri, and S. Roy, hep-th 9411225
[792] L. Landau and E. Lifshitz, Statistical physics, Permagon, 1982
[793] L. Landau and E. Lifshitz, Classical mechanics, Addison-Wesley, 1976
[794] L. Landau and I. Khalatnikov, Sov. Phys. Dokl., 96 (1954), 469-472
[795] G. Landi, An introduction to noncommutative spaces and their geometry, Springer, 1997
[796] P. Landsberg and V. Vedral, Phys. Lett. A, 247 (1998), 211
[797] N. Landsman, Mathematical topics between classical and quantum mechanics, Springer, 1998
[798] O. Lange and B. Schroers, nlin.PS 0201047
[799] M. Lapidus and M. van Frankenhuysen, Fractal geometry and number theory, Birkhäuser, 2000
[800] V. Latora, M. Baranger, A. Rapisarda, and C. Tsallis, Phys. Lett. A, 273 (2000), 97
[801] R. Laughlin, A different universe - Reinventing physics from the bottom down, Basic Books, 2005
[802] R. Laughlin, Phys. Rev. Lett., 60 (1988), 2677
[803] A. Lavagno, cond-mat 0207353
[804] P. Lax, C. Levermore, and S. Venakidis, in Important developments in soliton theory, Springer, 1993, pp. 205-241
[805] J. Lee and T. Parker, Bull. Amer. Math. Soc., 17 (1987), 37-91
[806] J. Lee, Thermal physics, World Scientific, 2004
[807] J.H. Lee and O. Pashaev, Teor. Mat. Fizika, 127 (2001), 432-443
[808] J.H. Lee, O. Pashaev, C. Rogers, and W. Schief, Jour. Plasma Phys., to appear
[809] B. Lehnert, Found. Phys. Lett., 15 (2002), 95
[810] T. Leistner, math.DG 0501239
[811] N. Lemos, Phys. Lett. A, 78 (1980), 237 and 239
[812] T. LePage, Acad. Roy. Belg. Bull. Cl. Sci., 22 (1936), 716-729 and 1036-1046; 27 (1941), 27-46; 28 (1942), 73-92 and 247-265
[813] J. van de Leur and R. Martini, solv-int 9808008
[814] M. Levin and X. Wen, cond-mat 0407140; hep-th 0507118
[815] C. Levit and J. Sarfatti, physics 9704029
[816] J. Lévy Lablond, Ann. Inst. H. Poincaré, 31 (1965), 1

[817] J. Lévy-Lablond, Phys. Lett. A, 125 (1987), 44
[818] M. Li and T. Yoneya, hep-th 9806240
[819] S. Liberati, S. Sonego, and M. Visser, Annal. Phys., 298 (2002), 167
[820] S. Liberati, gr-qc 0009050
[821] R. Libof, Kinetic theory, Springer, 2003
[822] A. Liddle, An introduction to cosmology, Wiley, 2004
[823] E. Lieb, Comm. Math. Phys., 62 (1978), 35
[824] E. Lieb and R. Seiringer, math-ph 0504042
[825] S. Lin, O. Pashaev, and S. Roan, hep-th 9709073
[826] L. Loday, Cyclic homology, Springer, 1998
[827] F. Lombardo and P. Villar, quant-ph 0412205
[828] F. London, Zeit. f. Phys., 42 (1927), 375
[829] J. Lott, math.DG 0211065
[830] S. Luo, Jour. Phys. A, 35 (2002), 5181-5187
[831] E. Madelung, Zeit. Phys., 40 (1926), 322
[832] J. Madore, An introduction to noncommutative geometry and its physical applications, Cambridge Univ. Press, 1995
[833] K. Maeda and M. Sakamoto, Phys. Lett. B, 344 (1995), 105
[834] G. Magnano, gr-qc 9511027
[835] G. Magnano and L. Sokolowski, Phys. Rev. D, 50 (1994), 5039-5059
[836] B. Mahomed and L. Stenflo, Jour. Phys. A, 24 (1991), L1149-L1153
[837] V. Majernik and L. Richerek, Euro. Jour. Phys., 18 (1997), 79-89
[838] S. Majid, Foundations of quantum group theory, Cambridge Univ. Press, 1995
[839] S. Majid, mathQA 0006150, 0006151, 0006152; hep-th 0006166, 0006167
[840] A. Majtey, P. Lamberti, M. Martin, and A. Plastino, quant-ph 0408082
[841] J. Makela, gr-qc 0701128
[842] J. Makela and A. Peltola, gr-qc 0612078
[843] V. Makhankov and O. Pashaev, Sov. Sci. Rev. Math. Phys., 9 (1992), 1
[844] V. Makhankov, R. Myrsakulov, and O. Pashaev, Lett. Math. Phys., 16 (1983), 83
[845] Z. Malik and C. Dewdney, quant-ph 9506026
[846] S. Manakov and V. Zakharov, Lett. Math. Phys., 5 (1981), 247
[847] L. Mandel and E. Wolf, Optical coherence and quantum optics, Cambridge Univ. Press, 1995
[848] L. Mandel, Phys. Rev., 144 (1966), 1071
[849] B. Mandelbrot, Gaussian self affinities and fractals, Springer, 2002
[850] B. Mandelbrot, The fractal geometry of nature, Freeman, 1982
[851] P. Mannheim, gr-qc 9810087
[852] L. Marchildon, quant-ph 0007068
[853] F. Markopoulou and L. Smolin, gr-qc 0311059
[854] F. Markopoulou, gr-qc 9601038; Class. Quant. Gravity, 13 (1996), 2577
[855] P. Markowich, G. Rein, and G. Wolavsky, math-ph 0101020
[856] H. Markum and R. Pullirsch, nlin.CD 0303057
[857] J. Marsden and T. Ratiu, Introduction to mechanics and symmetry, Springer, 1999
[858] O. Maroney, quant-ph 0311149 and 0411172
[859] L. Martina, O. Pashaev, and G. Soliani, Class. Quant. Gravity, 14 (1997), 3179-3186; Phys. Rev. D, 58 (1998), 084025; Phys. Rev. B, 48 (1993), 15787-15791; Jour. Phys. A, 27 (1994), 943-954; hep-th 9411120, 9506130, and 9603048; Teor. Mat. Fiz., 99 (1994), 462
[860] S. Martinez, F. Pennini, and A. Plastino, Physica A (2001), 246 and 416
[861] K. Maruyama, C. Brukner, and V. Vedral, quant-ph 0407151

[862] V. Maslov and M. Fedoriuk, Semi-classical approximations in quantum mechanics, Reidel, 1981
[863] S. Massen and C. Panos, nucl-th 0101040
[864] M. Matone, hep-th 0005274 and 0212260
[865] M. Matone, hep-th 0502134
[866] D. Mattingly and T. Jacobson, gr-qc 0112012
[867] R. Mauldin and C. Williams, Trans. Amer. Math. Soc., 295 (1986), 325-346
[868] K. McDonald, physics 0003062
[869] B. Mehlig and M. Wilkinson, Ann. Phys., 18 (2001), 541-555
[870] S. Mercuri and G. Montani, gr-qc 0310077, 0312077, and 0410077
[871] S. Merkulov, Class. Quant. Grav., 1 (1984), 349-354
[872] N. Mermin, quant-ph 0305088
[873] N. Mermin and T. Ho Phys. REv. Lett., 36 (1976), 594
[874] A. Messiah, Quantum mechanics, Dover, 1999
[875] R. Metzler, E. Barkai, and J. Klafter, Phys. Rev. Lett., 82 (1999), 3563-3567
[876] B. Mielnik, Comm. Math. Phys., 37 (1974), 221
[877] B. Mielnik, quant-ph 0112041
[878] P. Milgrom, Astrophys. Jour., 270 (1983), 365, 371, and 384; 333 (1988), 689; 302 (1986), 617; 287 (1984), 571
[879] P. Milonni, The quantum vacuum, Academic Press, 1994
[880] K. Milton, hep-th 0009173, 0401117, and 0406024
[881] D. Minic and C. Tze, Phys. Rev. D, 68 (2003), 061501(R); hep-th 0305193, 0309239, and 0401028
[882] J. Miritzis, gr-qc 0402039
[883] A. Mironov, A. Morozov, and L. Vinet, hep-th 9312213
[884] C. Misner, K. Thorne, and J. Wheeler, Gravitation, Freeman Press,1973
[885] G. Modense, hep-th 0009046, 0005038, and 0011250
[886] M. Monk and B. Hiley, Found. Phys. Lett., 11 (1998), 371-377
[887] S. Montangero, L. Fronzoni, and P. Grigolini, cond-mat 9911412
[888] G. Montani, Nucl. Phys. B, 634 (2002), 370-392
[889] J. Moret Bailly, physics 0112048 and 0203051
[890] C. Morette, Phys. Rev., 81 (1951), 848-852
[891] J. Morgan and G. Tian, math.DG 0607607
[892] E. Moro, cond-mat 0105044
[893] A. Moroianu, Lectures on Kähler geometry, math.DG 0402223
[894] A. Morozov, hep-th 0502010
[895] A. Morozov, hep-th 9311142
[896] A. Morozov and L. Vinet, hep-th 9309026
[897] R. Mosna, I. Hamilton, and L. Delle Site, quant-ph 0504124 and 0511068
[898] I. Moss, Quantum theory, black holes, and inflation, Wiley, 1996
[899] A. Mostafazadeh and F. Zamani, quant-ph 0312078
[900] A. Mostafazadeh, gr-qc 0306003
[901] H. Motavali, H. Salehi, and M. Golshani, hep-th 0011062
[902] H. Motavali and M. Golshani, hep-th 0011064
[903] E. Mottola, Jour. Math. Phys., 36 (1995), 2470-2511
[904] A. Mukhopadhyay and T. Padmanabhan, Phys. Rev. D, 74 (2006), 124023
[905] R. Müller, Differential Harnack inequalities and the Ricci flow, Eur. Math. Soc. Pub. House, 2006
[906] M. Nagasawa, Prob. Theory Rel. Fields, 82 (1089), 109-136; Chaos, Solitons, and Fractals, 7 (1996), 631-643

[907] M. Nagasawa, Stochastic processes, Physics, and Geometry II, World Scientific, 1995, pp. 545-556
[908] M. Nagasawa, Schrödinger equations and diffusion theory, Birkäuser, 1993
[909] M. Nagasawa, Diffusion processes and related problems in analysis, I, Birkhäuser, 1990, pp. 155-200
[910] M. Nagasawa, Stochastic processes in quantum physics, Birkhäuser, 2000
[911] C. Nair, physics 0506093
[912] M. El Naschie, O. Rössler, and I. Prigogine, Quantum mechanics, diffusions, and chaotic fractals, Elsevier, 1995
[913] M. El Naschie, Chaos, Solitons, and Fractals, 3 (1993), 89-98; 7 (1996), 499-518; 11 (1997), 1873-1886; 4 (1994), 403-409; 9 (1998), 517-529; 3 (1993), 675-685; 5 (1995), 661-684, 1503-1508; 6 (1995), 1031-1032; 7 (1996), 955-959, 1501-1506; 8 (1997), 753-759, 1865-1872; 9 (1998), 913-919, 2023-2030; 10 (1999), 567-580; 11 (2000), 453-464, 2391-2395; 12 (2001), 851-858; 13 (2002), 1935-1945; 18 (2003), 401-420; 19 (2004), 209-236 and 689-697; 20 (2004), 437-450
[914] M. El Naschie, Nuovo Cimento B, 107 (1992), 583-594 and 109 (1994), 149-157; Chaos, Solitons, and Fractals, 2 (1992), 91-94, 4 (1994), 177-179, 293-296, 2121-2132, and 2269-2272
[915] S. Nasiri, Symmetry, Integrability, and Geometry, 2 (2006), 12 pages
[916] S. Nasiri and H. Safari, quant-ph 0505147
[917] S. Nasiri, Y. Sobouti, and f. Taati, quant-ph 0605129
[918] P. Natterman and W. Scherer, quant-ph 9506033
[919] P. Natterman and R. Zhdanov, solv-int 9510001
[920] P. Natterman and R. Zhdanov, Jour. Phys. A, 24 (1996), 2869-2886
[921] P. Natterman, quant-ph 9703017 and 9709044
[922] J. Naudts, cond-mat 0203489, 0211444, 0311438, 0405508, 0407804, and 0412683
[923] J. Naudts, M. Kuna, and W. deRoeck, hep-th 0210188
[924] J. Naudts and E. Van der Straeten, cond-mat 0406050
[925] J. Needleman, R. Strichartz, A. Teplyaev, and P. Yung, math.GM 0312027
[926] E. Nelson, Quantum fluctuations, Princeton Univ. Press, 1985; Dynamical theory of Brownian motion, Princeton Univ. Press, 1967
[927] C. Neves and W. Oliveira, hep-th 0310064
[928] L. Ni, math.DG 0602337
[929] L. Nickisch and J. Mollere, physics 0205086
[930] H. Nikolić, Found. Phys., 17 (2004), 363-380 (quant-ph 0208185); 18 (2005), 123-138 (quant-ph 0302152); 18 (2005), 549-561 (quant-oh 0406173); hep-th 0103053, 0103251, and 0610138; quant-ph 0307179, and 0305131
[931] H. Nikolić, Euro. Phys. Jour. C, 421 (2005), 365-374 (hep-th 0407228); gr-qc 9909035 and 0111029; hep-th 0202204 and 0601027
[932] H. Nikolić, gr-qc 0312063; hep-th 0501046; quant-ph 0603207 and 0512065
[933] H. Nikoli, hep-th 0402145 and 0607140
[934] H. Nikolić, physics 9812006; gr-qc 9901045, 9901057, 0009068, 0307011, 0403121, and 0601037; quant-ph 0602024
[935] H. Nikolić, hep-th 0005240, 0105176, 0205022, and 0210307
[936] H. Nikolić, quant-ph 0609163
[937] H. Nikolić, quant-ph 0505143
[938] hep-th 0402145 and 0607140
[939] H. Nitta, T. Kudo, and H. Minowa, Amer. Jour. Phys., 67 (1999), 966-971
[940] J. Noldus, gr-qc 0508104

[941] L. Nottale, Fractal space-time and microphysics: Toward a theory of scale relativity, World Scientific, 1993

[942] L. Nottale, Chaos, Solitons, and Fractals, 7 (1996), 877-938; 12 (2000), 1577-1583

[943] L. Nottale, M. Célérier, and T. Lehner, hep-th 0307093

[944] L. Nottale, Chaos, solitons, and Fractals, 10 (1999), 459-468

[945] L. Nottale, La relativité dans tous ses états, Hachette, 2000

[946] L. Nottale, Chaos, Solitons, and Fractals, 16 (2003), 539-564

[947] L. Nottale and J. Schneider, Jour. Math. Phys., 25 (1984), 1296-1300

[948] S. Odintsov and R. Percacci, hep-th 9404020

[949] R. Oeckl, Comm. Math. Phys., 217 (2001), 451-473; hep-th 0003018

[950] H. Ohanian and R. Ruffini, Gravity and spacetime, Norton, 1994

[951] P. O'Hara, gr-qc 0502078 and 9701034; quant-ph 0310016

[952] M. Ohya and D. Petz, Quantum entropy and its use, Springer, 2004

[953] B. Oksendal, Stochastic differential equations, Springer, 2003

[954] L. Olavo, quant-ph 9503020, 9503021, 9503022, 9503024, 9503025, 9509012, 9509013, 9511028, 9511039, 9601002, 9607002, 9607003, 9609003,9609023, 9703006, 9704004

[955] L. Olavo, Physica A, 262 (1999), 197-214 and 271 (1999), 260-302; Phys. Rev. E, 64 (2001), 036125

[956] L. Olavo and A. Figueiredo, Physica A, 271 (1999), 181-196

[957] L. Olavo, A. Bakuzis, and R. Amilcar, Physica A, 271 (1999), 303-323

[958] A. Oliveira, Mod. Phys. Lett. A, 16 (2001), 541-555

[959] H. deOliveira, J. Salim, an S. Sautu, Class. Quant. Grav., 14 (1997), 2833-2843

[960] R. Omnes, the interpretation of quantum mechanics, Princeton Univ. Press, 1994

[961] N. Ó Murchadha, gr-qc 0502055

[962] L. O'Raifeartaigh, The dawning of gauge theory, Princeton Univ. Press, 1997

[963] L. O'Raifeartaigh and N. Straumann, hep-th 9810524

[964] G. Ord, Chaos, Solitons, and Fractals, 8 (1997), 727-741; 9 (1998), 1011-1029; Jour. Phys. A, 16 (1983), 1869-1884

[965] G. Ord, Chaos, Solitons, and Fractals, 11 (2000), 383-391; 17 (2003), 609-620

[966] G. Ord and J. Gualtieri, Chaos, Solitons, and Fractals, 14 (2002), 929-935

[967] G. Ord and R. Mann, quant-ph 0206095; 0208004

[968] G. Ord and A. Deakin, Phys. Rev. A, 54 (1996), 3772-3778

[969] S. Orfanidis, Phys. Rev. D, 21 (1980), 1513-1522

[970] A. Orlov, nlin.SI 0207030; math-ph 0210012 and 0302011; nlin.SI 0209063

[971] B. Osgood, R. Phillips, and P. Sarnak, Jour. Funct. Anal., 80 (1988), 148-211 and 212-234; Ann. Math., 129 (1989), 293-362

[972] E. O'Shea, gr-qc 0404117

[973] Y. Ozhigov, quant-ph 0702237

[974] T. Padmanabhan, gr-qc 0503107

[975] T. Padmanabhan, hep-th 9608122

[976] T. Padmanabhan, Inter. Jour. Mod. Phys. D, 14 (2005), 2263 (gr-qc 0408051); Gen. Relativ. Grav., 34 (2002), 2029-2103

[977] T. Padmanabhan, hep-th 0205278; gr-qc 0204019, 0412068, and 0510015; Class. Quant. Gravity, 21 (2004), 4485-4494

[978] T. Padmanabhan, Brazil. Jour. Phys., 35 (2005), 362 (gr-qc 0412068); gr-qc 0510015 and 0606061; astro-ph 0603114

[979] T. Padmanabhan, Phys. Reports, 406 (2005), 49 (gr-qc 0311036); gr-qc 0202078; Gen. Relativ. Grav., 35 (2003), 2097-2103

[980] T. Padmanabhan, gr-qc 0209088

[981] T. Padmanabhan, gr-qc 0609012
[982] T. Padmanabhan and A. Paranjape, gr-qc 0701003
[983] T. Padmanabhan and A. Patel, gr-qc 0309053; hep-th 0305165
[984] T. Padmanabhan, astro-ph 0603114 and 0602117
[985] D. Page, Phys. Rev. A, 36 (1987), 3479-3481; Phys. Rev. D, 25 (1982), 1499
[986] A. Paranjape, S. Sarkar, and T. Padmanabhan, hep-th 0607240
[987] M. Pardy, quant-ph 0111105
[988] R. Parentani, Class. Quant. Gravity, 17 (2000), 1527-1547
[989] L. Parker and D. Toms, Phys. Rev. D, 31 (1985), 953-956 and 2424-2438
[990] L. Parker, Phys. Rev. D, 19 (1979), 438-441
[991] R. Parmenter and A. DiRienzo, quant-ph 0305183
[992] R. Parmenter and R. Valentine, Phys. Lett. A, 201 (1995), 1
[993] M. Partovi, quant-ph 0107083; Phys. Rev. Lett., 82 (1999), 3424-3427 and 50 (1983), 1883-1885
[994] A. Parvate and A. Gangal, math-ph 0310047
[995] R. Parwani, quant-ph 0408185 and 0412192; hep-th 0401190
[996] R. Parwani, quant-ph 0506005 and 0508125
[997] R. Parwani and G. Tabia, quant-ph 0607222
[998] O. Pashaev and J. Lee, Chaos, Solitons, and Fractals, 11 (2000), 2193-2202; ANZIAM Jour., 44 (2002), 73-80; Mod. Phys. Lett. A, 17 (2002), 1601-1609; Jour. Nonlin. Math. Phys., 8 (2001), 230-234; hep-th 9906104
[999] O. Pashaev, Nucl. Phys. B, Supp. 57 (1997), 338-341; Jour. Math. Phys., 37 (1996), 4368-4387; hep-th 9505178 and 9511184
[1000] O. Passon, quant-ph 0404128 and 0412119
[1001] W. Pauli, Pauli lectures on physics, Vol. 6, Dover, 2000
[1002] N. Pauna and I. Vancea, gr-qc 9812009
[1003] M. Pavon, Jour. Math. Phys., 36 (1995), 6774-6800; quant-ph 0306052
[1004] L. de la Pena and A. Cetto, The quantum dice - An introduction to SED, Kluwer, 1996
[1005] L. de la Pena and A. Cetto, quant-ph 0501011
[1006] F. Pennini and A. Plastino, cond-mat 0405033, 0311586, 0312680, 0402467, and 0407110
[1007] F. Pennini, A. Plastino, and A.R. Plastino, cond-mat 0110135
[1008] R. Penrose, The road to reality, J. Cape Pub., 2004
[1009] G. Perelman, math.DG 0211159, 0303109, and 0307245
[1010] A. Perelomov, Generalized coherent states and their applications, Springer, 1986
[1011] A. Peres and D. Terno, Rev. Mod. Phys., 76 (2004), 93
[1012] A. Peres, Nuovo Cimento 26 (1962), 53
[1013] A. Peres, gr-qc 0409061
[1014] A. Peres, Quantum theory: Concepts and methods, Kluwer, 1995
[1015] V. Perlich, Class. Quantum Gravity, 8 (1991), 1369-1385
[1016] Y. Pesin, Dimension theory in dynamical systems, Univ. Chicago Press, 1997
[1017] M. Peskin and D. Schroeder, An introduction to quantum field theory, Addison-Wesley, 1995
[1018] D. Petz, Jour. Math. Phys., 35 (1994), 780-795; Jour. Phys. A, 35 (2002), 929-939 (quant-ph 0106125)
[1019] D. Petz and C. Sudar, Jour. Math. Phys., 37 (1996), 2662-2673; quant-ph 0102132
[1020] H. Pfeiffer and J. York, gr-qc 0207095
[1021] P. Pierce, Jour. Acoust. Soc. Amer., 87 (1990), 2292

[1022] N. Pinto-Neto and E. Santini, Phys. Rev. D, 59 (1999), 123517 (gr-qc 9811067); Gen. Relativ. Gravitation, 34 (2002), 505
[1023] N. Pinto-Neto, gr-qc 0410117
[1024] N. Pinto-Neto, A. Velasco, and R. Colistete Jr., gr-qc 0001074
[1025] N. Pinto-Neto and E. Santini, gr-qc 0009080 and 0302112; Gener. Relativ. Gravitation, 34 (2002), 505; Phys. Lett. A, 315 (2003), 36; Phys. Rev. D, 59 (1999), 123517 (gr-qc 0408185)
[1026] N. Pinto-Neto, hep-th 0410225 and 0410001
[1027] N. Pinto-Neto and R. Colistete Jr., gr-qc 0106063; Phys. Lett. A, 290 (2001), 219
[1028] J. Pissondes, Chaos, Solitons, and Fractals, 9 (1998), 1115-1142; Jour. Phys. A, 32 (1999), 2871
[1029] L. Pitaevskij and S. Stringari, Bose-Einstein condensation, Oxford Univ. Press, 2003
[1030] A.R. Plastino and A. Plastino, Phys. Rev. E, 54 (1996), 4423-4426
[1031] A. Plastino, A.R. Plastino, and H. Miller, Phys. Lett. A, 235 (1997), 129
[1032] A.R. Plastino, H. Miller, and A. Plastino, Phys. Rev. E, 56 (1997), 3927-3934
[1033] A. Plastino and E. Curado, cond-mat 0412336
[1034] A. Plastino and A.R. Plastino, Nonextensive statistical mechanics and thermodynamics, Brazil. Jour. Phys., 29 (1999), 50
[1035] A.R. Plastino and A. Plastino, Condensed matter theories, Nova Sci. Publ., 1996, 11, p. 327
[1036] E. Poisson, gr-qc 9912045
[1037] J. Polchinski, Phys. Rev. Lett., 66 (1991), 397-400
[1038] D. Pope, P. Drummond, and W. Munro, quant-ph 0003131
[1039] S. Popescu, A. Short, and A. Winter, quant-ph 0511225
[1040] F. Potiguar and U. Costa, cond-mat 0208357 and 0210525
[1041] G. Preparata, S. Rovelli, and S. Xue, gr-qc 9806044
[1042] I. Prigogine, Chaotic dynamics and transport in fluids and plasmas, AIP, 1993
[1043] W. Puszkarz, quant-ph 9912006
[1044] W. Puszkarz, quant-ph 9802001, 9903010, and 9905046
[1045] W. Puszkarz, quant-ph 9710007, 0710008, 9710009, 9710010, and 9710011
[1046] H. Puthoff, Phys. Rev. D, 35 (1987), 3266-3269; Phys. Rev. A, 39 (1989), 2333-2342; 40 (1989), 4857-4862
[1047] H. Puthoff, Phys. Rev. A, 47 (1993), 3454-3455
[1048] H. Puthoff, Found. Phys., 32 (2002), 927-943
[1049] S. Qian, M. Gong, and Q. Li, Phys. Rev. A, 70 (2004), 022105
[1050] L. Querella, gr-qc 9806005
[1051] L. Querella, gr-qc 9992044
[1052] I. Quiros, gr-qc 0004014, 0011056, and 9904004; hep-th 0009169 and 0010146
[1053] I. Quiros, R. Bonal, an R. Cardenas, gr-qc 9908075
[1054] I. Quiros, R. Cardenas, and R. Bonal, gr-qc 0002071
[1055] I. Quiros, R. Bonal, and R. Cardenas, gr-qc 9905071
[1056] T. Radozycki, quant-ph 0401115
[1057] G. Rainich, Trans. Amer. Math. Soc., 27 (1925), 106
[1058] A. Rajagopal and S. Abe, Phys. Rev. Lett., 83 (1999), 1711
[1059] J. Rankin, gr-qc 9404023 and 9408029; Class. Quant. Grav., 9 91992), 104-1067
[1060] F. Ravndal, hep-th 0009208
[1061] E. Recami and G. Salesi, quant-ph 9607025
[1062] M. Reginatto, quant-ph 9909065
[1063] M. Reginatto, Phys. Rev. A, 58 (1998), 1775-1778

[1064] M. Reginatto and F. Lenguel, cond-mat 9910039
[1065] M. Reginatto, gr-qc 0501030
[1066] J. Rehacek and Z. Hradil, Jour. Mod. Optics, 15 (2004), 979-982
[1067] F. Reif, Fundamentals of statistical and thermal physics, McGraw-Hill, 1965
[1068] M. Reuter, hep-th 9804036
[1069] D. Revuz and M. Yor, Continuous martingales and Brownian motion, Springer, 1999
[1070] K. Richardson, Jour. Fnl. Anal., 122 (1994), 52-83
[1071] C. Ridgely, physics 0010018 and 0103078
[1072] W. Rindler, Essential relativity, Springer, 1977
[1073] H. Risken, The Fokker Planck equation, Springer, 1984
[1074] F. Ritort, cond-mat 0405077
[1075] J. Robbin and D. Salmon, Math. Zeit., 221 (1996), 307-335
[1076] M. Roberts, hep-th 0012062
[1077] H. Robertson, Phys. Rev., 46 (1934), 794-801
[1078] D. Da Rocha and L. Nottale, astro-ph 0310036; Chaos, Solitons, and Fractals, 16 (2003), 565-595
[1079] L. Rogers and D. Williams, Diffusion, Markov processes, and martingales, Vol. 1, Cambridge Univ. Press, 2001
[1080] H. Roman and M. Giona, Jour. Phys. A, 25 (1992), 2107-2117
[1081] N. Romao and J. Speight, hep-th 0403215
[1082] N. Rosen, Found. Phys., 12 (1982), 213-248; 13 (1983), 363-372
[1083] G. Rousseaux, physics 0506203
[1084] G. Rousseaux and E. Guyon, Bull. Union Phys., 96 (2002), 107-135
[1085] C. Rovelli, Quantum gravity, Cambridge Univ. Press, 2004
[1086] C. Rovelli, gr-qc 0207043
[1087] C. Rovelli and L. Smolin, Nucl. Phys. B, 331 (1990), 80; Phys. Rev. Lett., 61 (1988), 1155; Phys. Rev. D, 52 (1995), 5743
[1088] S. Roy and V. Singh, quant-ph 9811041
[1089] S. Roy, quant-ph 9811047
[1090] V. Rubikov, classical theory of gauge fields, Princeton Univ. Press, 2002
[1091] A. Rueda and B. Haisch, Found. Phys., 28 (1998), 1057-1108; Phys. Lett. A, 240 (1998), 115-126
[1092] A. Rueda, B. Haisch, and R. Tung, gr-qc 0108026
[1093] A. Rueda, Phys. Rev. A, 23 (1981), 2020-2040
[1094] A. Rueda and B. Haisch, gr-qc 0504061; physics 9802030 and 9802031
[1095] H. Rund, The Hamilton Jacobi theory in the calculus of variations, Krieger-Huntington, 1973
[1096] L. Ryder, Quantum field theory, Cambridge Univ. Press, 1996
[1097] Y. Rylov, physics 9505044, 0210003, 0303065, and 0604111; Jour. Math. Phys., 40 (1999), 256-278 and 30 (1989), 2516-2520; Found. Physics, 28 (1998), 245-272;
[1098] C. Sabot, math-ph 0201041
[1099] P. Saffman, Vortex dynamics, Cambridge Univ. Press, 1992
[1100] T. Sakaguchi, gr-qc 9704039
[1101] H. Sakaguchi and t. Higashiuchi, nlin.PS 0607072
[1102] J. Sakurai, Advanced quantum mechanics, Addison-Wesley, 1967
[1103] J. Sakurai, Quantum mechanics, Addison-Wesley, 1994
[1104] H. Salehi, H. Motavali, and M. Golshani, hep-th 0011063
[1105] H. Salehi, hep-th 9909157, 9912115, and 0302178
[1106] H. Salehi, H. Sepangi, and F. Darabi, gr-qc 0002058

[1107] H. Salehi and S. Mirabotalebi, gr-qc 0402074
[1108] H. Salehi and H. Sepangi, hep-th 9810207
[1109] H. Salehi, H. Motavali, and M. Golshani, hep-th 0011063
[1110] H. Salehi, H. Sepangi, and F. Darabi, gr-qc 0002058
[1111] H. Salehi and Y. Bisabr, hep-th 0301208 and 0001095
[1112] G. Salesi, E. Recami, H. Hernandez, and L. Kretly, hep-th 9802106
[1113] G. Salesi, quant-ph 0412045 and 0112052
[1114] J. Salim and S. Sautu, Class. Quant. Grav., 13 (1996), 353-360
[1115] E. Santamato, Phys. Rev. D, 29 (1984), 216-222
[1116] E. Santamoto, Phys. Rev. D 32 (1985), 2615-2621; Jour. Math. Phys., 25 (1984), 2477-2480
[1117] E. Santamoto, Phys. Lett. A, 130 (1988), 199-202
[1118] D. Santiago and A. Silbergleit, gr-qc 9904003
[1119] E. Santini, gr-qc 0005092
[1120] R. Santos, Jour. Math. Phys., 38 (1997), 4104
[1121] E. Santos, quant-ph 0510059
[1122] A. Sanz, quant-ph 0412050
[1123] A. Sanz and F. Borondo, quant-ph 0310096
[1124] A. Sarkar and T. Padmanabhan, gr-qc 0607042
[1125] A. Scardicchio, Phys. Lett. A, 300 (2002), 7
[1126] A. Scarfone, cond-mat 0503684
[1127] L. Schiff, Quantum mechanics, McGraw-Hill, 1941
[1128] T. Schilling, Thesis, Penn. State Univ., 1996
[1129] I. Schmelzer, gr-qc 9811073, 0001101, and 0205035
[1130] R. Schoen, Lect. Notes Math., 1365 (1989), 120-184
[1131] E. Scholz, astro-ph 0403446 and 0409635
[1132] E. Scholz, Weyl geometry as an alternative framework for cosmology, book, 2004
[1133] J. Schouten, Ricci calculus, Springer, 1954
[1134] D. Schuch, Phys. Lett. A, 338 (2005), 225-231; Inter. Jour. Quant. Chem., 23 (1989), 59-72; 42 (1992), 663-383; 72 (1999), 537-547; Phys. Rev. A, 555 (1997), 935-940; Quantum theory and symmetries, World Scientific, 2000, pp. 152-157
[1135] D. Schuch, K. Chung, and H. Hartmann, Jour. Math. Phys., 24 (1983), 1652-1660; 25 (1984), 3062-3092
[1136] D. Schuch and K. Chung, Inter. Jour. Quant. Chem., 29 (1986), 1561
[1137] L. Schulman, Techniques and applications of path integration, Wiley, 1981
[1138] F. Schwabl, Statistical mechanics, Springer, 2002
[1139] H. Schwetlick and M. Struwe, Jour. Reine Angew. Math., 562 (2003), 59-100
[1140] J. Schwinger, Phys. Rev., 82 (1951), 914 and 91 (1953), 355
[1141] J. Schwinger, Phys. Rev., 74 (1948), 1439-1461
[1142] M. Scully and M. Zubairy, Quantum optics, Cambridge Univ. Press, 2002
[1143] R. Seligar and G. Whitham, Proc. Roy. Soc. Lond. A, 305 (1968), 1-25
[1144] O. Senatchin, physics 0101054 and 0105054
[1145] M. Serva, Ann. Inst. H. Poincaré, Phys. Theor., 49 (1988), 415-432
[1146] G. Sewell, Quantum mechanics and its emergent macrophysics, Princeton Univ. Press, 2002
[1147] A. Shalyt-Margolin and A. Tregubovich, gr-qc 0307018
[1148] C. Shannon, Bell Syst. Tech. Jour., 27 (1948), 379-423 and 623-656
[1149] V. Shchesnovich and R. Kraenkel, cond-mat 0306725
[1150] J. Shifflett, gr-qc 0310124
[1151] A. Shojai and M. Golshani, quant-ph 9612022 and 9612019

[1152] F. Shojai and M. Golshani, Inter. Jour. Mod. Phys. A, 13 (1998), 2135-2144 (gr-qc 9903047)
[1153] F. Shojai and M. Golshani, Int. Jour. Mod. Phys. A, 13 (1998), 677-693
[1154] F. Shojai, A. Shojai, and M. Golshani, Mod. Phys. Lett. A, 13 (1998), 2725-2729 and 2915-2922 (gr-qc 9903049 and 9903050)
[1155] A. Shojai, F. Shojai, and M. Golshani, Mod. Phys. Lett. A, 13 (1998), 2965-2969 (gr-qc 9903048)
[1156] A. Shojai, Inter. Jour. Mod. Phys. A, 15 (2000), 1757-1771 (gr-qc 0010013)
[1157] F. Shojai and A. Shojai, gr-qc 0409036, 0105102, and 0109052
[1158] F. Shojai and A. Shojai, Physica Scripta, 64 (2001), 413 (quant-ph 0109025)
[1159] F. Shojai and A. Shojai, Gravitation and Cosmology, 9 (2003), 163 (gr-qc 0306099)
[1160] F. Shojai and A. Shojai, gr-qc 0404102
[1161] F. Shojai and A. Shojai, Inter. Jour. Mod. Phys. A, 15 (2000), 1859-1868 (gr-qc 0010012)
[1162] A. Shojai and M. Golshani, quant-ph 9812019, 9612023, 9612020, and 9612021
[1163] F. and A. Shojai, Class. Quant. Grav., 21 (2004), 1-9 (gr-qc 0311076) and 0409035
[1164] A. and F. Shojai, gr-qc 0306100, 0311076, and 0105102; quant-ph 0609109 and 0109025
[1165] F. Shojai, gr-qc 9907093
[1166] A. and F. Shojai, astro-ph 0211272
[1167] A. and F. Shojai, gr-qc 0409020
[1168] A. and F. Shojai and N. Dadhich, gr-qc 0504137
[1169] F. Shojai, Phys. Rev. D, 60 (1999), 124001
[1170] Y. Shtanov, quant-ph 9705024; Phys. Rev. D, 54 (1996), 2564-2570
[1171] B. Sidharth, Chaos, Solitons, and Fractals, 12 (2001), 173-178, 613-616, 1371-1373; 14 (2002), 1325-1330
[1172] B. Sidharth, physics 0204007, 0204073, 0208045, and 0211012
[1173] B. Sidharth, Chaos, Solitons, and Fractals, 11 (2000), 1171-1174 and 1269-1278; 18 (2003), 197-201; 20 (2004), 701-703; physics 0202024, 0210057, 0312109, and 0404024; physics 0301015, 0311140, 0405157, and 0406147; quant-ph 9808031 and 990107; hep-th 9807190; physics 0006048, 0010026, 0106100, and 0110040; quant-ph 9803048; physics 0007020
[1174] G. Simeone, Gen. Relativ. Gravitation, 34 (2002), 1887
[1175] B. Simon, Phys. Rev. Lett., 51 (1983), 2167
[1176] T. Singh, Class. Quant. Gravity, 7 (1990), L149-L154
[1177] K. Sinha, C. Sivaram, and E. Sudarshan, Found. Phys., 6 (1976), 65 and 717; 8 (1978), 823
[1178] J. Sipe, Phys. Rev. A, 52 (1995), 1875-1883
[1179] L. Delle Site, quant-ph 0412128
[1180] A. Smailagic, E. Spallucci, and T. Padmanabhan, hep-th 0308122
[1181] L. Smolin, Phys. Lett. A, 113 (1986), 408-412
[1182] L. Smolin, Quo vadis quantum mechanics, Springer, 2005, pp. 121-152
[1183] L. Smolin, Three roads to quantum gravity, Oxford Univ. Press, 2000
[1184] L. Smolin, hep-th 0408048 and 0201031
[1185] L. Smolin and C. Rovelli, Nucl. Phys. B, 331 (1990), 80-152
[1186] L. Smolin and C. Soo, gr-qc 9405015
[1187] L. Smolin, quant-ph 0609109
[1188] Y. Sobouti and S. Nasiri, Inter. Jour. Mod. Phys. B, 7 (1993), 3255-3273
[1189] I. Sokolov, A. Chechkin, and J. Klafter, cond-mat 0401146
[1190] L. Sokolowski, gr-qc 9511073

[1191] D. Solomons, P. Dunsby, and G. Ellis, gr-qc 0103087
[1192] C. Soo, gr-qc 0512025 and 0109046; Class. Quant. Gravity, 19 (2002), 1051-1063
[1193] D. Sornette, Euro. Jour. Phys., 11 (1990), 334-337
[1194] T. Sotiriou and S. Liberati, gr-qc 0603096
[1195] A. Speliotopoulos, cond-mat 9504100
[1196] E. Squires, quant-ph 9508014
[1197] Y. Srivastava, G. Vitiello, and A. Widom, quant-ph 9810095
[1198] Z. Stachniak, Finite Alg. and AI, AISC 2004, LNAI 3249, pp. 8-14
[1199] H. Stephani, Relativity, Cambridge Univ. Press, 1982
[1200] P. Stevenson, hep-th 0109204
[1201] M. Stone, The physics of quantum fields, Springer, 2000
[1202] M. Stone, cond mat 0012316; Phys. Rev. E, 62 (2000), 1341-1350
[1203] A. Stotland, A. Pomeransky, E. Bachmat, and D. Cohen, quant-ph 0401021
[1204] N. Straumann, astro-ph 0006423
[1205] R. Streater, math-ph 0002050 and 0308037
[1206] R. Streater, Statistical dynamics, World Scientific, 1995
[1207] R. Streater, Lost causes in theoretical physics, Wikipedia
[1208] F. Strocchi, hep-th 0401143; Rev. Mod. Phys., 31 (1966), 36-40
[1209] D. Stroock, Markov processes from K. Ito's perspective, Princeton Univ. Press, 2003
[1210] C. Su, physics 0208082, 0208083, 0208084, 0208085, and 0211019
[1211] A. Sudberry, Jour. Phys. A, 19 (1986), L33-L36
[1212] G. Svetlichny, quant-ph 0410036
[1213] K. Svozil, quant-ph 000033 and 0110054; physics 0210091 and 0305048
[1214] T. Takabayashi, Prog. Theor. Phys., 8 (1952), 143-182; 9 (1953), 187-222
[1215] M. Tegmark, quant-ph 9709032 and 0302131
[1216] C. Teitelboim, Ann. Phys., 80 (1973), 542
[1217] C. Teitelboim, Phy. Rev. D, 25 (1982), 3159 and 28 (1983), 297
[1218] S. Teufel, K. Berndl, and D. Dürr, quant-ph 9609005
[1219] S. Teufel and R. Tumulka, math-ph 0406030
[1220] T. Thiemann, gr-qc 0110034
[1221] T. Thiemann, Lecture Notes Physics, 631 (2003), 41-135
[1222] T. Thiemann, Phys. Lett. B, 380 (1966), 257-264
[1223] T. Thiemann, Class. Quant. Grav., 15 (1998), 839-873; 875-905, 1249-1280, and 1281-1314
[1224] T. Thiemann, gr-qc 9606090, 9705017, and 9705019
[1225] T. Thiemann, gr-qc 0305080 and 0411031
[1226] F. Tian and J. Ye, Comm. Pure Appl. Math., 52 (1999), 0655-0692
[1227] F. Tipler, quant-ph 0003146 and 0611245
[1228] U. Tirnakli, G. Ananos, and C. Tsallis, cond-mat 0005210
[1229] S. Tiwari, quant-ph 0109048; Phys. Rev. A, 56 (1997), 157-161
[1230] S. Tiwari, Superluminal phenomena in modern perspective, Rinton Press, 2003
[1231] S. Tiwari, gr-qc 0307079
[1232] S. Tiwari, gr-qc 0612099
[1233] S. Tomonaga, Prog. Theor. Phys., 1 (1946), 27
[1234] D. Toms, hep-th 0411233
[1235] P. Topping, Lectures on the Ricci flow, Cambridge Univ. Press, 2006
[1236] A. de la Torre, quant-ph 0410171, 0410179, and 0503023
[1237] A. de la Torre and A. Daleo, quant-ph 9905032

[1238] G. Torres-Vega and J. Frederick, Jour. Chem. Phys., 93 (1990), 8862-8874 and 98 (1993), 176
[1239] C. Tricot, Courbes et dimension fractale, Springer, 1999
[1240] C. Tsallis, R. Mendes, and A.R. Plastino, Physica A, 261 (1998), 543-554
[1241] C. Tsallis, F. Baldovin, R. Cerbino, and P. Pierobon, cond-mat 0309093
[1242] C. Tsallis, Jour. Stat. Phys., 52 (1988), 479
[1243] C. Tsallis, R. Mendes, and A.R. Plastino, Physica A, 261 (1998), 534
[1244] N. Tsamis and R. Woodward, Phys. Rev. D, 36 (1987), 3641
[1245] R. Tumulka, math.PR 0312326; quant-ph 0408113, 0210207, and 0501167
[1246] J. Uffink and J. van Lith, Found. Phys., 29 (1999), 655-692
[1247] H. Umezawa, Advanced field theory, AIP Press, 1996
[1248] W. Unruh, Phys. Rev. Lett., 46 (1981), 1351; Phys. Rev. D, 51 (1995), 282 and 14 (1976), 870
[1249] N. Ustinov and Y. Brezhnev, nlin.SI 0012039
[1250] L. Vaidman, quant-ph 0609006, 0001057, and 0111072
[1251] A. Valentini, quant-ph 0104067, 0106098, 0309107, 0112151, and 0403034
[1252] A. Valentini, Phys. Lett. A, 156 (1991), 5
[1253] L. Valeri and G. Scharf, quant-ph 0502115
[1254] P. Van and T. Fülöp, quant-ph 0304062
[1255] P. Van, cond-mat 0112214 and 0210402
[1256] J. van Holten, physics 0107041
[1257] I. Vancea, gr-qc 9801072; hep-th 0399214
[1258] J. van Vleck, Bull. Natl. Sci. Res. Council, 10 (1926), 1-316; Proc. Nat'l Acad. Sci., 14 (1928), 176
[1259] J. Vigier, IEEE Trans. Plasma Sci., 18 (1990), 64
[1260] J. Vigier, Lett. Nuovo Cimento, 29 (1980), 467
[1261] J. Vigier, Phys. Lett. A, 135 (1989), 99
[1262] J. Vink, Phys. Rev. A, 48 (1993), 1808-1818; Nucl. Phys. B, 369 (1992), 707-728
[1263] M. Visser, Lorentz wormholes, AIP Press, 1996
[1264] M. Visser, Phys. Rev. D, 47 (1993), 2395-2402; gr-qc 9311026
[1265] M. Visser, Class. Quant. Grav., 15 (1998), 1767
[1266] A. Volkov and V. Kiselev, Sov. Phys. JETP, 30 (1979), 733
[1267] G. Volovik, The universe in a helium droplet, Oxford Univ. Press, 2003
[1268] G. Volovik, gr-qc 9809081, 0005091, 0101111, 0104046, 0212003, 0301043, and 0306011
[1269] G. Volovik, gr-qc 0505104
[1270] M. Wadati, K. Konno, and Y. Ichikawa, Jour. Phys. Soc. Japan, 46 (1979), 1965-1966
[1271] D. Wagenaar, Information geometry for neural networks, http://www.its.caltech. edu pinelab/wagenaar/infogeom/pdf
[1272] R. Wald, Quantum field theory in curved spacetime and black holes, Univ. Chicago Press, 1994
[1273] R. Wald, Phys. Rev. D, 48 (1993), R3427-R3431
[1274] R. Wald, General relativity, Univ. Chicago Press, 1984
[1275] D. Wallace, quant-ph 0112148, 0112149, and 0103092
[1276] T. Wallstrom, Phys. Rev. A, 49 (1994), 1613-1617
[1277] C. Wang, gr-qc 0304101, 0308026, 0308081, 0406079, and 0501024
[1278] C. Wang, gr-qc 0512023, 051124, 0603062, and 0605124; Phys. Rev. D, 72 (2005), 087501
[1279] C. Wang, S. Kessari, and E. Irvine, gr-qc 0310029

[1280] Q. Wang, cond-mat 0405373 and 0407570
[1281] Q. Wang, L. Nivaneu, and A. LeMéhauté, cond-mat 0311175
[1282] R. Ward, hep-th 0207100
[1283] G. Watanabe, S. Gifford, G. Bay, and C. Pethick, cond-mat 0605151
[1284] A. Wehner, Dissertation, Utah State Univ., 2001
[1285] A. Wehner and J. Wheeler, hep-th 9812099, 0001061, and 0001191
[1286] S. Weinberg, The quantum theory of fields, Vols. 1-3, Cambridge Univ. Press, 1995
[1287] S. Weinberg, Phys. Rev. Lett., 62 (1989), 485-488
[1288] S. Weinberg, Gravitation and cosmology, Wiley, 1972
[1289] Y. Weinstein, S. Lloyd, and C. Tsallis, cond-mat 0206039
[1290] X. Wen, cond-mat 0406441 and 0110397; hep-th 0109120
[1291] X. Wen, Quantum field theory of many body systems, Oxford Univ. Press, 2004
[1292] P. Werbos and L. Werbos, quant-ph 0309087
[1293] B. West, M. Bologna, and P. Grigolini, Physics of fractal operators, Springer, 2003
[1294] H. Weyl, Space, time, and matter, Dover, 1952
[1295] J. A. Wheeler, Ann. Phys., 2 (1957), 604
[1296] J.A. Wheeler, Proc. Third Internat. Sympos. on Foundations of Quantum Mechanics, Phys. Soc. Japan, 1990, p. 354
[1297] J.A. Wheeler, Battelle Rencontres, Benjamin, 1967, pp. 242-307
[1298] J.A. Wheeler, in Gilbert-Newton eds., Gordon-Breach, 1970, pp. 335-378
[1299] J. Wheeler, Phys. Rev. D, 41 (1990), 431
[1300] J. Wheeler, hep-th 0002068 and 0305017; gr-qc 9411030
[1301] J. Wheeler, Canad. Jour. Phys., (2002), 1-49
[1302] J. Wheeler, hep-th 9706215
[1303] J. Wheeler, hep-th 9708088
[1304] J. Wheeler, Jour. Math. Phys., 39 (1998), 299 (hep-th 9706214)
[1305] G. Whitham, Proc. Royal Soc. London A, 283 (1965), 238-261
[1306] G. Whitham, Linear and nonlinear waves, Wiley, 1999
[1307] Wikipedia, Many worlds
[1308] F. Wilczek, Phys. Rev. Lett., 49 (1982), 957
[1309] T. Willmore, Riemannian geometry, Oxford Univ. Press, 1993
[1310] H. Wiseman, quant-ph 0302080
[1311] D. Wisniacki and E. Pujals, quant-ph 0502108
[1312] D. Wisniacki, E. Vergini, R. Benito, and F. Borondo, nlin.CD 0402022
[1313] D. Wisniacki, F. Borondo, and R. Benito, quant-ph 0309083
[1314] D. Wisniacki, E. Pujals, and F. Borondo, nlin.CD 0507015
[1315] E. Witten, Comm. Math. Phys., 117 (1988), 353; 118 (1988), 411; 121 (1989), 351; Phys. Lett. B, 206 (1988), 601
[1316] D. Woicik, I. Bialynicki-Birula, and K. Zyczkowski, Phys. Rev. Lett., 85 (2000), 5022
[1317] M. Wong, Weyl transforms, Springer, 1998
[1318] W. Wood and G. Papini, gr-qc 9612042; Found. Phys. Lett. 6 (1993), 207-223
[1319] R. Woodward, Class. Quant. Gravity, 10 (1993), 483-496
[1320] W. Wootters, Phys. Rev. D, 23 (1981), 357-362; quant-ph 0406032
[1321] O. Wright, M. Forest and K. McLaughlin, Phys. Lett. A, 257 (1999), 170-174
[1322] C. Wu and L. Ford, quant-ph 0012144
[1323] T. Yamano, cond-mat 0009078 and 0010074
[1324] H. Yamasaki, Prog. Theor. Phys., 36 (1966), 72-85

[1325] F. Yndurain, Relativistic quantum mechanics and introduction to field theory, Springer, 1996

[1326] T. Yoneya, hep-th 0004074

[1327] J. York, Phys. Rev. Lett., 26 (1971), 1665-1668; 28 (1972), 1082-1085; 82 (1999), 1350-1353

[1328] J. York, Jour. Math. Phys., 14 (1973), 456-464 and 13 (1972), 125-130; gr-qc 0405005 and 9307022; Found. Phys., 16 (1986), 29-257

[1329] J. Yor, Phys. Rev. Lett., 82 (1999)

[1330] C. Zachos and T. Curtright, Prog. Theor. Phys. Suppl. 135 (1990), 244-258

[1331] C. Zachos, hep-th 0008010 and 0110114

[1332] C. Zachos and T. Curtright,hep-th 9903254

[1333] J. Zambrini, Phys. Rev. A, 38 (1987), 3631-3649; 33 (1986), 1532-1548; Jour. Math. Phys., 27 (1986), 2307-2330

[1334] T. Zastawnicak, Europhys. Lett., 13 (1990), 13-17

[1335] A. Zee, hep-th 0309032

[1336] A. Zee, Quantum field theory in a nutshell, Princeton Univ. Press, 2003

[1337] H. Zeh, Phys. Lett. A, 309 (2003), 329-334

[1338] T. Zlosnik, P. Ferreira, and G. Starkman, gr-qc 0606039

[1339] T. Zuyeva, math-ph 0211052

[1340] B. Zweibach, A first course in string theory, Cambridge Univ. Press, 2004

Index

www.ingramcontent.com/pod-product-compliance
Ingram Content Group UK Ltd.
Pitfield, Milton Keynes, MK11 3LW, UK
UKHW021315180426
11947UKWH00015B/1247

9 781845 492113